Introductory Tensorial Algorithms In Structural Mechanics

By:

FAYSAL ALKHALIL, PH.D.
Structural Engineering and Structural Mechanics
University of California at Berkeley, U.S.A. Earthquake Engineering Research Center.

Associate Professor 1974, Random Vibration & Stochastic Processes
University of Simon Bolivar, Graduate School.

Professor 1985, Structural Engineering
University of Damascus.

AuthorHouse™
1663 Liberty Drive
Bloomington, IN 47403
www.authorhouse.com
Phone: 1-800-839-8640

First published by AuthorHouse 06/17/2011

ISBN: 978-1-4567-8165-1 (sc)

Printed in the United States of America

This book is printed on acid-free paper.

INTRODUCTION:

Well established engineering research in theoretical and applied structural dynamics and continuum mechanics has directly rushed researchers into numerical programming. An equivalent research into tensorial upgrading of theoretical or applied structural dynamics and continuum mechanics is plausible.

Unfortunately, it became familiar, the use of computer programs as software closed boxes, with little concern to their frame of reference construction and proper or accurate use. Most users became satisfied by consumption of such programs!

Several work shops were executed to establish a methodology to fill up the gap between developed continuum theories or engineering practice and numerical computer programming.

Utilizing tensorial analysis and reference frame independence the subjects in this book are discussed

This book is designed for use as a textbook for courses in: Computer and Numerical Programming, Tensor Analysis, Continuum Mechanics, Structural Dynamics, Vibrations and Stochastic Processes; such might be offered in Civil Engineering, Engineering Mechanics and Analysis, Earthquake Engineering, Ocean Engineering, Aeronautics Engineering and Astronautics Engineering. It is also intended to be used as a reference for engineers and as a comprehensive guide lines or code and methodological approach for researchers in order to establish engineering programming algorithms. Algorithms are presented compactly, consequently and frame ordered. Computer programs are written in FORTRAN.

Vibration of structures is a distinctive subject that is in considerable focus of researchers in structural dynamics. In contrast with earlier practice, which was restricted to temporal or spectral structural dynamics, earthquake engineering requirements have emphasized that vibration resulting from such sources of ground shake, explosion, storms or any vibrating source introduces complexities that have not been familiar to those engaged in structural dynamics. Because such vibration exhibits random characteristics, it has become essential to device concepts and methods of analysis that are new in structural vibration, dynamic control and design.

2 Prof. Faysal Alkhalil, Ph.D. INTRODUCTORY TENSORIAL ALGORITHMS IN STRUCTURAL MECHANICS

CONTENTS:

1.
Structural Fixed Tensorials:

Bases of globally fixed coordinate system are unique, constant, orthonotmal and time invariant. Whereas bases of all other coordinate systems may *change in position and time.*

1.1.
Repeated Index Rule:

Given two structural variables P and Q:

P(I,J,K) Q(K,L)

P is function of 3 indices (I,J,K) Q is function of 2 indices (K,L)

Their Multiplication:

$$R(I,J,K,L) = P(I,J,\underline{K}) * Q(\underline{K},L)$$

A single index, $\underline{K}$ in common is *underlined* in both P and Q

Result, R is function of 4 indices (I,J,K,L), P 3 indices plus Q 2 indices minus *one*: 3 + 2 - *1* = 4

Sum over K of products:

$$S(I,J,L) = \Sigma_K R(I,J,K,L)$$

Contracting K and dropping symbol Σ, sum of products according to "Repeated Index Rule" is written:

$$S(I,J,L) = P(I,J,K) \; Q(K,L)$$

Result, S is a function of 3 indices (I,J,L), P 3 indices plus Q 2 indices minus *two*: 3 + 2 - *2* = 3

1.2.
Generalized Delta:

Given a structural variable R in terms of ordered N indices:

$$R = r(i1,i2,i3,..,iJ,\dots,iN)$$

(iJ) runs from 1 to space dimension E,

$$1 \leq (iJ) \leq E$$

(iJ) is in J-th order:

$$1 \leq (J) \leq N$$

In general N and E are independent.

1.2.1.
Transposition:

Given a structural variable R in terms of ordered N indices:

$$R = r(r1,r2,r3,\dots, rJ,\dots, rK,..,rN)$$

An interchange of only two indices rJ, rK, is a "Transpose in rJ, rK" of the given structural variable:

$$RT = r(r1,r2,r3,\dots,rK,\dots, rJ,..,rN)$$

Corollary:

There is no transposable R = r(r1), a singly indexed structural variable.

1.2.2.
Symmetry:

A structural variable R is "Symmetric in rJ, rK" if its value is *Unchanged* by an interchange of indices:

$$r(r1,r2,r3,\dots, rJ,\dots, rK,..,rN) = R = RT = r(r1,r2,r3,\dots, rK,\dots, rJ,..,rN)$$

R is "Symmetric" if its value is *Unchanged* by *any* permutation of indices.

1.2.3.
Skew-Symmetry:

A Skew-Symmetric structural variable:
Q = Q(i1,i2,i3,…, iJ,…,,iN):
Q is "Skew-Symmetric" if its value is *Unchanged* by any *Even* permutation of indices,
While, it *Reverses sign* by an *Odd* permutation of indices.
Two important Skew-Symmetric structural variables are:

Covariant	Contravariant
$Q_{Q1\ Q2\ Q3\ \ldots\ldots\ QN}$	$Q^{Q1\ Q2\ Q3\ \ldots\ldots\ QN}$

Corollary:

There is no skew-symmetric R = r(r1), a singly indexed structural variable.

Corollary:

A structural variable term of two-Like Indices must be Zero.

A *single*, that is an odd permutation of two like-indices must reverse sign, and zero is the only such value.

Q(i1,i2,i3,…<u>*lx*</u>, …, <u>*lx*</u>,…,iN) = Q(i1,i2,i3,…<u>*lx*</u>, …, <u>*lx*</u>,…,iN) = 0

This is valid in:

Covariant: $Q_{Q1\ Q2\ Q3\ \ldots \underline{k} \ldots \underline{k} \ldots\ QN} = 0$

Contravariant: $Q^{Q1\ Q2\ Q3\ \ldots \underline{k} \ldots \underline{k} \ldots\ QN} = 0$

1.2.4.
Isotropic Identity:

Skew-Symmetric *Isotropic Identity*,
R = e(i1,i2,i3,…, iJ,…,,iN):
(iJ) runs from 1 to space dimension E,
$1 \leq (iJ) \leq E$
(iJ) is in J-th order:
$1 \leq (J) \leq N$
But:
N and E are related, N is less or equal E:
$N \leq E$
Isotropiityc:
$1 \leq (J) \leq N \leq E$
And:
When (i1,i2,i3,…,iN) is an Even permutation of (1,2,3,..,N) then:
R = +1
When (i1,i2,i3,…,iN) is an Odd permutation of (1,2,3,..,N) then:
R = -1
Skew-Symmetric *Isotropic Identity* is two types:
First type:
Covariant *Isotropic Identity*:
$Q = e_{Q1\ Q2\ Q3\ \ldots\ldots\ QN}$
When (Q1,Q2,Q3,…,QN) is an Even permutation of (1,2,3,..,N) then: Q = +1
When (Q1,Q2,Q3,…,QN) is an Odd permutation of (1,2,3,..,N) then: Q = -1
Second type:
Contravariant *Isotropic Identity*:
$Q = e^{Q1\ Q2\ Q3\ \ldots\ldots\ QN}$
When (Q1,Q2,Q3,…,QN) is an Even permutation of (1,2,3,..,N) then: Q = +1
When (Q1,Q2,Q3,…,QN) is an Odd permutation of (1,2,3,..,N) then: Q = -1

According to past corollary *Isotropic Identity* two - like indices term must be Zero:

Covariant *Isotropic Identity*

$$e_{Q1\,Q2\,Q3\,\ldots\underline{Q}\ldots\underline{Q}\ldots\,QN} = 0$$

Contravariant *Isotropic Identity*

$$e^{Q1\,Q2\,Q3\,\ldots\underline{Q}\ldots\underline{Q}\ldots\,QN} = 0$$

1.2.4.1.
Isotropic Identity of Second Order:

For N = 2:

$e_{11} = 0$	$e_{12} = +1$
$e_{21} = -1$	$e_{22} = 0$

R has two like indices must be zero $e_{\underline{i}\,\underline{i}} = 0$, namely $e_{11} = e_{22} = 0$

$e_{11} = 0$	
	$e_{22} = 0$

And:

(1,2) is zero permutation of (1,2), so it is an even permutation of (1,2).

$$e_{12} = +1$$

(2,1) is one permutation of (1,2), so it is an odd permutation of (1,2).

$$e_{21} = -1$$

1.2.4.2.
Isotropic Identity of Third Order:

For N = 3:

First Level: I = 1:

$e_{111} = 0$	$e_{112} = 0$	$e_{113} = 0$
$e_{121} = 0$	$e_{122} = 0$	$e_{123} = +1$
$e_{131} = 0$	$e_{132} = -1$	$e_{133} = 0$

:

Second Level: I = 2:

$e_{211} = 0$	$e_{212} = 0$	$e_{213} = -1$
$e_{221} = 0$	$e_{222} = 0$	$e_{223} = 0$
$e_{231} = +1$	$e_{232} = 0$	$e_{233} = 0$

:

Third Level: I = 3:

$e_{311} = 0$	$e_{312} = +1$	$e_{313} = 0$
$e_{321} = -1$	$e_{322} = 0$	$e_{323} = 0$
$e_{331} = 0$	$e_{332} = 0$	$e_{333} = 0$

$i \neq j \neq k \neq i$	$e_{ijk} = e_{jki} = e_{kij} = -e_{kji} = -e_{ikj} = -e_{jik} = +1$
$e_{\underline{i}\,\underline{i}k} = e_{i\underline{j}\,\underline{j}} = e_{\underline{k}j\underline{k}} = 0$	For two like indices it must be zero: $e_{111} = e_{222} = e_{112} = e_{121} = e_{211} = e_{122} = e_{221} = \ldots = 0$

Ordered (i,j,k)

$e_{ijk} = e_{jki} = e_{kij} = 1$: i j k are even permutation of (1,2,3)
$e_{kji} = e_{ikj} = e_{jik} = -1$: i j k are odd permutation of (1,2,3)

1.2.5.
Forms of Generalized Delta or Isotropic Identity:

Define a structural variable with:

P_{Even} Indices are even permutation of (1,2,3,…,N)

$$P^{P1\,P2\,P3\,....\,PN} = \Pi^{P1\,P2\,P3\,....\,PN}{}_{1\,2\,3\,....\,N}$$

And:

R_{Odd} Indices are odd permutation of (1,2,3,…,N)

$$R_{Q1\,Q2\,Q3\,....\,QN} = \Theta^{1\,2\,3\,...N}{}_{Q1\,Q2\,Q3\,..\,QN}$$

1.2.5.1. Definition of Generalized Delta In Terms of Isotropic Identity:

Generalized Delta, Δ is defined in terms of previously defined Isotropic Identity :

$$\Delta^{P1\,P2\,P3\,......\,PN}{}_{Q1\,Q2\,Q3\,......\,QN} = (e^{P1\,P2\,P3\,......\,PN}) * (e_{Q1\,Q2\,Q3\,.......\,QN})$$

And two sets of indices:

(P1 P2 P3 …. PN) $1 \leq PI \leq E$ $1 \leq I \leq N \leq E$

(Q1 Q2 Q3 …. QN) $1 \leq QJ \leq E$ $1 \leq J \leq N \leq E$

Where each of PI or QJ runs from 1 to E.

A Skew-Symmetric Isotropic Identity in both:
(P1 P2 P3 …. PN) and (Q1 Q2 Q3 …. QN):
When:

If (P1 P2 …. PN) by an *even* permutation is equal (Q1 Q2 …. QN) then: $\Delta^{P1\,P2\,P3\,....\,PN}{}_{Q1\,Q2\,....\,QN} = +1$

If (P1 P2 …. PN) by an *odd* permutation is equal (Q1 Q2 …. QN) then: $\Delta^{P1\,P2\,P3\,....\,PN}{}_{Q1\,Q2\,....\,QN} = -1$

1.2.5.2. Isotropic Identity In Terms of Ordered Delta:

Isotropic Identity In Terms of the ordered generalized Delta:

$$e^{P1\,P2\,P3\,....\,PN} = \Delta^{P1\,P2\,P3\,....\,PN}{}_{1\,2\,3\,....\,N} \qquad \Big| \qquad e_{Q1\,Q2\,Q3\,....\,QN} = \Delta^{1\,2\,3\,...N}{}_{Q1\,Q2\,Q3\,..\,QN}$$

1.2.5.3. Definition of Generalized Delta In Terms of Two Ordered Delta:

Generalized Delta

$$\Delta^{P1\,P2\,P3\,......\,PN}{}_{Q1\,Q2\,Q3\,......\,QN}$$
$$=$$
$$(e^{P1\,P2\,P3\,......\,PN}) * (e_{Q1\,Q2\,Q3\,.......\,QN})$$
$$=$$
$$(\Delta^{P1\,P2\,P3\,....\,PN}{}_{1\,2\,3\,....\,N})\,(\Delta^{1\,2\,3\,...N}{}_{Q1\,Q2\,Q3\,..\,QN})$$

1.2.5.4. Illustrations:

N = 1 and E = 3:

$$\Delta^1{}_1 = \Delta^2{}_2 = \Delta^3{}_3 = +1$$
$$\Delta^1{}_2 = \Delta^2{}_1 = \Delta^3{}_1 = -1$$

$$\Delta^{P1}_{Q1} = (\Delta^{P1}_{1})(\Delta^{1}_{Q1})$$

N = 2 and E = 3:

$$\Delta^{11}_{IJ} = \Delta^{22}_{IJ} = \Delta^{33}_{IJ} = \Delta^{12}_{23} = \Delta^{32}_{21} = ... = 0$$
$$\Delta^{12}_{12} = \Delta^{21}_{21} = \Delta^{31}_{31} = ... = +1$$
$$\Delta^{12}_{21} = \Delta^{21}_{12} = \Delta^{31}_{13} = ... = -1$$

$$\Delta^{P1\,P2}_{Q1\,Q2} = (\Delta^{P1\,P2}_{12})(\Delta^{12}_{Q1\,Q2}) = (e^{P1\,P2}e_{12})(e^{12}e_{Q1\,Q2})$$

$$e^{11}e_{IJ} = e^{22}e_{IJ} = e^{33}e_{IJ} = e^{12}e_{23} = e^{32}e_{21} = ... = 0$$
$$e^{12}e_{12} = e^{21}e_{21} = e^{31}e_{31} = ... = +1$$
$$e^{12}e_{21} = e^{21}e_{12} = e^{31}e_{13} = ... = -1$$

N = 3 and E = 3:

$$\Delta^{11L}_{IJK} = \Delta^{22L}_{IJK} = \Delta^{33L}_{IJK} = \Delta^{12L}_{23K} = \Delta^{32L}_{21K} = ... = 0$$
$$\Delta^{123}_{123} = \Delta^{213}_{213} = \Delta^{312}_{312} = ... = +1$$
$$\Delta^{123}_{213} = \Delta^{213}_{123} = \Delta^{312}_{132} = ... = -1$$

$$\Delta^{P1\,P2\,P3}_{Q1\,Q2\,Q3} = (\Delta^{P1\,P2\,P3}_{123})(\Delta^{123}_{Q1\,Q2\,Q3}) = (e^{P1\,P2\,P3}e_{123})(e^{123}e_{Q1\,Q2\,Q3})$$

$$e^{11L}e_{IJK} = e^{22L}e_{IJK} = e^{33L}e_{IJK} = e^{12L}e_{23K} = e^{32L}e_{21K} = ... = 0$$
$$e^{123}e_{123} = e^{213}e_{213} = e^{312}e_{312} = ... = +1$$
$$e^{123}e_{213} = e^{213}e_{123} = e^{312}e_{132} = ... = -1$$

1.2.6. Contraction of Generalized Delta:

Generalized Delta Δ in terms of the defined Isotropic Identity :

In the General Case:

$$\Delta^{P1\,P2\,P3\,......\,PN}_{Q1\,Q2\,Q3\,......\,QN} = (e^{P1\,P2\,P3\,......\,PN}) * (e_{Q1\,Q2\,Q3\,......\,QN})$$

While Keeping J scripts uncontracted and it is wanted to contract the rest of *(K - J) scripts, from J + 1 to K*, that is:

$$\Delta^{P1P2P3...PJ}_{Q1Q2Q3...QJ}\ (N - J)! = (N - K)!\ \Delta^{P1P2P3...PJ\ PJ+1...PK}_{Q1Q2Q3...QJ\ PJ+1...PK}$$

J K N |

$$\Delta^{P1P2P3...PJ}_{Q1Q2Q3...QJ}$$
$$=$$
$$(N - K)!/(N - J)!\ \Delta^{P1\,P2\,P3...\,PJ\,PJ+1...PK}_{Q1Q2Q3...QJ\,PJ+1...PK}$$
$$=$$
$$(N - K)!/(N - J)!\ (e^{P1\,P2\,P3...\,PJ\,PJ+1...PK})(e_{Q1Q2Q3...QJ\,PJ+1...PK})$$

1.2.6.1. Contraction in Three Dimensioal Space:

J = 2, K = 3 and N = 3:

$$\Delta^{ABM}_{\quad M} = \Delta^{AB1}_{\quad 1} + \Delta^{AB2}_{\quad 2} + \Delta^{AB3}_{\quad 3} \equiv \Delta^{AB}$$

2 3 3 | $$\Delta^{AB} = (3 - 3)!/(3 - 2)!\ \Delta^{ABM}_{\quad M} = 1\ \Delta^{ABM}_{\quad M}$$

J = 1, K = 2 and N = 3: Kronecker Delta:

$$\Delta^{AM}_{\ BM} = \Delta^{A1}_{\ B1} + \Delta^{A2}_{\ B2} + \Delta^{A3}_{\ B3} = 2\Delta^{A}_{\ B}$$

1 2 3 | $$\Delta^{A}_{\ B} = (3 - 2)!/(3 - 1)!\ \Delta^{AM}_{\ BM} = 1/2\ \Delta^{AM}_{\ BM}$$

For A = B = 1

$$\Delta^{1M}_{\ 1M} = \Delta^{11}_{\ 11} + \Delta^{12}_{\ 12} + \Delta^{13}_{\ 13} = 0 + 1 + 1 = 2\Delta^{1}_{\ 1}$$

For A = 1 and B = 2

$$\Delta^{1M}_{\ 2M} = \Delta^{11}_{\ 21} + \Delta^{12}_{\ 22} + \Delta^{13}_{\ 23} = 0 + 0 + 0 = 2\Delta^{1}_{\ 2}$$

So:

$$\Delta^{A}_{\ B} = 1/2\ \Delta^{AM}_{\ BM} = 1/2(\Delta^{A1}_{\ B1} + \Delta^{A2}_{\ B2} + \Delta^{A3}_{\ B3})$$

If: A = 1 then:

$$\Delta^{1}_{\ B} = 1/2(\Delta^{11}_{\ B1} + \Delta^{12}_{\ B2} + \Delta^{13}_{\ B3})$$

$$\Delta^{1}_{\ B} = 1/2(0 + \Delta^{12}_{\ B2} + \Delta^{13}_{\ B3})$$

This vanishes unless B = 1

$$\Delta^{1}_{\ 1} = 0$$

Similarly:

If: A = 2 then:

$$\Delta^{2}_{\ B} = 1/2(\Delta^{21}_{\ B1} + \Delta^{22}_{\ B2} + \Delta^{23}_{\ B3})$$

$$\Delta^{2}_{\ B} = 1/2(\Delta^{21}_{\ B1} + 0 + \Delta^{23}_{\ B3})$$

This vanishes unless B = 2

$$\Delta^{2}_{\ 2} = 0$$

Similarly:

If: A = 3 then:

$$\Delta^{3}_{\ B} = 1/2(\Delta^{31}_{\ B1} + \Delta^{32}_{\ B2} + \Delta^{33}_{\ B3})$$

$$\Delta^{3}_{\ B} = 1/2(\Delta^{31}_{\ B1} + \Delta^{32}_{\ B2} + 0)$$

This vanishes unless B = 3

$$\Delta^{3}_{\ 3} = 0$$

So:

$$\Delta^{A}_{\ B} = 1 \text{ if } A = B$$

$$\Delta^{A}_{\ B} = 0 \text{ if } A \neq B$$

This Special Case Will represented By:

$$\delta^{A}_{\ B} = 1 \text{ if } A = B$$

$$\delta^{A}_{\ B} = 0 \text{ if } A \neq B$$

It is Called Kronecker Delta.

J = 0, K = 1 and N = 3:

$$\Delta^{M}_{\ M} = \Delta^{1}_{\ 1} + \Delta^{2}_{\ 2} + \Delta^{3}_{\ 3} = 3\Delta = 3$$

So:

0 1 3 | $\Delta = 1 = (3-1)!/3!\ \Delta^{M}_{M} = 1/3\ \Delta^{M}_{M}$

1.2.6.2. Contraction in Special Cases:

Special Case: K = N:

Keeping J scripts and Contraction the Rest of *(N - J) scripts, from J + 1 to N*:

$$\Delta^{P1\,P2\,P3\,...\,PJ}_{Q1\,Q2\,Q3\,...\,QJ}(N-J)! = (N-N)!\,\Delta^{P1\,P2\,P3\,..\,PJ\,PJ+1\,...\,PN}_{Q1\,Q2\,Q3\,...\,QJ\,PJ+1\,...\,PN}$$

Or:

$$\Delta^{P1\,P2\,P3\,...\,PJ}_{Q1\,Q2\,Q3\,...\,QJ}(N-J)! = \Delta^{P1\,P2\,P3\,...\,PJ\,PJ+1\,...\,PN}_{Q1\,Q2\,Q3\,...\,QJ\,PJ+1\,...\,PN}$$

J N N | $\Delta^{P1P2P3...PJ}_{Q1Q2Q3...QJ} = 1/(N-J)!\ \Delta^{P1P2P3...PJ\,PJ+1...PN}_{Q1Q2Q3...QJ\,PJ+1...PN}$

Special Case: J = 0:

Keeping None J = 0 scripts and Contraction all of *(K) scripts, from 1 to K*:

$$(N-0)! = (N-K)!\,\Delta^{P1\,P2\,P3\,...\,PK}_{P1\,P2\,P3\,...\,PK}$$

Or:

$$(N)! = (N-K)!\ \Delta^{P1\,P2\,P3\,...\,PK}_{P1\,P2\,P3\,...\,PK}$$

Or:

$$1 = (N-K)!/(N)!\ \Delta^{P1\,P2\,P3\,...\,PK}_{P1\,P2\,P3\,...\,PK}$$

0 K N | $1 = (N-K)!/(N)!\ \Delta^{P1\,P2\,P3\,...\,PK}_{P1\,P2\,P3\,...\,PK}$

Or:

0 K N | $(N)!/(N-K)! = \Delta^{P1\,P2\,P3\,...\,PK}_{P1\,P2\,P3\,...\,PK} = e^{P1\,P2\,P3\,......\,PK}\ e_{P1\,P2\,P3\,...\,PK}$

Special Case: J = 0 and K = N:

Keeping None J = 0 scripts and Contraction all of (N) scripts, from 1 to K = N:

$$(N-0)! = (N-N)!\,\Delta^{P1\,P2\,P3\,...\,PN}_{P1\,P2\,P3\,...\,PN}$$

Or:

$$1 = 1/(N)!\ \Delta^{P1\,P2\,P3\,...\,PN}_{P1\,P2\,P3\,...\,PN}$$

0 N N | $1 = 1/(N)!\ \Delta^{P1\,P2\,P3\,...\,PN}_{P1\,P2\,P3\,...\,PN}$

Or:

0 K N | $(N)! = \Delta^{P1\,P2\,P3\,...\,PN}_{P1\,P2\,P3\,...\,PN} = e^{P1\,P2\,P3\,.....PN}\ e_{P1\,P2\,P3\,.......\,PN}$

1.2.6.3. Summary of (Δ) Contraction:

J	K	N	$\Delta^{P1P2P3...PJ}_{Q1Q2Q3...QJ} = (N-K)!/(N-J)!\Delta^{P1P2P3...PJ\,PJ+1...PK}_{Q1Q2Q3...QJ\,PJ+1...PK}$
J	N	N	$\Delta^{P1P2P3...PJ}_{Q1Q2Q3...QJ} = 1/(N-J)!\ \Delta^{P1P2P3...PJ\,PJ+1...PN}_{Q1Q2Q3...QJ\,PJ+1...PN}$
0	K	N	$1 = (N-K)!/N!\ \Delta^{P1\,P2\,P3\,...\,PK}_{P1\,P2\,P3\,...\,PK}$
0	N	N	$1 = 1/N!\ \Delta^{P1\,P2\,P3\,...\,PN}_{P1\,P2\,P3\,...\,PN}$

$$\Delta^{KIJ}_{RIJ} = e^{K\ IJ}\ e_{R\ IJ} = 2\,\Delta^{K}_{R}$$

J	K	N	
J	K	3	$\Delta^{P1...PJ}_{Q1...QJ} = (3-K)!/(3-J)!\Delta^{...PJ\,PJ+1...PK}_{...QJ\,PJ+1...PK}$
J	3	3	$\Delta^{P1...PJ}_{Q1...QJ} = (3-3)!/(3-J)!\Delta^{...PJ\,PJ+1...P3}_{...QJ\,PJ+1...P3}$
2	3	3	$\Delta^{P1P2}_{Q1Q2} = (3-3)!/(3-2)!\ \Delta^{P1\,P2\,P3}_{Q1\,Q2\,P3}$
2	3	3	$\Delta^{AB}_{\ } = 1\ \Delta^{ABN}_{N}$
1	3	3	$\Delta^{P1}_{Q1} = (3-3)!/(3-1)!\ \Delta^{P1\,P2\,P3}_{Q1\,P2\,P3}$
1	3	3	$\Delta^{P}_{Q} = 1/2\ \Delta^{PMN}_{QMN}$
0	3	3	$\Delta = 1 = (3-3)!/(3-0)!\ \Delta^{P1\,P2\,P3}_{P1\,P2\,P3}$
0	3	3	$1 = 1/6\ \Delta^{LMN}_{LMN}$
J	2	3	$\Delta^{P1...PJ}_{Q1...QJ} = (3-2)!/(3-J)!\Delta^{...PJ\,PJ+1...P2}_{...QJ\,PJ+1...P2}$
1	2	3	$\Delta^{P1}_{Q1} = (3-2)!/(3-1)!\ \Delta^{P1\,P2}_{Q1\,P2}$
1	2	3	$\Delta^{A}_{B} = 1/2\ \Delta^{AM}_{BM}$
0	2	3	$\Delta = 1 = (3-2)!/(3-0)!\ \Delta^{P1\,P2}_{P1\,P2}$
0	2	3	$1 = 1/6\ \Delta^{MN}_{MN}$
J	1	3	$\Delta^{PJ}_{QJ} = (3-1)!/(3-J)!\ \Delta^{P1}_{P1}$
0	1	3	$\Delta = 1 = (3-1)!/(3-0)!\ \Delta^{P1}_{P1}$
0	1	3	$1 = 1/3\ \Delta^{M}_{M}$
1	3	3	$\Delta^{P}_{Q} = 1/2\ \Delta^{PMN}_{QMN} = 1/2\, e^{PMN}\ e_{QMN}$
1	3	3	$\Delta^{P}_{P} = 1/2\ \Delta^{PMN}_{PMN}$
0	1	3	$1 = 1/3\ \Delta^{P}_{P}$
0	3	3	$1 = 1/6\ \Delta^{LMN}_{LMN}$

$$\Delta^{KMN}_{RMN} = e^{K\ MN}\ e_{R\ MN} = 2\,\Delta^{K}_{R}$$

1.2.7. Product of Two Isotropic Identities:

N = 1 and E = 3:

$$\Delta^{1}_{1} = \Delta^{2}_{2} = \Delta^{3}_{3} = +1$$

$$\Delta^{1}_{2} = \Delta^{2}_{1} = \Delta^{3}_{1} = -1$$

N = 2 and E = 3:

$$\Delta^{11}_{IJ} = \Delta^{22}_{IJ} = \Delta^{33}_{IJ} = \Delta^{12}_{23} = \Delta^{32}_{21} = ... = 0$$

$$\Delta^{12}_{12} = \Delta^{21}_{21} = \Delta^{31}_{31} = ... = +1$$

$$\Delta^{12}_{21} = \Delta^{21}_{12} = \Delta^{31}_{13} = ... = -1$$

Equivalently:

$$e^{11} e_{IJ} = e^{22} e_{IJ} = e^{33} e_{IJ} = e^{12} e_{23} = e^{32} e_{21} = \dots = 0$$
$$e^{12} e_{12} = e^{21} e_{21} = e^{31} e_{31} = \dots = +1$$
$$e^{12} e_{21} = e^{21} e_{12} = e^{31} e_{13} = \dots = -1$$

N = 3 and E = 3:

$$\Delta^{11L}_{IJK} = \Delta^{22L}_{IJK} = \Delta^{33L}_{IJK} = \Delta^{12L}_{23K} = \Delta^{32L}_{21K} = \dots = 0$$
$$\Delta^{123}_{123} = \Delta^{213}_{213} = \Delta^{312}_{312} = \dots = +1$$
$$\Delta^{123}_{213} = \Delta^{213}_{123} = \Delta^{312}_{132} = \dots = -1$$

Equivalently:

$$e^{11L} e_{IJK} = e^{22L} e_{IJK} = e^{33L} e_{IJK} = e^{12L} e_{23K} = e^{32L} e_{21K} = \dots = 0$$
$$e^{123} e_{123} = e^{213} e_{213} = e^{312} e_{312} = \dots = +1$$
$$e^{123} e_{213} = e^{213} e_{123} = e^{312} e_{132} = \dots = -1$$

The Isotropic Identity in terms of the defined Generalized Delta:

$e^{P1\,P2\,P3} = \Delta^{P1\,P2\,P3}_{123}$	$e_{Q1\,Q2\,Q3\,\dots\,QN} = \Delta^{123}_{Q1\,Q2\,Q3}$

So:

Generalized Delta

$$\Delta^{P1\,P2\,P3}_{Q1\,Q2\,Q3} = (\Delta^{P1\,P2\,P3}_{123})(\Delta^{123}_{Q1\,Q2\,Q3}) = (e^{P1\,P2\,P3})(e_{Q1\,Q2\,Q3})$$

J K N	$\Delta^{P1P2P3\dots PJ}_{Q1Q2Q3\dots QJ}$ $=$ $(N-K)!/(N-J)!\ \Delta^{P1\,P2\,P3\dots\,PJ\,PJ+1\dots PK}_{Q1Q2Q3\dots QJ\,PJ+1\dots PK}$ $=$ $(N-K)!/(N-J)!\ (e^{P1\,P2\,P3\dots\,PJ\,PJ+1\dots PK})(e_{Q1Q2Q3\dots QJ\,PJ+1\dots PK})$
J 3 3	$\Delta^{P1P2P3\dots PJ}_{Q1Q2Q3\dots QJ}$ $=$ $1/(3-J)!\ \Delta^{\dots PJ\,PJ+1\dots P3}_{\dots QJ\,PJ+1\dots P3}$ $=$ $1/(3-J)!\ (e^{\dots PJ\,PJ+1\dots P3})(e_{\dots QJ\,PJ+1\dots P3})$

3 3 3	$\Delta^{P1\,P2\,P3}_{Q1\,Q2\,Q3} = 1/1\ e^{P1\,P2\,P3}\, e_{Q1\,Q2\,Q3}$
2 3 3	$\Delta^{P1\,P2}_{Q1\,Q2} = 1/1\ e^{P1\,P2\,P3}\, e_{Q1\,Q2\,P3}$
1 3 3	$\Delta^{P1}_{Q1} = 1/2\ e^{P1\,P2\,P3}\, e_{Q1\,P2\,P3}$
0 3 3	$\Delta = 1/6\ e^{P1\,P2\,P3}\, e_{P1\,P2\,P3}$

Isotropic Triplet Identity: N = 3

Zeros unless:

$e_{123} = e_{231} = e_{312} = -e_{321} = -e_{213} = -e_{132} = 1$	$e^{123} = e^{231} = e^{312} = -e^{321} = -e^{213} = -e^{132} = 1$

$e_{ijk}\, e^{pqr}$ is equal zeros unless:

	$e_{123} = 1$	$e_{231} = 1$	$e_{312} = 1$	$e_{321} = -1$	$e_{213} = -1$	$e_{132} = -1$
$e^{123} = 1$	+1	+1	+1	-1	-1	-1
$e^{231} = 1$	+1	+1	+1	-1	-1	-1
$e^{312} = 1$	+1	+1	+1	-1	-1	-1
$e^{321} = -1$	-1	-1	-1	+1	+1	+1

$e^{213} = -1$	-1	-1	-1	+1	+1	+1
$e^{132} = -1$	-1	-1	-1	+1	+1	+1

These consist of six groups:

First group of diagonally symmetric of (+ 1):

	$e_{123} = 1$	$e_{231} = 1$	$e_{312} = 1$	$e_{321} = -1$	$e_{213} = -1$	$e_{132} = -1$
$e^{123} = 1$	1					
$e^{231} = 1$		1				
$e^{312} = 1$			1			
$e^{321} = -1$				1		
$e^{213} = -1$					1	
$e^{132} = -1$						1

Similar Second group of diagonally symmetric of (- 1):

	$e_{123} = 1$	$e_{231} = 1$	$e_{312} = 1$	$e_{321} = -1$	$e_{213} = -1$	$e_{132} = -1$
$e^{123} = 1$				- 1		
$e^{231} = 1$					- 1	
$e^{312} = 1$						- 1
$e^{321} = -1$	- 1					
$e^{213} = -1$		- 1				
$e^{132} = -1$			- 1			

Third group of diagonally skew-Symmetric of (+ 1):

	$e_{123} = 1$	$e_{231} = 1$	$e_{312} = 1$	$e_{321} = -1$	$e_{213} = -1$	$e_{132} = -1$
$e^{123} = 1$			1			
$e^{231} = 1$	1					
$e^{312} = 1$		1				
$e^{321} = -1$					1	
$e^{213} = -1$						1
$e^{132} = -1$				1		

Similar Fourth group of diagonally skew-Symmetric of (+ 1):

	$e_{123} = 1$	$e_{231} = 1$	$e_{312} = 1$	$e_{321} = -1$	$e_{213} = -1$	$e_{132} = -1$
$e^{123} = 1$		1				
$e^{231} = 1$			1			
$e^{312} = 1$	1					
$e^{321} = -1$						1
$e^{213} = -1$				1		
$e^{132} = -1$					1	

Fifth group of diagonally symmetric of alternating (- 1):

	$e_{123} = 1$	$e_{231} = 1$	$e_{312} = 1$	$e_{321} = -1$	$e_{213} = -1$	$e_{132} = -1$
$e^{123} = 1$						- 1
$e^{231} = 1$				- 1		
$e^{312} = 1$					- 1	
$e^{321} = -1$		- 1				

$e^{213} = -1$			- 1			
$e^{132} = -1$	- 1					

Similar Sixth group of diagonally symmetric of alternating (- 1):

	$e_{123} = 1$	$e_{231} = 1$	$e_{312} = 1$	$e_{321} = -1$	$e_{213} = -1$	$e_{132} = -1$
$e^{123} = 1$					- 1	
$e^{231} = 1$						- 1
$e^{312} = 1$				- 1		
$e^{321} = -1$			- 1			
$e^{213} = -1$	- 1					
$e^{132} = -1$		- 1				

Some properties of Isotropic Triplet Identity: N = 3

$$e^{IJK} e_{IQR} = e^{1JK} e_{1QR} + e^{2JK} e_{2QR} + e^{3JK} e_{3QR}$$

1 3 3 | $$\Delta^{P1}_{Q1} = 1/2\ e^{P1\,P2\,P3} e_{Q1\,P2\,P3}$$

Or:

$$e^{IJK} e_{IJR} = e^{12K} e_{12R} + e^{23K} e_{23R} + e^{31K} e_{31R} + e^{21K} e_{21R} + e^{32K} e_{32R} + e^{13K} e_{13R} =$$
$$2\,(e^{12K} e_{12R} + e^{23K} e_{23R} + e^{31K} e_{31R}) = 2\,\Delta^{K}_{R}$$

0 3 3 | $$\Delta = 1/6\ e^{P1\,P2\,P3} e_{P1\,P2\,P3}$$

Or:

$$e^{ijk} e_{ijk} = 6$$

$e^{LMN} e_{LMN} = 6$ Is sum of diagonal values

	$e_{123} = 1$	$e_{231} = 1$	$e_{312} = 1$	$e_{321} = -1$	$e_{213} = -1$	$e_{132} = -1$
$e^{123} = 1$	1					
$e^{231} = 1$		1				
$e^{312} = 1$			1			
$e^{321} = -1$				1		
$e^{213} = -1$					1	
$e^{132} = -1$						1

Multiplied by their corresponding $\delta^{i}_{p}\,\delta^{j}_{q}\,\delta^{k}_{r}$:

	$e_{123} = 1$	$e_{231} = 1$	$e_{312} = 1$	$e_{321} = -1$	$e_{213} = -1$	$e_{132} = -1$
$e^{123} = 1$	1 * 1 * 1					
$e^{231} = 1$		1 * 1 * 1				
$e^{312} = 1$			1 * 1 * 1			
$e^{321} = -1$				1 * 1 * 1		
$e^{213} = -1$					1 * 1 * 1	
$e^{132} = -1$						1 * 1 * 1

Sum of products:

$$e_{ijk}\, e^{pqr}\, \delta^{i}_{p}\, \delta^{j}_{q}\, \delta^{k}_{r} = e_{ijk}\, e^{ijk} = 6$$

$$1/6 \quad e_{ijk} \; e^{pqr} \; \delta^{i}_{p} \; \delta^{j}_{q} \; \delta^{k}_{r} = 1$$

1.2.8.
Multiplication of Generalized Delta by Symmetric a Symmetric Variable:

If:

$$\Delta^{Q1\,Q2\,Q3\,\ldots\,QJ}_{R1\,R2\,R3\ldots\,RJ}$$

Multiplied by a symmetric A:

$A^{P1\,P2\,P3\,\ldots\ldots\,PJ}_{Q1\,Q2\,Q3\,\ldots\ldots\,QJ\,QJ+1\ldots QK}$ which is symmetric in two or more subscripts,

Then:

$$(\Delta^{Q1\,Q2\,Q3\,\ldots\,QJ}_{R1\,R2\,R3\ldots\,RJ})(A^{P1\,P2\,P3\,\ldots\ldots\,PJ}_{Q1\,Q2\,Q3\,\ldots\ldots\,QJ\,QJ+1\ldots QK}) = 0$$

If:

$$\Delta^{P1\,P2\,P3\,\ldots\ldots\,PJ}_{Q1\,Q2\,Q3\,\ldots\ldots\,QJ}$$

Multiplied by a symmetric A:

$A^{Q1\,Q2\,Q3\ldots QJ}_{R1\,R2\,R3\ldots\,RJ\,RJ+1\,\ldots RK}$ which is symmetric in two or more superscripts,

Then:

$$(\Delta^{P1\,P2\,P3\,\ldots\ldots\,PJ}_{Q1\,Q2\,Q3\,\ldots\ldots\,QJ})\;(A^{Q1\,Q2\,Q3\ldots QJ}_{R1\,R2\,R3\ldots\,RJ\,RJ+1\,\ldots RK}) = 0$$

1.2.9.
Isotropic Quatret Identity:

Isotropic 4 Identity:

$$e^{R\,PQ} \; e_{RIJ} = \Delta^{PQ}_{IJ} \equiv \delta^{P}_{I}\;\delta^{Q}_{J} - \delta^{Q}_{I}\;\delta^{P}_{J}$$

δ^{P}_{I}	δ^{P}_{J}
δ^{Q}_{I}	δ^{Q}_{J}

Quatret of Bases:

$\Delta^{PQ}_{IJ} = 0$ Except in what following cases:

$(P \neq Q)$ $\quad \Delta^{\underline{PQ}}_{QP} = -1$	
	$\Delta^{\underline{KL}}_{KL} = +1$ $\quad (K \neq L)$

1.2.9.1.
Three D Isotropic Quatret Identity:

$3^4 = 81$ values are detailed here:

(6 + 6 = 12:)

$\Delta^{12}_{21} = \Delta^{21}_{12} = \Delta^{23}_{32} = \Delta^{32}_{23} = \Delta^{31}_{13} = \Delta^{13}_{31} = -1$ (6 values of - 1)	
	$\Delta^{12}_{12} = \Delta^{21}_{21} = \Delta^{23}_{23} = \Delta^{32}_{32} = \Delta^{31}_{31} = \Delta^{13}_{13} = +1$ (6 values of + 1)

Plus:

(27 + 27 = 54:)

	$\Delta^{\underline{11}}_{IJ} = \Delta^{\underline{22}}_{IJ} = \Delta^{\underline{33}}_{IJ} = 0$ (3x9 values)

$\Delta^{PQ}{}_{\underline{11}} = \Delta^{PQ}{}_{\underline{22}} = \Delta^{PQ}{}_{\underline{33}} = 0$

(3x9 values)

Minus:

The following (9 in common values)

$(\Delta^{\underline{11}}{}_{11} = \Delta^{\underline{11}}{}_{22} = \Delta^{\underline{11}}{}_{33} = \Delta^{\underline{22}}{}_{11} = \Delta^{\underline{22}}{}_{22} = \Delta^{\underline{22}}{}_{33} = \Delta^{\underline{33}}{}_{11} = \Delta^{\underline{33}}{}_{22} = \Delta^{\underline{33}}{}_{33})$

Or:

$(\Delta^{11}{}_{\underline{11}} = \Delta^{11}{}_{\underline{22}} = \Delta^{11}{}_{\underline{33}} = \Delta^{22}{}_{\underline{11}} = \Delta^{22}{}_{\underline{22}} = \Delta^{22}{}_{\underline{33}} = \Delta^{33}{}_{\underline{11}} = \Delta^{33}{}_{\underline{22}} = \Delta^{33}{}_{\underline{33}})$

Plus:

One upper index and upper index are not alke: (24 + 24 = 48:)

$\Delta^{13}{}_{21} = \Delta^{31}{}_{21} = \Delta^{23}{}_{21} = \Delta^{32}{}_{21} \ldots = 0$ (6x4)

$\Delta^{13}{}_{12} = \Delta^{31}{}_{12} = \Delta^{23}{}_{12} = \Delta^{32}{}_{12} \ldots = 0$ (6x4)

Minus (24 in common values)

The total:

(12 + 54 - 9 + 48 – 24 = 81)

1.2.9.2. Two D Isotropic Quatret Identity:

$2^4 = 16$ values values are detailed here:

(2 + 2 = 4:)

$\Delta^{12}{}_{21} = \Delta^{21}{}_{12} = -1$

(2 values)

$\Delta^{12}{}_{12} = \Delta^{21}{}_{21} = +1$

(2 values)

Plus:

(8 + 8 = 16:)

$\Delta^{\underline{11}}{}_{IJ} = \Delta^{\underline{22}}{}_{IJ} = 0$

(2x4 values)

$\Delta^{PQ}{}_{\underline{11}} = \Delta^{PQ}{}_{\underline{22}} = 0$

(2x4 values)

Minus (4 in common values)

$(\Delta^{\underline{11}}{}_{11} = \Delta^{\underline{11}}{}_{22} = \Delta^{\underline{22}}{}_{11} = \Delta^{\underline{22}}{}_{22})$

The total:

(12 + 54 - 9 + 48 – 24 = 81)

$\Delta^{12}{}_{21} = \Delta^{21}{}_{12} = -1$

$\Delta^{11}{}_{11} = \Delta^{11}{}_{22} = \Delta^{22}{}_{11} = \Delta^{22}{}_{22} = \Delta^{11}{}_{12} = \Delta^{11}{}_{21} = \Delta^{22}{}_{12} = \Delta^{22}{}_{21} = 0$

$\Delta^{11}{}_{11} = \Delta^{11}{}_{22} = \Delta^{22}{}_{11} = \Delta^{22}{}_{22} = \Delta^{12}{}_{11} = \Delta^{21}{}_{11} = \Delta^{12}{}_{22} = \Delta^{21}{}_{22} = 0$

$\Delta^{12}{}_{12} = \Delta^{21}{}_{21} = +1$

1.2.9.3. One D Isotropic Quatret Identity:

One dimension:

$1^4 = 1$ value = 2 values - 1 common value

$\Delta^{11}{}_{11} = 0$

$\Delta^{11}{}_{11} = 0$

1.2.9.4. Applications:

Application 1:

$$\delta^J{}_K\delta^K{}_L\delta^L{}_M\delta^M{}_N =$$

Answer

$$\delta^J{}_K\delta^K{}_L\delta^L{}_M\delta^M{}_N = \delta^J{}_L\delta^L{}_M\delta^M{}_N = \delta^J{}_M\delta^M{}_N = \delta^J{}_N$$

Application 2:

$$\delta^M{}_M =$$

Answer

$$\delta^M{}_M = \delta^1{}_1 + \delta^2{}_2 + \delta^3{}_3 = 1 + 1 + 1 = 3$$

Application 3:

$$\delta^M{}_N\delta^M{}_N =$$

Answer

$$\delta^M{}_N\delta^M{}_N = \delta^M{}_N\delta^N{}_M = \delta^M{}_M = 3$$

Or

$$\delta^M{}_N\delta^M{}_N = \delta^N{}_M\,\delta^M{}_N = \delta^N{}_N = 3$$

Application 4:

$$\delta^J{}_K\delta^K{}_L\delta^L{}_M\delta^M{}_J =$$

Answer

$$\delta^J{}_K\delta^K{}_L\delta^L{}_M\delta^M{}_J = \delta^J{}_J = 3$$

Application 5:

$$\delta^J{}_K u^K =$$

Answer

$$\delta^J{}_K u^K = \delta^J{}_1 u^1 + \delta^J{}_2 u^2 + \delta^J{}_3 u^3 = u^J$$

Application 6:

$$\delta_{KM}\, u^K v^M =$$

Answer

$$\delta_{KM}\, u^K v^M = \delta_{11} u^1 v^1 + \delta_{22} u^2 v^2 + \delta_{33} u^3 v^3 = u^1 v^1 + u^2 v^2 + u^3 v^3$$

Application 7:

$$\Delta_{KM}\, u^K v^M w^Q =$$

Answer

$$\Delta_{KM}\, u^K v^M w^Q = (\Delta_{KM} u^K v^M)\, w^Q = (u^1 v^1 + u^2 v^2 + u^3 v^3)\, w^Q$$

Define $$\Delta^{PQ}{}_{IJ} \equiv \delta^{P}{}_{I}\,\delta^{Q}{}_{J} - \delta^{Q}{}_{I}\,\delta^{P}{}_{J}$$

Application 8:

Anti symmetric about contra-diagonal

$$\Delta^{PQ}{}_{IJ} = -\Delta^{QP}{}_{IJ}$$

Answer

$$\Delta^{PQ}{}_{IJ} = \delta^{P}{}_{I}\,\delta^{Q}{}_{J} - \delta^{Q}{}_{I}\,\delta^{P}{}_{J}$$

$$\Delta^{QP}{}_{IJ} = \delta^{Q}{}_{I}\,\delta^{P}{}_{J} - \delta^{P}{}_{I}\,\delta^{Q}{}_{J}$$

$$= -\Delta^{PQ}{}_{IJ}$$

So

$$\Delta^{PQ}{}_{IJ} = -\Delta^{QP}{}_{IJ}$$

Similarly

$$\Delta_{IJ}{}^{PQ} = -\Delta_{IJ}{}^{QP}$$

Answer

$$\Delta_{IJ}{}^{PQ} = \Delta_{I}{}^{P}\,\Delta_{J}{}^{Q} - \Delta_{I}{}^{Q}\,\Delta_{J}{}^{P}$$

$$\Delta_{IJ}{}^{QP} = \Delta_{I}{}^{Q}\,\Delta_{J}{}^{P} - \Delta_{I}{}^{P}\,\Delta_{J}{}^{Q}$$

$$= -\Delta_{IJ}{}^{PQ}$$

So

$$\Delta_{IJ}{}^{PQ} = -\Delta_{IJ}{}^{QP}$$

Such as

$$\Delta_{12}{}^{12} = -\Delta_{12}{}^{21}$$
$$\Delta_{21}{}^{21} = -\Delta_{21}{}^{12}$$
$$\Delta_{12}{}^{23} = -\Delta_{12}{}^{32}$$
$$\Delta_{21}{}^{32} = -\Delta_{21}{}^{23}$$
$$\Delta_{12}{}^{31} = -\Delta_{12}{}^{13}$$
$$\Delta_{21}{}^{13} = -\Delta_{21}{}^{31}$$

$$\Delta_{23}{}^{23} = -\Delta_{23}{}^{32}$$
$$\Delta_{32}{}^{32} = -\Delta_{32}{}^{23}$$
$$\Delta_{23}{}^{31} = -\Delta_{23}{}^{13}$$
$$\Delta_{32}{}^{13} = -\Delta_{32}{}^{31}$$
$$\Delta_{23}{}^{12} = -\Delta_{23}{}^{21}$$
$$\Delta_{32}{}^{21} = -\Delta_{32}{}^{12}$$

$$\Delta_{31}{}^{31} = -\Delta_{31}{}^{13}$$
$$\Delta_{13}{}^{13} = -\Delta_{13}{}^{31}$$
$$\Delta_{31}{}^{12} = -\Delta_{31}{}^{21}$$
$$\Delta_{13}{}^{21} = -\Delta_{13}{}^{12}$$
$$\Delta_{31}{}^{23} = -\Delta_{31}{}^{32}$$
$$\Delta_{13}{}^{32} = -\Delta_{13}{}^{23}$$

Application 9:

Anti symmetric about co-diagonal

$$\Delta^{PQ}{}_{IJ} = -\Delta^{PQ}{}_{JI}$$

Similarly

$$\Delta_{IJ}{}^{PQ} = -\Delta_{JI}{}^{PQ}$$

Answer

$$\Delta_{IJ}{}^{PQ} = \delta_{I}{}^{P}\,\delta_{J}{}^{Q} - \delta_{I}{}^{Q}\,\delta_{J}{}^{P}$$

$$\Delta_{JI}{}^{PQ} = \delta_{J}{}^{P}\,\delta_{I}{}^{Q} - \delta_{J}{}^{Q}\,\delta_{I}{}^{P}$$

$$= \delta_{I}{}^{Q}\,\delta_{J}{}^{P} - \delta_{I}{}^{P}\,\delta_{J}{}^{Q}$$

$$= -\Delta_{IJ}{}^{PQ}$$

So

$$\Delta_{IJ}{}^{PQ} = -\Delta_{JI}{}^{PQ}$$

Such as

$$\Delta_{12}{}^{12} = -\Delta_{21}{}^{12}$$

$$\Delta_{23}{}^{23} = -\Delta_{32}{}^{23}$$

$$\Delta_{31}{}^{31} = -\Delta_{13}{}^{31}$$

$$\Delta_{21}^{\ \ 21} = -\Delta_{12}^{\ \ 21}$$
$$\Delta_{12}^{\ \ 23} = -\Delta_{21}^{\ \ 23}$$
$$\Delta_{21}^{\ \ 32} = -\Delta_{12}^{\ \ 32}$$
$$\Delta_{12}^{\ \ 31} = -\Delta_{21}^{\ \ 31}$$
$$\Delta_{21}^{\ \ 13} = -\Delta_{12}^{\ \ 13}$$

$$\Delta_{32}^{\ \ 32} = -\Delta_{23}^{\ \ 32}$$
$$\Delta_{23}^{\ \ 31} = -\Delta_{32}^{\ \ 31}$$
$$\Delta_{32}^{\ \ 13} = -\Delta_{23}^{\ \ 13}$$
$$\Delta_{23}^{\ \ 12} = -\Delta_{32}^{\ \ 12}$$
$$\Delta_{32}^{\ \ 21} = -\Delta_{23}^{\ \ 21}$$

$$\Delta_{13}^{\ \ 13} = -\Delta_{31}^{\ \ 13}$$
$$\Delta_{31}^{\ \ 12} = -\Delta_{13}^{\ \ 12}$$
$$\Delta_{13}^{\ \ 21} = -\Delta_{31}^{\ \ 21}$$
$$\Delta_{31}^{\ \ 23} = -\Delta_{13}^{\ \ 23}$$
$$\Delta_{13}^{\ \ 32} = -\Delta_{31}^{\ \ 32}$$

Application 10:

Anti symmetric about contra-diagonal and about co-diagonal:
Symmetric: Anti symmetric * Anti symmetric

$$\Delta^{QP}_{\ \ IJ} = \Delta^{PQ}_{\ \ JI}$$

Similarly

$$\Delta_{IJ}^{\ \ QP} = \Delta_{JI}^{\ \ PQ}$$

Answer

Since

$$\Delta_{IJ}^{\ \ PQ} = -\Delta_{IJ}^{\ \ QP}$$
$$\Delta_{IJ}^{\ \ PQ} = -\Delta_{JI}^{\ \ PQ}$$

So

$$\Delta_{IJ}^{\ \ QP} = \Delta_{JI}^{\ \ PQ}$$

Such as

$$\Delta_{12}^{\ \ 12} = \Delta_{21}^{\ \ 21}$$
$$\Delta_{12}^{\ \ 21} = \Delta_{21}^{\ \ 12}$$
$$\Delta_{12}^{\ \ 23} = \Delta_{21}^{\ \ 32}$$
$$\Delta_{12}^{\ \ 32} = \Delta_{21}^{\ \ 23}$$
$$\Delta_{12}^{\ \ 31} = \Delta_{21}^{\ \ 13}$$
$$\Delta_{12}^{\ \ 13} = \Delta_{21}^{\ \ 31}$$

$$\Delta_{23}^{\ \ 23} = \Delta_{32}^{\ \ 32}$$
$$\Delta_{23}^{\ \ 32} = \Delta_{32}^{\ \ 23}$$
$$\Delta_{23}^{\ \ 31} = \Delta_{32}^{\ \ 13}$$
$$\Delta_{23}^{\ \ 13} = \Delta_{32}^{\ \ 31}$$
$$\Delta_{23}^{\ \ 12} = \Delta_{32}^{\ \ 21}$$
$$\Delta_{23}^{\ \ 21} = \Delta_{32}^{\ \ 12}$$

$$\Delta_{31}^{\ \ 31} = \Delta_{13}^{\ \ 13}$$
$$\Delta_{31}^{\ \ 13} = \Delta_{13}^{\ \ 31}$$
$$\Delta_{31}^{\ \ 12} = \Delta_{13}^{\ \ 21}$$
$$\Delta_{31}^{\ \ 21} = \Delta_{13}^{\ \ 12}$$
$$\Delta_{31}^{\ \ 23} = \Delta_{13}^{\ \ 32}$$
$$\Delta_{31}^{\ \ 32} = \Delta_{13}^{\ \ 23}$$

Application 11:

Symmetric about co-diagonal and contra-diagonal

$$\Delta^{PQ}_{\ \ IJ} = \Delta^{IJ}_{\ \ PQ}$$

Similarly

$$\Delta_{IJ}^{\ \ PQ} = \Delta_{PQ}^{\ \ IJ}$$

Answer

$$\Delta_{IJ}^{\ \ PQ} = \delta_I^{\ P}\,\delta_J^{\ Q} - \delta_I^{\ Q}\,\delta_J^{\ P}$$
$$\Delta_{PQ}^{\ \ IJ} = \delta_P^{\ I}\,\delta_Q^{\ J} - \delta_P^{\ J}\,\delta_Q^{\ I}$$

But δ is symmetric

$$\Delta_{PQ}^{\ \ IJ} = \delta_I^{\ P}\,\delta_J^{\ Q} - \delta_J^{\ P}\,\delta_I^{\ Q}$$
$$= \delta_I^{\ P}\,\delta_J^{\ Q} - \delta_I^{\ Q}\,\delta_J^{\ P}$$
$$= \Delta_{IJ}^{\ \ PQ}$$

So

$$\Delta_{IJ}^{\ \ PQ} = \Delta_{PQ}^{\ \ IJ}$$

Such as

$\Delta_{12}^{21} = \Delta_{21}^{12}$ $\Delta_{23}^{32} = \Delta_{32}^{23}$ $\Delta_{31}^{13} = \Delta_{13}^{31}$

$\Delta_{12}^{32} = \Delta_{32}^{12}$ $\Delta_{23}^{13} = \Delta_{13}^{23}$ $\Delta_{31}^{21} = \Delta_{21}^{31}$

$\Delta_{12}^{13} = \Delta_{13}^{12}$ $\Delta_{23}^{21} = \Delta_{21}^{23}$ $\Delta_{31}^{32} = \Delta_{32}^{31}$

Application 12:

$$\Delta^{PQ}_{IJ} = \Delta^{IQ}_{PJ} + \Delta^{PI}_{QJ} = \Delta^{JQ}_{IP} + \Delta^{PJ}_{IQ}$$

Similarly

$$\Delta_{IJ}^{PQ} = \Delta_{PJ}^{IQ} + \Delta_{QJ}^{PI} = \Delta_{IP}^{JQ} + \Delta_{IQ}^{PJ}$$

Divided into two parts:

First part:

$$\Delta_{IJ}^{PQ} = \Delta_{PJ}^{IQ} + \Delta_{QJ}^{PI}$$

Answer

$$\Delta_{PJ}^{IQ} = \delta_P^I \delta_J^Q - \delta_P^Q \delta_J^I$$

$$\Delta_{QJ}^{PI} = \delta_Q^P \delta_J^I - \delta_Q^I \delta_J^P$$

By Symmetry

$$\Delta_{PJ}^{IQ} + \Delta_{QJ}^{PI} = \delta_P^I \delta_J^Q - \delta_Q^I \delta_J^P$$
$$= \Delta_{IJ}^{PQ}$$

So

$$\Delta_{IJ}^{PQ} = \Delta_{PJ}^{IQ} + \Delta_{QJ}^{PI}$$

Second part:

$$\Delta_{IJ}^{PQ} = \Delta_{IP}^{JQ} + \Delta_{IQ}^{PJ}$$

Answer

$$\Delta_{IP}^{JQ} = \delta_I^J \delta_P^Q - \delta_I^Q \delta_P^J$$

$$\Delta_{IQ}^{PJ} = \delta_I^P \delta_Q^J - \delta_I^J \delta_Q^P$$

By Symmetry

$$\Delta_{IP}^{JQ} + \Delta_{IQ}^{PJ} = \delta_I^P \delta_J^Q - \delta_I^Q \delta_J^P$$
$$= \Delta_{IJ}^{PQ}$$

So

$$\Delta_{IJ}^{PQ} = \Delta_{IP}^{JQ} + \Delta_{IQ}^{PJ}$$

Application 13:

Sym * (Anti symmetric about contra-diagonal and about co-diagonal): sym * (sym)

$$\Delta^{PI}_{QJ} = \Delta^{JQ}_{IP}$$

Similarly

$$\Delta_{QJ}^{PI} = \Delta_{IP}^{JQ}$$

Since

$$\Delta_{IJ}^{PQ} = \Delta_{PJ}^{IQ} + \Delta_{QJ}^{PI} = \Delta_{IP}^{JQ} + \Delta_{IQ}^{PJ}$$

and

$$\Delta_{PJ}^{IQ} = \Delta_{IQ}^{PJ}$$

So

$$\Delta_{QJ}^{PI} = \Delta_{IP}^{JQ}$$

Or

$$\Delta_{QJ}^{PI} = - \Delta_{JQ}^{PI} = - (- \Delta_{JQ}^{IP}) = \Delta_{JQ}^{IP} = \Delta_{IP}^{JQ}$$

So

$$\Delta_{QJ}^{PI} = \Delta_{IP}^{JQ}$$

Application 14:

Diagonal co vanish

If:(I = J no sum on: repeated I or repeated J) $\Delta^{..}_{II} = \Delta^{PQ}_{II} = 0$

Similarly

If:(I = J no sum on: repeated I or repeated J) $\Delta_{II}^{..} = \Delta_{II}^{PQ} = 0$

Answer

$$\Delta_{IJ}^{PQ} = \delta_I^P \delta_J^Q - \delta_I^Q \delta_J^P$$

$$\Delta_{II}^{PQ} = \delta_I^P \delta_I^Q - \delta_I^Q \delta_I^P = 0$$

Application 15:

Diagonal contra vanish

If:(P = Q no sum on: repeated P or repeated Q) $\Delta^{PP}_{..} = \Delta^{PQ}_{IJ} = 0$ Similarly

If:(P = Q no sum on: repeated P or repeated Q) $\Delta^{PP}_{..} = \Delta_{IJ}^{PQ} = 0$

Answer

$$\Delta_{IJ}^{PQ} = \delta_I^P \delta_J^Q - \delta_I^Q \delta_J^P$$

$$\Delta_{IJ}^{PP} = \delta_I^P \delta_J^P - \delta_I^P \delta_J^P = 0$$

Application 16:

$\Delta^{IJ}_{IJ} = 1$ if:(I ≠ J) and (P ≠ Q) and(I J = P Q) (I = P) and(J = Q)

Similarly

$\Delta_{IJ}^{IJ} = 1$ if:(I ≠ J) and (P ≠ Q) and(I J = P Q) (I = P) and(J = Q)

Answer

$$\Delta_{IJ}^{PQ} = \delta_I^P \delta_J^Q - \delta_I^Q \delta_J^P$$

$$\Delta_{IJ}^{IJ} = \delta_I^I \delta_J^J - \delta_I^J \delta_J^I = 1 * 1 - 0 * 0 = 1$$

Application 17:

$\Delta^{JI}_{IJ} = -1$ if:(I ≠ J) and (P ≠ Q) and(I J = Q P) (I = Q) and(J = P)

Similarly

$\Delta_{IJ}^{JI} = -1$ if:(I ≠ J) and (P ≠ Q) and(I J = Q P) (I = Q) and(J = P)

Answer

$$\Delta_{IJ}^{PQ} = \delta_I^P \delta_J^Q - \delta_I^Q \delta_J^P$$

$$\Delta_{IJ}^{JI} = \delta_I^J \delta_J^I - \Delta_I^I \Delta_J^J = 0 * 0 - 1 * 1 = - 1$$

Application 18:

$$\text{Sum of } \Delta^{IJ}_{IJ} = 6$$

Similarly

$$\text{Sum of } \Delta_{IJ}^{IJ} = 6$$

Answer

$$\Delta_{IJ}^{PQ} = \delta_I^P \delta_J^Q - \delta_I^Q \delta_J^P$$

$$\text{Sum of } \Delta_{IJ}^{IJ} = \Delta_{12}^{12} + \Delta_{23}^{23} + \Delta_{31}^{31} + \Delta_{21}^{21} + \Delta_{32}^{32} + \Delta_{13}^{13} = 6$$

Application 19:

$$\text{Sum of } \Delta^{JI}_{IJ} = -6$$

Similarly

$$\text{Sum of } \Delta_{IJ}^{JI} = -6$$

Answer

$$\Delta_{IJ}^{PQ} = \delta_I^P \delta_J^Q - \delta_I^Q \delta_J^P$$

$$\text{Sum of } \Delta_{IJ}^{JI} = \Delta_{12}^{21} + \Delta_{23}^{32} + \Delta_{31}^{13} + \Delta_{21}^{12} + \Delta_{32}^{23} + \Delta_{13}^{31} = -6$$

1.3.
Matrices:

1.3.1.
Operation on Matrices:

Trace of a matrix:

$$T_k^k$$

Product of matrices:

$(R)(S) = R_M^k S_k^N = T_M^N$	$(R)(S) = R^M_k S^k_N = T^M_N$
$(R\,S^S)_p^i = R_p^k S_i^k$ row p * row i	$(R\,S)_p^i = R_p^k S_k^i$ row p * col i
$(R^S S^S)_p^i = R_k^i S_p^k$ col i * row p	$(R^S S)_p^i = R_k^p S_k^i$ col p * col i

Product of identical matrices:

$(T\,T^T)_P^J = T_P^k T_J^k$ row P * row J	$(T\,T)_P^J = T_P^k T_k^J$ row P * col J
$(T^T T^T)_P^J = T_k^J T_P^k$ col J * row P	$(T^T T)_P^J = T_k^P T_k^J$ col P * col J

Squares sum and Crosses sum

$tr(T\,T^T) = T_P^K T_P^K$ squares sum	$tr(T\,T) = T_P^K T_K^P$ crosses sum
$tr(T^T T^T) = T_K^P T_P^K$ crosses sum	$tr(T^T T) = T_K^P T_K^P$ squares sum

Perpendicular matrices:

$(s)(t) = s_m^i t_i^n = \delta_m^n$	$(s)(t) = s^m_i t^i_n = \delta^m_n$
$\delta_N^M = 1$ if:$(N = M)$	$\delta^M_N = 1$ if:$(N = M)$
$\delta_N^M = 0$ if:$(N \neq M)$	$\delta^M_N = 0$ if:$(N \neq M)$

Determinant of E raws by E columns matrix s:

$RE \equiv |R|$

Inverse of R is R^{-1}:

$R_M^{\ i} (R^{-1})_i^{\ N} = \delta_M^{\ N}$ | $R^M_{\ i} (R^{-1})^i_{\ N} = \delta^M_{\ N}$

Multiply both sides by RE:

$RE \ R_m^{\ i} (R^{-1})_i^{\ n} = RE \ \delta_m^{\ n}$ | $RE \ R^m_{\ i} (R^{-1})^i_{\ n} = RE \ \delta^m_{\ n}$

Cofactor of matrix R:

$S_i^{\ n} = RE \ (R^{-1})_i^{\ n}$ | $S^i_{\ n} = RE \ (R^{-1})^i_{\ n}$

R^{-1}:

$(R^{-1})_i^{\ n} = S_i^{\ n} / |R| \equiv S_i^{\ n} / RE$ | $(R^{-1})^i_{\ n} = S^i_{\ n} / |R| \equiv S^i_{\ n} / RE$

Then:

$R_m^{\ i} S_i^{\ n} / RE \equiv R_m^{\ i} S_i^{\ n} / |R| = \delta_m^{\ n}$ | $R^m_{\ i} S^i_{\ n} / RE \equiv R^m_{\ i} S^i_{\ n} / |R| = \delta^m_{\ n}$

S cofactor of R

Cofactor (transposed signed minor) of $R_i^{\ j}$ is $S_j^{\ i}$ such that:

$R_k^{\ n} S_n^{\ i} = RE \ \delta_k^{\ i}$ expand $R_k^{\ n}$ fixed k | $R^k_{\ l} S^l_{\ i} = RE \ \delta^k_{\ i}$ expand $R^k_{\ l}$ fixed k

Notice:

$S_l^{\ m} R_m^{\ j} = RE \ \delta_l^{\ j}$ expand $R_m^{\ j}$ fixed j | $S^l_{\ m} R^m_{\ j} = RE \ \delta^l_{\ j}$ expand $R^m_{\ j}$ fixed j

But:

$\delta_n^{\ n} = N$ | $\delta^n_{\ n} = N$

RE can be written in two forms

$RE = R_K^{\ \underline{M}} S_{\underline{M}}^{\ K} = 1/N \ R_K^{\ M} S_M^{\ K}$ | $RE = R^{\underline{M}}_{\ K} S^K_{\ \underline{M}} = 1/N \ R^M_{\ K} S^K_{\ M}$

1.3.2. Determinant of 3 by 3 Matrix:

The concept of sum over (e) indices is very important

$e_{ijk} e^{pqr} R_p^{\ i} R_q^{\ j} R_r^{\ k}$ | $e_{ijk} e^{pqr} R^i_{\ p} R^j_{\ q} R^k_{\ r}$

All components are detailed consequentally, on different levels:

I, J, K, P, Q, R

Dtailed sum over I

$$= + e_{1JK} e^{PQR} R_P^{\ 1} R_Q^{\ J} R_R^{\ K} + e_{2JK} e^{PQR} R_P^{\ 2} R_Q^{\ J} R_R^{\ K} + e_{3JK} e^{PQR} R_P^{\ 3} R_Q^{\ J} R_R^{\ K}$$

$$= + e_{1JK} e^{PQR} R^1_{\ P} R^J_{\ Q} R^K_{\ R} + e_{2JK} e^{PQR} R^2_{\ P} R^J_{\ Q} R^K_{\ R} + e_{3JK} e^{PQR} R^3_{\ P} R^J_{\ Q} R^K_{\ R}$$

Dtailed sum over I , J

$$= + e_{11K} e^{PQR} R_P^{\ 1} R_Q^{\ 1} R_R^{\ K} + e_{21K} e^{PQR} R_P^{\ 2} R_Q^{\ 1} R_R^{\ K} + e_{31K} e^{PQR} R_P^{\ 3} R_Q^{\ 1} R_R^{\ K}$$
$$+ e_{12K} e^{PQR} R_P^{\ 1} R_Q^{\ 2} R_R^{\ K} + e_{22K} e^{PQR} R_P^{\ 2} R_Q^{\ 2} R_R^{\ K} + e_{32K} e^{PQR} R_P^{\ 3} R_Q^{\ 2} R_R^{\ K}$$

$$= + e_{11K} e^{PQR} R^1_{\ P} R^1_{\ Q} R^K_{\ R} + e_{21K} e^{PQR} R^2_{\ P} R^1_{\ Q} R^K_{\ R} + e_{31K} e^{PQR} R^3_{\ P} R^1_{\ Q} R^K_{\ R}$$
$$+ e_{12K} e^{PQR} R^1_{\ P} R^2_{\ Q} R^K_{\ R} + e_{22K} e^{PQR} R^2_{\ P} R^2_{\ Q} R^K_{\ R} + e_{32K} e^{PQR} R^3_{\ P} R^2_{\ Q} R^K_{\ R}$$

$$
\begin{aligned}
&+\\
&+ e_{13K}\, e^{PQR} R^{1}_{P} R^{3}_{Q} R^{K}_{R}\\
&+ e_{23K}\, e^{PQR} R^{2}_{P} R^{3}_{Q} R^{K}_{R}\\
&+ e_{33K}\, e^{PQR} R^{3}_{P} R^{3}_{Q} R^{K}_{R}
\end{aligned}
$$

Dtailed sum over I , J, K

$$
\begin{aligned}
&=\\
&+ e_{211}\, e^{PQR} R^{2}_{P} R^{1}_{Q} R^{1}_{R}\\
&+ e_{311}\, e^{PQR} R^{3}_{P} R^{1}_{Q} R^{1}_{R}\\
&+ e_{121}\, e^{PQR} R^{1}_{P} R^{2}_{Q} R^{1}_{R}\\
&+ e_{321}\, e^{PQR} R^{3}_{P} R^{2}_{Q} R^{1}_{R}\\
&+ e_{131}\, e^{PQR} R^{1}_{P} R^{3}_{Q} R^{1}_{R}\\
&+ e_{231}\, e^{PQR} R^{2}_{P} R^{3}_{Q} R^{1}_{R}\\
&+\\
&+ e_{212}\, e^{PQR} R^{2}_{P} R^{1}_{Q} R^{2}_{R}\\
&+ e_{312}\, e^{PQR} R^{3}_{P} R^{1}_{Q} R^{2}_{R}\\
&+ e_{122}\, e^{PQR} R^{1}_{P} R^{2}_{Q} R^{2}_{R}\\
&+ e_{322}\, e^{PQR} R^{3}_{P} R^{2}_{Q} R^{2}_{R}\\
&+ e_{132}\, e^{PQR} R^{1}_{P} R^{3}_{Q} R^{2}_{R}\\
&+ e_{232}\, e^{PQR} R^{2}_{P} R^{3}_{Q} R^{2}_{R}\\
&+\\
&+ e_{213}\, e^{PQR} R^{2}_{P} R^{1}_{Q} R^{3}_{R}\\
&+ e_{313}\, e^{PQR} R^{3}_{P} R^{1}_{Q} R^{3}_{R}\\
&+ e_{123}\, e^{PQR} R^{1}_{P} R^{2}_{Q} R^{3}_{R}\\
&+ e_{323}\, e^{PQR} R^{3}_{P} R^{2}_{Q} R^{3}_{R}\\
&+ e_{133}\, e^{PQR} R^{1}_{P} R^{3}_{Q} R^{3}_{R}\\
&+ e_{233}\, e^{PQR} R^{2}_{P} R^{3}_{Q} R^{3}_{R}
\end{aligned}
$$

$$
\begin{aligned}
&+\\
&+ e_{13K}\, e^{PQR} R^{1}_{P} R^{3}_{Q} R^{K}_{R}\\
&+ e_{23K}\, e^{PQR} R^{2}_{P} R^{3}_{Q} R^{K}_{R}\\
&+ e_{33K}\, e^{PQR} R^{3}_{P} R^{3}_{Q} R^{K}_{R}
\end{aligned}
$$

$$
\begin{aligned}
&=\\
&+ e_{211}\, e^{PQR} R^{2}_{P} R^{1}_{Q} R^{1}_{R}\\
&+ e_{311}\, e^{PQR} R^{3}_{P} R^{1}_{Q} R^{1}_{R}\\
&+ e_{121}\, e^{PQR} R^{1}_{P} R^{2}_{Q} R^{1}_{R}\\
&+ e_{321}\, e^{PQR} R^{3}_{P} R^{2}_{Q} R^{1}_{R}\\
&+ e_{131}\, e^{PQR} R^{1}_{P} R^{3}_{Q} R^{1}_{R}\\
&+ e_{231}\, e^{PQR} R^{2}_{P} R^{3}_{Q} R^{1}_{R}\\
&+\\
&+ e_{212}\, e^{PQR} R^{2}_{P} R^{1}_{Q} R^{2}_{R}\\
&+ e_{312}\, e^{PQR} R^{3}_{P} R^{1}_{Q} R^{2}_{R}\\
&+ e_{122}\, e^{PQR} R^{1}_{P} R^{2}_{Q} R^{2}_{R}\\
&+ e_{322}\, e^{PQR} R^{3}_{P} R^{2}_{Q} R^{2}_{R}\\
&+ e_{132}\, e^{PQR} R^{1}_{P} R^{3}_{Q} R^{2}_{R}\\
&+ e_{232}\, e^{PQR} R^{2}_{P} R^{3}_{Q} R^{2}_{R}\\
&+\\
&+ e_{213}\, e^{PQR} R^{2}_{P} R^{1}_{Q} R^{3}_{R}\\
&+ e_{313}\, e^{PQR} R^{3}_{P} R^{1}_{Q} R^{3}_{R}\\
&+ e_{123}\, e^{PQR} R^{1}_{P} R^{2}_{Q} R^{3}_{R}\\
&+ e_{323}\, e^{PQR} R^{3}_{P} R^{2}_{Q} R^{3}_{R}\\
&+ e_{133}\, e^{PQR} R^{1}_{P} R^{3}_{Q} R^{3}_{R}\\
&+ e_{233}\, e^{PQR} R^{2}_{P} R^{3}_{Q} R^{3}_{R}
\end{aligned}
$$

So on:

Dtailed sum over I , J, K , P

Dtailed sum over I , J, K , P, Q

Dtailed sum over I , J, K , P, Q, R

Collecting non - zero components:

$$
\begin{aligned}
&e_{IJK}\, e^{PQR} R^{I}_{P} R^{J}_{Q} R^{K}_{R}\\
&=\\
&+ R^{3}_{3} R^{2}_{2} R^{1}_{1}\\
&+ (-1) R^{2}_{3} R^{3}_{2} R^{1}_{1}\\
&+ (-1) R^{3}_{3} R^{1}_{2} R^{2}_{1}\\
&+ R^{1}_{3} R^{3}_{2} R^{2}_{1}\\
&+ R^{2}_{3} R^{1}_{2} R^{3}_{1}\\
&+ (-1) R^{1}_{3} R^{2}_{2} R^{3}_{1}\\
&+ (-1) R^{3}_{2} R^{2}_{3} R^{1}_{1}\\
&+ R^{2}_{2} R^{3}_{3} R^{1}_{1}\\
&+ R^{3}_{2} R^{1}_{3} R^{2}_{1}
\end{aligned}
$$

$$
\begin{aligned}
&e_{IJK}\, e^{PQR} R^{I}_{P} R^{J}_{Q} R^{K}_{R}\\
&=\\
&+ R^{3}_{3} R^{2}_{2} R^{1}_{1}\\
&+ (-1) R^{2}_{3} R^{3}_{2} R^{1}_{1}\\
&+ (-1) R^{3}_{3} R^{1}_{2} R^{2}_{1}\\
&+ R^{1}_{3} R^{3}_{2} R^{2}_{1}\\
&+ R^{2}_{3} R^{1}_{2} R^{3}_{1}\\
&+ (-1) R^{1}_{3} R^{2}_{2} R^{3}_{1}\\
&+ (-1) R^{3}_{2} R^{2}_{3} R^{1}_{1}\\
&+ R^{2}_{2} R^{3}_{3} R^{1}_{1}\\
&+ R^{3}_{2} R^{1}_{3} R^{2}_{1}
\end{aligned}
$$

$+(-1)\ R_2^1 R_3^3 R_1^2$	$+(-1)\ R^1_{\ 2} R^3_{\ 3} R^2_{\ 1}$
$+(-1)\ R_2^2 R_3^1 R_1^3$	$+(-1)\ R^2_{\ 2} R^1_{\ 3} R^3_{\ 1}$
$+\ R_2^1 R_3^2 R_1^3$	$+\ R^1_{\ 2} R^2_{\ 3} R^3_{\ 1}$
$+(-1)\ R_3^3 R_1^2 R_2^1$	$+(-1)\ R^3_{\ 3} R^2_{\ 1} R^1_{\ 2}$
$+\ R_3^2 R_1^3 R_2^1$	$+\ R^2_{\ 3} R^3_{\ 1} R^1_{\ 2}$
$+\ R_3^3 R_1^1 R_2^2$	$+\ R^3_{\ 3} R^1_{\ 1} R^2_{\ 2}$
$+(-1)\ R_3^1 R_1^3 R_2^2$	$+(-1)\ R^1_{\ 3} R^3_{\ 1} R^2_{\ 2}$
$+(-1)\ R_3^2 R_1^1 R_2^3$	$+(-1)\ R^2_{\ 3} R^1_{\ 1} R^3_{\ 2}$
$+\ R_3^1 R_1^2 R_2^3$	$+\ R^1_{\ 3} R^2_{\ 1} R^3_{\ 2}$
$+\ R_1^3 R_3^2 R_2^1$	$+\ R^3_{\ 1} R^2_{\ 3} R^1_{\ 2}$
$+(-1)\ R_1^2 R_3^3 R_2^1$	$+(-1)\ R^2_{\ 1} R^3_{\ 3} R^1_{\ 2}$
$+(-1)\ R_1^3 R_3^1 R_2^2$	$+(-1)\ R^3_{\ 1} R^1_{\ 3} R^2_{\ 2}$
$+\ R_1^1 R_3^3 R_2^2$	$+\ R^1_{\ 1} R^3_{\ 3} R^2_{\ 2}$
$+\ R_1^2 R_3^1 R_2^3$	$+\ R^2_{\ 1} R^1_{\ 3} R^3_{\ 2}$
$+(-1)\ R_1^1 R_3^2 R_2^3$	$+(-1)\ R^1_{\ 1} R^2_{\ 3} R^3_{\ 2}$
$+\ R_2^3 R_1^2 R_3^1$	$+\ R^3_{\ 2} R^2_{\ 1} R^1_{\ 3}$
$+(-1)\ R_2^2 R_1^3 R_3^1$	$+(-1)\ R^2_{\ 2} R^3_{\ 1} R^1_{\ 3}$
$+(-1)\ R_2^3 R_1^1 R_3^2$	$+(-1)\ R^3_{\ 2} R^1_{\ 1} R^2_{\ 3}$
$+\ R_2^1 R_1^3 R_3^2$	$+\ R^1_{\ 2} R^3_{\ 1} R^2_{\ 3}$
$+\ R_2^2 R_1^1 R_3^3$	$+\ R^2_{\ 2} R^1_{\ 1} R^3_{\ 3}$
$+(-1)\ R_2^1 R_1^2 R_3^3$	$+(-1)\ R^1_{\ 2} R^2_{\ 1} R^3_{\ 3}$
$+(-1)\ R_1^3 R_2^2 R_3^1$	$+(-1)\ R^3_{\ 1} R^2_{\ 2} R^1_{\ 3}$
$+\ R_1^2 R_2^3 R_3^1$	$+\ R^2_{\ 1} R^3_{\ 2} R^1_{\ 3}$
$+\ R_1^3 R_2^1 R_3^2$	$+\ R^3_{\ 1} R^1_{\ 2} R^2_{\ 3}$
$+(-1)\ R_1^1 R_2^3 R_3^2$	$+(-1)\ R^1_{\ 1} R^3_{\ 2} R^2_{\ 3}$
$+(-1)\ R_1^2 R_2^1 R_3^3$	$+(-1)\ R^2_{\ 1} R^1_{\ 2} R^3_{\ 3}$
$+\ R_1^1 R_2^2 R_3^3$	$+\ R^1_{\ 1} R^2_{\ 2} R^3_{\ 3}$

Construct the following 36 values of given matrix $R^P_{\ Q}$:

$R^1_{\ 1} R^2_{\ 2} R^3_{\ 3}$	$R^1_{\ 2} R^2_{\ 3} R^3_{\ 1}$	$R^1_{\ 3} R^2_{\ 1} R^3_{\ 2}$	$R^1_{\ 3} R^2_{\ 2} R^3_{\ 1}$	$R^1_{\ 2} R^2_{\ 1} R^3_{\ 3}$	$R^1_{\ 1} R^2_{\ 3} R^3_{\ 2}$
$R^2_{\ 1} R^3_{\ 2} R^1_{\ 3}$	$R^2_{\ 2} R^3_{\ 3} R^1_{\ 1}$	$R^2_{\ 3} R^3_{\ 1} R^1_{\ 2}$	$R^2_{\ 3} R^3_{\ 2} R^1_{\ 1}$	$R^2_{\ 2} R^3_{\ 1} R^1_{\ 3}$	$R^2_{\ 1} R^3_{\ 3} R^1_{\ 2}$
$R^3_{\ 1} R^1_{\ 2} R^2_{\ 3}$	$R^3_{\ 2} R^1_{\ 3} R^2_{\ 1}$	$R^3_{\ 3} R^1_{\ 1} R^2_{\ 2}$	$R^3_{\ 3} R^1_{\ 2} R^2_{\ 1}$	$R^3_{\ 2} R^1_{\ 1} R^2_{\ 3}$	$R^3_{\ 1} R^1_{\ 3} R^2_{\ 2}$
$R^3_{\ 1} R^2_{\ 2} R^1_{\ 3}$	$R^3_{\ 2} R^2_{\ 3} R^1_{\ 1}$	$R^3_{\ 3} R^2_{\ 1} R^1_{\ 2}$	$R^3_{\ 3} R^2_{\ 2} R^1_{\ 1}$	$R^3_{\ 2} R^2_{\ 1} R^1_{\ 3}$	$R^3_{\ 1} R^2_{\ 3} R^1_{\ 2}$
$R^2_{\ 1} R^1_{\ 2} R^3_{\ 3}$	$R^2_{\ 2} R^1_{\ 3} R^3_{\ 1}$	$R^2_{\ 3} R^1_{\ 1} R^3_{\ 2}$	$R^2_{\ 3} R^1_{\ 2} R^3_{\ 1}$	$R^2_{\ 2} R^1_{\ 1} R^3_{\ 3}$	$R^2_{\ 1} R^1_{\ 3} R^3_{\ 2}$
$R^1_{\ 1} R^3_{\ 2} R^2_{\ 3}$	$R^1_{\ 2} R^3_{\ 3} R^2_{\ 1}$	$R^1_{\ 3} R^3_{\ 1} R^2_{\ 2}$	$R^1_{\ 3} R^3_{\ 2} R^2_{\ 1}$	$R^1_{\ 2} R^3_{\ 1} R^2_{\ 3}$	$R^1_{\ 1} R^3_{\ 3} R^2_{\ 2}$

Multiply these 36 values by $e_{ijk}\, e^{pqr}$ to get six groups:

First group:

	$e^{123} = 1$	$e^{231} = 1$	$e^{312} = 1$	$e^{321} = -1$	$e^{213} = -1$	$e^{132} = -1$
$e_{123} = 1$	$R^1_{\ 1} R^2_{\ 2} R^3_{\ 3}$					

$e_{231}=1$		$R^2_2R^3_3R^1_1$				
$e_{312}=1$			$R^3_3R^1_1R^2_2$			
$e_{321}=-1$				$R^3_3R^2_2R^1_1$		
$e_{213}=-1$					$R^2_2R^1_1R^3_3$	
$e_{132}=-1$						$R^1_1R^3_3R^2_2$

Second group:

	$e^{123}=1$	$e^{231}=1$	$e^{312}=1$	$e^{321}=-1$	$e^{213}=-1$	$e^{132}=-1$
$e_{123}=1$				$-R^1_3R^2_2R^3_1$		
$e_{231}=1$					$-R^2_2R^3_1R^1_3$	
$e_{312}=1$						$-R^3_1R^1_3R^2_2$
$e_{321}=-1$	$-R^3_1R^2_2R^1_3$					
$e_{213}=-1$		$-R^2_2R^1_3R^3_1$				
$e_{132}=-1$			$-R^1_3R^3_1R^2_2$			

Third group:

	$e^{123}=1$	$e^{231}=1$	$e^{312}=1$	$e^{321}=-1$	$e^{213}=-1$	$e^{132}=-1$
$e_{123}=1$			$R^1_3R^2_1R^3_2$			
$e_{231}=1$	$R^2_1R^3_2R^1_3$					
$e_{312}=1$		$R^3_2R^1_3R^2_1$				
$e_{321}=-1$					$R^3_2R^2_1R^1_3$	
$e_{213}=-1$						$R^2_1R^1_3R^3_2$
$e_{132}=-1$				$R^1_3R^3_2R^2_1$		

Fourth group:

	$e_{123}=1$	$e_{231}=1$	$e_{312}=1$	$e_{321}=-1$	$e_{213}=-1$	$e_{132}=-1$
$e_{123}=1$		$R^1_2R^2_3R^3_1$				
$e_{231}=1$			$R^2_3R^3_1R^1_2$			
$e_{312}=1$	$R^3_1R^1_2R^2_3$					
$e_{321}=-1$						$R^3_1R^2_3R^1_2$
$e_{213}=-1$				$R^2_3R^1_2R^3_1$		
$e_{132}=-1$					$R^1_2R^3_1R^2_3$	

Fifth group:

	$e^{123}=1$	$e^{231}=1$	$e^{312}=1$	$e^{321}=-1$	$e^{213}=-1$	$e^{132}=-1$
$e^{123}=1$						$-R^1_1R^2_3R^3_2$
$e^{231}=1$				$-R^2_3R^3_2R^1_1$		
$e^{312}=1$					$-R^3_2R^1_1R^2_3$	
$e^{321}=-1$		$-R^3_2R^2_3R^1_1$				
$e^{213}=-1$			$-R^2_3R^1_1R^3_2$			
$e^{132}=-1$	$-R^1_1R^3_2R^2_3$					

Sixth group:

$e^{123}=1$	$e^{231}=1$	$e^{312}=1$	$e^{321}=-1$	$e^{213}=-1$	$e^{132}=-1$

$e_{123} = 1$					$-R^1_2 R^2_1 R^3_3$	
$e_{231} = 1$						$-R^2_1 R^3_3 R^1_2$
$e_{312} = 1$				$-R^3_3 R^1_2 R^2_1$		
$e_{321} = -1$			$-R^3_3 R^2_1 R^1_2$			
$e_{213} = -1$	$-R^2_1 R^1_2 R^3_3$					
$e_{132} = -1$		$-R^1_2 R^3_3 R^2_1$				

Determinant of 3 by 3 Matrix:

Collect and sum all components of past groups:

	$e^{123} - 1$	$e^{231} = 1$	$e^{312} = 1$	$e^{321} = -1$	$e^{213} = -1$	$e^{132} = -1$
$e^{123} = 1$	$R^1_1 R^2_2 R^3_3$	$R^1_2 R^2_3 R^3_1$	$R^1_3 R^2_1 R^3_2$	$-R^1_3 R^2_2 R^3_1$	$-R^1_2 R^2_1 R^3_3$	$-R^1_1 R^2_3 R^3_2$
$e^{231} = 1$	$R^2_1 R^3_2 R^1_3$	$R^2_2 R^3_3 R^1_1$	$R^2_3 R^3_1 R^1_2$	$-R^2_3 R^3_2 R^1_1$	$-R^2_2 R^3_1 R^1_3$	$-R^2_1 R^3_3 R^1_2$
$e^{312} = 1$	$R^3_1 R^1_2 R^2_3$	$R^3_2 R^1_3 R^2_1$	$R^3_3 R^1_1 R^2_2$	$-R^3_3 R^1_2 R^2_1$	$-R^3_2 R^1_1 R^2_3$	$-R^3_1 R^1_3 R^2_2$
$e^{321} = -1$	$-R^3_1 R^2_2 R^1_3$	$-R^3_2 R^2_3 R^1_1$	$-R^3_3 R^2_1 R^1_2$	$R^3_3 R^2_2 R^1_1$	$R^3_2 R^2_1 R^1_3$	$R^3_1 R^2_3 R^1_2$
$e^{213} = -1$	$-R^2_1 R^1_2 R^3_3$	$-R^2_2 R^1_3 R^3_1$	$-R^2_3 R^1_1 R^3_2$	$R^2_3 R^1_2 R^3_1$	$R^2_2 R^1_1 R^3_3$	$R^2_1 R^1_3 R^3_2$
$e^{132} = -1$	$-R^1_1 R^3_2 R^2_3$	$-R^1_2 R^3_3 R^2_1$	$-R^1_3 R^3_1 R^2_2$	$R^1_3 R^3_2 R^2_1$	$R^1_2 R^3_1 R^2_3$	$R^1_1 R^3_3 R^2_2$

Equal:

$$+6 R^1_1 R^2_2 R^3_3 + 6 R^1_2 R^2_3 R^3_1 + 6 R^1_3 R^2_1 R^3_2$$
$$-6 R^1_3 R^2_2 R^3_1 - 6 R^1_2 R^2_1 R^3_3 - 6 R^1_1 R^2_3 R^3_2$$

Divide by 6 to get Determinant of R^P_Q:

(Each component is a factored product of three R values)

$$R3 = 1/6\, e_{IJK}\, e^{UVW} R^I_P R^J_Q R^K_W$$

$$=$$

$$+R^1_1 R^2_2 R^3_3 + R^1_2 R^2_3 R^3_1 + R^1_3 R^2_1 R^3_2$$
$$-R^1_3 R^2_2 R^3_1 - R^1_2 R^2_1 R^3_3 - R^1_1 R^2_3 R^3_2$$

$$=$$

$$1/6\, [\, R^I_I R^J_J R^K_K + 2\,(R^I_J R^J_K R^K_I) - 3\, R^I_I (R^J_K R^K_J)\,]$$

And:

$$R^P_I S^I_P = 3\, R^P_I S^I_{\underline{P}} = 3\ S3 \equiv 3\ ||R||$$

3 by 3 Matrix Determinant:

$$R3 = (1/6)\, e_{ijk}\, e^{pqr} R^i_p R^j_q R^k_r \qquad R3 = (1/6)\, e_{ijk}\, e^{pqr} R^i_p R^j_q R^k_r$$

To be substituted in:

$$R_k{}^n S_n{}^i = R3\, \delta_k{}^i \qquad R^k{}_l S^l{}_i = R3\ \delta^k{}_i$$

To get:

$$R_m{}^i S_i{}^n = (1/6)\, e_{ijk}\, e^{pqr} R^i_p R^j_q R^k_r\, \delta_m{}^n \qquad R^m{}_i S^i{}_n = (1/6) e_{ijk}\, e^{pqr} R^i_p R^j_q R^k_r\, \delta^m{}_n$$

Since:

$$R_m{}^i S_i{}^m = (1/6)\, e_{ijk}\, e^{pqr} R^i_p R^j_q R^k_r\, \delta_m{}^m \qquad R^m{}_i S^i{}_m = (1/6) e_{ijk}\, e^{pqr} R^i_p R^j_q R^k_r\, \delta^m{}_m$$

Where:

$$\delta_m{}^m = 3 \qquad \delta^m{}_m = 3$$

So:

$$R_m{}^i S_i{}^m = (1/2)\, e_{ijk}\, e^{pqr} R_p{}^i R_q{}^j R_r{}^k \qquad\Big|\qquad R^m{}_i S^i{}_m = (1/2) e_{ijk}\, e^{pqr} R^i{}_p R^j{}_q R^k{}_r$$

And:

$$R_m{}^i S_i{}^m = R_1{}^i S_i{}^1 + R_2{}^i S_i{}^2 + R_3{}^i S_i{}^3 = 3\, R_{\underline{m}}{}^i S_i{}^{\underline{m}} \qquad\Big|\qquad R^m{}_i S^i{}_m = R^1{}_i S^i{}_1 + R^2{}_i S^i{}_2 + R^3{}_i S^i{}_3 = 3R^{\underline{m}}{}_i S^i{}_{\underline{m}}$$

This is detailed more:

$$R_1{}^i S_i{}^1 = R_1{}^1 S_1{}^1 + R_1{}^2 S_2{}^1 + R_1{}^3 S_3{}^1 \qquad\Big|\qquad R^1{}_i S^i{}_1 = R^1{}_1 S^1{}_1 + R^1{}_2 S^2{}_1 + R^1{}_3 S^3{}_1$$

$$R_2{}^i S_i{}^2 = R_2{}^1 S_1{}^2 + R_2{}^2 S_2{}^2 + R_2{}^3 S_3{}^2 \qquad\Big|\qquad R^2{}_i S^i{}_2 = R^2{}_1 S^1{}_2 + R^2{}_2 S^2{}_2 + R^2{}_3 S^3{}_2$$

$$R_3{}^i S_i{}^3 = R_3{}^1 S_1{}^3 + R_3{}^2 S_2{}^3 + R_3{}^3 S_3{}^3 \qquad\Big|\qquad R^3{}_i S^i{}_3 = R^3{}_1 S^1{}_3 + R^3{}_2 S^2{}_3 + R^3{}_3 S^3{}_3$$

:

3 by 3 Matrix Determinant:

$$R3 = R_{\underline{m}}{}^i S_i{}^{\underline{m}} = 1/3\, R_m{}^i S_i{}^m = (1/6)\, e_{ijk}\, e^{pqr} R_p{}^i R_q{}^j R_r{}^k \qquad\Big|\qquad R3 = R^{\underline{m}}{}_i S^i{}_{\underline{m}} = 1/3\, R^m{}_i S^i{}_m = (1/6)\, e_{ijk}\, e^{pqr} R^i{}_p R^j{}_q R^k{}_r$$

:

3 by 3 Matrix Determinant:

First row cofactors::

$R_1{}^i$ * cofactors | $R^1{}_i$ * cofactors

$$R3 = R_1{}^i S_i{}^1 = R_1{}^1 S_1{}^1 + R_1{}^2 S_2{}^1 + R_1{}^3 S_3{}^1 \qquad\Big|\qquad R3 = R^1{}_i S^i{}_1 = R^1{}_1 S^1{}_1 + R^1{}_2 S^2{}_1 + R^1{}_3 S^3{}_1$$

Second row cofactors:

$R_2{}^i$ * cofactors | $R^2{}_i$ * cofactors

$$R3 = R_2{}^i S_i{}^2 = R_2{}^1 S_1{}^2 + R_2{}^2 S_2{}^2 + R_2{}^3 S_3{}^2 \qquad\Big|\qquad R3 = R^2{}_i S^i{}_2 = R^2{}_1 S^1{}_2 + R^2{}_2 S^2{}_2 + R^2{}_3 S^3{}_2$$

Third row cofactors:

$R_3{}^i$ * cofactors | $R^3{}_i$ * cofactors

$$R3 = R_3{}^i S_i{}^3 = R_3{}^1 S_1{}^3 + R_3{}^2 S_2{}^3 + R_3{}^3 S_3{}^3 \qquad\Big|\qquad R3 = R^3{}_i S^i{}_3 = R^3{}_1 S^1{}_3 + R^3{}_2 S^2{}_3 + R^3{}_3 S^3{}_3$$

In general:

$$R3 = R_{\underline{P}}{}^i S_i{}^{\underline{P}} \qquad\Big|\qquad R3 = R^{\underline{P}}{}_i S^i{}_{\underline{P}}$$

$$3 * R3 = R_P{}^i S_i{}^P\, 3 = R_{\underline{P}}{}^i S_i{}^{\underline{P}} \qquad\Big|\qquad 3 * R3 = R^P{}_i S^i{}_P = 3\, R^{\underline{P}}{}_i S^i{}_{\underline{P}}$$

1.3.2.1. Third Invariant:

Define third invariant, R3 is the determinant of $R^P{}_Q$:

Each component is a factored of three R values product:

1/6 * Sum of all components in:

	$e_{123} = 1$	$e_{231} = 1$	$e_{312} = 1$	$e_{321} = -1$	$e_{213} = -1$	$e_{132} = -1$
$e^{123} = 1$	$R^1{}_1 R^2{}_2 R^3{}_3$	$R^1{}_2 R^2{}_3 R^3{}_1$	$R^1{}_3 R^2{}_1 R^3{}_2$	$-R^1{}_3 R^2{}_2 R^3{}_1$	$-R^1{}_2 R^2{}_1 R^3{}_3$	$-R^1{}_1 R^2{}_3 R^3{}_2$
$e^{231} = 1$	$R^2{}_1 R^3{}_2 R^1{}_3$	$R^2{}_2 R^3{}_3 R^1{}_1$	$R^2{}_3 R^3{}_1 R^1{}_2$	$-R^2{}_3 R^3{}_2 R^1{}_1$	$-R^2{}_2 R^3{}_1 R^1{}_3$	$-R^2{}_1 R^3{}_3 R^1{}_2$
$e^{312} = 1$	$R^3{}_1 R^1{}_2 R^2{}_3$	$R^3{}_2 R^1{}_3 R^2{}_1$	$R^3{}_3 R^1{}_1 R^2{}_2$	$-R^3{}_3 R^1{}_2 R^2{}_1$	$-R^3{}_2 R^1{}_1 R^2{}_3$	$-R^3{}_1 R^1{}_3 R^2{}_2$
$e^{321} = -1$	$-R^3{}_1 R^2{}_2 R^1{}_3$	$-R^3{}_2 R^2{}_3 R^1{}_1$	$-R^3{}_3 R^2{}_1 R^1{}_2$	$R^3{}_3 R^2{}_2 R^1{}_1$	$R^3{}_2 R^2{}_1 R^1{}_3$	$R^3{}_1 R^2{}_3 R^1{}_2$
$e^{213} = -1$	$-R^2{}_1 R^1{}_2 R^3{}_3$	$-R^2{}_2 R^1{}_3 R^3{}_1$	$-R^2{}_3 R^1{}_1 R^3{}_2$	$R^2{}_3 R^1{}_2 R^3{}_1$	$R^2{}_2 R^1{}_1 R^3{}_3$	$R^2{}_1 R^1{}_3 R^3{}_2$
$e^{132} = -1$	$-R^1{}_1 R^3{}_2 R^2{}_3$	$-R^1{}_2 R^3{}_3 R^2{}_1$	$-R^1{}_3 R^3{}_1 R^2{}_2$	$R^1{}_3 R^3{}_2 R^2{}_1$	$R^1{}_2 R^3{}_1 R^2{}_3$	$R^1{}_1 R^3{}_3 R^2{}_2$

$$R3 = (1/6)\, e^{IJK} e^{UVW} (R_{IU} \quad R_{JV}\, R_{KW}) = 1/6\, [R_{II} R_{JJ} R_{KK} + 2\,(R_{IJ} R_{JK} R_{KI}) - 3\, R_{II} (R_{JK} R_{KJ})]$$

$$=$$

$$+ R_{11} R_{22} R_{33} + R_{12} R_{23} R_{31} + R_{13} R_{21} R_{32} - R_{11} R_{23} R_{32} - R_{13} R_{22} R_{31} - R_{12} R_{21} R_{33}$$

1.3.2.2.
Second Invariant:

Second invariant R2:
Each component is a factored product of two R values:
1/6 * Sum of all components in:

Replaced third R by δ:

	$e_{123} = 1$	$e_{231} = 1$	$e_{312} = 1$	$e_{321} = -1$	$e_{213} = -1$	$e_{132} = -1$
$e^{123} = 1$	$R^1_1 R^2_2 \delta^3_3$	$R^1_2 R^2_3 \delta^3_1$	$R^1_3 R^2_1 \delta^3_2$	$-R^1_3 R^2_2 \delta^3_1$	$-R^1_2 R^2_1 \delta^3_3$	$-R^1_1 R^2_3 \delta^3_2$
$e^{231} = 1$	$R^2_1 R^3_2 \delta^1_3$	$R^2_2 R^3_3 \delta^1_1$	$R^2_3 R^3_1 \delta^1_2$	$-R^2_3 R^3_2 \delta^1_1$	$-R^2_2 R^3_1 \delta^1_3$	$-R^2_1 R^3_3 \delta^1_2$
$e^{312} = 1$	$R^3_1 R^1_2 \delta^2_3$	$R^3_2 R^1_3 \delta^2_1$	$R^3_3 R^1_1 \delta^2_2$	$-R^3_3 R^1_2 \delta^2_1$	$-R^3_2 R^1_1 \delta^2_3$	$-R^3_1 R^1_3 \delta^2_2$
$e^{321} = -1$	$-R^3_1 R^2_2 \delta^1_3$	$-R^3_2 R^2_3 \delta^1_1$	$-R^3_3 R^2_1 \delta^1_2$	$R^3_3 R^2_2 \delta^1_1$	$R^3_2 R^2_1 \delta^1_3$	$R^3_1 R^2_3 \delta^1_2$
$e^{213} = -1$	$-R^2_1 R^1_2 \delta^3_3$	$-R^2_2 R^1_3 \delta^3_1$	$-R^2_3 R^1_1 \delta^3_2$	$R^2_3 R^1_2 \delta^3_1$	$R^2_2 R^1_1 \delta^3_3$	$R^2_1 R^1_3 \delta^3_2$
$e^{132} = -1$	$-R^1_1 R^3_2 \delta^2_3$	$-R^1_2 R^3_3 \delta^2_1$	$-R^1_3 R^3_1 \delta^2_2$	$R^1_3 R^3_2 \delta^2_1$	$R^1_2 R^3_1 \delta^2_3$	$R^1_1 R^3_3 \delta^2_2$

Showing only non-zero values:

	$e_{123} = 1$	$e_{231} = 1$	$e_{312} = 1$	$e_{321} = -1$	$e_{213} = -1$	$e_{132} = -1$
$e^{123} = 1$	$R^1_1 R^2_2$				$-R^1_2 R^2_1$	
$e^{231} = 1$		$R^2_2 R^3_3$		$-R^2_3 R^3_2$		
$e^{312} = 1$			$R^3_3 R^1_1$			$-R^3_1 R^1_3$
$e^{321} = -1$		$-R^3_2 R^2_3$		$R^3_3 R^2_2$		
$e^{213} = -1$	$-R^2_1 R^1_2$				$R^2_2 R^1_1$	
$e^{132} = -1$			$-R^1_3 R^3_1$			$R^1_1 R^3_3$

$$R2 = (1/6)\, 3\, e^{IJK} e^{UVW} (R_{IU} \quad R_{JV} . \delta_{KW}) = 1/2\, e^{IJK} e^{UVK} R_{IU} R_{IV} = 1/2\, (R_{II} R_{JJ} - R_{IK} R_{KI})$$

$$=$$

$$1/2\{\, [(R_{11}{}^2 + R_{22}{}^2 + R_{33}{}^2) + 2(R_{11} R_{22} + R_{22} R_{33} + R_{33} R_{11})] - [(R_{11}{}^2 + R_{22}{}^2 + R_{33}{}^2) + 2(R_{12} R_{21} + R_{23} R_{32} + R_{31} R_{13})]\, \}$$

$$=$$

$$(R_{11} R_{22} + R_{22} R_{33} + R_{33} R_{11}) - (R_{12} R_{21} + R_{23} R_{32} + R_{31} R_{13})$$

$R2 = (1/2\, R_{II} R_{JJ} - 1/2\, R_{IK} R_{KI})$ Difference of two positive values.

1.3.2.3.
First Invariant:

First invariant R1:
Each component is a factored product of one R values:
1/6 * Sum of all components in:

Replaced second R and third R by δ:

	$e_{123} = 1$	$e_{231} = 1$	$e_{312} = 1$	$e_{321} = -1$	$e_{213} = -1$	$e_{132} = -1$
$e^{123} = 1$	$R^1_1 \delta^2_2 \delta^3_3$	$R^1_2 \delta^2_3 \delta^3_1$	$R^1_3 \delta^2_1 \delta^3_2$	$-R^1_3 \delta^2_2 \delta^3_1$	$-R^1_2 \delta^2_1 \delta^3_3$	$-R^1_1 \delta^2_3 \delta^3_2$
$e^{231} = 1$	$R^2_1 \delta^3_2 \delta^1_3$	$R^2_2 \delta^3_3 \delta^1_1$	$R^2_3 \delta^3_1 \delta^1_2$	$-R^2_3 \delta^3_2 \delta^1_1$	$-R^2_2 \delta^3_1 \delta^1_3$	$-R^2_1 \delta^3_3 \delta^1_2$

$e^{312} = 1$	$R^3_1\delta^1_2\delta^2_3$	$R^3_2\delta^1_3\delta^2_1$	$R^3_3\delta^1_1\delta^2_2$	$-R^3_3\delta^1_2\delta^2_1$	$-R^3_2\delta^1_1\delta^2_3$	$-R^3_1\delta^1_3\delta^2_2$
$e^{321} = -1$	$-R^3_1\delta^2_2\delta^1_3$	$-R^3_2\delta^2_3\delta^1_1$	$-R^3_3\delta^2_1\delta^1_2$	$R^3_3\delta^2_2\delta^1_1$	$R^3_2\delta^2_1\delta^1_3$	$R^3_1\delta^2_3\delta^1_2$
$e^{213} = -1$	$-R^2_1\delta^1_2\delta^3_3$	$-R^2_2\delta^1_3\delta^3_1$	$-R^2_3\delta^1_1\delta^3_2$	$R^2_3\delta^1_2\delta^3_1$	$R^2_2\delta^1_1\delta^3_3$	$R^2_1\delta^1_3\delta^3_2$
$e^{132} = -1$	$-R^1_1\delta^3_2\delta^2_3$	$-R^1_2\delta^3_3\delta^2_1$	$-R^1_3\delta^3_1\delta^2_2$	$R^1_3\delta^3_2\delta^2_1$	$R^1_2\delta^3_1\delta^2_3$	$R^1_1\delta^3_3\delta^2_2$

Showing only non-zero values:

	$e_{123} = 1$	$e_{231} = 1$	$e_{312} = 1$	$e_{321} = -1$	$e_{213} = -1$	$e_{132} = -1$
$e^{123} = 1$	R^1_1					
$e^{231} = 1$		R^2_2				
$e^{312} = 1$			R^3_3			
$e^{321} = -1$				R^3_3		
$e^{213} = -1$					R^2_2	
$e^{132} = -1$						R^1_1

$$R1 = (1/6)\,3\,e^{IJK}e^{UVW}R_{IU}\delta_{JV}\delta_{KW} = 1/2\,e^{IJK}e^{UJK}R_{IU} = \delta^{IU}R_{IU} = R_{KK}$$

$$= R_{11} + R_{22} + R_{33}$$

1.3.2.4.
Zero Invariant:

Replaced first second and third R by δ:

	$e_{123} = 1$	$e_{231} = 1$	$e_{312} = 1$	$e_{321} = -1$	$e_{213} = -1$	$e_{132} = -1$
$e^{123} = 1$	δ^1_1					
$e^{231} = 1$		δ^2_2				
$e^{312} = 1$			δ^3_3			
$e^{321} = -1$				δ^3_3		
$e^{213} = -1$					δ^2_2	
$e^{132} = -1$						δ^1_1

$$R0 = 1/6\,e^{IJK}e^{UVW}\delta_{IU}\delta_{JV}\delta_{KW} = e^{IJK}e^{IJK} = (1/6)\,\delta_{KK}$$

$$= 1$$

1.3.2.5.
List of Invariant:

$$R3 = 1/6\,e^{IJK}e^{UVW}(R_{IU}\;R_{JV}\;R_{KW})$$

$$= 1/6\,[R_{II}R_{JJ}R_{KK} + 2(R_{IJ}R_{JK}R_{KI}) - 3R_{II}(R_{JK}\;R_{KJ})]$$

$$= +R_{11}R_{22}R_{33} + R_{12}R_{23}R_{31} + R_{13}R_{21}R_{32} - R_{11}R_{23}R_{32} - R_{13}R_{22}R_{31} - R_{12}R_{21}R_{33}$$

$$R2 = (1/6)\,3\,e^{IJK}e^{UVW}(R_{IU}\;R_{JV}\;.\;\delta_{KW})$$

$$= 3/6\,e^{IJK}e^{UVK}R_{IU}R_{IV} = 3/6\,(R_{II}\;R_{JJ} - R_{IK}R_{KI})$$

$$= +R_{11}R_{22} + R_{22}R_{33} + R_{33}R_{11} - R_{12}R_{21} - R_{23}R_{32} - R_{31}R_{13}$$

$$R1 = (1/6)\,3\, e^{IJK} e^{UVW} R_{IU}\,\delta_{JV}\,\delta_{KW}$$
$$=$$
$$1/2\, e^{IJK} e^{UJK} R_{IU} = \delta^{IU} R_{IU} = R_{KK}$$
$$=$$
$$R_{11} + R_{22} + R_{33}$$

$$R0 = 1/6\, e^{IJK} e^{UVW} \delta_{IU}\,\delta_{JV}\,\delta_{KW}$$
$$=$$
$$1/6\, e^{IJK} e^{IJK}$$
$$=$$
$$1$$

1.3.2.6.
Generalized Invariants:

$$RUV\ldots\mathbf{N} \equiv (R^{U}_{\;IJ} R^{V}_{\;JK} \ldots)^{N}$$

:

$R1\mathbf{N} \equiv (R_{MM})^{N}$	$R11 \equiv (R_{MM})^{1}$	$R12 \equiv (R_{MM})^{2}$	$R13 \equiv (R_{MM})^{3}$
$R2\mathbf{N} \equiv (R_{PM} R_{MP})^{N}$	$R21 \equiv (R_{PM} R_{MP})^{1}$	$R22 \equiv (R_{PM} R_{MP})^{2}$	$R23 \equiv (R_{PM} R_{MP})^{3}$
$R3\mathbf{N} \equiv (R_{IJ} R_{JK} R_{KI})_{N}$	$R31 \equiv (R_{IJ} R_{JK} R_{KI})^{1}$	$R32 \equiv (R_{IJ} R_{JK} R_{KI})^{2}$	$R33 \equiv (R_{IJ} R_{JK} R_{KI})^{2}$

List of Invariant of 3 by 3 Matrix become:

$R3 = 1/6\,[R_{II} R_{JJ} R_{KK} + 2(R_{IJ} R_{JK} R_{KI}) - 3 R_{II} (R_{JK} R_{KJ})]$ $=$ $1/6\,[R13 + 2(R31) - 3\,R11\,(R21)]$ $=$ $+R_{11}R_{22}R_{33} + R_{12}R_{23}R_{31} + R_{13}R_{21}R_{32} - R_{11}R_{23}R_{32} - R_{13}R_{22}R_{31} - R_{12}R_{21}R_{33}$
$R2 = 3/6\,(R_{II} R_{JJ} - R_{IK} R_{KI})$ $=$ $1/2\,(R12 - R21)$ $=$ $+R_{11}R_{22} + R_{22}R_{33} + R_{33}R_{11} - R_{12}R_{21} - R_{23}R_{32} - R_{31}R_{13}$
$R1$ $=$ $R11$ $=$ R_{KK}

Square of Sum of Diagonals R12, Sum of Squares R21 and Invariant R2

Square of Sum of Diagonals:

$$R12 = [(R_{11} + R_{22} + R_{33})^2 = [(R_{11}^{\;2} + R_{22}^{\;2} + R_{33}^{\;2}) + 2(R_{11} R_{22} + R_{22} R_{33} + R_{33} R_{11})]$$

And

Sum of Squares

$$R21 = R_{IK} R_{KI} = [(R_{11}^{\;2} + R_{22}^{\;2} + R_{33}^{\;2}) + 2(R_{12} R_{21} + R_{23} R_{32} + R_{31} R_{13})]$$

So:

Invariant R2:

$$R2 = 1/2\; R12 - 1/2\; R21 = (R_{11} R_{22} + R_{22} R_{33} + R_{33} R_{11}) - (R_{12} R_{21} + R_{23} R_{32} + R_{31} R_{13})$$

Square of Sum of Diagonals R12 Minus Second Invariant R2:

R21 - (R2)

=

R21 - (1/2 R12 - 1/2 R21) = 3/2 R21 - 1/2 R12

If **Square of Sum of Diagonals, R12 and Sum of Squares,** R21 are substituted in

R21 - (R2) = 3/2 R21 - 1/2 R12

=

$$3/2\,(R_{11}^{\,2}+R_{22}^{\,2}+R_{33}^{\,2}) + 3/2 * 2\,(R_{12}R_{21}+R_{23}R_{32}+R_{31}R_{13})$$

−

$$[\ 1/2\,[(R_{11}^{\,2}+R_{22}^{\,2}+R_{33}^{\,2}) + 1/2 * 2\,(R_{11}R_{22}+R_{22}R_{33}+R_{33}R_{11})\ ]$$

=

$$(R_{11}^{\,2}+R_{22}^{\,2}+R_{33}^{\,2}) + 3(R_{12}R_{21}+R_{23}R_{32}+R_{31}R_{13}) - (R_{11}R_{22}+R_{22}R_{33}+R_{33}R_{11})$$

Add

Zero relation:

$$1/2\ [(R_{11}-R_{22})^2 + (R_{22}-R_{33})^2 + (R_{33}-R_{11})^2] - (R_{11}^{\,2}+R_{22}^{\,2}+R_{33}^{\,2}) - (R_{11}R_{22}+R_{22}R_{33}+R_{33}R_{11}) = 0$$

To get:

R21 - (R2) = 3/2 R21 - 1/2 R12

=

$$(R_{11}^{\,2}+R_{22}^{\,2}+R_{33}^{\,2}) + 3\,(R_{12}R_{21}+R_{23}R_{32}+R_{31}R_{13}) - (R_{11}R_{22}+R_{22}R_{33}+R_{33}R_{11})$$

+

$$+1/2 * [(R_{11}-R_{22})^2 + (R_{22}-R_{33})^2 + (R_{33}-R_{11})^2] - (R_{11}^{\,2}+R_{22}^{\,2}+R_{33}^{\,2}) + (R_{11}R_{22}+R_{22}R_{33}+R_{33}R_{11})$$

Or:

$$R21 - R2 = 3/2\ R21 - 1/2\ R12 = 1/2[(R_{11}-R_{22})^2+(R_{22}-R_{33})^2+(R_{33}-R_{11})^2] + 3(R_{12}R_{21}+R_{23}R_{32}+R_{31}R_{13})]$$

1.3.3.
Symmetric Matrix Invariants:

s_{11}	s_{12}	s_{13}
s_{21}	s_{22}	s_{23}
s_{31}	s_{32}	s_{33}

Eigenvalue 3 by 3 Matrix Invariants:

F as function of a scalar s:

$$F(s) \equiv \det(s_{MN} - \delta_{MN}\,.\,s) \equiv 1/6\ e^{IJK}\ e^{UVW}\ (s_{IU} - \delta_{IU}\,.\,s)\,(s_{JV} - \delta_{JV}\,.\,s)\,(s_{KW} - \delta_{KW}\,.\,s)$$

This is Expanded:

$$F(s) = 1/6\ e^{IJK}\ e^{UVW}\ (s_{IU}s_{JV}s_{KW} - 3\,s_{IU}s_{JV}\,.\,\delta_{KW}\,.\,s + 3\,s_{IU}\,.\,\delta_{JV}\delta_{KW}\,.\,s^2 - \delta_{IU}\ \delta_{JV}\ \delta_{KW}\,.\,s^3)$$

F as function of a scalar s:

$$F(s) = (s3) - (s2)\,.\,s + (s1)\,.\,s^2 - s^3$$

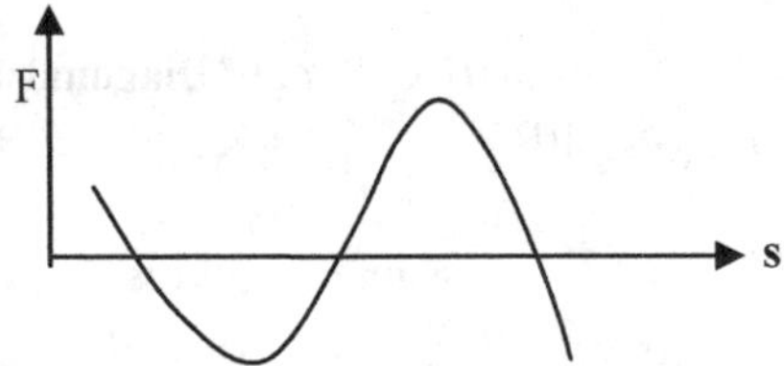

Invariants: s0, s1, s2, s3:

$$s3 \equiv \det(s_{IJ}) \equiv 1/6\ e^{IJK}\ e^{UVW}\ (s_{IU}\ \ s_{JV}\ s_{KW}) = 1/6\ [s_{II}s_{JJ}s_{KK} + 2\,(s_{IJ}s_{JK}s_{KI}) - 3\,s_{II}(s_{JK}s_{KJ})\,]$$

=

1/6 [s13 + 2 (s31) – 3 s11 (s21)]

$$=$$

$$+s_{11}s_{22}s_{33}+s_{12}s_{23}s_{31}+s_{13}s_{21}s_{32}-s_{11}s_{23}s_{32}-s_{13}s_{22}s_{31}-s_{12}s_{21}s_{33}$$

$$s2 \equiv (1/6)\,3\,e^{IJK}e^{UVW}(s_{IU}\;s_{JV}.\delta_{KW}) = 1/2\,e^{IJK}e^{UVK}s_{IU}s_{IV} = 1/2\,(s_{II}s_{JJ} - s_{IK}s_{KI})$$

$$=$$

$$1/2\;(s1^2 - s2^1)$$

$$=$$

$$+s_{11}s_{22}+s_{22}s_{33}+s_{33}s_{11}-s_{12}s_{21}-s_{23}s_{32}-s_{31}s_{13}$$

$$s1 \equiv 1/2\,e^{IJK}e^{UVW}s_{IU}\delta_{JV}\delta_{KW} = 1/2\,e^{IJK}e^{UJK}s_{IU} = \delta^{IU}s_{IU} = s_{KK}$$

$$=$$

$$s_{11}+s_{22}+s_{33}$$

$$s0 \equiv 1/6\,e^{IJK}e^{UVW}\delta_{IU}\delta_{JV}\delta_{KW} = 1/6\,e^{IJK}e^{IJK}$$

$$=$$

$$1$$

1.3.3.1. Diagonal Matrix Invariants:

$$s_{ij} = 0 \text{ if } i \neq j$$

s_{11}		
	s_{22}	
		s_{33}

Third Invariant:

$$s3 \equiv \det(s_{ij}) = s_{11}\,s_{22}\,s_{33}$$

Second Invariant:

$$s2 = s_{11}\,s_{22} + s_{22}\,s_{33} + s_{33}\,s_{11} - (s_{12}^2 + s_{23}^2 + s_{31}^2) = s_{11}\,s_{22} + s_{22}\,s_{33} + s_{33}\,s_{11}$$

First Invariant:

$$s1 = s_{11} + s_{22} + s_{33}$$

Where:

$$s_{ik}\,s_{ki} = s_{11}s_{11} + s_{12}s_{21} + s_{13}s_{31} + s_{21}s_{12} + s_{22}s_{22} + s_{23}s_{32} + s_{31}s_{13} + s_{32}s_{23} + s_{33}\,s_{33}$$

$$=$$

$$s_{11}^2 + s_{22}^2 + s_{33}^2$$

Distortion Property:

$$s1^2 - 3\,s2 = ½\,[(s_{11} - s_{22})^2 + (s_{22} - s_{33})^2 + (s_{33} - s_{11})^2] + 3(s_{12}^2 + s_{23}^2 + s_{31}^2)$$

$$=$$

$$½\,[(s_{11} - s_{22})^2 + (s_{22} - s_{33})^2 + (s_{33} - s_{11})^2]$$

So:

$$F \equiv (s_{11}\,s_{22}\,s_{33}) - (s_{11}\,s_{22} + s_{22}\,s_{33} + s_{33}\,s_{11})\,s + (s_{11} + s_{22} + s_{33})\,s^2 - s^3$$

1.3.3.2. Equal Diagonals Matrix Invariants (Spherical Invariants):

$$s_{11} = s_{22} = s_{33} = sP \qquad s_{IJ} = 0 \text{ if } I \neq J$$

sP		
	sP	
		sP

Third Invariant:

$$s3 \equiv \det(s) = sP^3$$

Second Invariant:

$$s2 = s_{11}\, s_{22} + s_{22}\, s_{33} + s_{33}\, s_{11} - (s_{12}^2 + s_{23}^2 + s_{31}^2) = sP\, sP + sP\, sP + sP\, sP = 3\, sP^2$$

First variant:

$$s1 \equiv s_{ii} = s_{11} + s_{22} + s_{33} = sP + sP + sP \equiv 3 * sP$$

Distortion Property:

$$s1^2 - 3\, s2 = \tfrac{1}{2}\,[(s_{11} - s_{22})^2 + (s_{22} - s_{33})^2 + (s_{33} - s_{11})^2] + 3(s_{12}^2 + s_{23}^2 + s_{31}^2) = 0$$

F can be looked at as if it is written as a function of $(3sP^1)$, $(3sP^2)$ and (sP^3):

3D: $(3sP)$, $(3sP^2)$, (sP^3) $\quad F \equiv \det(sP - s\,\delta_{mn}) = (sP - s)^3 = (sP^3) - s\,(3\, sP^2) + s^2\,(3\, sP) - s^3$

3D: F is a function of: $(3sP)$, $(3sP^2)$, (sP^3), following graph is for $0 < ss$:

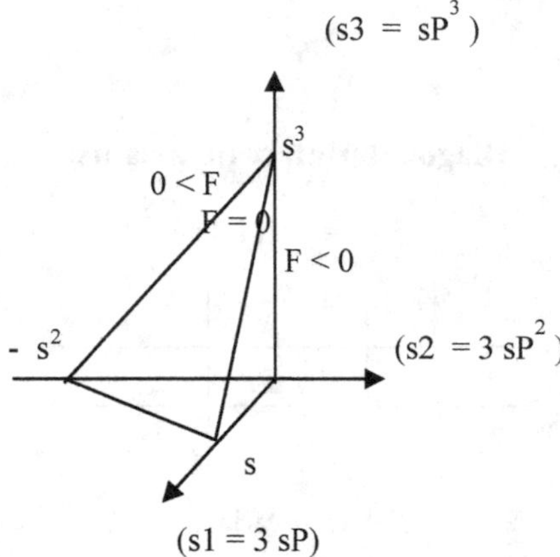

F Graph For $0 < s$

1.3.3.3.
Zero Diagonal Matrix Invariants:

$$s_{JK} = 0 \text{ if } J = K$$

	s_{12}	s_{13}
s_{21}		s_{23}
s_{31}	s_{32}	

Third Invariant:

$$s3 = 1/6\,(s_{JJ})^3 + 1/3\, s_{JK}\, s_{KL}\, s_{LJ} - 1/2\,(s_{JJ})\,(s_{KL} s_{LK}) = 0 + 1/3\, s_{JK}\, s_{KL}\, s_{LJ} - 0$$

Second Invariant:

$$s2 = s_{11}\, s_{22} + s_{22}\, s_{33} + s_{33}\, s_{11} - (s_{12}^2 + s_{23}^2 + s_{31}^2) = -(s_{12}^2 + s_{23}^2 + s_{31}^2)$$

First Invariant:

$$s1 = 0$$

Distortion Property

$$s1^2 - 3s2 = \tfrac{1}{2}[(s_{11} - s_{22})^2 + (s_{22} - s_{33})^2 + (s_{33} - s_{11})^2] + 3(s_{12}^2 + s_{23}^2 + s_{31}^2) = 3(s_{12}^2 + s_{23}^2 + s_{31}^2)$$

So:

$$F \equiv (1/3\; s_{ij}\, s_{jk}\, s_{ki}) + (s_{12}^2 + s_{23}^2 + s_{31}^2)\, s - s^3$$

F as a function of only two invariants (r2) and (r3):

2D: s2, s3 $F \equiv \det(s_{mn} - s\,\delta_{mn}) \equiv (s3) - s\ (s2) + 0 - s^3$

2D: F is a Function of (s2), (s3), following graph For 0 < s:

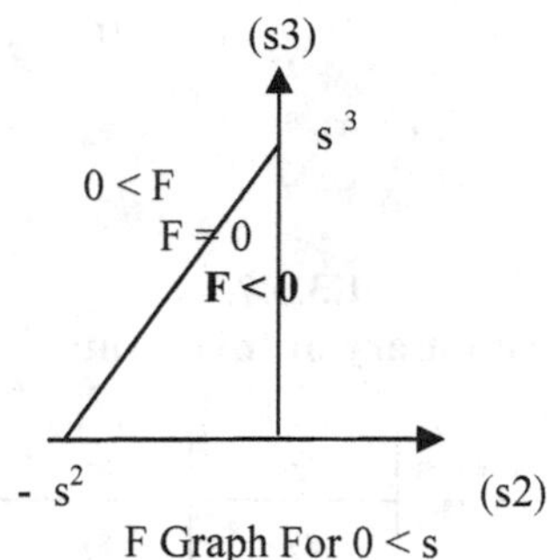

F Graph For 0 < s

1.3.3.4.
General Deviatory Matrix Invariants:

If:

$r_{mn} = s_{mn} - \delta_{mn}\, s_{kk}/3$

s_{11}	s_{12}	s_{13}
s_{21}	s_{22}	s_{23}
s_{31}	s_{32}	s_{33}

r_{11}	r_{12}	r_{13}
r_{21}	r_{22}	r_{23}
r_{31}	r_{32}	r_{33}

$$r1 = r_{kk} = r_{11} + r_{22} + r_{33} = s_{11} - s_{kk}/3 + s_{22} - s_{kk}/3 + s_{33} - s_{kk}/3 = s_{jj} - 3\, s_{kk}/3 = 0$$

$$F \equiv (r3) - (r2)\, r + 0 - r^3$$

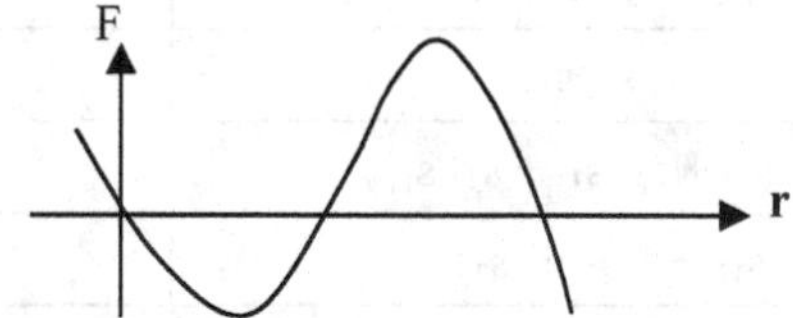

F as a function of only two invariants (r2) and (r3):

2D: r2, r3 $F \equiv \det(r_{mn} - r\,\delta_{mn}) \equiv (r3) - r\ (r2) + 0 - r^3$

2D: F is a Function of (r2), (r3): Graph For 0 < r:

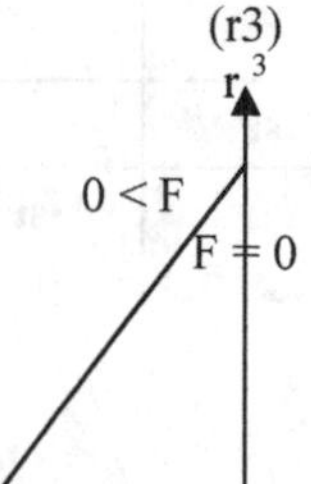

F < 0

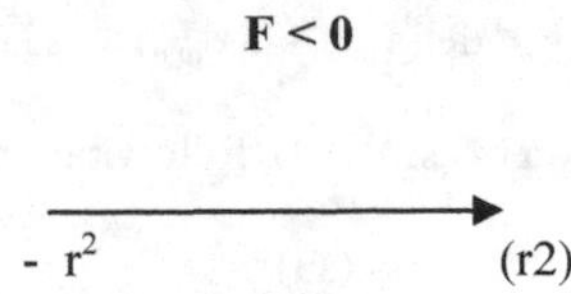

1.3.3.7.
Summary of Invariants:

s_{11}	s_{12}	s_{13}
s_{21}	s_{22}	s_{23}
s_{31}	s_{32}	s_{33}

F	General $F \equiv s3 - s2 * s + s^2 - s^3$	Deviatory $F \equiv r3 - r2\, r + 0 - r^3$
s_{Lm}	s_{mn}	$r_{mn} = s_{mn} - \delta_{mn} s_{LL}/3$
s3	$1/6(s_{JJ})^3 + 1/3 s_{JK} s_{KL} s_{LJ} - 1/2(s_{JJ})(s_{KL} s_{LK})$	$1/3\, r_{JK} r_{KL} r_{LJ}$
s2	$s_{11}s_{22} + s_{22}s_{33} + s_{33}s_{11} - (s_{12}^2 + s_{23}^2 + s_{31}^2)$	$r_{11}r_{22} + r_{22}r_{33} + r_{33}r_{11} - (r_{12}^2 + r_{23}^2 + r_{31}^2) \le 0$
s1	$s_{11} + s_{22} + s_{33}$	0
$s21 = s_{JL}\, s_{LJ}$	$s_{11}^2 + s_{22}^2 + s_{33}^2 + 2(s_{12}^2 + s_{23}^2 + s_{31}^2)$	$r_{11}^2 + r_{22}^2 + r_{33}^2 + 2(r_{12}^2 + r_{23}^2 + r_{31}^2)$
s12-3s2	$½[(s_{11} - s_{22})^2 + (s_{22} - s_{33})^2 + (s_{33} - s_{11})^2] + 3(s_{12}^2 + s_{23}^2 + s_{31}^2)$	$½[(r_{11} - r_{22})^2 + (r_{22} - r_{33})^2 + (r_{33} - r_{11})^2] + 3(r_{12}^2 + r_{23}^2 + r_{31}^2)$

$s_{JK} = 0 \text{ Jf } J \neq K$

s_{11}		
	s_{22}	
		s_{33}

s_{Lm}	$s_{JK} = 0 \text{ Jf } J \neq K$	$r_{JK} = 0 \text{ Jf } J \neq K$
s3	$s_{11}\, s_{22}\, s_{33}$	$r_{11}\, r_{22}\, r_{33}$
s2	$s_{11}\, s_{22} + s_{22}\, s_{33} + s_{33}\, s_{11}$	$r_{11}\, r_{22} + r_{22}\, r_{33} + r_{33}\, r_{11}$
s1	$s_{11} + s_{22} + s_{33}$	0
$s21 = s_{JL}\, s_{LJ}$	$s_{11}^2 + s_{22}^2 + s_{33}^2 + 2(s_{12}^2 + s_{23}^2 + s_{31}^2)$	$(r_{11}^2 + r_{22}^2 + r_{33}^2)$
s12-3s2	$½[(s_{11} - s_{22})^2 + (s_{22} - s_{33})^2 + (s_{33} - s_{11})^2]$	$½[(r_{11} - r_{22})^2 + (r_{22} - r_{33})^2 + (r_{33} - r_{11})^2]$

:

Equal Diagonals MatrJx

$s_{11} = s_{22} = s_{33} = sP$ $\quad s_{IJ} = 0 \text{ Jf } I \neq J$

sP		
	sP	
		sP

s_{Lm}	$s_{11} = s_{22} = s_{33} = sP$	0
s3	sP^3	0
s2	$3\ sP^2$	0
s1	$s_{11} + s_{22} + s_{33}$	0
s21= $s_{JL}\ s_{LJ}$	$3\ sP^2$	0
s12-3s2	0	0

$$s_{JK} = 0 \text{ Jf } J=K$$

	s_{12}	s_{13}
s_{21}		s_{23}
s_{31}	s_{32}	

s_{Lm}	$s_{JK} = 0$ Jf $J = K$	$r_{JK} = 0$ Jf $J = K$
s3	$1/3\ s_{JK}\, s_{KL}\, s_{LJ}$	$1/3\ r_{JK}\, r_{KL}\, r_{LJ}$
s2	$-(s_{12}^2 + s_{23}^2 + s_{31}^2)$	$-(r_{12}^2 + r_{23}^2 + r_{31}^2) \leq 0$
s1	$s_{11} + s_{22} + s_{33}$	0
s21= $s_{JL}\ s_{LJ}$	$2(s_{12}^2 + s_{23}^2 + s_{31}^2)$	$2(r_{12}^2 + r_{23}^2 + r_{31}^2)$
s12-3s2	$3(s_{12}^2 + s_{23}^2 + s_{31}^2)$	$3(r_{12}^2 + r_{23}^2 + r_{31}^2)$

1.3.4.
Equal-Angle Octahedral Plane:

Octahedral Plane makes equal angles with each of the axes:

$n_1 = 1/3^{1/2}$ $\quad n_2 = 1/3^{1/2}$ $\quad n_3 = 1/3^{1/2}$

The Normal Vector:

$$sN = \begin{bmatrix} 1/3^{1/2} & 1/3^{1/2} & 1/3^{1/2} \end{bmatrix} \begin{bmatrix} s_{11} & & \\ & s_{22} & \\ & & s_{33} \end{bmatrix} \begin{bmatrix} 1/3^{1/2} \\ 1/3^{1/2} \\ 1/3^{1/2} \end{bmatrix}$$

Or:

$$sN = s_{11}\, n_1^2 + s_{22}\, n_2^2 + s_{33}\, n_3^2 = 1/3\, (s_{11} + s_{22} + s_{33}) = sP$$

The Square of Tangent Vector:

$$sT^2 = (s_{11}\, n_1)^2 + (s_{22}\, n_2)^2 + (s_{33}\, n_3)^2 - (sN)^2$$

$$sT^2 = s_{11}^2\ 1/3 + s_{22}^2\ 1/3 + s_{33}^2\ 1/3 - [1/3\,(s_{11} + s_{22} + s_{33})]^2$$

$$sT^2 = s_{11}^2\ 3/9 + s_{22}^2\ 3/9 + s_{33}^2\ 3/9 - 1/9(s_{11}^2 + s_{22}^2 + s_{33}^2 + 2s_{11}\, s_{22} + 2s_{22}\, s_{33} + 2s_{33}2s_{11})$$

$$sT^2 = 1/9\,[s_{11}^2\ 3 + s_{22}^2\ 3 + s_{33}^2\ 3 - (s_{11}^2 + s_{22}^2 + s_{33}^2) - (2s_{11}\, s_{22} + 2s_{22}\, s_{33} + 2s_{33}2s_{11})]$$

$$sT^2 = 1/9\,[2(s_{11}^2 + s_{22}^2 + s_{33}^2) - (2s_{11}\, s_{22} + 2s_{22}\, s_{33} + 2s_{33}\, s_{11})]$$

$$sT^2 = 1/9\,[(s_{11} - s_{22})^2 + (s_{22} - s_{33})^2 + (s_{33} - s_{11})^2]$$

If Compared With:

Diagonal Deviatory $r_{ii} = 0$: and $r_{ij} = 0$ if $i \neq j$: Half Squares Sum:

$$0 \leq -r2 = 1/6\,[(s_{11} - s_{22})^2 + (s_{22} - s_{33})^2 + (s_{33} - s_{11})^2]$$

To get:

$$0 \leq sT^2/(-r_2) = 2/3$$

Trigonometric Paragraph

$$\cos\alpha - 4\cos\alpha\sin^2\alpha = \cos\alpha\,(1 - 2\sin^2\alpha)\sin\alpha\,(2\sin\alpha\cos\alpha), \text{ equals:}$$

$$\cos\alpha - 4\cos\alpha\sin^2\alpha = \cos\alpha\cos 2\alpha - \sin\alpha\sin 2\alpha, \text{ equals:}$$

$$4\cos\alpha\,(1 - \sin^2\alpha) - 3\cos\alpha = \cos(\alpha + 2\alpha), \text{ equals:}$$

$$4\cos^3\alpha - 3\cos\alpha = \cos 3\alpha, \text{ equals:}$$

$$+ 1/4\cos 3\alpha + 3/4\cos\alpha - \cos^3\alpha = 0$$

Let:

$$r \equiv \rho\cos\alpha$$

In:

$$F \equiv r3 - r2\,r + \quad - r^3 = 0$$

To Get:

$$r3/\rho^3 - r2/\rho^2\cos\alpha \quad - \cos^3\alpha = 0$$

Compared terms with:

$$+ 1/4\cos 3\alpha + 3/4\cos\alpha - \cos^3\alpha = 0$$

To Get:

$$r3/\rho^3 = 1/4\cos 3\alpha$$

$$- r2/\rho^2 = 3/4$$

Second Leads To:

$$\rho = 2(-r2/3)^{1/2}$$

$$\rho^2 = 4(-r2/3)$$

$$\rho^3 = 8(-r2/3)^{3/2}$$

First and Second Lead To:

$$\cos 3\alpha = 4\,r3/\rho^3 = ½\,r3/(-r2/3)^{3/2}$$

First Root:

$$0 \leq 3\alpha_1 \leq \pi$$

All Roots:

$$0 \leq \alpha_1 \leq \pi/3$$

$$\alpha_2 = \alpha_1 + 2\pi/3$$

$$\alpha_3 = \alpha_2 + 2\pi/3 = \alpha_1 + 4\pi/3$$

$$\alpha_1 = \alpha_3 + 2\pi/3 = \alpha_2 + 4\pi/3 = \alpha_1 + 6\pi/3$$

As Defined Earlier:

$$r_{\underline{K}K} \equiv \rho\cos\alpha_K$$

And Found:

$$\rho = 2(-r2/3)^{1/2}$$

Then:

$$r_{\underline{K}K} \equiv \rho\cos\alpha_K = 2(-r2/3)^{1/2}\cos\alpha_K$$

Values:

$$s_{11} - sP \equiv r_{11} = 2(-r2/3)^{1/2}\cos\alpha_1$$

$$s_{22} - sP \equiv r_{22} = 2(-r2/3)^{1/2}\cos\alpha_2$$

$$s_{33} - sP \equiv r_{33} = 2(-r2/3)^{1/2}\cos\alpha_3$$

$$r_{33} \leq r_{22} \leq r_{11}$$

New Invariants are: α_1, s1 and r2 , Where:

$$0 \leq \alpha_1 \leq \pi/3$$

$$sP = s_{mm}/3 = s1/3$$

$$(-r2/3)^{1/2}$$

Criteria:

$$F(\alpha_1 , s1 , r2) = 0$$

1.3.5.
General N by N Matrix Determinant: ||a|| and ||b||

$$\|a^{P}{}_{Q}\| = e_{P1\,P2\,P3\ldots PN}\, a^{P1}{}_{1}\, a^{P2}{}_{2}\, a^{P3}{}_{3}\ldots a^{PN}{}_{N} \qquad \|a_{P}{}^{Q}\| = e^{P1\,P2\,P3\ldots PN}\, a_{P1}{}^{1}\, a_{P2}{}^{2}\, a_{P3}{}^{3}\ldots a_{PN}{}^{N}$$

$$\|b^{Q}{}_{R}\| = e_{Q1Q2Q3\ldots QN}\, b^{Q1}{}_{1}\, b^{Q2}{}_{2}\, b^{Q3}{}_{3}\ldots b^{QN}{}_{N} \qquad \|b_{Q}{}^{R}\| = e^{Q1Q2Q3\ldots QN}\, b_{Q1}{}^{1}\, b_{Q2}{}^{2}\, b_{Q3}{}^{3}\ldots b_{QN}{}^{N}$$

Generalize to:

$$e_{Q1Q2Q3\ldots QN}\,\|a^{P}{}_{Q}\| = e_{P1P2P3\ldots PN}\, a^{P1}{}_{Q1}\, a^{P2}{}_{Q2}\, a^{P3}{}_{Q3}\ldots a^{PN}{}_{QN} \qquad e^{Q1Q2Q3\ldots QN}\,\|a_{P}{}^{Q}\| = e^{P1P2P3\ldots PN}\, a_{P1}{}^{Q1}\, a_{P2}{}^{Q2}\, a_{P3}{}^{Q3}\ldots a_{PN}{}^{QN}$$

$$e_{R1R2R3\ldots RN}\,\|b^{Q}{}_{R}\| = e_{Q1Q2Q3\ldots QN}\, b^{Q1}{}_{R1}\, b^{Q2}{}_{R2}\, b^{Q3}{}_{R3}\ldots b^{QN}{}_{RN} \qquad e^{R1R2R3\ldots RN}\,\|b_{Q}{}^{R}\| = e^{Q1Q2Q3\ldots QN}\, b_{Q1}{}^{R1}\, b_{Q2}{}^{R2}\, b_{Q3}{}^{R3}\ldots b_{QN}{}^{RN}$$

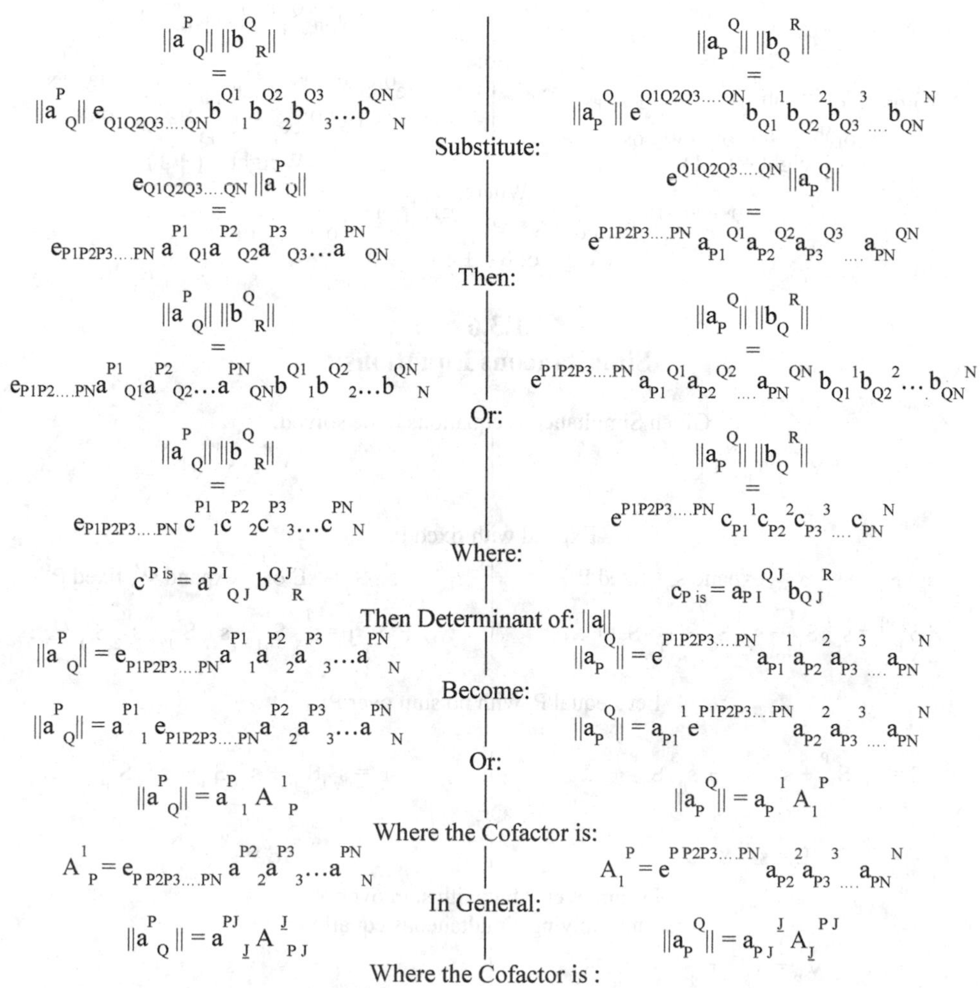

Multiplication of Matrices:

$$\|a^{P}{}_{Q}\|\,\|b^{Q}{}_{R}\| = \|a^{P}{}_{Q}\|\, e_{Q1Q2Q3\ldots QN}\, b^{Q1}{}_{1}\, b^{Q2}{}_{2}\, b^{Q3}{}_{3}\ldots b^{QN}{}_{N} \qquad \|a_{P}{}^{Q}\|\,\|b_{Q}{}^{R}\| = \|a_{P}{}^{Q}\|\, e^{Q1Q2Q3\ldots QN}\, b_{Q1}{}^{1}\, b_{Q2}{}^{2}\, b_{Q3}{}^{3}\ldots b_{QN}{}^{N}$$

Substitute:

$$e_{Q1Q2Q3\ldots QN}\,\|a^{P}{}_{Q}\| = e_{P1P2P3\ldots PN}\, a^{P1}{}_{Q1}\, a^{P2}{}_{Q2}\, a^{P3}{}_{Q3}\ldots a^{PN}{}_{QN} \qquad e^{Q1Q2Q3\ldots QN}\,\|a_{P}{}^{Q}\| = e^{P1P2P3\ldots PN}\, a_{P1}{}^{Q1}\, a_{P2}{}^{Q2}\, a_{P3}{}^{Q3}\ldots a_{PN}{}^{QN}$$

Then:

$$\|a^{P}{}_{Q}\|\,\|b^{Q}{}_{R}\| = e_{P1P2\ldots PN}\, a^{P1}{}_{Q1}\, a^{P2}{}_{Q2}\ldots a^{PN}{}_{QN}\, b^{Q1}{}_{1}\, b^{Q2}{}_{2}\ldots b^{QN}{}_{N} \qquad \|a_{P}{}^{Q}\|\,\|b_{Q}{}^{R}\| = e^{P1P2P3\ldots PN}\, a_{P1}{}^{Q1}\, a_{P2}{}^{Q2}\ldots a_{PN}{}^{QN}\, b_{Q1}{}^{1}\, b_{Q2}{}^{2}\ldots b_{QN}{}^{N}$$

Or:

$$\|a^{P}{}_{Q}\|\,\|b^{Q}{}_{R}\| = e_{P1P2P3\ldots PN}\, c^{P1}{}_{1}\, c^{P2}{}_{2}\, c^{P3}{}_{3}\ldots c^{PN}{}_{N} \qquad \|a_{P}{}^{Q}\|\,\|b_{Q}{}^{R}\| = e^{P1P2P3\ldots PN}\, c_{P1}{}^{1}\, c_{P2}{}^{2}\, c_{P3}{}^{3}\ldots c_{PN}{}^{N}$$

Where:

$$c^{P}{}_{is} = a^{PI}{}_{QJ}\, b^{QJ}{}_{R} \qquad c_{P\,is} = a_{PI}{}^{QJ}\, b_{QJ}{}^{R}$$

Then Determinant of: ||a||

$$\|a^{P}{}_{Q}\| = e_{P1P2P3\ldots PN}\, a^{P1}{}_{1}\, a^{P2}{}_{2}\, a^{P3}{}_{3}\ldots a^{PN}{}_{N} \qquad \|a_{P}{}^{Q}\| = e^{P1P2P3\ldots PN}\, a_{P1}{}^{1}\, a_{P2}{}^{2}\, a_{P3}{}^{3}\ldots a_{PN}{}^{N}$$

Become:

$$\|a^{P}{}_{Q}\| = a^{P1}{}_{1}\, e_{P1P2P3\ldots PN}\, a^{P2}{}_{2}\, a^{P3}{}_{3}\ldots a^{PN}{}_{N} \qquad \|a_{P}{}^{Q}\| = a_{P1}{}^{1}\, e^{P1P2P3\ldots PN}\, a_{P2}{}^{2}\, a_{P3}{}^{3}\ldots a_{PN}{}^{N}$$

Or:

$$\|a^{P}{}_{Q}\| = a^{P}{}_{1}\, A^{1}{}_{P} \qquad \|a_{P}{}^{Q}\| = a_{P}{}^{1}\, A_{1}{}^{P}$$

Where the Cofactor is:

$$A^{1}{}_{P} = e_{P\,P2P3\ldots PN}\, a^{P2}{}_{2}\, a^{P3}{}_{3}\ldots a^{PN}{}_{N} \qquad A_{1}{}^{P} = e^{P\,P2P3\ldots PN}\, a_{P2}{}^{2}\, a_{P3}{}^{3}\ldots a_{PN}{}^{N}$$

In General:

$$\|a^{P}{}_{Q}\| = a^{PJ}{}_{\underline{J}}\, A^{\underline{J}}{}_{PJ} \qquad \|a_{P}{}^{Q}\| = a_{PJ}{}^{\underline{J}}\, A_{\underline{J}}{}^{PJ}$$

Where the Cofactor is :

$$A^{J}_{\ PJ} = e_{P1\,P2P3...\,PJ\,...\,PN}\ a^{P1}_{\ 1} a^{P2}_{\ 2} a^{P2}_{\ 2} a^{P3}_{\ 3} .. a^{PJ-1}_{\ J-1} a^{PJ+1}_{\ J+1} ... a^{PN}_{\ N}$$

$$A_{J}^{\ PJ} = e^{P1\,P2P3...PJ....PN\,PN}\ a_{P1}^{\ 1} a_{P2}^{\ 2} a_{P3}^{\ 3} ... a_{PJ-1}^{\ J-1} a_{PJ+1}^{\ J+1} ... a_{PN}^{\ N}$$

Derivative of ||a||:

$$\partial\|a^{P}_{\ Q}\|/\partial b^{K} = \partial a^{P1}_{\ 1}/\partial b^{K} A^{1}_{\ P1} + \partial a^{P2}_{\ 2}/\partial b^{K} A^{2}_{\ P2} + \ldots + \partial a^{PJ}_{\ J}/\partial b^{K} A^{J}_{\ PJ} + \ldots + \partial a^{PN}_{\ N}/\partial b^{K} A^{N}_{\ PN}$$

$$\partial\|a_{P}^{\ Q}\|/\partial b^{J} = \partial a_{P1}^{\ 1}/\partial b^{K} A_{1}^{\ P1} + \partial a_{P2}^{\ 2}/\partial b^{K} A_{2}^{\ P2} + \ldots + \partial a_{PJ}^{\ J}/\partial b^{K} A_{J}^{\ PJ} + \ldots + \partial a_{PN}^{\ N}/\partial b^{K} A_{N}^{\ PN}$$

Or:

$$\partial\|a^{P}_{\ Q}\| /\partial b^{K} = \partial a^{P}_{\ J}/\partial b^{K} A^{J}_{\ P}$$

$$\partial\|a_{P}^{\ Q}\|/\partial b^{J} = \partial a_{P}^{\ J}/\partial b^{K} A_{J}^{\ P}$$

Transformation of e:

Let:

$$\partial a^{P}/\partial b^{Q} = ab^{P}_{\ Q}$$

$$\partial b_{P}/\partial a_{Q} = ba_{P}^{\ Q}$$

In Generalized Determinant:

$$e_{Q1Q2Q3....QN} \|ab^{P}_{\ Q}\| = e_{P1P2P3....PN}\ ab^{P1}_{\ Q1} ab^{P2}_{\ Q2} ab^{P3}_{\ Q3} \ldots ab^{PN}_{\ QN}$$

$$e^{Q1Q2Q3....QN} \|ba_{P}^{\ Q}\| = e^{P1P2P3....PN}\ ba_{P1}^{\ Q1} ba_{P2}^{\ Q2} ba_{P3}^{\ Q3} .... ba_{PN}^{\ QN}$$

Or:

$$e_{Q1Q2Q3....QN} = \|ab^{P}_{\ Q}\|^{(-1)} e_{P1P2P3....PN}\ ab^{P1}_{\ Q1} ab^{P2}_{\ Q2} ab^{P3}_{\ Q3} \ldots ab^{PN}_{\ QN}$$

$$e^{Q1Q2Q3....QN} = \|ba_{P}^{\ Q}\|^{(-1)} e^{P1P2P3....PN}\ ba_{P1}^{\ Q1} ba_{P2}^{\ Q2} ba_{P3}^{\ Q3} .... ba_{PN}^{\ QN}$$

But:

$$\|ba_{P}^{\ Q}\|^{(-1)} = \|ab^{P}_{\ Q}\|^{(+1)}$$

So:

$$e_{Q1Q2Q3....QN} = \|ab^{P}_{\ Q}\|^{(-1)} e_{P1P2P3....PN}\ ab^{P1}_{\ Q1} ab^{P2}_{\ Q2} ab^{P3}_{\ Q3} \ldots ab^{PN}_{\ QN}$$

Weight = (- 1)

$$e^{Q1Q2Q3....QN} = \|ab^{P}_{\ Q}\|^{(+1)} e^{P1P2P3....PN}\ ba_{P1}^{\ Q1} ba_{P2}^{\ Q2} ba_{P3}^{\ Q3} .... ba_{PN}^{\ QN}$$

Weight = (+ 1)

Where:

$$e^{P1P2P3....PN} e_{Q1Q2Q3....QN} = \delta^{P1\,P2\,P3\,......\,PJ}_{Q1\,Q2\,Q3\,......\,QJ}$$

Weight of δ = + 1 – 1 = 0

1.3.6.
Simultaneous Equations:

Given Simultaneous equations to be solved:

$$v_{P} = s_{P}^{\ N} r_{N} \qquad\qquad v^{P} = s^{P}_{\ N} r^{N}$$

Expand with fixed P:

$$s_{P}^{\ N} S_{N}^{\ J} = sE\,\delta_{P}^{\ J} \text{ expand } s_{P}^{\ N} \text{ fixed P} \qquad\qquad s^{P}_{\ 1} S^{1}_{\ J} = sE\,\delta^{P}_{\ J} \text{ expand } s^{P}_{\ 1} \text{ fixed P}$$

$$sE\,\delta_{P}^{\ J} = s_{P}^{\ 1} S_{1}^{\ J} + s_{P}^{\ 2} S_{2}^{\ J} + s_{P}^{\ 3} S_{3}^{\ J} + \ldots \qquad\qquad sE\,\delta^{P}_{\ J} = s^{P}_{\ 1} S^{1}_{\ J} + s^{P}_{\ 2} S^{2}_{\ J} + s^{P}_{\ 2} S^{2}_{\ J} + \ldots$$

Let J equal P, with no sum over P:

$$sE = s_{\underline{P}}^{\ 1} S_{1}^{\ P} + s_{\underline{P}}^{\ 2} S_{2}^{\ P} + s_{\underline{P}}^{\ 3} S_{3}^{\ P} + \ldots \qquad\qquad sE = s^{\underline{P}}_{\ 1} S^{1}_{\ P} + s^{\underline{P}}_{\ 2} S^{2}_{\ P} + s^{\underline{P}}_{\ 3} S^{3}_{\ P} + \ldots$$

Or:

$$sE = s_{\underline{P}}^{\ N} S_{N}^{\ P} \qquad\qquad sE = s^{\underline{P}}_{\ N} S^{N}_{\ P}$$

No sum over P but with sum over N.

Pre - multiplying simultaneous equations:

$$v_{P} = s_{P}^{\ N} r_{N} \qquad\qquad v^{P} = s^{P}_{\ N} r^{N}$$

by:

$S_1^{\ P}$	To get:	$S_P^{\ 1}$
$S_1^{\ P} \quad v_P = S_1^{\ P} s_P^{\ N} r_N =$ $sE\, \delta_1^{\ N} r_N = sE\, r_1$		$S_P^{\ 1} v^P = S_P^{\ 1} s^P_{\ N} r^N =$ $sE\, \delta^1_{\ N} r^N = sE\, r^1$
$r_1 = 1/sE\, S_1^{\ P} v_P$	Or:	$r^1 = 1/sE\, S_P^{\ 1} v^P$

1.3.7.
Dyadic Circle (M, R):

1.3.7.1.
Circle (M, R)

Assume coordinates system XYZ

Area * Y*Y	Area * Y*Z
Area * Z*Z Y	Area * Z*Z

Stated, simply:

YY	YZ
ZY	ZZ

Eigenvalues:

YY - VV	YZ
ZY	ZZ - VV

Its Determinant:

$$(YY * ZZ - YZ * ZY) - (YY + ZZ)\, VV + VV * VV = 0$$

$$VV_{min} = M - R \quad | \quad VV_{max} = M + R$$

(M, R):

$$M = (YY + ZZ)/2 \quad | \quad C = (YY - ZZ)/2 \quad | \quad R^2 = [(YY + ZZ)/2]^2 - (YY * ZZ - YZ * ZY)$$

$$R^2 = [(YY + ZZ)/2]^2 - (2/2\, YY * ZZ - YZ * ZY)$$

$$R^2 = [YY^2/4 + ZZ^2/4 + 2 * YY * ZZ/4] + (-2 * YY * ZZ/2 + YZ * ZY)$$

$$R^2 = YY^2/4 + ZZ^2/4 - 2 * YY * ZZ/4 + 2 * YZ * ZY/2 = [(YY - ZZ)/2]^2 + (YZ)^2$$

$$R^2 = C^2 + (YZ)^2$$

$$Y = V(\beta = 0) \quad | \quad Z = W(\beta = 0)$$

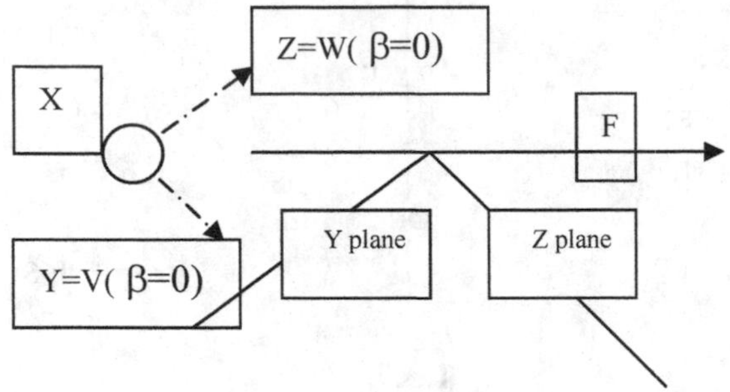

Q:(ZZ, ZY)
Circle (M, R)

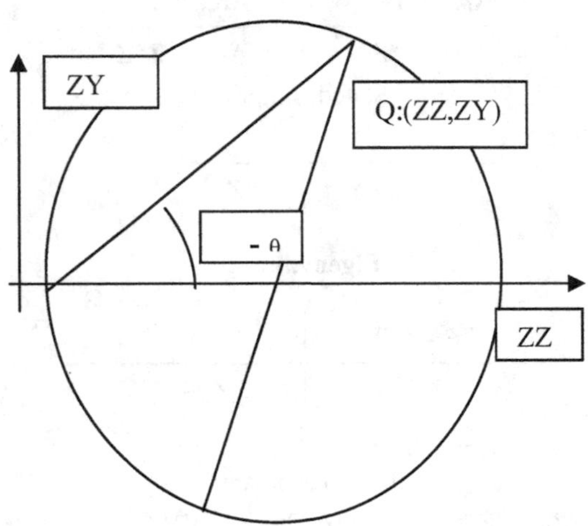

P:(YY,YZ)
Circle (M, R)

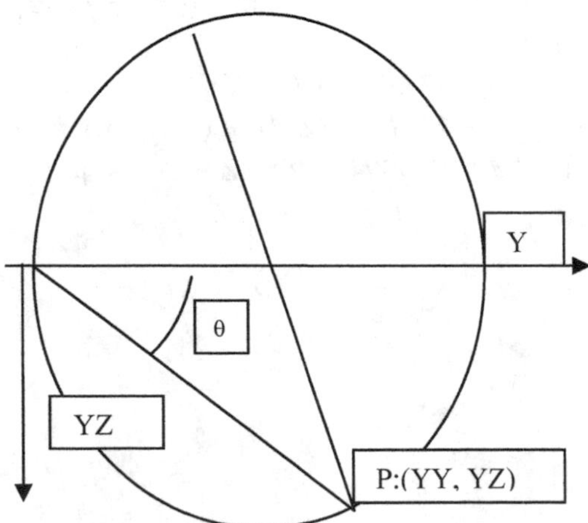

$M = (YY + ZZ)/2$ | $R^2 = [(YY + ZZ)/2]^2 - (YY * ZZ - YZ * ZY)$

50 Prof. Faysal Alkhalil, Ph.D. INTRODUCTORY TENSORIAL ALGORITHMS IN STRUCTURAL MECHANICS

M + C = YY	M - C = ZZ

1.3.7.2.
Rotation of a Dyadic Point Y:

When YZ coordinates are rotated by β so as to coincide with a new VW coordinates:

$$2 * \beta \equiv 2 * \theta + 2 * \alpha$$

A point [P(YY,YZ)] with a central angle (2 θ) is rotated to [PP(VV,VW)] with a central angle (2 β)
Point Q is similarly is rotate to QQ

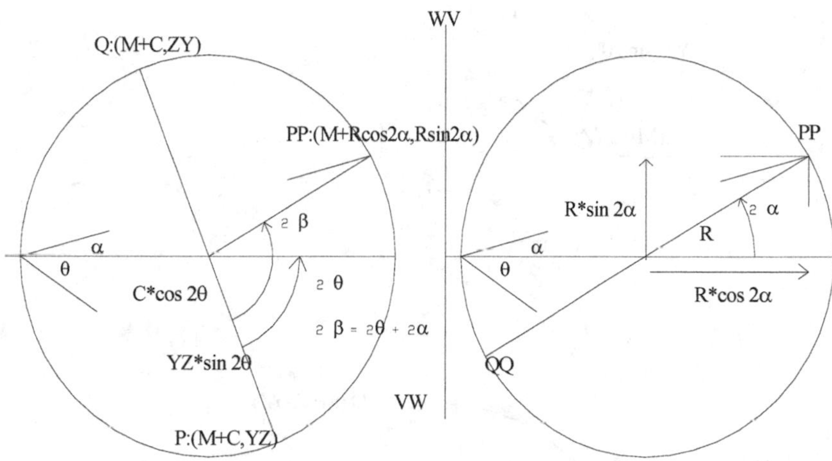

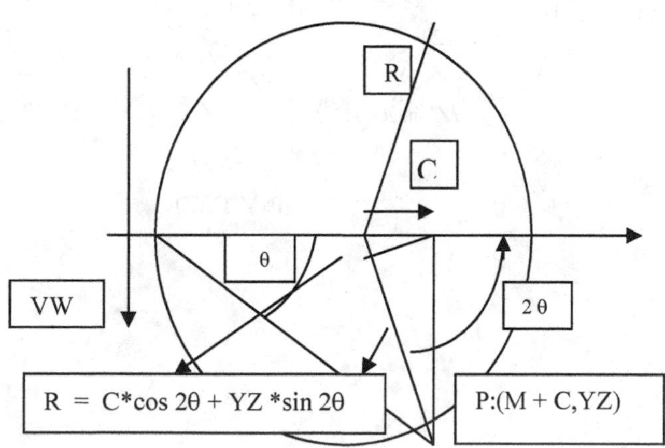

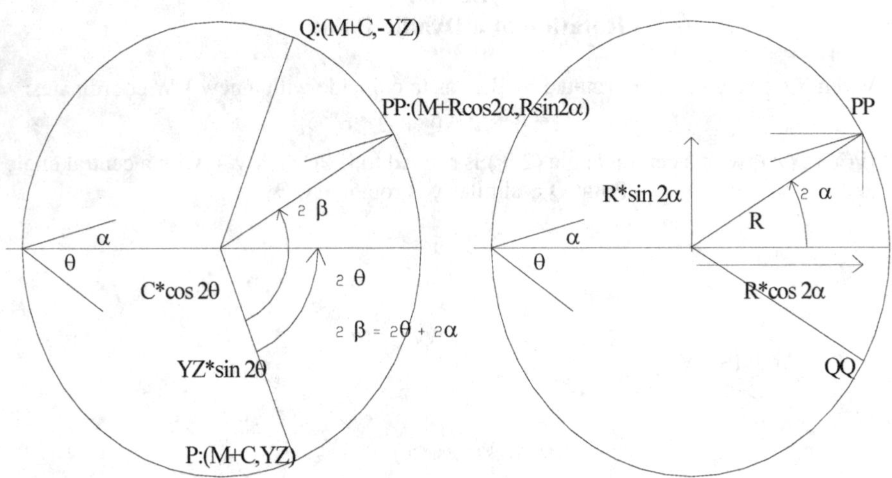
Q:(M+C,-YZ)
PP:(M+Rcos2α,Rsin2α)
2 β
α
θ
C*cos 2θ
2 θ
2 β = 2θ + 2α
YZ*sin 2θ
P:(M+C,YZ)
PP
R*sin 2α
2 α
R
α
θ
R*cos 2α
QQ

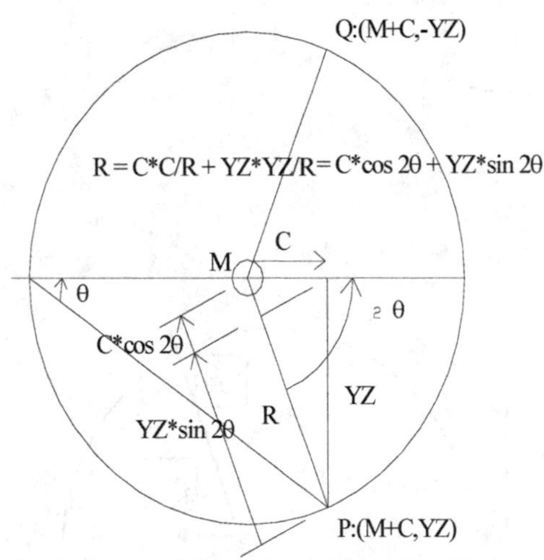
Q:(M+C,-YZ)
R = C*C/R + YZ*YZ/R= C*cos 2θ + YZ*sin 2θ
C
M
θ
2 θ
C*cos 2θ
YZ
R
YZ*sin 2θ
P:(M+C,YZ)

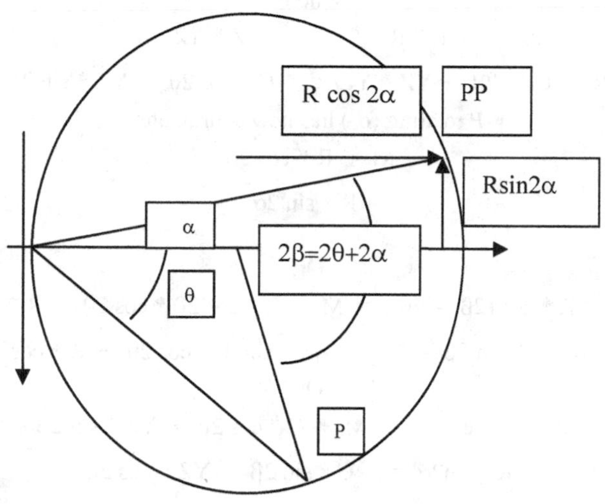

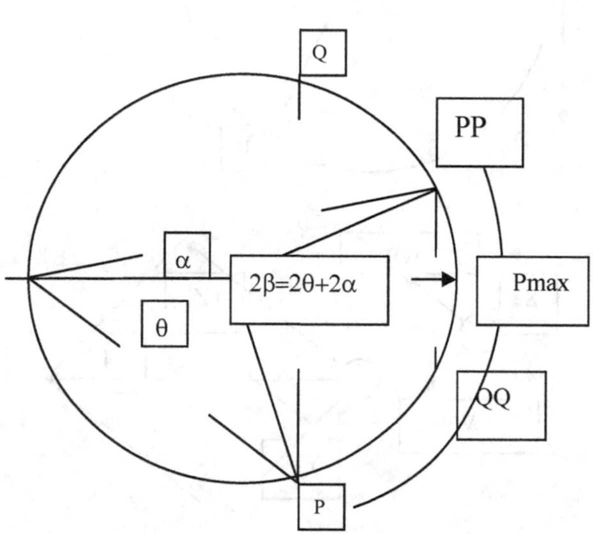

As point P is rotating (2θ) through, it passes to an extreme point [Pmax(M+R, 0) and continues rotating additional angle (2α) to reach point PP

$$0 < \beta \equiv \theta + \alpha < 90°$$

$$\tan 2\theta \equiv YZ / C$$

$$0 < +2 * \alpha < 2 * 90^{\circ} = 180^{\circ}$$

Where:

$$R * R \equiv C * C + YZ * YZ$$

$$R = C * C/R + YZ * YZ/R = C * \cos 2\theta + YZ * \sin 2\theta$$

P rotating (α) has new coordinate:

$$M + R * \cos 2\alpha$$

$$- R * \sin 2\alpha$$

Or:

$$M + R * \cos 2\alpha \equiv R * \cos (2\beta - 2\theta) = M + R * \cos 2\beta * \cos 2\theta + R * \sin 2\beta * \sin 2\theta$$

$$- R * \sin 2\alpha \equiv - R * \sin(2\beta - 2\theta) = - R * \sin 2\beta * \cos 2\theta + R * \cos 2\beta * \sin 2\theta$$

Or:

$$M + R * \cos 2\alpha = M + C * \cos 2\beta + YZ * \sin 2\beta$$

$$- R * \sin 2\alpha = - C * \sin 2\beta + YZ * \cos 2\beta$$

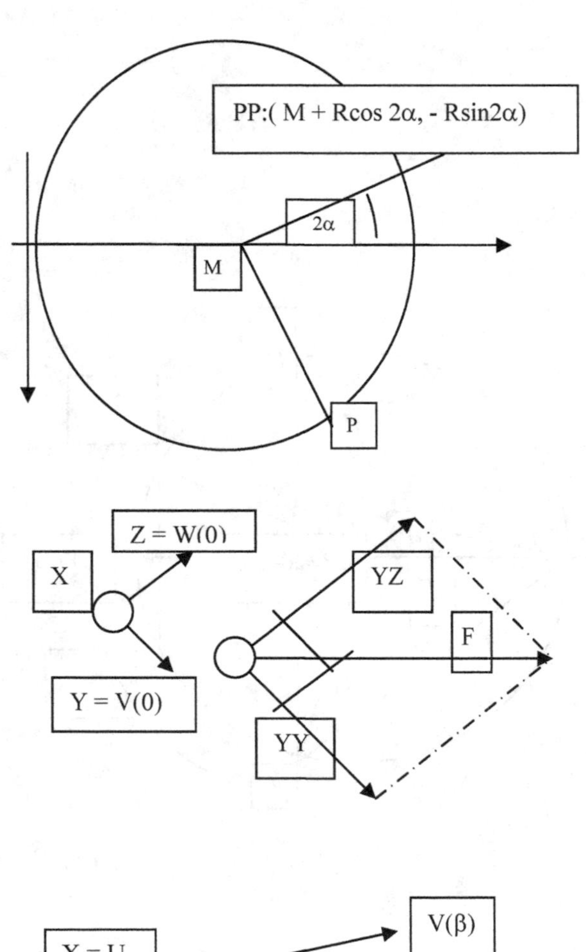

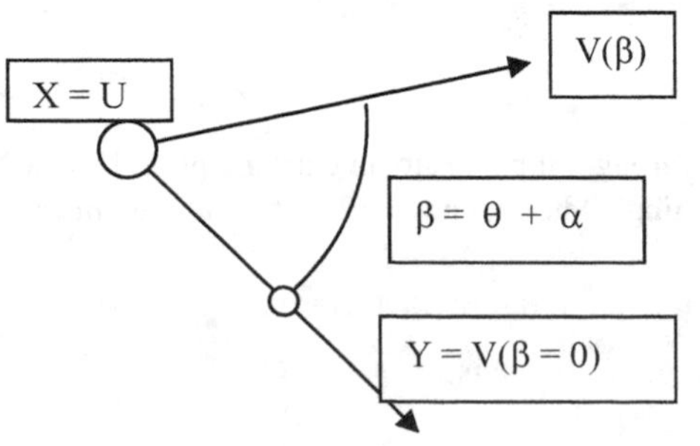

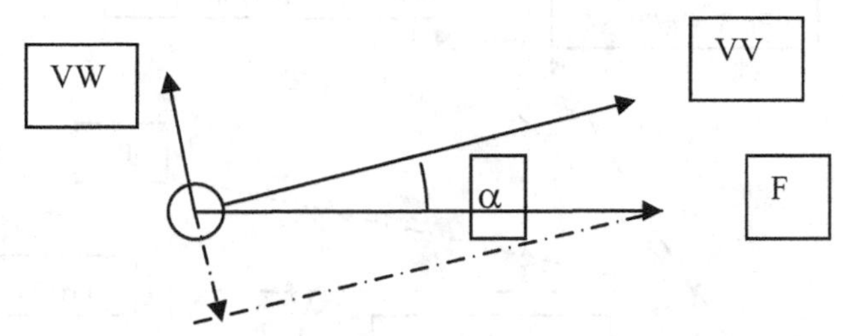
VW
VV
α
F

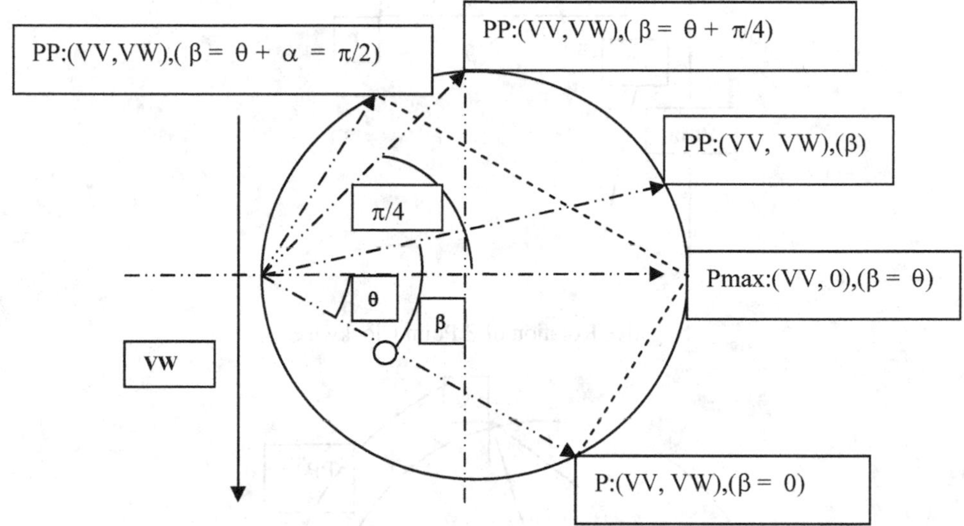
PP:(VV,VW),(β = θ + α = π/2)
PP:(VV,VW),(β = θ + π/4)
PP:(VV, VW),(β)
π/4
Pmax:(VV, 0),(β = θ)
θ
β
VW
P:(VV, VW),(β = 0)

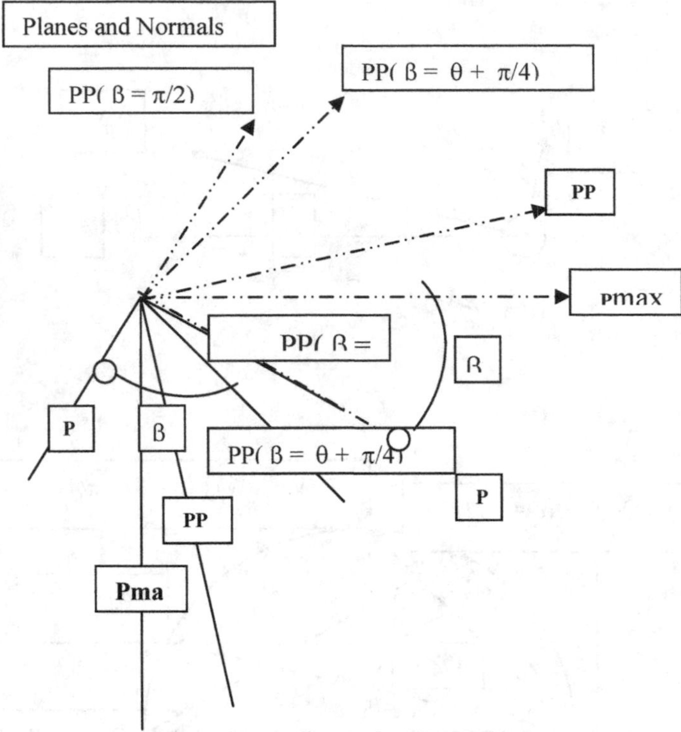

Negative Rotation of Z Point Clockwise:

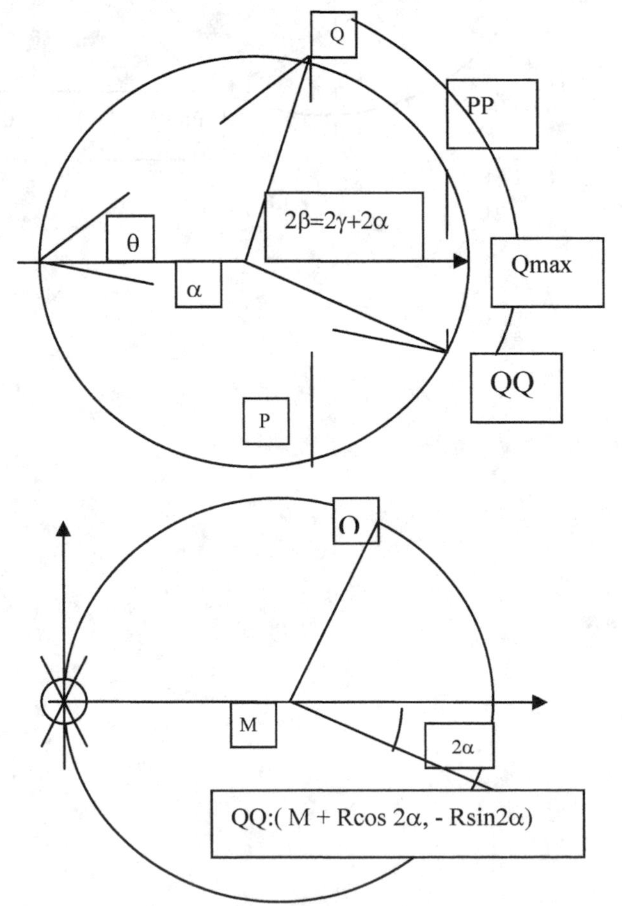

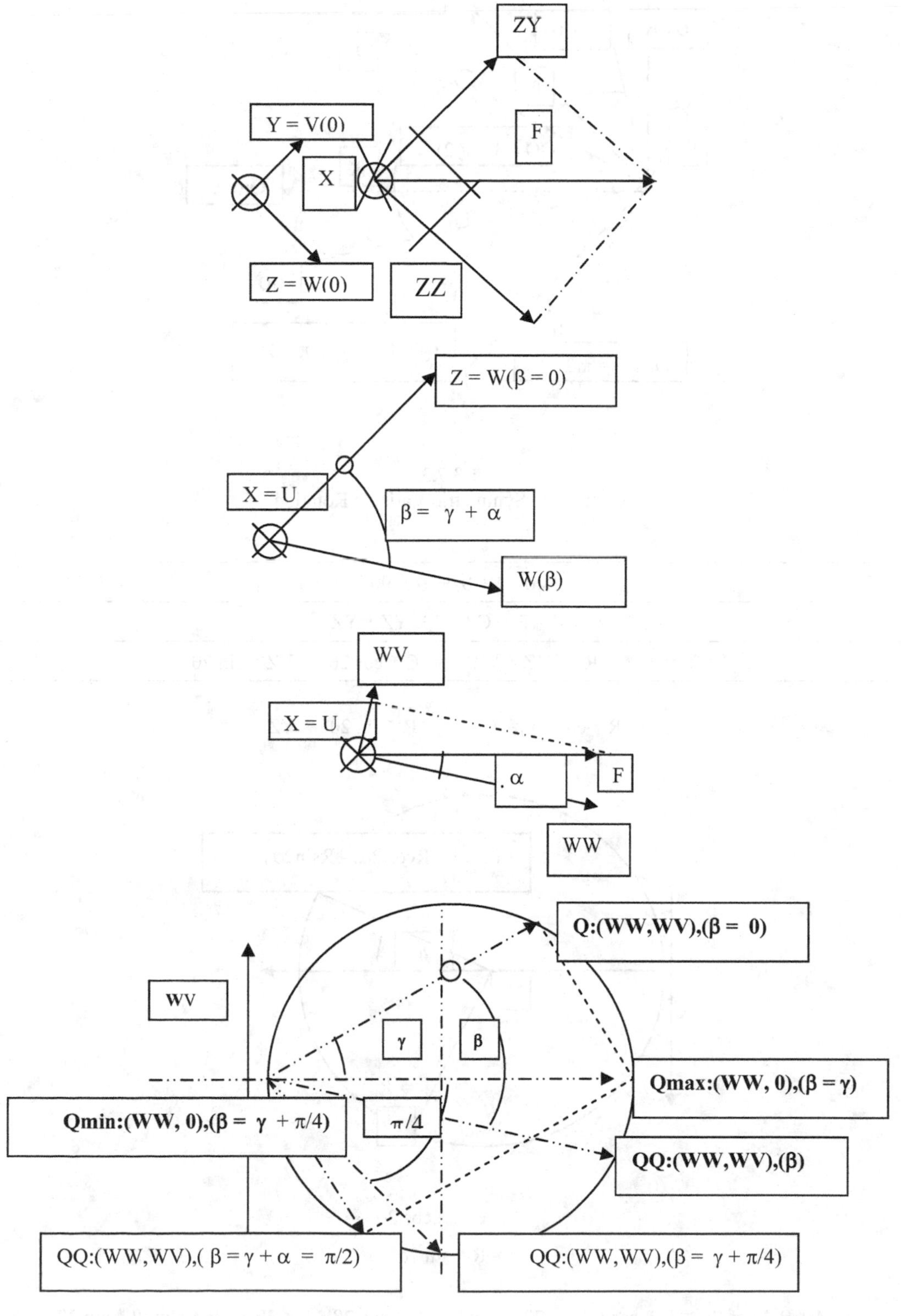
ZY
Y = V(0)
F
X
Z = W(0)
ZZ
Z = W(β = 0)
X = U
β = γ + α
W(β)
WV
X = U
α
F
WW
Q:(WW,WV),(β = 0)
WV
γ
β
Qmax:(WW, 0),(β = γ)
Qmin:(WW, 0),(β = γ + π/4)
π/4
QQ:(WW,WV),(β)
QQ:(WW,WV),(β = γ + α = π/2)
QQ:(WW,WV),(β = γ + π/4)

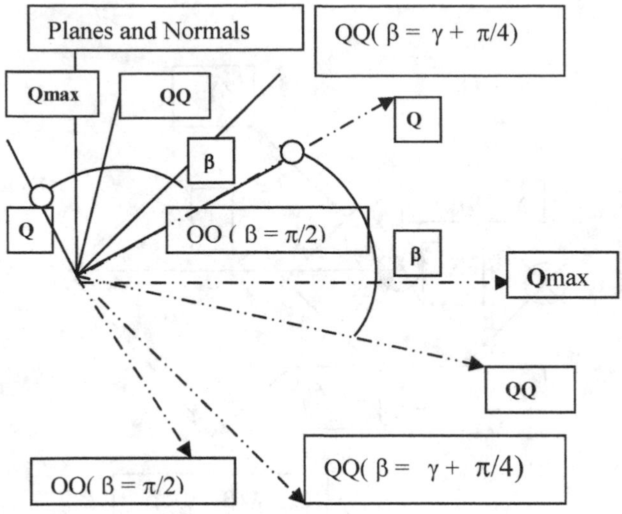

1.3.7.3.
Coordinate Symmetric Variable Rotation:

$0 < \beta \equiv \theta + \alpha < 90^\circ$
$R * R \equiv C * C + YZ * YZ$
$R = C * C/R + YZ * YZ/R = C * \cos 2\theta + YZ * \sin 2\theta$

$R * \cos 2\theta = C$ | $R * \sin 2\theta = YZ$

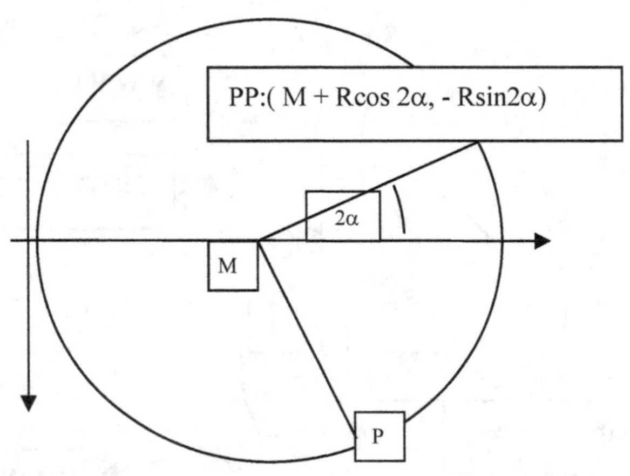

$M + R * \cos 2\alpha$

$- R * \sin 2\alpha$

$$M + R * \cos 2\alpha \equiv R * \cos (2\beta - 2\theta) = M + R * \cos 2\beta * \cos 2\theta + R * \sin 2\beta * \sin 2\theta$$

$$- R * \sin 2\alpha \equiv - R * \sin(2\beta - 2\theta) = - R * \sin 2\beta * \cos 2\theta + R * \cos 2\beta * \sin 2\theta$$

That is

$$VV - M = C * \cos 2\beta + YZ * \sin 2\beta$$

$$VW = -C * \sin 2\beta + YZ * \cos 2\beta$$

$v = y \cos\beta + z \sin\beta$ | $w = y(-1)\sin\beta + z\cos\beta$

$$\begin{bmatrix} v \\ w \end{bmatrix} = \begin{bmatrix} \cos\beta & \sin\beta \\ -\sin\beta & \cos\beta \end{bmatrix} \begin{bmatrix} y \\ z \end{bmatrix}$$

$$\begin{bmatrix} y \\ z \end{bmatrix} = \begin{bmatrix} \cos\beta & -\sin\beta \\ \sin\beta & \cos\beta \end{bmatrix} \begin{bmatrix} v \\ w \end{bmatrix}$$

$$\begin{bmatrix} v & w \end{bmatrix} = \begin{bmatrix} j & k \end{bmatrix} \begin{bmatrix} \cos\beta & -\sin\beta \\ \sin\beta & \cos\beta \end{bmatrix}$$

$$\begin{bmatrix} j & k \end{bmatrix} = \begin{bmatrix} v & w \end{bmatrix} \begin{bmatrix} \cos\beta & \sin\beta \\ -\sin\beta & \cos\beta \end{bmatrix}$$

Multiplying:

$$\begin{bmatrix} \cos\beta & \sin\beta \\ -\sin\beta & \cos\beta \end{bmatrix} \begin{bmatrix} YY & YZ \\ ZY & ZZ \end{bmatrix} \begin{bmatrix} \cos\beta & -\sin\beta \\ \sin\beta & \cos\beta \end{bmatrix}$$

Or:

$$\begin{bmatrix} \cos\beta & \sin\beta \\ -\sin\beta & \cos\beta \end{bmatrix} \begin{bmatrix} YY\cos\beta + YZ\sin\beta & -YY\sin\beta + YZ\cos\beta \\ YZ\cos\beta + ZZ\sin\beta & -YZ\sin\beta + ZZ\cos\beta \end{bmatrix}$$

Resulting:

$$\begin{array}{c|c} +YY * \cos\beta * \cos\beta + YZ * \sin\beta * \cos\beta + & -YY * \sin\beta * \cos\beta + YZ * \cos\beta * \cos\beta \\ +YZ * \cos\beta * \sin\beta + ZZ * \sin\beta * \sin\beta & -YZ * \sin\beta * \sin\beta + ZZ * \cos\beta * \sin\beta \\ \hline -YY * \cos\beta * \sin\beta - YZ * \sin\beta * \sin\beta + & +YY * \sin\beta * \sin\beta - YZ * \cos\beta * \sin\beta + \\ +YZ * \cos\beta * \cos\beta + ZZ * \sin\beta * \cos\beta & -YZ * \sin\beta * \cos\beta + ZZ * \cos\beta * \cos\beta + \end{array}$$

But:

$$\begin{array}{c|c} YY = M + C & ZZ = M - C \\ \hline \cos^2\beta = (1 + \cos 2\beta)/2 & \sin^2\beta = (1 - \cos 2\beta)/2 \end{array}$$

And:

$$YY * \cos^2\beta + ZZ * \sin^2\beta = YY * (1 + \cos 2\beta)/2 + ZZ * (1 - \cos 2\beta)/2$$
$$= (YY + ZZ)/2 + (YY - ZZ)/2 * \cos 2\beta = M + C * \cos 2\beta$$

$$YY * \sin^2\beta + ZZ * \cos^2\beta = YY * (1 - \cos 2\beta)/2 + Z * (1 + \cos 2\beta)/2$$
$$= (YY + ZZ)/2 - (YY - ZZ)/2 * \cos 2\beta = M - C * \cos 2\beta$$

$$-YY/2 * \sin 2\beta + ZZ/2 * \sin 2\beta = -(YY - ZZ)/2 * \sin 2\beta = -C * \sin 2\beta$$

So:

$$\begin{array}{c|c} VV - M = C * \cos 2\beta + YZ * \sin 2\beta & -C * \sin 2\beta + YZ * \cos 2\beta \\ \hline -C * \sin 2\beta + YZ * \cos 2\beta & WW - M = -C * \cos 2\beta - YZ * \sin 2\beta \end{array}$$

$$\begin{bmatrix} VV - M & VW \end{bmatrix} \quad \begin{bmatrix} \cos 2\beta & \sin 2\beta \end{bmatrix} \quad \begin{bmatrix} C & YZ \end{bmatrix}$$

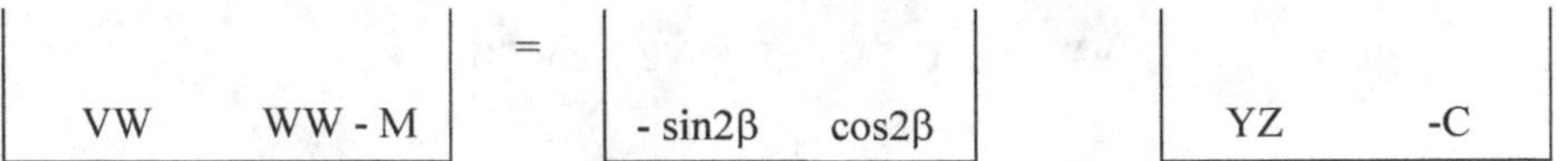

1.3.7.4.
Extreme Values:

Extreme Values VV_{min} , VV_{max} of first component:

min	max
$\alpha_{min} = 180^{\circ}$	$\alpha_{max} = 0$
$VV_{min} = M - R$	$VV_{max} = M + R$
$VW_{min} = 0$	$VW_{max} = 0$

Same values of eigenvalue problem

Extreme Values of $\mathbf{WW_{min}}$, $\mathbf{WW_{max}}$ second component:

min	max
$\boldsymbol{\alpha_{min} = 270^{\circ}}$	$\boldsymbol{\alpha_{max} = 90^{\circ}}$
$\mathbf{VV_{min} = M}$	$\mathbf{VV_{max} = M}$
$\mathbf{WW_{min} = - R}$	$\mathbf{WW_{max} = R}$

1.3.7.5.
Triaxial Test Circle:

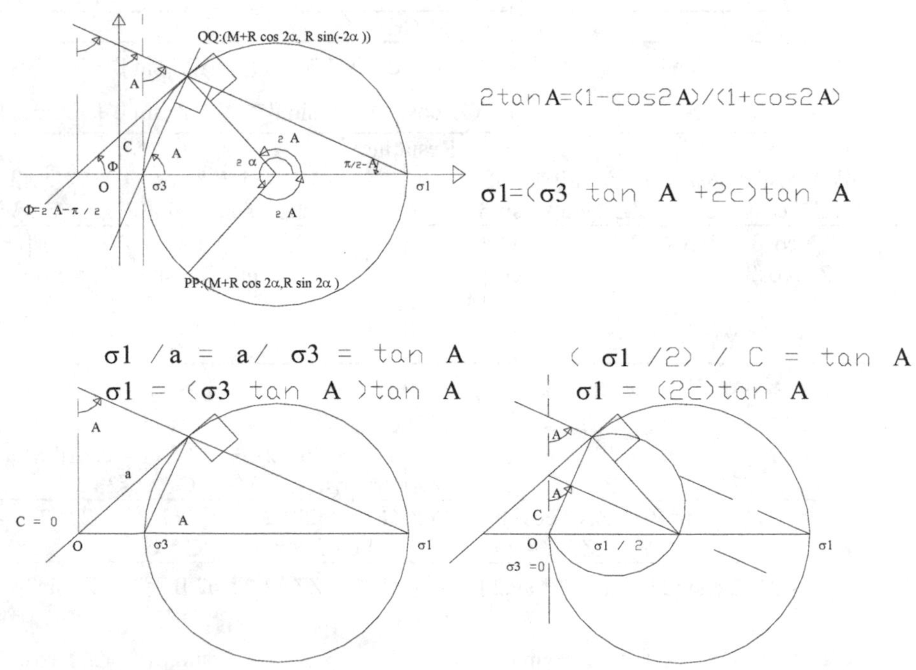

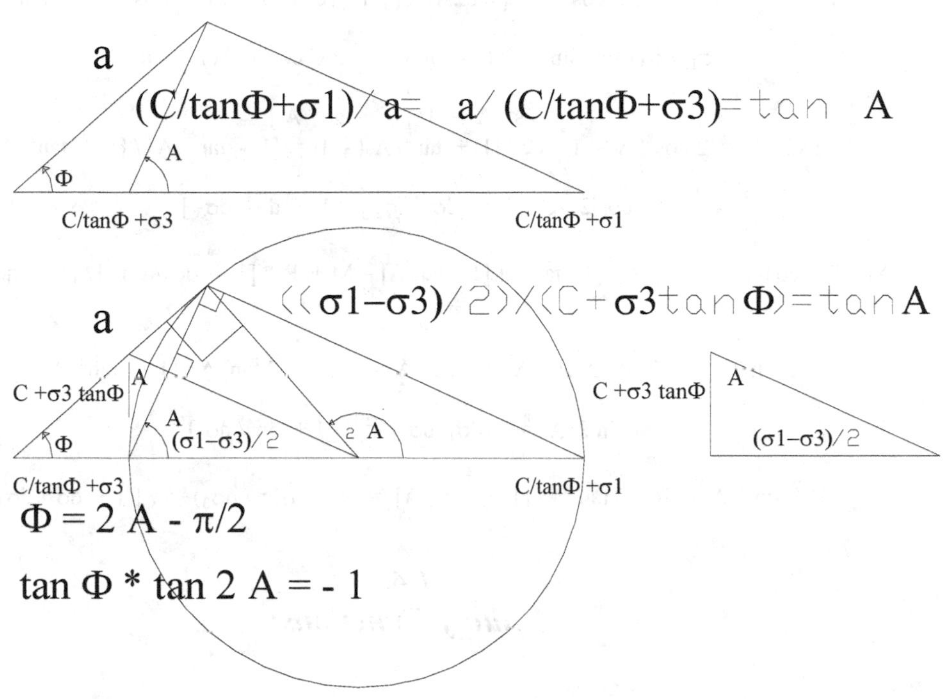

Substitute:

$$\sigma = [(\sigma_1 + \sigma_3)/2 + (\sigma_1 - \sigma_3)/2 \cos 2A]$$

$$\tau = (\sigma_1 - \sigma_3)/2 \sin 2A$$

In:

$$\tau = C + \sigma \tan\Phi$$

To get:

$$(\sigma_1 - \sigma_3)/2 \sin 2A = C + [(\sigma_1 + \sigma_3)/2 + (\sigma_1 - \sigma_3)/2 \cos 2A] \tan\Phi$$

But:

$$\Phi = 2A - \pi/2$$

And:

$$\tan\Phi \tan 2A = -1$$

So: multiply by: tan 2A : to get:

$$(\sigma_1 - \sigma_3)/2 \sin 2A \tan 2A = C \tan 2A - [(\sigma_1 + \sigma_3)/2 + (\sigma_1 - \sigma_3)/2 \cos 2A]$$

$$(\sigma_1/2 - \sigma_3/2) \sin 2A \tan 2A = C \tan 2A - [\sigma_1/2 + \sigma_3/2 + (\sigma_1/2 - \sigma_3/2) \cos 2A]$$

$$\sigma_1/2 [\sin 2A \tan 2A + \cos 2A + 1] = C \tan 2A - \sigma_3/2[1 - \cos 2A - \sin 2A \tan 2A]$$

$$\sigma_1 = 2C \tan 2A/[\sin 2A \tan 2A + \cos 2A + 1] + \sigma_3[\sin 2A \tan 2A + \cos 2A - 1]/[\sin 2A \tan 2A + \cos 2A + 1]$$

But:

$$1/[\sin 2A \tan 2A + \cos 2A + 1] = [1 - 1/(2\cos^2 A)]$$

$$\tan 2A / [\sin 2A \tan 2A + \cos 2A + 1] = \tan 2A [1 - 1/(2\cos^2 A)] = \tan A$$

$$[\sin 2A \tan 2A + \cos 2A - 1]/[\sin 2A \tan 2A + \cos 2A + 1] = 1 - 2[1 - 1/(2\cos^2 A)] = \tan^2 A$$

So:

$$\sigma_1 = 2C \tan A + \sigma_3 \tan^2 A$$

$$d\sigma_1/d\sigma_3 = \tan^2 A$$

$$2A = \pi/2 + \Phi$$

$$\cos 2A = -\sin\Phi$$

$$2\cos^2 A = 1 + \cos 2A = 1 - \sin\Phi$$

$$2\sin^2 A = 1 - \cos 2A = 1 + \sin\Phi$$

$$\tan^2 A = \sin^2 A / \cos^2 A = [(1 - \cos 2A)/2]/[(1 + \cos 2A)/2] = (1 + \sin\Phi)/(1 - \sin\Phi)$$

$$\tan A = \sin\alpha \cos A / \cos^2 A = [\sin 2A/2]/[(1 + \cos 2A)/2] = \cos\Phi/(1 - \sin\Phi)$$

:

$$\sigma_1 = \sigma_3 (1 + \sin\Phi)/(1 - \sin\Phi) + 2C\cos\Phi/(1 - \sin\Phi)$$

:

$$\cos^2 A = 1/[1 + \tan^2 A]$$

$$\cos 2A = 2\cos^2 A - 1 = 2/[1 + \tan^2 A] - 1 = [1 - \tan^2 A]/[1 + \tan^2 A]$$

$$\cos 2A = [1 - d\sigma_1/d\sigma_3]/[1 + d\sigma_1/d\sigma_3]$$

:

$$M + R * \cos 2A = M + R * [1 - \tan^2 A]/[1 + \tan^2 A] = M + R * [1 - d\sigma_1/d\sigma_3]/[1 + d\sigma_1/d\sigma_3]$$

:

$$\sin 2A = 2\sin A\cos A = 2\tan A\cos^2 A = 2\tan A/(1 + \tan^2 A)$$

$$\sin 2A = 2(d\sigma_1/d\sigma_3)^{1/2} / [1 + d\sigma_1/d\sigma_3]$$

:

$$R * \sin 2A = R * 2\tan A/[1 + \tan^2 A] = R * 2(d\sigma_1/d\sigma_3)^{1/2} / [1 + d\sigma_1/d\sigma_3]$$

1.4.
Unitary Functions:

Define:

For $0 \le m$:Unit Functions: $\eta^{(0)} = 1$

$\eta^{(-m)} = 0$ if $(s - S < 0)$ $\quad\quad$ $\eta^{(-m)} = (s - S)^m/m!$ if $(0 < s - S)$

Also Define:

Only For $-1 = m$: Unit Impulse function:

$\eta^{(1)} = \delta(s - S) = 0$ if $(s - S < 0)$ $\quad\quad$ $\eta^{(1)} = \delta(s - S) = 0$ if $(0 < s - S)$

:

Such that for m = 0

$$\eta^{(0)} = (s - S)^0/0! = 1 \text{ if } (0 < s - S)$$

$$\eta^{(0)} = \int\delta(s - S)\,ds = 1 \text{ if } (0 < s - S)$$

:

For $(0 < s - S)$:

$$\eta^{[-(m)]} = d\eta^{[-(m+1)]}/ds = d[(s - S)^{m+1}/(m + 1)!]/ds = (s - S)^m/m!$$

$$\eta^{[-(m+1)]} = \int \eta^{[-(m)]}\,ds = \int d\eta^{[-(m+1)]} = \int d[(s - S)^{m+1}/(m + 1)!] = (s - S)^{m+1}/(m + 1)!$$

$$\eta^{[-(m-1)]} \equiv d\eta^{[-(m)]}/ds = d[(s - S)^m/(m)!]/ds \equiv (s - S)^{(m-1)}/(m - 1)!$$

$$\eta^{[-(m)]} = \int \eta^{[-(m-1)]}\,ds = \int d\eta^{[-(m)]} = \int d[(s - S)^m/(m)!] = (s - S)^m/(m)!$$

Take m = - 1:

$$\eta^{[1]} = d\eta^{[-(-1+1)]}/ds = d[(s - S)^{-1+1}/(-1 + 1)!]/ds = (s - S)^{-1}/-1! \text{ Not Applicable}$$

$$\eta^{[0]} = \int \eta^{[-(-1)]}\,ds = \int d\eta^{[-(-1+1)]} = \int d[(s - S)^{-1+1}/(-1 + 1)!] = (s - S)^{-1+1}/(-1 + 1)! = 1$$

$$\eta^{[2]} \equiv d\eta^{[-(-1)]}/ds = d[(s - S)^{-1}/(-1)!]/ds \equiv (s - S)^{(-1-1)}/(-1 - 1)! \text{ Not Applicable}$$

$$\eta^{[1]} = \int \eta^{[-(-1-1)]}\,ds = \int d\eta^{[-(-1)]} = \int d[(s - S)^{-1}/(-1)!] = (s - S)^{-1}/(-1)! \text{ Not Applicable}$$

Take m = 0:

$$\eta^{[0]} = \eta^{[-0]} = d\eta^{[-(0+1)]}/ds = d[(s - S)^{0+1}/(0+1)!]/ds = (s - S)^0/0! = 1$$
$$\eta^{[-1]} = \int \eta^{[-0]} ds = \int d\eta^{[-(0+1)]} = \int d[(s - S)^{0+1}/(0+1)!] = (s - S)$$

$$\eta^{[1]} \equiv d\eta^{[-(0)]}/ds = d[(s - S)^0/(0)!]/ds \equiv (s - S)^{(0-1)}/(0-1)! \text{ Not Applicable}$$
$$\eta^{[-(0)]} = \int \eta^{[-(0-1)]} ds = \int d\eta^{[-(0)]} = \int d[(s - S)^0/(0)!] = (s - S)^0/(0)! = 1$$

Take m = 1:

$$\eta^{[-1]} = d\eta^{[-(1+1)]}/ds = d[(s - S)^{1+1}/(1+1)!]/ds = (s - S)^1/1! = (s - S)$$
$$\eta^{[-2]} = \int \eta^{[-1]} ds = \int d\eta^{[-(1+1)]} = \int d[(s - S)^{1+1}/(1+1)!] = (s - S)^2/2!$$

$$\eta^{[0]} \equiv d\eta^{[-(1)]}/ds = d[(s - S)^1/(1)!]/ds \equiv (s - S)^{(1-1)}/(1-1)! = 1$$
$$\eta^{[-1]} = \int \eta^{[-(1-1)]} ds = \int d\eta^{[-(1)]} = \int d[(s - S)^1/(1)!] = (s - S)^1/(1)!$$

Unit Impulse function:

Only For: $-\varepsilon < s - S < \varepsilon$ where $\varepsilon \rightarrow +0$

Only For m = - 1

:

$\eta^{(1)} \equiv \delta(s - S) \equiv 0$ if $(s \neq S)$ $\qquad$ $\eta^{(1)} \equiv \delta(s - S) \rightarrow \infty$ if $(s \rightarrow S)$

So:

Only For m = 0

$$\eta^{(0)} \equiv \int \eta^{(1)} ds \equiv \int \delta\, ds \equiv 1$$

For m = - 1

$$\eta^{(-1)} \equiv \int \eta^{(0)} ds \equiv \int_{-\varepsilon}^{+\varepsilon} 1\, ds = 0$$

And:

For m = 0 :

$$\eta^{(0)} \equiv \int \eta^{(-1)} ds \equiv \int_{-\varepsilon}^{+\varepsilon} 0\, ds = 0$$

So:

For any $0 \leq m$:

$$\eta^{(-m)} \equiv \int \eta^{(0)}(s - S)\, ds \equiv \int_{-\varepsilon}^{+\varepsilon} 0\, ds = 0$$

So:

For $0 \leq m$: if $(0 < s - S)$:

$$\eta^{(-m)} \equiv (s - S)^m/m!$$

Where:

$$\eta^{(-m)} \equiv d\eta^{-(m+1)}/ds = d[(s - S)^{m+1}/(m+1)!]/ds \equiv (s - S)^m/m!$$

Particularly For m = 0

$$\eta \equiv \eta^{(0)} \equiv d\eta^{(-1)}/ds = d[(s - S)^1/(1)!]/ds \equiv \eta$$

Then Only For m = 0

$$\eta \equiv \eta^{(0)} \equiv \int \eta^{(1)} ds \equiv \int \delta(s - S)\, ds \equiv 1$$

Or:

$$\eta^{(1)} \equiv d\eta^{(0)}/ds \equiv d\eta/ds \equiv \delta(s - S) \rightarrow \infty \text{ if } (s \rightarrow S)$$

$$\eta^{(-m)} = (s - S)^m / (m!)$$

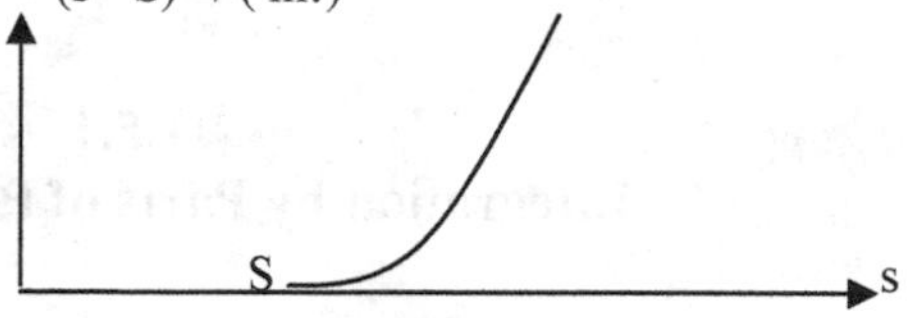

S s

$$\eta^{(-1)} = (s - S)^1 / (1!)$$

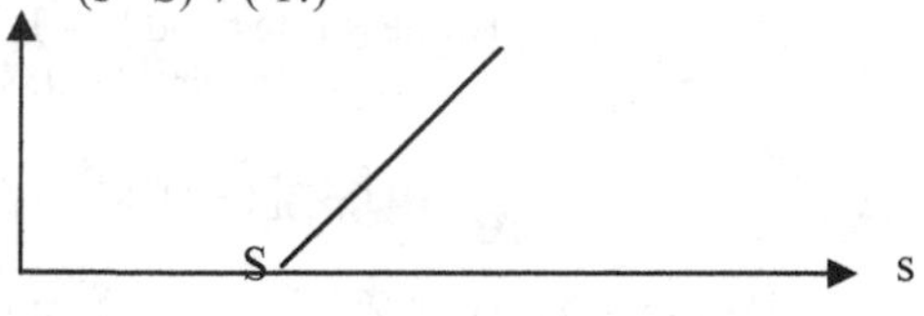

S s

$\eta^{(0)}$

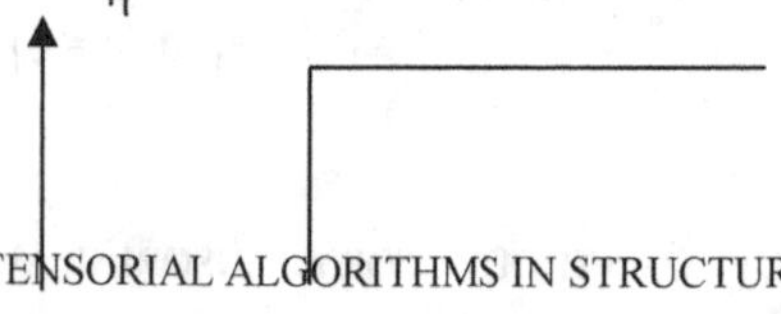

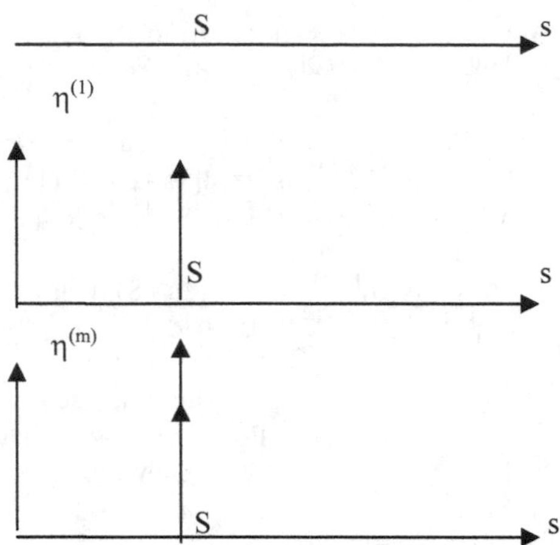

For $0 \leq m$: if $(0 < s - S)$: unit: Time ^{m} :

$\eta^{(-m)} \equiv (s - S)^{m}/m!$: unit: Time ^{m} :

Such that:

$\eta^{[-(m+1)]} \equiv \int \eta^{[-m]} ds$ $\qquad$ $d\eta^{[-(m+1)]} \equiv \eta^{[-m]} ds$ $\qquad$ $\eta^{[-m]} \equiv d\eta^{[-(m+1)]}/ds$

If:

$\phi(s - S) \equiv f * \eta(s - S)$

Then:

$f\eta^{[-(m+1)]} \equiv \int f\eta^{[-m]} ds$ $\qquad$ $df\eta^{[-(m+1)]} \equiv f\eta^{[-m]} ds$ $\qquad$ $f\eta^{[-m]} \equiv d f\eta^{[-(m+1)]}/ds$

$\phi^{[-(m+1)]} \equiv \int \phi^{[-m]} ds$ $\qquad$ $d\phi^{[-(m+1)]} \equiv \phi^{[-m]} ds$ $\qquad$ $\phi^{[-m]} \equiv d\phi^{[-(m+1)]}/ds$

If $m = 0$:

$f\eta^{[-1]} \equiv \int f\eta\, ds$ $\qquad$ $df\eta^{[-1]} \equiv f\eta^{[} ds$ $\qquad$ $f\eta \equiv d f\eta^{[-1]}/ds$

If $m = -1$:

$f\eta \equiv \int f\eta^{[-1]} ds$ $\qquad$ $d f\eta \equiv f\eta^{[1]} ds$ $\qquad$ $f\eta^{[1]} \equiv d f\eta/ds$

$f\eta \equiv \int f\delta\, ds$ $\qquad$ $d f\eta \equiv f\delta\, ds$ $\qquad$ $f\delta \equiv d f\eta/ds$

1.5.
Integration by Parts:

$$\int_{v=v(s1)}^{v=v(s2)} u(s) * dv(s) = u(s) * v(s)\,\Big|_{v=v(s1)}^{v=v(s2)} - \int_{u=u(s1)}^{u=u(s2)} v(s)\, du(s)$$

$$\int_{s=s1}^{s=s2} u(s) * dv(s)/ds * ds = [u(s) * v(s)]\,\Big|_{s=s1}^{s=s2} - \int_{s=s1}^{s=s2} [v(s) * du(s)/ds * ds$$

1.5.1.
Integration by Parts of Power Function:

Unitless v:

Unitless Integrand $[(-1)^{n} s^{n} * v(s)^{(n)}]$:

Integral Unit: s^{1} :

:

$$\int_{s1}^{s2} [(-1)^{n} s^{n} * v^{(n)}] * ds$$

=

$$[\, n!/n!\,(-1)^{n} s^{n} v^{(n-1)} + n!/(n-1)!\,(-1)^{n-1} s^{n-1} v^{(n-2)} + \ldots$$
$$+ n!/4!\,(-1)^{4} s^{4} v^{(3)} + n!/3!\,(-1)^{3} s^{3} v^{(2)} + n!/2!(-1)^{2} s^{2} v^{(1)} + n!/1!\,(-1)^{1} s\, v\,]\,\Big|_{s1}^{s2} +$$
$$+ \int_{s1}^{s2} [\, n!/0! * v] * ds$$

$$n = 1$$
$$\int_{s1}^{s2} [(-1)^1 s * v]^{\bullet}\, ds$$
$$=$$
$$[(-1)^1 s\, v]|_{s1}^{s2} + \int_{s1}^{s2} [v] * ds$$
$$=$$
$$[1!/1!\,(-1)^1 s^1 v^{(0)}]|_{s1}^{s2} + \int_{s1}^{s2} [1!/0! * v] ds$$
$$n = 2$$
$$(-1)^2 \int_{s1}^{s2} s^2 v^{(2)} ds = s^2 v^{(1)}|_{s1}^{s2} - \int_{s1}^{s2} 2\, s\, v^{(1)} ds$$
$$(-1)^2 \int_{s1}^{s2} s^2 v^{(2)} ds = s^2 v^{(1)}|_{s1}^{s2} + 2\{(-1)^1 \int_{s1}^{s2} s\, v^{(1)} ds\}$$
$$(-1)^2 \int_{s1}^{s2} s^2 v^{(2)} ds = s^2 v^{(1)}|_{s1}^{s2} + 2\{[(-1)^1 s\, v]|_{s1}^{s2} + \int_{s1}^{s2} v(s)\, ds\}$$
$$(-1)^2 \int_{s1}^{s2} s^2 v^{(2)} ds = \{[(-1)^2 s^2 v^{(1)} + 2(-1)^1 s\, v]|_{s1}^{s2} + 2\int_{s1}^{s2} v(s)\, ds\}$$
$$(-1)^2 \int_{s1}^{s2} s^2 v^{(2)} ds = \{[(-1)^2 s^2 v^{(1)} + 2!\,(-1)^1 s\, v]|_{s1}^{s2} + 2!\int_{s1}^{s2} v(s)\, ds\}$$
$$(-1)^2 \int_{s1}^{s2} s^2 v^{(2)} ds = 2!/2!\,(-1)^2 s^2 v^{(2-1)} + 2!/1!\,(-1)^{2-1} s^{2-1} v^{(2-2)} + 2!/0! \int_{s1}^{s2} v\, ds\}$$
$$n = 3$$
$$(-1)^3 \int_{s1}^{s2} s^3 v^{(3)} ds = (-1)^3 s^3 v^{(2)}|_{s1}^{s2} + (-1)^2 \int_{s1}^{s2} 3\, s^2 v^{(2)} ds$$
$$(-1)^3 \int_{s1}^{s2} s^3 v^{(3)} ds = (-1)^3 s^3 v^{(2)}|_{s1}^{s2} + 3(-1)^2 \int_{s1}^{s2} s^2 v^{(2)} ds$$
$$(-1)^3 \int_{s1}^{s2} s^3 v^{(3)} ds = (-1)^3 s^3 v^{(2)}|_{s1}^{s2} + 3\{[(-1)^2 s^2 v^{(1)} + 2(-1)^1 s\, v]|_{s1}^{s2} + 2\int_{s1}^{s2} v\, ds\}$$
$$(-1)^3 \int_{s1}^{s2} s^3 v^{(3)} ds = \{[(-1)^3 s^3 v^{(2)} + 3(-1)^2 s^2 v^{(1)} + 3 * 2(-1)^1 s\, v]|_{s1}^{s2} + 3 * 2\int_{s1}^{s2} v\, ds\}$$
$$(-1)^3 \int_{s1}^{s2} s^3 v^{(3)} ds = \{(-1)^3 s^3 v^{(2)} + 3(-1)^2 s^2 v^{(1)} + 3!\,(-1)^1 s\, v]|_{s1}^{s2} + 3!\int_{s1}^{s2} v\, ds\}$$
$$(-1)^3 \int_{s1}^{s2} s^3 v^{(3)} ds = \{(-1)^3 s^3 v^{(2)} + 3!/2!\,(-1)^2 s^2 v^{(1)} + 3!\,(-1)^1 s\, v]|_{s1}^{s2} + 3!\int_{s1}^{s2} v\, ds\}$$
$$= 3!/3!\,(-1)^3 s^3 v^{(2)} + 3!/2!\,(-1)^2 s^2 v^{(1)} + 3!/1!\,(-1)^1 s^0 v^{(0)} + 3!/0! \int_{s1}^{s2} v\, ds\}$$
$$n = 4$$
$$(-1)^4 \int_{s1}^{s2} s^4 v^{(4)} ds = (-1)^4 s^4 v^{(3)}|_{s1}^{s2} + (-1)^3 \int_{s1}^{s2} 4\, s^3 v^{(3)} ds$$
$$(-1)^4 \int_{s1}^{s2} s^4 v^{(4)} ds = (-1)^4 s^4 v^{(3)}|_{s1}^{s2} + 4\{(-1)^3 s^3 v^{(2)} + 3!/2!\,(-1)^2 s^2 v^{(1)} + 3!\,(-1)^1 s\, v]|_{s1}^{s2} + 3! \int_{s1}^{s2} v\, ds\}$$
$$(-1)^4 \int_{s1}^{s2} s^4 v^{(4)} ds = \{(-1)^4 s^4 v^{(3)} + 4!/3!(-1)^3 s^3 v^{(2)} + 4!/2!(-1)^2 s^2 v^{(1)} + 4!\,(-1)^1 s\, v]|_{s1}^{s2} + 4! \int_{s1}^{s2} v\, ds\}$$
$$= [4!/4!\,(-1)^4 s^4 v^{(3)} + 4!/3!\,(-1)^3 s^3 v^{(2)} + 4!/2!(-1)^2 s^2 v^{(1)} + 4!/1!\,(-1)^1 s\, v]|_{s1}^{s2} + + \int_{s1}^{s2} [4!/0!\, v] * ds$$

1.5.2.
Integration by Parts of Exponential Function:

$$\int_{s1}^{s2} [\exp(s/t)]\,[dv(s)/ds]\, ds = [\exp(s/t)]\,[v(s)]|_{s1}^{s2} - \int_{s1}^{s2} [v(s)]\,[1/t \exp(s/t)]\, ds$$

$$\int_{s1}^{s2} [u(s)]\,[\exp(s/t)]\, ds = [u(s)]\,[t \exp(s/t)]|_{s1}^{s2} - \int_{s1}^{s2} [t \exp(s/t)]\,[du(s)/ds\, ds]\, ds$$

1.5.3.
Integration by Parts of Unit Functions :

$$\eta^{[-(m+1)]} \equiv \int \eta^{[-m]} ds \qquad d\eta^{[-(m+1)]} \equiv \eta^{[-m]} ds \qquad \eta^{[-m]} \equiv d\eta^{[-(m+1)]}/ds$$

$$\int_{s=s1}^{s=s2} \eta^{(-m)} (s - S)\, dv(s)/ds\, ds = \eta^{(-m)} v(s)|_{s=s1}^{s=s2} - \int_{s=s1}^{s=s2} [v(s)]\,[d\eta^{(-m)}/ds]\, ds$$
$$\int_{s=s1}^{s=s2} \eta^{(-m)} (s - S)\, dv(s)/ds\, ds = \eta^{(-m)} v(s)|_{s=s1}^{s=s2} - \int_{s=s1}^{s=s2} [v(s)]\,[\eta^{[-(m)+1]}]\, ds$$

$$\int_{s=s1}^{s=s2} u(s)\, d\eta^{[-m]}/ds\, ds = [u(s)]\,[\eta^{[-m]}]|_{s=s1}^{s=s2} - \int_{s=s1}^{s=s2} \eta^{[-m]}\,[du(s)/ds]\, ds$$
$$\int_{s=s1}^{s=s2} u(s)\, \eta^{[-m+1]} ds = [u(s)]\,[\eta^{(m)}]|_{s=s1}^{s=s2} - \int_{s=s1}^{s=s2} \eta^{[-m]}\,[du(s)/ds]\, ds$$

But:

$$f\eta \equiv \int f\delta\, du \qquad d\, f\eta \equiv f\delta\, du \qquad f\delta \equiv d\, f\eta/du$$

Then:

$$\int_{s=s1}^{s=s2} \eta(s - S)\, dv(s)/ds\, ds = \eta(s - S)\, v(s)\Big|_{s=s1}^{s=s2} - \int_{s=s1}^{s=s2} [v(s)]\, [\delta(s - S)]\, ds$$
$$\int_{s=s1}^{s=s2} \eta(s - S)\, dv(s)/ds\, ds = [\, \eta(s_2 - S)\, v(s_2) - \eta(s_1 - S)\, v(s_1)\,] - [v(s_2 - S)]$$

$$\int_{s=s1}^{s=s2} u(s)\, \delta(s - S)\, ds = [u(S)\,]$$

1.6.
Probability Density Funtions:

Expectations and Derivatives Expectations are Dependent on Probabity Density p(X) :

{p(X)]}

1.6.1.
First Order Probability:

X

:

X^1

X^2

...

X^R

...

X^m

:

Expectations are Dependent on Probabity Density p($\underline{X}$) :

[p(X)]

:

First Order Probability:

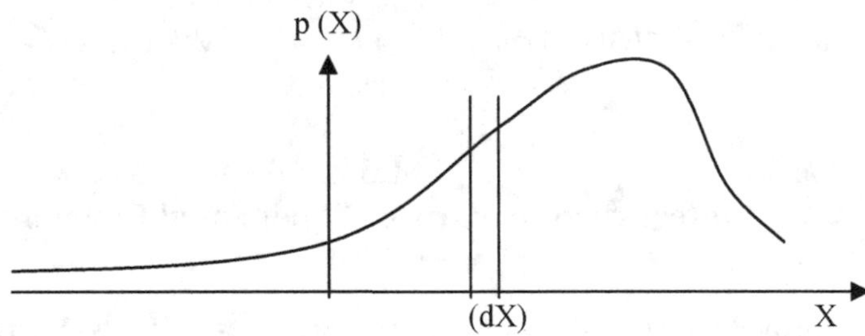

Probability Distribution of X in the Region:

$$a < X < b$$
$$P(a < X < b) = \int_a^b p(X)\ dX$$
$$P(-\infty < X < \infty) = \int_{-\infty}^{+\infty} p(X)\ dX = 1$$

Units:

$$[P(a < X < b)\,] = [p(X)\,] * [X\,] = 1$$
$$[p(X)\,] = 1/\,[X\,]$$

1.6.2.
Second Order Joint Probability:

Two Dimensional Probability Space:

X_I X_J

:

Expectations are Dependent on Probabity Density p($\underline{X}$) :

[p(X)]

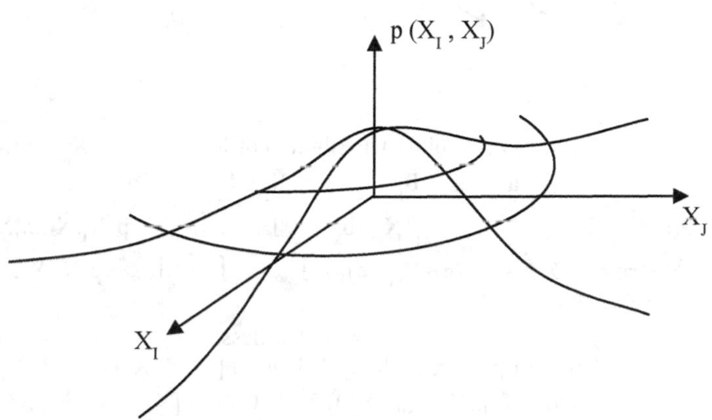

Joint Two Probability Distribution of X_I and X_J in the Region:

$$a_I < X_I < b_I \qquad a_J < X_J < b_J$$

$$P(a_I < X_I < b_I, a_J < X_J < b_J) = \int_{aJ}^{bJ} \int_{aI}^{bI} p(X_I, X_J)\, dX_I\, dX_J$$

$$P(-\infty < X_I < \infty, -\infty < X_J < -\infty) = \int_{-\infty}^{+\infty} \int_{-\infty}^{+\infty} p(X_I, X_J)\, dX_I\, dX_J = 1$$

Units:

Unit of [$p(X_I, X_J)$] * Unit of [X_I * X_J] = 1

Unit of [$p(X_I, X_J)$] = 1 / Unit of [X_I * X_J]

1.6.3.
M - th Order Joint Probability:

(M - th Order) Dimensional Probability Space:

X_1	X_2	...	X_M
$X_1^{\ 1}$	$X_2^{\ 1}$	...	$X_M^{\ 1}$
$X_1^{\ 2}$	$X_2^{\ 2}$	...	$X_M^{\ 2}$
$X_1^{\ 2}$	$X_2^{\ 2}$	...	$X_M^{\ 2}$
$X_1^{\ m}$	$X_2^{\ m}$	...	$X_M^{\ m}$

:

Expectations are Dependent on Probabity Density p($\underline{X}$) :

[p(X)]

Symbolic (M) Dimensional Probability Space

$$p(X_1, X_2, ..., X_M)$$

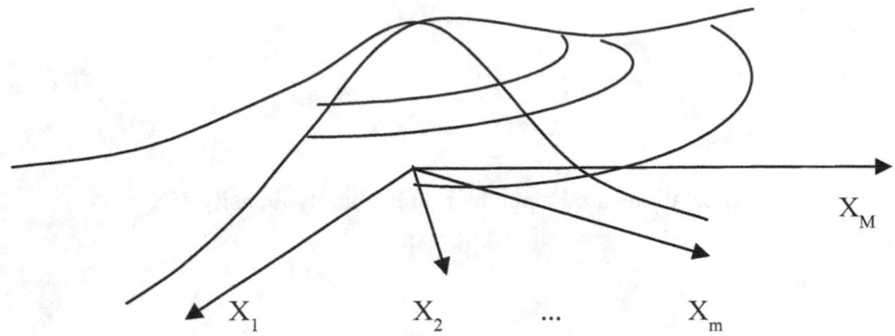

Joint M th Probability Distribution of X_1 X_2 ... and X_M in the Region:

$$a_1 < X_1 < b_1 \qquad a_2 < X_2 < b_2 \;\ldots\qquad a_M < X_M < b_M$$

$$P(a_1<X_1<b_1,a_2<X_2<b_2,\ldots a_M<X_M<b_M) = \int_{aM}^{bM}\ldots\int_{a2}^{b2}\int_{a1}^{b1} p(X_1,X_2,\ldots,X_M)dX_1dX_2\ldots dX_M$$

$$P(-\infty<X_1<\infty,-\infty<X_2<\infty,\ldots -\infty<X_M<\infty) = \int_{-\infty}^{+\infty}\ldots\int_{-\infty}^{+\infty}\int_{-\infty}^{+\infty} p(X_1,X_2,\ldots,X_M)dX_1dX_2\ldots dX_M = 1$$

:

Unitless:

$$\text{Unit of}[\,p(X_1,X_2,\ldots,X_M)] * \text{Unit of}[\,X_1 * X_2 * \ldots * X_M)] = 1$$

$$\text{Unit of}[p(X_1,X_2,\ldots,X_M)] = 1/\,\text{Unit of}[X_1 * X_2 * \ldots * X_M)]$$

Probability Distribution is Unitless

$$[P(a_1<X_1<b_1,a_2<X_2<b_2,\ldots a_M<X_M<b_M)\,] = [p(X_1,X_2,\ldots,X_M)] * [X_1 * X_2 * \ldots X_M)] = 1$$

And Probability Density Unit is Inverse Unit Variable Products

$$\text{Unit of}[p(X_1,X_2,\ldots,X_M)] = 1/\,\text{Unit of}[X_1 * X_2 * \ldots * X_M)]$$

1.6.4.
Gaussian Distribution:

Expectations are Dependent on Probabity Density $p(\underline{X})$:

$[p(X)]$

:

K only :

$$(p(X_K)) = 1/\,(2\pi)^{1/2}\,/\,(\sigma_{\underline{K}\,\underline{K}}) \quad \exp\{-1/2[(X_K - E(X_K))]^2/\sigma_{\underline{K}\,\underline{K}}{}^2\}$$

$$(p(X_K)) = 1/\,(2\pi)^{1/2}\,/\,(\sigma_{\underline{K}\,\underline{K}}{}^2)^{1/2} \quad \exp\{-1/2[(X_K - E(X_K))]^2/\sigma_{\underline{K}\,\underline{K}}{}^2\}$$

$$(p(X_K)) = 1/\,(2\pi)^{1/2}/|\mu 2_{\underline{K}\,\underline{K}}|^{1/2} \exp\{-1/2[(X_{\underline{K}} - E(X_{\underline{K}}))]\;\mu 2_{\underline{K}\,\underline{K}}{}^{-1}\,[(X_{\underline{K}} - E(X_{\underline{K}}))]\}$$

$$(p(X_K)) = 1/\,(2\pi)^{1/2}\;\nu 2_{\underline{K}\,\underline{K}}{}^{1/2} \quad \exp\{-1/2[(X_{\underline{K}} - E(X_{\underline{K}}))]\;\nu 2_{\underline{K}\,\underline{K}}\,[(X_{\underline{K}} - E(X_{\underline{K}}))]\}$$

K and L:

$$(p(X_K, X_L)) = 1/(2\pi)^{1/2}\;|\,\underline{\nu 2}\,|^{1/2} \exp\{-1/2[(X_K - E(X_K))]\;\nu 2_{KL}\,[\,(X_L - E(X_L))]\}$$

$$(x) = \{X - E(X)\}$$

$$(p(x)) = 1/\,(2\pi)^{1/2} \quad * \quad \exp\{-1/2\,(x/\sigma)^2\}/\sigma$$

:

Commonly Occurring Integrals

$$\int x^N \;\exp\{-1/2\,(x/\sigma)^2\}\,dx/\sigma^{N+1} = (2)^{(N-1)/2} \quad \Gamma\{(N+1)/2\}$$

:

N is Even: N/2 is Integer:

$2^{(N-1)/2}\Gamma\{(N+1)/2\}$

$\int x^N \exp\{-1/2(x/\sigma)^2\}\, dx / \sigma^{N+1}$

:

$\int x^0 \exp\{-1/2(x/\sigma)^2\}\, dx / \sigma^{0+1}$

$=$

$(2\pi)^{1/2}/2 \;*\; (1)$

$\int x^2 \exp\{-1/2(x/\sigma)^2\}\, dx / \sigma^{2+1}$

$=$

$(2\pi)^{1/2}/2 \;*\; (1)$

$\int x^4 \exp\{-1/2(x/\sigma)^2\}\, dx / \sigma^{4+1}$

$=$

$(2\pi)^{1/2}/2 \;*\; (3)$

N is Odd: N/2 is Not Integer:

$2^{(N-1)/2}\Gamma\{(N+1)/2\}$

$=$

$(N+1)/2$

$\int x^N \exp\{-1/2(x/\sigma)^2\}\, dx / \sigma^{N+1}$

$=$

$(N+1)/2$

$\int x^1 \exp\{-1/2(x/\sigma)^2\}\, dx / \sigma^{1+1}$

$=$

$(1+1)/2$

$\int x^3 \exp\{-1/2(x/\sigma)^2\}\, dx / \sigma^{3+1}$

$=$

$(3+1)/2$

$\int x^5 \exp\{-1/2(x/\sigma)^2\}\, dx / \sigma^{5+1}$

$=$

$(5+1)/2$

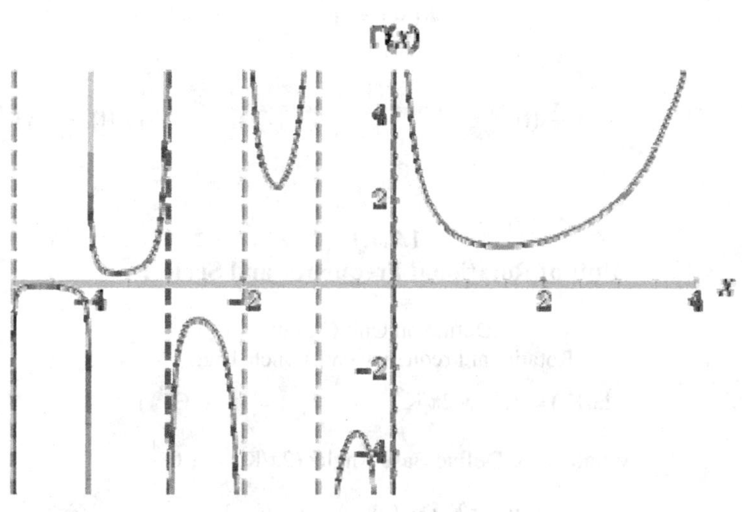

Incomplete Integral:

$\int_0^X \exp\{-1/2\,(x/\sigma)^2\}\, dx/\sigma = (2\pi)^{1/2}/2 \;*\; \mathrm{erf}\{1/2^{1/2}(X/\sigma)\}$

:

(X/σ) is small

$\int_0^X \exp\{-1/2(x/\sigma)^2\}dx/\sigma = (2\pi)^{1/2}/2 \;*$

$\mathrm{erf}\{1/2^{1/2}(X/\sigma)\}$

$\approx$

Value at Zero

Plus

$(X/\sigma)^1/(1) - (X/\sigma)^3/(3 * 2) + (X/\sigma)^5/(5 * 2^2 * 2!)$

$\int_0^X \exp\{-1/2\,(x/\sigma)^2\}\, dx/\sigma$

$\approx$

(0)

Plus

$(X/\sigma)^1/(1) - (X/\sigma)^3/(3 * 2) + (X/\sigma)^5/(5 * 2^2 * 2!)$

(X/σ) is large

$\int_0^X \exp\{-1/2(x/\sigma)^2\}dx/\sigma = (2\pi)^{1/2}/2 \;*$

$\mathrm{erf}\{1/2^{1/2}(X/\sigma)\}$

$\approx$

Value at Infinity

Minus

$\exp\{-1/2(x/\sigma)^2\} * \{(\sigma/X)^1 - (\sigma/X)^3 + 3(\sigma/X)^5\}$

$\int_0^X \exp\{-1/2\,(x/\sigma)^2\}\, dx/\sigma$

$\approx$

$(2\pi)^{1/2}/2$

Minus

$\exp\{-1/2(x/\sigma)^2\} * \{(\sigma/X)^1 - (\sigma/X)^3 + 3(\sigma/X)^5\}$

1.6.5.
Generalized Gaussian Distribution:

$$(p(\underline{X})) = 1/(2\pi)^{M/2} / |\underline{\mu 2}|^{1/2} \exp\{ - 1/2 \ [(\underline{X} - E(\underline{X}))]^{T} \underline{\mu 2}^{-1} [(\underline{X} - E(\underline{X}))]\}$$

$$(p(\underline{X})) = 1/(2\pi)^{M/2} * |\underline{v2}|^{1/2} \exp\{ - 1/2[(X_m - E(X_m))] \ v2_{mn} [(X_n - E(X_n))]\}$$

1.7.
Exponential Transformation Pairs:

1.7.1.
Exponential Series:

:

Given Phenomenon Continuous Variable Pairs:

(t) **(w)**

Such that their Product is One:

(t w) = 1

→ 1 / (w) = (t) —(1)→ 1 / (t) = (w) →

1.7.1.1.
Unit of Rotational Frequency and Series:

Define on Unit Circle:

Rotational Frequency (w_K) Such That:

$$(2\pi/K) \equiv (t \quad w\, 2\pi/K) \equiv (t \quad w_K) \equiv (t \ w \ \Theta^{1})$$

Whatever K Define Base Angle: $(2\pi/K) \equiv (\Theta^{1})$

$$(w\, 2\pi/K) \equiv (w_K) \equiv (w\,\Theta^{1})$$

$$\{\exp(j\, 2\pi/K)\} \equiv \{\exp(j \ t\, w\, 2\pi/K)\} \equiv \{\exp(j \ t\, w_K)\} \equiv \{\exp(j \ t\, w\, \Theta^{1})\}$$

If $t \rightarrow \infty$ then $w \rightarrow 0$:

If $t \rightarrow \infty$ then $w \rightarrow 0$: For any K, $(2\pi\, w/K) \rightarrow 0$ Equivalently $w_K \rightarrow 0$ and $w\,\Theta^{1} \rightarrow 0$)

Increased K

K	1	2	3	4	5	6	…	.
$2\pi/K \equiv (tw2\pi/K) = tw_K$	$2\pi/1=360$	$2\pi/2=180$	$2\pi/3=120$	$2\pi/4=90$	$2\pi/5=72$	$2\pi/6=60$	→	0

Decreased Values as K is Increased

On Unit Circle:

Define Unit of rotational frequency $\{ I_K \}$:

Unit, $\{ I_K \}$of rotational frequency:

$$\{ I_K \} \equiv \{\exp(j\, 2\pi \ 1/K)\} \equiv \{\exp(j\, 2\pi \ t\, w/K)\} \equiv \{\exp(j \ t\, w_K)\}$$

For any I_K of Constant Amplitude (Radius) of 1:

$$\| I_K \| = 1$$

$[\exp(j\,2\pi/K)] = [\exp(j\,t\,w_K)]$ On Unit Circle Increased K

K	1	2	3	4	5	6	…	.
$\exp(j\,t\,w_K)$	1	- 1	$-1/2 + j\sqrt{3}/2$	$+j$	$0.31 + j\,0.95$	$1/2 + j\,\sqrt{3}/2$	$\rightarrow$	1
$\lVert I_K \rVert$	1	1	1	1	1	1	1	1

:

K = 1

$$\{I_1\} \equiv \{\exp(j\,2\pi/1)\} \equiv \{\exp(j\,2\pi\;t\,w/1)\} \equiv \{\exp(j\;t\,w_1)\} = \{\exp(j\,t\;w\;\Theta^1)\}$$

K = 1 is $\{\exp(j\,2\pi)\}$ w_1 = Contiuous w

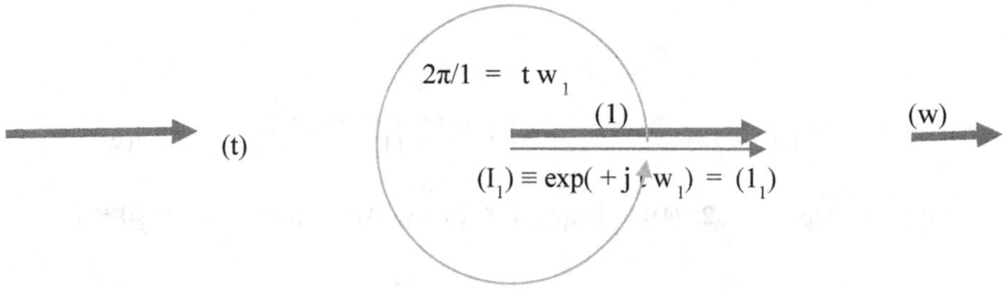

$2\pi/1 = t\,w_1$

(1)

(w)

(t)

$(I_1) \equiv \exp(+j\,t\,w_1) = (1_1)$

$$\{I_1\}^1 \equiv \{\exp(j\,1 * 2\pi/1)\} \equiv \{\exp(j\,1 * 2\pi\;t\,w/1)\} \equiv \{\exp(j\;1 *\;t\,w_1)\} \equiv 1$$

K = 2

$$\{I_2\} \equiv \{\exp(j\,2\pi/2)\} \equiv \{\exp(j\,2\pi\;t\,w/2)\} \equiv \{\exp(j\;t\,w_2)\}$$

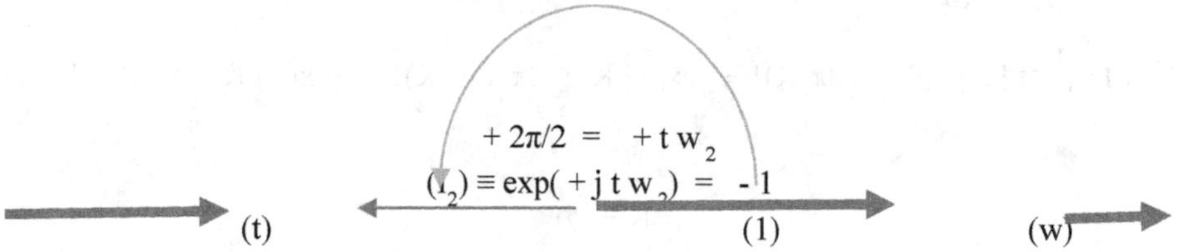

$+2\pi/2 = +t\,w_2$

$(I_2) \equiv \exp(+j\,t\,w_2) = -1$

(t)

(1)

(w)

$$\{I_2\}^2 \equiv \{\exp(j\,2 * 2\pi/2)\} \equiv \{\exp(j\,2 * 2\pi\;t\,w/2)\} \equiv \{\exp(j\,2\;t\,w_2)\} \equiv 1$$

K = 3

$$\{I_3\} \equiv \{\exp(j\,2\pi/3)\} \equiv \{\exp(j\,2\pi\;t\,w/3)\} \equiv \{\exp(j\;t\,w_3)\}$$

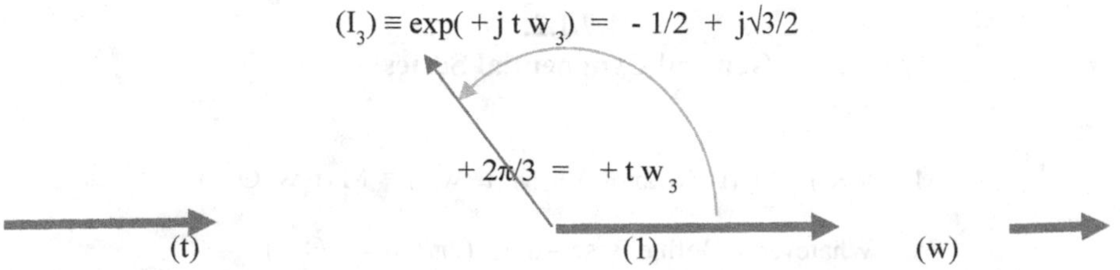

$(I_3) \equiv \exp(+j\,t\,w_3) = -1/2 + j\sqrt{3}/2$

$+2\pi/3 = +t\,w_3$

(t)

(1)

(w)

$$\{ I_3 \}^3 \equiv \{\exp(j\,3 \;*\; 2\pi /3)\} \equiv \{\exp(j\,3 \;*\; 2\pi \;\; t\,w /3)\} \equiv \{\exp(j\;\; 3\;\; t\,w_3)\} \equiv 1$$

K = 4

$$\{ I_4 \} \equiv \{\exp(j\,2\pi /4)\} \equiv \{\exp(j\,2\pi \;\; t\,w /4)\} \equiv \{\exp(j \;\; t\,w_4)\}$$

This is Known ($I_4 = j$)

$$(I_4)^{+1} = j = \exp(+j\,t\,w_4) = (+j)$$

$+ 2\pi/4 = +t\,w_4$

(t) (1) (w)

$$\{ I_4 \}^4 \equiv \{\exp(j\,4 \;*\; 2\pi /4)\} \equiv \{\exp(j\,4 \;*\; 2\pi \;\; t\,w /4)\} \equiv \{\exp(j\;\; 4\;\; t\,w_4)\} \equiv 1$$

So on ...

$$\{ I_K \} \equiv \{\exp(j\,2\pi /K)\} \equiv \{\exp(j\,2\pi \;\; t\,w /K)\} \equiv \{\exp(j \;\; t\,w_K)\}$$

$$(I_K) \equiv \exp(+j\,t\,w_K) = \cos(+t\,w_K) + \sin(+t\,w_K)$$

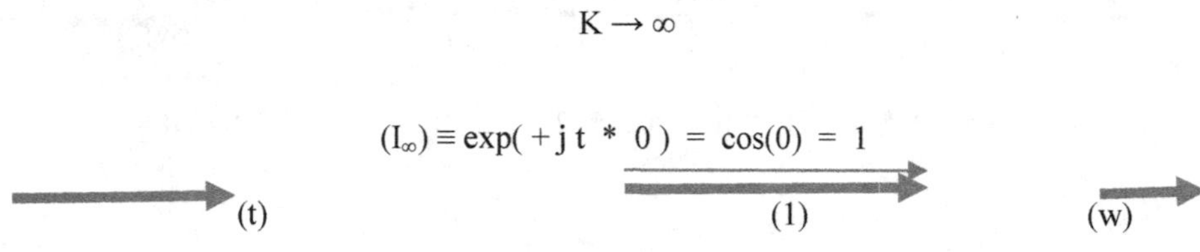

$$\{ I_K \}^K \equiv \{\exp(j\,K \;*\; 2\pi /K)\} \equiv \{\exp(j\,K \;*\; 2\pi \;\; t\,w /K)\} \equiv \{\exp(j\,K \;\; t\,w_K)\} \equiv 1$$

$$K \rightarrow \infty$$

$$(I_\infty) \equiv \exp(+j\,t \;*\; 0) = \cos(0) = 1$$

(t) (1) (w)

1.7.1.2.
General Exponential Series:

$$M\;(2\pi/K) \equiv M\;(t \quad w\,2\pi/K) \equiv M\,(t \;\; w_K) \equiv M\,(t\;\; w\;\; \Theta^1)$$

Whatever K Define Base Angle: $(2\pi/K) \equiv (\Theta^1)$

$$M(\ w\,2\pi/K\ ) \equiv M\ (\ w_K\ ) \equiv\ M\,(w\,\Theta^1\ )$$

$$\{\exp(\ j\ 2\pi/\ K)\} \equiv \{\exp(\ j\ \ t\,w\ 2\pi\ /K)\} \equiv\ \{\exp(\ j\ \ \ t\,w_K)\} \equiv\ \{\exp(\ j\ \ \ t\,w\ \Theta^1\ )\}$$

:

Whatever K Define Base Angle: $(\ w\,2\pi/K\ ) \equiv\ (\ w_K\ ) \equiv\ (\ w\,\Theta^1\ )$

$$\{\ (I_K)^M\ \} \equiv \{\{\exp(\ j\,M\ 2\pi/\ K)\} \equiv \{\exp(\ j\ \ M\,t\,w\ 2\pi\ /K)\} \equiv\ \{\exp(\ j\,M\ \ t\,w_K)\} \equiv\ \{\exp(\ j\,M\ \ t\ \ w\ \ \Theta^1\ )\}$$

:

For any M

$$(I_K)^M \equiv \{\exp(\ j\ \ t\,M\,w_K)\} \equiv\ \{\exp(\ j\,t\,M\ \ \Theta^1\ )\}$$

$$2\pi\ M/K\ =\ \ +t\,M\,w_K\ =\ \ +t\,M\Theta^1$$

(t) (1) (w)

If $t\to\infty$ then $w\to 0$:

If $t\to\infty$ then $w\to 0$: For any K or M: $(2\pi\ w/K)\to(\ 0\ )$, $(w_K)\to(0)$ and $M\,\Theta^1\to(\ 0\ )$ and $\exp(\ j\,t\,M\Theta^1\ )\to 1$

If $t\to\infty$ then $w\to 0$: For any K or M: $\{\ (I_K)^M\ \}\to(\ 1\ )$

Where:

$$\{\ (I_K)^M\ \} \equiv \{\{\exp(\ j\,M\ 2\pi/\ K)\} \equiv \{\exp(\ j\ \ M\,t\,w\ 2\pi\ /K)\} \equiv\ \{\exp(\ j\,M\ \ t\,w_K)\} \equiv\ \{\exp(\ j\,M\ \ t\,\Theta^1\ )\}$$

:

If: K = 1

$$\{\ (I_1)^M\ \} \equiv \{\{\exp(\ j\,M\ 2\pi/\ 1)\} \equiv \{\exp(\ j\ \ M\,t\,w\ 2\pi\ /1)\} \equiv\ \{\exp(\ j\,M\ \ t\,w_1)\} \equiv\ \{\exp(\ j\,M\ \ t\,\Theta^1\ )\}$$

If: K = 2

$$\{\ (I_2)^M\ \} \equiv \{\{\exp(\ j\,M\ 2\pi/\ 2)\} \equiv \{\exp(\ j\ \ M\,t\,w\ 2\pi\ /2)\} \equiv\ \{\exp(\ j\,M\ \ t\,w_2)\} \equiv\ \{\exp(\ j\,M\ \ t\,\Theta^1\ )\}$$

Met: K = 3

$$\{\ (I_3)^M\ \} \equiv \{\{\exp(\ j\,M\ 2\pi/\ 3)\} \equiv \{\exp(\ j\ \ M\,t\,w\ 2\pi\ /3)\} \equiv\ \{\exp(\ j\,M\ \ t\,w_3)\} \equiv\ \{\exp(\ j\,M\ \ t\,\Theta^1\ )\}$$

Met: K = 4

$$\{\ (I_4)^M\ \} \equiv \{\{\exp(\ j\,M\ 2\pi/\ 4)\} \equiv \{\exp(\ j\ \ M\,t\,w\ 2\pi\ /4)\} \equiv\ \{\exp(\ j\,M\ \ t\,w_4)\} \equiv\ \{\exp(\ j\,M\ \ t\,\Theta^1\ )\}$$

…

Any K:

$$\{\ (I_K)^M\ \} \equiv \{\{\exp(\ j\,M\ 2\pi/\ K)\} \equiv \{\exp(\ j\ \ M\,t\,w\ 2\pi\ /K)\} \equiv\ \{\exp(\ j\,M\ \ t\,w_K)\} \equiv\ \{\exp(\ j\,M\ \ t\,\Theta^1\ )\}$$

Let:

$$(\ \Theta^1\ ) \equiv\ 1\ \text{Radian}$$

Then:

Any K:

$$\{\ (I_K)^M\ \} \equiv \{\{\exp(\ j\,M\ 2\pi/\ K)\} \equiv \{\exp(\ j\ \ M\,t\,w\ 2\pi\ /K)\} \equiv\ \{\exp(\ j\,M\ \ t\,w_K)\} \equiv\ \{\exp(\ j\,t\ \ w\,M)\}$$

Consider in what follows M = 1, unit circle rotational frequency (w_L) such that:

$\{ (I_L) \} \equiv \{\{\exp(j\ 2\pi/ L)\} \equiv \{\exp(j\ \ t\, w\, 2\pi /L)\} \equiv \{\exp(j\ \ t\, w_L)\}$

If L is dropped:

$\{ (I) \} \equiv \{\{\exp(j\ 2\pi)\} \equiv \{\exp(j\ \ t\, w\, 2\pi)\} \equiv \{\exp(j\ \ t\, w)\}$

1.7.1.3. Exponential Integral and Series:

:

Exponential Integrals:

Integral on t Real Line

$1/(2\pi) \int_{-\infty}^{+\infty} R(t) \exp(-\ j\, t\, w_L)\, dt = S(w_L)$

$[R]\,[t] = [\,S\,]$

Integral on w Real Line

$\int_{-\infty}^{+\infty} S(w_L) \exp(+\ j\, t\, w_L)\, d\, w = R(t)$

$[S]\ \ [w] = [R]$

Exponential Series:

Integral on t Real Line

$\text{Lim}_{T\to\infty}\ 1/T \int_{-T/2}^{T/2} \{ R(t) \}\ \exp(-j\, t\, w_L)\, dt = \{ S_{,T}(w_L)\}$

$[R] = [\,S_{,T}\,]$

Sum on Unit Circle

$\sum_{L=-\infty}^{+\infty} S_{,T}(w_L)\ \ \exp(+j\ \ t\, w_L) = \{ R(t) \}$

$[S_{,T}] = [R]$

:

Exponential Integrals and Series:

$\int_{-\infty}^{+\infty} S(w_L) \exp(+\ j\, t\, w_L)\, d\, w = R(t)$

$\sum_{L=-\infty}^{\infty}\ \ S_{,T}(w_L)\ \ \exp(+j\ \ t\, w_L) = \{ R(t) \}$

Units

$[R]\,[t] = [\,S\,]$ $\qquad$ $[S] * [w] = [R]$

$[R] = [\,S_{,T}\,]$ $\qquad$ $[S_{,T}] = [R]$

Exponential Integrals and Series At 0:

$(w_L) \to 0$

$1/(2\pi) \int_{-\infty}^{\infty} R(t) \exp(-\, j\, t\, 0)\, dt$

$=$

$1/(2\pi)\ \text{Lim}_{T\to\infty} T/T \int_{-\infty}^{\infty} R(t)\ 1\ dt$

$=$

$1/(2\pi)\ \text{Lim}_{T\to\infty}\ 1/T \int_{-T/2}^{+T/2}\ T\ \ \ R(t)\ \ dt$

$=$

$S(w_L = 0\,)$

$(\,t\,) = 0$

$\int_{-\infty}^{+\infty} S(w_L) \exp(+\ j\, 0\, w_L)\, d\, w = R(t = 0)$

$\text{Lim}_{T\to\infty}\ 1/T \int_{-T/2}^{T/2}\ 1\ \ \{ R(t) \}\ \ \ dt \equiv \{ S_{,T}(w_L = 0)\}$

$\sum_{L=-\infty}^{+\infty} S_{,T}(w_L)\ \exp(+\ j\, 0\, w_L) \equiv \{ R_{,W}(t = 0) \}$

Here Only mention Different Forms of Exponential Integrals and Series:

$1/(2\pi) \int_{-T/2}^{+T/2} R(t) \exp(-\ j\, t\, w)\, dt = S(w)$ $\qquad$ $\int_{-\infty}^{+\infty} S(w) \exp(+\ j\, t\, w)\, dw = w\ *\ R_{,W}(t)$

:

$1/(2\pi)\ \mathrm{Lim}_{T\to\infty}\ 1/T \int_{-T/2}^{+T/2} T\, R(t) \exp(-j\, t\, w_L)\, dt$

$=$

$S(w_L)$

$\mathrm{Lim}_{T\to\infty}\ 1/T \int_{-T/2}^{T/2} 1 \{ R(t) \} \exp(-j\, t\, w_L)\, dt$

$=$

$\{ S_{,T}(w_L) \}$

$\mathrm{Lim}_{T\to\infty}\ 1/T \sum_{L=-\infty}^{\infty} T\, S(w_L) \exp(+j\, t w_L)\, \Delta w$

$\approx$

$R(t)$

$\mathrm{Lim}_{T\to\infty}\ 1/T \sum_{L=-\infty}^{\infty} T\, S_{,T}(w_L) \exp(+j\, t\, w_L)$

$=$

$\{ R(t) \}$

Units

$[R]\,[t] = [S]$

$[R] = [S_{,T}]$

$[S] * [w] = [R]$

$[S_{,T}] = [R]$

1.7.1.4.
Second Average:

Exponential Integrals and Series:

$\int_{-\infty}^{+\infty} S(w_L) \exp(+\, j\, t\, w_L)\, dw$

$=$

$R(t)$

$\sum_{L=-\infty}^{\infty} S_{,T}(w_L) \exp(+j\, t\, w_L)$

$=$

$\{ R(t) \}$

:

The Inner Product:

$\langle R, T \rangle = \mathrm{Lim}_{T\to\infty}\ 1/T \int_{-T/2}^{T/2} R(t)\ T^C(t)\ dt$

$\langle R, R \rangle = \mathrm{Lim}_{T\to\infty}\ 1/T \int_{-T/2}^{T/2} R(t)\ R^C(t)\ dt$

Exponential Integrals Inner Product and Series Inner Product:

$\langle R, R \rangle = \langle \|R(t)\|^2 \rangle$

$=$

$\mathrm{Lim}_{T\to\infty}\ 1/T \int_{-T/2}^{T/2}$

$\int_{-\infty}^{+\infty} S(w_M) \exp(+\, j\, t\, w_M)\, dw$

$\int_{-\infty}^{+\infty} S(w_N) \exp(-\, j\, t\, w_N)\, dw$

(dt)

$\langle \|R(t)\|^2 \rangle$

$=$

$\mathrm{Lim}_{T\to\infty}\ 1/T \int_{-T/2}^{T/2}$

$\sum_{M=-\infty}^{\infty} S_{,T}(w_M) \exp(+j\, t w_M)$

$\sum_{N=-\infty}^{\infty} S_{,T}(w_N) \exp(-j\, t w_N)$

(dt)

Integrate on real (t) First and Since:

$$\int_{-T/2}^{+T/2} \{ \exp(+ j t M) \} \{ \exp(- j t N) \} dt = 0$$

Unless:

$$(+ j t M - j t N \} = 0$$

Then:

$$1/T \int_{-T/2}^{T/2} (1) \, dt = 1$$

And:

$$\lim_{T \to \infty} 1/T \int_{-T/2}^{T/2} (1) \, dt = 1$$

Then Exponential Integrals Inner Product and Series Inner Product:

$$< \|R(t)\|^2 > = \{ \int_{-\infty}^{+\infty} S(w_L) \, dw \}^2$$

$$< \|R(t)\|^2 > = \sum_{L=-\infty}^{\infty} \|S_{,T}(w_L)\|^2$$

1.7.1.5.
Periodic Rectangular Wave:

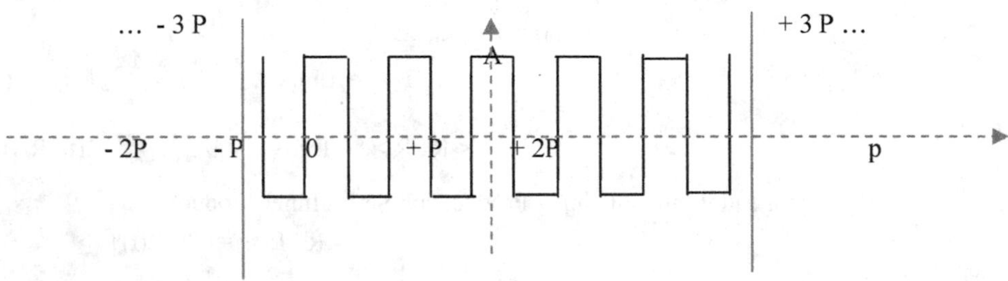

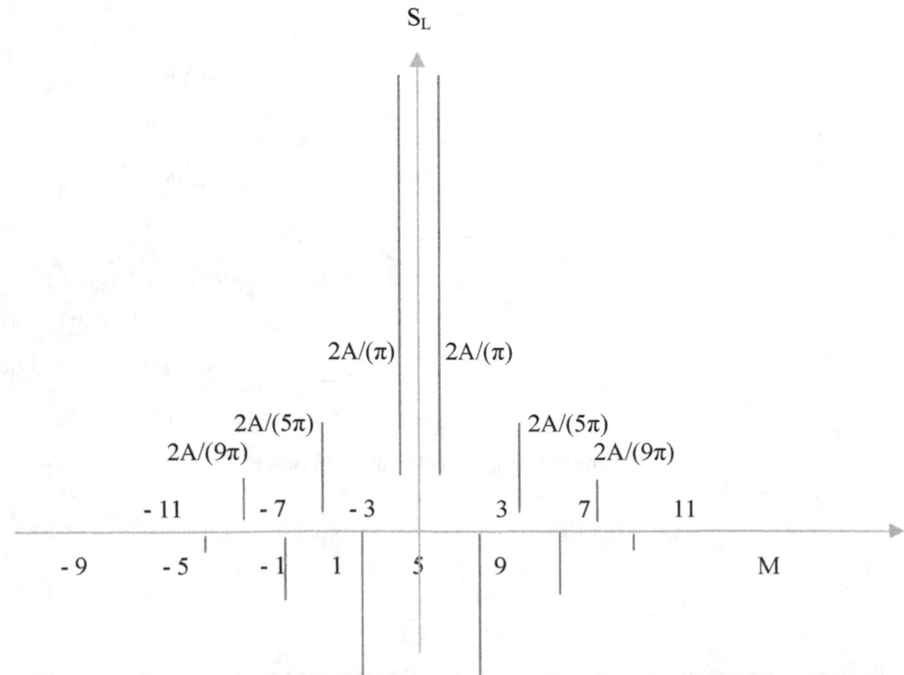

$2A/(11\pi)$ $2A/(11\pi)$

$2A/(7\pi)$ $2A/(7\pi)$

$2A/(3\pi)$ $2A/(3\pi)$

Exponential Series:

Integral on t Real Line

$1/T \int_{-T/2}^{T/2} R_{,W}(t) \exp(-j\, t\, w\, L)\, dt$

=

$S_{L,T}(w) = S(w)\,(2\pi)/T$

$S_{L,T}(w)$ is Function of (L, T and w)

Sum on Unit Circle

$\sum_{L=-\infty}^{+\infty} S_{L,T}(w) \exp(+j\, t\, w\, L)$

=

$R_{,W}(t) = R(t) / w$

$R_{,W}(t)$ is Function of (w and t)

:

Applied to Periodic Rectangular Wave:

$A/T \int_{-T/2}^{+T/2} SIGN(A)\, \{\cos(L\, w\, t) - j * \sin(L * w * t)\}\, dt$

=

$S_{L,T}(w)$

Even L:

$S_{L,T}(w) = 0$

Odd L:

$S_{L,T}(w) = 2A / (L\pi) * (-1) ** \{(L - 1)/2\}$

A^2 is the Mean $S_{L,T}$, Square:

$$A^2 = 8A^2/\pi^2 (1/1^2 + 1/3^2 + 1/5^2 + 1/7^2 + \ldots)$$

$$1 = 8/\pi^2 (1/1^2 + 1/3^2 + 1/5^2 + 1/7^2 + \ldots)$$

This is the Well Known Identity:

$$\sum_{L=1}^{+\infty} 1/(2L-1)^2/ = (\pi^2)/8$$

1.7.2. Exponential Integrals

Transformation Pair Rates:

Given Phenomenon Function Pairs:

(R) (S)

Exponential Integrals:

$S(w,\Pi,\Theta)$ is Independent of t.

$\int_{-\infty}^{+\infty}$

$R(t,\Pi,\Theta)$

$R(t,\Pi,\Theta)$ is Independent of w .

$\int_{-\infty}^{+\infty}$

$S(w,\Pi,\Theta)$

$$*$$
$$\{\exp(-j * 2\pi\, t\, \Theta)\, dt\}$$
$$=$$
$$S(w,\Pi,\Theta)$$

$S(w,\Pi,\Theta)$ is Independent of t

$$*$$
$$\{\exp(+j * 2\pi\, t\, w) * d\, w\}$$
$$=$$
$$R(t,\Pi,\Theta)$$

$R(t,\Pi,\Theta)$ is Independent of w

1.7.3
Product of Exponential Integral Pairs:

$$S(w) * R(t)$$
$$=$$
$$\{\int_{-\infty}^{+\infty} R(t) * \exp[(j * 2\pi)(-t * w)] * dt\} * \{\int_{-\infty}^{+\infty} S(w) * \exp[(j * 2\pi)(+w * t)] * d\, w\}$$

Not necessarily Equal

$$\int_{-\infty}^{+\infty}\int_{-\infty}^{+\infty} R(t) * S(w) * \{\exp[(j * 2\pi)(-t * w + w * t = 0)] = 1\} * dt * d\, w$$
$$=$$
$$\int_{-\infty}^{+\infty}\int_{-\infty}^{+\infty} R(t) * S(w) * dt * d\, w$$

Independent of t and Independent of w

But $S(w = 0) * R(t = 0)$:

At 0:

$$S(w = 0) * R(t = 0)$$
$$=$$
$$\{\int_{-\infty}^{+\infty} R(t) * [1] * dt\} * \{\int_{-\infty}^{+\infty} S(w) * [1] * d\, w\}$$

Equal

$$\{\int_{-\infty}^{+\infty}\int_{-\infty}^{+\infty} R(t) * S(w) * dt * d\, w\}$$

Independent of t and Independent of w

Units

$$[S] * [R]$$
$$=$$
$$[R] * [S] * [dt] * [d\, w]$$

1.7.4.
Delta Function:

:

Cross Substitution to get $S(\Pi,\Theta)$:

$$\int_{-\infty}^{+\infty}\{\int_{-\infty}^{+\infty} S(\upsilon,\Pi,\Theta)\, \exp(+j * 2\pi\, \tau\, \upsilon)\, d\upsilon\} * \{\exp(-j * 2\pi\, \tau\, \Theta)\, d\tau\}$$

=

S(Π,Θ)

:

$$\int_{-\infty}^{\infty} \{\int_{-\infty}^{\infty} \exp(-j * 2\pi\ \tau\ (\Theta - \upsilon))\ d\tau\ \} * S(\upsilon,\Pi,\Theta)\ d\upsilon$$

=

S(Π,Θ)

Compared with Sieve Property:

$$\int_{-\infty}^{+\infty} \{\ \delta(\Theta - \upsilon)\ \} * S(\upsilon,\Pi,\Theta)\ d\upsilon$$

=

S(Π,Θ)

That is:

$$\int_{-\infty}^{\infty} \{\int_{-\infty}^{\infty} \exp(-j * 2\pi\ \tau\ (\Theta - \upsilon))\ d\tau\ \} * S(\upsilon,\Pi,\Theta)\ d\upsilon$$

=

$$\int_{-\infty}^{+\infty} \{\ \delta(\Theta - \upsilon)\ \} * S(\upsilon,\Pi,\Theta)\ d\upsilon$$

So:

$$\{\int_{-\infty}^{\infty} \exp(-j * 2\pi\ \tau\ (\Theta - \upsilon))\ d\tau\ \}$$

=

$$\{\ \delta(\Theta - \upsilon)\ \}$$

:

Similarly

$$\{\int_{-\infty}^{\infty} \exp(-j * 2\pi\ (\Pi - \tau)\ \upsilon)\ d\tau\ \}$$

=

$$\{\ \delta(\Pi - \tau)\ \}$$

Units

[R] * [τ] = [S]　　　　[S] * [υ] = [R]

Product Units

[R] * [τ] * [S] * [υ] = [S] * [R] = 1

Cross Substitution Units

[S] * [dυ] * [d τ] = [S]　　　　[R] * [d τ] * [dυ] = [R]

Unitless:

[1] = [δ(Θ - υ) dυ] * [1]　　　　[1] = [δ(Π - τ) d τ] * [1]

:

Cross Substitution to get R(Θ,Π):

$$\int_{-\infty}^{+\infty} \{\int_{-\infty}^{+\infty} R(\tau,\Theta,\Pi)\ \exp(-j * 2\pi\ \upsilon\ \tau)\ d\tau\} * \{\exp(+j * 2\pi\ \upsilon\ \Pi)\ d\upsilon\ \}$$

=

R(Θ,Π)

:

$$\int_{-\infty}^{\infty} \{\int_{-\infty}^{\infty} \exp(+j * 2\pi\ \upsilon\ (\Pi - \tau))\ d\upsilon\ \} * R(\tau,\Theta,\Pi)\ d\tau$$

=

R(Θ,Π)

Compared with Rieve Property:

$$\int_{-\infty}^{+\infty} \{\ \delta(\Pi - \tau)\ \} * R(\tau,\Theta,\Pi)\ d\tau$$

$$=$$

$$R(\Theta,\Pi)$$

That is:

$$\int_{-\infty}^{\infty} \{\int_{-\infty}^{\infty} \exp(+j * 2\pi \upsilon (\Pi - \tau))\, d\upsilon \} * R(\tau,\Theta,\Pi)\, d\tau$$

$$=$$

$$\int_{-\infty}^{+\infty} \{ \delta(\Pi - \tau) \} * R(\tau,\Theta,\Pi)\, d\tau$$

So:

$$\{ \int_{-\infty}^{\infty} \exp(+j * 2\pi \upsilon (\Pi - \tau))\, d\upsilon \}$$

$$=$$

$$\{ \delta(\Pi - \tau) \}$$

:

Similarly

$$\{ \int_{-\infty}^{\infty} \exp(+j * 2\pi (\Theta - \upsilon) \tau)\, d\upsilon \}$$

$$=$$

$$\{ \delta(\Theta - \upsilon) \}$$

Units

$$[R] * [\upsilon] = [R] \qquad [R] * [\tau] = [R]$$

Product Units

$$[R] * [\upsilon] * [R] * [\tau] = [R] * [R] = 1$$

Cross Substitution Units

$$[R] * [d\tau] * [d\upsilon] = [R] \qquad [R] * [d\upsilon] * [d\tau] = [R]$$

Unitless:

$$[1] = [\delta(\Pi - \tau)\, d\tau] * [1] \qquad [1] = [\delta(\Theta - \upsilon)\, d\upsilon] * [1]$$

1.7.5.
Incremental Variables and Functions:

Phenomenon Rates:

$$\int_{-\infty}^{+\infty} R(p) * \exp[(j * 2\pi)(-p * q)] * dp = S(q) \qquad \int_{-\infty}^{+\infty} S(q) * \exp[(j * 2\pi)(+q * p)] * dq = R(p)$$

:

$$\sum_{M=-\infty}^{+\infty} \qquad \sum_{N=-\infty}^{+\infty}$$

$$R(M * \Delta p) \qquad S(N * \Delta q)$$

$$* \qquad *$$

$$\{ \exp(j * 2\pi) \} * * (- M * \Delta p * N * \Delta q) \qquad \{ \exp(j * 2\pi) \} * * (+N * \Delta q * M * \Delta p)$$

$$* \qquad *$$

$$\Delta p \qquad \Delta q$$

$$= \qquad =$$

$$S(N * \Delta q) = S(q) \qquad R(M * \Delta p) = R(p)$$

:

Phenomenon Metric (Product of Phenomenon Rates)

$$\{ \sum_{m=-\infty}^{+\infty} R(m * \Delta p) * \exp[(j * 2\pi)(-m * \Delta p * N * \Delta q)] * \Delta p \}$$

$$*$$

$$\{ \sum_{n=-\infty}^{+\infty} S(n * \Delta q) * \exp[(j * 2\pi)(+n * \Delta q * M * \Delta p)] * \Delta q \}$$

$$=$$

$$S(N * \Delta q) * R(M * \Delta p)$$

Not necessarily Equal

$$\{\sum_{m=-\infty}^{+\infty} \sum_{n=-\infty}^{+\infty}$$

$$R(m * \Delta p) * S(n * \Delta q) * \{\exp[(j * 2\pi)[(-m * \Delta p * n * \Delta q) + (+n * \Delta q * m * \Delta p) = 0] = 1\} * \Delta p * \Delta q$$

$$=$$

$$\sum_{m=-\infty}^{+\infty} \sum_{n=-\infty}^{+\infty} R(p) * S(q) * \Delta p * \Delta q$$

But $S(q = 0) * R(p = 0)$:

Phenomenon Metric (Product of Phenomenon Rates) at 0:

$$\sum_{m=-\infty}^{+\infty} R(m * \Delta p) * \Delta p = S(q = 0) \qquad \sum_{n=-\infty}^{+\infty} S(n * \Delta q) * \Delta q = R(p = 0)$$

$$\{ \sum_{m=-\infty}^{+\infty} R(m * \Delta p) * \Delta p$$

$$*$$

$$\{ \sum_{n=-\infty}^{+\infty} S(n * \Delta q) * \Delta q \}$$

$$=$$

$$S(q = 0) * R(p = 0)$$

EQUAL

$$\sum_{m=-\infty}^{+\infty} \sum_{n=-\infty}^{+\infty} R(m * \Delta p) * S(n * \Delta q) * \Delta p * \Delta q$$

Assume Unitless:

$$[(-M * \Delta p * N * \Delta q)] = 1 = [(+N * \Delta q * M * \Delta p)]$$

$$[\{\exp(i * 2\pi)\} * * (-M * \Delta p * N * \Delta q)] = 1 = [\{\exp(i * 2\pi)\} * * (+N * \Delta q * M * \Delta p)]$$

Units:

$$[x(p)] * [p] = [X(q)] \qquad [X(q)] * [q] = [x(p)]$$

Cross Substitution:

$$[X(q)] * [q] * [p] = [X(q)] \qquad [x(p)] * [p] * [q] = [x(p)]$$

Or:

$$[q] * [p] = 1 = [p] * [q]$$

$$\sum_{M=1}^{+\infty} \{ A(M) * \cos(2\pi * M) + (i) * B(M) * \sin(2\pi * M)\}$$

1.7.6.
Constant Product of Independent Variables:

:

Define Phenomenon Pairs:

[R(u)] [S(v)]

Let:

$(u) = A * p$ $(v) = 1/A * q$

Bands:

$(du) = A * dp$ $(dv) = 1/A * dq$

Where:

$P = a p$ $P Q = p q$ $Q = q/a$

$1/(2\pi)\int_{-\infty}^{\infty} R(Ap) \exp(-j(Ap)(q/A))\, d(Ap) = S(q/A)$

$\int_{-\infty}^{\infty} S(q/A) \exp(+ j A p q/A)\, d(q/A) = R(Ap)$

Then:

$A\ 1/(2\pi)\int_{-\infty}^{+\infty} R(Ap) * \exp(-j p q)\, dp = S(q/A)$

$1/A \int_{-\infty}^{+\infty} X \exp(+ j p q)\, dq = R(Ap)$

Compared with:

$S(q) = 1/(2\pi) \int_{-\infty}^{+\infty} R \exp(-jpq)\, dp$ $R(p) = \int_{-\infty}^{+\infty} X * \exp(+ ipq)\, dq$

Then:

$A\, S(q) = S(q/A)$ $1/A\, R(p) = R(Ap)$

At Origin:

$1/(2\pi)\int_{-\infty}^{+\infty} R(\tau) \exp(-j\tau\omega)\, d\tau = 1/(2\pi) \int_{-\infty}^{+\infty} R(\tau)\, d\tau = S(\omega = 0)$

$\int_{-\infty}^{+\infty} S(\omega) \exp(+ j \tau \omega)\, d\omega = \int_{-\infty}^{+\infty} S(\omega)\, d\omega = R(t = 0)$

Phenomenon Metric (Product of Phenomenon Rates)

$S(v)\ R(u) =$

$\{ \int_{-\infty}^{+\infty} R(u)\exp[(j * 2\pi)(-uv)]\, A\, du \} \{\int_{-\infty}^{+\infty} S(v)\exp[(j * 2\pi)(+vu)]\, 1/A\, dv \}$

$=$

$\{ \int_{-\infty}^{+\infty} R(u)\exp[(j * 2\pi)(-uv)]\, du \} \{\int_{-\infty}^{+\infty} S(v)\exp[(j * 2\pi)(+vu)]\, dv \}$

Not necessarily Equal

$\int_{-\infty}^{+\infty} \int_{-\infty}^{+\infty} R(u)\, S(v) \{ \exp[(j * 2\pi)(-uv + vu = 0)] = 1\}\, du\, dv$

$=$

$\int_{-\infty}^{+\infty} \int_{-\infty}^{+\infty} R(u)\, S(v)\, du\, dv$

But $S(q = 0) * R(p = 0)$:

Phenomenon Metric (Product of Phenomenon Rates) at 0:

$S(v = 0)\ R(u = 0)$

$=$

$\{ \int_{-\infty}^{+\infty} R(u)[1]A\, du \} \{\int_{-\infty}^{+\infty} S(v)[1]\, 1/A\, dv \}$

=

$\{ \int_{-\infty}^{+\infty} R(u)\, du \} \{\int_{-\infty}^{+\infty} S(v)\, dv \}$

Equal

$\{\int_{-\infty}^{+\infty} \int_{-\infty}^{+\infty} R(u)\ S(v)\ du\, dv \}$

Cross Substitution of:

Phenomenon Rates:

$\int_{-\infty}^{+\infty} R(u) \exp[(j * 2\pi)(- uv)]\, A\, du =$ $[S(v)]$

$\int_{-\infty}^{+\infty} S(v) \exp[(j * 2\pi)(+ vu)]\, 1/A\, dv =$ $[R(u)]$

$\int_{-\infty}^{+\infty} R(U) \exp[(j * 2\pi)(- Uv)]\, A\, dU =$ $[S(v)]$

$\int_{-\infty}^{+\infty} S(V)\exp[(j * 2\pi)(+ Vu)]\, 1/A\, dV =$ $[R(u)]$

To Get:

$S(v) = \int_{-\infty}^{+\infty}\int_{-\infty}^{+\infty} S(V)\exp[(j * 2\pi)(+ Vu)]$

$\exp[(j * 2\pi)(- uv)]A\, du\ * 1/A\, dv =$

$\int_{-\infty}^{\infty}\int_{-\infty}^{\infty} S(V)\exp[(j * 2\pi)(- 1)u(v - V)]du dv =$

$S(v)$

$S(v) = \int_{-\infty}^{+\infty} S(V)\ \delta(v - V)\ dv$

$= \int_{-\infty}^{+\infty} \int_{-\infty}^{+\infty}$

$S(V) \exp[(j * 2\pi)[- u (v - V)]\, du\ dv$

$R(u] = \int_{-\infty}^{+\infty} \int_{-\infty}^{+\infty} [R(U)] \exp[(j * 2\pi)(- Uv)]$

$\exp[(j * 2\pi)(+ vu)]\, A\, du\, 1/A\, dv =$

$\int_{-\infty}^{+\infty}\int_{-\infty}^{+\infty} R(U) \exp[(j * 2\pi)\, v(u - U)]\, du\, dv$

$[R(u]) = \int_{-\infty}^{+\infty} R(U)\ \delta(u - U)\ du$

$= \int_{-\infty}^{+\infty}\int_{-\infty}^{+\infty}$

$[R(U)] \exp[(j * 2\pi)[+ v (u - U)]\, du\ dv$

So:

$\delta(v - V) = \int_{-\infty}^{+\infty} \exp[(j * 2\pi)[- u (v - V)]\, du$

$\delta(u - U) = \int_{-\infty}^{+\infty} \exp[(j * 2\pi)[+ v(u - U)]\ dv$

1.7.7. Variable Merge Transformation:

1.7.7.1. Left - Sided Merge Transformation:

$(2\pi) * (p) = P$ Left - Sided Merge Transformation:

Phenomenon Rates:

$\int_{-\infty}^{+\infty} R(p) \exp(- j\, 2\pi\, p\, q)\, dp = S(q)$

$\int_{-\infty}^{+\infty} S(q) \exp(+ j\, 2\pi\, p\, q)\, dq = R(p)$

:

$(2\pi) * (p) = T$

$(q) = \Omega$

$(2\pi) * p * q = T * \Omega$

:

$(dp) = 1/(2\pi)\, dT$

$(dq) = d\Omega$

$(dp)(dq) = 1/(2\pi)\, dT\ d\Omega$

Then:

Phenomenon Rates:

$1/(2\pi) \int_{-\infty}^{+\infty} R(T) \exp(- j\, T\, \Omega)\, dT = S(\Omega)$

$\int_{-\infty}^{+\infty} S(\Omega) \exp(+ j\, T\, \Omega)\, d\Omega = R(T)$

:

Phenomenon Rates at 0:

$1/(2\pi)\int_{-\infty}^{+\infty} R(T)\ dT = S(\Omega = 0)$

$\int_{-\infty}^{+\infty} S(\Omega)\ d\Omega = R(T = 0)$

Phenomenon Metric (Product of Phenomenon Rates) at 0:

$$1/(2\pi)\int_{-\infty}^{+\infty}\int_{-\infty}^{+\infty} R(T)\ \ S(\Omega)\ dT\ \ d\Omega = (R(T=0)\ S(\Omega=0))$$

Delta Function:

$$S(\omega) = \int_{-\infty}^{+\infty} S(\Omega) * \delta(\omega-\Omega)\, d\Omega \qquad (R(t)) = \int_{-\infty}^{+\infty} R(T) * \delta(t-T)\, dT$$

$$= \int_{-\infty}^{+\infty}\int_{-\infty}^{+\infty} \qquad = \int_{-\infty}^{+\infty}\int_{-\infty}^{+\infty}$$

$$S(\Omega) * (\exp(-j * t(\omega-\Omega)) * dt * d\Omega/(2\pi) \qquad (R(T)) * (\exp(+j * \omega(t-T)) * dT * d\omega/(2\pi)$$

$$\delta(\omega-\Omega) = \int_{-\infty}^{+\infty}(\exp(-j * t(\omega-\Omega))\, dt/(2\pi) \qquad \delta(t-T) = \int_{-\infty}^{+\infty}(\exp(+j * \omega(t-T)) * d\omega/(2\pi)$$

Since this left merge will be used:

Including:

Keeping in mind:

$$(-j)^{2K} = (-1)^{K} = (+j)^{2K}$$

Exponential Pair: :

$$1/(2\pi)\int_{-\infty}^{\infty} R(\tau)\ \exp(-j\tau\omega)\, d\tau = S(\omega) \qquad \int_{-\infty}^{+\infty} S(\omega)\ \exp(+j\tau\omega)\, d\omega = R(\tau)$$

Unit of [R] Unit of [τ] ≡ Unit of [S] Unit of [S] Unit of [ω] ≡ Unit of [R]

Density when independent variables are 0:

$$1/(2\pi)\int_{-\infty}^{\infty} R(\tau)\ \ * 1 *\ \ d\tau = S(\omega=0) \qquad \int_{-\infty}^{+\infty} S(\omega)\ \ * 1 *\ \ d\omega = R(\tau=0)$$

$$1/(2\pi)\, R(\tau) = d\{S(\omega=0)\}/\, d\tau \qquad S(\omega) = d\{R(0)\}/\, d\omega$$

Such That:

$R(\tau)$		If $0 < S(\omega)$ is Positive
Generally it is	◄	Then
Not Necessarily Positive		$0 < \int_{-\infty}^{\infty} S(\omega)\, d\omega = R(0)$ is Positive

And:

If $0 < 1/(2\pi)\, R(\tau)$ is Positive		$S(\omega)$
Then	►	Generally it is
$0 < 1/(2\pi)\int_{-\infty}^{\infty} R(\tau)\, d\tau = S(0)$ is Positive		Not Necessarily Positive

:

Derivatives:

:

Since:

$$[d^{2K}\{\exp(-j\tau\omega)\}/\, d\omega^{2K}] = (-j\tau)^{2K}\ \exp(-j\tau\omega) = (-1)^{K}\ (\tau)^{2K}\ \exp(-j\tau\omega)$$

$$[d^{2K}\{\exp(+j\tau\omega)\}/\, d\tau^{2K}] = (+j\omega)^{2K}\ \exp(+j\tau\omega) = (-1)^{K}\ (\omega)^{2K}\ \exp(+j\tau\omega)$$

:

Derivative of Exponential Pair:

$$(-1)^{K}/(2\pi)\int_{-\infty}^{\infty} \tau^{2K} R(\tau)\exp(-j\tau\omega)\, d\tau = \{d^{2K} S(\omega)/(d\omega)^{2K}\}$$

$$(-1)^{K}\int_{-\infty}^{+\infty} \omega^{2K}\ S(\omega)\ \exp(+j\tau\omega)\ d\omega = \{d^{2K} R(\tau)/(d\tau)^{2K}\}$$

Density when independent variables are 0:

$$(-1)^{K}/(2\pi)\int_{-\infty}^{\infty} \tau^{2K}\ R(\tau)\ \ * 1 *\ d\tau \qquad (-1)^{K}\int_{-\infty}^{+\infty} \omega^{2K}\ S(\omega)\ \ * 1 *\ \ d\omega$$

= =

$\{d^{2K}S(\omega = 0)/(d\omega)^{2K}$ $\{d^{2K}R(\tau = 0)/(d\tau)^{2K}\}$

Such That

If $0 < (-1)^K \omega^{2K} S(\omega)$ is Positive

Then

$0 < (-1)^K \int_{-\infty}^{\infty} \omega^{2K} S(\omega)\, d\omega$ is Positive

=

$0 < \{d^{2K}R(\tau = 0)/(d\tau)^{2K}\}$ is Positive

◀

$(-1)^K/(2\pi)\, \tau^{2K} R(\tau)$

Generally it Is

Not Necessarily Positive

And:

If $0 < 1/(2\pi)\, R(\tau)$ is Positive

Then

$0 < 1/(2\pi)\int_{-\infty}^{\infty} R(\tau)\, d\tau = S(0)$ is Positive

▶

$(-1)^K \omega^{2K} S(\omega)$

Generally it Is

Not Necessarily Positi

:

Exponential Pair of Derivatives $U^{(M)} * V^{(N)} \ldots$:

$(-1)^K/(2\pi)\int_{-\infty}^{\infty} R_{U(M)V(N)\ldots}(\tau) \exp(-j\tau\omega)\, d\tau$ $(-1)^K \int_{-\infty}^{+\infty} S_{U(M)V(N)\ldots}(\omega) \exp(+j\tau\omega)\, d\omega$

= =

$S_{U(M)V(N)\ldots}(\omega)$ $R_{U(M)V(N)\ldots}(\tau)$

Density when independent variables are 0:

$1/(2\pi)\int_{-\infty}^{\infty} R_{U(M)V(N)\ldots}(\tau)\, d\tau = S_{U(M)V(N)\ldots}(\omega = 0)$ $\int_{-\infty}^{+\infty} S_{U(M)V(N)\ldots}(\omega)\, d\omega = R_{U(M)V(N)\ldots}(\tau = 0)$

$1/(2\pi)\, R_{U(M)V(N)\ldots}(\tau) = d\{S_{U(M)V(N)\ldots}(\omega = 0)\}/d\tau$ $S_{U(M)V(N)\ldots}(\omega) = d\{R_{U(M)V(N)\ldots}(0)\}/d\omega$

Such That:

If $0 < S_{U(M)V(N)\ldots}(\omega)$ is Positive

Then

$0 < \int_{-\infty}^{\infty} S_{U(M)V(N)\ldots}(\omega)\, d\omega = R_{U(M)V(N)\ldots}(0)$ is Positive

◀

$R_{U(M)V(N)\ldots}(\tau)$

Generally it is

Not Necessarily Positive

And:

If $0 < 1/(2\pi)\, R_{U(M)V(N)\ldots}(\tau)$ is Positive

Then

$0 < 1/(2\pi)\int_{-\infty}^{\infty} R_{U(M)V(N)\ldots}(\tau)\, d\tau = S_{U(M)V(N)\ldots}(0)$ is Positive

▶

$S_{U(M)V(N)\ldots}(\omega)$

Generally it is

Not Necessarily Positive

:

Derivative of Exponential Pair of Derivatives $U^{(M)} * V^{(N)} \ldots$:

$(-1)^K/(2\pi)\int_{-\infty}^{\infty} \tau^{2K} R_{U(M)V(N)\ldots}(\tau) \exp(-j\tau\omega)\, d\tau$ $(-1)^K \int_{-\infty}^{+\infty} \omega^{2K} S_{U(M)V(N)\ldots}(\omega) \exp(+j\tau\omega)\, d\omega$

= =

$\{d^{2K} S_{U(M)V(N)\ldots}(\omega)/(d\omega)^{2K}\}$ $\{d^{2K} P_{U(M)V(N)\ldots}(\tau)/(d\tau)^{2K}\}$

Density when independent variables are 0:

$(-1)^K/(2\pi)\int_{-\infty}^{\infty} \tau^{2K} R_{U(M)V(N)...}(\tau) * 1 * d\tau$ = $\{d^{2K} S_{U(M)V(N)...}(\omega = 0)/(d\omega)^{2K}$

$(-1)^K \int_{-\infty}^{+\infty} \omega^{2K} S_{U(M)V(N)...}(\omega) * 1 * d\omega$ = $\{d^{2K} R_{U(M)V(N)...}(\tau = 0)/(d\tau)^{2K}\}$

Such That

If $0 < (-1)^K \omega^{2K} S_{U(M)V(N)...}(\omega)$ is Positive

Then

$0 < (-1)^K \int_{-\infty}^{\infty} \omega^{2K} S_{U(M)V(N)...}(\omega)\ d\omega$ is Positive

◄ $(-1)^K/(2\pi)\ \tau^{2K} R_{U(M)V(N)...}(\tau)$

Generally it Is

Not Necessarily Positive

And:

If $0 < (-1)^K/(2\pi)\ \tau^{2K} R_{U(M)V(N)...}(\tau)$ is Positive

Then

$0 < 1/(2\pi)\int_{-\infty}^{\infty} R(\tau)\ d\tau = S(0)$ is Positive

► $(-1)^K \omega^{2K} S_{U(M)V(N)...}(\omega)$

Generally it is

Not Necessarily Positive

1.7.7.2. Symmetric Merge Transformation:

(2π) Symmetric Merge Transformation

Phenomenon Rates:

$$\int_{-\infty}^{+\infty} R(p) \exp(-j\, 2\pi\, p\, q)\, dp = S(q) \qquad \int_{-\infty}^{+\infty} S(q) \exp(+j\, 2\pi\, p\, q)\, dq = R(p)$$

:

$$\sqrt{(2\pi)} * (p) = T$$

$$\sqrt{(2\pi)} * (q) = \Omega$$

$$(2\pi) * p * q = T * \Omega$$

:

$$(dp) = 1/\sqrt{(2\pi)}\ dT$$

$$(dq) = 1/\sqrt{(2\pi)}\ d\Omega$$

$$(dp)(dq) = 1/(2\pi)\ dT\ d\Omega$$

Then:

Phenomenon Rates:

$$\int_{-\infty}^{+\infty} R(T) \exp(-j\, T\, \Omega)\, dT/\sqrt{(2\pi)} = S(\Omega) \qquad \sqrt{(2\pi)} \int_{-\infty}^{+\infty} S(\Omega) \exp(+j\, T\, \Omega)\, d\Omega = R(T)$$

:

Phenomenon Rates at 0:

$$S(\Omega = 0) = \int_{-\infty}^{+\infty} R(T)\ dT/\sqrt{(2\pi)} \qquad (R(T = 0)) = \int_{-\infty}^{+\infty} S(\Omega)\ d\Omega/\sqrt{(2\pi)}$$

Phenomenon Metric (Product of Phenomenon Rates) at 0:

$$1/(2\pi) \int_{-\infty}^{+\infty}\int_{-\infty}^{+\infty} R(T)\ S(\Omega)\ dT\ d\Omega = (R(T = 0)\ S(\Omega = 0))$$

:

Delta Function:

$$S(\omega) = \int_{-\infty}^{+\infty} S(\Omega) * \delta(\omega - \Omega)\, d\Omega = \int_{-\infty}^{+\infty}\int_{-\infty}^{+\infty} S(\Omega) * (\exp(-j * t(\omega - \Omega)) * dt * d\Omega/(2\pi)$$

$$\delta(\omega - \Omega) = \int_{-\infty}^{+\infty} (\exp(-j * t(\omega - \Omega))\, dt/(2\pi)$$

$$(R(t)) = \int_{-\infty}^{+\infty} R(T) * \delta(t - T)\, dT = \int_{-\infty}^{+\infty}\int_{-\infty}^{+\infty} (R(T)) * (\exp(+j * \omega(t - T)) * dT * d\omega/(2\pi)$$

$$\delta(t - T) = \int_{-\infty}^{+\infty} (\exp(+j * \omega(t - T)) * d\omega/(2\pi)$$

Define asymmetric R(τ) and S(ω):

$SQRT\{|B|/(2\pi)^{1+A}\}$ $\int_{-\infty}^{+\infty} (R(\tau)) \exp(+jB\omega\tau)\, d\tau = S(\omega)$

$SQRT\{|B|/(2\pi^{1-A})\}$ $\int_{-\infty}^{+\infty} S(\omega) \exp(-jB\omega\tau)\, d\omega = R(\tau)$

When A = 1 and B = - 1:

$SQRT\{|-1|/(2\pi^{1+1})\}$ $\int_{-\infty}^{+\infty} (R(\tau)) \exp(+j(-1)\omega\tau)\, d\tau = S(\omega)$

$SQRT\{|-1|/(2\pi)^{1-1}\}$ $\int_{-\infty}^{+\infty} S(\omega) \exp(-j(-1)\omega\tau)\, d\omega = R(\tau)$

When A = 0 and B = - 2 π:

$SQRT\{|-2\pi|/(2\pi)^{1+0}\}$ $\int_{-\infty}^{\infty} (R(\tau)) \exp(+j(-2\pi)\omega\tau)\, d\tau = S(\omega)$

$SQRT\{|-2\pi|/(2\pi^{1-0})\}$ $\int_{-\infty}^{+\infty} S(\omega) \exp(-j(-2\pi)\omega\tau)\, d\omega = R(\tau)$

$1/(2\pi)\int_{-\infty}^{+\infty} R(\tau) \exp(-j\omega\tau)\, d\tau = S(\omega)$

$1 \int_{-\infty}^{+\infty} S(\omega) \exp(+j\omega\tau)\, d\omega = (R(\tau))$

1.7.7.3.
Right - Sided Merge Transformation:

$(2\pi) * (q) = Q$ Right - Sided Merge Transformation

Phenomenon Rates:

$\int_{-\infty}^{+\infty} R(p) \exp(-j\, 2\pi\, p\, q)\, dp = S(q)$

$\int_{-\infty}^{+\infty} S(q) \exp(+j\, 2\pi\, p\, q)\, dq = R(p)$

:

$(p) = T$

$(2\pi) * (q) = \Omega$

$(2\pi) * p * q = T * \Omega$

:

$(dp) = dT$

$(dq) = 1/(2\pi)\, d\Omega$

$(dp)(dq) = 1/(2\pi)\, dT\, d\Omega$

Then:

Phenomenon Rates:

$\int_{-\infty}^{+\infty} R(T) \exp(-j\, T\, \Omega)\, dT = S(\Omega)$

$1/(2\pi) \int_{-\infty}^{+\infty} S(\Omega) \exp(+j\, T\, \Omega)\, d\Omega = R(T)$

:

Phenomenon Rates at 0:

$\int_{-\infty}^{+\infty} R(T)\; dT = S(\Omega = 0)$

$\int_{-\infty}^{+\infty} S(\Omega)\; d\Omega / (2\pi) = R(T = 0)$

Phenomenon Metric (Product of Phenomenon Rates) at 0:

$1/(2\pi) \int_{-\infty}^{+\infty}\int_{-\infty}^{+\infty} R(T)\; S(\Omega)\; dT\; d\Omega = (R(T=0)\, S(\Omega=0))$

Delta Function:

$S(\omega) = \int_{-\infty}^{+\infty} S(\Omega) * \delta(\omega - \Omega)\, d\Omega$
$= \int_{-\infty}^{+\infty} \int_{-\infty}^{+\infty}$
$S(\Omega) * (\exp(-j * t(\omega - \Omega)) * dt * d\Omega / (2\pi)$
$\delta(\omega - \Omega) = \int_{-\infty}^{+\infty} (\exp(-j * t(\omega - \Omega))\, dt / (2\pi)$

$(R(t)) = \int_{-\infty}^{+\infty} R(T) * \delta(t - T)\, dT$
$= \int_{-\infty}^{+\infty} \int_{-\infty}^{+\infty}$
$(R(T)) * (\exp(+j * \omega(t - T)) * dT * d\omega / (2\pi)$
$\delta(t - T) = \int_{-\infty}^{+\infty} (\exp(+j * \omega(t - T)) * d\omega / (2\pi)$

1.7.8.
Applications of Variable Merge Transformation:

Left:

$1/(2\pi)\int_{-\infty}^{+\infty} R(T)\,\{\exp(-j\,T\Omega)\}\,dT = S(\Omega)$ $\int_{-\infty}^{+\infty} S(\Omega)\,\{\exp(+j\,T\Omega)\}\,d\Omega = R(T)$

Symmetric:

$1/\sqrt{(2\pi)}\int_{-\infty}^{+\infty} R(T)\,\{\exp(-j\,T\Omega)\}\,dT = S(\Omega)$ $1/\sqrt{(2\pi)}\int_{-\infty}^{+\infty} S(\Omega)\,\{\exp(+j\,T\Omega)\}\,d\Omega = R(T)$

Right:

$\int_{-\infty}^{+\infty} R(T)\,\{\exp(-j\,T\Omega)\}\,dT = S(\Omega)$ $1/(2\pi)\int_{-\infty}^{+\infty} S(\Omega)\,\{\exp(+j\,T\Omega)\}\,d\Omega = R(T)$

At 0:

Left:

$1/(2\pi)\int_{-\infty}^{+\infty} R(T)\,dT = S(0)$ $\int_{-\infty}^{+\infty} S(\Omega)\,d\Omega = R(0)$

Symmetric:

$1/\sqrt{(2\pi)}\int_{-\infty}^{+\infty} R(T)\,dT = S(0)$ $1/\sqrt{(2\pi)}\int_{-\infty}^{+\infty} S(\Omega)\,d\Omega = R(0)$

Right:

$\int_{-\infty}^{+\infty} R(T)\,dT = S(0)$ $1/(2\pi)\int_{-\infty}^{+\infty} S(\Omega)\,d\Omega = R(0)$

Phenomenon Metric (Product of Phenomenon Rates) at 0:

$1/(2\pi)\ \int_{-\infty}^{+\infty}\int_{-\infty}^{+\infty} R(T)\ S(\Omega)\ dT\,d\Omega = (R(0)\,S(0))$

1.7.8.1. Exponent Transformation of Delta:

Exponent Transformation of Delta Function:

$\int_{-\infty}^{\infty} \delta(v)\exp(j\,v)\,dv = \lim_{V\to\infty}\{1/V * \int_{-V/2}^{V/2} \delta(v)\exp(j\,v)\,dv\} = \lim_{V\to\infty}\{1/V * V\} = 1$

Left:

Left = Symmetric * { $1/\sqrt{(2\pi)}$ } Right = Symmetric / { $1/\sqrt{(2\pi)}$ }

$1/(2\pi)\int_{-\infty}^{\infty}\delta(\sigma) * \exp(-j\upsilon\,\sigma)\,d\sigma = 1/(2\pi)$ ► $\int_{-\infty}^{+\infty}1/(2\pi) * \exp(+j\,\sigma\upsilon)\,d\upsilon = \delta(\sigma)$

$1/(2\pi)\int_{-\infty}^{+\infty} 1 * \exp(+j\upsilon\,\sigma)\,d\sigma = \delta(\upsilon)$ ◄ $\int_{-\infty}^{+\infty}\delta(\upsilon) * \exp(+j\sigma\,\upsilon)\,d\upsilon = 1$

Symmetric:

Symmetric: Symmetric:

$1/\sqrt{(2\pi)}\int_{-\infty}^{\infty}\delta(\sigma) * \exp(-j\upsilon\,\sigma)d\tau = 1/\sqrt{(2\pi)}$ ► $1/\sqrt{(2\pi)}\int_{-\infty}^{+\infty}1/\sqrt{(2\pi)} * \exp(+j\,\sigma\upsilon)d\upsilon = \delta(\sigma)$

$1/\sqrt{(2\pi)}\int_{-\infty}^{+\infty}1/\sqrt{(2\pi)} * \exp(+j\upsilon\,\sigma)d\sigma = \delta(\upsilon)$ ◄ $1/\sqrt{(2\pi)}\int_{-\infty}^{+\infty}\delta(\upsilon) * \exp(+j\sigma\,\upsilon)d\upsilon = 1/\sqrt{(2\pi)}$

Right:

Left = Symmetric / { $1/\sqrt{(2\pi)}$ } Right = Symmetric * { $1/\sqrt{(2\pi)}$ }

$\int_{-\infty}^{\infty}\delta(\sigma) * \exp(-j\upsilon\,\sigma)\,d\sigma = 1$ ► $1/(2\pi)\int_{-\infty}^{+\infty} 1 * \exp(+j\,\sigma\upsilon)d\upsilon = \delta(\sigma)$

$\int_{-\infty}^{+\infty}1/(2\pi) * \exp(+j\upsilon\,\sigma)\,d\sigma = \delta(\upsilon)$ ◄ $1/(2\pi)\int_{-\infty}^{+\infty}\delta(\upsilon) * \exp(+j\sigma\,\upsilon)d\upsilon = 1/(2\pi)$

:

◄► ► ◄►

$\delta(\sigma) \rightarrow 1/(2\pi)$ Left	$1/(2\pi)$	$1/(2\pi) \rightarrow \delta(\sigma)$
$\delta(\sigma) \rightarrow 1/\sqrt{(2\pi)}$ Symmetric	$1/\sqrt{(2\pi)}$	Symmetric $1/\sqrt{(2\pi)} \rightarrow \delta(\sigma)$
$\delta(\sigma) \rightarrow 1$	1	Right $1 \rightarrow \delta(\sigma)$
◀▶	◀	◀▶
$1 \rightarrow \delta(\upsilon)$ Left	1	$\delta(\sigma) \rightarrow 1$
$1/\sqrt{(2\pi)} \rightarrow \delta(\upsilon)$ Symmetric	$1/\sqrt{(2\pi)}$	Symmetric $\delta(\sigma) \rightarrow 1/\sqrt{(2\pi)}$
$1/(2\pi) \rightarrow \delta(\upsilon)$	$1/(2\pi)$	Right $\delta(\sigma) \rightarrow 1/(2\pi)$

At 0:
Left:

Left = Symmetric * $\{ 1/\sqrt{(2\pi)} \}$		Right = Symmetric / $\{ 1/\sqrt{(2\pi)} \}$
$1/(2\pi)\int_{-\infty}^{\infty} \delta(\sigma)\, d\sigma = 1/(2\pi)$	▶	$\int_{-\infty}^{+\infty} 1/(2\pi)\, d\upsilon = \delta(0)$
$1/(2\pi)\int_{-\infty}^{+\infty} 1\, d\sigma = \delta(0)$	◀	$\int_{-\infty}^{+\infty} \delta(\upsilon)\, d\upsilon = 1$

Phenomenon Metric (Product of Phenomenon Rates) at 0:

$$1/(2\pi) \int_{-\infty}^{+\infty}\int_{-\infty}^{+\infty} \delta(\sigma) * 1/(2\pi)\, d\sigma\, d\Omega = 1/(2\pi)\, \delta(0)$$

$$1/(2\pi) \int_{-\infty}^{+\infty}\int_{-\infty}^{+\infty} 1 * \delta(\sigma)\, d\sigma\, d\Omega = \delta(0) * 1$$

Symmetric:

Symmetric:		Symmetric:
$1/\sqrt{(2\pi)}\int_{-\infty}^{\infty} \delta(\sigma)\, d\tau = 1/\sqrt{(2\pi)}$	▶	$1/\sqrt{(2\pi)}\int_{-\infty}^{+\infty} 1/\sqrt{(2\pi)}\, d\upsilon = \delta(0)$
$1/\sqrt{(2\pi)}\int_{-\infty}^{+\infty} 1/\sqrt{(2\pi)}\, d\sigma = \delta(0)$	◀	$1/\sqrt{(2\pi)}\int_{-\infty}^{+\infty} \delta(\upsilon)\, d\upsilon = 1/\sqrt{(2\pi)}$

Phenomenon Metric (Product of Phenomenon Rates) at 0:

$$1/(2\pi) \int_{-\infty}^{+\infty}\int_{-\infty}^{+\infty} \delta(\sigma) * 1/\sqrt{(2\pi)}\, d\sigma\, d\Omega = 1/\sqrt{(2\pi)}\ \delta(0)$$

$$1/(2\pi) \int_{-\infty}^{+\infty}\int_{-\infty}^{+\infty} 1/\sqrt{(2\pi)} * \delta(\sigma)\, d\sigma\, d\Omega = \delta(0) * 1/\sqrt{(2\pi)}$$

Right:

Left = Symmetric / $\{ 1/\sqrt{(2\pi)} \}$		Right = Symmetric * $\{ 1/\sqrt{(2\pi)} \}$
$\int_{-\infty}^{\infty} \delta(\sigma)\, d\sigma = 1$	▶	$1/(2\pi)\int_{-\infty}^{+\infty} 1\, d\upsilon = \delta(0)$
$\int_{-\infty}^{+\infty} 1/(2\pi)\, d\sigma = \delta(0)$	◀	$1/(2\pi)\int_{-\infty}^{+\infty} \delta(\upsilon)\, d\upsilon = 1/(2\pi)$

Phenomenon Metric (Product of Phenomenon Rates) at 0:

$$1/(2\pi) \int_{-\infty}^{+\infty}\int_{-\infty}^{+\infty} \delta(\sigma) * 1\, d\sigma\, d\Omega = 1\ \delta(0)$$

$$1/(2\pi) \int_{-\infty}^{+\infty}\int_{-\infty}^{+\infty} 1/(2\pi) * \delta(\sigma)\, d\sigma\, d\Omega = \delta(0) * 1/(2\pi)$$

1.7.8.2.
Exponent Transformation of Exponent:

Ω_n Discret: { exp(j Ω_n T)}

$1/(2\pi)\int_{-\infty}^{+\infty}\{\exp(j\Omega_n T)\}\{\exp(-jT\Omega)\}dT$

=

$1/(2\pi)\int_{-\infty}^{+\infty}\exp\{j(\Omega_n - \Omega)T\}dT$

=

$\{\delta(\Omega_n - \Omega)\}$

=

$S(\Omega)$

$\int_{-\infty}^{+\infty} S(\Omega)\{\exp(+jT\Omega)\,d\Omega$

=

$R(T)$

T_n Discret : { exp(j Ω T_n)}

$\int_{-\infty}^{+\infty}\{\exp(j\,\Omega T_n)\}\{\exp(+jT\Omega)\,d\Omega$

=

$\int_{-\infty}^{+\infty}\exp\{j(T_n + T)\Omega\}d\Omega$

=

$(2\pi)\{\delta(T_n + T)\}$

=

$R(T)$

$1/(2\pi)\int_{-\infty}^{+\infty} R(T)\{\exp(-jT\Omega)\}\,dT$

=

$S(\Omega)$

{ exp(j Ω_n σ}

Left:

Left = Symmetric * { $1/\sqrt{(2\pi)}$ }

Right = Symmetric / { $1/\sqrt{(2\pi)}$ }

$1/(2\pi)\int_{-\infty}^{\infty}\{\exp(j\Omega_n\sigma)\} * \exp(-j\upsilon\sigma)\,d\sigma$ ► $\int_{-\infty}^{+\infty}\{\delta(\Omega_n - \upsilon)\} * \exp(+j\sigma\upsilon)\,d\upsilon$

=

$\{\delta(\Omega_n - \upsilon)\}$

$\{\exp(+j\sigma\Omega_n)\}$

$\int_{-\infty}^{+\infty}\{\exp(j\,T_n\upsilon)\} * \exp(+j\sigma\upsilon)\,d\upsilon$

$1/(2\pi)\int_{-\infty}^{+\infty} 1 * \exp(+j\upsilon\sigma)\,d\sigma = \delta(\upsilon)$ ◄

=

$(2\pi)\{\delta(T_n + \sigma)\}$

Phenomenon Metric (Product of Phenomenon Rates) at 0:

$1/(2\pi)\int_{-\infty}^{+\infty}\int_{-\infty}^{+\infty}\delta(\sigma)\;1/(2\pi)\;dT\,d\Omega = \{\delta(\Omega_n)\}\{1\}$

Symmetric:

Symmetric:

Symmetric:

$1/\sqrt{(2\pi)}\int_{-\infty}^{\infty}\delta(\sigma) * \exp(-j\upsilon\sigma)d\tau = 1/\sqrt{(2\pi)}$ ► $1/\sqrt{(2\pi)}\int_{-\infty}^{+\infty}1/\sqrt{(2\pi)} * \exp(+j\sigma\upsilon)d\upsilon = \delta(\sigma)$

$1/\sqrt{(2\pi)}\int_{-\infty}^{+\infty}1/\sqrt{(2\pi)} * \exp(+j\upsilon\sigma)d\sigma = \delta(\upsilon)$ ◄ $1/\sqrt{(2\pi)}\int_{-\infty}^{+\infty}\delta(\sigma) * \exp(+j\sigma\upsilon)d\upsilon = 1/\sqrt{(2\pi)}$

Right:

Left = Symmetric / { $1/\sqrt{(2\pi)}$ }

Right = Symmetric * { $1/\sqrt{(2\pi)}$ }

$\int_{-\infty}^{\infty}\delta(\sigma) * \exp(-j\upsilon\,\sigma)\,d\tau = 1$	►	$1/(2\pi)\int_{-\infty}^{+\infty} 1 * \exp(+j\,\sigma\upsilon)d\upsilon = \delta(\sigma)$
$\int_{-\infty}^{+\infty}1/(2\pi) * \exp(+j\upsilon\,\sigma)\,d\sigma = \delta(\upsilon)$	◄	$1/(2\pi)\int_{-\infty}^{+\infty}\delta(\sigma) * \exp(+j\sigma\,\upsilon)d\upsilon = 1/(2\pi)$
	:	
◄►	►	◄►
$\delta(\sigma) \rightarrow 1/(2\pi)$ Left	$1/(2\pi)$	$1/(2\pi) \rightarrow \delta(\sigma)$
$\delta(\sigma) \rightarrow 1/\sqrt{(2\pi)}$ Symmetric	$1/\sqrt{(2\pi)}$	Symmetric $1/\sqrt{(2\pi)} \rightarrow \delta(\sigma)$
$\delta(\sigma) \rightarrow 1$	1	Right $1 \rightarrow \delta(\sigma)$
◄►	◄	◄►
$1 \rightarrow \delta(\upsilon)$ Left	1	$\delta(\sigma) \rightarrow 1$
$1/\sqrt{(2\pi)} \rightarrow \delta(\upsilon)$ Symmetric	$1/\sqrt{(2\pi)}$	Symmetric $\delta(\sigma) \rightarrow 1/\sqrt{(2\pi)}$
$1/(2\pi) \rightarrow \delta(\upsilon)$	$1/(2\pi)$	Right $\delta(\sigma) \rightarrow 1/(2\pi)$

1.7.8.3.
Exponent Transformation of Cosine:

Ω_n Discret: $\{\exp(+j\Omega_n T) + \exp(-j\Omega_n T)\}/2$

$1/(2\pi)\int_{-\infty}^{+\infty}$

$\{\exp(+j\,\Omega_n\,T) + \exp(-j\,\Omega_n\,T)\}/2$

$\{\exp(-j\,T\Omega)\}\,dT$

=

$\{\delta(\Omega_n - \Omega) + \delta(-\Omega_n - \Omega)\}/2$

=

$S(\Omega)$

$\int_{-\infty}^{+\infty}$

$S(\Omega)$

$\{\exp(+j\,T\Omega)\,d\Omega$

=

$R(T)$

$1/(2\pi)\int_{-\infty}^{+\infty}$

$R(T)$

$\{\exp(-j\,T\Omega)\}\,dT$

=

$S(\Omega)$

Ω_n Discret: $\{\exp(+j\Omega_n T) + \exp(-j\Omega_n T)\}/2$

$\int_{-\infty}^{+\infty}$

$\{\exp(+j\,\Omega_n\,T) + \exp(-j\,\Omega_n\,T)\}/2$

$\{\exp(+j\,T\Omega)\,d\Omega$

=

$(2\pi)\{\delta(\Omega_n + \Omega) + \delta(-\Omega_n + \Omega)\}/2$

=

$R(T)$

1.7.8.4.

Exponent Transformation of Sine:

Ω_n Discret: $\{\exp(+j\Omega_n T) - \exp(-j\Omega_n T)\}/(2i)$

$$1/(2\pi) \int_{-\infty}^{+\infty} \{\exp(+j\Omega_n T) - \exp(-j\Omega_n T)\}/(2i) \{\exp(-jT\Omega)\}\, dT = \{\delta(\Omega_n - \Omega) - \delta(-\Omega_n - \Omega)\}/(2i) = S(\Omega)$$

$$\int_{-\infty}^{+\infty} S(\Omega) \{\exp(+jT\Omega)\, d\Omega = R(T)$$

$$1/(2\pi) \int_{-\infty}^{+\infty} R(T) \{\exp(-jT\Omega)\}\, dT = S(\Omega)$$

Ω_n Discret: $\{\exp(+j\Omega_n T) - \exp(-j\Omega_n T)\}/(2i)$

$$\int_{-\infty}^{+\infty} \{\exp(+j\Omega_n T) - \exp(-j\Omega_n T)\}/(2i) \{\exp(+jT\Omega)\, d\Omega = (2\pi)\{\delta(\Omega_n + \Omega) - \delta(-\Omega_n + \Omega)\}/(2i) = R(T)$$

1.7.9. Derivatives:

Keeping in Mind Delta Transformations:

$1/(2\pi) \int_{-\infty}^{\infty} \delta(\tau) \exp(-j\omega\tau)\, d\tau = 1/(2\pi)$ ► $\int_{-\infty}^{+\infty} 1/(2\pi) \exp(+j\tau\omega)\, d\omega = \delta(\tau)$

$1/(2\pi) \int_{-\infty}^{+\infty} (1) \exp(+j\omega\tau)\, d\tau = \delta(\omega)$ ◄ $\int_{-\infty}^{+\infty} \delta(\omega) \exp(+j\tau\omega)\, d\omega = (1)$

Transformations:

$1/(2\pi)\int_{-\infty}^{\infty} R(\tau) \exp(-j\tau\omega)\, d\tau = S(\omega)$ | $\int_{-\infty}^{+\infty} S(\omega) \exp(+j\tau\omega)\, d\omega = R(\tau)$

$1/(2\pi)\int_{-\infty}^{\infty} R(\tau)\, d\tau = S(0)$ | $\int_{-\infty}^{+\infty} S(\omega)\, d\omega = R(0)$

$1/(2\pi) R(\tau) = d\{S(0)\}/d\tau$ | $S(\omega) = d\{R(0)\}/d\omega$

$R(\tau)$ Temporal Density of $S(0)$ | $S(\omega)$ Spectral Density of $R(0)$

Unit of [R] Unit of [τ] ≡ Unit of [S] | Unit of [S] Unit of [ω] ≡ Unit of [R]

Such That:

If $0 < S(\omega)$ Positive $S(\omega)$ Mean Square Spectral Density of $R(0)$ ◄ Then $R(\tau)$ Not Necessarily Greeater than 0

If

0 < R(τ) Positive
R(τ)
Mean Square Spectral Density of
S(0)

► Then
S(ω) Not Necessarily Greeater than
0

Exponential Transformation Derivatives

$1/(2\pi)\int_{-\infty}^{\infty} R(\tau)\ \exp(-j\ \tau\ \omega)\ d\tau = S(\omega)$

$\int_{-\infty}^{+\infty} S(\omega)\ \exp(+\ j\ \tau\ \omega)\ d\omega = R(\tau)$

$1/(2\pi)\int_{-\infty}^{\infty} (-\ j\tau)^{M}\ R(\tau)\ \exp(-j\ \tau\omega)\ d\tau = d^{N}S(\omega)/(d\omega)^{N}$

$\int_{-\infty}^{+\infty} (j\omega)^{N}\ S(\omega)\ \exp(+j\tau\omega)\ d\omega = d^{N}R(\tau)/(d\tau)^{N}$

$1/(2\pi)\int_{-\infty}^{\infty} (-\ j\tau)^{M}\ R(\tau)\ d\tau = d^{N}S(0)/(d\omega)^{N}$

$\int_{-\infty}^{+\infty} (j\omega)^{N}\ S(\omega)\ d\omega = d^{N}R(0)/(d\tau)^{N}$

$1/(2\pi)(-j\tau)^{M} R(\tau)$ Temporal Density of $d^{N}S(0)/(d\omega)^{N}$

$(j\omega)^{N} S(\omega)$ Temporal Density of $d^{N}R(0)/(d\tau)^{N}$

1.7.10.
Some Properties:

$1/(2\pi)\int_{-\infty}^{+\infty} R(\tau)\ \exp(-j\omega\tau)\ d\tau = S(\omega)$

$1\int_{-\infty}^{+\infty} S(\omega)\ \exp(+\ j\ \omega\ \tau)\ d\omega = (R(\tau))$

$1/(2\pi)\int_{-\infty}^{+\infty} \delta(\tau)\ \exp(-j\omega\tau)d\tau = 1/(2\pi)$

$\int_{-\infty}^{+\infty} 1/(2\pi)\ \exp(+j\omega\tau)d\omega = \delta(\tau)$

$1/(2\pi)\int_{-\infty}^{+\infty} (1)\ \exp(+j\omega\tau)d\omega = \delta(\tau)$

$\int_{-\infty}^{+\infty} \delta(\tau)\ \exp(+j\omega\tau)d\omega = (1)$

$1/(2\pi)\int_{-\infty}^{+\infty} \{\cos(q_x\ p)\} * \exp(-jpq)\ dp = (2\pi)/2\ (\delta(q - q_x)) + \delta(q + q_x))$

$\int_{-\infty}^{+\infty} (2\pi)/2\ (\delta(q - q_x)) + \delta(q + q_x)) * \exp(+\ ipq)\ dq = (\cos(q_x\ p))$

$1/(2\pi)\int_{-\infty}^{+\infty} \{\sin(q_x\ p)\} * \exp(-jpq)\ dp = (2\pi)/(2i)\ (\delta(q - q_x)) - \delta(q + q_x))$

$\int_{-\infty}^{+\infty} (2\pi)/(2i)(\delta(q - q_x)) - \delta(q + q_x)) * \exp(+\ ipq)\ dq = (\sin(q_x\ p))$

$1/(2\pi)\int_{-\infty}^{+\infty} R(\tau)\ \exp(-j\omega\tau)\ d\tau = S(\omega)$

$\int_{-\infty}^{+\infty} S(\omega)\ \exp(+\ j\ \omega\ \tau)\ d\omega = R(\tau)$

$(i/\tau) = S(\omega)$ This is independent of ω ►

$\int_{-\infty}^{+\infty} i/\tau\ \exp(+\ j\ \omega\ \tau)\ d\omega = i/\tau\ (2\pi) * \delta(\tau) = R(\tau)$

$1/(2\pi)\int_{-\infty}^{\infty} i/\tau\ (2\pi)\ \delta(\tau)\ \exp(-j\omega\tau)\ d\tau = i/\tau = S(\omega)$ ◄

$\int_{-\infty}^{+\infty} i/\tau\ \exp(+\ j\ \omega\ \tau)\ d\omega = i/\tau\ (2\pi) * \delta(\tau) = R(\tau)$

$1/(2\pi)\int_{-\infty}^{+\infty} R(\tau)\, d\tau = S(\omega = 0)$ $\quad$ $\int_{-\infty}^{+\infty} S(\omega)\, d\omega = R(t = 0)$

$1/(2\pi)\, R(\tau) = dS(\omega = 0)/d\tau$ $\quad$ $S(\omega) = dR(\tau = 0)/d\omega$

$(i/\tau) = S(\omega)$ ► $\int_{-\infty}^{+\infty} i/\tau \; d\omega = R(\tau)$

$\int_{-\infty}^{\infty} i/\tau\; \delta(\tau)\, d\tau = i/\tau = S(\omega)$ ◄ $\int_{-\infty}^{+\infty} i/\tau \; d\omega = R(\tau)$

:

$$\int u(\sigma)\,(dv/d\sigma)\, d\sigma + \int v(\sigma)\,(du/d\sigma)\, d\sigma = (u * v)$$

:

$$\int R(\tau)\,(dS/d\tau)\, d\tau + \int S(\tau)\,(dR/d\tau)\, d\tau = (R * S)$$

$$\int R(\omega)\,(dS/d\omega)\, d\omega + \int S(\omega)\,(dR/d\omega)\, d\omega = (R * S)$$

$S(-\omega) = S(\omega)^C$	**$R(s)$**
$S(-\omega) = S(\omega)$	**$R(s)$ is even**
$S(-\omega) = -S(\omega)$	**$R(s)$ is odd**
$S(\omega/a)$	**$1/a$**
$(a * S(\omega/a))$	**$(R(at))$**
$1/\|a\|\, S(\omega/a)$	**$(R(at))$**
$S(\omega)\exp(j\,\omega\, t_0)$	**$(R(s - t_0))$**
$S(\omega - \omega_0)$	**$(\exp(-j\,\omega\, t_0) * R(s))$**
$S(\omega)\exp(j\,\omega\, t_0)$	**$(R(s - t_0))$**
$(j\,\omega)^N\, S(\omega)$	**$(d^N (R(s))/dt^N)$**
$\{d(S(\omega))/dt\}$	**$(-j\, t)\, R(s)$**
$S(\omega)/(-j\,\omega)$	**$\int R(s)\, dt$**
$\int S(\omega)\, d\omega$	**$-1/(-j\, t)$**

1.7.11. Rectangular Pulse:

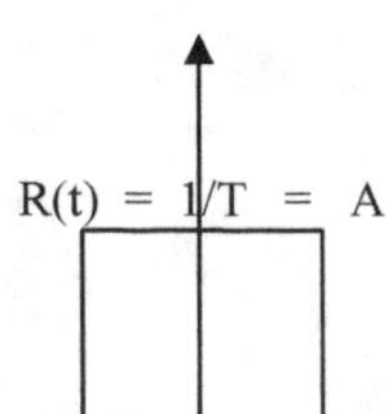

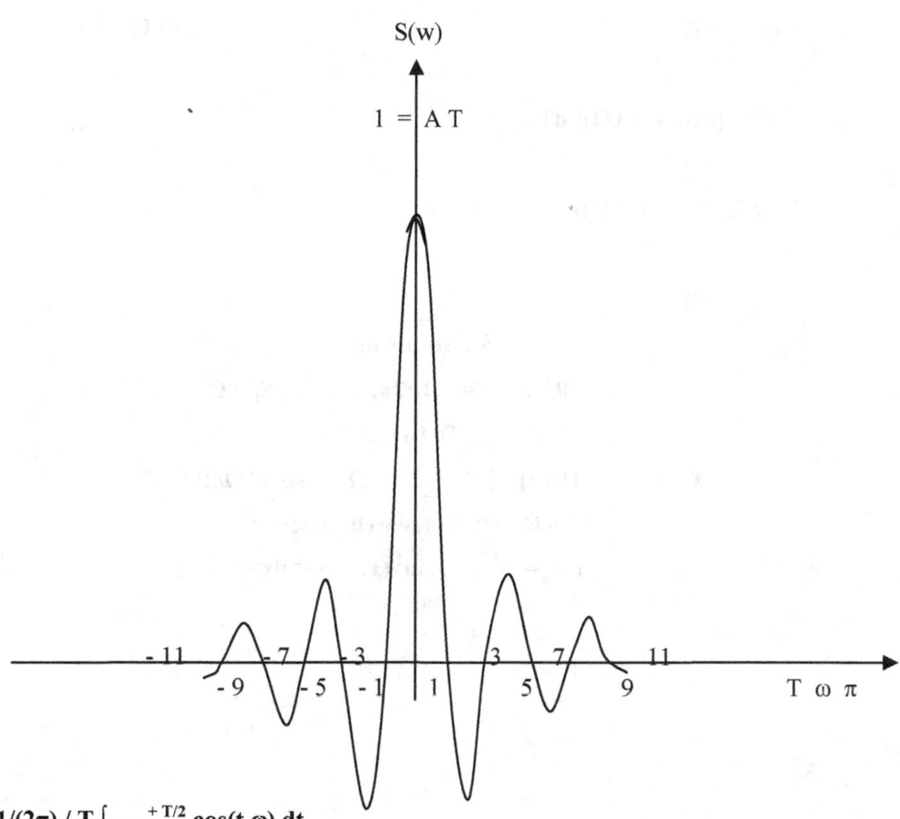

$$1/(2\pi) / T \int_{-T/2}^{+T/2} \cos(t\ \omega)\ dt$$

=

$$1/(2\pi) / T\ *\ 2/\omega\ \{\sin(t\ \omega)\}|_0^{+T/2}$$

=

$$1/(2\pi) /(T/2\ *\ \omega) \sin (T/2\ \omega)$$

=

$$S(\omega)$$

:

$$1/(2\pi) \int_{-\infty}^{+\infty} \{ R(t) \}^2 dt = \int_{-\infty}^{+\infty} |S(\omega)|^2 d\omega$$

$$1/(2\pi)\ 1/T^2\ T = \int_{-\infty}^{+\infty} |1/(\pi * T * \omega) \sin (T/2\ \omega) |^2 d\omega$$

$$1/(2\pi)\ 1/T = \int_{-\infty}^{+\infty} |1/ (2\ \pi\ \omega) \sin (\ \omega) |^2\ 2/T\ \ d\omega$$

$$1/(2\pi)\ 1/T = \int_{-\infty}^{+\infty} 1/ (2\ \pi\ )^2 | (1/\omega) \sin (\ \omega) |^2\ 2/T\ \ d\omega$$

$$1 = 1/\pi \int_{-\infty}^{+\infty} \{ \sin(\omega) / \omega \}^2\ \ d\omega$$

:

2 A * sin (T w /2) / w

Where:

$$\mathbf{1/(2\pi)\int_{-\infty}^{+\infty} R(T)\,\{exp(-\ j\ T\Omega)\}\ dT = S(\Omega)}$$

$$\mathbf{\int_{-\infty}^{+\infty} S(\Omega)\,\{exp(+\ j\ T\Omega)\}\ d\Omega = R(T)}$$

$$\mathbf{\int_{-T/2}^{+T/2} A\,\{exp(-\ j\ T\Omega)\}\ dT = 2\,A/\Omega\ *\ sin\,(T\Omega/2) = S(\Omega)}$$

Substitute in:

$$\mathbf{\int_{-\infty}^{+\infty} R^2\ dT = 1/(2\pi)\ \int_{-\infty}^{+\infty} |\,S|^2\ d\Omega}$$

To Get:

$$\mathbf{A^2\ T = 1/(2\pi)\ \int_{-\infty}^{+\infty} |\,2\,A/\Omega\ *\ sin\,(T\Omega/2)\,|^2\ d\Omega}$$

This Results in the well - known:

$$\mathbf{(\pi) = \int_{-\infty}^{+\infty} sin^2\,(x)\,/\ (x)^2\ dx}$$

2.
Structural Coordinate Tensorials:

Bases of fixed coordinate system are unique, constant, orthonotmal and time invariant.
Whereas bases of all other coordinate systems may *change in position and time.*
Bases of coordinate system are underlined.
These bases are defined relative to bases of fixed coordinates..
A form of structure is defined by its own coordinate system bases.

Coordinate systems may be curved and depend on structural forms and may vary due to:
Structural own dead loads and imposed live loads, temperature, static or dynamic.

Structural coordinate systems are general, orthogonal or orthonormal.
General coordinate system is represented by Letters: a, b ...
Orthogonal coordinate system is represented by Letters: r, s ...
Orthonormal coordinate system is represented by Letters: x, y, z.
Fixed coordinate system is represented by Letter o.

Structural variables may be scalars, otherwise any structural variable is defined by its indices.
Index location, low or high, indicates its type:
Covariant (Low Index) Contravariant (Upper Index)

Time dependent structural variables are function of Position Vector

Postion vector itself is a structural variable and its bases and components are defined as:
An unfeformed postion vector, $\underline{p}(t)$:

$$\underline{a}_i\, a^i = a_i\, \underline{a}^i = \underline{f}_i\, f^i = f_i\, \underline{f}^i = \underline{r}_i\, r^i = r_i\, \underline{r}^i = \underline{x}_i\, x^i = x_i\, \underline{x}^i$$

Also a defeformed postion vector, $\underline{q}(t)$:

$$\underline{b}_i\, b^i = b_i\, \underline{b}^i = \underline{g}_i\, g^i = g_i\, \underline{g}^i = \underline{s}_i\, s^i = s_i\, \underline{s}^i = \underline{y}_i\, y^i = y_i\, \underline{y}^i$$

	$\underline{p}(t)$	$\underline{q}(t)$	
$\underline{a}_i$	a^i	$\underline{b}_i$	b^i
$\underline{f}_i$	f^i	$\underline{g}_i$	g^i
$\underline{r}_i$	r^i	$\underline{s}_i$	s^i
$\underline{x}_i$	x^i	$\underline{y}_i$	y^i

A postion component tensor is of the same order as bases order.

2.1.
Fixed Bases:

Fixed bases are *three orthonormal constant and time invariant vectors.*
All other coordinate bases are referenced to these bases.

2.1.1
Bases, Components and Position Vector:

Bases of fixed System are underlined

$\underline{o}_i$ Low $\underline{o}^i$ Up

covariant — contravariant

System Basis — Conjugate System Basis

Fixed covariant basis is equal its corresponding contravariant basis with unitary norm

$$\underline{o}_i = \underline{o}^i$$

$$\| \underline{o}_i \| = 1 = \| \underline{o}^i \|$$

Covariant component is equal its corresponding contravariant component

$$o^i \text{ Up} = o_i \text{ Low}$$

contravariant = covariant

A contravariant component is associated with a covariant basis, and vice-versa

A postion tensor is of three components associated with three bases:

$\underline{o} = \underline{o}_i o^i$ — $\underline{o} = \underline{o}^i o_i$

covariant $\underline{o}_i$ * contravariant o^i — contravariant $\underline{o}^i$ * covariant o_i

Position vector:

$$\underline{o}_N o^N \equiv (\underline{o}) \equiv \underline{o}^N o_N$$

2.1.2. Function of Position Vector:

Structural variables are Function of Position Vector

$$f(\underline{o}_i o^i) = f(\underline{o}) = f(\underline{o}^i o_i)$$

2.1.2.1. Basis Differential:

Since bases are fixed the differential of a position basis is zero

$$d\underline{o}_i \equiv 0 \qquad d\underline{o}^i \equiv 0$$

2.1.2.2. Component Differential:

Whereas components have differentials

$$do^n \qquad do_n$$

2.1.2.3. Partial Derivative with respect to a Coordinate:

$$\partial [\,]/\partial o^n \equiv \partial/\partial o^n[\,] \equiv [\,]_n \qquad \partial [\,]/\partial o_n \equiv \partial/\partial o_n[\,] \equiv [\,]^n$$

Partial Derivative of a basis with respect to o is null:

$$\partial \underline{o}^m / \partial o^n = \underline{o}^m{}_,{}^n = 0 \qquad \partial \underline{o}_m / \partial o_n = \underline{o}_{m},{}_n = 0$$

Partial Derivative of a Scalar s with respect to o:

$$\partial s / \partial o^n \equiv so_n \qquad \partial s / \partial o_n \equiv so^n$$

Partial Derivative of o with respect to a Scalar s:

$$\partial o_n / \partial s \equiv os_n \qquad \partial o^n / \partial s \equiv os^n$$

Partial Derivative of o with respect to o:

$$\partial o_m / \partial o_n = \delta^n{}_m \qquad \partial o^m / \partial o^n = \delta^m{}_n$$

2.1.2.4.
Partial Derivative of Positin Vector:

Differential of Position Vector:

$d\underline{o} = d(\underline{o}_i \, o^i) = \underline{o}_i \, do^i$ $\qquad$ $d\underline{o} = d(\underline{o}^i \, o_i) = \underline{o}^i \, do_i$

Partial Derivative of a Position Vector with respect to o:

$$\partial\underline{o}/\partial o_n = \partial(\underline{o}^i o_i)/\partial o_n = \partial(\underline{o}^i)/\partial o_n \, o_i + \underline{o}^i \, \partial o_i/\partial o_n = \underline{o}^{i}{}_{,}{}^{n} o_i + \underline{o}^i \, \delta_i^n = 0 + \underline{o}^n$$

$$\partial(\underline{o}^i o_i)/\partial o_n = \underline{o}^n$$

$$\partial\underline{o}/\partial o^n = \partial(\underline{o}_i o^i)/\partial o^n = \partial(\underline{o}_i)/\partial o^n \, o^i + \underline{o}_i \, \partial o^i/\partial o^n = \underline{o}_{i,n} o^i + \underline{o}_i \, \delta^i_n = 0 + \underline{o}_n$$

$$\partial(\underline{o}_i o^i)/\partial o^n = \underline{o}_n$$

2.1.3.
Outer and Inner Products:

2.1.3.1.
Outer Product of Fixed Bases:

Bases Outer Product:

$$\underline{o}_I \wedge \underline{o}_J = e_{IJK} \, \underline{o}^K$$

$$\underline{o}^I \wedge \underline{o}^J = e^{IJK} \, \underline{o}_K$$

2.1.3.2.
Inner Product of Fixed Bases:

Direction Cosine or Inner Product of Bases:

Inner Product oo^{IK} Upper and oo_{IK} Lower are Called Metric

$$\cos(\underline{o}^I, \underline{o}^K) = \underline{o}^I \cdot \underline{o}^K = oo^{IK} = \delta^{IK}$$

$$\cos(\underline{o}_I, \underline{o}_K) = \underline{o}_I \cdot \underline{o}_K = oo_{IK} = \delta_{IK}$$

2.1.3.3.
Outer Product of Vectors:

Vectors Outer product:

$$\underline{u} \wedge \underline{v} = u^I \underline{o}_I \wedge v^J \underline{o}_J = e_{IJK} \, u^I v^J \underline{o}^K$$

$$\underline{u} \wedge \underline{v} = u_I \underline{o}^I \wedge v_J \underline{o}^J = e^{IJK} \, u_I v_J \underline{o}_K$$

2.1.3.4.
Inner Product of Vectors:

$$\underline{u} \cdot \underline{v}$$

$$\underline{o}_i \, u^i \cdot \underline{o}_k \, v^k = u^i \, v^k \, \delta_{ik} = u^i \, v^i$$

$$\underline{o}^i u_i \cdot \underline{o}^k v_k = u_i v_k \delta^{ik} = u_i v_i$$

2.1.3.5.
Mixed Product of Fixed Bases:

$$\underline{o}_I \wedge \underline{o}_J \cdot \underline{o}_K$$

$$\underline{o}_I \wedge \underline{o}_J \cdot \underline{o}_K = e_{IJp}\underline{o}^p \cdot \underline{o}_K = e_{IJK}$$

$$\underline{o}^I \wedge \underline{o}^J \cdot \underline{o}^K = e^{IJp}\underline{o}_p \cdot \underline{o}^K = e^{IJK}$$

2.1.3.6.
Mixed Product of Vectors:

$$\underline{u} \wedge \underline{v} \cdot \underline{w}$$

$$\underline{u} \wedge \underline{v} \cdot \underline{w} = u^I \underline{o}_I \wedge v^J \underline{o}_J \cdot w^K \underline{o}_K = e_{IJK} u^I v^J w^K$$

$$\underline{u} \wedge \underline{v} \cdot \underline{w} = u_I \underline{o}^I \wedge v_J \underline{o}^J \cdot w_K \underline{o}^K = e^{IJK} u_I v_J w_K$$

2.2.
Projection and Bases:

2.2.1.
Position Vector Projection:

For a position vector there are two projection types: perpendicular projection and parallel projection.
Knowig:
There is one plane perpendicular to an axis whereas there are infinite planes parallel to an axis.
And
Two planes intersect in a line. And a line intersect a plane in a point.
Projections are constructed.

In Plane System

Perpendicular projection: Parallel projection:

Vector Components Bases Bases Vector Components

▼

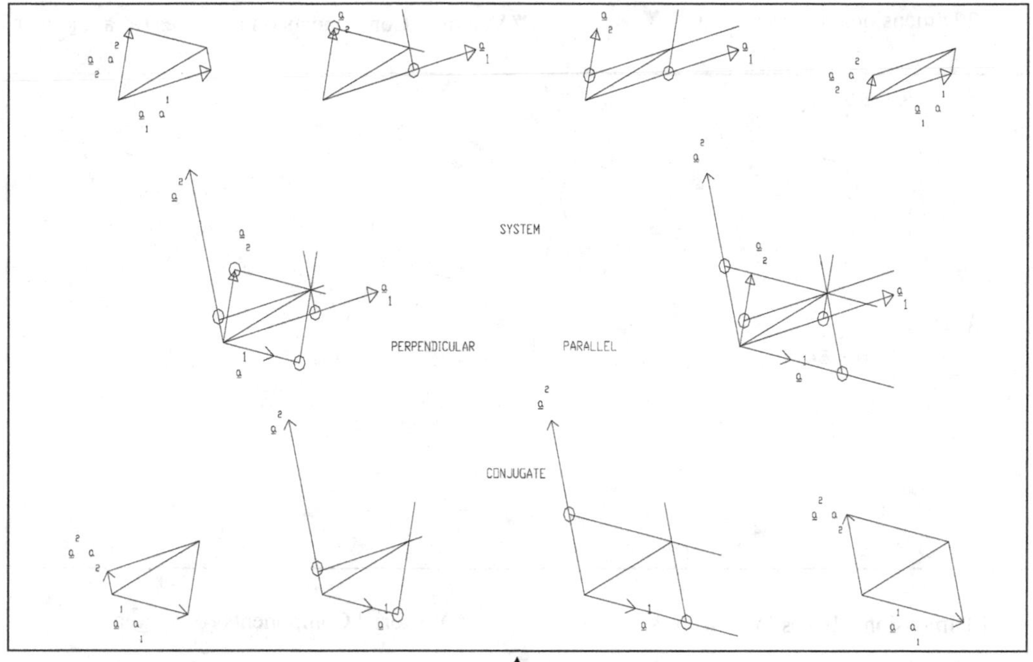

▲

Vector Components   Bases   Bases   Vector Components

Perpendicular projection: Parallel projection

In Plane Conjugate System :

In Plane Since $\underline{a}^J$ is Perpendicular to $\underline{a}_I$ then a Perpendicular line to $\underline{a}^J$ is $\underline{a}_I$ parallel

In Plane Since $\underline{a}_J$ is Perpendicular to $\underline{a}^I$ then a Perpendicular line to $\underline{a}_J$ is $\underline{a}^I$ parallel

In Plane $\underline{a}^J \perp \underline{a}_I$ *In Plane* $\underline{a}^J \perp \underline{a}$ $\underline{a}_I \,//\, \underline{a}$

In Plane $\underline{a}_J \perp \underline{a}^I$ *In Plane* $\underline{a}_J \perp \underline{a}$ $\underline{a}^I \,//\, \underline{a}$

Parallel and Perpendicular Projection:

▼

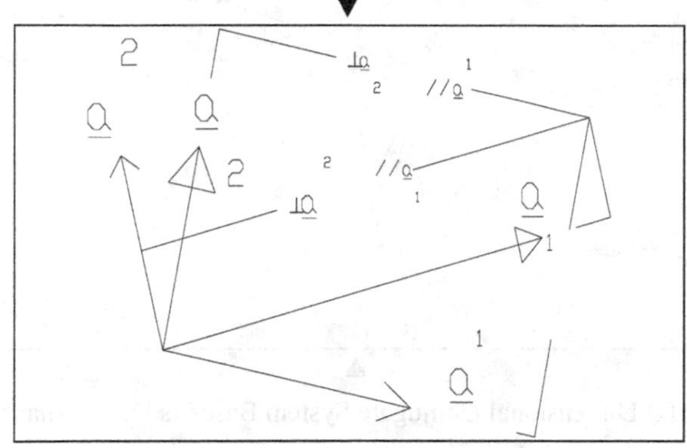

***Parallel Projecton* will be Considered:**

Parallel Projecton Conjugates of a postion vector:

$$\underline{a}_i a^i = a_i \underline{a}^i$$

$$(\underline{a}_1 a^1 + \underline{a}_2 a^2 + \underline{a}_3 a^3) = (\underline{a}^1 a_1 + \underline{a}^2 a_2 + \underline{a}^3 a_3)$$

Parallel: $(\underline{a}_1 a^1 + \underline{a}_2 a^2)$

▼

2 Dimensional Bases ($\underline{a}_1$, $\underline{a}_2$) ▼ ▼Vectorial Components on its Bases ($\underline{a}_1 a^1$, $\underline{a}_2 a^1$)

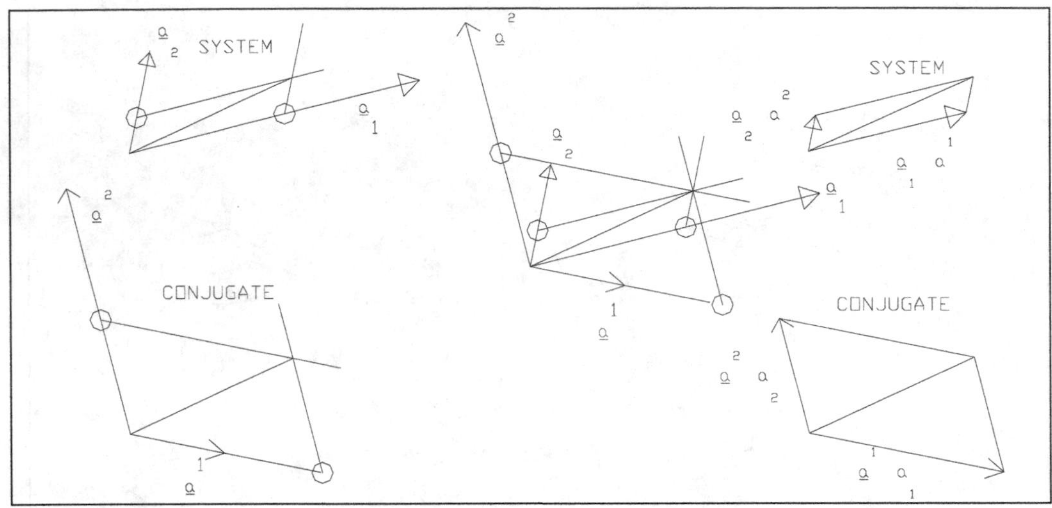

2 Dimensional Bases ($\underline{a}^1$, $\underline{a}^2$) ▲ ▲Vectorial Components ($\underline{a}^1 a_1$, $\underline{a}^2 a_2$)

▲

Parallel *Conjugate*: ($\underline{a}^1 a_1 + \underline{a}^2 a_2$)

2.2.2.
Bases:

Generally there are two bases types: system bases and conjugate system bases.

Contained Volume of 3 Dimensional System Bases Is Determinant of ($\underline{a}_1$, $\underline{a}_2$, $\underline{a}_3$)

▼

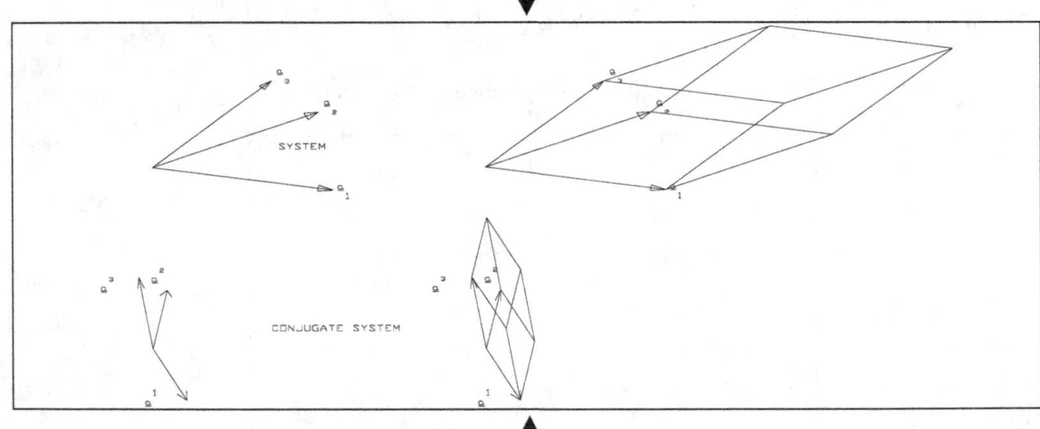

▲

Contained Volume of 3 Dimensional Conjugate System Bases Is Determinant of ($\underline{a}^1$, $\underline{a}^2$, $\underline{a}^3$)

Adjoint of two bases is perpendicular to them (conjugate of the third).
Outer product of two bases of the same system is:
Equal their adjoint multiplied by all bases determinant
Outer product of two bases of the same system is:
Less or equal their adjoint multiplied by the norms of these two bases

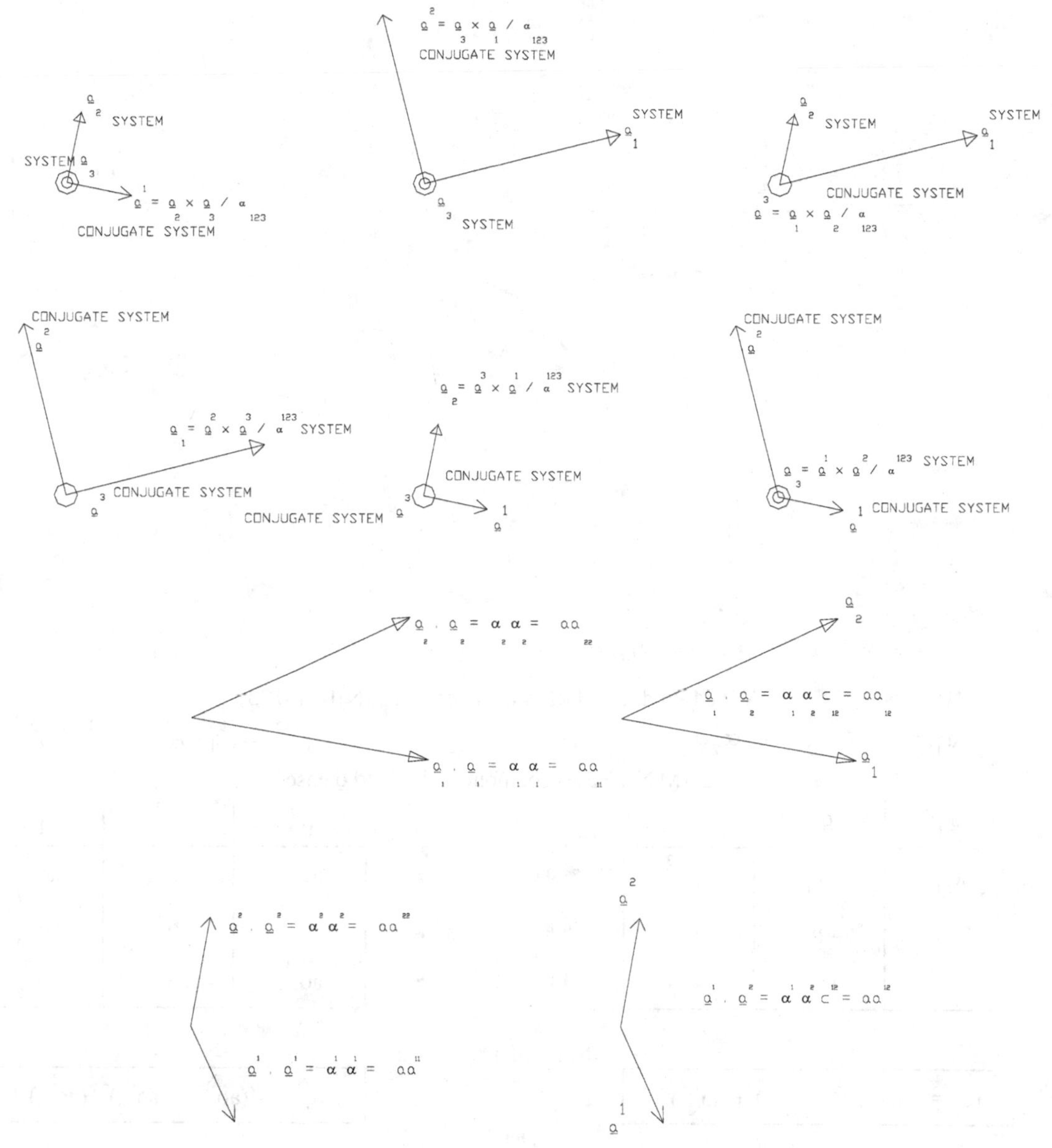
CONJUGATE SYSTEM
SYSTEM
SYSTEM
SYSTEM
CONJUGATE SYSTEM
SYSTEM
SYSTEM
SYSTEM
CONJUGATE SYSTEM
CONJUGATE SYSTEM
CONJUGATE SYSTEM
CONJUGATE SYSTEM
SYSTEM
CONJUGATE SYSTEM
SYSTEM
CONJUGATE SYSTEM
CONJUGATE SYSTEM
SYSTEM
CONJUGATE SYSTEM

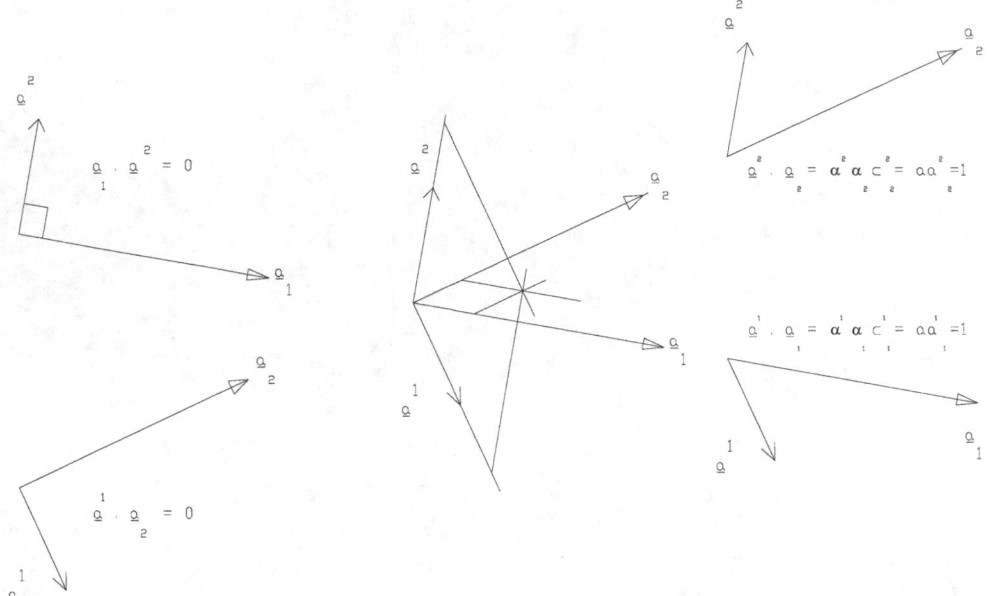

Where:

Mixed Product is Determinant of [ao(M,N)] ≡ (a123):

$a_{123} \equiv \|_a\| \le \alpha_{123} \equiv \alpha_1 \alpha_2 \alpha_3$ $\qquad$ $a^{123} \equiv \|^a\| \le \alpha^{123} \equiv \alpha^1 \alpha^2 \alpha^3$

[ao(M,N)] Bases components in fixed o bases:

o_1	o_2	o_3	
ao^1_1	ao^2_1	ao^3_1	◀ a_1
ao^1_2	ao^2_2	ao^3_2	◀ a_2
ao^1_3	ao^2_3	ao^3_3	◀ a_3

	o^1	o^2	o^3
a^1 ▶	ao^1_1	ao^1_2	ao^1_3
a^2 ▶	ao^2_1	ao^2_2	ao^2_3
a^3 ▶	ao^3_1	ao^3_2	ao^3_3

Norm of a basis is α

$$\alpha_M = \sqrt{[(ao_M^1)^2+(ao_M^2)^2+(ao_M^3)^2]} \qquad \alpha^N = \sqrt{[(ao^N_1)^2+(ao^N_2)^2+(ao^N_3)^2]}$$

Detailed:

$$\alpha_1 = \sqrt{[(ao_1^1)^2+(ao_1^2)^2+(ao_1^3)^2]} \qquad \alpha^1 = \sqrt{[(ao^1_1)^2+(ao^1_2)^2+(ao^1_3)^2]}$$

$$\alpha_2 = \sqrt{[(ao_2^1)^2+(ao_2^2)^2+(ao_2^3)^2]} \qquad \alpha^2 = \sqrt{[(ao^2_1)^2+(ao^2_2)^2+(ao^2_3)^2]}$$

$$\alpha_3 = \sqrt{[(ao_3^1)^2+(ao_3^2)^2+(ao_3^3)^2]} \qquad \alpha^3 = \sqrt{[(ao^3_1)^2+(ao^3_2)^2+(ao^3_3)^2]}$$

Such that:

No Sum over M: $\boldsymbol{\alpha_{\underline{M}} \alpha^{\underline{M}} = 1}$

Detailed:

$$\alpha_{\underline{1}} \alpha^{\underline{1}} = 1$$

$$\alpha_{\underline{2}} \alpha^{\underline{2}} = 1$$

$$\alpha_{\underline{3}} \alpha^{\underline{3}} = 1$$

In terms of bases angles cosines C and D

$$\alpha_N = \sqrt{\{[C^1_N\alpha_N]^2+[C^2_N\alpha_N]^2+[C^3_N\alpha_N]^2\}} \quad \Big| \quad \alpha^M = \sqrt{\{[C_1^M\alpha^M]^2+[C_2^M\alpha^M]^2+[C_3^M\alpha^M]^2\}}$$

$\alpha_N = \sqrt{\{[\alpha_N D_N^{1}]^2 + [\alpha_N D_N^{2}]^2 + [\alpha_N D_N^{3}]^2\}}$ | $\alpha^M = \sqrt{\{[\alpha^M D^M{}_1]^2 + [\alpha^M D^M{}_2]^2 + [\alpha^M D^M{}_1]^2\}}$

Define in terms of α:

$\underline{a}_K \equiv \alpha_{\underline{K}}\ \underline{A}_{\underline{K}}$ $\qquad$ $\underline{a}^K \equiv \alpha^{\underline{K}}\ \underline{A}^K$

$(ao_K{}^M) \equiv \alpha_{\underline{K}}\ AO_{\underline{K}}{}^M$ $\qquad$ $(ao^K{}_M) \equiv \alpha^{\underline{K}}\ AO^{\underline{K}}{}_M$

And resolution of bases into fixed cootdinates:

$\underline{a}_K = (ao_K{}^M)(\underline{o}_M)$ $\qquad$ $\underline{a}^K = (ao^K{}_M)(\underline{o}^M)$

Then:

$\underline{a}_K \equiv \alpha_{\underline{K}}\ \underline{A}_{\underline{K}} = (\alpha_{\underline{K}}\ AO_{\underline{K}}{}^M)(\underline{o}_M)$ $\qquad$ $\underline{a}^K \equiv \alpha^{\underline{K}}\ \underline{A}^{\underline{K}} = (\alpha^{\underline{K}} AO^{\underline{K}}{}_M)(\underline{o}^M)$

Or:

$\underline{A}_{\underline{K}} = (AO_{\underline{K}}{}^M)(\underline{o}_M)$ $\qquad$ $\underline{A}^{\underline{K}} = (AO^{\underline{K}}{}_M)(\underline{o}^M)$

And amplitudes:

$1 = ||\underline{A}_K|| = ||AO_{\underline{K}}{}^M||\ \ ||\underline{o}_M||$ $\qquad$ $1 = ||\underline{A}^K|| = ||AO_M{}^K||\ \ ||\underline{o}^M||$

$1 = ||AO_{\underline{K}}{}^M||$ $\qquad$ $1 = ||AO_M{}^K||$

2.2.2.1.
Outer Product of Bases:

M≠P≠Q≠M Ordered with no sum

$\underline{o}_P \wedge \underline{o}_Q = e_{PQM}\ \underline{o}^M$ | $\underline{o}^P \wedge \underline{o}^Q = e^{PQM}\ \underline{o}_M$

Sine Oriantation:

Composition of b:
α : positive angle between a and b
$\Sigma = \sin\alpha$

Resolution into b:
β : positive angle between b and a
$\sin\beta = T$

Outer Product Of Bases

$\underline{a}^M \wedge \underline{b}^N$ $\qquad$ $\underline{a}^M \wedge \underline{b}_N$ $\qquad$ $\underline{b}_N \wedge \underline{a}^M$ $\qquad$ $\underline{b}^N \wedge \underline{a}^M$

$\underline{a}_M \wedge \underline{b}^N$ $\qquad$ $\underline{a}_M \wedge \underline{b}_N$ $\qquad$ $\underline{b}_N \wedge \underline{a}_M$ $\qquad$ $\underline{b}^N \wedge \underline{a}_M$

M≠P≠Q≠M Ordered with no sum

$\underline{a}_P \wedge \underline{a}_Q = \underline{o}_i\, a^i{}_P \wedge \underline{o}_j\, a^j{}_P = e_{ijM}\ a^i{}_P\, a^j{}_P\, \underline{o}^M$ | $\underline{a}^P \wedge \underline{a}^Q = \underline{o}^i\, a_i{}^P \wedge \underline{o}^j\, a_j{}^Q = e^{ijM}\ a_i{}^P\ a_j{}^Q\ \underline{o}_M$

$\underline{a}_P \wedge \underline{a}_Q\ /\alpha_{123} \le \underline{a}_P \wedge \underline{a}_Q / ||_a|| = \underline{a}^M$ | $\underline{a}^P \wedge \underline{a}^Q\ /\alpha^{123} \le \underline{a}^P \wedge \underline{a}^Q / ||^a|| = \underline{a}_M$

$\underline{a}_P \wedge \underline{a}_Q = a_{123}\ \underline{a}^M \le \alpha_{123}\ \underline{a}^M = \alpha_p\, \alpha_q\, \underline{A}^M$ | $\underline{a}^P \wedge \underline{a}^Q = a^{123}\ \underline{a}_M \le \alpha^{123}\ \underline{a}_{\underline{M}} = \alpha^p\, \alpha^q\, \underline{A}_M$

A notice:
Knowing:
Ordered $i \neq j \neq K \neq i$

$\underline{a}_I \wedge \underline{a}_J = a_{123}\ \underline{a}^K \le \alpha_{123}\ \underline{a}^K$ $\qquad$ $\underline{a}^I \wedge \underline{a}^J = a^{123}\ \underline{a}_K \le \alpha^{123}\ \underline{a}_K$

Then:
Inner Product By K Basis with No Sum:

$\underline{a}_I \wedge \underline{a}_J \cdot \underline{a}_K = a_{123}\, \underline{a}^K \cdot \underline{a}_{\underline{K}} \le \alpha_{123}\, \underline{a}^K \cdot \underline{a}_K$ $\qquad$ $\underline{a}^I \wedge \underline{a}^J \cdot \underline{a}^K = a^{123}\, \underline{a}_{\underline{K}} \cdot \underline{a}^K \le \alpha^{123}\, \underline{a}_{\underline{K}} \cdot \underline{a}^K$

But:

$aa^I{}_K \equiv \underline{a}^I \cdot \underline{a}_K = \delta^I{}_{\underline{K}}$

$aa_I{}^K \equiv \underline{a}_I \cdot \underline{a}^K = \delta^K_{\underline{I}}$

So:

$0 < \underline{a}_1 \wedge \underline{a}_2 \cdot \underline{a}_3 \equiv a_{123} \le \alpha_{123}$ $\qquad$ $0 < \underline{a}^I \wedge \underline{a}^J \cdot \underline{a}^K \equiv a^{123} \le \alpha^{123}$

Norms Of Outer Product Of Bases:

$\| \underline{a}_i \wedge \underline{a}_j \| = a_{123}\, \| \underline{a}^K \| = a_{123}\, \alpha^K \le \alpha_{123}\, \alpha^K$ $\qquad$ $\| \underline{a}^i \wedge \underline{a}^j \| = a^{123}\, \| \underline{a}_{\underline{K}} \| = a^{123}\, \alpha_K \le \alpha^{123}\, \alpha_K$

$\|\underline{a}^M \wedge \underline{b}^N\|$ = $\alpha^{\underline{M}}\,{}^M\Sigma^{\underline{N}}\,\beta^{\underline{N}}$ = $-\beta^{\underline{N}}\,{}^N T^M\,\alpha^{\underline{M}}$	$\|\underline{a}^M \wedge \underline{b}_N\|$ = $\alpha^{\underline{M}}\,{}^M\Sigma_{\underline{N}}\,\beta_{\underline{N}}$ = $-\beta_{\underline{N}}\,{}_N T^M\,\alpha^{\underline{M}}$	$-\alpha^{\underline{M}}\,{}^M\Sigma_{\underline{N}}\,\beta_{\underline{N}}$ = $\beta_{\underline{N}}\,{}_N T^M\,\alpha^{\underline{M}}$ = $\|\underline{b}_N \wedge \underline{a}^M\|$	$-\alpha^{\underline{M}}\,{}^M\Sigma^{\underline{N}}\,\beta^{\underline{N}}$ = $\beta^{\underline{N}}\,{}^{\underline{N}}T^M\,\alpha^{\underline{M}}$ = $\|\underline{b}^N \wedge \underline{a}^M\|$
$\|\underline{a}_M \wedge \underline{b}^N\|$ = $\alpha_{\underline{M}}\,{}_{\underline{M}}\Sigma^{\underline{N}}\,\beta^{\underline{N}}$ = $-\beta^{\underline{N}}\,{}^{\underline{N}}T_M\,\alpha_{\underline{M}}$	$\|\underline{a}_M \wedge \underline{b}_N\|$ = $\alpha_{\underline{M}}\,{}_M\Sigma_{\underline{N}}\,\beta_{\underline{N}}$ = $-\beta_{\underline{N}}\,{}_N T_M\,\alpha_{\underline{M}}$	$-\alpha_{\underline{M}}\,{}_M\Sigma_{\underline{N}}\,\beta_{\underline{N}}$ = $\beta_{\underline{N}}\,{}_N T_M\,\alpha_{\underline{M}}$ = $\|\underline{b}_N \wedge \underline{a}_M\|$	$-\alpha_{\underline{M}}\,{}_M\Sigma^{\underline{N}}\,\beta^{\underline{N}}$ = $\beta^{\underline{N}}\,{}^{\underline{N}}T_M\,\alpha_{\underline{M}}$ = $\|\underline{b}^N \wedge \underline{a}_M\|$

So:

$\alpha M * \underline{M}\Sigma\underline{N} * \beta N = -\beta N * \underline{N}T\underline{M} * \alpha M$

	bNM	gNM	sNM	yNM	oNM
aMN	αM*MΣN* βN = -βN*NTM*αM	αM*MΣN*1 = -1*NTM*αM	αM*MΣN* σN = -σN*NTM*αM	αM*MΣN* 1 = -1*NTM*αM	αM*MΣN*1 = -1*NTM*αM
fMN	1*MΣN* βN = -βN*NTM*1	1*MΣN*1 = -1*NTM*1	1*MΣN* σN = -σN* NTM*1	1*MΣN* 1 = -1*NTM*1	1*MΣN*1 = -1*NTM*1
rMN	ρM*MΣN* βN = -βN*NTM* ρM	ρM*MΣN*1 = -1*NTM* ρM	ρM*MΣN* σN = -σN* NTM*ρM	ρM*MΣN* 1 = -1*NTM*ρM	ρM*MΣN*1 = -1*NTM*ρM
xMN	1*MΣN*βN = -βN*NTM*1	1*MΣN*1 = -1*NTM*1	1*MΣN* σN = -σN*NTM*1	1*MΣN* 1 = -1*NTM*1	1*MΣN*1 = -1*NTM*1
oMN	1*MΣN*βN = –βN*NTM*1	1*MΣN *1 = -1*NTM*1	1*MΣN * σN = –σN*NTM*1	1*MΣN*1 = –1*NTM*1	1*MΣN*1 = –1*NTM*1

$\|\underline{o}_m \wedge \underline{a}^n\| = S_m{}^n\,\alpha^n$ $\qquad$ $\|\underline{a}_n \wedge \underline{o}^m\| = \alpha_n\,T_n{}^m$ $\qquad$ $\|\underline{o}^n \wedge \underline{a}_m\| = S^n{}_m\,\alpha_m$ $\qquad$ $\|\underline{a}^m \wedge \underline{o}_n\| = \alpha^m\,T^m{}_n$

Where

$S_m{}^n = S(\underline{o}_m, \underline{a}^n)$ $\qquad$ $T_n{}^m = S(\underline{a}_n, \underline{o}^m)$ $\qquad$ $S^n{}_m = S(\underline{o}^n, \underline{a}_m)$ $\qquad$ $T^m{}_n = S(\underline{a}^m, \underline{o}_n)$

$o{\wedge}a_m{}^n = S_m{}^n\,\alpha^n$ $\qquad$ $a{\wedge}o_n{}^m = \alpha_n\,T_n{}^m$ $\qquad$ $o{\wedge}a^n{}_m = S^n{}_m\,\alpha_m$ $\qquad$ $a{\wedge}o^m{}_n = \alpha^m\,T^m{}_n$

$$-S_m^{\ n} = T^n_{\ m}$$

$$T_n^{\ m} = -S^m_{\ n}$$

$$S_m^{\ n}\alpha^n = \alpha^n T^n_{\ m}$$

$$\alpha_n T_n^{\ m} = S^m_{\ n}\alpha_n$$

2.2.2.2.
InnerProduct of Bases

m≠p≠q≠m

$\underline{a}^I . \underline{a}^K \equiv aa^{IK} \le \alpha^I\alpha^K$	$\underline{a}^m . \underline{a}_{\underline{m}} = \underline{a}_p \wedge \underline{a}_q . \underline{a}_m / \Vert\ \Vert = \Vert\ \Vert / \Vert\ \Vert \equiv 1$
$\underline{a}_{\underline{m}} . \underline{a}^m = \underline{a}^p \wedge \underline{a}^q . \underline{a}^m / a^{123} = a^{123} / a^{123} = 1$	$\underline{a}_I . \underline{a}_K \equiv aa_{IK} \le \alpha_I\alpha_K$
$\underline{a}^I . \underline{a}^K = aa^{IK} = AA^{\underline{IK}}\alpha^{\underline{I}}\alpha^{\underline{K}}$	$a_{123} \equiv \Vert_a\Vert \equiv (\underline{a}_1 \wedge \underline{a}_2 . \underline{a}_3) \le \alpha_1\alpha_2\alpha_3$ $a_{123} \equiv a_{123}\underline{a}^{\underline{m}} . \underline{a}_{\underline{m}} = a_{123} AA^{\underline{m}}_{\underline{m}}\alpha^{\underline{m}}\alpha_{\underline{m}}$ $1 = \underline{a}^{\underline{m}} . \underline{a}_{\underline{m}} = AA^{\underline{m}}_{\underline{m}}\alpha^{\underline{m}}\alpha_{\underline{m}}$
$a^{123} \equiv \Vert^a\Vert \equiv (\underline{a}^1 \wedge \underline{a}^2 . \underline{a}^3) \le \alpha^1\alpha^2\alpha^3$ $a^{123} \equiv \Vert^a\Vert = a^{123}\underline{a}_{\underline{m}} . \underline{a}^{\underline{m}} = a^{123} AA_{\underline{m}}^{\underline{m}}\alpha_{\underline{m}}\alpha^{\underline{m}}$ $1 = \underline{a}_{\underline{m}} . \underline{a}^{\underline{m}} = AA_{\underline{m}}^{\underline{m}}\alpha_{\underline{m}}\alpha^{\underline{m}}$	$\underline{a}_I . \underline{a}_K = aa_{IK} = AA_{\underline{IK}}\alpha_{\underline{I}}\alpha_{\underline{K}}$

$\underline{A}$: Unit Vector:

$$\underline{a}_K \equiv \underline{A}_{\underline{K}}\alpha_{\underline{K}} \qquad \Vert\underline{A}_K\Vert = 1 = \Vert\underline{A}^K\Vert \qquad \underline{a}^K \equiv \underline{A}^{\underline{K}}\alpha^{\underline{K}}$$

and:

$$aa_{IK} \equiv \underline{a}_I . \underline{a}_K = \underline{A}_{\underline{I}} \cdot \underline{A}_{\underline{K}}\alpha_{\underline{I}}\alpha_{\underline{K}}$$

$$aa^{IK} \equiv \underline{a}^I . \underline{a}^K = \underline{A}^{\underline{I}} . \underline{A}^{\underline{K}}\alpha^{\underline{I}}\alpha^{\underline{K}}$$

Direction cosines:

$$-1 \le AA_{\underline{LK}} \equiv \underline{A}_{\underline{L}} \cdot \underline{A}_{\underline{K}} \le 1$$

$$-1 \le AA^{\underline{LK}} \equiv \underline{A}^{\underline{l}} . \underline{A}^{\underline{K}} \le 1$$

$$-\alpha_I\alpha_K \le aa_{IK} = AA_{\underline{IK}}\alpha_{\underline{I}}\alpha_{\underline{K}} \le \alpha_I\alpha_K$$

$$aa^I_{\ K} \equiv \underline{a}^I . \underline{a}_K = \delta^I_{\underline{K}}$$

$$aa_I^{\ K} \equiv \underline{a}_I . \underline{a}^K = \delta^K_{\underline{I}}$$

$$-\alpha^I\alpha^K \le aa^{IK} = AA^{\underline{IK}}\alpha^{\underline{I}}\alpha^{\underline{K}} \le \alpha^I\alpha^K$$

p≠q = r Ordered with no sum

$$aa^r_{\ q} = \underline{a}_p \wedge \underline{a}_q . \underline{a}_q / \Vert_a\Vert = 0 / \Vert_a\Vert = 0$$

$$aa_r^{\ q} = \underline{a}^p \wedge \underline{a}^q . \underline{a}^q / \Vert^a\Vert = 0 / \Vert^a\Vert = 0$$

and: r≠p≠q≠r Ordered with no sum

$$aa^{\underline{r}}_{\ \underline{r}} = \underline{a}_p \wedge \underline{a}_q . \underline{a}_r / \Vert_a\Vert = \Vert_a\Vert / \Vert_a\Vert \equiv 1$$

$aa_{\underline{r}}^{\underline{r}} = \underline{a}^{p} \wedge \underline{a}^{q} . \underline{a}^{r} / ||^{a}|| = ||^{a}||/ ||^{a}|| \equiv 1$

2.2.2.3. Mixed Product of Bases:

Mixed product (volume of formed bases) is equal or less products of vector norms:

Mixed product of same system bases:

$a_{123} \equiv \|\|_{a}\|\| \equiv \underline{a}_p \wedge \underline{a}_q . \underline{a}_r = (1/6) e_{ijk} e^{pqr} a^i_p a^j_q a^k_r$	$a^{123} \equiv \|\|^{a}\|\| \equiv \underline{a}^p \wedge \underline{a}^q . \underline{a}^r = (1/6) e^{ijk} e_{pqr} a^p_i a^q_j a^r_k$
$\underline{a}_K \wedge \underline{a}_L . \underline{a}_M = e_{\alpha\beta\Gamma}\, oa^{\alpha}_{K}\, oa^{\beta}_{L}\, oa^{\Gamma}_{M}$	$\underline{a}^K \wedge \underline{a}^L . \underline{a}^M = e^{\alpha\beta\Gamma}\, oa_{\alpha}^{K}\, oa_{\beta}^{L}\, oa_{\Gamma}^{M}$
$a_{123} \equiv \|\|_{a}\|\| \equiv \underline{o}_i a^i_p \wedge \underline{o}_j a^j_q . \underline{o}_k a^k_r = \underline{o}_i \wedge \underline{o}_j . \underline{o}_k a^i_p a^j_q a^k_r$	$a^{123} \equiv \|\|^{a}\|\| \equiv \underline{o}^i a^p_i \wedge \underline{o}^j a^q_j . \underline{o}^k a^r_k = \underline{o}^i \wedge \underline{o}^j . \underline{o}^k a^p_i a^q_j a^r_k$

Covariant Bases	Contravariant Bases
$a_{123} \equiv (\underline{a}_1 \wedge \underline{a}_2 . \underline{a}_3) = \|oa(+n,-m)\| \equiv oa[+n,-m]$ $\leq$ $\alpha_{123} \equiv \alpha_1 \alpha_2 \alpha_3$	$a^{123} \equiv (\underline{a}^1 \wedge \underline{a}^2 . \underline{a}^3) = \| oa(-m,+n) \| \equiv oa[-m ,+n]$ $\leq \alpha^{123} \equiv \alpha^1 \alpha^2 \alpha^3$
$f_{123} \equiv (\underline{f}_1 \wedge \underline{f}_2 . \underline{f}_3) = \|of(+n,-m)\| \equiv of[+n,-m]$ $\leq$ $1 = \phi_{123} \equiv \phi_1 \phi_2 \phi_3$	$f^{123} \equiv (\underline{f}^1 \wedge \underline{f}^2 . \underline{f}^3) = \| of(-m,+n) \| \equiv of[-m ,+n]$ $\leq$ $1 = \phi^{123} \equiv \phi^1 \phi^2 \phi^3$
$r_{123} \equiv (\underline{r}_1 \wedge \underline{r}_2 . \underline{r}_3) = \|or(+n,-m)\| \equiv or[+n,-m]$ $=$ $\rho_{123} \equiv \rho_1 \rho_2 \rho_3$	$r^{123} \equiv (\underline{r}^1 \wedge \underline{r}^2 . \underline{r}^3) = \| or(-m,+n) \| \equiv or[-m ,+n]$ $=$ $\rho^{123} \equiv \rho^1 \rho^2 \rho^3$
$x_{123} \equiv (\underline{x}_1 \wedge \underline{x}_2 . \underline{x}_3) = \|ox(+n,-m)\| \equiv ox[+n,-m] = 1$	$x^{123} \equiv (\underline{x}^1 \wedge \underline{x}^2 . \underline{x}^3) = \|ox(-m,+n)\| \equiv ox[-m ,+n] = 1$
$o_{123} \equiv (\underline{o}_1 \wedge \underline{o}_2 . \underline{o}_3) = \|oo(+n,-m)\| \equiv oo[+n,-m] = 1$	$o^{123} \equiv (\underline{o}^1 \wedge \underline{o}^2 . \underline{o}^3) = \|oo(-m,+n)\| \equiv oo[-m ,+n] = 1$

If a basis is repeated:

$\underline{a}_I \wedge \underline{a}_J . \underline{a}_I = 0 = \underline{a}_I \wedge \underline{a}_J . \underline{a}_J$	$\underline{a}^I \wedge \underline{a}^J . \underline{a}^I = 0 = \underline{a}^I \wedge \underline{a}^J . \underline{a}^J$
$\underline{a}_1 \wedge \underline{a}_2 . \underline{a}_1 = \underline{a}_1 \wedge \underline{a}_2 . (\underline{o}_m\, oa^m_{\ 1})$ $=$ $oa^{\alpha}_{\ 1}\, oa^{\beta}_{\ 2}\, e_{\alpha\beta m}\, oa^m_{\ 1} = 0$	$\underline{a}^1 \wedge \underline{a}^2 . \underline{a}^1 = \underline{a}^1 \wedge \underline{a}^2 . (\underline{o}^m\, oa_m^{\ 1})$ $=$ $oa_{\alpha}^{\ 1}\, oa_{\beta}^{\ 2}\, e^{\alpha\beta m}\, oa_m^{\ 1} = 0$
$\underline{a}_1 \wedge \underline{a}_2 . \underline{a}_2 = \underline{a}_1 \wedge \underline{a}_2 . (\underline{o}_m\, oa^m_{\ 2})$ $=$ $oa^{\alpha}_{\ 1}\, oa^{\beta}_{\ 2}\, e_{\alpha\beta m}\, oa^m_{\ 2} = 0$	$\underline{a}^1 \wedge \underline{a}^2 . \underline{a}^2 = \underline{a}^1 \wedge \underline{a}^2 . (\underline{o}^m\, oa_m^{\ 2})$ $=$ $oa_{\alpha}^{\ 1}\, oa_{\beta}^{\ 2}\, e^{\alpha\beta m}\, oa_m^{\ 2} = 0$

Mixed product of two bases outer product and inner a fixed basis:

$\underline{a}_i \wedge \underline{a}_j . \underline{o}_m = oa^{\alpha}_{\ i}\, oa^{\beta}_{\ j}\, e_{\alpha\beta k}\, \underline{o}^k . \underline{o}_m$ $=$ $oa^{\alpha}_{\ i}\, oa^{\beta}_{\ j}\, e_{\alpha\beta m}$	$\underline{a}^i \wedge \underline{a}^j . \underline{o}^m = oa_{\alpha}^{\ i}\, oa_{\beta}^{\ j}\, e^{\alpha\beta k}\, \underline{o}_k . \underline{o}^m$ $=$ $oa_{\alpha}^{\ i}\, oa_{\beta}^{\ j}\, e^{\alpha\beta m}$
$\underline{a}_i \wedge \underline{a}_j . \underline{o}_m = oa^{\alpha}_{\ i}\, oa^{\beta}_{\ j}\, e_{\alpha\beta m}$ $\underline{f}_i \wedge \underline{f}_j . \underline{o}_m = of^{\alpha}_{\ i}\, of^{\beta}_{\ j}\, e_{\alpha\beta m}$ $\underline{r}_i \wedge \underline{r}_j . \underline{o}_m = or^{\alpha}_{\ i}\, or^{\beta}_{\ j}\, e_{\alpha\beta m}$	$\underline{a}^i \wedge \underline{a}^j . \underline{o}^m = oa_{\alpha}^{\ i}\, oa_{\beta}^{\ j}\, e^{\alpha\beta m}$ $\underline{f}^i \wedge \underline{f}^j . \underline{o}^m = of_{\alpha}^{\ i}\, of_{\beta}^{\ j}\, e^{\alpha\beta m}$ $\underline{r}^i \wedge \underline{r}^j . \underline{o}^m = or_{\alpha}^{\ i}\, or_{\beta}^{\ j}\, e^{\alpha\beta m}$

$$\underline{x}_i \wedge \underline{x}_j \cdot \underline{o}_m = ox^{\alpha}_{\ i}\, ox^{\beta}_{\ j}\; e_{\alpha\beta m}$$
$$\underline{o}_i \wedge \underline{o}_j \cdot \underline{o}_m = oo^{\alpha}_{\ i}\, oo^{\beta}_{\ j}\; e_{\alpha\beta m}$$

$$\underline{x}^i \wedge \underline{x}^j \cdot \underline{o}^m = ox_{\alpha}^{\ i}\, ox_{\beta}^{\ j}\; e^{\alpha\beta m}$$
$$\underline{o}^i \wedge \underline{o}^j \cdot \underline{o}^m = oo_{\alpha}^{\ i}\, oo_{\beta}^{\ j}\; e^{\alpha\beta m}$$

2.2.2.4. Operations on Bases:

Knowing:

Ordered $i \neq j \neq K \neq i$

$$\underline{a}_I \wedge \underline{a}_J = a_{123}\ \underline{a}^K \qquad\qquad \underline{a}^I \wedge \underline{a}^J = a^{123}\ \underline{a}_K$$

Then:

$$(a_{123}\ \underline{a}^I) \wedge (a_{123}\ \underline{a}^J) = a_{123}\ a_{123}\ a^{123}\ \underline{a}_K \qquad\qquad (a^{123}\ \underline{a}_I) \wedge (a^{123}\ \underline{a}_J) = a^{123}\ a^{123}\ a_{123}\ \underline{a}^K$$

Or:

$$(\underline{a}_J \wedge \underline{a}_K) \wedge (\underline{a}_K \wedge \underline{a}_I) = a_{123}\ a_{123}\ a^{123}\ \underline{a}_K \qquad\qquad (\underline{a}^I \wedge \underline{a}^J) \wedge (\underline{a}^J \wedge \underline{a}^K) = a^{123}\ a^{123}\ a_{123}\ \underline{a}^K$$

But

$$(\underline{A} \wedge \underline{B}) \wedge \underline{C} = (\underline{C} \cdot \underline{A})\,\underline{B} - (\underline{B} \cdot \underline{C})\,\underline{A}$$

$$(\underline{A} \wedge \underline{B}) \wedge (\underline{C} \wedge \underline{A}) = (\underline{C} \wedge \underline{A} \cdot \underline{A})\,\underline{B} - (\underline{B} \cdot \underline{C} \wedge \underline{A})\,\underline{A} = (\underline{A} \cdot \underline{B} \wedge \underline{C})\,\underline{A}$$

So:

$$a_{123}\ a_{123}\ a^{123}\ \underline{a}_K = (\underline{a}_J \wedge \underline{a}_K) \wedge (\underline{a}_K \wedge \underline{a}_I) = a_{123}\ \underline{a}_K = \|_a\|\ \underline{a}_K$$

$$a^{123}\ a^{123}\ a_{123}\ \underline{a}^K = (\underline{a}^J \wedge \underline{a}^K) \wedge (\underline{a}^K \wedge \underline{a}^I) = a^{123}\ \underline{a}^K = \|^a\|\ \underline{a}^K$$

Or:

$$(\underline{a}_1 \wedge \underline{a}_2 \cdot \underline{a}_3)(\underline{a}^1 \wedge \underline{a}^2 \cdot \underline{a}^3) \equiv \|_a\|\ \|^a\| = 1 \qquad\qquad (\underline{a}^1 \wedge \underline{a}^2 \cdot \underline{a}^3)(\underline{a}_1 \wedge \underline{a}_2 \cdot \underline{a}_3) \equiv \|^a\|\ \|_a\| = 1$$

$$\underline{\nabla} \wedge \underline{\nabla}\, U = 0$$

$$\underline{\nabla} \cdot \underline{\nabla}\, U = \underline{\nabla}\ \underline{\nabla} \cdot \underline{V}$$

$$\underline{\nabla} \cdot \underline{\nabla} \wedge \underline{V} = 0$$

Since:

$$\underline{\nabla} \wedge (B\ \underline{C}) = (\underline{\nabla}\, B) \wedge \underline{C} - B\,(\underline{\nabla} \wedge \underline{C})$$

Then:

$$\underline{\nabla} \wedge (a^J\ \underline{\nabla} a^K) = \underline{\nabla} a^J \wedge \underline{\nabla} a^K - a^J\,(\underline{\nabla} \wedge \underline{\nabla} a^K) = \underline{\nabla} a^J \wedge \underline{\nabla} a^K - a^J(0) = \underline{\nabla} a^J \wedge \underline{\nabla} a^K$$

$$\underline{\nabla} \wedge (a_J\ \underline{\nabla} a_K) = \underline{\nabla} a_J \wedge \underline{\nabla} a_K - a_J\,(\underline{\nabla} \wedge \underline{\nabla} a_K) = \underline{\nabla} a_J \wedge \underline{\nabla} a_K - a_J(0) = \underline{\nabla} a_J \wedge \underline{\nabla} a_K$$

Because:

$$\underline{\nabla} \wedge (\underline{\nabla}\, B) = 0$$

So:

$i \neq j \neq K \neq i$ Ordered with no sum

$$\underline{\nabla} \wedge (a^J\ \underline{a}^K) = \underline{a}^J \wedge \underline{a}^K = a^{123}\ \underline{a}_L \qquad\qquad \underline{\nabla} \wedge (a_J\ \underline{a}_K) = \underline{a}_J \wedge \underline{a}_K = a_{123}\ \underline{a}^L$$

and Since:

$$\underline{\nabla} \cdot (\underline{\nabla} \wedge \underline{C}) = 0$$

Then:

$\underline{\nabla} \cdot (\underline{\nabla} \wedge (a^{J}\ \underline{a}^{K})) = \underline{\nabla} \cdot (a^{123}\ \underline{a}_{L}) = 0$ | $\underline{\nabla} \cdot (\underline{\nabla} \wedge (a_{J}\ \underline{a}_{K})) = \underline{\nabla} \cdot (\|_{a}\|\ \underline{a}^{L}) = 0$

2.2.2.5. Covariant-Contravariant Basis Change:

Multiply Position Vector:

$$\underline{a}_{i}\ a^{i} = \underline{a}^{i}\ a_{i}$$

$$\underline{o}_{k}\ oa^{k}_{\ i}\ a^{i} = \underline{o}^{j}\ oa_{j}^{\ m}\ a_{m}$$

by:

$$\underline{o}^{i} = \underline{o}_{i}$$

To Get:

$$\underline{o}^{i} \cdot \underline{o}_{k}\ oa^{k}_{\ n}\ a^{n} = \underline{o}_{i} \cdot \underline{o}^{j}\ oa_{j}^{\ m}\ a_{m}$$

$$oa^{i}_{\ n}\ a^{n} = oa^{nm}\ a_{m} = oa_{i}^{\ m}\ a_{m} = o_{i}$$

Multiply by:

$$oa^{i}_{\ I} = ao^{i}_{I}$$

To Get:

$$oa^{i}_{\ I}\ oa^{i}_{\ n}\ a^{n} = ao^{i}_{I}\ o_{i} = a_{I}$$

Or:

$$oa^{i}_{\ I}\ oa^{i}_{\ n}\ a^{n} = a_{I}$$

Or:

$$aa_{In}\ a^{n} = a_{I}$$

Where it is defined:

$$oa^{i}_{\ I}\ oa^{i}_{\ n} \equiv aa_{In}$$

In a similar way get:

$$aa^{In}\ a_{n} = a^{I}$$

Both results:

$$aa_{In}\ a^{n} = a_{I} \qquad\qquad aa^{In}\ a_{n} = a^{I}$$

2.2.3. Systems Bases

Align drawings

2.2.3.1. General Bases:

$(\underline{a}_{1}, \underline{a}_{2}, \underline{a}_{3})$ coordinates | $(\underline{a}_{1}, \underline{a}_{2})$ coordinates | $(\underline{a}^{3})$ conjugate of $(\underline{a}_{1}, \underline{a}_{2})$

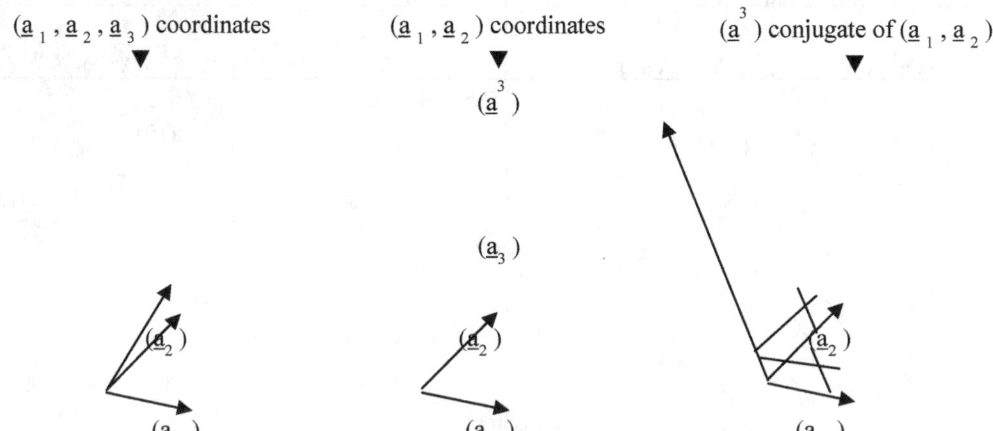

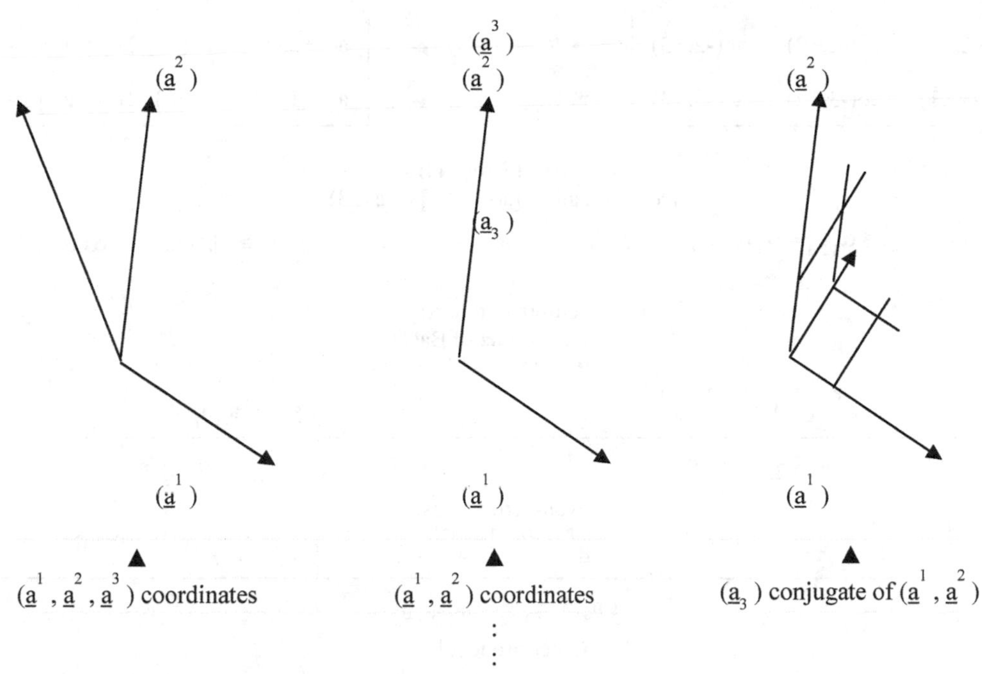

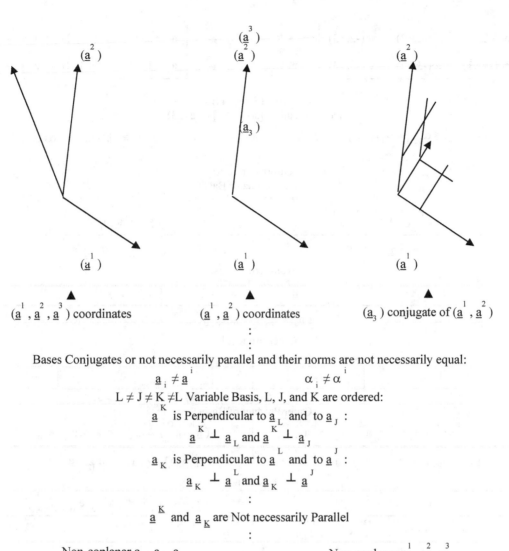

$(\underline{a}^1, \underline{a}^2, \underline{a}^3)$ coordinates $\quad$ $(\underline{a}^1, \underline{a}^2)$ coordinates $\quad$ $(\underline{a}_3)$ conjugate of $(\underline{a}^1, \underline{a}^2)$

:

:

Bases Conjugates or not necessarily parallel and their norms are not necessarily equal:

$$\underline{a}_i \neq \underline{a}^i \qquad\qquad \alpha_i \neq \alpha^i$$

$L \neq J \neq K \neq L$ Variable Basis, L, J, and K are ordered:

$\underline{a}^K$ is Perpendicular to $\underline{a}_L$ and to $\underline{a}_J$:

$$\underline{a}^K \perp \underline{a}_L \text{ and } \underline{a}^K \perp \underline{a}_J$$

$\underline{a}_K$ is Perpendicular to $\underline{a}^L$ and to $\underline{a}^J$:

$$\underline{a}_K \perp \underline{a}^L \text{ and } \underline{a}_K \perp \underline{a}^J$$

:

$\underline{a}^{\underline{K}}$ and $\underline{a}_{\underline{K}}$ are Not necessarily Parallel

:

Non-coplanar $\underline{a}_1$, $\underline{a}_2$, $\underline{a}_3$

$$\underline{a}_K \equiv \underline{A}_{\underline{K}}\, \alpha_{\underline{K}}$$

$$1 = \|\underline{A}_K\|$$

$$\alpha_{\underline{K}}$$

Non-coplanar $\underline{a}^1$, $\underline{a}^2$, $\underline{a}^3$

$$\underline{a}^K \equiv \underline{A}^{\underline{K}}\, \alpha^{\underline{K}}$$

$$1 = \|\underline{A}^K\|$$

$$\alpha^{\underline{K}}$$

$(\underline{a}_K \equiv \underline{A}_{\underline{K}}\, \alpha_{\underline{K}})$ perpendicular to $(\underline{a}^I \equiv \underline{A}^I\, \alpha^I)$ $\quad$ $(\underline{a}_K \equiv \underline{A}_{\underline{K}}\, \alpha_{\underline{K}})$ perpendicular to $(\underline{a}^J \equiv \underline{A}^J\, \alpha^J)$

$$\underline{a}_K = (\,ao_K^{\;M}\,)\,(\underline{o}_M)$$

$$\alpha_{\underline{M}} \equiv \|\,\underline{a}_{\underline{M}}\| = \sqrt{(\underline{a}_{\underline{M}} \cdot \underline{a}_{\underline{M}})}$$

$$\underline{a}^K = (\,ao^K_{\;M}\,)\,(\underline{o}^K)$$

$$\alpha^N \equiv \|\,\underline{a}^N\| = \sqrt{(\underline{a}^{\underline{N}} \cdot \underline{a}^{\underline{N}})}$$

Norm of a basis is α

$$\alpha_M = \sqrt{[(ao_M^{\;1})^2 + (ao_M^{\;2})^2 + (ao_M^{\;3})^2]}$$

$$\alpha^N = \sqrt{[(ao^N_{\;1})^2 + (ao^N_{\;2})^2 + (ao^N_{\;3})^2]}$$

Such that:

$$\boxed{\alpha_{\underline{M}}\, \alpha^{\underline{M}} = 1}$$

ao(-1,+1)	ao(-1,+2)	ao(-1,+3)	◄ α_1	α^1 ►	ao(+1,-1)	ao(+1,-2)	ao(+1,-3)

ao(-2,+1)	ao(-2,+2)	ao(-2,+3)	◄ α_2	α^2 ►	ao(+2,-1)	ao(+2,-2)	ao(+2,-3)
ao(-3,+1)	ao(-3,+2)	ao(-3,+3)	◄ α_3	α^3 ►	ao(+3,-1)	ao(+3,-2)	ao(+3,-3)

Mixed Product:

Determinant of [ao(M,N)] ≡ (a123):

$a_{123} \equiv \|_a\| \leq \alpha_{123} \equiv \alpha_1 \alpha_2 \alpha_3$ $\qquad$ $a^{123} \equiv \|^a\| \leq \alpha^{123} \equiv \alpha^1 \alpha^2 \alpha^3$

Outer Product:

Outer Product of Bases:

$P \neq Q \neq M \neq P$

$(\underline{a}^P \wedge \underline{a}^Q) = a^{123}\, \underline{a}_M$ $\qquad$ $(\underline{a}^P \wedge \underline{a}_Q) = 0$

$(\underline{a}_P \wedge \underline{a}^Q) = 0$ $\qquad$ $(\underline{a}_P \wedge \underline{a}_Q) = a_{123}\, \underline{a}^M$

Non-zero values:

$(\underline{a}^P \wedge \underline{a}^Q) = a^{123}\, \underline{a}_M$

$(\underline{a}_P \wedge \underline{a}_Q) = a_{123}\, \underline{a}^M$

Outer product by:

$\underline{a}^N$ $\qquad$ $\underline{a}_N$

To get:

Outer Product of Bases:

$(\underline{a}^P \wedge \underline{a}^Q) \wedge \underline{a}^N = a^{123}\, \underline{a}_M \wedge \underline{a}^N$ $\qquad$ $(\underline{a}^P \wedge \underline{a}^Q) \wedge \underline{a}_N = a^{123}\, \underline{a}_M \wedge \underline{a}_N$

$(\underline{a}_P \wedge \underline{a}_Q) \wedge \underline{a}^N = a_{123}\, \underline{a}^M \wedge \underline{a}^N$ $\qquad$ $(\underline{a}_P \wedge \underline{a}_Q) \wedge \underline{a}_N = a_{123}\, \underline{a}^M \wedge \underline{a}_N$

Knowing:

$\| \underline{a}_M \wedge \underline{a}_N \| \leq \|\alpha_{123}\, e_{MNK}\, \underline{a}^K\|$

$\| \underline{a}^M \wedge \underline{a}^N \| \leq \|\alpha^{123}\, e^{MNK}\, \underline{a}_K\|$

Then:

$\|(\underline{a}^P \wedge \underline{a}^Q) \wedge \underline{a}_N\| = a^{123} \|\underline{a}_M \wedge \underline{a}_N\| \leq \|\alpha^{123}\, e^{PQM}\, \underline{a}_M \wedge \underline{a}_N\|$

$\|(\underline{a}_P \wedge \underline{a}_Q) \wedge \underline{a}^N\| = a_{123} \| \underline{a}^M \wedge \underline{a}^N \| \leq \|\alpha_{123} e_{PQM} \underline{a}^M \wedge \underline{a}^N \|$

:

Outer Product of Bases:

$\underline{a}_M \wedge \underline{a}^N = (\underline{a}^P \wedge \underline{a}^Q) \wedge \underline{a}^N / a^{123}$ $\qquad$ $\underline{a}_M \wedge \underline{a}_N = (\underline{a}^P \wedge \underline{a}^Q) \wedge \underline{a}_N / a^{123}$

$\underline{a}^M \wedge \underline{a}^N = (\underline{a}_P \wedge \underline{a}_Q) \wedge \underline{a}^N / a_{123}$ $\qquad$ $\underline{a}^M \wedge \underline{a}_N = (\underline{a}_P \wedge \underline{a}_Q) \wedge \underline{a}_N / a_{123}$

$\underline{a}_P \wedge \underline{a}_Q = \underline{o}_\alpha\, oa^\alpha{}_P \wedge \underline{o}_\beta\, oa^\beta{}_Q = oa^\alpha{}_P\, oa^\beta{}_Q\, e_{\alpha\beta k}\, \underline{o}^k$

$\underline{a}^P \wedge \underline{a}^Q = \underline{o}^\alpha\, oa_\alpha{}^P \wedge \underline{o}^\beta\, oa_\beta{}^Q = oa_\alpha{}^P\, oa_\beta{}^Q\, e^{\alpha\beta k}\, \underline{o}_k$

$\| \underline{a}_P \wedge \underline{a}_Q \| \leq \|\alpha_{123}\, e_{PQk}\, \underline{a}^k\|$ $\qquad$ $\| \underline{a}^P \wedge \underline{a}^Q \| \leq \|\alpha^{123}\, e^{PQk}\, \underline{a}_k\|$

Inner Product of Bases:

$$aa^{IK} \equiv \underline{a}^{I} . \underline{a}^{K} = ao^{I}_{J}\, ao^{K}_{J} \qquad aa^{I}_{K} \equiv \underline{a}^{I} . \underline{a}_{K} = \delta^{I}_{\underline{K}}$$

$$aa_{I}^{K} \equiv \underline{a}_{I} . \underline{a}^{K} = \delta_{\underline{I}}^{K} \qquad aa_{IK} \equiv \underline{a}_{I} . \underline{a}_{K} = ao_{I}^{J}\, ao_{K}^{J}$$

2.2.3.2.
General Unitary Bases, a Special Case of General Unitary Bases:

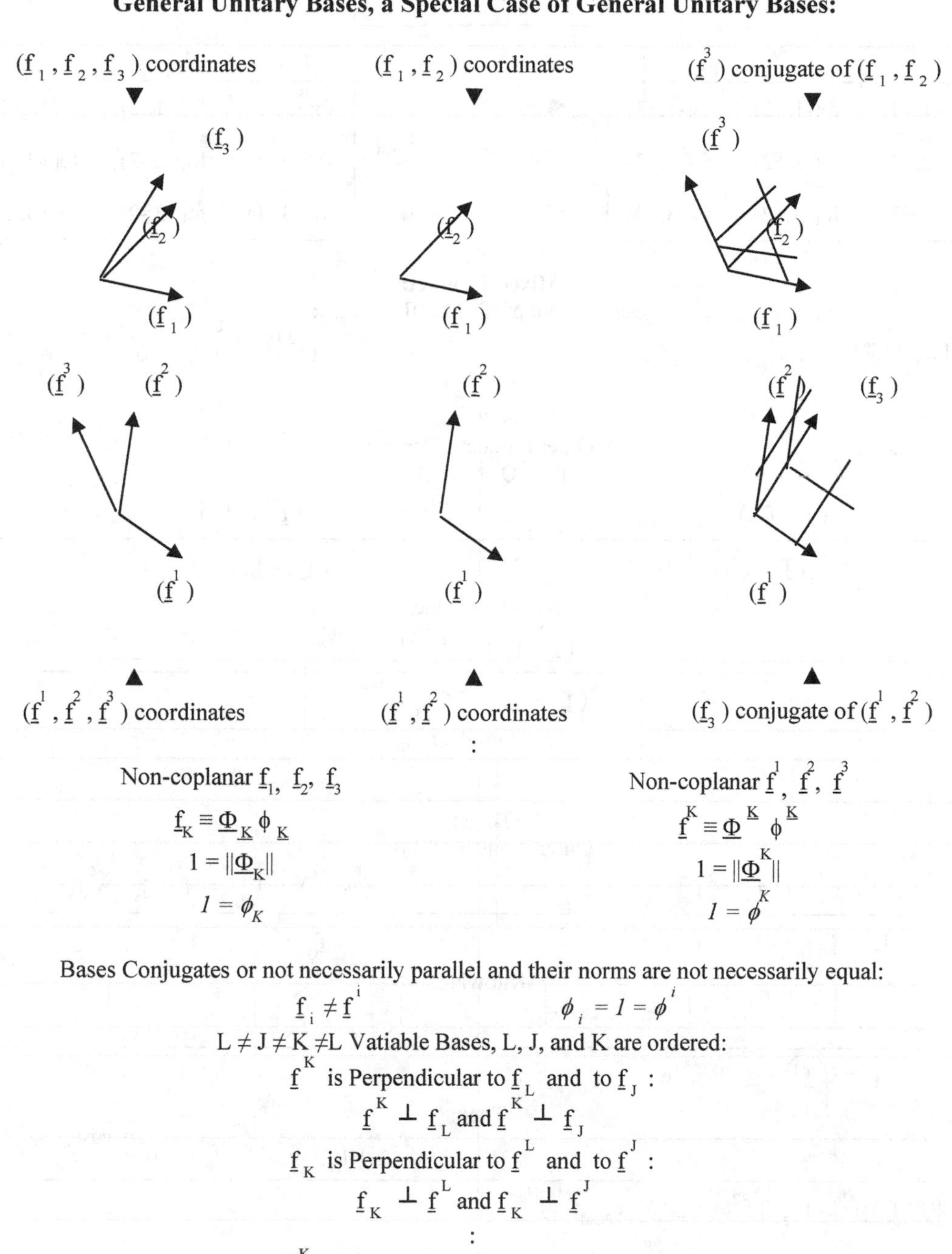

:

Non-coplanar $\underline{f}_1$, $\underline{f}_2$, $\underline{f}_3$

$$\underline{f}_K \equiv \underline{\Phi}_{\underline{K}}\, \phi_{\underline{K}}$$

$$1 = \|\underline{\Phi}_K\|$$

$$1 = \phi_K$$

Non-coplanar $\underline{f}^1$, $\underline{f}^2$, $\underline{f}^3$

$$\underline{f}^K \equiv \underline{\Phi}^{\underline{K}}\, \phi^{\underline{K}}$$

$$1 = \|\underline{\Phi}^K\|$$

$$1 = \phi^K$$

Bases Conjugates or not necessarily parallel and their norms are not necessarily equal:

$$\underline{f}_i \neq \underline{f}^i \qquad \phi_i = 1 = \phi^i$$

$L \neq J \neq K \neq L$ Vatiable Bases, L, J, and K are ordered:

$\underline{f}^K$ is Perpendicular to $\underline{f}_L$ and to $\underline{f}_J$:

$$\underline{f}^K \perp \underline{f}_L \text{ and } \underline{f}^K \perp \underline{f}_J$$

$\underline{f}_K$ is Perpendicular to $\underline{f}^L$ and to $\underline{f}^J$:

$$\underline{f}_K \perp \underline{f}^L \text{ and } \underline{f}_K \perp \underline{f}^J$$

:

$\underline{f}^{\underline{K}}$ and $\underline{f}_{\underline{K}}$ are Not necessarily Parallel

:

$(\underline{f}_K \equiv \underline{\Phi}_{\underline{K}}\, \phi_{\underline{K}})$ Perpendicular to $(\underline{f}^I \equiv \underline{\Phi}^{\underline{I}}\, \phi^{\underline{I}})$ $\qquad$ $(\underline{f}_K \equiv \underline{\Phi}_{\underline{K}}\, \phi_{\underline{K}})$ Perpendicular to $(\underline{f}^J \equiv \underline{\Phi}^{\underline{J}}\, \phi^{\underline{J}})$

$$\underline{f}_K = (fo_K^{\ M})(\underline{o}_M) \qquad \underline{f}^K = (fo^K_{\ M})(\underline{o}^K)$$

$\phi_M \equiv \| \underline{f}_M \| = \sqrt{(\underline{f}_M \cdot \underline{f}_M)}$ $\qquad$ $\phi^N \equiv \| \underline{f}^N \| = \sqrt{(\underline{f}^N \cdot \underline{f}^N)}$

Norm of f Basis is ϕ

$\phi_M = \sqrt{[(fo_M^{\ 1})^2 + (fo_M^{\ 2})^2 + (fo_M^{\ 3})^2]}$ $\qquad$ $\phi^N = \sqrt{[(fo^N{}_1)^2 + (fo^N{}_2)^2 + (fo^N{}_3)^2]}$

Such that:

$$\phi_M = 1 = \phi^M$$

fo(-1,+1)	fo(-1,+2)	fo(-1,+3)	◄ ϕ_1	ϕ^1 ►	fo(+1,-1)	fo(+1,-2)	fo(+1,-3)
fo(-2,+1)	fo(-2,+2)	fo(-2,+3)	◄ ϕ_2	ϕ^2 ►	fo(+2,-1)	fo(+2,-2)	fo(+2,-3)
fo(-3,+1)	fo(-3,+2)	fo(-3,+3)	◄ ϕ_3	ϕ^3 ►	fo(+3,-1)	fo(+3,-2)	fo(+3,-3)

Mixed Product:

Determinant of [fo(M,N)] ≡ (f123):

$f_{123} \equiv \|_f\| \leq \underline{1} = \phi_{123} \equiv \phi_1 \phi_2 \phi_3$ $\qquad$ $f^{123} \equiv \|^f\| \leq \underline{1} = \phi^{123} \equiv \phi^1 \phi^2 \phi^3$

Outer Product:

Outer Product of Bases:

$P \neq Q \neq M \neq P$

$(\underline{f}^P \wedge \underline{f}^Q) = f^{123}\ \underline{f}_M$	$(\underline{f}^P \wedge \underline{f}_Q) = 0$
$(\underline{f}_P \wedge \underline{f}^Q) = 0$	$(\underline{f}_P \wedge \underline{f}_Q) = f_{123}\ \underline{f}^M$

Non-zero values:

$$(\underline{f}^P \wedge \underline{f}^Q) = f^{123}\ \underline{f}_M$$

$$(\underline{f}_P \wedge \underline{f}_Q) = f_{123}\ \underline{f}^M$$

Outer product by:

$\underline{f}^N$	$\underline{f}_N$

To get:

Outer Product of Bases:

$(\underline{f}^P \wedge \underline{f}^Q) \wedge \underline{f}^N = f^{123}\ \underline{f}_M \wedge \underline{f}^N$	$(\underline{f}^P \wedge \underline{f}^Q) \wedge \underline{f}_N = f^{123}\ \underline{f}_M \wedge \underline{f}_N$
$(\underline{f}_P \wedge \underline{f}_Q) \wedge \underline{f}^N = f_{123}\ \underline{f}^M \wedge \underline{f}^N$	$(\underline{f}_P \wedge \underline{f}_Q) \wedge \underline{f}_N = f_{123}\ \underline{f}^M \wedge \underline{f}_N$

Knowing:

	$\| \underline{f}_M \wedge \underline{f}_N \| \leq \|\phi_{123}\ e_{MNK}\ \underline{f}^K\|$
$\| \underline{f}^M \wedge \underline{f}^N \| \leq \|\phi^{123}\ e^{MNK}\ \underline{f}_K\|$	

Then:

	$\|(\underline{f}^P \wedge \underline{f}^Q) \wedge \underline{f}_N\| = f^{123}\ \|\underline{f}_M \wedge \underline{f}_N\| \leq \|\phi^{123}\ e^{PQM}\ \underline{f}_M \wedge \underline{f}_N\|$
$\|(\underline{f}_P \wedge \underline{f}_Q) \wedge \underline{f}^N\| = f_{123}\ \|\underline{f}^M \wedge \underline{f}^N\| \leq \|\phi_{123} e_{PQM} \underline{f}^M \wedge \underline{f}^N\|$	

:

Outer Product of Bases:

$\underline{f}_M \wedge \underline{f}^N = (\underline{f}^P \wedge \underline{f}^Q) \wedge \underline{f}^N / f^{123}$	$\underline{f}_M \wedge \underline{f}_N = (\underline{f}^P \wedge \underline{f}^Q) \wedge \underline{f}_N / f^{123}$
$\underline{f}^M \wedge \underline{f}^N = (\underline{f}_P \wedge \underline{f}_Q) \wedge \underline{f}^N / f_{123}$	$\underline{f}^M \wedge \underline{f}_N = (\underline{f}_P \wedge \underline{f}_Q) \wedge \underline{f}_N / f_{123}$

$$\underline{f}_P \wedge \underline{f}_Q = \underline{o}_\phi\, of^\phi{}_P \wedge \underline{o}_\beta\, of^\beta{}_Q = of^\phi{}_P\, of^\beta{}_Q\, e_{\phi\beta k}\, \underline{o}^k$$

$$\underline{f}^P \wedge \underline{f}^Q = \underline{o}^\phi\, of_\phi{}^P \wedge \underline{o}^\beta\, of_\beta{}^Q = of_\phi{}^P\, of_\beta{}^Q\, e^{\phi\beta k}\, \underline{o}_k$$

$$\| \underline{f}_P \wedge \underline{f}_Q \| \le \|\phi_{123}\, e_{PQk}\, \underline{f}^k\|$$

$$\| \underline{f}^P \wedge \underline{f}^Q \| \le \|\phi^{123}\, e^{PQk}\, \underline{f}_k\|$$

Inner Product of Bases:

$$ff^{IK} \equiv \underline{f}^I . \underline{f}^K = fo^I{}_J\, fo^K{}_J$$

$$ff_I{}^K \equiv \underline{f}_I . \underline{f}^K = \delta_I^K$$

$$ff^I{}_K \equiv \underline{f}^I . \underline{f}_K = \delta^I_K$$

$$ff_{IK} \equiv \underline{f}_I . \underline{f}_K = fo_I{}^J\, fo_K{}^J$$

2.2.3.3.
Orthogonal Bases:

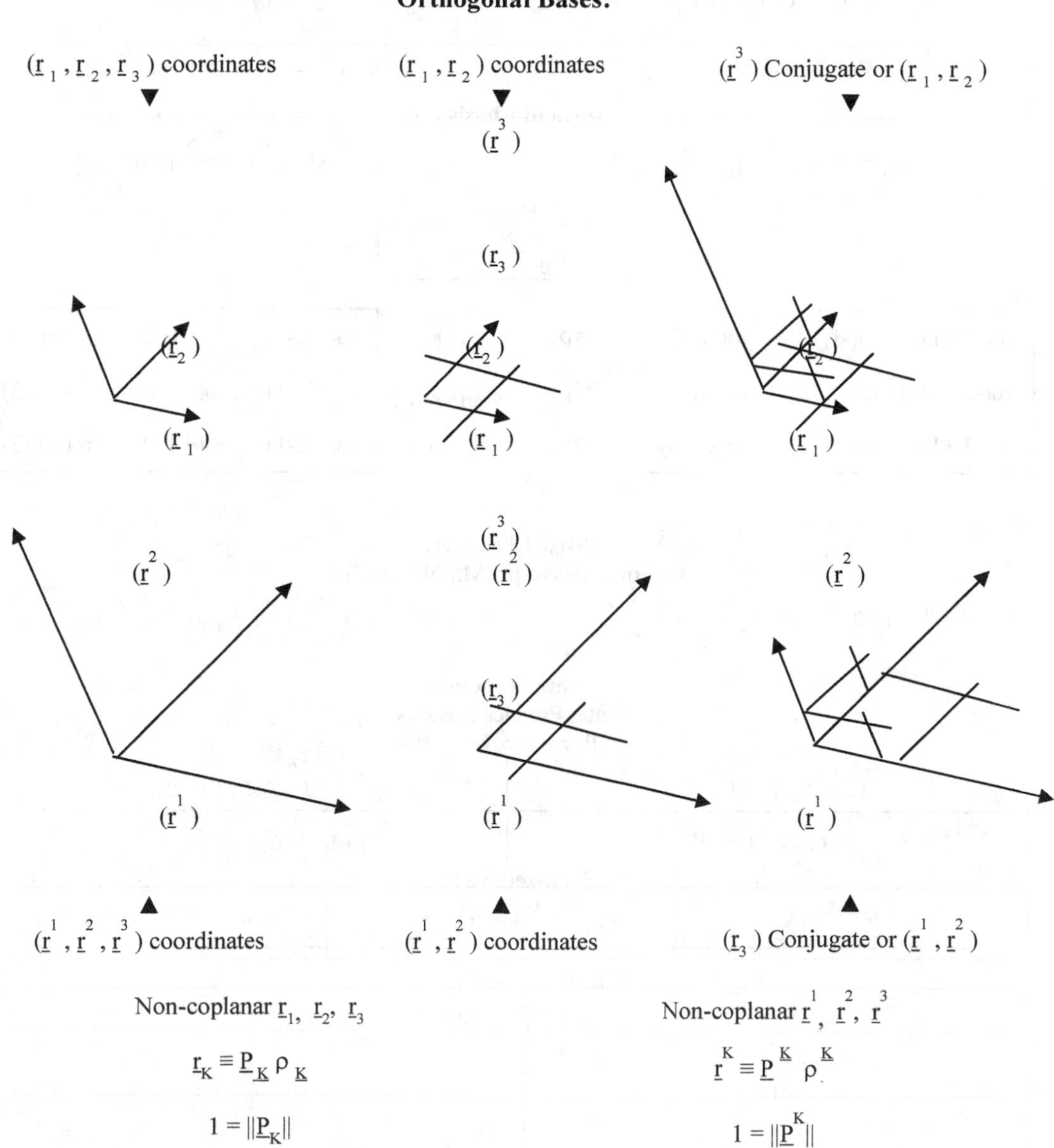

Non-coplanar $\underline{r}_1,\ \underline{r}_2,\ \underline{r}_3$

$$\underline{r}_K \equiv \underline{P}_K\, \rho_K$$

$$1 = \|\underline{P}_K\|$$

Non-coplanar $\underline{r}^1,\ \underline{r}^2,\ \underline{r}^3$

$$\underline{r}^K \equiv \underline{P}^K\, \rho^K$$

$$1 = \|\underline{P}^K\|$$

$\rho_{\underline{K}}$ $\rho^{\underline{K}}$

:

Bases Conjugates or not necessarily parallel and their norms are not necessarily equal:

$\underline{r}_i \neq \underline{r}^i$ $\rho_i \neq \rho^i$

$L \neq J \neq K \neq L$ Variable Basis, L, J, and K are ordered:

$\underline{r}^K$ is Perpendicular to $\underline{r}_L$ and to $\underline{r}_J$:

$\underline{r}^K \perp \underline{r}_L$ and $\underline{r}^K \perp \underline{r}_J$

$\underline{r}_K$ is Perpendicular to $\underline{r}^L$ and to $\underline{r}^J$:

$\underline{r}_K \perp \underline{r}^L$ and $\underline{r}_K \perp \underline{r}^J$

:

$(\underline{r}_K \equiv \underline{P}_{\underline{K}} \rho_{\underline{K}})$ perpendicular to $(\underline{r}^I \equiv \underline{P}^{\underline{I}} \rho^{\underline{I}})$ $(\underline{r}_K \equiv \underline{P}_{\underline{K}} \rho_{\underline{K}})$ perpendicular to $(\underline{r}^J \equiv \underline{P}^{\underline{J}} \rho^{\underline{J}})$

$\underline{r}_K = (ro_K^{\ M})(\underline{o}_M)$ $\underline{r}^K = (ro^K_{\ M})(\underline{o}^K)$

$\rho_{\underline{M}} \equiv \| \underline{r}_{\underline{M}} \| = \sqrt{(\underline{r}_{\underline{M}} \cdot \underline{r}_{\underline{M}})}$ $\rho^N \equiv \| \underline{r}^N \| = \sqrt{(\underline{r}^{\underline{N}} \cdot \underline{r}^{\underline{N}})}$

Norm of r basis is ρ

$\rho_M = \sqrt{[(ro_M^{\ 1})^2 + (ro_M^{\ 2})^2 + (ro_M^{\ 3})^2]}$ $\rho^N = \sqrt{[(ro^N_{\ 1})^2 + (ro^N_{\ 2})^2 + (ro^N_{\ 3})^2]}$

Such that:

$$\rho_{\underline{M}}\, \rho^{\underline{M}} = 1$$

ro(-1,+1)	ro(-1,+2)	ro(-1,+3)	◄ρ_1	ρ^1►	ro(+1,-1)	ro(+1,-2)	ro(+1,-3)
ro(-2,+1)	ro(-2,+2)	ro(-2,+3)	◄ρ_2	ρ^2►	ro(+2,-1)	ro(+2,-2)	ro(+2,-3)
ro(-3,+1)	ro(-3,+2)	ro(-3,+3)	◄ρ_3	ρ^3►	ro(+3,-1)	ro(+3,-2)	ro(+3,-3)

Mixed Product:

Determinants of [ro(M,N)] ≡ (r123):

$r_{123} \equiv \|_r\|$ *Equal* $\rho_{123} \equiv \rho_1\ \rho_2\ \rho_3$ $r^{123} \equiv \|^r\|$ *Equal* $\rho^{123} \equiv \rho^1 \rho^2 \rho^3$

Outer Product:

Outer Product of Bases:

$P \neq Q \neq M \neq P$

$(\underline{r}^P \wedge \underline{r}^Q) = r^{123}\ \underline{r}_M$	$(\underline{r}^P \wedge \underline{r}_Q) = 0$
$(\underline{r}_P \wedge \underline{r}^Q) = 0$	$(\underline{r}_P \wedge \underline{r}_Q) = r_{123}\ \underline{r}^M$

Non-zero vaues:

$$(\underline{r}^P \wedge \underline{r}^Q) = r^{123}\ \underline{r}_M$$

$$(\underline{r}_P \wedge \underline{r}_Q) = r_{123}\ \underline{r}^M$$

Outer product by:

$\underline{r}^N$	$\underline{r}_N$

To get:

Outer Product of Bases:

$(\underline{r}^{P} \wedge \underline{r}^{Q}) \wedge \underline{r}^{N} = r^{123}\, \underline{r}_{M} \wedge \underline{r}^{N}$	$(\underline{r}^{P} \wedge \underline{r}^{Q}) \wedge \underline{r}_{N} = r^{123}\, \underline{r}_{M} \wedge \underline{r}_{N}$
$(\underline{r}_{P} \wedge \underline{r}_{Q}) \wedge \underline{r}^{N} = r_{123}\, \underline{r}^{M} \wedge \underline{r}^{N}$	$(\underline{r}_{P} \wedge \underline{r}_{Q}) \wedge \underline{r}_{N} = r_{123}\, \underline{r}^{M} \wedge \underline{r}_{N}$

Knowing:

	$\| \underline{r}_{M} \wedge \underline{r}_{N} \|$ *Equal* $\| \rho_{123}\, e_{MNK}\, \underline{r}^{K} \|$
$\| \underline{r}^{M} \wedge \underline{r}^{N} \|$ *Equal* $\| \rho^{123}\, e^{MNK}\, \underline{r}_{K} \|$	

Then:

	$\|(\underline{r}^{P} \wedge \underline{r}^{Q}) \wedge \underline{r}_{N}\| = r^{123}\, \|\underline{r}_{M} \wedge \underline{r}_{N}\|$ *Equal* $\|\rho^{123}\, e^{PQM}\, \underline{r}_{M} \wedge \underline{r}_{N}\|$
$\|(\underline{r}_{P} \wedge \underline{r}_{Q}) \wedge \underline{r}^{N}\| = r_{123}\, \|\underline{r}^{M} \wedge \underline{r}^{N}\|$ *Equal* $\|\rho_{123}\, e_{PQM}\, \underline{r}^{M} \wedge \underline{r}^{N}\|$	

:

Outer Product of Bases:

$\underline{r}_{M} \wedge \underline{r}^{N} = (\underline{r}^{P} \wedge \underline{r}^{Q}) \wedge \underline{r}^{N} / r^{123}$	$\underline{r}_{M} \wedge \underline{r}_{N} = (\underline{r}^{P} \wedge \underline{r}^{Q}) \wedge \underline{r}_{N} / r^{123}$
$\underline{r}^{M} \wedge \underline{r}^{N} = (\underline{r}_{P} \wedge \underline{r}_{Q}) \wedge \underline{r}^{N} / r_{123}$	$\underline{r}^{M} \wedge \underline{r}_{N} = (\underline{r}_{P} \wedge \underline{r}_{Q}) \wedge \underline{r}_{N} / r_{123}$

$\underline{r}_{P} \wedge \underline{r}_{Q} = \underline{o}_{\alpha}\, or^{\alpha}{}_{P} \wedge \underline{o}_{\beta}\, or^{\beta}{}_{Q} = or^{\alpha}{}_{P}\, or^{\beta}{}_{Q}\, e_{\alpha\beta k}\, \underline{o}^{k}$	$\underline{r}^{P} \wedge \underline{r}^{Q} = \underline{o}^{\alpha}\, or_{\alpha}{}^{P} \wedge \underline{o}^{\beta}\, or_{\beta}{}^{Q} = or_{\alpha}{}^{P}\, or_{\beta}{}^{Q}\, e^{\alpha\beta k}\, \underline{o}_{k}$
$\| \underline{r}_{P} \wedge \underline{r}_{Q} \|$ *Equal* $\| \rho_{123}\, e_{PQk}\, \underline{r}^{k} \|$	$\| \underline{r}^{P} \wedge \underline{r}^{Q} \|$ *Equal* $\| \rho^{123}\, e^{PQk}\, \underline{r}_{k} \|$

Inner Product of Bases:

$rr^{IK} \equiv \underline{r}^{I} . \underline{r}^{K} = \rho^{I} \rho^{K}$	$rr^{I}{}_{K} \equiv \underline{r}^{I} . \underline{r}_{K} = \delta^{I}{}_{K}$
$rr_{I}{}^{K} \equiv \underline{r}_{I} . \underline{r}^{K} = \delta_{I}{}^{K}$	$rr_{IK} \equiv \underline{r}_{I} . \underline{r}_{K} = \rho_{I}\, \rho_{K}$

2.2.3.4.
Orthonormal Bases, a Special Case of Orthogonal Unitary Bases:

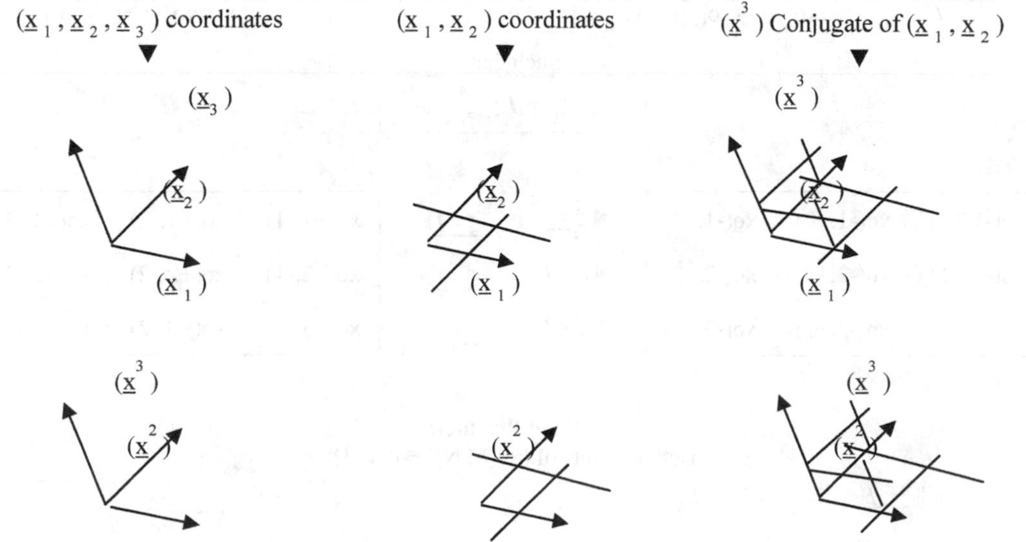

$(\underline{x}^1)$	$(\underline{x}^1)$	$(\underline{x}^1)$
▲	▲	▲
$(\underline{x}^1, \underline{x}^2, \underline{x}^3)$ coordinates	$(\underline{x}^1, \underline{x}^2)$ coordinates	$(\underline{x}_3)$ Conjugate of $(\underline{x}^1, \underline{x}^2)$

:

Non-coplanar $\underline{x}_1, \underline{x}_2, \underline{x}_3$	Non-coplanar $\underline{x}^1, \underline{x}^2, \underline{x}^3$
$\underline{x}_K \equiv \underline{\Xi}_K \xi_{\underline{K}}$	$\underline{x}^K \equiv \underline{\Xi}^K \xi^K$
$1 = \|\underline{\Xi}_K\|$	$1 = \|\underline{\Xi}^K\|$
$1 = \xi_{\underline{K}}$	$1 = \xi^{\underline{K}}$

:

Bases Conjugates or not necessarily parallel and their norms are not necessarily equal:

x_i *Equal* x^i $\qquad \xi_i = 1 = \xi^j$

$L \neq J \neq K \neq L$ Variable Basis, L, J, and K are ordered:

$\underline{x}^K$ is Perpendicular to $\underline{x}_L$ and to $\underline{x}_J$:

$\underline{x}^K \perp \underline{x}_L$ and $\underline{x}^K \perp \underline{x}_J$

$\underline{x}_K$ is Perpendicular to $\underline{x}^L$ and to $\underline{x}^J$:

$\underline{x}_K \perp \underline{x}^L$ and $\underline{x}_K \perp \underline{x}^J$

:

$\underline{x}^{\underline{K}}$ and $\underline{x}_{\underline{K}}$ are Not necessxrily Parallel

:

$(\underline{x}_K \equiv \underline{\Xi}_{\underline{K}} \xi_{\underline{K}})$ perpendicular to $(\underline{x}^I \equiv \underline{\Xi}^I \xi^I)$ $\qquad$ $(\underline{x}_K \equiv \underline{\Xi}_{\underline{K}} \xi_{\underline{K}})$ perpendicular to $(\underline{x}^J \equiv \underline{\Xi}^J \xi^J)$

$$\underline{x}_K = (xo_K^{\ M})(\underline{o}_M) \qquad \underline{x}^K = (xo^K_{\ M})(\underline{o}^K)$$

$$1 = \xi_M \equiv \|\underline{x}_{\underline{M}}\| = \sqrt{(\underline{x}_{\underline{M}} \cdot \underline{x}_{\underline{M}})} \qquad 1 = \xi^N \equiv \|\underline{x}^N\| = \sqrt{(\underline{x}^{\underline{N}} \cdot \underline{x}^{\underline{N}})}$$

Norm of x basis is ξ

$$1 = \xi_M = \sqrt{[(xo_M^{\ 1})^2 + (xo_M^{\ 2})^2 + (xo_M^{\ 3})^2]} \qquad 1 = \xi^N = \sqrt{[(xo^N_{\ 1})^2 + (xo^N_{\ 2})^2 + (xo^N_{\ 3})^2]}$$

Such that:

$$\boxed{\xi_M = 1 = \xi^M}$$

xo(-1,+1)	xo(-1,+2)	xo(-1,+3)	◄ $\underline{\xi_1 = 1}$	$\underline{\xi^1 = 1}$ ►	xo(+1,-1)	xo(+1,-2)	xo(+1,-3)
xo(-2,+1)	xo(-2,+2)	xo(-2,+3)	◄ $\underline{\xi_2 = 1}$	$\underline{\xi^2 = 1}$ ►	xo(+2,-1)	xo(+2,-2)	xo(+2,-3)
xo(-3,+1)	xo(-3,+2)	xo(-3,+3)	◄ $\underline{\xi_3 = 1}$	$\underline{\xi^3 = 1}$ ►	xo(+3,-1)	xo(+3,-2)	xo(+3,-3)

Mixed Product:

Determinants of [xo(M,N)] ≡ (x123):

$$x_{123} \equiv \|_x\| \underline{= 1 =} \xi_{123} \equiv \xi_1 \xi_2 \xi_3 \qquad x^{123} \equiv \|^x\| \underline{= 1 =} \xi^{123} \equiv \xi^1 \xi^2 \xi^3$$

Outer Product:

Outer Product of Bases:

$P \neq Q \neq M \neq P$

$(\underline{x}^{P} \wedge \underline{x}^{Q}) = x^{123}\ \underline{x}_{M} = 1\ \underline{x}_{M}$	$(\underline{x}^{P} \wedge \underline{x}_{Q}) = 0$
$(\underline{x}_{P} \wedge \underline{x}^{Q}) = 0$	$(\underline{x}_{P} \wedge \underline{x}_{Q}) = x_{123}\ \underline{x}^{M} = 1\ \underline{x}^{M}$

Non-zero values:

$$(\underline{x}^{P} \wedge \underline{x}^{Q}) = x^{123}\ \underline{x}_{M} = 1\ \underline{x}_{M}$$

$$(\underline{x}_{P} \wedge \underline{x}_{Q}) = x_{123}\ \underline{x}^{M} = 1\ \underline{x}^{M}$$

Outer product by:

$\underline{x}^{N}$	$\underline{x}_{N}$

To get:

Outer Product of Bases:

$(\underline{x}^{P} \wedge \underline{x}^{Q}) \wedge \underline{x}^{N} = x^{123}\ \underline{x}_{M} \wedge \underline{x}^{N}$	$(\underline{x}^{P} \wedge \underline{x}^{Q}) \wedge \underline{x}_{N} = x^{123}\ \underline{x}_{M} \wedge \underline{x}_{N}$
$(\underline{x}_{P} \wedge \underline{x}_{Q}) \wedge \underline{x}^{N} = x_{123}\ \underline{x}^{M} \wedge \underline{x}^{N}$	$(\underline{x}_{P} \wedge \underline{x}_{Q}) \wedge \underline{x}_{N} = x_{123}\ \underline{x}^{M} \wedge \underline{x}_{N}$

Knowing:

	$\|\| x_{M} \wedge x_{N} \|\| = \|\|\xi_{123}\ e_{MNK}\ \underline{x}^{K}\|\|$
$\|\| \underline{x}^{M} \wedge \underline{x}^{N} \|\| = \|\|\xi^{123}\ e^{MNK}\ \underline{x}_{K}\|\|$	

Then:

	$\|\|(\underline{x}^{P} \wedge \underline{x}^{Q}) \wedge \underline{x}_{N}\|\| = x^{123}\ \|\|\underline{x}_{M} \wedge \underline{x}_{N}\|\| = \|\|\xi^{123}\ e^{PQM}\ \underline{x}_{M} \wedge \underline{x}_{N}\|\|$
$\|\|(\underline{x}_{P} \wedge \underline{x}_{Q}) \wedge \underline{x}^{N}\|\| = x_{123}\ \|\|\underline{x}^{M} \wedge \underline{x}^{N}\|\| = \|\|\xi_{123} e_{PQM} \underline{x}^{M} \wedge \underline{x}^{N}\|\|$	

:

Outer Product of Bases:

$\underline{x}_{M} \wedge \underline{x}^{N} = (\underline{x}^{P} \wedge \underline{x}^{Q}) \wedge \underline{x}^{N} / x^{123}$	$\underline{x}_{M} \wedge \underline{x}_{N} = (\underline{x}^{P} \wedge \underline{x}^{Q}) \wedge \underline{x}_{N} / x^{123}$
$\underline{x}^{M} \wedge \underline{x}^{N} = (\underline{x}_{P} \wedge \underline{x}_{Q}) \wedge \underline{x}^{N} / x_{123}$	$\underline{x}^{M} \wedge \underline{x}_{N} = (\underline{x}_{P} \wedge \underline{x}_{Q}) \wedge \underline{x}_{N} / x_{123}$

$\underline{x}_{P} \wedge \underline{x}_{Q} = \underline{x}^{k}$	$\underline{x}^{P} \wedge \underline{x}^{Q} = \underline{x}_{k}$
$=$	$=$
$\underline{o}_{\alpha}\ ox^{\alpha}_{P} \wedge \underline{o}_{\beta}\ ox^{\beta}_{Q}$	$\underline{o}^{\alpha}\ ox_{\alpha}^{P} \wedge \underline{o}^{\beta}\ ox_{\beta}^{Q}$
$=$	$=$
$ox^{\alpha}_{P}\ ox^{\beta}_{Q}\ e_{\alpha\beta k}\ \underline{o}^{k}$	$ox_{\alpha}^{P}\ ox_{\beta}^{Q}\ e^{\alpha\beta k}\ \underline{o}_{k}$
$\|\| \underline{x}_{P} \wedge \underline{x}_{Q} \|\|$ <u>*Equal*</u> $1 = \|\|\underline{x}^{k}\|\|$	$\|\| \underline{x}^{P} \wedge \underline{x}^{Q} \|\|$ <u>*Equal*</u> $1 = \|\|\underline{x}_{k}\|\|$

Inner Product of Bases:

$xx^{IK} \equiv \underline{x}^{I} . \underline{x}^{K} = \delta^{IK}$	$xx^{I}_{K} \equiv \underline{x}^{I} . \underline{x}_{K} = \delta^{I}_{K}$
$xx_{I}^{K} \equiv \underline{x}_{I} . \underline{x}^{K} = \delta_{I}^{K}$	$xx_{IK} \equiv \underline{x}_{I} . \underline{x}_{K} = \delta_{IK}$

2.2.3.5.
Two Sets of Bases in All Coordinate Systems:

General: bN

aM	$ba_N{}^M = ab^M{}_N$	$\neq$	$ba^N{}_M = ab_M{}^N$	$ab^M{}_N = ba_N{}^M$	$\neq$	$ab_M{}^N = ba^N{}_M$
fM	$bf_N{}^M = fb^M{}_N$	$\neq$	$bf^N{}_M = fb_M{}^N$	$fb^M{}_N = bf_N{}^M$	$\neq$	$fb_M{}^N = bf^N{}_M$
rM	$br_N{}^M = rb^M{}_N$	$\neq$	$br^N{}_M = rb_M{}^N$	$rb^M{}_N = bb_N{}^M$	$\neq$	$rb_M{}^N = bb^N{}_M$
xM	$bx_N{}^M = xb^M{}_N$	$\neq$	$bx^N{}_M = xb_M{}^N$	$xb^M{}_N = bx_N{}^M$	$\neq$	$xb_M{}^N = bx^N{}_M$
oM	$bo_N{}^M = ob^M{}_N$	$\neq$	$bo^N{}_M = ob_M{}^N$	$ob^M{}_N = bo_N{}^M$	$\neq$	$ob_M{}^N = bo^N{}_M$

General Unitary: gN

aM	$ga_N{}^M = ag^M{}_N$	$\neq$	$ga^N{}_M = ag_M{}^N$	$ag^M{}_N = ga_N{}^M$	$\neq$	$ag_M{}^N = ga^N{}_M$
fM	$gf_N{}^M = fg^M{}_N$	$\neq$	$gf^N{}_M = fg_M{}^N$	$fg^M{}_N = gf_N{}^M$	$\neq$	$fg_M{}^N = gf^N{}_M$
rM	$gr_N{}^M = rg^M{}_N$	$\neq$	$gr^N{}_M = rg_M{}^N$	$rg^M{}_N = gg_N{}^M$	$\neq$	$rg_M{}^N = gg^N{}_M$
xM	$gx_N{}^M = xg^M{}_N$	$\neq$	$gx^N{}_M = xg_M{}^N$	$xg^M{}_N = gx_N{}^M$	$\neq$	$xg_M{}^N = gx^N{}_M$
oM	$go_N{}^M = og^M{}_N$	$\neq$	$go^N{}_M = og_M{}^N$	$og^M{}_N = go_N{}^M$	$\neq$	$og_M{}^N = go^N{}_M$

Orthogonal: sN

aM	$sa_N{}^M = as^M{}_N$	$\neq$	$sa^N{}_M = as_M{}^N$	$as^M{}_N = sa_N{}^M$	$\neq$	$as_M{}^N = sa^N{}_M$
fM	$sf_N{}^M = fs^M{}_N$	$\neq$	$sf^N{}_M = fs_M{}^N$	$fs^M{}_N = sf_N{}^M$	$\neq$	$fs_M{}^N = sf^N{}_M$
rM	$sr_N{}^M = rs^M{}_N$	$\neq$	$sr^N{}_M = rs_M{}^N$	$rs^M{}_N = sr_N{}^M$	$\neq$	$rs_M{}^N = sr^N{}_M$
xM	$sx_N{}^M = xs^M{}_N$	$\neq$	$sx^N{}_M = xs_M{}^N$	$xs^M{}_N = sx_N{}^M$	$\neq$	$xs_M{}^N = sx^N{}_M$
oM	$so_N{}^M = os^M{}_N$	$\neq$	$so^N{}_M = os_M{}^N$	$os^M{}_N = so_N{}^M$	$\neq$	$os_M{}^N = so^N{}_M$

Orthonormal: yN

aM	$ya_N{}^M = ay^M{}_N$	$\neq$	$ya^N{}_M = ay_M{}^N$	$ay^M{}_N = ya_N{}^M$	$\neq$	$ay_M{}^N = ya^N{}_M$
fM	$yf_N{}^M = fy^M{}_N$	$\neq$	$yf^N{}_M = fy_M{}^N$	$fy^M{}_N = yf_N{}^M$	$\neq$	$fy_M{}^N = yf^N{}_M$
rM	$yr_N{}^M = ry^M{}_N$	$\neq$	$yr^N{}_M = ry_M{}^N$	$ry^M{}_N = yr_N{}^M$	$\neq$	$ry_M{}^N = yr^N{}_M$
xM	$yx_N{}^M = xy^M{}_N$	$\neq$	$yx^N{}_M = xy_M{}^N$	$xy^M{}_N = yx_N{}^M$	$\neq$	$xy_M{}^N = yx^N{}_M$
oM	$yo_N{}^M = oy^M{}_N$	$\neq$	$yo^N{}_M = oy_M{}^N$	$oy^M{}_N = yo_N{}^M$	$\neq$	$oy_M{}^N = yo^N{}_M$

Fixed: oN

aM	$oa_N{}^M = ao^M{}_N$	$\neq$	$oa^N{}_M = ao_M{}^N$	$ao^M{}_N = oa_N{}^M$	$\neq$	$ao_M{}^N = oa^N{}_M$
fM	$of_N{}^M = fo^M{}_N$	$\neq$	$of^N{}_M = fo_M{}^N$	$fo^M{}_N = of_N{}^M$	$\neq$	$fo_M{}^N = of^N{}_M$
rM	$or_N{}^M = ro^M{}_N$	$\neq$	$or^N{}_M = ro_M{}^N$	$ro^M{}_N = or_N{}^M$	$\neq$	$ro_M{}^N = or^N{}_M$
xM	$ox_N{}^M = xo^M{}_N$	$\neq$	$ox^N{}_M = xo_M{}^N$	$xo^M{}_N = ox_N{}^M$	$\neq$	$xo_M{}^N = ox^N{}_M$
oM	$oo_N{}^M = oo^M{}_N$	$\neq$	$oo^N{}_M = oo_M{}^N$	$oo^M{}_N = oo_N{}^M$	$\neq$	$oo_M{}^N = oo^N{}_M$

Sine Oriantation:

$\Sigma^{MN} \equiv \sin(\underline{a}^M, \underline{b}^N)$	$\Sigma^M{}_N \equiv \sin(\underline{a}^M, \underline{b}_N)$	$\sin(\underline{b}_N, \underline{a}^M) \equiv -\Sigma_N{}^M$	$\sin(\underline{b}^N, \underline{a}^M) \equiv -\Sigma^{NM}$
$\Sigma_M{}^N \equiv \sin(\underline{a}_M, \underline{b}^N)$	$\Sigma_{MN} \equiv \sin(\underline{a}_M, \underline{b}_N)$	$\sin(\underline{b}_N, \underline{a}_M) \equiv -\Sigma_{NM}$	$\sin(\underline{b}^N, \underline{a}_M) \equiv -\Sigma^N{}_M$

Cosine Oriantation:

$X^{MN} \equiv \cos(\underline{a}^{M},\underline{b}^{N})$	$X^{M}_{N} \equiv \cos(\underline{a}^{M},\underline{b}_{N})$	$\cos(\underline{b}_{N},\underline{a}^{M}) \equiv X_{N}^{M}$	$\cos(\underline{b}^{N},\underline{a}^{M}) \equiv X^{NM}$
$X_{M}^{N} \equiv \cos(\underline{a}_{M},\underline{b}^{N})$	$X_{MN} \equiv \cos(\underline{a}_{M},\underline{b}_{N})$	$\cos(\underline{b}_{N},\underline{a}_{M}) \equiv X_{NM}$	$\cos(\underline{b}^{N},\underline{a}_{M}) \equiv X^{N}_{M}$

2.2.3.6. Outer Product of Vectors in Different Systems:

$$u \wedge v$$

$$u_i \underline{a}^i \wedge v_j \underline{a}^j = a^{123} \; e^{ijk} u_i v_j \underline{a}_k$$
$$u_i \underline{f}^i \wedge v_j \underline{f}^j = f^{123} \; e^{ijk} u_i v_j \underline{f}_k$$
$$u_i \underline{r}^i \wedge v_j \underline{r}^j = r^{123} \; e^{ijk} u_i v_j \underline{r}_k$$
$$u_i \underline{x}^i \wedge v_j \underline{x}^j = x^{123} \; e^{ijk} u_i v_j \underline{x}_k$$
$$u_i \underline{o}^i \wedge v_j \underline{o}^j = o^{123} \; e^{ijk} u_i v_j \underline{o}_k$$

$$u^i \underline{a}_i \wedge v^j \underline{a}_j = a_{123} \; e_{ijk} u^i v^j \underline{a}^k$$
$$u^i \underline{f}_i \wedge v^j \underline{f}_j = f_{123} \; e_{ijk} u^i v^j \underline{f}^k$$
$$u^i \underline{r}_i \wedge v^j \underline{r}_j = r_{123} \; e_{ijk} u^i v^j \underline{r}^k$$
$$u^i \underline{x}_i \wedge v^j \underline{x}_j = x_{123} \; e_{ijk} u^i v^j \underline{x}^k$$
$$u^i \underline{o}_i \wedge v^j \underline{o}_j = o_{123} \; e_{ijk} u^i v^j \underline{o}^k$$

2.2.4. Function of Position Vector:

Bases of all other coordinate systems may *change in position and time.*
These bases are defined relative to fixed coordinates..
These general coordinate system may be called curved coordinate system.

$f = f(\underline{a}_i \, a^i)$ $\qquad$ $f = f(\underline{a}^i \, a_i)$

2.2.4.1. Bases Differential:

$d\underline{a}_i$ $\qquad$ $d\underline{a}^i$

2.2.4.2. Component Differential:

da^n $\qquad$ da_n

$$\underline{a}^n \; da_n$$

$dx_k \; xy^k{}_l = dy_l$	Function of: : x: To y:	$\underline{y}^j = yx^j{}_i \;\; \underline{x}^i$
$dy_k \; yz^k{}_l = dz_l$	Function of: : y To z:	$\underline{z}^j = zy^j{}_i \;\; \underline{y}^i$
$do_k \;\; oz^k{}_n = dz_n$	Function of: : o To z:	$\underline{z}^l = zo^l{}_i \;\; \underline{o}^i$
$ox^k{}_l \; xy^l{}_m \; yz^m{}_n = oz^k{}_n$	o onto x to y to z	$zy^l{}_k \; yx^k{}_j \; xo^j{}_i = zo^l{}_i$
$do_{kp} \; ox^k{}_l \; ox^p{}_q = dx_{lq}$	Function of: : o To: x	$\underline{x}^{jn} = xo^j{}_i \; xo^n{}_m \; \underline{o}^{im}$

General Composition (L) First Index operation Examples:

$dx^{k}\, xy_{k}{}^{l} = dy^{l}$	Function of: : x To: y	$\underline{y}_{j} = yx_{j}{}^{i}\, \underline{x}_{i}$
$dy^{k}\, yz_{k}{}^{l} = dz^{l}$	Function of: : y To: z	$\underline{z}_{j} = zy_{j}{}^{i}\, \underline{y}_{i}$
$do^{k}\, oz_{k}{}^{n} = dz^{n}$	Function of: : o To: z	$\underline{z}_{j} = zo_{j}{}^{i}\, \underline{o}_{i}$
$ox_{k}{}^{l}\, xy_{l}{}^{m}\, yz_{m}{}^{n} = oz_{k}{}^{n}$	o onto x to y to z	$zy_{l}{}^{k}\, yx_{k}{}^{j}\, xo_{j}{}^{i} = zo_{l}{}^{i}$
$do^{kp}\, ox_{k}{}^{l}\, ox_{p}{}^{q} = dx^{lq}$	Function of: : o To: x	$\underline{x}_{jn} = xo_{j}{}^{i}\, xo_{m}{}^{n}\, \underline{o}_{im}$

do oa = da :Composition	oa	Resolution: $\underline{o} = \underline{a}$ ao
Resolution: da ao = do	ao	$\underline{a} = \underline{o}$ oa :Composition
do oa ao = do	oa ao = δ	$\underline{o} = \underline{o}$ oa ao
da ao oa = da	ao oa = δ	$\underline{a} = \underline{a}$ ao oa
$\|oa_{i}{}^{j} = ao^{j}{}_{i}\| \le (\alpha^{1}\alpha^{2}\alpha^{3})$	$\|oa^{i}{}_{j}\|\,\|ao^{j}{}_{i}\| = 1 = \|ao_{j}{}^{i}\|\,\|oa_{i}{}^{j}\|$	$\|oa^{i}{}_{j} = ao_{j}{}^{i}\| \le (\alpha_{1}\alpha_{2}\alpha_{3})$
	$\alpha_{1}\alpha_{2}\alpha_{3}\,\alpha^{1}\alpha^{2}\alpha^{3}\,C_{1}{}^{1}C_{2}{}^{2}C_{3}{}^{3} = 1$	

2.2.4.3.
Partial Derivative with respect to a Component:

$\partial[\;]/\partial a^{n} \equiv \partial/\partial a^{n}[] \equiv []_{n}$ $\qquad$ $\partial[]/\partial a_{n} \equiv \partial/\partial a_{n}[] \equiv []^{n}$

Partial Derivative of a Scalar s with respect to a:

$\partial f/\partial a^{n} \equiv fa_{n}$ $\qquad$ $\partial f/\partial a_{n} \equiv fa^{n}$

Partial Derivative of a with respect to a Scalar, s:

$\partial a_{n}\,/\,\partial f \equiv af_{n}$ $\qquad$ $\partial a^{n}/\partial f \equiv af^{n}$

Partial Derivative of a with respect to a:

$\partial a_{m}\,/\,\partial a_{n} = \delta^{n}{}_{m}$ $\qquad$ $\partial a^{m}/\partial a^{n} = \delta^{m}{}_{n}$

Partial Derivative of a basis with respect to a

$\partial\underline{a}^{m}/\partial a_{n} = \underline{a}^{m}{}_{,}{}^{n}$ $\qquad$ $\partial\underline{a}_{m}\,/\,\partial a^{n} = \underline{a}_{m,n}$

2.2.4.4.
Partial Derivative of a Position Vector:

$$\partial\underline{a}/\partial a_{n} = \partial(\underline{a}^{i}\, a_{i})/\partial a_{n} \qquad \partial\underline{a}/\partial a^{n} = \partial(\underline{a}_{i}\, a^{i})/\partial a^{n}$$

$$= \qquad\qquad =$$

$$\partial(\underline{a}^{i})/\partial a_{n}\, a_{i} + \underline{a}^{i}\,\partial a_{i}/\partial a_{n} \qquad \partial(\underline{a}_{i})/\partial a^{n}\, a^{i} + \underline{a}_{i}\,\partial a^{i}/\partial a^{n}$$

Or:

$$\partial\underline{a}/\partial a_{n} = \underline{a}^{i}{}_{,}{}^{n}\, a_{i} + \underline{a}^{i}\,\delta_{i}{}^{n} \qquad \partial\underline{a}/\partial a^{n} = \underline{a}_{i,n}\, a^{i} + \underline{a}_{i}\,\delta^{i}{}_{n}$$

Or:

$$\partial\underline{a}/\partial a_{n} = \underline{a}^{i}{}_{,}{}^{n}\, a_{i} + \underline{a}^{n} \qquad \partial\underline{a}/\partial a^{n} = \underline{a}_{i,n}\, a^{i} + \underline{a}_{n}$$

Differential of Position Vector:

$$d\underline{a} = \underline{a}^{i}{}_{,}{}^{n}\, a_{i}\, da_{n} + \underline{a}^{n}\, da_{n} \qquad d\underline{a} = \underline{a}_{i,n}\, a^{i}\, da^{n} + \underline{a}_{n}\, da^{n}$$

Partial derivative of a position vector of *fixed bases*:

$\partial \underline{a} / \partial a_n = \underline{a}^n$ $\qquad$ $\partial \underline{a} / \partial a^n = \underline{a}_n$

Partial derivative of a position vector of bases:
Recall:

$d\,\underline{a} = d\,(\underline{o}^i a_i) = \underline{o}^i\ d(a_i)$ $\qquad$ $d\,\underline{a} = d\,(\underline{o}_i a^i) = \underline{o}_i\ d(a^i)$

So:

$\partial\underline{a} / \partial o_n = \underline{o}^i\ \partial a_i / \partial o_n = \underline{o}^i\ ao_i^{\ n}$ $\qquad$ $\partial\underline{a} / \partial o^n = \underline{o}_i\ \partial a^i / \partial o_n = \underline{o}_i\ ao^i_{\ n}$

Differential of Position Vector:

$d\,\underline{a} = \underline{o}^i\ ao_i^{\ n}\ do_n$ $\qquad$ $d\,\underline{a} = \underline{o}_i\ ao^i_{\ n}\ do^n$

2.3.
Initial and Current Description Differentials and Partial Derivatives:

Partial Derivatives of Structural Coordinates of
Lower Position w.t. to Upper Position,
And
Partial Derivatives of Structural Coordinates of
Upper Position w.t. to Lower Position:

2.3.1.
Orthonormality and Transpose:

Orthonormality of Function of Two Indices:

$$CN\,({}^{Q}\backslash_{p}\ {}_{p}/^{q})\,C\,({}^{P}\backslash_{q}\ {}_{q}/^{Q}) \equiv C\,({}^{P}\backslash_{q}\ {}_{p}/^{Q})\,CN\,({}^{Q}\backslash_{p}\ {}_{q}/^{P}) = \delta\,({}^{Q}\backslash_{q}\ {}_{q}/^{Q})$$

$CN({}^{Q}\backslash_{p})C({}^{P}\backslash_{q}) \equiv C({}^{P}\backslash_{q})CN({}^{Q}\backslash_{p}) = \delta({}^{Q}\backslash_{q})$ $\qquad$ $CN({}_{p}/^{q})C({}_{q}/^{Q}) \equiv C({}_{p}/^{Q})CN({}_{q}/^{P}) = \delta({}_{p}/^{Q})$

Transpose of Function of Two Indices:
Transpose with respect to two indices is inter replacement of these two indices

$$CT\,({}^{q}\backslash_{p}\ {}_{q}/^{P}) = C\,({}^{P}\backslash_{q}\ {}_{p}/^{q})$$

$CT\,({}^{q}\backslash_{p}\ {}_{q}/^{P}) = C\,({}^{P}\backslash_{q}\ {}_{p}/^{q})$ $\qquad$ $CT\,({}^{q}\backslash_{p}\ {}_{q}/^{P}) = C\,({}^{P}\backslash_{q}\ {}_{p}/^{q})$

FORTRAN $CT^{I}\backslash_{J} = C^{J}\backslash_{I}$:
EQUIVALENT (CT(+I,-J) , C(+J,-I))

2.3.2.
Covariant and Contravariant Partial Derivatives:

Differential in terms of partial derivatives:
(f) = (f(p(q(c…))))

$\partial f^F/\partial p^P\ \partial p^P/\partial q^q\ \partial q^q/\partial c^c \ldots = \partial f^F/\partial c^c$ $\qquad$ $\partial f_F/\partial p_p\ \partial p_p/\partial q_q\ \partial q_q/\partial c_c \ldots = \partial f_F/\partial c_c$

$(fp)^F\backslash_P(pq)^P\backslash_Q(qc)^Q\backslash_C(cf)^C\backslash_f = \delta^F\backslash_f$ $\qquad$ $(fp)_f/^P(pq)_P/^Q(qc)_Q/^C(cf)_C/^F = \delta_f/^F$

$[p^P] = [p^P(q)] \equiv [pq^P] \equiv P^P$ $\qquad$ $[p_P] = [p_P(q)] \equiv [pq_P] \equiv P_P$

$[q^Q] = q^Q(p)] \equiv [qp^Q] \equiv Q^Q$ $\qquad$ $[q_Q] = [q_Q(p)] \equiv [pq_Q] \equiv P_Q$

$$dp^{P} = \partial p^{P}/\partial q^{Q}\ dq^{Q} \equiv P^{P}{}_{Q}\ dq^{Q} \quad \vdots \quad dp_{p} = \partial p_{p}/\partial q_{Q}\ dq_{Q} \equiv P_{p}{}^{Q}\ dq_{Q}$$

$$dq^{Q} = \partial q^{Q}/\partial p^{p}\ \ dp^{p} = Q^{Q}{}_{p}\ \ dp^{p} \quad \vdots \quad dq_{Q} = \partial q_{Q} / \partial p_{P}\ dp_{P} = Q_{Q}{}^{P}\ \ dp_{P}$$

$$dp^{P} = P^{P}{}_{Q}\ Q^{Q}{}_{p}\ dp^{p} \quad \vdots \quad dp_{p} = P_{p}{}^{Q}\ Q_{Q}{}^{P}\ dp_{P}$$

$$dp^{P}/dp^{p} \equiv pp^{P}\backslash_{p} = \delta^{P}\backslash_{p} \quad \vdots \quad dp_{p}/dp_{P} \equiv pp_{p}/^{P} = \delta_{p}/^{P}$$

$$(pq)^{P}\backslash_{Q}\ (qp)^{Q}\backslash_{p} = P^{P}\backslash_{q}\ Q^{q}\backslash_{p} = \delta^{P}\backslash_{p} \quad \vdots \quad (pq)_{p}/^{q}\ (qp)_{q}/^{P} = P_{p}/^{q}\ Q_{q}/^{P} = \delta_{p}/^{P}$$

Orthogonality of System Jacobeans and their Conjugate System Jacobeans

$$P(\,^{P}\backslash_{q}\ {}_{p}/^{Q})\ \ Q(\,^{Q}\backslash_{p}\ {}_{q}/^{P}) \quad = \delta(\,^{Q}\backslash_{q}\ {}_{q}/^{Q})$$

$$P(\,^{P}\backslash_{q})\,Q(\,^{Q}\backslash_{p}) = \delta(\,^{Q}\backslash_{q}\,{}_{q}/^{Q}) \qquad P(\,_{p}/^{Q})\,Q(\,_{q}/^{P}) = \delta(\,_{q}/^{Q})$$

:
:

Since Orthonormality:

:

$$P(\,^{P}\backslash_{q}\ {}_{p}/^{Q})\ \ PN(\,^{Q}\backslash_{p}\ {}_{q}/^{P}) \quad = \delta(\,^{Q}\backslash_{q}\,{}_{q}/^{Q})$$

$$P(\,^{P}\backslash_{q})\,PN(\,^{Q}\backslash_{p}) = \delta(\,^{Q}\backslash_{q}) \quad \vdots \quad P(\,_{p}/^{P})\ PN(\,_{q}/^{P}) = \delta(\,_{q}/^{Q})$$

Then

Partial Derivatives:

$$Q(\,^{Q}\backslash_{p}\ {}_{p}/^{Q}) = PN(\,^{Q}\backslash_{p}\ {}_{p}/^{Q})$$

$$Q(\,^{Q}\backslash_{p}) = PN(\,^{Q}\backslash_{p}) \qquad Q(\,_{q}/^{P}) \ = PN(\,_{q}/^{P})$$

FORTRAN instruction for $Q^{J}\backslash_{I} = PN^{J}\backslash_{I}$:

EQUIVALENT (Q(+J,-I) , PN(+J,-I))

Differential Series

Serial Indices (I J) (=) (I J) are Identical in: Q=PN

:

Pre-Transformer	$P\,\vert_{C} = QN\,\vert_{C}$	$P\backslash_{C} = QN\ \backslash_{C}$	$P_{C}/ = QN_{C}/$
Pre-Transformer	$Q\,\vert_{C} = PN\,\vert_{C}$	$Q\backslash_{C} = PN\ \backslash_{C}$	$Q_{C}/ = PN_{C}/$

:

Symmetry Remark:

QN:

Partial Derivatives:

$$QN(\,_{q}/^{P}\,^{P}\backslash_{q}\,) = P(\,_{q}/^{P}\,^{P}\backslash_{q}\,)$$

$$QN(\,_{q}/^{P}) = P(\,_{q}/^{P}) \qquad QN(\,^{P}\backslash_{q}) = P(\,^{P}\backslash_{q})$$

FORTRAN instruction for $QN_{I}/^{J} = P_{I}/^{J}$:

EQUIVALENT (QN(-I,+J) , P(-I,+J))

Transpose of Q:

$$QT(\,^{q}\backslash_{p}\ \text{Or}\ {}_{p}/^{q}) = Q(\,^{P}\backslash_{q}\ \text{Or}\ {}_{q}/^{P})$$

$$QT(\,^{q}\backslash_{p}\ {}_{q}/^{P}) = Q(\,^{P}\backslash_{q}\ {}_{p}/^{q}) \qquad QT(\,^{q}\backslash_{p}\ {}_{q}/^{P}) = Q(\,^{P}\backslash_{q}\ {}_{p}/^{q})$$

So:

:

$$QT(\,^{q}\backslash_{p}\ \text{Or}\ {}_{p}/^{q}) = Q(\,^{P}\backslash_{q}\ {}_{q}/^{P}) \quad = PN(\,^{P}\backslash_{q}\ {}_{q}/^{P})$$

$QT(^{q}\backslash_{p}) = Q(^{p}\backslash_{q}) = PN(^{p}\backslash_{q})$ $\qquad$ $QT(_{p}/^{q}) = Q(_{q}/^{p}) = PN(_{q}/^{p})$

FORTRAN instruction for $QT^{I}\backslash_{J} = Q^{J}\backslash_{I} = PN^{J}\backslash_{I}$:
EQUIVALENT (QT(+I,-J) , Q(+J,-I) , PN(+J,-I))

2.3.3.
The Squared Differential Distance:

Position Differential Duality:

$$\underline{b}_M \, db^M = \underline{a}_M \, da^M$$
$$\underline{b}^N \, db_N = \underline{a}^N \, da_N$$

$$(\underline{b}^N \, db_N) \cdot \underline{b}_M \, db^M = (\underline{a}^N \, da_N) \cdot \underline{a}_M \, da^M$$

$$(db_N \, db^M) \, \delta^N_M = (da_N \, da^M) \, \delta^N_M$$

$$(db_N \, db^N) = (da_M \, da^M)$$

The Squared Differential Distances

$$d\underline{o} \cdot d\underline{o} = d\underline{a} \cdot d\underline{a} = d\underline{r} \cdot d\underline{r} = d\underline{f} \cdot d\underline{f} = d\underline{x} \cdot d\underline{x} = d\underline{o} \cdot d\underline{o} = d\underline{q} \cdot d\underline{q} = d\underline{p} \cdot d\underline{p}$$
$$do_i \, do^i = da_i \, da^i = dr_i \, dr^i = df_i \, df^i = dx_i \, dx^i = do_i \, do^i = dq_i \, dq^I = dp^j \, dp_i$$

2.3.3.1.
Symmetry:

Given
The Squared Differential Distances

$$dq_i \, dq^I = dp^j \, dp_j$$
$$P^p\backslash_q \equiv (dp^p / dq^q) = (dq_q / dp_p) \equiv Q_q/^p$$

$$P^p\backslash_q \, _q/^p = Q_q/^{pp}\backslash_q$$

$P^p\backslash_q = Q_q/^p$ $\qquad$ $P_q/^p = Q^p\backslash_q$

FORTRAN:
EQUIVALENT (P(-J,+I) , Q(+I,-J))

Symmetry :
Symmetric Q=P Indices about the Equal Sign: (I J) (=) (J I):
Q=P Indices Far to the Left or Right from the Equal Sign are Identical.
Q=P Indices Close to the Equal Sign are Identical.
Identical Q=P Indices are On the Same Level, Covariant or Contravariant.

Inverseses are also equal:

$$QN(_q/^{pp}\backslash_q) = PN(^p\backslash_q \, _q/^p)$$

$QN(_q/^p) = PN(^p\backslash_q)$ $\qquad$ $QN(^p\backslash_q) = PN(_q/^p)$

FORTRAN instruction for $QN_{I}/^{J} = PN^{J}\backslash_{I}$:

EQUIVALENT (QN(-I,+J) , PN(+J,-I))

:

2.3.3.2. Orthonormality

:

$$P({}_{q}/^{p\,p}\backslash_{q}) = Q({}^{p}\backslash_{q\,q}/^{p}) = PN({}^{p}\backslash_{q\,q}/^{p})$$

$$P({}_{q}/^{p}) = Q({}^{p}\backslash_{q}) = PN({}^{p}\backslash_{q}) \qquad P({}^{p}\backslash_{q}) = Q({}_{q}/^{p}) = PN({}_{q}/^{p})$$

:

FORTRAN:

EQUIVALENT (P(-I,+J) , Q(+J,-I) , PN(+J,-I))

Symmetry:

$$P^{p}\backslash_{q\,q}/^{p} = Q_{q}/^{p\,p}\backslash_{q}$$

:

PN:

Partial Derivatives:

$$Q({}^{p}\backslash_{q\,q}/^{p}) = PN({}^{p}\backslash_{q\,q}/^{p})$$

:

:

Orthonormality

$$P({}_{q}/^{p\,p}\backslash_{q}) = Q({}^{p}\backslash_{q\,q}/^{p}) = PN({}^{p}\backslash_{q\,q}/^{p})$$

Equivilantly Partial Derivatives:

$$QN({}_{q}/^{p\,p}\backslash_{q}) = P({}_{q}/^{p\,p}\backslash_{q})$$

:

Orthonormality of

$$QN({}_{q}/^{p\,p}\backslash_{q}) = P({}_{q}/^{p\,p}\backslash_{q}) = Q({}^{p}\backslash_{q\,q}/^{p})$$

:

$$QN({}_{q}/^{p\,p}\backslash_{q}) = P({}_{q}/^{p\,p}\backslash_{q}) = Q({}^{p}\backslash_{q\,q}/^{p})$$

$$QN({}_{q}/^{p}) = P({}_{q}/^{p}) = Q({}^{p}\backslash_{q}) \qquad QN({}^{p}\backslash_{q}) = P({}^{p}\backslash_{q}) = Q({}_{q}/^{p})$$

:

FORTRAN:

EQUIVALENT (QN(-I,+J) , P(-I,+J) , Q(+J,-I))

:

Orthonormality:

$$QN({}_{q}/^{p\,p}\backslash_{q}) = P({}_{q}/^{p\,p}\backslash_{q}) = Q({}^{p}\backslash_{q\,q}/^{p}) = PN({}^{p}\backslash_{q\,q}/^{p})$$

$$QN({}_{q}/^{p}) = P({}_{q}/^{p}) = Q({}^{p}\backslash_{q}) = PN({}^{p}\backslash_{q}) \qquad QN({}^{p}\backslash_{q}) = P({}^{p}\backslash_{q}) = Q({}_{q}/^{p}) = PN({}_{q}/^{p})$$

:

FORTRAN:

EQUIVALENT (QN(-I,+J) , P(-I,+J) , Q(+J,-I) , PN(+J,-I))

Orthogonality of System Jacobeans and their Conjugate System Jacobeans

$$P({}^{p}\backslash_{q\,p}/^{Q})\ Q({}^{Q}\backslash_{p\,q}/^{p}) = \delta({}^{Q}\backslash_{q\,q}/^{Q})$$

$$P({}^{p}\backslash_{q})\,Q({}^{Q}\backslash_{p}) = \delta({}^{Q}\backslash_{q\,q}/^{Q}) \qquad P({}_{p}/^{Q})\,Q({}_{q}/^{p}) = \delta({}_{q}/^{Q})$$

Orthonormality:

$$Q\left({}^{Q}\backslash_{p}\,{}_{p}/^{Q}\right) = PN\left({}^{Q}\backslash_{p}\,{}_{p}/^{Q}\right)$$

$$Q({}^{Q}\backslash_{p}) = PN({}^{Q}\backslash_{p}) \qquad\qquad Q({}_{q}/^{p}) = PN({}_{q}/^{p})$$

Or:

$$\ldots\ldots\ldots\ldots\ldots P_{I}/\backslash_{I} = Q\backslash_{II}/ = PN\backslash_{II}/$$

$$P^{p}\backslash_{q}\,{}_{q}/^{p} = Q_{q}/^{pp}\backslash_{q}$$

$$P^{p}\backslash_{q} = Q_{q}/^{p} \qquad\qquad P_{q}/^{p} = Q^{p}\backslash_{q}$$

Or:

$$QN_{I}/\backslash_{I} = P_{I}/\backslash_{I} = \ldots\ldots\ldots\ldots\ldots$$

All Combined:

$$\ldots\ldots\ldots\ldots\ldots P_{I}/\backslash_{I} = Q\backslash_{II}/ = PN\backslash_{II}/$$

$$QN_{I}/\backslash_{I} = P_{I}/\backslash_{I} = \ldots\ldots\ldots\ldots\ldots$$

So:

$$QN_{I}/\backslash_{I} = P_{I}/\backslash_{I} = Q\backslash_{II}/ = PN\backslash_{II}/$$

Or:

$$Q\backslash_{II}/ = QN_{I}/\backslash_{I} = PN\backslash_{II}/ = P_{I}/\backslash_{I}$$

Putting parentheses Including Transposed Tensors:

$$QT^{M}\backslash_{N} = \{\,[Q^{N}\backslash_{M}] = (QN_{M}/^{N}\,]\,\} = \{\,(PN^{N}\backslash_{M}) = [P_{M}/^{N}\,]\,\} = PT_{N}/^{M}$$

QT(+M,-N)= { [Q(+N,-M)]= (QN(-M,+N)) } = { (PN(+N,-M)) =[P(-M,+N)] } =PT(-N,+M)

Detailed:

(QN(-M,+N)) = (PN(+N,-M))
[Q(+N,-M)] … … … … = … … … … [P(-M,+N)]
{ [Q(+N,-M)] = (QN(-M,+N)) } = { (PN(+N,-M)) = [P(-M,+N)] }
QT(+M,-N) … … … … … … … … = … … … … … … … … PT(-N,+M)

2.3.4.
Initial and Current Descriptions of Continuum:

Let:

(p)	(q)
Initial Material Vector Description, Green Lagrangian Solid.	Current Spatial Vector Description, Almansi Eulerian Fluid.

Differential Components are of Two Conjugates:

(dp_{I}) $\qquad$ (dq_{I})

(dp^{I}) $\qquad$ (dq^{I})

:

Pre-Transformer	$P\,|_C = QN\,|_C$	$P\backslash_C = QN\,\backslash_C$	$P_C/ = QN_C/$
Pre-Transformer	$Q\,|_C = PN\,|_C$	$Q\backslash_C = PN\,\backslash_C$	$Q_C/ = PN_C/$

:

Pre-Transformer and Metric	$Q_Q/^P\ Q^Q\backslash_P = \delta_P/^P$	$P_Q/^P\ P^Q\backslash_P = \delta_P/^P$

:

Pre-Transformer and Metric	$P\backslash_{C\,C}/ = PN_C/\backslash_C$	$P\backslash_C = PN_C/$	$P_C/ = PN\backslash_C$
Pre-Transformer and Metric	$Q\backslash_{C\,C}/ = QN_C/\backslash_C$	$Q\backslash_C = QN_C/$	$Q_C/ = QN\backslash_C$

:

$$QT^{C}\backslash/^{C} \neq QN\backslash_{CC}/ = P\backslash_{CC}/ = P_C/\backslash_C = PN_C/\backslash_C \neq PT/^{CC}\backslash$$

$$PT^{C}\backslash/^{C} \neq PN\backslash_{CC}/ = Q\backslash_{CC}/ = Q_C/\backslash_C = QN_C/\backslash_C \neq QT/^{CC}\backslash$$

:

$$QN\backslash_C = P\backslash_C = PN_C/$$

Putting parentheses Including Transposed Tensors:

QT(+J,-I) = { [Q(+I,-J)] = (QN(-J,+I)) } = { (PN(+I,-J)) = [P(-J,+I)] } = PT(-I,+J)

Detailed:

:

(QN(-J,+I)) = (PN(+I,-J))

[Q(+I,-J)] … … … … = … … … … [P(-J,+I)]

{ [Q(+I,-J)] = (QN(-J,+I)) } = { (PN(+I,-J)) = [P(-J,+I)] }

QT(+J,-I) … … … … … … … … = … … … … … … … … PT(-I,+J)

In words by following the indices:

Q is The Iinverse of P	The Iinverse of Q is P
The Iinverse of P is Not Its Transpose	The Transpose of Q is Not Its Iinverse

If Q is Symmetric:

Trqnspose of Q or Q is The Iinverse of P	The Iinverse of Q is P
The Iinverse of P is Not Its Transpose	The Transpose of Q is Not Its Iinverse

If P is Symmetric:

Q is The Iinverse of P	The Iinverse of Q is P or Transpose of P
The Iinverse of P is Not Its Transpose	The Transpose of Q is Not Its Iinverse

If Q and P Are Symmetric:

Transpose of Q or Q is The Iinverse of P	The Iinverse of Q is P Or Transpose of P
The Iinverse of P is Not Its Transpose	The Transpose of Q is Not Its Iinverse

2.3.5.
Affine Initial and Current Descriptions of Continuum:

:

Affine:

:

QT J I = Q I J = QN J I = PN I J = P J I = PT I J

:

Q J I ≠ QT I J [=] QT J I = Q I J = QN J I = PN I J = P J I = PT I J [=] PT J I ≠ P I J

:

An Affine Transformation Component: Conjugates are Real and Equal:

$dp^{I} = dp_{I}$ $dq^{I} = dq_{I}$

Symmetric Affine:

$Q^{I}_{J} = Q^{IJ} = Q_{IJ} = Q_{I}^{J}$ = $P_{J}^{I} = P^{JI} = P_{JI} = P^{J}_{I}$

Briefly:

Q(I,J) = P(J,I)

Order of Affine Q=P Indices:

$^{I}_{J} = {}^{IJ} = {}_{IJ} = {}^{J}_{I}$ = $^{I}_{J} = {}^{JI} = {}_{JI} = {}^{J}_{I}$

Briefly:

(I,J) = (J,I)

:

Affine In General:

:

Q is The Iinverse of P
The Iinverse of P is Its Transpose

The Iinverse of Q is P
The Transpose of Q is Its Iinverse

Affine and Q is Symmetric:

Transpose of A Or A is The Iinverse of P
The Iinverse of P is Its Transpose

The Iinverse of A is P
The Transpose of A is Its Iinverse

Affine and P is Symmetric:

:

:

Q is The Iinverse of P
The Inverse of P is Its Transpose

The Iinverse of Q is P or Transpose of P
The Transpose of Q is Its Iinverse

Affine and Q and P are Symmetric:

:

Transpose of Q or Q is The Iinverse of P
The Iinverse of P is Its Transpose

The Iinverse of Q is P or Transpose of P
The Transpose of Q is Its Iinverse

2.4.
Similarity Transformations:

2.4.1.
Product of Bases and Differentials:

Define Two Forms of Transformation:

$\underline{b}$ db ◀ $\underline{a}$ da | da / a ▶ db / b

And:

ab ≡ da/db

Then:

$\underline{b}_{L}\, ba^{L}_{K} = \underline{a}_{K}$ $da^{M} = ab^{M}_{N}\, db^{N}$

$\underline{b}^{L}\, ba_{L}{}^{K} = \underline{a}^{K}$ $da_{M} = ab_{M}{}^{N}\, db_{N}$

But:

$$ba^{L}{}_{K} = ab_{K}{}^{L} \qquad ab^{M}{}_{N} = ba_{N}{}^{M}$$
$$ba_{L}{}^{K} = ab^{K}{}_{L} \qquad ab_{M}{}^{N} = ba^{N}{}_{M}$$

Then:

$$ab_{K}{}^{L}\,\underline{b}_{L} = \underline{a}_{K} \qquad da^{M} = db^{N}\,ba_{N}{}^{M}$$
$$ab^{K}{}_{L}\,\underline{b}^{L} = \underline{a}^{K} \qquad da_{M} = db_{N}\,ba^{N}{}_{M}$$

(ab)	(ba)
Pre Basis $\underline{b}$ Rotates $\underline{b}$ to $\underline{a}$	Post Differential db Rotates db to da

Transformations:

:

$$\underline{b}_{L}\,ba^{L}{}_{K} = \underline{a}_{K} \qquad \underline{b}_{L}\,ba^{L}{}_{K}\,\delta^{K}{}_{M}\,ab^{M}{}_{N}\,db^{N} = \underline{a}_{K}\,\delta^{K}{}_{M}\,da^{M} \qquad da^{M} = ab^{M}{}_{N}\,db^{N}$$

$$\underline{b}^{L}\,ba_{L}{}^{K} = \underline{a}^{K} \qquad \underline{b}^{L}\,ba_{L}{}^{K}\,\delta_{K}{}^{M}\,ab_{M}{}^{N}\,db_{N} = \underline{a}^{K}\,\delta_{K}{}^{M}\,da_{M} \qquad da_{M} = ab_{M}{}^{N}\,db_{N}$$

:

$$ab_{K}{}^{L}\,\underline{b}_{L} = \underline{a}_{K} \qquad db^{N}\,ba_{N}{}^{M}\,\delta_{M}{}^{K}\,ab_{K}{}^{L}\,\underline{b}_{L} = da^{M}\,\delta_{M}{}^{K}\,\underline{a}_{K} \qquad da^{M} = db^{N}\,ba_{N}{}^{M}$$

$$ab^{K}{}_{L}\,\underline{b}^{L} = \underline{a}^{K} \qquad db_{N}\,ba^{N}{}_{M}\,\delta^{M}{}_{K}\,ab^{K}{}_{L}\,\underline{b}^{L} = da_{M}\,\delta^{M}{}_{K}\,\underline{a}^{K} \qquad da_{M} = db_{N}\,ba^{N}{}_{M}$$

2.4.2. Replacement of Isotropic Identity:

Let:

$$\underline{a}_{K}\,T(a)^{K}{}_{M}\,da^{M}$$
$$\underline{a}^{K}\,T(a)_{K}{}^{M}\,da_{M}$$

Then:

$$\underline{b}_{L}\,ba^{L}{}_{K}\,T(a)^{K}{}_{M}\,ab^{M}{}_{N}\,db^{N}$$
$$\underline{b}^{L}\,ba_{L}{}^{K}\,T(a)_{K}{}^{M}\,ab_{M}{}^{N}\,db_{N}$$

(ba)	(ab)
Post Basis $\underline{b}$ = Pre function of a	Pre Differential db = Post function of a

Transformations:

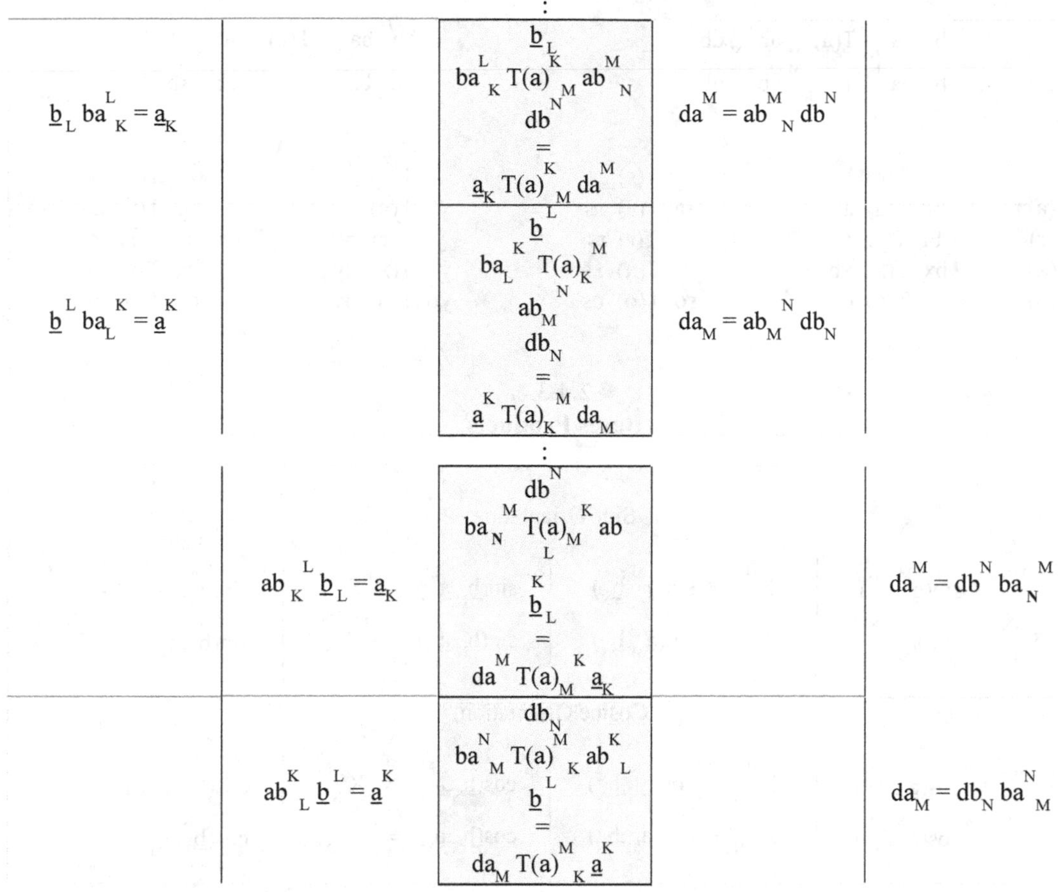

2.4.3. Applications in Different Systems:

2.4.3.1. Bases Transformation:

	b	s	y	o
a	b ba = a	s sa = a	y ya = a	o oa = a
r	b br = r	s sr = r	y yr = r	o or = r
x	b bx = x	s sx = x	y yx = x	o ox = x
o	b bo = o	s so = o	y yo = o	o oo = o

2.4.3.2. Differential Transformation:

	db	ds	dy	do
da	da = ab db	da = as ds	da = ay dy	da = ao do
dr	dr = rb db	dr = rs ds	dr = ry dy	dr = ro do
dx	dx = xb db	dx = xs ds	dx = xy dy	dx = xo do
do	do = ob db	do = os ds	do = oy dy	do = oo do

2.4.3.3. Similarity Transformation:

:

a T(a) da	da T(a) a

$\underline{b}_L\, ba^L_K\, T(a)^K_M\, ab^M_N\, db^N$	$db^N\, ba_N^M\, T(a)_M^K\, ab_K^L\, \underline{b}_L$
$\underline{b}^L\, ba_L^K\, T(a)_K^M\, ab_M^N\, db_N$	$db_N\, ba^N_M\, T(a)^M_K\, ab^K_L\, \underline{b}^L$

:

	T(b)	T(s)	T(y)	T(o)
T(a)	ba T(a) ab	sa T(a) as	ya T(a) ay	oa T(a) ao
T(r)	br T(r) rb	sr T(r) rs	yr T(r) ry	or T(r) ro
T(x)	bx T(x) xb	sx T(x) xs	yx T(x) xy	ox T(x) xo
T(o)	bo T(o) ob	so T(o) os	yo T(o) oy	oo T(o) oo

2.4.3.4.
Bases Products:

Sine Oriantation:

$\Sigma^{MN} \equiv \sin(\underline{a}^M,\underline{b}^N)$	$\Sigma^M_N \equiv \sin(\underline{a}^M,\underline{b}_N)$	$\sin(\underline{b}_N,\underline{a}^M) \equiv -\Sigma^M_N$	$\sin(\underline{b}^N,\underline{a}^M) \equiv -\Sigma^{NM}$
$\Sigma_M^N \equiv \sin(\underline{a}_M,\underline{b}^N)$	$\Sigma_{MN} \equiv \sin(\underline{a}_M,\underline{b}_N)$	$\sin(\underline{b}_N,\underline{a}_M) \equiv -\Sigma_{NM}$	$\sin(\underline{b}^N,\underline{a}_M) \equiv -\Sigma^N_M$

Cosine Oriantation:

$X^{MN} \equiv \cos(\underline{a}^M,\underline{b}^N)$	$X^M_N \equiv \cos(\underline{a}^M,\underline{b}_N)$	$\cos(\underline{b}_N,\underline{a}^M) \equiv X^M_N$	$\cos(\underline{b}^N,\underline{a}^M) \equiv X^{NM}$
$X_M^N \equiv \cos(\underline{a}_M,\underline{b}^N)$	$X_{MN} \equiv \cos(\underline{a}_M,\underline{b}_N)$	$\cos(\underline{b}_N,\underline{a}_M) \equiv X_{NM}$	$\cos(\underline{b}^N,\underline{a}_M) \equiv X^N_M$

	Variable bN	Orthogonal sN	Orthonormal yN	Fixed oN
aM	BNM = AMN	ΣNM = AMN	ΨNM = AMN	ONM = AMN
rM	BNM = PMN	ΣNM = PMN	ΨNM = PMN	ONM = PMN
xM	BNM = ΞMN	ΣNM = ΞMN	ΨNM = ΞMN	ONM = ΞMN
oM	BNM = OMN	ΣNM = OMN	ΨNM = OMN	ONM = OMN

AFFINE Transformation

$[ox^m{}_n]^{(-1)} = [ox^m{}_n]^{(T)} = ox^n{}_m$	$[ox_m{}^n]^{(-1)} = [ox_m{}^n]^{(T)} = ox_n{}^m$
$[xo^n{}_m]^{(-1)} = [xo^n{}_m]^{(T)} = xo^m{}_n$	$[xo_n{}^m]^{(-1)} = [xo_n{}^m]^{(T)} = xo_m{}^n$

Where:

$$\begin{bmatrix} \cos\alpha & -\sin\alpha \\ \sin\alpha & \cos\alpha \end{bmatrix} = \begin{bmatrix} ox^1{}_1 & ox^1{}_2 \\ ox^2{}_1 & ox^2{}_2 \end{bmatrix}$$

Resolution Transformation To x
β : positive angle between x and o

$\underline{x}^M$
$\underline{o}^M$

$\underline{y}^N$
$x_M{}^N = \Xi_{\underline{M}}{}^{\underline{N}}$
$o_M{}^N = O_{\underline{M}}{}^{\underline{N}}$

$\underline{o}^N$
$x_M{}^N = \Xi_{\underline{M}}{}^{\underline{N}}$

Where:

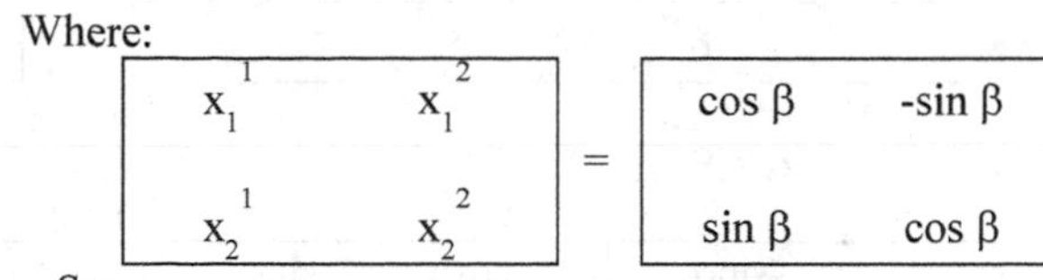

$$\begin{bmatrix} x_1^{\,1} & x_1^{\,2} \\ x_2^{\,1} & x_2^{\,2} \end{bmatrix} = \begin{bmatrix} \cos\beta & -\sin\beta \\ \sin\beta & \cos\beta \end{bmatrix}$$

So:

α : positive angle between o and x Composition Transformation Of x α = θ:	β : positive angle between x and o Resolution Transformation To x: :β = - θ

$$\begin{bmatrix} \cos\alpha & -\sin\alpha \\ \sin\alpha & \cos\alpha \end{bmatrix} = \begin{bmatrix} o^{1}_{\ 1} & o^{1}_{\ 2} \\ o^{2}_{\ 1} & o^{2}_{\ 2} \end{bmatrix} \quad \Bigg| \quad \begin{bmatrix} x_1^{\,1} & x_1^{\,2} \\ x_2^{\,1} & x_2^{\,2} \end{bmatrix} = \begin{bmatrix} \cos\beta & -\sin\beta \\ \sin\beta & \cos\beta \end{bmatrix}$$

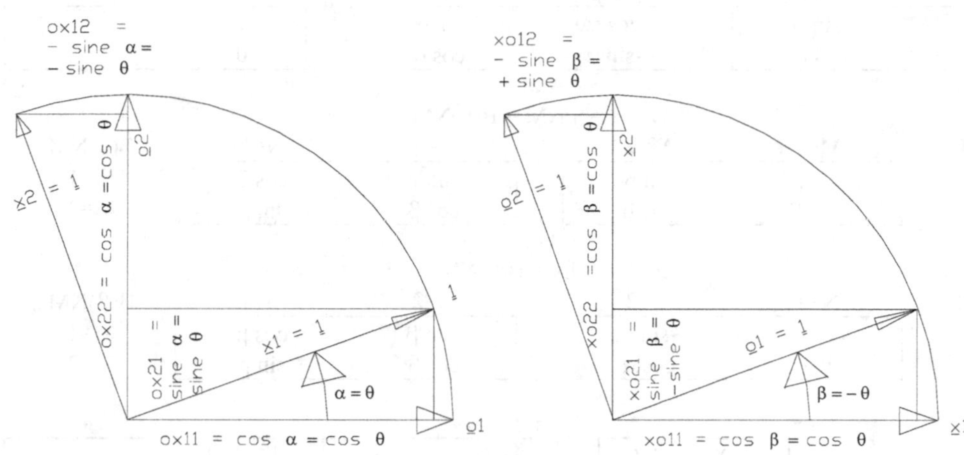

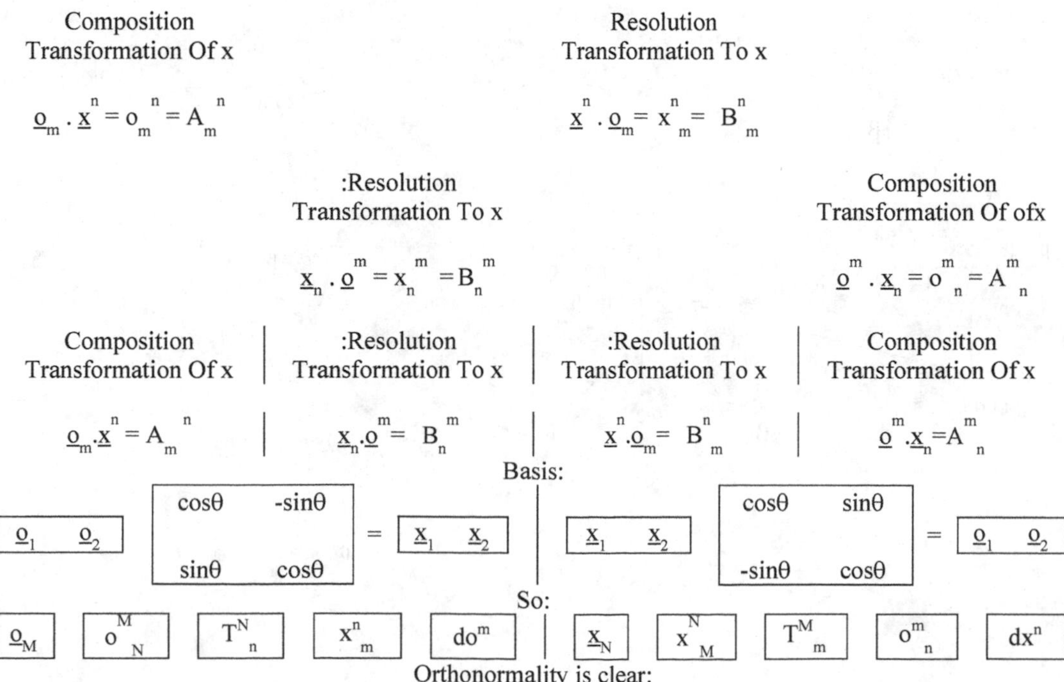

Composition Transformation Of x	Resolution Transformation To x
$\underline{o}_m \cdot \underline{x}^n = o^n_{\ m} = A^n_{\ m}$	$\underline{x}^n \cdot \underline{o}_m = x^n_{\ m} = B^n_{\ m}$

:Resolution Transformation To x	Composition Transformation Of ofx
$\underline{x}_n \cdot \underline{o}^m = x^m_{\ n} = B^m_{\ n}$	$\underline{o}^m \cdot \underline{x}_n = o^m_{\ n} = A^m_{\ n}$

Composition Transformation Of x	:Resolution Transformation To x	:Resolution Transformation To x	Composition Transformation Of x
$\underline{o}_m . \underline{x}^n = A^n_{\ m}$	$\underline{x}_n . \underline{o}^m = B^m_{\ n}$	$\underline{x}^n . \underline{o}_m = B^n_{\ m}$	$\underline{o}^m . \underline{x}_n = A^m_{\ n}$

Basis:

$$\begin{bmatrix} \underline{o}_1 & \underline{o}_2 \end{bmatrix} \begin{bmatrix} \cos\theta & -\sin\theta \\ \sin\theta & \cos\theta \end{bmatrix} = \begin{bmatrix} \underline{x}_1 & \underline{x}_2 \end{bmatrix} \quad \Bigg| \quad \begin{bmatrix} \underline{x}_1 & \underline{x}_2 \end{bmatrix} \begin{bmatrix} \cos\theta & \sin\theta \\ -\sin\theta & \cos\theta \end{bmatrix} = \begin{bmatrix} \underline{o}_1 & \underline{o}_2 \end{bmatrix}$$

So:

$$\boxed{\underline{o}_M} \ \boxed{o^M_{\ N}} \ \boxed{T^N_{\ n}} \ \boxed{x^n_{\ m}} \ \boxed{do^m} \quad \Big| \quad \boxed{\underline{x}_N} \ \boxed{x^N_{\ M}} \ \boxed{T^M_{\ m}} \ \boxed{o^m_{\ n}} \ \boxed{dx^n}$$

Orthonormality is clear:

$$o^{M}_{N}\,\delta^{N}_{n}\,x^{n}_{m} = \delta^{M}_{m} \qquad\qquad x^{N}_{M}\,\delta^{M}_{m}\,o^{m}_{n} = \delta^{N}_{n}$$

Since:

$\alpha = \theta$ $\qquad\qquad$ $\beta = -\theta$

Then:

cosα	-sinα
sinα	cosα

cosα	sinα
-sinα	cosα

cosβ	- sinβ
sinβ	cosβ

cosβ	sinβ
-sinβ	cosβ

Become:

cosθ	-sinθ
sinθ	cosθ

cosθ	sinθ
-sinθ	cosθ

cosθ	sinθ
-sinθ	cosθ

cosθ	-sinθ
sinθ	cosθ

Orthonormality:

A(α)MN = ΞMN = OMN

A(β)MN	N=1	N=2	M=2	M=1	ΞMN
M=1	cos α	-sin α	sin α	cos α	N=1
M=2	sin α	cos α	cos α	-sin α	N=2

N=1	N=2	M=1	M=2			
cos α	-sin α	cos α	sin α	=	1	0
sin α	cos α	-sin α	cos α		0	1

And:

ONM = ΨNM = B(β)NM

ΨNM	M=1	M=2	N=2	N=1	B(β)NM
N=1	cos β	sin β	-sin β	cos β	M=1
N=2	-sin β	cos β	cos β	sin β	M=2

Or:

A(α)MN = B(β)NM

A(β)MN	N=1	N=2	N=2	N=1	B(β)NM
M=1	cos α	-sin α	-sin β	cos β	M=1
M=2	sin α	cos α	cos β	sin β	M=2

Metrics:

$$\begin{bmatrix} \underline{x}_1 \\ \underline{x}_2 \end{bmatrix} \begin{bmatrix} \underline{x}_1 & \underline{x}_2 \end{bmatrix} = \begin{bmatrix} \cos\theta & \sin\theta \\ -\sin\theta & \cos\theta \end{bmatrix} \begin{bmatrix} \cos\theta & -\sin\theta \\ \sin\theta & \cos\theta \end{bmatrix} = \begin{bmatrix} 1 & \\ & 1 \end{bmatrix}$$

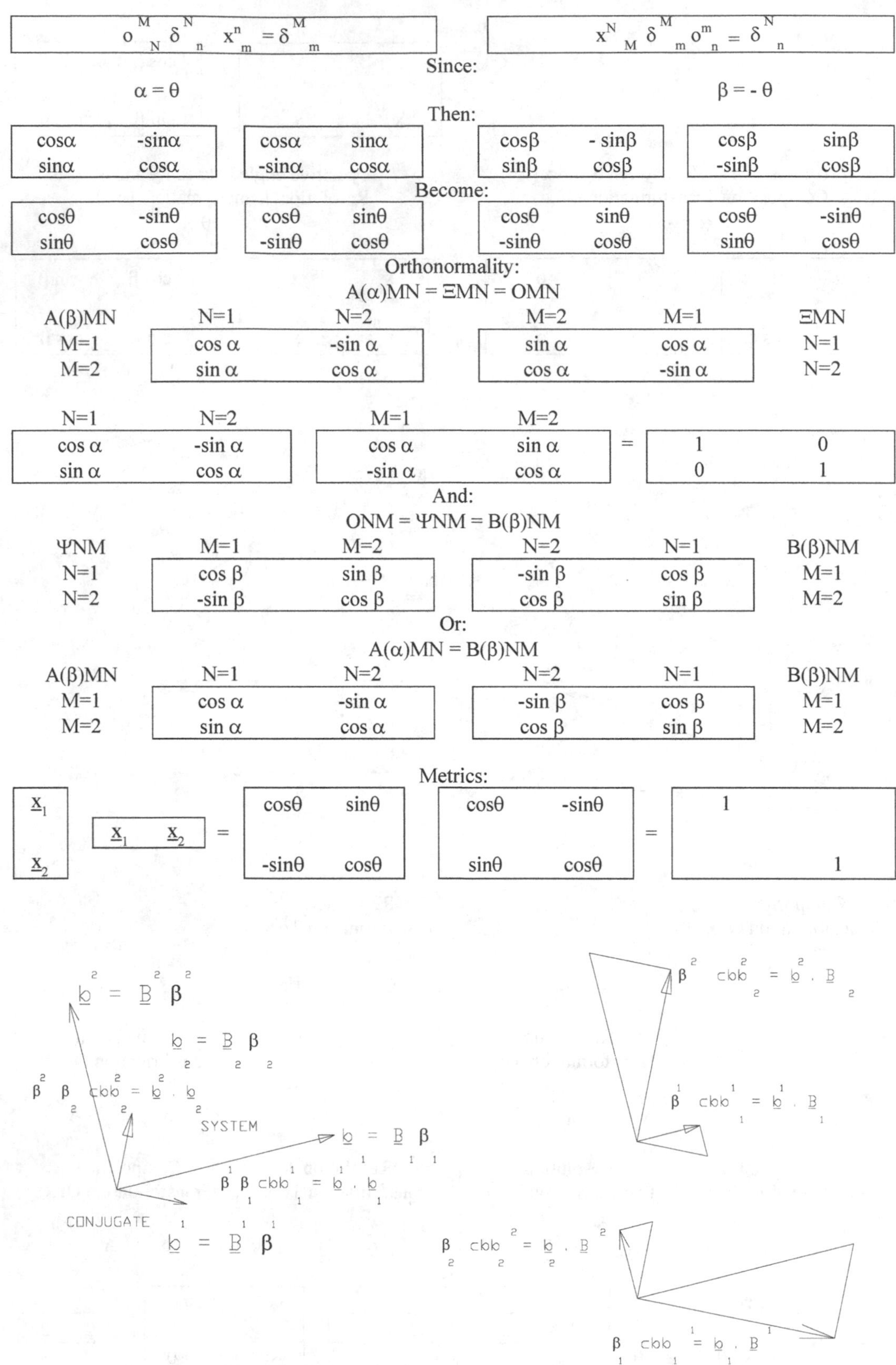

134 Prof. Faysal Alkhalil, Ph.D. INTRODUCTORY TENSORIAL ALGORITHMS IN STRUCTURAL MECHANICS

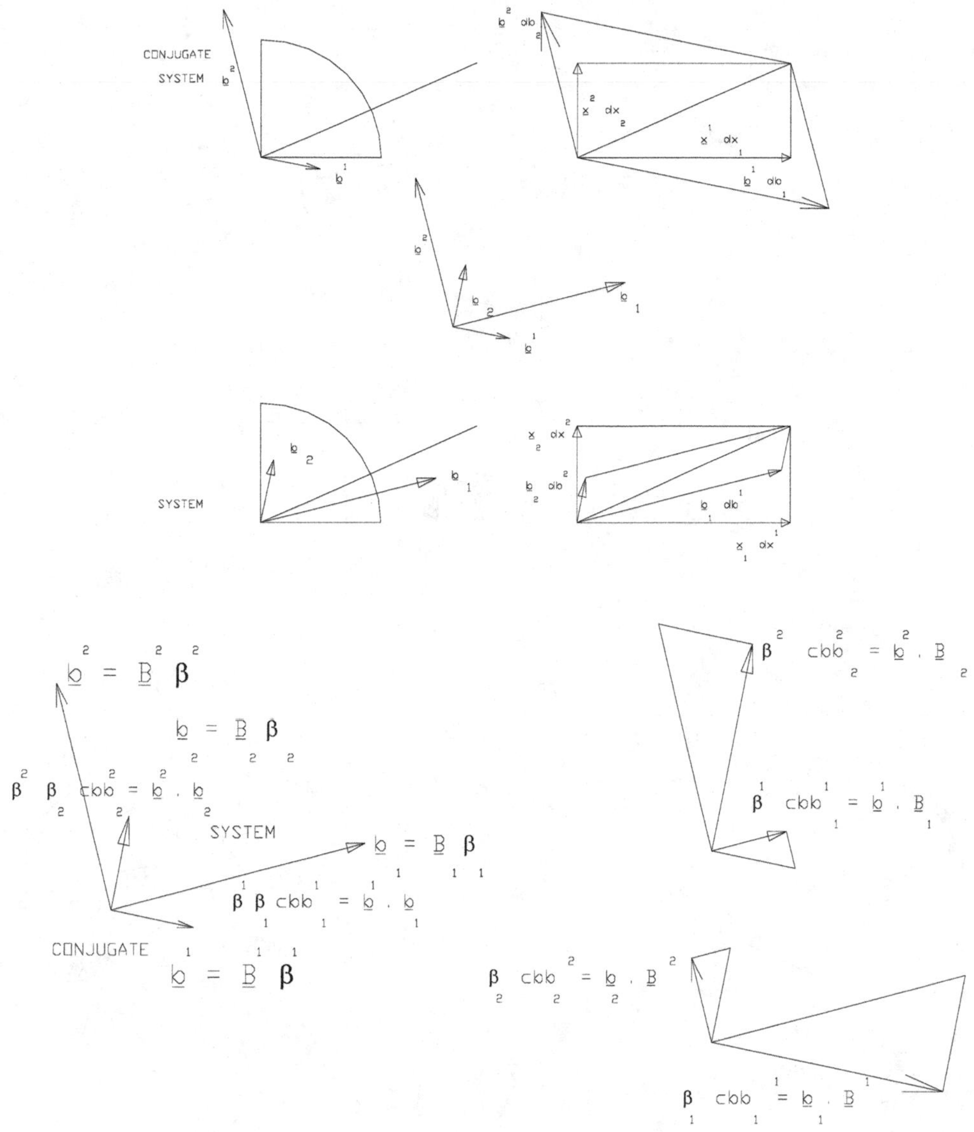
CONJUGATE
SYSTEM
SYSTEM
SYSTEM
CONJUGATE

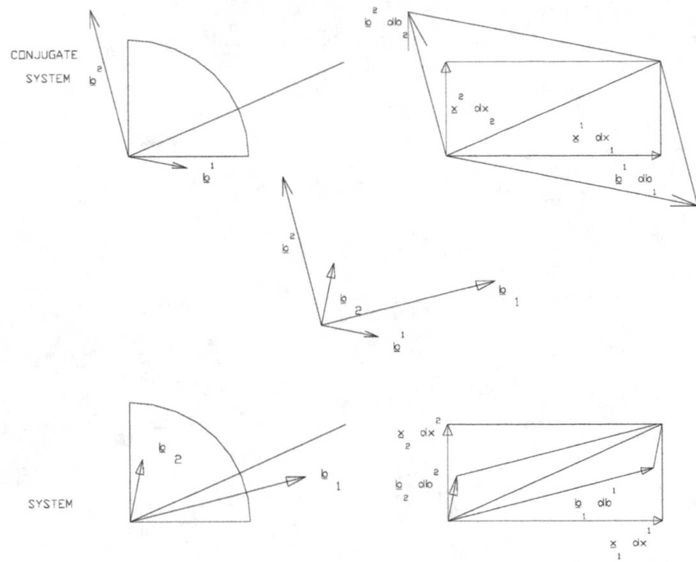

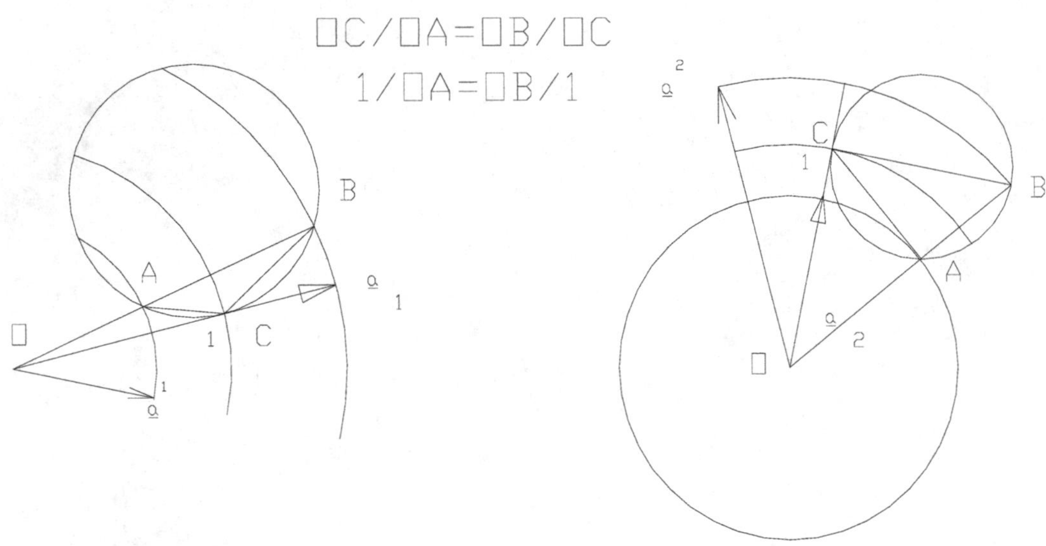

Orthogonal Basis And Fixed:

Composition Transformation Of r	o onto r :Resolution Transformation To:

Composition Transformation Of b onto x	x :Resolution Transformation To: b

2.4.4.
Second Order Transformation:

Second Partial Derivatives of
Lower and *Upper* w.t. to *Upper Position*:
Equivalently:
First Partial Derivatives of Transformation Of these
Lower and *Upper* with respect to *Upper Position*:

:

Lower First Order:

$$T(b)_M = T(a)_\mu \; ab^\mu{}_M$$

:

A tensor which is a product of two functions

$$T(b)_M = T(a)_\mu \; ab^\mu{}_M$$

Partial Derivative of a tensor which is a product of two functions with respect to *Upper Position*

$$\partial [T(b)_M]/\partial b^N = \partial [T(a)_\mu]/\partial b^N * ab^\mu{}_M + T(a)_\mu * \partial [ab^\mu{}_M]/\partial b^N$$

Substituting:

$$\partial [T(a)_\mu]/\partial b^N = \partial [T(a)_\mu]/\partial a^k \quad \partial a^k/\partial b^N \qquad \partial [ab^\mu{}_M]/\partial b^N$$

$$T(b)_{M,N} = \quad T(a)_{\mu,k} \quad ab^k{}_N \qquad [ab^\mu{}_M]_{,N}$$

Then:

$$T(b)_{M,N} = T(a)_\mu \; [ab^\mu{}_M]_{,N} + T(a)_{\mu,k} \; ab^\mu{}_M \, ab^k{}_N$$

:

Upper First Order:

$$T(b)^M = ba^M{}_\mu \; T(a)^\mu$$

A tensor which is a product of two functions

Partial Derivative of a tensor which is a product of two functions with respect to *Upper Position*

$$\partial [T(b)^M]/\partial b^N = \partial [ba^M{}_\mu]/\partial b^N \; T(a)^\mu + ba^M{}_\mu \; \partial [T(a)^\mu]/\partial b^N$$

Substituting:

$$\partial [ba^M{}_\mu]/\partial b^N \qquad \partial [T(a)^\mu]/\partial b^N = \partial [T(a)^\mu]/\partial a^K \quad ab^k{}_N$$

$$[ba^M{}_\mu]_{,N} \qquad T(a)^\mu{}_{,N} = T(a)^\mu{}_{,k} \; ab^k{}_N$$

Then:

$$T(b)^M{}_{,N} = [ba^M{}_\mu]_{,N} \; T(a)^\mu + ba^M{}_\mu \; T(a)^\mu{}_{,k} \; ab^k{}_N$$

:

Transformation Of P.D. of Lower and Upper with respect to Upper Position

$$T(b)_{M,N} = [ba_M{}^\mu]_{,N} \; T(a)_\mu + ba_M{}^\mu \; ba_N{}^k \; T(a)_{\mu,k}$$

$$T(b)^M{}_{,N} = T(a)^\mu \; [ab_\mu{}^M]_{,N} + T(a)^\mu{}_{,k} \; ab_\mu{}^M \; ab^k{}_N$$

:

If:

$$[ba^M{}_\mu]_{,N} = 0 = [ab^\mu{}_M]_{,N}$$

Then:

$$T(b)_{M,N} = ba_M^{\ \mu}\ ba_N^{\ k}\ T(a)_{\mu,k} \quad \dots \dots \dots \dots$$
$$T(b)_{M,N} = \dots \dots \dots \dots \dots\ T(a)_{\mu,k}\ ab^{\mu}_{\ M}\ ab^{k}_{\ N}$$

$$T(b)^{M}_{\ ,N} = \dots \dots \dots \dots \dots\ T(a)^{\mu}_{\ ,k}\ ab^{M}_{\ \mu}\ ab^{k}_{\ N}$$
$$T(b)^{M}_{\ ,N} = ba^{M}_{\ \mu}\ ba_N^{\ k}\ T(a)^{\mu}_{\ ,k} \quad \dots \dots \dots \dots$$

The following analogy of metrics and matrices will be generalized to any order:

:

$$aa^{\alpha\beta}\ ab_{\alpha}^{\ a}\ ab_{\beta}^{\ b} = bb^{ab} \qquad T(b)_{ab} = ba_{a}^{\ \alpha}\ ba_{b}^{\ \beta}\ T(a)_{\alpha\beta}$$
$$aa^{\alpha\beta}\ ba^{a}_{\ \alpha}\ ba^{b}_{\ \beta} = bb^{ab} \qquad T(b)_{ab} = ab^{\alpha}_{\ a}\ ab^{\beta}_{\ b}\ T(a)_{\alpha\beta}$$

:

$$aa^{\alpha\beta}\ ab_{\alpha}^{\ a}\ ab_{\beta}^{\ b} = bb^{ab} \qquad T(b)_{ab} = ba_{a}^{\ \alpha}\ ba_{b}^{\ \beta}\ T(a)_{\alpha\beta}$$
$$aa^{\alpha\beta}\ ba^{a}_{\ \alpha}\ ba^{b}_{\ \beta} = bb^{ab} \qquad T(b)_{ab} = ab^{\alpha}_{\ a}\ ab^{\beta}_{\ b}\ T(a)_{\alpha\beta}$$

Applied To Different Systems:

General: bN

$$aa^{\alpha\beta}\ ab_{\alpha}^{\ a}\ ab_{\beta}^{\ b} = bb^{ab} \qquad T(b)_{ab} = ba_{a}^{\ \alpha}\ ba_{b}^{\ \beta}\ T(a)_{\alpha\beta}$$
$$aa^{\alpha\beta}\ ba^{a}_{\ \alpha}\ ba^{b}_{\ \beta} = bb^{ab} \qquad T(b)_{ab} = ab^{\alpha}_{\ a}\ ab^{\beta}_{\ b}\ T(a)_{\alpha\beta}$$

$$rr^{\alpha\beta}\ rb_{\alpha}^{\ r}\ rb_{\beta}^{\ b} = bb^{rb} \qquad T(b)_{rb} = br_{r}^{\ \alpha}\ br_{b}^{\ \beta}\ T(r)_{\alpha\beta}$$
$$rr^{\alpha\beta}\ br^{r}_{\ \alpha}\ br^{b}_{\ \beta} = bb^{rb} \qquad T(b)_{rb} = rb^{\alpha}_{\ r}\ rb^{\beta}_{\ b}\ T(r)_{\alpha\beta}$$

$$xx^{\alpha\beta}\ xb_{\alpha}^{\ x}\ xb_{\beta}^{\ b} = bb^{xb} \qquad T(b)_{xb} = bx_{x}^{\ \alpha}\ bx_{b}^{\ \beta}\ T(x)_{\alpha\beta}$$
$$xx^{\alpha\beta}\ bx^{x}_{\ \alpha}\ bx^{b}_{\ \beta} = bb^{xb} \qquad T(b)_{xb} = xb^{\alpha}_{\ x}\ xb^{\beta}_{\ b}\ T(x)_{\alpha\beta}$$

$$oo^{\alpha\beta}\ ob_{\alpha}^{\ o}\ ob_{\beta}^{\ b} = bb^{ob} \qquad T(b)_{ob} = bo_{o}^{\ \alpha}\ bo_{b}^{\ \beta}\ T(o)_{\alpha\beta}$$
$$oo^{\alpha\beta}\ bo^{o}_{\ \alpha}\ bo^{b}_{\ \beta} = bb^{ob} \qquad T(b)_{ob} = ob^{\alpha}_{\ o}\ ob^{\beta}_{\ b}\ T(o)_{\alpha\beta}$$

Orthogonal: sN

$$aa^{\alpha\beta}\ as_{\alpha}^{\ a}\ as_{\beta}^{\ s} = ss^{as} \qquad T(s)_{as} = sa_{a}^{\ \alpha}\ sa_{s}^{\ \beta}\ T(a)_{\alpha\beta}$$
$$aa^{\alpha\beta}\ sa^{a}_{\ \alpha}\ sa^{s}_{\ \beta} = ss^{as} \qquad T(s)_{as} = as^{\alpha}_{\ a}\ as^{\beta}_{\ s}\ T(a)_{\alpha\beta}$$

$$rr^{\alpha\beta}\ rs_{\alpha}^{\ r}\ rs_{\beta}^{\ s} = ss^{rs} \qquad T(s)_{rs} = sr_{r}^{\ \alpha}\ sr_{s}^{\ \beta}\ T(r)_{\alpha\beta}$$
$$rr^{\alpha\beta}\ sr^{r}_{\ \alpha}\ sr^{s}_{\ \beta} = ss^{rs} \qquad T(s)_{rs} = rs^{\alpha}_{\ r}\ rs^{\beta}_{\ s}\ T(r)_{\alpha\beta}$$

$$xx^{\alpha\beta}\ xs_{\alpha}^{\ x}\ xs_{\beta}^{\ s} = ss^{xs} \qquad T(s)_{xs} = sx_{x}^{\ \alpha}\ sx_{s}^{\ \beta}\ T(x)_{\alpha\beta}$$
$$xx^{\alpha\beta}\ sx^{x}_{\ \alpha}\ sx^{s}_{\ \beta} = ss^{xs} \qquad T(s)_{xs} = xs^{\alpha}_{\ x}\ xs^{\beta}_{\ s}\ T(x)_{\alpha\beta}$$

$$oo^{\alpha\beta}\ os_{\alpha}^{\ o}\ os_{\beta}^{\ s} = ss^{os} \qquad T(s)_{os} = so_{o}^{\ \alpha}\ so_{s}^{\ \beta}\ T(o)_{\alpha\beta}$$
$$oo^{\alpha\beta}\ so^{o}_{\ \alpha}\ so^{s}_{\ \beta} = ss^{os} \qquad T(s)_{os} = os^{\alpha}_{\ o}\ os^{\beta}_{\ s}\ T(o)_{\alpha\beta}$$

Orthonormal: yN

$$aa^{\alpha\beta}\, ay_{\alpha}^{\ a}\, ay_{\beta}^{\ y} = yy^{ay} \qquad T(y)_{ay} = ya_{a}^{\ \alpha}\, ya_{y}^{\ \beta}\, T(a)_{\alpha\beta}$$

$$aa^{\alpha\beta}\, ya^{a}_{\ \alpha}\, ya^{y}_{\ \beta} = yy^{ay} \qquad T(y)_{ay} = ay^{\alpha}_{\ a}\, ay^{\beta}_{\ y}\, T(a)_{\alpha\beta}$$

$$rr^{\alpha\beta}\, ry_{\alpha}^{\ r}\, ry_{\beta}^{\ y} = yy^{ry} \qquad T(y)_{ry} = yr_{r}^{\ \alpha}\, yr_{y}^{\ \beta}\, T(r)_{\alpha\beta}$$

$$rr^{\alpha\beta}\, yr^{r}_{\ \alpha}\, yr^{y}_{\ \beta} = yy^{ry} \qquad T(y)_{ry} = ry^{\alpha}_{\ r}\, ry^{\beta}_{\ y}\, T(r)_{\alpha\beta}$$

$$xx^{\alpha\beta}\, xy_{\alpha}^{\ x}\, xy_{\beta}^{\ y} = yy^{xy} \qquad T(y)_{xy} = yx_{x}^{\ \alpha}\, yx_{y}^{\ \beta}\, T(x)_{\alpha\beta}$$

$$xx^{\alpha\beta}\, yx^{x}_{\ \alpha}\, yx^{y}_{\ \beta} = yy^{xy} \qquad T(y)_{xy} = xy^{\alpha}_{\ x}\, xy^{\beta}_{\ y}\, T(x)_{\alpha\beta}$$

$$oo^{\alpha\beta}\, oy_{\alpha}^{\ o}\, oy_{\beta}^{\ y} = yy^{oy} \qquad T(y)_{oy} = yo_{o}^{\ \alpha}\, yo_{y}^{\ \beta}\, T(o)_{\alpha\beta}$$

$$oo^{\alpha\beta}\, yo^{o}_{\ \alpha}\, yo^{y}_{\ \beta} = yy^{oy} \qquad T(y)_{oy} = oy^{\alpha}_{\ o}\, oy^{\beta}_{\ y}\, T(o)_{\alpha\beta}$$

Fixed: oN

$$aa^{\alpha\beta}\, ao_{\alpha}^{\ a}\, ao_{\beta}^{\ o} = oo^{ao} \qquad T(o)_{ao} = oa_{a}^{\ \alpha}\, oa_{o}^{\ \beta}\, T(a)_{\alpha\beta}$$

$$aa^{\alpha\beta}\, oa^{a}_{\ \alpha}\, oa^{o}_{\ \beta} = oo^{ao} \qquad T(o)_{ao} = ao^{\alpha}_{\ a}\, ao^{\beta}_{\ o}\, T(a)_{\alpha\beta}$$

$$rr^{\alpha\beta}\, ro_{\alpha}^{\ r}\, ro_{\beta}^{\ o} = oo^{ro} \qquad T(o)_{ro} = or_{r}^{\ \alpha}\, or_{o}^{\ \beta}\, T(r)_{\alpha\beta}$$

$$rr^{\alpha\beta}\, or^{r}_{\ \alpha}\, or^{o}_{\ \beta} = oo^{ro} \qquad T(o)_{ro} = ro^{\alpha}_{\ r}\, ro^{\beta}_{\ o}\, T(r)_{\alpha\beta}$$

$$xx^{\alpha\beta}\, xo_{\alpha}^{\ x}\, xo_{\beta}^{\ o} = oo^{xo} \qquad T(o)_{xo} = ox_{x}^{\ \alpha}\, ox_{o}^{\ \beta}\, T(x)_{\alpha\beta}$$

$$xx^{\alpha\beta}\, ox^{x}_{\ \alpha}\, ox^{o}_{\ \beta} = oo^{xo} \qquad T(o)_{xo} = xo^{\alpha}_{\ x}\, xo^{\beta}_{\ o}\, T(x)_{\alpha\beta}$$

$$oo^{\alpha\beta}\, oo_{\alpha}^{\ o}\, oo_{\beta}^{\ o} = oo^{oo} \qquad T(o)_{oo} = oo_{o}^{\ \alpha}\, oo_{o}^{\ \beta}\, T(o)_{\alpha\beta}$$

$$oo^{\alpha\beta}\, oo^{o}_{\ \alpha}\, oo^{o}_{\ \beta} = oo^{oo} \qquad T(o)_{oo} = oo^{\alpha}_{\ o}\, oo^{\beta}_{\ o}\, T(o)_{\alpha\beta}$$

2.5.
Summary of Coordinate Transformations

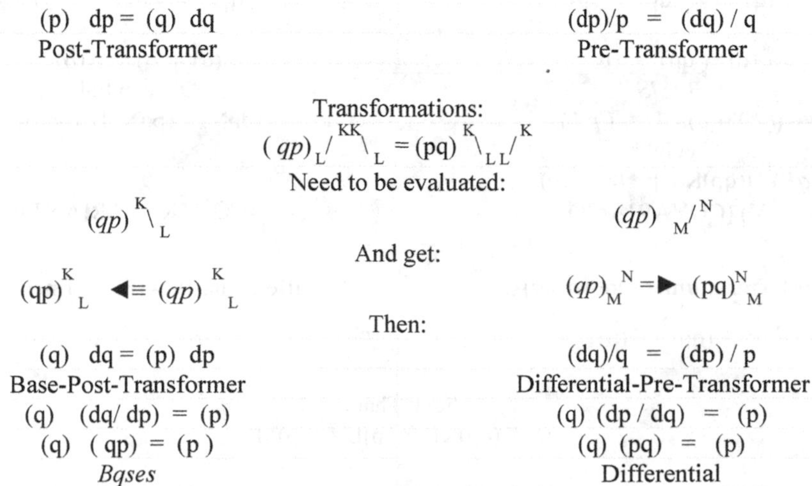

(p) dp = (q) dq — Post-Transformer

(dp)/p = (dq) / q — Pre-Transformer

Transformations:

$$(qp)_L/^{KK}\backslash_L = (pq)^{K}\backslash_{LL}/^{K}$$

Need to be evaluated:

$$(qp)^{K}\backslash_L \qquad (qp)_M/^{N}$$

And get:

$$(qp)^{K}_{\ L} \blacktriangleleft\equiv (qp)^{K}_{\ L} \qquad (qp)_M^{\ N} =\blacktriangleright (pq)^{N}_{\ M}$$

Then:

(q) dq = (p) dp	(dq)/q = (dp) / p
Base-Post-Transformer	Differential-Pre-Transformer
(q) (dq/ dp) = (p)	(q) (dp / dq) = (p)
(q) (qp) = (p)	(q) (pq) = (p)
Bqses	Differential

$(\underline{q}_K)\,(qp)^K{}_L = (\underline{p})_L$ $\qquad$ $(dp)^N = (pq)^N{}_M\,(dq)^M$

Weight

$(q)_K\ [(qp)^K{}_L]^{**}W = (p_L)$

(∇) (covariant) $\qquad$ (contravariant)

Bases-Post-Transformer (pre-tensor) $\qquad$ Differential-Pre-Transformer (post-tensor)

$T(q)^q = (qp)^q{}_p\,T(p)^p$ $\qquad$.\ $\qquad$ $T(p)_p = T(q)_q\ (qp)^q{}_p$

Such That:

$$(\underline{q}_K)\,(qp)^K{}_L\,[pTp]^L{}_M\,(dp)^M$$
$$=$$
$$(\underline{q}_K)\,[qTp]^K{}_L\,(dq)^L = (\underline{p})_K\ [pTp]^k{}_l\ (dp)^l$$
$$=$$
$$(\underline{p})_k\ [pTp]^k{}_l\,(pq)^l{}_L\ (dq)^L$$

So:

$$[qTq]^K{}_L = (qp)^K{}_k\,[pTp]^k{}_l\ (pq)^l{}_L$$

Mixed Order:

$$(\underline{qqq})_{ABD}\,(qp)^A{}_q\,(qp)^B{}_p\,(qp)^D{}_d\ [pppTp]^{qpd}{}_c\ (dp)^D$$
$$= (\underline{qqq}_{ABD})\,[qqqTp]^{AB\ D}{}_C\ (dq)^C = (\underline{ppp})_{qpd}\ [pppTp]^{qpd}{}_c\ (dp)^c =$$
$$(\underline{ppp})_{qpd}\ [pppTp]^{qpd}{}_c\ (pq)^c{}_C\ (dq)^C$$

So:

$$[qqqTq]^{AB\ D}{}_C = (qp)^A{}_q\,(qp)^B{}_p\,(qp)^D{}_d\ [pppTp]^{qpd}{}_c\ (pq)^c{}_C$$

:

Affine Transformations:

(*qp*) K L $\quad$ = (pq) LK

Transformations:

Need to be Evaluated, only:

(*qp*) K L

Then its Transpose:

(*qp*) M N ◄= [(*qp*) N M]T

And Get:

(qp)KL ◄≡ (*qp*) KL $\qquad$ (*qp*) M N ≡► (pq) N M

Then:

(q) dq = (p) dp $\qquad$ (dq)/q = (dp) / p

Post-Transformer $\qquad$ Pre-Transformer

(q) (dq/ dp) = (p) $\qquad$ (q) (dp / dq) = (p)

(q) (qp) = (p) $\qquad$ (q) (pq) = (p)

BASIS $\qquad$ Differentiql

(*q*K) (qp)K L = (*p* L) $\qquad$ (dpN) = (pqNM) (dqM)

Weight

(*qK*) [(qp)K L]***W* = (*p*)*L*

(∇) (COVARIANT) $\qquad$ (CONTRAVARIANT)

Bases-Post-Transformer (pre-tensor) $\qquad$ Differential-Pre-Transformer (post-tensor)

$T(q)^q = (qp)^q{}_p\,T(p)^p$ $\qquad$.\ $\qquad$ $T(p)_p = T(q)_q\ (qp)^q{}_p$

Such That:

(*q*)*K* (qp)KL [pTp]LM (dp)M

=

(<u>q</u>)*K* [qTp]AC (dq)C = (<u>p</u>)*q* [pTp]qc (dp)c

=

(<u>p</u>)*L* [pTp]LM (pq)MN (dq)N

So:

[qTq]AC = (qp)Aq [pTp]qc (pq)cC

Mixed order

(<u>qqq</u>)*ABD* (qp)Aq (qp)Bp (qp)Dd [pppTp]ABCD (dp)C

=

(<u>qqq</u>)*ABD* [qqqTp]ABCD (dqC)

=

(<u>ppp</u>)*qpd* [pppTp]ABCD (dp)C

=

(<u>ppp</u>) *qpd* [pppTp]qpc d (pq)cC (dqC)

So:

[qqqTq]ABCD

=

(qp)Aq (qp)Bp (qp)Dd [pppTp]qpcd (pq)cC

3.
Structural Coordinate Metrics:

3.1.
Direction Cosines and Inner Products:

Unit Bases:

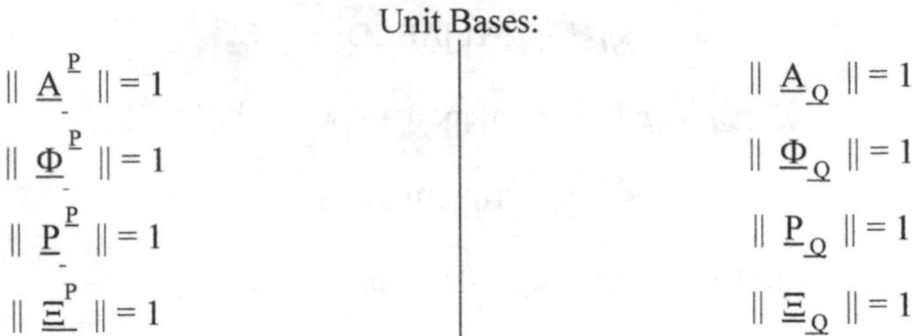

$\Vert \underline{A}^{\underline{P}} \Vert = 1$	$\Vert \underline{A}_{\underline{Q}} \Vert = 1$
$\Vert \underline{\Phi}^{\underline{P}} \Vert = 1$	$\Vert \underline{\Phi}_{\underline{Q}} \Vert = 1$
$\Vert \underline{P}^{\underline{P}} \Vert = 1$	$\Vert \underline{P}_{\underline{Q}} \Vert = 1$
$\Vert \underline{\Xi}^{\underline{P}} \Vert = 1$	$\Vert \underline{\Xi}_{\underline{Q}} \Vert = 1$

Bases:

$\underline{a}^{P} = \alpha^{\underline{P}} \underline{A}^{\underline{P}}$	$\alpha^{\underline{K}} = 1/\alpha_{\underline{K}}$	$\underline{a}_{Q} = \alpha_{\underline{Q}} \underline{A}_{\underline{Q}}$
$\underline{f}^{P} = \phi^{\underline{P}} \underline{\Phi}^{\underline{P}} = \underline{\Phi}^{\underline{P}}$	$\phi^{\underline{K}} = 1 = \phi_{\underline{K}}$	$\underline{f}_{Q} = \phi_{\underline{Q}} \underline{\Phi}_{\underline{Q}} = \underline{\Phi}_{\underline{Q}}$
$\underline{r}^{P} = \rho^{\underline{P}} \underline{P}^{\underline{P}}$	$\rho_{\underline{K}} \rho^{\underline{K}} = 1$	$\underline{r}_{Q} = \rho_{\underline{Q}} \underline{P}_{\underline{Q}}$
$\underline{x}^{P} = \xi^{\underline{P}} \underline{\Xi}^{\underline{P}} = \underline{\Xi}^{\underline{P}}$	$\xi^{\underline{K}} = 1 = \xi_{K}$	$\underline{x}_{Q} = \xi_{\underline{Q}} \underline{\Xi}_{\underline{Q}} = \underline{\Xi}_{\underline{Q}}$

Direction cosines of two bases:

$\Vert \cos(\underline{a}^{I}, \underline{a}^{K}) \equiv AA^{IK} \Vert \leq 1$ and $AA^{\underline{KK}} = 1$	$\Vert \cos(\underline{a}_{I}, \underline{a}_{K}) \equiv AA_{IK} \Vert \leq 1$ and $AA_{\underline{KK}} = 1$
$\Vert \cos(\underline{f}^{I}, \underline{f}^{K}) \equiv \Phi\Phi^{IK} \Vert \leq 1$ and $\Phi\Phi^{\underline{KK}} = 1$	$\Vert \cos(\underline{f}_{I}, \underline{f}_{K}) \equiv \Phi\Phi_{IK} \Vert \leq 1$ and $\Phi\Phi_{\underline{KK}} = 1$
$\cos(\underline{r}^{I}, \underline{r}^{K}) = \delta^{IK}$	$\cos(\underline{r}_{I}, \underline{r}_{K}) = \delta_{IK}$
$\cos(\underline{x}^{I}, \underline{x}^{K}) = \delta^{IK}$	$\cos(\underline{x}_{I}, \underline{x}_{K}) = \delta_{IK}$

Inner products of two bases:

$aa^{IK} = \alpha^{I} \alpha^{K} AA^{IK}$	$aa_{IK} = \alpha_{I} \alpha_{K} AA_{IK}$
$ff^{IK} = \Phi\Phi^{IK}$	$ff_{IK} = \Phi\Phi_{IK}$
$rr^{IK} = \rho^{I} \rho^{K} \delta^{IK}$	$rr_{IK} = \rho_{I} \rho_{K} \delta_{IK}$
$xx^{IK} = \delta^{IK}$	$xx_{IK} = \delta_{IK}$

Bases are related to fixed coordinates:

$\underline{a}^{J} = ao^{J}_{n} \underline{o}^{n} = \underline{o}^{n} oa^{J}_{n}$	$\underline{a}_{I} = ao_{I}^{m} \underline{o}_{m} = \underline{o}_{m} oa^{m}_{I}$
$\underline{f}^{J} = fo^{J}_{n} \underline{o}^{n} = \underline{o}^{n} of^{J}_{n}$	$\underline{f}_{I} = fo_{I}^{m} \underline{o}_{m} = \underline{o}_{m} of^{m}_{I}$
$\underline{r}^{J} = ro^{J}_{n} \underline{o}^{n} = \underline{o}^{n} or^{J}_{n}$	$\underline{r}_{I} = ro_{I}^{m} \underline{o}_{m} = \underline{o}_{m} or^{m}_{I}$
$\underline{x}^{J} = xo^{J}_{n} \underline{o}^{n} = \underline{o}^{n} ox^{J}_{n}$	$\underline{x}_{I} = xo_{I}^{m} \underline{o}_{m} = \underline{o}_{m} ox^{m}_{I}$

Reciprocal Metrics:

$$\alpha_{\underline{K}} \alpha^{\underline{K}} = \sqrt{aa_{\underline{KK}}} \sqrt{aa^{\underline{KK}}} = 1$$

Cosine of angle between two bases of a

aa_{KL} / √($aa_{\underline{KK}}$ $aa_{\underline{LL}}$) = cos θKL = aa^{KL} / √(aa^{KK} aa^{LL})

3.1.1.
General Bases Direction Cosines and Inner Products:

A postion vector:

$\underline{a}_i\, a^i = a_i\, \underline{a}^i$

Bases:

$\underline{a}^P = \alpha^{\underline{P}}\, \underline{A}^P$ | $\underline{a}_Q = \alpha_{\underline{Q}}\, \underline{A}_{\underline{Q}}$

Direction cosine of two bases:

$\| \cos(\underline{a}^I, \underline{a}^K) \equiv AA^{IK} \| \leq 1$ and $AA^{\underline{KK}} = 1$ | $\| \cos(\underline{a}_I, \underline{a}_K) \equiv AA_{IK} \| \leq 1$ and $AA_{\underline{KK}} = 1$

Inner products of two bases:

$aa^{IK} = \alpha^{\underline{I}}\, \alpha^{\underline{K}}\, AA^{\underline{IK}}$ | $aa_{IK} = \alpha_{\underline{I}}\, \alpha_{\underline{K}}\, AA_{\underline{IK}}$

Bases are related to fixed coordinates:

$\underline{a}^J = ao^J_{\ n}\, \underline{o}^n = \underline{o}^n\, oa^J_{\ n}$ | $\underline{a}_I = ao^m_I\, \underline{o}_m = \underline{o}_m\, oa^m_{\ I}$

3.1.1.1.
General Bases Direction Cosines:

$\| \cos(\underline{a}^I, \underline{a}^K) \equiv AA^{IK} \| \leq 1$ and $AA^{\underline{KK}} = 1$ | $\| \cos(\underline{a}_I, \underline{a}_K) \equiv AA_{IK} \| \leq 1$ and $AA_{\underline{KK}} = 1$

If: I = K

$AA^{\underline{II}} = 1$ | $AA_{\underline{II}} = 1$

If: I ≠ K

$0 < \| AA^{IK} \| \leq 1$ | $0 < \| AA_{IK} \| \leq 1$

3.1.1.2.
General Bases Inner Products:

$\| aa^{IK} = \alpha^I \alpha^K AA^{Ik} \| \leq \alpha^I \alpha^K$ | $\| aa_{IK} = \alpha_I \alpha_K AA_{IK} \| \leq \alpha_I \alpha_K$

If: I = K

$aa^{\underline{II}} = \alpha^{\underline{I}} \alpha^{\underline{I}}$ | $aa_{\underline{II}} = \alpha_{\underline{I}} \alpha_{\underline{I}}$

If: I ≠ K

$\| aa^{IK} \| \leq \alpha^I \alpha^K$ | $\| aa_{IK} \| \leq \alpha_I \alpha_K$

Inner products of bases:

$aa^{IK} \equiv \underline{a}^I . \underline{a}^K = \underline{A}^{\underline{I}}\, \alpha^{\underline{I}} . \underline{A}^{\underline{K}}\, \alpha^{\underline{K}} = \alpha^{\underline{I}}\, \alpha^{\underline{K}}\, AA^{\underline{IK}}$ | $aa_{IK} \equiv \underline{a}_I . \underline{a}_K = \underline{A}_{\underline{I}}\, \alpha_{\underline{I}} . \underline{A}_{\underline{K}}\, \alpha_{\underline{K}} = \alpha_{\underline{I}}\, \alpha_{\underline{K}}\, AA_{\underline{IK}}$

Since:

$AA^{II} = \cos 0 = 1$ | $AA_{II} = \cos 0 = 1$

Then:

$\alpha^I = \text{sqrt}\, | aa^{II} |$ | $\alpha_I = \text{sqrt}\, | aa_{II} |$

$\underline{a}^I . \underline{A}^K = \underline{A}^{\underline{I}}\, \alpha^{\underline{I}} . \underline{A}^{\underline{K}} = \alpha^{\underline{I}}\, AA^{\underline{IK}}$ | $\underline{a}_I . \underline{A}_K = \underline{A}_{\underline{I}}\, \alpha_{\underline{I}} . \underline{A}_{\underline{K}} = \alpha_{\underline{I}}\, AA_{\underline{IK}}$

3.1.1.3.

Direction Cosines of General Bases and Fixed Bases:

$OA_m^{\ n} = \cos(\underline{o}_m, \underline{a}^n)$	$AO_n^{\ m} = \cos(\underline{a}_n, \underline{o}^m)$	$OA^n_{\ m} = \cos(\underline{o}^n, \underline{a}_m)$	$AO^m_{\ n} = \cos(\underline{a}^m, \underline{o}_n)$
$OA_m^{\ n}$	$=$	$OA^n_{\ m}$	
	$AO_n^{\ m}$	$=$	$AO^m_{\ n}$
$oa_m^{\ n} = OA_m^{\ n}\ \alpha^n$	$=$	$ao^n_{\ m} = \alpha^n\ OA^n_{\ m}$	
	$ao_n^{\ m} = \alpha_n\ AO_n^{\ m}$	$=$	$oa^m_{\ n} = AO^m_{\ n}\ \alpha_n$

aa(-m,-n) ≡ ba(-m).ba(-n) = bo(-j) oa(+j,-m). bo(-k) oa(+k,-n) = oa(+k,-m)oa(+k,-n):
aa(-m,-n) = oa(+k,-m) oa(+k,-n)

m	aa(-m,-1) n = 1	aa(-m,-2) n = 2	aa(-m,-3) n = 3
1	oa(+m,-1) oa(+m,-1)	oa(+m,-1) oa(+m,-2)	oa(+m,-1) oa(+m,-3)
2	oa(+m,-2) oa(+m,-1)	oa(+m,-2) oa(+m,-2)	oa(+m,-2) oa(+m,-3)
3	oa(+m,-3) oa(+m,-1)	oa(+m,-3) oa(+m,-2)	oa(+m,-3) oa(+m,-3)

:

Sum of product of column by column: aa(-m,-n) = oa(+k,-m)oa(+k,-n)

m	aa(-m,-1) n = 1	aa(-m,-2) n = 2	aa(-m,-3) n = 3
1	first * first columns sum	first * second columns Sum	first * third columns Sum
2	Symmetric	second * second columns sum	second * third columns sum
3	Symmetric	Symmetric	third * third columns sum

3.1.1.4.
Inner Products of General Bases and Fixed Bases:

Inner products of general bases by fixed bases:

$\underline{o}_m \cdot \underline{a}^n = OA_m^{\ n}\ \alpha^n$	$\underline{a}_n \cdot \underline{o}^m = \alpha_n\ AO_n^{\ m}$	$\underline{o}^n . \underline{a}_m = OA^n_{\ m}\ \alpha_m$	$\underline{a}^m . \underline{o}_n = \alpha^m\ AO^m_{\ n\,n}$

Inner Products:

$aa_{IJ} \equiv \underline{a}_I \cdot \underline{a}_J$	$aa^{IJ} \equiv \underline{a}^I . \underline{a}^J$
$=$	$=$
$\underline{o}_m\, oa^m_{\ I} \cdot \underline{o}_n\, oa^n_{\ J}$	$\underline{o}^m\, oa_m^{\ I} . \underline{o}^n\, oa_n^{\ J}$
$=$	$=$
$\delta_{mn}\, oa^m_{\ I}\, oa^n_{\ J}$	$\delta^{mn}\, oa_m^{\ I}\, oa_n^{\ J}$
$=$	$=$
$oa^M_{\ I}\, oa^M_{\ J}$	$oa_M^{\ I}\, oa_M^{\ J}$

Sum of products of corresponding inner products:

$$aa_{IJ}\, aa^{IJ} = oa^m_{\ I}\, oa^m_{\ J}\, oa_p^{\ I}\, oa_p^{\ J} = oa_p^{\ I}\, oa^m_{\ I}\, oa^m_{\ J}\, oa_p^{\ J}$$

But:

$oa_p^{\ I}\, oa^m_{\ I} = \delta^m_p$	$oa^m_{\ J}\, oa_p^{\ J} = \delta^m_p$

So:

$$aa_{IJ}\, aa^{IJ} = \delta^m_p\, \delta^m_p = 3$$

3.1.2.
General Unitary Bases Direction Cosines and Inner Products:

A postion vector:

$$f_i f^i = f_i \underline{f}^i$$

Bases:

$\underline{f}^P = \phi^{\underline{P}} \underline{\Phi}^{\underline{P}} = \underline{\Phi}^{\underline{P}}$	$\underline{f}_Q = \phi_{\underline{Q}} \underline{\Phi}_{\underline{Q}} = \underline{\Phi}_{\underline{Q}}$

Direction cosine of two bases:

$\Vert \cos(\underline{f}^I, \underline{f}^K) \equiv \Phi\Phi^{IK} \Vert \leq 1$ and $\Phi\Phi^{\underline{KK}} = 1$	$\Vert \cos(\underline{f}_I, \underline{f}_K) \equiv \Phi\Phi_{IK} \Vert \leq 1$ and $\Phi\Phi_{\underline{KK}} = 1$

Inner products of two bases:

$ff^{IK} = \Phi\Phi^{\underline{IK}}$	$ff_{IK} = \Phi\Phi_{\underline{IK}}$

Bases are related to fixed coordinates:

$\underline{f}^J = fo^J_n \underline{o}^n = \underline{o}^n of^J_n$	$\underline{f}_I = fo^m_I \underline{o}_m = \underline{o}_m of^m_I$

3.1.2.1.
General Unitary Bases Direction Cosines:

$\Vert \cos(\underline{f}^I, \underline{f}^K) \equiv \Phi\Phi^{IK} \Vert \leq 1$ and $\Phi\Phi^{\underline{KK}} = 1$	$\Vert \cos(\underline{f}_I, \underline{f}_K) \equiv \Phi\Phi_{IK} \Vert \leq 1$ and $\Phi\Phi_{\underline{KK}} = 1$

If: I = K

$\Phi\Phi^{\underline{II}} = 1$	$\Phi\Phi_{\underline{II}} = 1$

If: $I \neq K$

$\Vert \Phi\Phi^{IK} \Vert \leq 1$	$\Vert \Phi\Phi_{IK} \Vert < 1$

3.1.2.2.
General Unitary Bases Inner Products:

$\Vert ff^{IK} = \underline{\Phi}^I \phi^{\underline{I}} . \underline{\Phi}^K \phi^{\underline{K}} = \Phi\Phi^{Ik} \Vert \leq 1$	$\Vert ff_{IK} = \underline{\Phi}_{\underline{I}} \phi_{\underline{I}} . \underline{\Phi}_{\underline{K}} \phi_{\underline{K}} = \Phi\Phi_{IK} \Vert \leq 1$

If: I = K

$ff^{\underline{II}} = 1$	$ff_{\underline{II}} = 1$

If: $I \neq K$

$\Vert ff^{IK} \Vert \leq 1$	$\Vert ff^{IK} \Vert \leq 1$

Inner products of bases:

$ff^{IK} = \underline{\Phi}^I \phi^{\underline{I}} . \underline{\Phi}^K \phi^{\underline{K}} = \phi^I \phi^K \Phi\Phi^{\underline{IK}} = \Phi\Phi^{\underline{IK}}$	$ff_{IK} = \underline{\Phi}_{\underline{I}} \phi_{\underline{I}} . \underline{\Phi}_{\underline{K}} \phi_{\underline{K}} = \phi_I \phi_K \Phi\Phi_{\underline{IK}} = \Phi\Phi_{IK}$

Since:

$\Phi\Phi^{II} = \cos 0 = 1$	$\Phi\Phi_{II} = \cos 0 = 1$

Then:

$1 \equiv \phi^I = \text{sqrt} \mid ff^{II} \mid = 1$	$1 \equiv \phi_I = \text{sqrt} \mid ff_{II} \mid = 1$
$\underline{f}^I . \underline{\Phi}^K = \underline{\Phi}^I \phi^{\underline{I}} . \underline{\Phi}^K = \Phi\Phi^{\underline{IK}}$	$\underline{f}_I . \underline{\Phi}_K = \underline{\Phi}_{\underline{I}} \phi_{\underline{I}} . \underline{\Phi}_{\underline{K}} = \Phi\Phi_{\underline{IK}}$

3.1.2.3.
Direction Cosines of General Unitary Bases and Fixed Bases:

$OF_m^{\ n} = \cos(\underline{o}_m, \underline{f}^n)$	$FO_n^{\ m} = \cos(\underline{f}_n, \underline{o}^m)$	$OF^n_{\ m} = \cos(\underline{o}^n, \underline{f}_m)$	$FO^m_{\ n} = \cos(\underline{f}^m, \underline{o}_n)$
$OF_m^{\ n}$	=	$OF^n_{\ m}$	
	$FO_n^{\ m}$	=	$FO^m_{\ n}$
$of_m^{\ n} = OF_m^{\ n} \phi^n = OF_m^{\ n}$	=	$fo^n_{\ m} = \phi^n OF^n_{\ m} = OF^n_{\ m}$	

$$\left| fo_n^{\ m} = \phi_n FO_n^{\ m} = FO_n^{\ m} \right| \quad = \quad \left| of^m_{\ n} = FO^m_{\ n} \phi_n = FO^m_{\ n} \right.$$

ff(-m,-n) ≡ bf(-m).bf(-n) = bo(-j) of(+j,-m). bo(-k) of(+k,-n) = of(+k,-m)of(+k,-n):
ff(-m,-n) = of(+k,-m) of(+k,-n)

m	ff(-m,-1) n = 1	ff(-m,-2) n = 2	ff(-m,-3) n = 3
1	of(+m,-1) of(+m,-1) = 1	of(+m,-1) of(+m,-2)	of(+m,-1) of(+m,-3)
2	of(+m,-2) of(+m,-1)	of(+m,-2) of(+m,-2) = 1	of(+m,-2) of(+m,-3)
3	of(+m,-3) of(+m,-1)	of(+m,-3) of(+m,-2)	of(+m,-3) of(+m,-3) = 1

:

Sum of product of column by column: ff(-m,-n) = of(+k,-m)of(+k,-n)

m	ff(-m,-1) n = 1	ff(-m,-2) n = 2	ff(-m,-3) n = 3
1	first * first columns sum	first * second columns Sum	first * third columns Sum
2	*Symmetric*	second * second columns sum	second * third columns sum
3	*Symmetric*	*Symmetric*	third * third columns sum

3.1.2.4.
Inner Products of General Unitary Bases and Fixed Bases:

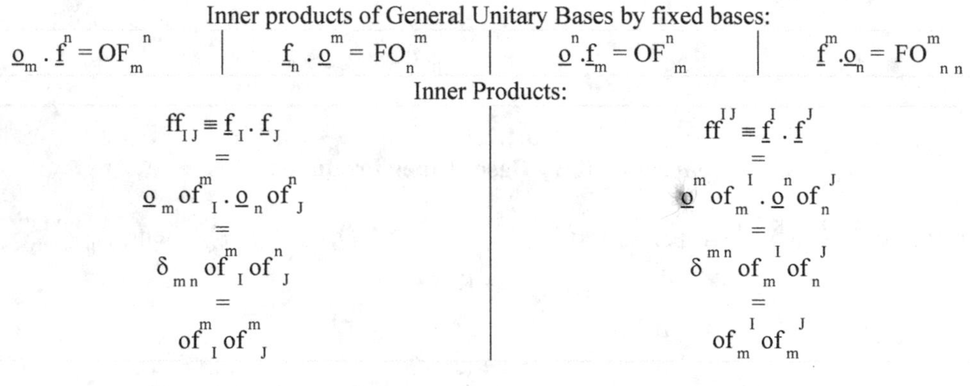

Inner products of General Unitary Bases by fixed bases:

$$\underline{o}_m \cdot \underline{f}^n = OF_m^{\ n} \quad \Big| \quad \underline{f}_n \cdot \underline{o}^m = FO_n^{\ m} \quad \Big| \quad \underline{o}^n \cdot \underline{f}_m = OF^n_{\ m} \quad \Big| \quad \underline{f}^m \cdot \underline{o}_n = FO^m_{\ n\,n}$$

Inner Products:

$$ff_{IJ} \equiv \underline{f}_I \cdot \underline{f}_J = \underline{o}_m of^m_{\ I} \cdot \underline{o}_n of^n_{\ J} = \delta_{mn} of^m_{\ I} of^n_{\ J} = of^m_{\ I} of^m_{\ J}$$

$$ff^{IJ} \equiv \underline{f}^I \cdot \underline{f}^J = \underline{o}^m of_m^{\ I} \cdot \underline{o}^n of_n^{\ J} = \delta^{mn} of_m^{\ I} of_n^{\ J} = of_m^{\ I} of_m^{\ J}$$

Sum of products of corresponding inner products:

$$ff_{IJ}\, ff^{IJ} = of^m_{\ I} of^m_{\ J} of_p^{\ I} of_p^{\ J} = of_p^{\ I} of^m_{\ I} of^m_{\ J} of_p^{\ J}$$

But:

$$of_p^{\ I} of^m_{\ I} = \delta_p^{\ m} \quad \Big| \quad of^m_{\ J} of_p^{\ J} = \delta^m_{\ p}$$

So:

$$ff_{IJ}\, ff^{IJ} = \delta_p^{\ m} \delta^m_{\ p} = 3$$

3.1.3.
Orthogonal Bases Direction Cosines and Inner Products:

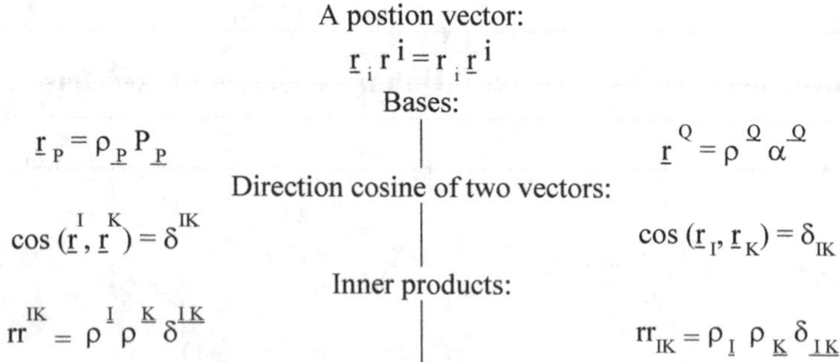

A postion vector:

$$\underline{r}_i r^i = r_i \underline{r}^i$$

Bases:

$$\underline{r}_P = \rho_{\underline{P}} P_{\underline{P}} \quad \Big| \quad \underline{r}^Q = \rho^{\underline{Q}} \alpha^{\underline{Q}}$$

Direction cosine of two vectors:

$$\cos(\underline{r}^I, \underline{r}^K) = \delta^{IK} \quad \Big| \quad \cos(\underline{r}_I, \underline{r}_K) = \delta_{IK}$$

Inner products:

$$rr^{IK} = \rho^{\underline{I}} \rho^{\underline{K}} \delta^{\underline{IK}} \quad \Big| \quad rr_{IK} = \rho_{\underline{I}} \rho_{\underline{K}} \delta_{\underline{IK}}$$

Bases are related to fixed coordinates:

$\underline{r}_I = ro_I^m\,\underline{o}_m = \underline{o}_m\,or_I^m$	$\underline{r}^J = ro_n^J\,\underline{o}^n = \underline{o}^n\,or_n^J$

3.1.3.1.
Orthogonal Bases Direction Cosines:

$\cos(\underline{r}^I, \underline{r}^K) \equiv \delta^{IK}$	$\cos(\underline{r}_I, \underline{r}_K) \equiv \delta_{IK}$
If: I = K	
$\delta^{\underline{II}} = 1$	$\delta_{\underline{II}} = 1$
If: I ≠ K	
$\delta^{IK} = 0$	$\delta_{IK} = 0$

3.1.3.2.
Orthogonal Bases Inner Products:

$rr^{IK} = \rho^{\underline{I}}\,\rho^{\underline{K}}\,\delta^{\underline{IK}}$	$rr_{IK} \equiv \rho_{\underline{I}}\,\rho_{\underline{K}}\,\delta_{\underline{IK}}$
If: I = K	
$rr^{\underline{II}} = \rho^I\,\rho^I$	$rr_{\underline{II}} = \rho_{\underline{I}}\,\rho_{\underline{I}}$
If: I ≠ K	
$rr^{IK} = 0$	$rr_{IK} = 0$

3.1.3.3.
Direction Cosines Of Orthogonal Bases and Fixed Bases:

$OR_m^n = \cos(\underline{o}_m, \underline{r}^n)$	$RO_n^m = \cos(\underline{r}_n, \underline{o}^m)$	$OR_m^n = \cos(\underline{o}^n, \underline{r}_m)$	$RO_n^m = \cos(\underline{r}^m, \underline{o}_n)$
OR_m^n	=	OR_m^n	
	RO_n^m	=	RO_n^m
$or_m^n = OR_m^n\,\rho^n$	=	$ro_m^n = \rho^n\,OR_m^n$	
	$ro_n^m = \rho_n\,RO_n^m$	=	$or_n^m = RO_n^m\,\rho_n$

rr(-m,-n) ≡ br(-m).br(-n) = bo(-j) or(+j,-m). bo(-k) or(+k,-n) = or(+k,-m)or(+k,-n):
rr(-m,-n) = or(+k,-m) or(+k,-n)

m	rr(-m,-1) n = 1	rr(-m,-2) n = 2	rr(-m,-3) n = 3
1	or(+m,-1) or(+m,-1)	or(+m,-1) or(+m,-2)	or(+m,-1) or(+m,-3)
2	or(+m,-2) or(+m,-1)	or(+m,-2) or(+m,-2)	or(+m,-2) or(+m,-3)
3	or(+m,-3) or(+m,-1)	or(+m,-3) or(+m,-2)	or(+m,-3) or(+m,-3)

Sum of product of column by column: xx(-m,-n) = ox(+k,-m) ox(+k,-n):

m	rr(-m,-1) n = 1	rr(-m,-2) n = 2	rr(-m,-3) n = 3
1	first * first sum = $(\rho^1)^2$	0	0
2	0	second * second sum = $(\rho^2)^2$	0
3	0	0	third * third sum = $(\rho^3)^2$

3.1.3.4.

Inner Products of Orthogonal Bases and Fixed Bases:

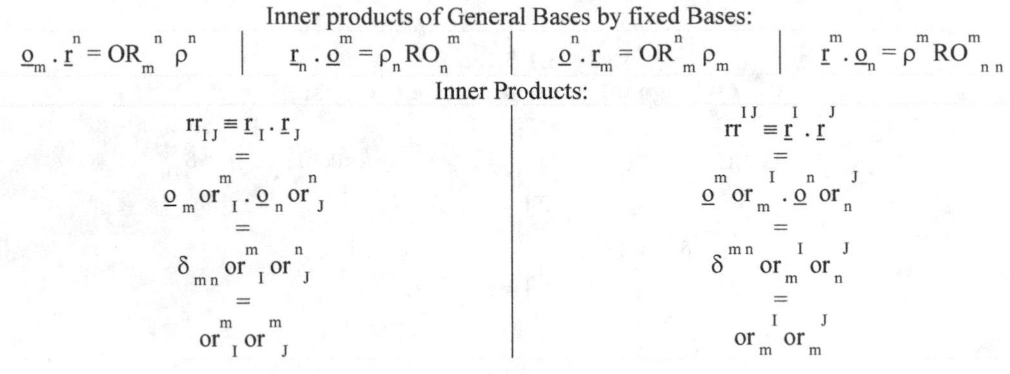

Inner products of General Bases by fixed Bases:

$\underline{o}_m \cdot \underline{r}^n = OR_m^{\ n}\, \rho^n$ | $\underline{r}_n \cdot \underline{o}^m = \rho_n RO_n^{\ m}$ | $\underline{o}^n \cdot \underline{r}_m = OR^n_{\ m}\, \rho_m$ | $\underline{r}^m \cdot \underline{o}_n = \rho^m RO^m_{\ n\,n}$

Inner Products:

$$rr_{IJ} \equiv \underline{r}_I \cdot \underline{r}_J = \underline{o}_m or^m_{\ I} \cdot \underline{o}_n or^n_{\ J} = \delta_{mn}\, or^m_{\ I}\, or^n_{\ J} = or^m_{\ I}\, or^m_{\ J}$$

$$rr^{IJ} \equiv \underline{r}^I \cdot \underline{r}^J = \underline{o}^m or_m^{\ I} \cdot \underline{o}^n or_n^{\ J} = \delta^{mn}\, or_m^{\ I}\, or_n^{\ J} = or_m^{\ I}\, or_m^{\ J}$$

Sum of products of corresponding inner products:

$$rr_{IJ}\, rr^{IJ} = or^m_{\ I}\, or^m_{\ J}\, or_p^{\ I}\, or_p^{\ J} = or_p^{\ I}\, or^m_{\ I}\, or^m_{\ J}\, or_p^{\ J}$$

But:

$or_p^{\ I}\, or^m_{\ I} = \delta^m_{\ p}$ | $or^m_{\ J}\, or_p^{\ J} = \delta^m_{\ p}$

So:

$$rr_{IJ}\, rr^{IJ} = \delta^m_{\ p}\, \delta^m_{\ p} = 3$$

3.1.4.
Orthonormal Bases Direction Cosines and Inner Products:

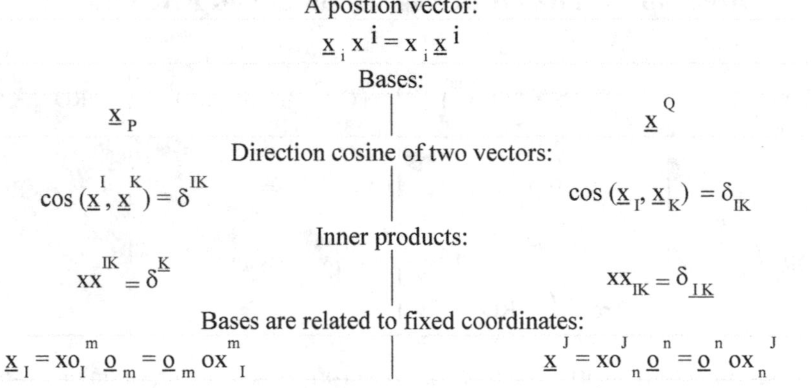

A postion vector:

$$\underline{x}_i\, x^i = x_i\, \underline{x}^i$$

Bases:

$\underline{x}_P$ | $\underline{x}^Q$

Direction cosine of two vectors:

$\cos(\underline{x}^I, \underline{x}^K) = \delta^{IK}$ | $\cos(\underline{x}_I, \underline{x}_K) = \delta_{IK}$

Inner products:

$xx^{IK} = \delta^{K}$ | $xx_{IK} = \delta_{IK}$

Bases are related to fixed coordinates:

$\underline{x}_I = xo_I^{\ m}\, \underline{o}_m = \underline{o}_m\, ox^m_{\ I}$ | $\underline{x}^J = xo^J_{\ n}\, \underline{o}^n = \underline{o}^n\, ox_n^{\ J}$

3.1.4.1.
Orthonormal Bases Direction Cosines:

$\cos(\underline{x}^I, \underline{x}^K) \equiv \delta^{IK}$ | $\cos(\underline{x}_I, \underline{x}_K) \equiv \delta_{IK}$

3.1.4.2.
Orthonormal Inner Products:

$xx^{IK} = \delta^{IK}$ | $xx_{IK} = \delta_{IK}$

3.1.4.3.
Direction Cosines of Orthonormal Bases and Fixed Bases:

$OX_m^{\ n} = \cos(\underline{o}_m, \underline{x}^n)$	$XO_n^{\ m} = \cos(\underline{x}_n, \underline{o}^m)$	$OX^n_{\ m} = \cos(\underline{o}^n, \underline{x}_m)$	$XO^m_{\ n} = \cos(\underline{x}^m, \underline{o}_n)$
$OX_m^{\ n}$	$=$	$OX^n_{\ m}$	
	$XO_n^{\ m}$	$=$	$XO^m_{\ n}$
$ox_m^{\ n\,n} = OX_m^{\ n}$	$=$	$xo^n_{\ m} = OX^n_{\ m}$	
	$xo_n^{\ m} = XO_n^{\ m}$	$=$	$ox^m_{\ n} = XO^m_{\ n}$

xx(-m,-n) ≡ bx(-m).bx(-n) = bo(-j) ox(+j,-m). bo(-k) ox(+k,-n) = ox(+k,-m)ox(+k,-n):
xx(-m,-n) = ox(+k,-m) ox(+k,-n)

	xx(-m,-1)	xx(-m,-2)	xx(-m,-3)
m	n = 1	n = 2	n = 3
1	ox(+m,-1) ox(+m,-1)	ox(+m,-1) ox(+m,-2)	ox(+m,-1) ox(+m,-3)
2	ox(+m,-2) ox(+m,-1)	ox(+m,-2) ox(+m,-2)	ox(+m,-2) ox(+m,-3)
3	ox(+m,-3) ox(+m,-1)	ox(+m,-3) ox(+m,-2)	ox(+m,-3) ox(+m,-3)

Sum of product of column by column: xx(-m,-n) = ox(+k,-m) ox(+k,-n):

	xx(-m,-1)	xx(-m,-2)	xx(-m,-3)
m	n = 1	n = 2	n = 3
1	first * first columns sum = 1	0	0
2	0	second*second columns sum = 1	0
3	0	0	third * third columns sum=1

3.1.4.4.
Inner Products of Orthonormal Bases and Fixed Bases:

Inner pxoducts of Genexal Bases by fixed Bases:

$\underline{o}_m \cdot \underline{x}^n = OX_m^{\ n}$ $\underline{x}_n \cdot \underline{o}^m = XO_n^{\ m}$ $\underline{o}^n . \underline{x}_m = OX^n_{\ m}$ $\underline{x}^m . \underline{o}_n = XO^m_{\ n\,n}$

Inner Products:

$xx_{IJ} \equiv \underline{x}_I \cdot \underline{x}_J$	$xx^{IJ} \equiv \underline{x}^I \cdot \underline{x}^J$
$=$	$=$
$\underline{o}_m\, ox^m_{\ I} \cdot \underline{o}_n\, ox^n_{\ J}$	$\underline{o}^m\, ox_m^{\ I} \cdot \underline{o}^n\, ox_n^{\ J}$
$=$	$=$
$\delta_{mn}\, ox^m_{\ I}\, ox^n_{\ J}$	$\delta^{mn}\, ox_m^{\ I}\, ox_n^{\ J}$
$=$	$=$
$ox^m_{\ I}\, ox^m_{\ J}$	$ox_m^{\ I}\, ox_m^{\ J}$

Sum of products of corresponding inner products:

$$xx_{IJ}\, xx^{IJ} = ox^m_{\ I}\, ox^m_{\ J}\, ox_p^{\ I}\, ox_p^{\ J} = ox_p^{\ I}\, ox^m_{\ I}\, ox^m_{\ J}\, ox_p^{\ J}$$

But:

$ox_p^{\ I}\, ox^m_{\ I} = \delta^m_{\ p}$ | $ox^m_{\ J}\, ox_p^{\ J} = \delta^m_{\ p}$

So:

$$xx_{IJ}\, xx^{IJ} = \delta^m_{\ p}\, \delta^m_{\ p} = 3$$

3.1.5.
Inner Product of Vectors:

u . v

$\underline{u} \cdot \underline{v} = metric^{ik}\, u_i\, v_k$	$\underline{u} \cdot \underline{v} = u^i\, v_i$ (conjugates)
$\underline{u} \cdot \underline{v} = u^i\, v_i$ (conjugates)	$\underline{u} \cdot \underline{v} = metric_{ik}\, u^i\, v^k$

Different coordinate systems

$\underline{a}^i u_i \cdot \underline{a}^k v_k = aa^{ik} u_i v_k$	$\underline{a}^i u_i \cdot \underline{a}_k v^k = u_i v^k \delta_i{}^k = u_i v^i$
$\underline{a}_i u^i \cdot \underline{a}^k v_k = u^i v_k \delta_i{}^k = u^i v_i$	$\underline{a}_i u^i \cdot \underline{a}_k v^k = aa_{ik} u^i v^k$
$\underline{f}^i u_i \cdot \underline{f}^k v_k = ff^{ik} u_i v_k$	$\underline{f}^i u_i \cdot \underline{f}_k v^k = u_i v^k \delta_i{}^k = u_i v^i$
$\underline{f}_i u^i \cdot \underline{f}^k v_k = u^i v_k \delta_i{}^k = u^i v_i$	$\underline{f}_i u^i \cdot \underline{f}_k v^k = ff_{ik} u^i v^k$
$\underline{r}^i u_i \cdot \underline{r}^k v_k = rr^{ik} u_i v_k$	$\underline{r}^i u_i \cdot \underline{r}_k v^k = u_i v^k \delta_i{}^k = u_i v^i$
$\underline{r}_i u^i \cdot \underline{r}^k v_k = u^i v_k \delta_i{}^k = u^i v_i$	$\underline{r}_i u^i \cdot \underline{r}_k v^k = rr_{ik} u^i v^k$
$\underline{x}^i u_i \cdot \underline{x}^k v_k = xx^{ik} u_i v_k = 1\, u_k v_k$	$\underline{x}^i u_i \cdot \underline{x}_k v^k = u_i v^k \delta_i{}^k = u_i v^i$
$\underline{x}_i u^i \cdot \underline{x}^k v_k = u^i v_k \delta_i{}^k = u^i v_i$	$\underline{x}_i u^i \cdot \underline{x}_k v^k = xx_{ik} u^i v^k = 1\, u^k v^k$

3.2. Metrics:

3.2.1. Bases Inner Products are Metrics:

Bases Inner Products are Metrics of Coordinate Bases:
aaMK = aM . aK :

Two conjugate Bases Metrics

$$aa_K{}^M = \underline{a}_K \cdot \underline{a}^M = \alpha_{\underline{K}} \alpha^{\underline{M}} \delta_{\underline{K}}{}^{\underline{M}} = AA_K{}^M = \delta_K{}^M$$

$$aa^M{}_K = \underline{a}^M \cdot \underline{a}_K = \alpha_{\underline{M}} \alpha^{\underline{K}} \delta^{\underline{K}}{}_{\underline{M}} = AA_M{}^K = \delta_M{}^K$$

If: M = K

$$aa_K{}^M = \underline{a}_K \cdot \underline{a}^M = 1 \qquad aa^M{}_{\underline{M}} = \underline{a}^M \cdot \underline{a}_{\underline{M}} = 1$$

If: M ≠ K

$$aa_K{}^M = \underline{a}_K \cdot \underline{a}^M = 0 \qquad aa^M{}_K = \underline{a}^M \cdot \underline{a}_K = 0$$

A Metric of a single *General* basis is *Square of its norm, other* Metrics:

$$\| aa_{MK} \| = \alpha_M \alpha_K \, \| AA_{MK} \| \leq \alpha_M \alpha_K \qquad \| aa^{MK} \| = \alpha^M \alpha^K \, \| AA^{MK} \| \leq \alpha^M \alpha^K$$

Same System:
Two Bases Inner Products of a system are Metrics:

$aa_{PQ} \equiv \underline{a}_P \cdot \underline{a}_Q = oa^m{}_P\, oa^m{}_Q$	$aa^{PQ} \equiv \underline{a}^P \cdot \underline{a}^Q = oa_m{}^P\, oa_m{}^Q$
$ff_{\underline{P}\underline{P}} \equiv \underline{f}_P \cdot \underline{f}_P = of^m{}_P\, of^m{}_{\underline{P}}$	$ff^{PP} \equiv \underline{f}^{\underline{P}} \cdot \underline{f}^{\underline{P}} = of_{\underline{P}}{}^m\, of^m{}_{\underline{P}}$
$rr_{\underline{P}\underline{P}} \equiv \underline{r}_P \cdot \underline{r}_P = or^m{}_{\underline{P}}\, or^m{}_{\underline{P}}$	$rr^{PP} \equiv \underline{r}^{\underline{P}} \cdot \underline{r}^{\underline{P}} = or^m{}_{\underline{P}}\, or^m{}_{\underline{P}}$
$xx_{\underline{P}\underline{P}} \equiv \underline{x}_P \cdot \underline{x}_P = ox^m{}_{\underline{P}}\, ox^m{}_P$	$xx^{PP} \equiv \underline{x}^{\underline{P}} \cdot \underline{x}^{\underline{P}} = ox^m{}_{\underline{P}}\, ox^m{}_P$

FORTRAN Algorithmic Forms:

aa(-i,-j) ≡ oa(+n, -i) * oa(+n, -j)	aa(+i,+j) ≡ oa(-m, +i) * oa(-m,+j)
ff(-i,-j) ≡ of(+n, -i) * of(+n, -j)	ff(+i,+j) ≡ of(-m, +i) * of(-m,+j)
rr(-i,-j) ≡ or(+n, -i) * or(+n, -j)	rr(+i,+j) ≡ or(-m, +i) * or(-m,+j)
xx(-i,-j) ≡ ox(+n, -i) * ox(+n, -j)	xx(+i,+j) ≡ ox(-m, +i) * ox(-m,+j)

General $\underline{a}$	Unitary Bases $\underline{f}$	Orthogonal $\underline{r}$	Orthonormal $\underline{x}$
If: I = K			
$aa^{II} = (\alpha^I)^2$	$ff^{II} = 1$	$rr^{II} = (\rho^I)^2$	$xx^{II} = 1$
$aa_{II} = (\alpha_I)^2$	$ff_{II} = 1$	$rr_{II} = (\rho_I)^2$	$xx_{II} = 1$
If: I ≠ K			
$\lVert aa^{IK} \rVert \leq \alpha^I \alpha^K$	$\lVert ff^{IK} \rVert \leq 1$	$rr^{IK} = 0$	$xx^{IK} = 0$
$\lVert aa_{IK} \rVert \leq \alpha_I \alpha_K$	$\lVert ff_{IK} \rVert \leq 1$	$rr_{IK} = 0$	$xx_{IK} = 0$

$$(dC\ dC) = (do^k\ do^k)$$

=

da(+m) * ∂o(+k)/∂a(+m) * da(+n) * ∂o(+k) /∂a(+n)

=

da(+m) * da(+n) * aa(-m,-n)

Where:

aa(-m,-n)

=

oaoa(-m,-n)

≡

oa(+k,-m) oa(+k,-n)

and:

aa(+m,+n) * aa(-k,-n) = δ(+m,-k)

aa(+m,+n) * aa(-k,-n) = aaaa(+m,-k)

aa(+m,+n) is the cofactor of aa(-k,-n) and vice-versa

aa[-m,-n] ≡ ||aa(-m,-n)|| = ||oa(+k,-m)|| ||oa(+k,-n)|| = || oa(+k,-m)||**2

So:

aa[-m,-n] is of weight 2

Transformation of (w) weight : qa(+k,-j) = oa[+s,-r]**w * ao(+k,-n) * qo(+n,-m) * oa(+m,-j)

Bases-Post-Transformer (pre-tensor)		Differential-Pre-Transformer (post-tensor)
$T(b)^b = (ba)^b{}_a\, T(a)^a$	. \ | \	$T(a)_a = T(b)_b\ (ba)^b{}_a$

fo(-j) = X[+s-r]**t fx(-m) X(+m - j)

fx(-j) = O[+s-r]**t fo(- m) O(+m -j)

$dg^n = (gf^n{}_m)^t\ df^m$	$df^m = (fg^m{}_n)^t\ dg^n$
$dg_n = (gf_n{}^m)^t\ df_m$	$df_m = (fg_m{}^n)^t\ dg_n$

Coordinates Metrics:

$$da_m\, da^m = do_i\, oa^i{}_m\, do^j\, oa_j{}^m = do_i\ \delta^i{}_j\, do^j = do_i\, do^i$$

=

$$do^j\, do_j = da^n\, ao_n{}^j\, da_m\, ao^m{}_j = da^n\ \delta_n{}^m\, da_m = da^n\, da_n$$

Inner Product of Bases

$\underline{a}^j \cdot \underline{a}^n$	$\underline{a}^j \cdot \underline{a}_n$
=	=
$\alpha^j \alpha^n \cos(j,n)$	$ao^j{}_i \underline{o}^i \cdot ao_n{}^m \underline{o}_m$
=	=
$ao^j{}_i \underline{o}^i \cdot ao^n{}_m \underline{o}^m$	$ao^j{}_i\ ao_n{}^m\ \delta^i{}_m$
=	=
$ao^j{}_i\ ao^n{}_m\ \delta^{im}$	$ao^j{}_i\ ao_n{}^i$
=	=
$ao^j{}_i\ ao^{ni} = ao^j{}_i\ ao^n{}_i$	$\delta^j{}_n$
$\underline{a}_j \cdot \underline{a}^n$	$\underline{a}_j \cdot \underline{a}_n$
=	=
$ao_j{}^i \underline{o}_i \cdot ao^n{}_m \underline{o}^m$	$\alpha_j\ \alpha_n\ \cos(\underline{a}_j,\underline{a}_n)$
=	=
$ao_j{}^i\ ao^n{}_m\ \delta_i{}^m$	$ao_j{}^i \underline{o}_i \cdot ao_n{}^m \underline{o}_m$
=	=
$ao_j{}^i\ ao^n{}_I$	$ao_j{}^i\ ao_n{}^m\ \delta_{im}$
=	=
$\delta_j{}^n$	$ao_j{}^i\ ao_n{}^i = ao_j{}^i\ ao_n{}^i$

$$\begin{bmatrix} ao^1{}_1 & ao^1{}_2 & ao^1{}_3 \\ ao^2{}_1 & ao^2{}_2 & ao^2{}_3 \\ ao^3{}_1 & ao^3{}_2 & ao^3{}_3 \end{bmatrix} \ast_{\text{Row By Row}} \begin{bmatrix} ao^1{}_1 & ao^1{}_2 & ao^1{}_3 \\ ao^2{}_1 & ao^2{}_2 & ao^2{}_3 \\ ao^3{}_1 & ao^3{}_2 & ao^3{}_3 \end{bmatrix} = \begin{bmatrix} aa^{11} & aa^{12} & aa^{13} \\ aa^{21} & aa^{22} & aa^{23} \\ aa^{31} & aa^{32} & aa^{33} \end{bmatrix}$$

$$\begin{bmatrix} ao^1{}_1 & ao^1{}_2 & ao^1{}_3 \\ ao^2{}_1 & ao^2{}_2 & ao^2{}_3 \\ ao^3{}_1 & ao^3{}_2 & ao^3{}_3 \end{bmatrix} \ast_{\text{Row By Row}} \begin{bmatrix} ao_1{}^1 & ao_1{}^2 & ao_1{}^3 \\ ao_2{}^1 & ao_2{}^2 & ao_2{}^3 \\ ao_3{}^1 & ao_3{}^2 & ao_3{}^3 \end{bmatrix} = \begin{bmatrix} aa^1{}_1 & aa^1{}_2 & aa^1{}_3 \\ aa^2{}_1 & aa^2{}_2 & aa^2{}_3 \\ aa^3{}_1 & aa^3{}_2 & aa^3{}_3 \end{bmatrix}$$

$$\begin{bmatrix} ao_1{}^1 & ao_1{}^2 & ao_1{}^3 \\ ao_2{}^1 & ao_2{}^2 & ao_2{}^3 \\ ao_3{}^1 & ao_3{}^2 & ao_3{}^3 \end{bmatrix} \ast_{\text{Row By Row}} \begin{bmatrix} ao^1{}_1 & ao^1{}_2 & ao^1{}_3 \\ ao^2{}_1 & ao^2{}_2 & ao^2{}_3 \\ ao^3{}_1 & ao^3{}_2 & ao^3{}_3 \end{bmatrix} = \begin{bmatrix} aa_1{}^1 & aa_1{}^2 & aa_1{}^3 \\ aa_2{}^1 & aa_2{}^2 & aa_2{}^3 \\ aa_3{}^1 & aa_3{}^2 & aa_3{}^3 \end{bmatrix}$$

$$\begin{bmatrix} ao_1{}^1 & ao_1{}^2 & ao_1{}^3 \\ ao_2{}^1 & ao_2{}^2 & ao_2{}^3 \\ ao_3{}^1 & ao_3{}^2 & ao_3{}^3 \end{bmatrix} \ast_{\text{Row By Row}} \begin{bmatrix} ao_1{}^1 & ao_1{}^2 & ao_1{}^3 \\ ao_2{}^1 & ao_2{}^2 & ao_2{}^3 \\ ao_3{}^1 & ao_3{}^2 & ao_3{}^3 \end{bmatrix} = \begin{bmatrix} aa_{11} & aa_{12} & aa_{13} \\ aa_{21} & aa_{22} & aa_{23} \\ aa_{31} & aa_{32} & aa_{33} \end{bmatrix}$$

3.2.1.1. General Bases Metrics:

:

$aaIK = \underline{a}I \cdot \underline{a}K$:

$\Vert aa_{IK} \Vert = \alpha_I \alpha_K\ \Vert AA_{IK} \Vert \leq \alpha_I \alpha_K$	$\Vert aa^{IK} \Vert = \alpha^I \alpha^K\ \Vert AA^{IK} \Vert\ \leq \alpha^I \alpha^K$

If: I = K :

$aa_{II} = (\alpha_I)^2$	$aa^{II} = (\alpha^I)^2$

If: $I \neq K$

$\Vert aa_{IK} \Vert \leq \alpha_I \alpha_K$	$\Vert aa^{IK} \Vert\ \leq \alpha^I \alpha^K$

Sum of products of corresponding Metrics:

$$aa_{IJ}\, aa^{IJ} = oa^{m}_{I}\, oa^{m}_{J}\, oa^{I}_{p}\, oa^{J}_{p} = oa^{I}_{p}\, oa^{m}_{I}\, oa^{m}_{J}\, oa^{J}_{p}$$

But:

$$oa^{I}_{p}\, oa^{m}_{I} = \delta^{m}_{p} \qquad oa^{m}_{J}\, oa^{J}_{p} = \delta^{m}_{p}$$

So:

$$aa_{IJ}\, aa^{IJ} = \delta^{m}_{p}\, \delta^{m}_{p} = 3$$

3.2.1.2.
General Unitary Bases Metrics:

:

$ffIK = \underline{f}I \,.\, \underline{f}K$:

$$\| ff_{IK} \| = \| \Phi\Phi_{IK} \| \leq 1 \qquad \| ff^{IK} \| = \| \Phi\Phi^{IK} \| \leq 1$$

If: I = K :

$$ff_{\underline{II}} = 1 \qquad ff^{\underline{II}} = 1$$

If: $I \neq K$

$$\| ff_{IK} \| \leq 1 \qquad \| ff^{IK} \| \leq 1$$

Sum of products of corresponding Metrics:

$$ff_{IJ}\, ff^{IJ} = of^{m}_{I}\, of^{m}_{J}\, of^{I}_{p}\, of^{J}_{p} = of^{I}_{p}\, of^{m}_{I}\, of^{m}_{J}\, of^{J}_{p}$$

But:

$$of^{I}_{p}\, of^{m}_{I} = \delta^{m}_{p} \qquad of^{m}_{J}\, of^{J}_{p} = \delta^{m}_{p}$$

So:

$$ff_{IJ}\, ff^{IJ} = \delta^{m}_{p}\, \delta^{m}_{p} = 3$$

3.2.1.3.
Orthogonal Bases Metrics:

Orthogonal Metrics Changes from General:

$$\| aa_{IK} \| = \alpha_I \alpha_K \; \| AA_{IK} \| \leq \alpha_I \alpha_K \qquad \| aa^{IK} \| = \alpha^I \alpha^K \; \| AA^{IK} \| \leq \alpha^I \alpha^K$$

Changed to | Changed to

$$r_{IK} = \rho_I\, \rho_K\; \delta_{\underline{IK}} \qquad rr_{IK} = \rho_I\, \rho_K\; \delta_{\underline{IK}}$$

$rr\ IK = \underline{r}I \,.\, \underline{r}K$:

$$rr_{IK} = \rho_I\, \rho_K\; \delta_{\underline{IK}} \qquad rr^{IK} = \rho^I \rho^K\; \delta^{\underline{IK}}$$

If: I = K:

$$\text{No change } rr_{\underline{II}} = (\rho_I)^2 \qquad \text{No change } rr^{\underline{II}} = (\rho^I)^2$$

If: $I \neq K$:

$$rr_{IK} = 0 \qquad rr^{IK} = 0$$

3.2.1.4.
Orthonormal Bases Metrics:

Orthonormal Metrics Changes from General:

$xxIK = \underline{x}I \,.\, \underline{x}K$:

$$Changed\ xx_{IK} = \delta_{IK} \qquad Changed\ xx^{IK} = \delta^{IK}$$

	If: I = K	
Changed $xx_{II} = 1$		*Changed* $xx^{II} = 1$
	If: I ≠ K	
$xx_{IK} = 0$		$xx^{IK} = 0$

3.2.2.
The Squared Differential Distance and Partial Derivatives:

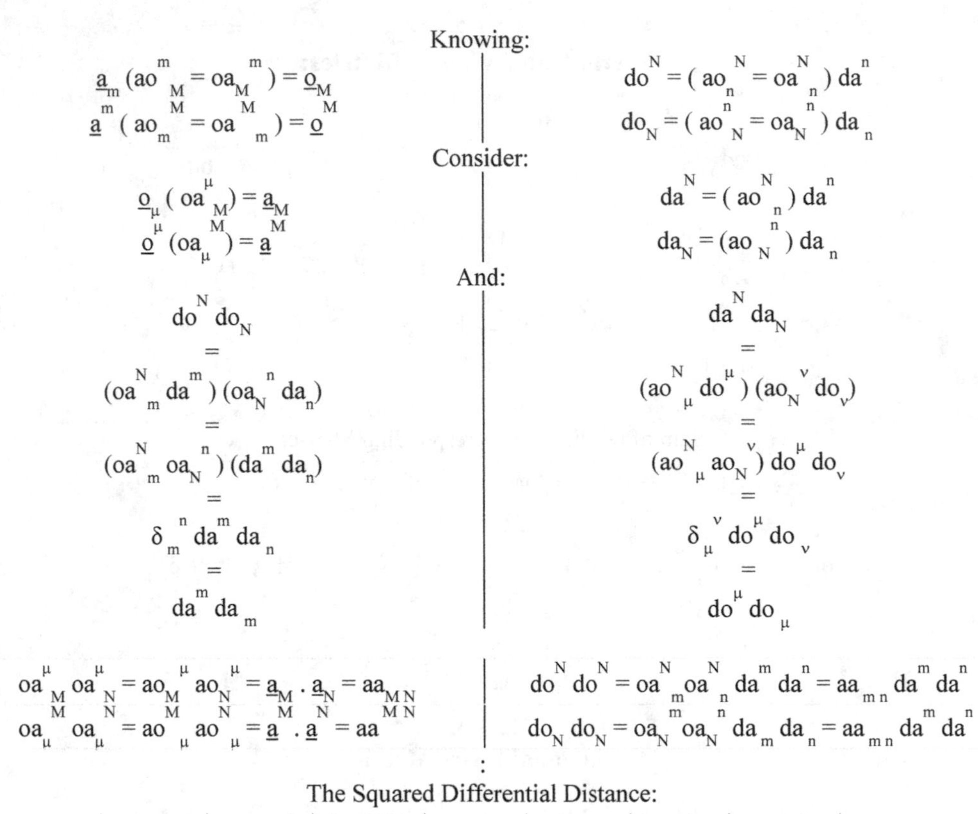

The Squared Differential Distance:

$$da^j . da_j = df^j . df_j = dr^j . dr_j = dx^j dx_j = do^i do_i = dy^j dy_j = ds^j . ds_j = db^j . db_j$$

:

Apply:

$$da^j da_j = db^i db_i$$

To Systems:

	General b^N	General Unitary Bases g^N	Orthogonal s^N	Orthonormal y^N	Fixed o^N
a^M	$da_M da^M = db_N db^N$	$da_M da^M = dg_N dg^N$	$da_M da^M = ds_N ds^N$	$da_M da^M = dy_N dy^N$	$da_M da^M = do_N do^N$
f^M	$df_M df^M = db_N db^N$	$df_M df^M = dg_N dg^N$	$df_M df^M = ds_N ds^N$	$df_M df^M = dy_N dy^N$	$df_M df^M = do_N do^N$
r^M	$dr_M dr^M = db_N db^N$	$dr_M dr^M = dg_N dg^N$	$dr_M dr^M = ds_N ds^N$	$dr_M dr^M = dy_N dy^N$	$dr_M dr^M = do_N do^N$
x^M	$dx_M dx^M = db_N db^N$	$dx_M dx^M = dg_N dg^N$	$dx_M dx^M = ds_N ds^N$	$dx_M dx^M = dy_N dy^N$	$dx_M dx^M = do_N do^N$
o^M	$do_M do^M = db_N db^N$	$do_M do^M = dg_N dg^N$	$do_M do^M = ds_N ds^N$	$do_M do^M = dy_N dy^N$	$do_M do^M = do_N do^N$

Partial derivatives:

$ab^{MN} \equiv \partial a^M / \partial b_N$	$ab^M{}_N \equiv \partial a^M / \partial b^N$	$\partial b_N / \partial a_M \equiv ba_N{}^M$	$\partial b^N / \partial a_M \equiv ba^{NM}$

$ab_M^{\ N} \equiv \partial a_M/\partial b_N$	$ab_{MN} \equiv \partial a_M/\partial b^N$	$\partial b_N/\partial a^M \equiv ba_{NM}$	$\partial b^N/\partial a^M \equiv ba^N_{\ M}$

Orthonormality:

baNM δMm abmn = δNn	abMN δNn banm = δMm

abMN ≡ aM.bN = ∂aM/∂bN = αM*ABMN*βN = βN*BANM*αM = ∂bN/∂aM = bN.aM ≡ baNM

Apply:
To Systems:

	General b^N	General Unitary Bases g^N	Orthogonal s^N	Orthonormal y^N	Fixed o^N
a^M	$ab_M^{\ N} \equiv \underline{a}_M \cdot \underline{b}^N$ = $\alpha_{\underline{M}}\ AB_{\underline{M}}^{\ \underline{N}}\ \beta^{\underline{N}}$ = $\partial a_M/\partial b_N$ = $\partial b^N/\partial a^M$ = $\beta^{\underline{N}}\ BA^{\underline{N}}_{\ \underline{M}}\ \alpha_{\underline{M}}$ = $\underline{b}^N \cdot \underline{a}_M \equiv ba^N_{\ M}$	$ag_M^{\ N} \equiv \underline{a}_M \cdot \underline{g}^N$ = $\alpha_{\underline{M}}\ A\Gamma_{\underline{M}}^{\ \underline{N}}\ 1$ = $\partial a_M/\partial g_N$ = $\partial g^N/\partial a^M$ = $1\ \Phi A^{\underline{N}}_{\ \underline{M}}\ \alpha_{\underline{M}}$ = $\underline{g}^N \cdot \underline{a}_M \equiv ga^N_{\ M}$	$as_M^{\ N} \equiv \underline{a}_M \cdot \underline{s}^N$ = $\alpha_{\underline{M}}\ A\Sigma_{\underline{M}}^{\ \underline{N}}\ \sigma^{\underline{N}}$ = $\partial a_M/\partial s_N$ = $\partial s^N/\partial a^M$ = $\sigma^{\underline{N}}\ \Sigma A^{\underline{N}}_{\ \underline{M}}\ \alpha_{\underline{M}}$ = $\underline{s}^N \cdot \underline{a}_M \equiv sa^N_{\ M}$	$ay_M^{\ N} \equiv \underline{a}_M \cdot \underline{y}^N$ = $\alpha_{\underline{M}}\ A\Psi_{\underline{M}}^{\ \underline{N}}\ 1$ = $\partial a_M/\partial y_N$ = $\partial y^N/\partial a^M$ = $1\ \Psi A^{\underline{N}}_{\ \underline{M}}\ \alpha_{\underline{M}}$ = $\underline{y}^N \cdot \underline{a}_M \equiv ya^N_{\ M}$	$ao_M^{\ N} \equiv \underline{a}_M \cdot \underline{o}^N$ = $\alpha_{\underline{M}}\ AO_{\underline{M}}^{\ \underline{N}}\ 1$ = $\partial a_M/\partial o_N$ = $\partial o^N/\partial a^M$ = $1\ OA^{\underline{N}}_{\ \underline{M}}\ \alpha_{\underline{M}}$ = $\underline{o}^N \cdot \underline{a}_M \equiv oa^N_{\ M}$
f^M	$fb_M^{\ N} \equiv \underline{f}_M \cdot \underline{b}^N$ = $1\ \Phi B_{\underline{M}}^{\ \underline{N}}\ \beta^{\underline{N}}$ = $\partial f_M/\partial b_N$ = $\partial b^N/\partial f^M$ = $\beta^{\underline{N}}\ \Phi B^{\underline{N}}_{\ \underline{M}}\ 1$ = $\underline{b}^N \cdot \underline{f}_M \equiv bf^N_{\ M}$	$fg_M^{\ N} \equiv \underline{f}_M \cdot \underline{g}^N$ = $1\ \Phi\Gamma_{\underline{M}}^{\ \underline{N}}\ 1$ = $\partial f_M/\partial g_N$ = $\partial g^N/\partial f^M$ = $1\ \Phi\Phi^{\underline{N}}_{\ \underline{M}}\ 1$ = $\underline{g}^N \cdot \underline{f}_M \equiv gf^N_{\ M}$	$fs_M^{\ N} \equiv \underline{f}_M \cdot \underline{s}^N$ = $1\ \Phi\Sigma_{\underline{M}}^{\ \underline{N}}\ \sigma^{\underline{N}}$ = $\partial f_M/\partial s_N$ = $\partial s^N/\partial f^M$ = $\sigma^{\underline{N}}\ \Phi\Sigma^{\underline{N}}_{\ \underline{M}}\ 1$ = $\underline{s}^N \cdot \underline{f}_M \equiv sf^N_{\ M}$	$fy_M^{\ N} \equiv \underline{f}_M \cdot \underline{y}^N$ = $1\ \Phi\Psi_{\underline{M}}^{\ \underline{N}}\ 1$ = $\partial f_M/\partial y_N$ = $\partial y^N/\partial f^M$ = $1\ \Phi\Psi^{\underline{N}}_{\ \underline{M}}\ 1$ = $\underline{y}^N \cdot \underline{f}_M \equiv yf^N_{\ M}$	$fo_M^{\ N} \equiv \underline{f}_M \cdot \underline{o}^N$ = $1\ \Phi O_{\underline{M}}^{\ \underline{N}}\ 1$ = $\partial f_M/\partial o_N$ = $\partial o^N/\partial f^M$ = $1\ \Phi O^{\underline{N}}_{\ \underline{M}}\ 1$ = $\underline{o}^N \cdot \underline{f}_M \equiv of^N_{\ M}$
r^M	$rb_M^{\ N} \equiv \underline{r}_M \cdot \underline{b}^N$ = $\rho_{\underline{M}}\ PB_{\underline{M}}^{\ \underline{N}}\ \beta^{\underline{N}}$ = $\partial r_M/\partial b_N$ = $\partial b^N/\partial r^M$ = $\beta^{\underline{N}}\ PB^{\underline{N}}_{\ \underline{M}}\ \rho_{\underline{M}}$ = $\underline{b}^N \cdot \underline{r}_M \equiv br^N_{\ M}$	$rg_M^{\ N} \equiv \underline{r}_M \cdot \underline{g}^N$ = $\rho_{\underline{M}}\ P\Gamma_{\underline{M}}^{\ \underline{N}}\ 1$ = $\partial r_M/\partial g_N$ = $\partial g^N/\partial r^M$ = $1\ P\Phi^{\underline{N}}_{\ \underline{M}}\ \rho_{\underline{M}}$ = $\underline{g}^N \cdot \underline{r}_M \equiv gr^N_{\ M}$	$rs_M^{\ N} \equiv \underline{r}_M \cdot \underline{s}^N$ = $\rho_{\underline{M}}\ P\Sigma_{\underline{M}}^{\ \underline{N}}\ \sigma^{\underline{N}}$ = $\partial r_M/\partial s_N$ = $\partial s^N/\partial r^M$ = $\sigma^{\underline{N}}\ P\Sigma^{\underline{N}}_{\ \underline{M}}\ \rho_{\underline{M}}$ = $\underline{s}^N \cdot \underline{r}_M \equiv sr^N_{\ M}$	$ry_M^{\ N} \equiv \underline{r}_M \cdot \underline{y}^N$ = $\rho_{\underline{M}}\ P\Psi_{\underline{M}}^{\ \underline{N}}\ 1$ = $\partial r_M/\partial y_N$ = $\partial y^N/\partial r^M$ = $1\ P\Psi^{\underline{N}}_{\ \underline{M}}\ \rho_{\underline{M}}$ = $\underline{y}^N \cdot \underline{r}_M \equiv yr^N_{\ M}$	$ro_M^{\ N} \equiv \underline{r}_M \cdot \underline{o}^N$ = $\rho_{\underline{M}}\ PO_{\underline{M}}^{\ \underline{N}}\ 1$ = $\partial r_M/\partial o_N$ = $\partial o^N/\partial r^M$ = $1\ PO^{\underline{N}}_{\ \underline{M}}\ \rho_{\underline{M}}$ = $\underline{o}^N \cdot \underline{r}_M \equiv or^N_{\ M}$

x^M	$xb_M{}^N \equiv \underline{x}_M \cdot \underline{b}^N$ = $1\ \Xi B_{\underline{M}}{}^{\underline{N}}\ \beta^{\underline{N}}$ = $\partial x_M/\partial b_N$ = $\partial b^N/\partial x^M$ = $\beta^{\underline{N}}\ \Xi B^{\underline{N}}{}_{\underline{M}}$ = $\underline{b}^N \cdot \underline{x}_M \equiv bx^N{}_M$	$xg_M{}^N \equiv \underline{x}_M \cdot \underline{g}^N$ = $1\ \Xi\Gamma_{\underline{M}}{}^{N}\ 1$ = $\partial x_M/\partial g_N$ = $\partial g^N/\partial x^M$ = $1\ \Xi\Phi^{\underline{N}}{}_{\underline{M}}\ 1$ = $\underline{g}^N \cdot \underline{x}_M \equiv gx^N{}_M$	$xs_M{}^N \equiv \underline{x}_M \cdot \underline{s}^N$ = $1\ \Xi\Sigma_{\underline{M}}{}^{\underline{N}}\ \sigma^{\underline{N}}$ = $\partial x_M/\partial s_N$ = $\partial s^N/\partial x^M$ = $\sigma^{\underline{N}}\ \Xi\Sigma^{\underline{N}}{}_{\underline{M}}\ 1$ = $\underline{s}^N \cdot \underline{x}_M \equiv sx^N{}_M$	$xy_M{}^N \equiv \underline{x}_M \cdot \underline{y}^N$ = $1\ \Xi\Psi_{\underline{M}}{}^{\underline{N}}\ 1$ = $\partial x_M/\partial y_N$ = $\partial y^N/\partial x^M$ = $1\ \Xi\Psi^{\underline{N}}{}_{\underline{M}}\ 1$ = $\underline{y}^N \cdot \underline{x}_M \equiv yx^N{}_M$	$xo_M{}^N \equiv \underline{x}_M \cdot \underline{o}^N$ = $1\ \Xi O_{\underline{M}}{}^{\underline{N}}\ 1$ = $\partial x_M/\partial o_N$ = $\partial o^N/\partial x^M$ = $1\ \Xi O^{\underline{N}}{}_{\underline{M}}\ 1$ = $\underline{o}^N \cdot \underline{x}_M \equiv ox^N{}_M$
o^M	$ob_M{}^N \equiv \underline{o}_M \cdot \underline{b}^N$ = $OB_{\underline{M}}{}^{\underline{N}}\ \beta^{\underline{N}}$ = $\partial o_M/\partial b_N$ = $\partial b^N/\partial o^M$ = $\beta^{\underline{N}}\ OB^{\underline{N}}{}_{\underline{M}}$ = $\underline{b}^N \cdot \underline{o}_M \equiv bo^N{}_M$	$og_M{}^N \equiv \underline{o}_M \cdot \underline{g}^N$ = $1\ O\Gamma_{\underline{M}}{}^{N}\ 1$ = $\partial o_M/\partial g_N$ = $\partial g^N/\partial o^M$ = $1\ O\Phi^{\underline{N}}{}_{\underline{M}}\ 1$ = $\underline{g}^N \cdot \underline{o}_M \equiv go^N{}_M$	$os_M{}^N \equiv \underline{o}_M \cdot \underline{s}^N$ = $1\ O\Sigma_{\underline{M}}{}^{\underline{N}}\ \sigma^{\underline{N}}$ = $\partial o_M/\partial s_N$ = $\partial s^N/\partial o^M$ = $\sigma^{\underline{N}}\ O\Sigma^{\underline{N}}{}_{\underline{M}}$ = $\underline{s}^N \cdot \underline{o}_M \equiv so^N{}_M$	$oy_M{}^N \equiv \underline{o}_M \cdot \underline{y}^N$ = $1\ O\Psi_{\underline{M}}{}^{\underline{N}}\ 1$ = $\partial o_M/\partial y_N$ = $\partial y^N/\partial o^M$ = $1\ O\Psi^{\underline{N}}{}_{\underline{M}}\ 1$ = $\underline{y}^N \cdot \underline{o}_M \equiv yo^N{}_M$	$oo_M{}^N \equiv \underline{o}_M \cdot \underline{o}^N$ = $1\ OO_{\underline{M}}{}^{\underline{N}}\ 1$ = $\partial o_M/\partial o_N$ = $\partial o^N/\partial o^M$ = $1\ OO^{\underline{N}}{}_{\underline{M}}\ 1$ = $\underline{o}^N \cdot \underline{o}_M \equiv oo^N{}_M$

Taking into Consideration:

$$\alpha^{\underline{M}}\alpha_{\underline{M}} = \beta^{\underline{N}}\beta_{\underline{N}} = \rho^{\underline{N}}\rho_{\underline{N}} = \sigma^{\underline{N}}\sigma_{\underline{N}} = \xi^{\underline{N}}\xi_{\underline{N}} = \psi^{\underline{N}}\psi_{\underline{N}} = o^{\underline{N}}o_{\underline{N}} = 1$$

And pre-multiplying and post- multiplying by applicable norms to get direction cosines for all systems:

Direction Cosines For System:

	General b^N	General Unitary Bases g^N	Orthogonal s^N	Orthonormal y^N	Fixed o^N
a^M	$\alpha^{\underline{M}}\ ab_M{}^N\ \beta_{\underline{N}}$ = $AB_{\underline{M}}{}^{\underline{N}} = BA^{\underline{N}}{}_{\underline{M}}$ = $\beta_{\underline{N}}\ ba^N{}_M\ \alpha^{\underline{M}}$	$\alpha^{\underline{M}}\ ag_M{}^N\ 1$ = $A\Gamma_{\underline{M}}{}^{\underline{N}} = \Gamma A^{\underline{N}}{}_{\underline{M}}$ = $1\ ga^N{}_M\ \alpha^{\underline{M}}$	$\alpha^{\underline{M}}\ as_M{}^N\ \sigma_{\underline{N}}$ = $A\Sigma_{\underline{M}}{}^{\underline{N}} = \Sigma A^{\underline{N}}{}_{\underline{M}}$ = $\sigma_{\underline{N}}\ sa^N{}_M\ \alpha^{\underline{M}}$	$\alpha^{\underline{M}}\ ay_M{}^N\ 1$ = $A\Psi_{\underline{M}}{}^{\underline{N}} = \Psi A^{\underline{N}}{}_{\underline{M}}$ = $1\ ya^N{}_M\ \alpha^{\underline{M}}$	$\alpha^{\underline{M}}\ ao_M{}^N\ 1$ = $AO_{\underline{M}}{}^{\underline{N}} = OA^{\underline{N}}{}_{\underline{M}}$ = $1\ oa^N{}_M\ \alpha^{\underline{M}}$
f^M	$1\ fb_M{}^N\ \beta_{\underline{N}}$ = $\Phi B_{\underline{M}}{}^{\underline{N}} = B\Phi^{\underline{N}}{}_{\underline{M}}$ = $\beta_{\underline{N}}\ bf^N{}_M\ 1$	$1\ fg_M{}^N\ 1$ = $\Phi\Gamma_{\underline{M}}{}^{\underline{N}} = \Gamma\Phi^{\underline{N}}{}_{\underline{M}}$ = $1\ gf^N{}_M\ 1$	$1\ fs_M{}^N\ \sigma_{\underline{N}}$ = $\Phi\Sigma_{\underline{M}}{}^{\underline{N}} = \Sigma\Phi^{\underline{N}}{}_{\underline{M}}$ = $\sigma_{\underline{N}}\ sf^N{}_M\ 1$	$1\ fy_M{}^N\ 1$ = $\Phi\Psi_{\underline{M}}{}^{\underline{N}} = \Psi\Phi^{\underline{N}}{}_{\underline{M}}$ = $1\ yf^N{}_M\ 1$	$1\ fo_M{}^N\ 1$ = $\Phi O_{\underline{M}}{}^{\underline{N}} = O\Phi^{\underline{N}}{}_{\underline{M}}$ = $1\ of^N{}_M\ 1$
r^M	$\rho^{\underline{M}}\ rb_M{}^N\ \beta_{\underline{N}}$ = $PB_{\underline{M}}{}^{\underline{N}} = BP^{\underline{N}}{}_{\underline{M}}$ = $\beta_{\underline{N}}\ br^N{}_M\ \rho^{\underline{M}}$	$\rho^{\underline{M}}\ rg_M{}^N\ 1$ = $P\Gamma_{\underline{M}}{}^{\underline{N}} = \Gamma P^{\underline{N}}{}_{\underline{M}}$ = $1\ gr^N{}_M\ \rho^{\underline{M}}$	$\rho^{\underline{M}}\ rs_M{}^N\ \sigma_{\underline{N}}$ = $P\Sigma_{\underline{M}}{}^{\underline{N}} = \Sigma P^{\underline{N}}{}_{\underline{M}}$ = $\sigma_{\underline{N}}\ br^N{}_M\ \rho^{\underline{M}}$	$\rho^{\underline{M}}\ ry_M{}^N\ 1$ = $P\Psi_{\underline{M}}{}^{\underline{N}} = \Psi P^{\underline{N}}{}_{\underline{M}}$ = $1\ yr^N{}_M\ \rho^{\underline{M}}$	$\rho^{\underline{M}}\ ro_M{}^N\ 1$ = $PO_{\underline{M}}{}^{\underline{N}} = OP^{\underline{N}}{}_{\underline{M}}$ = $1\ or^N{}_M\ \rho^{\underline{M}}$

x^M	$1\,xb_M{}^N\,\beta_{\underline{N}} = \Xi B_{\underline{M}}{}^{\underline{N}} = B\Xi^{\underline{N}}{}_{\underline{M}} = \beta_N\,bx^N{}_M 1$	$1\,xg_M{}^N\,1 = \Xi\Gamma_{\underline{M}}{}^{\underline{N}} = \Gamma\Xi^{\underline{N}}{}_{\underline{M}} = 1\,gx^N{}_M 1$	$1\,xs_M{}^N\,\sigma_{\underline{N}} = \Xi\Sigma_{\underline{M}}{}^{\underline{N}} = \Sigma\Xi^{\underline{N}}{}_{\underline{M}} = \sigma_N\,sx^N{}_M 1$	$1\,xy_M{}^N\,1 = \Xi\Psi_{\underline{M}}{}^{\underline{N}} = \Psi\Xi^{\underline{N}}{}_{\underline{M}} = 1\,yx^N{}_M 1$	$1\,xo_M{}^N\,1 = \Xi O_{\underline{M}}{}^{\underline{N}} = O\Xi^{\underline{N}}{}_{\underline{M}} = 1\,ox^N{}_M 1$
o^M	$1\,ob_M{}^N\,\beta_{\underline{N}} = OB_{\underline{M}}{}^{\underline{N}} = BO^{\underline{N}}{}_{\underline{M}} = \beta_N\,bo^N{}_M 1$	$1\,og_M{}^N\,1 = O\Gamma_{\underline{M}}{}^{\underline{N}} = \Gamma O^{\underline{N}}{}_{\underline{M}} = 1\,go^N{}_M 1$	$1\,os_M{}^N\,\sigma_{\underline{N}} = O\Sigma_{\underline{M}}{}^{\underline{N}} = \Sigma O^{\underline{N}}{}_{\underline{M}} = \sigma_N\,so^N{}_M 1$	$1\,oy_M{}^N\,1 = O\Psi_{\underline{M}}{}^{\underline{N}} = \Psi O^{\underline{N}}{}_{\underline{M}} = 1\,yo^N{}_M 1$	$1\,oo_M{}^N\,1 = OO_{\underline{M}}{}^{\underline{N}} = OO^{\underline{N}}{}_{\underline{M}} = 1\,oo^N{}_M 1$

3.2.2.1.
General Bases and Fixed Systems:

The Squared Differential Distance:

$$da^j \,.\, da_j = do^i\, do_i = db^j \,.\, db_j$$

$$aa_{ab} = \underline{a}_a \,.\, \underline{a}_b = \alpha_a\, \alpha_b \cos(_{ab}) \quad\Big|\quad aa^{ab} = \underline{a}^a \,.\, \underline{a}^b = \alpha^a\, \alpha^b \cos(^{ab})$$

$$(oa^m{}_M = ao_M{}^m)\,(oa^m{}_N = ao_N{}^m) = aa_{MN} \quad\Big|\quad aa^{MN} = (oa_n{}^M = ao^M{}_n)\,(oa_n{}^N = ao^N{}_n)$$

$$(oa_m{}^M = ao^M{}_m)\,(oa_m{}^N = ao^N{}_m) = aa^{MN} \quad\Big|\quad aa_{MN} = (oa^n{}_M = ao_M{}^n)\,(oa^n{}_N = ao_N{}^n)$$

$$oa^m{}_M\, oa^m{}_N = aa_{MN} \quad\Big|\quad aa^{MN} = ao^M{}_n\, ao^N{}_n$$

$$oa_m{}^M\, oa_m{}^N = aa^{MN} \quad\Big|\quad aa_{MN} = ao_M{}^n\, ao_N{}^n$$

$$ao_M{}^m\, ao_N{}^m = aa_{MN} \quad\Big|\quad aa^{MN} = oa_n{}^M\, oa_n{}^N$$

$$ao^M{}_m\, ao^N{}_m = aa^{MN} \quad\Big|\quad aa_{MN} = oa^n{}_M\, oa^n{}_N$$

$$(oo_{\alpha\beta} \equiv \delta_{\alpha\beta})\, oa^\alpha{}_a\, oa^\beta{}_b = aa_{ab} \quad\Big|\quad aa^{ab} = ao^a{}_\alpha\, ao^b{}_\beta\, (oo^{\alpha\beta} \equiv \delta^{\alpha\beta})$$

$$(oo^{\alpha\beta} \equiv \delta^{\alpha\beta})\, oa_\alpha{}^a\, oa_\beta{}^b = aa^{ab} \quad\Big|\quad aa_{ab} = ao_a{}^\alpha\, ao_b{}^\beta\, (oo_{\alpha\beta} \equiv \delta_{\alpha\beta})$$

$$\|\delta_{\alpha\beta}\|\ \|oa^\alpha{}_a\|\ \|oa^\beta{}_b\| \equiv \|aa_{ab}\| \quad\Big|\quad \|aa^{ab}\| \equiv \|ao^a{}_\alpha\|\ \|ao^b{}_\beta\|\ \|\delta^{\alpha\beta}\|$$

$$\|\delta^{\alpha\beta}\|\ \|oa_\alpha{}^a\|\ \|oa_\beta{}^b\| \equiv \|aa^{ab}\| \quad\Big|\quad \|aa_{ab}\| \equiv \|ao_a{}^\alpha\|\ \|ao_b{}^\beta\|\ \|\delta_{\alpha\beta}\|$$

$$\|o^\alpha{}_a \equiv \partial o^\alpha / \partial a^a\|^2 = \|aa_{ab}\| \quad\Big|\quad \|aa^{ab}\| = \|a^a{}_\alpha \equiv \partial a^a / \partial o^\alpha\|^2$$

$$\|o_\alpha{}^a \equiv \partial o_\alpha / \partial a_a\|^2 = \|aa^{ab}\| \quad\Big|\quad \|aa_{ab}\| = \|a_a{}^\alpha \equiv \partial a_a / \partial o_\alpha\|^2$$

$$do^0\, do_0 = oo^{ab}\, do_a\, do_b \quad\Big|\quad da^0\, da_0 = aa^{ab}\, da_a\, da_b$$

$$do^0\, do_0 = oo_{ab}\, do^a\, do^b \quad\Big|\quad da^0\, da_0 = aa_{ab}\, da^a\, da^b$$

Knowing:

$$do^s = do_s = oa_s{}^b\, da_b \quad\Big|\quad do^s = do_s = ob_s{}^b\, db_b$$

$$ao^a{}_s = oa_s{}^a \quad\Big|\quad bo^b{}_s = ob_s{}^b$$

$da^a = ao^a_s\, do^s = oa_s{}^a\, do_s$	$db^b = bo^b_s\, do^s = ob_s{}^b\, do_s$
$da^a = oa_s{}^a\, oa_s{}^b\, da_b = aa^{ab}\, da_b$	$db^b = ob_s{}^b\, ob_s{}^a\, db_a = bb^{ba}\, db_b$
Multiply by:	
da_a	db_a
To get:	
$da^a\, da_a = aa^{ab}\, da_a\, da_b$	$db^a\, db_a = bb^{ab}\, db_a\, db_b$
Similarly: Get:	
$da^a\, da_a = aa_{ab}\, da^a\, da^b$	$db^a\, db_a = bb_{ab}\, db^a\, db^b$
:	
So:	
$da^a\, da_a = aa^{ab}\, da_a\, da_b$	$db^a\, db_a = bb^{ab}\, db_a\, db_b$
$da^a\, da_a = aa_{ab}\, da^a\, da^b$	$db^a\, db_a = bb_{ab}\, db^a\, db^b$

3.2.2.2. General Unitary Bases and Fixed Systems:

The Squared Differential Distance:

$$df^j \,.\, df_j = do^i\, do_i = db^j \,.\, db_j$$

$ff_{fb} = \underline{f}_f \,.\, \underline{f}_b = \cos(_{fb})$	$ff^{fb} = \underline{f}^f \,.\, \underline{f}^b = \cos(^{fb})$
$(of^m_M = fo_M{}^m)\,(of^m_N = fo_N{}^m) = ff_{MN}$	$ff^{MN} = (of_n^M = fo^M_n)\,(of_n^N = fo^N_n)$
$(of_m^M = fo^M_m)\,(of_m^N = fo^N_m) = ff^{MN}$	$ff_{MN} = (of^n_M = fo_M{}^n)\,(of^n_N = fo_N{}^n)$
$of^m_M\, of^m_N = ff_{MN}$	$ff^{MN} = fo^M_n\, fo^N_n$
$of_m^M\, of_m^N = ff^{MN}$	$ff_{MN} = fo_M{}^n\, fo_N{}^n$
:	
$fo_M{}^m\, fo_N{}^m = ff_{MN}$	$ff^{MN} = of_n^M\, of_n^N$
$fo^M_m\, fo^N_m = ff^{MN}$	$ff_{MN} = of^n_M\, of^n_N$
$(oo_{\alpha\beta} \equiv \delta_{\alpha\beta})\, of^\alpha_f\, of^\beta_b = ff_{fb}$	$ff^{fb} = fo^f_\alpha\, fo^b_\beta\, (oo^{\alpha\beta} \equiv \delta^{\alpha\beta})$
$(oo^{\alpha\beta} \equiv \delta^{\alpha\beta})\, of_\alpha^f\, of_\beta^b = ff^{fb}$	$ff_{fb} = fo_f{}^\alpha\, fo_b{}^\beta\, (oo_{\alpha\beta} \equiv \delta_{\alpha\beta})$
:	
$\|\delta_{\alpha\beta}\|\; \|of^\alpha_f\|\; \|of^\beta_b\| \equiv \|ff_{fb}\|$	$\|ff^{fb}\| \equiv \|fo^f_\alpha\|\; \|fo^b_\beta\|\; \|\delta^{\alpha\beta}\|$
$\|\delta^{\alpha\beta}\|\; \|of_\alpha^f\|\; \|of_\beta^b\| \equiv \|ff^{fb}\|$	$\|ff_{fb}\| \equiv \|fo_f{}^\alpha\|\; \|fo_b{}^\beta\|\; \|\delta_{\alpha\beta}\|$
$\|o^\alpha_f \equiv \partial o^\alpha / \partial f^f\|^2 = \|ff_{fb}\|$	$\|ff^{fb}\| = \|f^f_\alpha \equiv \partial f^f / \partial o^\alpha\|^2$
$\|o_\alpha^f \equiv \partial o_\alpha / \partial f_f\|^2 = \|ff^{fb}\|$	$\|ff_{fb}\| = \|f_f{}^\alpha \equiv \partial f_f / \partial o_\alpha\|^2$
$do^0\, do_0 = oo^{fb}\, do_f\, do_b$	$df^0\, df_0 = ff^{fb}\, df_f\, df_b$
$do^0\, do_0 = oo_{fb}\, do^f\, do^b$	$df^0\, df_0 = ff_{fb}\, df^f\, df^b$

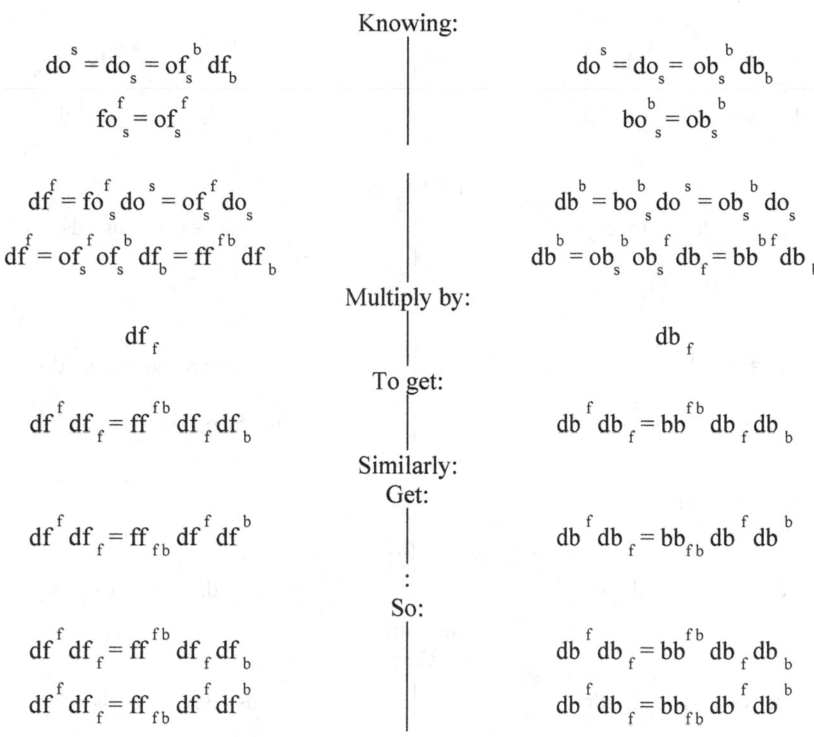

Knowing:

$do^s = do_s = of_s^b\ df_b$	$do^s = do_s = ob_s^b\ db_b$
$fo_s^f = of_s^f$	$bo_s^b = ob_s^b$
$df^f = fo_s^f\ do^s = of_s^f\ do_s$	$db^b = bo_s^b\ do^s = ob_s^b\ do_s$
$df^f = of_s^f\ of_s^b\ df_b = ff^{fb}\ df_b$	$db^b = ob_s^b\ ob_s^f\ db_f = bb^{bf}\ db_b$

Multiply by:

df_f	db_f

To get:

$df^f\ df_f = ff^{fb}\ df_f\ df_b$	$db^f\ db_f = bb^{fb}\ db_f\ db_b$

Similarly:
Get:

$df^f\ df_f = ff_{fb}\ df^f\ df^b$	$db^f\ db_f = bb_{fb}\ db^f\ db^b$

:

So:

$df^f\ df_f = ff^{fb}\ df_f\ df_b$	$db^f\ db_f = bb^{fb}\ db_f\ db_b$
$df^f\ df_f = ff_{fb}\ df^f\ df^b$	$db^f\ db_f = bb_{fb}\ db^f\ db^b$

3.2.2.3.
Orthogonal Bases and Fixed Systems:

The Squared Differential Distance:

$$dr^j\ .\ dr_j = do^i\ do_i = ds^j\ .\ ds_j$$

$ss_{sb} = \underline{s}_s\ .\ \underline{s}_b = \sigma_s\ \sigma_b\ \cos(_{sb})$	$ss^{sb} = \underline{s}^s\ .\ \underline{s}^b = \sigma^s\ \sigma^b\ \cos(^{sb})$
$os^m_M\ (= so^m_M)\ os^m_N\ (= so^m_N) = ss_{MN}$	$ss^{MN} = (\ os^M_n =)\ so^M_n\ (\ os^N_n =)\ so^N_n$
$os^M_m\ (= so^M_m)\ os^N_m\ (= so^N_m) = ss^{MN}$	$ss_{MN} = (\ os^n_M =)\ so^n_M\ (\ os^n_N =)\ so^n_N$
$os^m_M\ os^m_N = ss_{MN}$	$ss^{MN} = so^M_n\ so^N_n$
$os^M_m\ os^N_m = ss^{MN}$	$ss_{MN} = so^n_M\ so^n_N$
$so^m_M\ so^m_N = ss_{MN}$	$ss^{MN} = os^M_n\ os^N_n$
$so^M_m\ so^N_m = ss^{MN}$	$ss_{MN} = os^n_M\ os^n_N$
$(oo_{\alpha\beta} \equiv \delta_{\alpha\beta})\ os^\alpha_a\ os^\beta_b = ss_{ab}$	$ss^{ab} = so^a_\alpha\ so^b_\beta\ (oo^{\alpha\beta} \equiv \delta^{\alpha\beta})$
$(oo^{\alpha\beta} \equiv \delta^{\alpha\beta})\ os^a_\alpha\ os^b_\beta = ss^{ab}$	$ss_{ab} = so^\alpha_a\ so^\beta_b\ (oo_{\alpha\beta} \equiv \delta_{\alpha\beta})$
$\lVert\delta_{\alpha\beta}\rVert\ \lVert os^\alpha_a\rVert\ \lVert os^\beta_b\rVert \equiv \lVert ss_{ab}\rVert$	$\lVert ss^{ab}\rVert \equiv \lVert so^a_\alpha\rVert\ \lVert so^b_\beta\rVert\ \lVert\delta^{\alpha\beta}\rVert$
$\lVert\delta^{\alpha\beta}\rVert\ \lVert os^a_\alpha\rVert\ \lVert os^b_\beta\rVert \equiv \lVert ss^{ab}\rVert$	$\lVert ss_{ab}\rVert \equiv \lVert so^\alpha_a\rVert\ \lVert so^\beta_b\rVert\ \lVert\delta_{\alpha\beta}\rVert$
$\lVert o^\alpha_a \equiv \partial o^\alpha / \partial s^a \rVert^2 = \lVert ss_{ab}\rVert$	$\lVert ss^{ab}\rVert = \lVert s^a_\alpha \equiv \partial s^a / \partial o^\alpha \rVert^2$
$\lVert o^a_\alpha \equiv \partial o_\alpha / \partial s_a \rVert^2 = \lVert ss^{ab}\rVert$	$\lVert ss_{ab}\rVert = \lVert s^\alpha_a \equiv \partial s_a / \partial o_\alpha \rVert^2$

$do^{0}\, do_{0} = oo^{sb}\, do_{s}\, do_{b}$	$ds^{0}\, ds_{0} = ss^{sb}\, ds_{s}\, ds_{b}$
$do^{0}\, do_{0} = oo_{sb}\, do^{s}\, do^{b}$	$ds^{0}\, ds_{0} = ss_{sb}\, ds^{s}\, ds^{b}$

Knowing:

$do^{s} = do_{s} = or^{s}_{s}\, dr_{s}$	$do^{s} = do_{s} = os^{s}_{s}\, ds_{s}$
$ro^{r}_{s} = or^{r}_{s}$	$so^{s}_{s} = os^{s}_{s}$
$dr^{r} = ro^{r}_{s}\, do^{s} = or^{r}_{s}\, do_{s}$	$ds^{s} = so^{s}_{s}\, do^{s} = os^{s}_{s}\, do_{s}$
$dr^{r} = or^{r}_{s}\, or^{s}_{s}\, dr_{s} = rr^{rs}\, dr_{s}$	$ds^{s} = os^{s}_{s}\, os^{r}_{s}\, ds_{r} = ss^{sr}\, ds_{s}$

Multiply by:

dr_{r}	ds_{r}

To get:

$dr^{r}\, dr_{r} = rr^{rs}\, dr_{r}\, dr_{s}$	$ds^{r}\, ds_{r} = ss^{rs}\, ds_{r}\, ds_{s}$

Similarly:
Get:

$dr^{r}\, dr_{r} = rr_{rs}\, dr^{r}\, dr^{s}$	$ds^{r}\, ds_{r} = ss_{rs}\, ds^{r}\, ds^{s}$

:

So:

$dr^{r}\, dr_{r} = rr^{rs}\, dr_{r}\, dr_{s}$	$ds^{r}\, ds_{r} = ss^{rs}\, ds_{r}\, ds_{s}$
$dr^{r}\, dr_{r} = rr_{rs}\, dr^{r}\, dr^{s}$	$ds^{r}\, ds_{r} = ss_{rs}\, ds^{r}\, ds^{s}$

3.2.2.4.
Orthonormal Bases and Fixed Systems:

The Squared Differential Distance:

$$dx^{j}\, dx_{j} = do^{i}\, do_{i} = dy^{j}\, dy_{j}$$

$oy^{m}_{M}\,(= yo_{M}^{m})\; oy^{m}_{N}\,(= yo_{N}^{m}) = yy_{MN}$	$yy^{MN} = (\,oy^{M}_{n} =)\; yo^{M}_{n}\,(\,oy^{N}_{n} =)\; yo^{N}_{n}$
$oy_{m}^{M}\,(= yo^{M}_{m})\; oy_{m}^{N}\,(= yo^{N}_{m}) = yy^{MN}$	$yy_{MN} = (\,oy^{n}_{M} =)\; yo_{M}^{n}\;(\,oy^{n}_{N} =)\; yo_{N}^{n}$
$oy^{m}_{M}\, oy^{m}_{N} = yy_{MN}$	$yy^{MN} = yo^{M}_{n}\, yo^{N}_{n}$
$oy_{m}^{M}\, oy_{m}^{N} = yy^{MN}$	$yy_{MN} = yo_{M}^{n}\, yo^{n}_{N}$
:	:
$yo_{M}^{m}\, yo_{N}^{m} = yy_{MN}$	$yy^{MN} = oy_{n}^{M}\, oy_{n}^{N}$
$yo^{M}_{m}\, yo^{N}_{m} = yy^{MN}$	$yy_{MN} = oy^{n}_{M}\, oy^{n}_{N}$
$(oo_{\alpha\beta} \equiv \delta_{\alpha\beta})\; oy^{\alpha}_{a}\, oy^{\beta}_{b} = yy_{ab}$	$yy^{ab} = yo^{a}_{\alpha}\, yo^{b}_{\beta}\, (oo^{\alpha\beta} \equiv \delta^{\alpha\beta})$
$(oo^{\alpha\beta} \equiv \delta^{\alpha\beta})\; oy_{\alpha}^{a}\, oy_{\beta}^{b} = yy^{ab}$	$yy_{ab} = yo^{\alpha}_{a}\, yo^{\beta}_{b}\, (oo_{\alpha\beta} \equiv \delta_{\alpha\beta})$
:	:
$\Vert\delta_{\alpha\beta}\Vert\; \Vert oy^{\alpha}_{a}\Vert\; \Vert oy^{\beta}_{b}\Vert \equiv \Vert yy_{ab}\Vert = 1$	$1 = \Vert yy^{ab}\Vert \equiv \Vert yo^{a}_{\alpha}\Vert\; \Vert yo^{b}_{\beta}\Vert\; \Vert\delta^{\alpha\beta}\Vert$
$\Vert\delta^{\alpha\beta}\Vert\; \Vert oy_{\alpha}^{a}\Vert\; \Vert oy_{\beta}^{b}\Vert \equiv \Vert yy^{ab}\Vert = 1$	$1 = \Vert yy_{ab}\Vert \equiv \Vert yo_{a}^{\alpha}\Vert\; \Vert yo_{b}^{\beta}\Vert\; \Vert\delta_{\alpha\beta}\Vert$

$$\| o^{\alpha}_{a} \equiv \partial o^{\alpha} / \partial y^{a} \|^{2} = \| yy_{ab} \| = 1 \qquad 1 = \| yy^{ab} \| = \| y^{a}_{\alpha} \equiv \partial y^{a} / \partial o^{\alpha} \|^{2}$$

$$\| o^{a}_{\alpha} \equiv \partial o_{\alpha} / \partial y_{a} \|^{2} = \| yy^{ab} \| = 1 \qquad 1 = \| yy_{ab} \| = \| y^{\alpha}_{a} \equiv \partial y_{a} / \partial o_{\alpha} \|^{2}$$

$$do^{0}\, do_{0} = oo^{yb}\, do_{y}\, do_{b} \qquad dy^{0}\, dy_{0} = yy^{yb}\, dy_{y}\, dy_{b}$$

$$do^{0}\, do_{0} = oo_{yb}\, do^{y}\, do^{b} \qquad dy^{0}\, dy_{0} = yy_{yb}\, dy^{y}\, dy^{b}$$

Knowing:

$$do^{s} = do_{s} = ox^{y}_{s}\, dx_{y} \qquad do^{s} = do_{s} = oy^{y}_{s}\, dy_{y}$$

$$xo^{x}_{s} = ox^{x}_{s} \qquad yo^{y}_{s} = oy^{y}_{s}$$

$$dx^{x} = xo^{x}_{s}\, do^{s} = ox^{x}_{s}\, do_{s} \qquad dy^{y} = yo^{y}_{s}\, do^{s} = oy^{y}_{s}\, do_{s}$$

$$dx^{x} = ox^{x}_{s}\, ox^{y}_{s}\, dx_{y} = xx^{xy}\, dx_{y} \qquad dy^{y} = oy^{y}_{s}\, oy^{x}_{s}\, dy_{x} = yy^{yx}\, dy_{y}$$

Multiply by:

$$dx_{x} \qquad dy_{x}$$

To get:

$$dx^{x}\, dx_{x} = xx^{xy}\, dx_{x}\, dx_{y} \qquad dy^{x}\, dy_{x} = yy^{xy}\, dy_{x}\, dy_{y}$$

Similarly:
Get:

$$dx^{x}\, dx_{x} = xx_{xy}\, dx^{x}\, dx^{y} \qquad dy^{x}\, dy_{x} = yy_{xy}\, dy^{x}\, dy^{y}$$

:

So:

$$dx^{x}\, dx_{x} = xx^{xy}\, dx_{x}\, dx_{y} \qquad dy^{x}\, dy_{x} = yy^{xy}\, dy_{x}\, dy_{y}$$

$$dx^{x}\, dx_{x} = xx_{xy}\, dx^{x}\, dx^{y} \qquad dy^{x}\, dy_{x} = yy_{xy}\, dy^{x}\, dy^{y}$$

3.2.3. Metric Determinant:

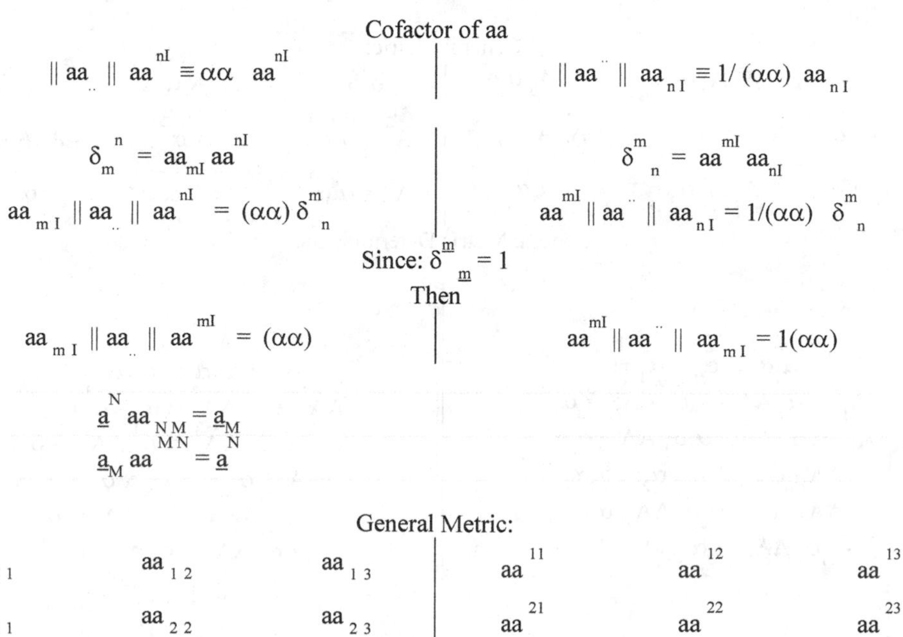

Cofactor of aa

$$\| aa_{..} \|\, aa^{nI} \equiv \alpha\alpha\ \ aa^{nI} \qquad \| aa^{..} \|\, aa_{nI} \equiv 1/(\alpha\alpha)\ aa_{nI}$$

$$\delta^{n}_{m} = aa_{mI}\, aa^{nI} \qquad \delta^{m}_{n} = aa^{mI}\, aa_{nI}$$

$$aa_{mI}\, \| aa_{..} \|\, aa^{nI} = (\alpha\alpha)\, \delta^{m}_{n} \qquad aa^{mI}\, \| aa^{..} \|\, aa_{nI} = 1/(\alpha\alpha)\ \delta^{m}_{n}$$

Since: $\delta^{\underline{m}}_{\underline{m}} = 1$
Then

$$aa_{mI}\, \| aa_{..} \|\, aa^{mI} = (\alpha\alpha) \qquad aa^{mI}\, \| aa^{..} \|\, aa_{mI} = 1(\alpha\alpha)$$

$$\underline{a}^{N}\, aa_{NM} = \underline{a}_{M}$$

$$\underline{a}_{M}\, aa^{MN} = \underline{a}^{N}$$

General Metric:

aa_{11}	aa_{12}	aa_{13}	aa^{11}	aa^{12}	aa^{13}
aa_{21}	aa_{22}	aa_{23}	aa^{21}	aa^{22}	aa^{23}

aa_{31} aa_{32} aa_{33} | aa^{31} aa^{32} aa^{33}

General Metric Determinants:

$$\|_{aa}\| \equiv e^{IJK} aa_{I1} aa_{J2} aa_{K3} = e^{IJK} oa^{u}_{I} oa^{u}_{1} \; oa^{v}_{J} oa^{v}_{2} \; oa^{w}_{K} oa^{w}_{3}$$

$$\|^{aa}\| \equiv e_{IJK} aa^{I1} aa^{J2} aa^{K3} = e_{IJK} oa^{I}_{u} oa^{1}_{u} \; oa^{J}_{v} oa^{2}_{v} \; oa^{K}_{w} oa^{3}_{w}$$

Or:

$$\|_{aa}\| \equiv (1/6)\, e^{IJK} e^{LMN} aa_{IL} aa_{JM} aa_{KN} = (1/6) e^{IJK} e^{LMN} oa^{u}_{I} oa^{u}_{L} oa^{v}_{J} oa^{v}_{M} oa^{w}_{K} oa^{w}_{N}$$

$$\|^{aa}\| \equiv (1/6)\, e_{IJK} e_{LMN} aa^{IL} aa^{JM} aa^{KN} = (1/6) e_{IJK} e_{LMN} oa^{I}_{u} oa^{L}_{u} oa^{J}_{v} oa^{M}_{v} oa^{K}_{w} oa^{N}_{w}$$

:

$$aa_{IK} \equiv \underline{a}_I . \underline{a}_K = \underline{A}_{\underline{I}} . \underline{A}_{\underline{K}} \; \alpha_{\underline{I}} \alpha_{\underline{K}} \qquad aa^{IK} \equiv \underline{a}^{I} . \underline{a}^{K} = \underline{A}^{\underline{I}} . \underline{A}^{\underline{K}} \; \alpha^{\underline{I}} \alpha^{\underline{K}}$$

$$\alpha_{\underline{I}} \alpha^{\underline{I}} = 1$$

Reciprocal Metrics

$$(aa^{\underline{I}}_{\underline{I}})^2 \quad (aa_{\underline{I}}^{\underline{I}})^2 = \alpha^{I} \alpha_{\underline{I}} \quad \alpha_{\underline{I}} \alpha^{\underline{I}} = (\alpha_{\underline{I}})^2 \quad (\alpha^{\underline{I}})^2 = (\alpha_{\underline{I}} \quad \alpha^{\underline{I}})^2 = 1$$

Metrics

aaIK = aI . aK :

$0 \le /aa_{IK}/ = \alpha_I \alpha_K / AA_{IK}/ \le \alpha_I \alpha_K$	$aa^{I}_{K} = \underline{a}^{I} . \underline{a}_{K} = \alpha_{I} \alpha^{K} = AA_{I}^{K} = \delta^{K}_{I}$
$aa^{I}_{K} = \underline{a}^{I} . \underline{a}_{K} = \alpha^{I} \alpha_{K} = AA^{I}_{K} = \delta^{I}_{K}$	$0 \le /aa^{IK}/ = \alpha^{I} \alpha^{K} / AA^{IK}/ \le \alpha^{I} \alpha^{K}$

3.2.3.1.
General Metric Determinant:

General Metric:

$\alpha_1\alpha_1$	$A_1.A_2\,\alpha_1\alpha_2$	$A_1.A_3\,\alpha_1\alpha_3$	$\alpha^1\alpha^1$	$A^1.A^2\alpha^1\alpha^2$	$A^1.A^3\alpha^1\alpha^3$
$A_2.A_1\,\alpha_2\alpha_1$	$\alpha_2\alpha_2$	$A_2.A_3\,\alpha_2\alpha_3$	$A^2.A^1\alpha^2\alpha^1$	$\alpha^2\alpha^2$	$A^2.A^3\alpha^2\alpha^3$
$A_3.A_1\,\alpha_3\alpha_1$	$A_3.A_2\,\alpha_3\alpha_2$	$\alpha_3\alpha_3$	$A^3.A^1\alpha^3\alpha^1$	$A^3.A^2\alpha^3\alpha^2$	$\alpha^3\alpha^3$

General Metric Determinants:

$$\|_{aa}\| = \alpha_1\alpha_1\,\alpha_2\alpha_2\,\alpha_3\alpha_3 + AA_{12}\,\alpha_1\alpha_2\,AA_{23}\,\alpha_2\alpha_3\,AA_{31}\,\alpha_3\alpha_1 + AA_{13}\,\alpha_1\alpha_3\,AA_{21}\,\alpha_2\alpha_1\,AA_{32}\,\alpha_3\alpha_2 + -AA_{12}\,\alpha_1\alpha_2\,AA_{21}\,\alpha_2\alpha_1\,\alpha_3\alpha_3 -AA_{13}\,\alpha_1\alpha_3\,\alpha_2\alpha_2\,AA_{31}\,\alpha_3\alpha_1 -\alpha_1\alpha_1\,AA_{23}\,\alpha_2\alpha_3\,AA_{32}\,\alpha_3\alpha_2$$

$$\|^{aa}\| = \alpha^1\alpha^1\,\alpha^2\alpha^2\,\alpha^3\alpha^3 + AA^{12}\,\alpha^1\alpha^2\,AA^{23}\,\alpha^2\alpha^3\,AA^{31}\,\alpha^3\alpha^1 + AA^{13}\,\alpha^1\alpha^3\,AA^{21}\,\alpha^2\alpha^1\,AA^{32}\,\alpha^3\alpha^2 + -AA^{12}\,\alpha^1\alpha^2\,AA^{21}\,\alpha^2\alpha^1\,\alpha^3\alpha^3 -AA^{13}\,\alpha^1\alpha^3\,\alpha^2\alpha^2\,AA^{31}\,\alpha^3\alpha^1 -\alpha^1\alpha^1\,AA^{23}\,\alpha^2\alpha^3\,AA^{32}\,\alpha^3\alpha^2$$

:

$$\|_{aa}\| \; \|^{aa}\| = \|_{aa}\| * (1 / \|_{aa}\|) = 1$$

3.2.3.2.
General Unitary Bases Metric Determinant:

General Unitary Bases Metrics:

1	$\Phi_1 . \Phi_2$	$\Phi_1 . \Phi_3$	1	$\Phi^1 . \Phi^2$	$\Phi^1 . \Phi^3$
$\Phi_2 . \Phi_1$	1	$\Phi_2 . \Phi_3$	$\Phi^2 . \Phi^1$	1	$\Phi^2 . \Phi^3$
$\Phi_3 . \Phi_1$	$\Phi_3 . \Phi_2$	1	$\Phi^3 . \Phi^1$	$\Phi^3 . \Phi^2$	1

General Unitary Bases Metrics Determinants:

$$\|_{ff}\| = 1 + 2(\Phi\Phi_{12}\ \Phi\Phi_{23}\ \Phi\Phi_{31}) + -(\Phi\Phi_{12})^2 - (\Phi\Phi_{23})^2 - (\Phi\Phi_{31})^2$$

$$\|^{ff}\| = 1 + 2(\Phi\Phi^{12}\ \Phi\Phi^{23}\ \Phi\Phi^{31})^2 + -(\Phi\Phi^{12})^2 - (\Phi\Phi^{23})^2 - (\Phi\Phi^{31})^2$$

:

$$\|_{ff}\| \; \|^{ff}\| = \|_{ff}\| * (1 / \|_{ff}\|) = 1$$

3.2.3.3.
Orthogonal Metric Determinant:

Orthogonal Metrics:

$\rho_1 \rho_1$	0	0	$\rho^1 \rho^1$	0	0
0	$\rho_2 \rho_2$	0	0	$\rho^2 \rho^2$	0
0	0	$\rho_3 \rho_3$	0	0	$\rho^3 \rho^3$

Orthogonal Metrics Determinants:

$$\|_{rr}\| = (\rho_1 \rho_1 \rho_2 \rho_2 \rho_3 \rho_3) \qquad \|^{rr}\| = (\rho^1 \rho^1 \rho^2 \rho^2 \rho^3 \rho^3)$$

:

$$\|_{rr}\| \; \|^{rr}\| = (\rho_1 \rho_1 \rho_2 \rho_2 \rho_3 \rho_3)(\rho^1 \rho^1 \rho^2 \rho^2 \rho^3 \rho^3) = 1$$

3.2.3.4.
Orthonormal Metric Determinant:

Orthonormal Metrics:

1	0	0	1	0	0
0	1	0	0	1	0
0	0	1	0	0	1

Orthonormal Metrics Determinants:

$$\|_{xx}\| = (\xi_1 \xi_1 \xi_2 \xi_2 \xi_3 \xi_3) = (1) \qquad \|^{xx}\| = (\xi^1 \xi^1 \xi^2 \xi^2 \xi^3 \xi^3) = (1)$$

:

$$\|_{xx}\| \; \|^{xx}\| = (\xi_1 \xi_1 \xi_2 \xi_2 \xi_3 \xi_3)(\xi^1 \xi^1 \xi^2 \xi^2 \xi^3 \xi^3) = (1) * (1) = 1$$

3.3.

Transformation of Metrics:

3.3.1. Metric Transformation, (aa) Description:

$\underline{a}_m \quad da^n$	\A $\quad$ $\underline{a}_m \quad$ A \T\ $\quad da^n$	\R $\quad \underline{a}_m \quad da^n$	$\underline{a}_m \quad da^n$
$AB^m{}_n = A^m{}_K B^K{}_n$	$AA^{mn\backslash} = A^m{}_K A^n{}_K$	$AB^m{}_n = A^m{}_K B^K{}_n$	$AA^{m\backslash}{}_{n/} = A^m{}_K A^K{}_n$
$BB_{\backslash mn} = B^K{}_m B^K{}_n$	$AB^n{}_m = B^K{}_m A^n{}_K$	$BB^{/m}{}_{\backslash n} = B_K{}^m B^K{}_n$	$BA^{/m}{}_{n/} = B_K{}^m A^K{}_n$

$\underline{a}^m \quad da_n$	/A $\quad \underline{a}^m \quad$ A /C/ $\quad da_n$	/R $\quad \underline{a}^m \quad da_n$	$\underline{a}^m \quad da_n$
$AB_m{}^n = A_m{}^K B_K{}^n$	$AA_{mn/} = A_m{}^K A_n{}^K$	$AB_m{}^n = A_m{}^K B_K{}^n$	$AA_m{}^n = A_m{}^K A^n{}_K$
$BB^{/mn} = B_K{}^m B_K{}^n$	$AB_n{}^m = B_K{}^m A_n{}^K$	$BB_{\backslash m}{}^{/n} = B^K{}_m B_K{}^n$	$BA_{\backslash m}{}^{n\backslash} = B^K{}_m A^n{}_K$

3.3.1.1. Upper-Lower (bb\) Metric Transformation to (aa\) Description:

\R $\quad \underline{a}_m \quad A^m{}_M (= B_M{}^m) \quad bb^M{}_N \quad (A_n{}^N =)\, B^N{}_n \quad da^n$

\

\R 00	$A^m{}_M \, bb^M{}_N \, B^N{}_n = \partial a^m/\partial b^M \, bb^M{}_N \, \partial b^N/\partial a^n = aa^m{}_n$
\R 01	$A^m{}_M \, bb^M{}_N \, A_n{}^N = \partial a^m/\partial b^M \, bb^M{}_N \, \partial a_n/\partial b_N = aa^m{}_n$
\R 10	$B_M{}^m \, bb^M{}_N \, B^N{}_n = \partial b_M/\partial a_m \, bb^M{}_N \, \partial b^N/\partial a^n = aa^m{}_n$
\R 11	$B_M{}^m \, bb^M{}_N \, A_n{}^N = \partial b_M/\partial a_m \, bb^M{}_N \, \partial a_n/\partial b_N = aa^m{}_n$

Affine :

\Affine $\quad \underline{a}_m \quad A^m{}_M = B^M{}_m \quad bb^M{}_N \quad A^n{}_N = B^N{}_n \quad da^n$

\

\A 00	$A^m{}_M \, bb^M{}_N \, B^N{}_n = \partial a^m/\partial b^M \, bb^M{}_N \, \partial b^M/\partial a^n = aa^m{}_n$
\A 10	$B^M{}_m \, bb^M{}_N \, B^N{}_n = \partial b^M/\partial a^m \, bb^M{}_N \, \partial b^N/\partial a^n = aa^m{}_n$
\A 01	$A^m{}_M \, bb^M{}_N \, A^n{}_N = \partial a^m/\partial b^M \, bb^M{}_N \, \partial a^n/\partial b^N = aa^m{}_n$
\A 11	$B^M{}_m \, bb^M{}_N \, A^n{}_N = \partial b^M/\partial a^m \, bb^M{}_N \, \partial a^n/\partial b^N = aa^m{}_n$

3.3.1.2. Lower-Upper (bb/) Metric Transformation to (aa/) Description:

/R $\quad \underline{a}^m \quad A_m{}^M (= B^M{}_m) \quad bb_M{}^N \quad (A^n{}_N =)\, B_N{}^n \quad da_n$

/

/R 00	$A_m{}^M \, bb_M{}^N \, B_N{}^n = \partial a_m/\partial b_M \, bb_M{}^N \, \partial b_N/\partial a_n = aa_m{}^n$
/R 01	$A_m{}^M \, bb_M{}^N \, A^n{}_N = \partial a_m/\partial b_M \, bb_M{}^N \, \partial a^n/\partial b^N = aa_m{}^n$
/R 10	$B^M{}_m \, bb_M{}^N \, B_N{}^n = \partial b^M/\partial a^m \, bb_M{}^N \, \partial b_N/\partial a_n = aa_m{}^n$
/R 11	$B^M{}_m \, bb_M{}^N \, A^n{}_N = \partial b^M/\partial a^m \, bb_M{}^N \, \partial a^n/\partial b^N = aa_m{}^n$

Affine :

/A	$\underline{a}^m$	$A_m^M = B_M^m$	bb_M^N	$A_n^N = B_N^n$	da_n

/

/A 00	$A_m^M \; bb_M^N \; B_N^n = \partial a_m/\partial b_M \; bb_M^N \; \partial b_N/\partial a_n = aa_m^n$
/A 10	$B_M^m \; bb_M^N \; B_N^n = \partial b_M/\partial a_m \; bb_M^N \; \partial b_N/\partial a_n = aa_m^n$
/A 01	$A_m^M \; bb_M^N \; A_n^N = \partial a_m/\partial b_M \; bb_M^N \; \partial a_n/\partial b_N = aa_m^n$
/A 11	$B_M^m \; bb_M^N \; A_n^N = \partial b_M/\partial a_m \; bb_M^N \; \partial a_n/\partial b_N = aa_m^n$

3.3.2.
Metric Transformation (bb) Description:

\R: $\underline{b}_m$ … db^n	B \T\: $\underline{b}_m$ … db^n	\A: $\underline{b}_m$ … db^n	$\underline{b}_m$ … db^n
$BA^m_n = B^m_K A^K_n$	$BB^{m\backslash}_{n/} = B^m_K B^K_n$	$BA^m_n = B^m_K A^K_n$	$BB^{mn\backslash} = B^m_K B^n_K$
$AA^{/m}_{\backslash n} = A_K^m A^K_n$	$AB^{/m}_{n/} = A_K^m B^K_n$	$AA_{\backslash mn} = A^K_m A^K_n$	$BA^n_m = A^K_m B^n_K$

/R: $\underline{b}^m$ … db_n	B /C/: $\underline{b}^m$ … db_n	/A: $\underline{b}^m$ … db_n	$\underline{b}^m$ … db_n
$BA_m^n = B^K_m A_K^n$	$BB_{m/}^{n\backslash} = B^K_m B^n_K$	$BA_m^n = B^K_m A^n_K$	$BB_{mn/} = B^K_m B^K_n$
$AA_{\backslash m}^{/n} = A^K_m A_K^n$	$AB_{\backslash m}^{n\backslash} = A^K_m B^n_K$	$AA^{/mn} = A_K^m A_K^n$	$BA_n^m = A_K^m B_n^K$

3.3.2.1.
Upper-Lower (aa\) Metric Transformation to (bb\) Description:

\R	$\underline{b}_m$	$B^m_M (= A_M^m)$	aa^M_N	$(B_n^N =) A^N_n$	db^n

\

\R 00	$B^m_M \; aa^M_N \; A^N_n = \partial b^m/\partial a^M \; aa^M_N \; \partial a^N/\partial b^n = bb^m_n$
\R 01	$B^m_M \; aa^M_N \; B_n^N = \partial b^m/\partial a^M \; aa^M_N \; \partial b_n/\partial a_N = bb^m_n$
\R 10	$A_M^m \; aa^M_N \; A^N_n = \partial a_M/\partial b_m \; aa^M_N \; \partial a^N/\partial b^n = bb^m_n$
\R 11	$A_M^m \; aa^M_N \; B_n^N = \partial a_M/\partial b_m \; aa^M_N \; \partial b_n/\partial a_N = bb^m_n$

Affine:

\Affine	$\underline{b}_m$	$B^m_M = A^M_m$	aa^M_N	$B^n_N = A^N_n$	db^n

\

\A 00	$B^m_M \; aa^M_N \; A^N_n = \partial b^m/\partial a^M \; aa^M_N \; \partial a^M/\partial b^n = bb^m_n$
\A 10	$A^M_m \; aa^M_N \; A^N_n = \partial a^M/\partial b^m \; aa^M_N \; \partial a^N/\partial b^n = bb^m_n$
\A 01	$B^m_M \; aa^M_N \; B^n_N = \partial b^m/\partial a^M \; aa^M_N \; \partial b^n/\partial a^N = bb^m_n$
\A 11	$A^M_m \; aa^M_N \; B^n_N = \partial a^M/\partial b^m \; aa^M_N \; \partial b^n/\partial a^N = bb^m_n$

:

(00)

Transformation $B^m_M \; aa^M_N \; A^N_n = \partial b^m/\partial a^M \; aa^M_N \; \partial a^N/\partial b^n = bb^m_n$

:

$[B^1_1 \;\; B^1_2 \;\; B^1_3] \; [aa^1_1 \;\; aa^1_2 \;\; aa^1_3] \; [A^1_1 \;\; A^1_2 \;\; A^1_3] \; [db^1]$

$$\begin{bmatrix} \underline{b}_1 & \underline{b}_2 & \underline{b}_3 \end{bmatrix} \quad \begin{bmatrix} B^2_1 & B^2_2 & B^2_3 \\ B^3_1 & B^3_2 & B^3_3 \end{bmatrix} \begin{bmatrix} aa^2_1 & aa^2_2 & aa^2_3 \\ aa^3_1 & aa^3_2 & aa^3_3 \end{bmatrix} \begin{bmatrix} A^2_1 & A^2_2 & A^2_3 \\ A^3_1 & A^3_2 & A^3_3 \end{bmatrix} \begin{bmatrix} db^2 \\ db^3 \end{bmatrix}$$

Or:

$$\begin{bmatrix} \underline{b}_1 & \underline{b}_2 & \underline{b}_3 \end{bmatrix} \begin{bmatrix} B^1_M\, aa^M_N\, A^N_1 & B^1_M\, aa^M_N\, A^N_2 & B^1_M\, aa^M_N\, A^N_3 \\ B^2_M\, aa^M_N\, A^N_1 & B^2_M\, aa^M_N\, A^N_2 & B^2_M\, aa^M_N\, A^N_3 \\ B^3_M\, aa^M_N\, A^N_1 & B^3_M\, aa^M_N\, A^N_2 & B^3_M\, aa^M_N\, A^N_3 \end{bmatrix} \begin{bmatrix} db^1 \\ db^2 \\ db^3 \end{bmatrix}$$

Where:

Orthonormal

$$\begin{bmatrix} \underline{b}_1 & \underline{b}_2 & \underline{b}_3 \end{bmatrix} \begin{bmatrix} B^1_M A^M_1 = 1 & B^1_M A^M_2 = 0 & B^1_M A^M_3 = 0 \\ B^2_M A^M_1 = 0 & B^2_M A^M_2 = 1 & B^2_M A^M_3 = 0 \\ B^3_M A^M_1 = 0 & B^3_M A^M_2 = 0 & B^3_M A^M_3 = 1 \end{bmatrix} \begin{bmatrix} db^1 \\ db^2 \\ db^3 \end{bmatrix}$$

(01)

Affine $\quad B^m_M\, aa^M_N\, B^n_N = \partial b^m/\partial a^M \;\; aa^M_N\, \partial b^n/\partial a^M \; = bb^m_n$

:

$$\begin{bmatrix} \underline{b}_1 & \underline{b}_2 & \underline{b}_3 \end{bmatrix} * \begin{bmatrix} B^1_1 & B^1_2 & B^1_3 \\ B^2_1 & B^2_2 & B^2_3 \\ B^3_1 & B^3_2 & B^3_3 \end{bmatrix} * \begin{bmatrix} aa^1_1 & aa^1_2 & aa^1_3 \\ aa^2_1 & aa^2_2 & aa^2_3 \\ aa^3_1 & aa^3_2 & aa^3_3 \end{bmatrix} * \begin{bmatrix} B^1_1 & B^2_1 & B^3_1 \\ B^1_2 & B^2_2 & B^3_2 \\ B^1_3 & B^2_3 & B^3_3 \end{bmatrix} * \begin{bmatrix} db^1 \\ db^2 \\ db^3 \end{bmatrix}$$

(10)

Affine $\quad A^M_m\, aa^M_N\, A^N_n \; = \; \partial a^M/\partial b^m \;\; aa^M_N \;\; \partial a^N/\partial b^n \; = bb^m_n$

:

$$\begin{bmatrix} \underline{b}_1 & \underline{b}_2 & \underline{b}_3 \end{bmatrix} * \begin{bmatrix} A^1_1 & A^2_1 & A^3_1 \\ A^1_2 & A^2_2 & A^3_2 \\ A^1_3 & A^2_3 & A^3_3 \end{bmatrix} * \begin{bmatrix} aa^1_1 & aa^1_2 & aa^1_3 \\ aa^2_1 & aa^2_2 & aa^2_3 \\ aa^3_1 & aa^3_2 & aa^3_3 \end{bmatrix} * \begin{bmatrix} A^1_1 & A^1_2 & A^1_3 \\ A^2_1 & A^2_2 & A^2_3 \\ A^3_1 & A^3_2 & A^3_3 \end{bmatrix} * \begin{bmatrix} db^1 \\ db^2 \\ db^3 \end{bmatrix}$$

(11)

Affine $\quad A^M_m\, aa^M_N\, B^n_N = \partial a^M/\partial b^m \;\; aa^M_N \;\; \partial b^n/\partial a^M \; = bb^m_n$

:

$$\begin{bmatrix} \underline{b}_1 & \underline{b}_2 & \underline{b}_3 \end{bmatrix} * \begin{bmatrix} A^1_1 & A^2_1 & A^3_1 \\ A^1_2 & A^2_2 & A^3_2 \\ A^1_3 & A^2_3 & A^3_3 \end{bmatrix} * \begin{bmatrix} aa^1_1 & aa^1_2 & aa^1_3 \\ aa^2_1 & aa^2_2 & aa^2_3 \\ aa^3_1 & aa^3_2 & aa^3_3 \end{bmatrix} * \begin{bmatrix} B^1_1 & B^2_1 & B^3_1 \\ B^1_2 & B^2_2 & B^3_2 \\ B^1_3 & B^2_3 & B^3_3 \end{bmatrix} * \begin{bmatrix} db^1 \\ db^2 \\ db^3 \end{bmatrix}$$

If (aa) is symmetric:

$$\begin{bmatrix} \underline{b}_1 & \underline{b}_2 & \underline{b}_3 \end{bmatrix} * \begin{bmatrix} B^1_1 & B^1_2 & B^1_3 \\ B^2_1 & B^2_2 & B^2_3 \\ B^3_1 & B^3_2 & B^3_3 \end{bmatrix} * \begin{bmatrix} 1a1 & a3a & a2a \\ a3a & 2a2 & a1a \\ a2a & a1a & 3a3 \end{bmatrix} * \begin{bmatrix} A^1_1 & A^1_2 & A^1_3 \\ A^2_1 & A^2_2 & A^2_3 \\ A^3_1 & A^3_2 & A^3_3 \end{bmatrix} * \begin{bmatrix} db^1 \\ db^2 \\ db^3 \end{bmatrix}$$

Result is:

$$\begin{bmatrix} \underline{b}_1 & \underline{b}_2 & \underline{b}_3 \end{bmatrix} * \begin{bmatrix} 1b1 & b3b & a2a \\ b3b & 2b2 & b1b \\ b2b & b1b & 3b3 \end{bmatrix} * \begin{bmatrix} db^1 \\ db^2 \\ db^3 \end{bmatrix}$$

Then (bb) is also symmetric

3.3.2.2.
Lower-Upper (aa/) Metric Transformation to (bb/) Description:

/R	$\underline{b}^m$	$B_m^M (= A_m^M)$	aa_M^N /	$(B_N^n =) A_N^n$	db_n
/R 00		$B_m^M \; aa_M^N \; A_N^n = \partial b_m/\partial a_M \; aa_M^N \; \partial a_N/\partial b_n = bb_m^n$			
/R 01		$B_m^M \; aa_M^N \; B_N^n = \partial b_m/\partial a_M \; aa_M^N \; \partial b^n/\partial a^N = bb_m^n$			
/R 10		$A_m^M \; aa_M^N \; A_N^n = \partial a^M/\partial b^m \; aa_M^N \; \partial a_N/\partial b_n = bb_m^n$			
/R 11		$A_m^M \; aa_M^N \; B_N^n = \partial a^M/\partial b^m \; aa_M^N \; \partial b^n/\partial a^N = bb_m^n$			

(00)
Transformation:
$$ba_m^M \; aa_M^N \; ab_N^n = bb_m^n$$
(01)
Affine Transformation:
$$ba_m^M \; aa_M^N \; (ba_n^N) = bb_{mn}$$
(10)
Affine Transformation:
$$(ab_M^m) \; aa_M^N \; ab_N^n = bb^{mn}$$
(11)
Affine Transformation:
$$(ab_M^m) \; aa_M^N \; (ba_n^N) = bb_n^m$$
If aa is symmetric so is bb
Affine:

/Affine	$\underline{b}^m$	$B_m^M = A_M^m$	aa_M^N /	$B_n^N = A_N^n$	db_n
/A 00		$B_m^M \; aa_M^N \; A_N^n = \partial b_m/\partial a_M \; aa_M^N \; \partial a_N/\partial b_n = bb_m^n$			
/A 10		$A_M^m \; aa_M^N \; A_N^n = \partial a_M/\partial b_m \; aa_M^N \; \partial a_N/\partial b_n = bb_m^n$			
/A 01		$B_m^M \; aa_M^N \; B_n^N = \partial b_m/\partial a_M \; aa_M^N \; \partial b_n/\partial a_N = bb_m^n$			
/A 11		$A_M^m \; aa_M^N \; B_n^N = \partial a_M/\partial b_m \; aa_M^N \; \partial b_n/\partial a_N = bb_m^n$			

3.3.3.
Delta Transformation:

(00)
Orthonormality of Transformation :
$$ba_m^M \; \delta_M^N \; ab_N^n = ba_m^M \; ab_M^n = \partial b_m/ \partial a_M \;\; \partial a_M / \partial b_n = \delta_m^n$$
(01)
Orthonormality of Transformation :
$$ba_m^M \; \delta_M^N \; (ba_n^N) = ba_m^M \; ba_n^M = \partial b_m/ \partial a_M \;\; \partial b_n / \partial a_M = \delta_m^n$$
(10)
Orthonormality of Transformation :
$$(ab_M^m) \; \delta_M^N \; ab_N^n = ab_M^m \; ab_M^n = \partial a_M/ \partial b_m \;\; \partial a_M / \partial b_n = \delta_m^n$$
(11)
Orthonormality of Transformation :
$$(ab_M^m) \; \delta_M^N \; (ba_n^N) = ab_M^m \; ba_n^M = \partial a_M/ \partial b_m \;\; \partial b_n / \partial a_M = \delta_m^n$$

3.3.4.

Different Systems Metric Transformations:

$aa_{\alpha\beta}\,(ab^{\alpha}_{a}=ba^{\alpha}_{a})\,(ab^{\beta}_{b}=ba^{\beta}_{b})=\underline{b}_a\cdot\underline{b}_b=bb_{ab}$		$bb^{ab}=\underline{b}^a\cdot\underline{b}^b=(ba^a_\alpha=ab^a_\alpha)\,(ba^b_\beta=ab^b_\beta)\,aa^{\alpha\beta}$
$bb_{ab}\,ba^a_\alpha\,ba^b_\beta=\underline{a}_\alpha\cdot\underline{a}_\beta=aa_{\alpha\beta}$ Post-Transformer up (low Post-Transformer)	\=/ /=\ \=/ /=\	$\underline{a}^\alpha\cdot\underline{a}^\beta=aa^{\alpha\beta}=a^\alpha_a\,a^\beta_b\,bb^{ab}$ (Post-Transformer low) up Post-Transformer
$bb_{ab}\,(ab^a_\alpha\,ab^b_\beta)=\underline{a}_\alpha\cdot\underline{a}_\beta=aa_{\alpha\beta}$ Post-Transformer low (up Post-Transformer)	\=/ /=\ /=\ \=/	$\underline{a}^\alpha\cdot\underline{a}^\beta=aa^{\alpha\beta}=(ba^\alpha_a\,ba^\beta_b)\,bb^{ab}$ (Post-Transformer up) low Post-Transformer
$aa_{\mu\theta}ab^\mu_M\,(=ba^\mu_M)\,ab^\theta_Q(=ba^\theta_Q)=bb_{MQ}$ $aa^{\mu\theta}ab^M_\mu\,(=ba^M_\mu)\,ab^Q_\theta(=ba^Q_\theta)=bb^{MQ}$		$bb^{NP}=(ab^N_\nu=)\,ba^N_\nu\,(ab^P_\pi=)\,ba^P_\pi\,aa^{\nu\pi}$ $bb_{NP}=(ab^\nu_N=)\,ba^\nu_N\,(ab^\pi_P=)ba^\pi_P\,aa_{\nu\pi}$
$aa_{\mu\theta}\,ab^\mu_M\,ab^\theta_Q=bb_{MQ}$ $aa^{\mu\theta}\,ab^M_\mu\,ab^Q_\theta=bb^{MQ}$		$bb^{NP}=ba^N_\nu\,ba^P_\pi\,aa^{\nu\pi}$ $bb_{NP}=ba^\nu_N\,ba^\pi_P\,aa_{\nu\pi}$
$aa_{\mu\theta}\,ba^\mu_M\,ba^\theta_Q=bb_{MQ}$ $aa^{\mu\theta}\,ba^M_\mu\,ba^Q_\theta=bb^{MQ}$		$bb^{NP}=ab^N_\nu\,ab^P_\pi\,aa^{\nu\pi}$ $bb_{NP}=ab^\nu_N\,ab^\pi_P\,aa_{\nu\pi}$
$aa_{\alpha\beta}\,ab^\alpha_a\,ab^\beta_b=bb_{ab}$ $aa_{\alpha\beta}\,ba^\alpha_a\,ba^\beta_b=bb_{ab}$		$bb^{ab}=ba^a_\alpha\,ba_\beta\,aa^{\alpha\beta}$ $bb^{ab}=ab^a_\alpha\,ab^b_\beta\,aa^{\alpha\beta}$
$\underline{bb}_{\mu\theta}\,[ba^\mu_M ba^\theta_Q=ab^\mu_M ab^\theta_Q]=\underline{aa}_{MQ}$ $\underline{bb}^{\mu\theta}\,[ba^M_\mu ba^Q_\theta=ab^M_\mu ab^Q_\theta]=\underline{aa}^{MQ}$		$aa^{NP}=[ba^N_\nu ba^P_\pi=ab^N_\nu ab^P_\pi]\,bb^{\nu\pi}$ $aa_{NP}=[ba^\nu_N ba^\pi_P=ab^\nu_N ab^\pi_P]\,bb_{\nu\pi}$
$bb_{\mu\theta}$ and aa_{MQ} $bb^{\mu\theta}$ and aa^{MQ}		aa^{NP} and $bb^{\nu\pi}$ aa_{NP} and $bb_{\nu\pi}$
$bb_{ab}\,ba^a_\alpha\,ba^b_\beta=aa_{\alpha\beta}$ $bb_{ab}\,ab^a_\alpha\,ab^b_\beta=aa_{\alpha\beta}$		$aa^{\alpha\beta}=ab^\alpha_a\,ab^\beta_b\,bb^{ab}$ $aa^{\alpha\beta}=ba^\alpha_a\,ba^\beta_b\,bb^{ab}$
$bb^{ab}\,ba^\alpha_a\,ba^\beta_b=aa^{\alpha\beta}$ $bb^{ab}\,ab^\alpha_a\,ab^\beta_b=aa^{\alpha\beta}$		$aa_{\alpha\beta}=ab^a_\alpha\,ab^b_\beta\,bb_{ab}$ $aa_{\alpha\beta}=ba^a_\alpha\,ba_\beta\,bb_{ab}$
$aa_{\alpha\beta}\,ab^\alpha_a\,ab^\beta_b=bb_{ab}$ $aa_{\alpha\beta}\,ba^\alpha_a\,ba^\beta_b=bb_{ab}$		$bb^{ab}=ba^a_\alpha\,ba^b_\beta\,aa^{\alpha\beta}$ $bb^{ab}=ab^a_\alpha\,ab^b_\beta\,aa^{\alpha\beta}$

Different Systems:

3.3.4.1.

General System Metric Transformations:

General: bN

$aa_{\alpha\beta}\, ab^{\alpha}{}_{a}\, ab^{\beta}{}_{b} = bb_{ab}$	$aa_{\alpha}{}^{\beta}\, ab^{\alpha}{}_{a}\, ab_{\beta}{}^{b} = bb^{b}{}_{a}$	$aa^{\alpha\beta}\, ab_{\alpha}{}^{a}\, ab_{\beta}{}^{b} = bb^{ab}$
$ff_{\alpha\beta}\, fb^{\alpha}{}_{f}\, fb^{\beta}{}_{b} = bb_{fb}$	$ff_{\alpha}{}^{\beta}\, fb^{\alpha}{}_{f}\, fb_{\beta}{}^{b} = bb^{b}{}_{f}$	$ff^{\alpha\beta}\, fb_{\alpha}{}^{f}\, fb_{\beta}{}^{b} = bb^{fb}$
$rr_{\alpha\beta}\, rb^{\alpha}{}_{r}\, rb^{\beta}{}_{b} = bb_{rb}$	$rr_{\alpha}{}^{\beta}\, rb^{\alpha}{}_{r}\, rb_{\beta}{}^{b} = bb^{b}{}_{r}$	$rr^{\alpha\beta}\, rb_{\alpha}{}^{r}\, rb_{\beta}{}^{b} = bb^{rb}$
$xx_{\alpha\beta}\, xb^{\alpha}{}_{x}\, xb^{\beta}{}_{b} = bb_{xb}$	$xx_{\alpha}{}^{\beta}\, xb^{\alpha}{}_{x}\, xb_{\beta}{}^{b} = bb^{b}{}_{x}$	$xx^{\alpha\beta}\, xb_{\alpha}{}^{x}\, xb_{\beta}{}^{b} = bb^{xb}$
$oo_{\alpha\beta}\, ob^{\alpha}{}_{o}\, ob^{\beta}{}_{b} = bb_{ob}$	$oo_{\alpha}{}^{\beta}\, ob^{\alpha}{}_{o}\, ob_{\beta}{}^{b} = bb^{b}{}_{o}$	$oo^{\alpha\beta}\, ob_{\alpha}{}^{o}\, ob_{\beta}{}^{b} = bb^{ob}$

3.3.4.2. General Unitary System Metric Transformations:

General Unitary: gN

$aa_{\alpha\beta}\, ag^{\alpha}{}_{a}\, ag^{\beta}{}_{g} = gg_{ag}$	$aa_{\alpha}{}^{\beta}\, ag^{\alpha}{}_{a}\, ag_{\beta}{}^{g} = gg^{g}{}_{a}$	$aa^{\alpha\beta}\, ag_{\alpha}{}^{a}\, ag_{\beta}{}^{g} = gg^{ag}$
$ff_{\alpha\beta}\, fg^{\alpha}{}_{f}\, fg^{\beta}{}_{g} = gg_{fg}$	$ff_{\alpha}{}^{\beta}\, fg^{\alpha}{}_{f}\, fg_{\beta}{}^{g} = gg^{g}{}_{f}$	$ff^{\alpha\beta}\, fg_{\alpha}{}^{f}\, fg_{\beta}{}^{g} = gg^{fg}$
$rr_{\alpha\beta}\, rg^{\alpha}{}_{r}\, rg^{\beta}{}_{g} = gg_{rg}$	$rr_{\alpha}{}^{\beta}\, rg^{\alpha}{}_{r}\, rg_{\beta}{}^{g} = gg^{g}{}_{r}$	$rr^{\alpha\beta}\, rg_{\alpha}{}^{r}\, rg_{\beta}{}^{g} = gg^{rg}$
$xx_{\alpha\beta}\, xg^{\alpha}{}_{x}\, xg^{\beta}{}_{g} = gg_{xg}$	$xx_{\alpha}{}^{\beta}\, xg^{\alpha}{}_{x}\, xg_{\beta}{}^{g} = gg^{g}{}_{x}$	$xx^{\alpha\beta}\, xg_{\alpha}{}^{x}\, xg_{\beta}{}^{g} = gg^{xg}$
$oo_{\alpha\beta}\, og^{\alpha}{}_{o}\, og^{\beta}{}_{g} = gg_{og}$	$oo_{\alpha}{}^{\beta}\, og^{\alpha}{}_{o}\, og_{\beta}{}^{g} = gg^{g}{}_{o}$	$oo^{\alpha\beta}\, og_{\alpha}{}^{o}\, og_{\beta}{}^{g} = gg^{og}$

3.3.4.3. Orthogonal System Metric Transformations:

Orthogonal: sN

$aa_{\alpha\beta}\, as^{\alpha}{}_{a}\, as^{\beta}{}_{s} = ss_{as}$	$aa_{\alpha}{}^{\beta}\, as^{\alpha}{}_{a}\, as_{\beta}{}^{s} = ss^{s}{}_{a}$	$aa^{\alpha\beta}\, as_{\alpha}{}^{a}\, as_{\beta}{}^{s} = ss^{as}$
$ff_{\alpha\beta}\, fs^{\alpha}{}_{f}\, fs^{\beta}{}_{s} = ss_{fs}$	$ff_{\alpha}{}^{\beta}\, fs^{\alpha}{}_{f}\, fs_{\beta}{}^{s} = ss^{s}{}_{f}$	$ff^{\alpha\beta}\, fs_{\alpha}{}^{f}\, fs_{\beta}{}^{s} = ss^{fs}$
$rr_{\alpha\beta}\, rs^{\alpha}{}_{r}\, rs^{\beta}{}_{s} = ss_{rs}$	$rr_{\alpha}{}^{\beta}\, rs^{\alpha}{}_{r}\, rs_{\beta}{}^{s} = ss^{s}{}_{r}$	$rr^{\alpha\beta}\, rs_{\alpha}{}^{r}\, rs_{\beta}{}^{s} = ss^{rs}$
$xx_{\alpha\beta}\, xs^{\alpha}{}_{x}\, xs^{\beta}{}_{s} = ss_{xs}$	$xx_{\alpha}{}^{\beta}\, xs^{\alpha}{}_{x}\, xs_{\beta}{}^{s} = ss^{s}{}_{x}$	$xx^{\alpha\beta}\, xs_{\alpha}{}^{x}\, xs_{\beta}{}^{s} = ss^{xs}$
$oo_{\alpha\beta}\, os^{\alpha}{}_{o}\, os^{\beta}{}_{s} = ss_{os}$	$oo_{\alpha}{}^{\beta}\, os^{\alpha}{}_{o}\, os_{\beta}{}^{s} = ss^{s}{}_{o}$	$oo^{\alpha\beta}\, os_{\alpha}{}^{o}\, os_{\beta}{}^{s} = ss^{os}$

3.3.4.4. Orthonormal System Metric Transformations:

Orthonormal: yN

$aa_{\alpha\beta}\, ay^{\alpha}{}_{a}\, ay^{\beta}{}_{y} = yy_{ay}$	$aa_{\alpha}{}^{\beta}\, ay^{\alpha}{}_{a}\, ay_{\beta}{}^{y} = yy^{y}{}_{a}$	$aa^{\alpha\beta}\, ay_{\alpha}{}^{a}\, ay_{\beta}{}^{y} = yy^{ay}$
$ff_{\alpha\beta}\, fy^{\alpha}{}_{f}\, fy^{\beta}{}_{y} = yy_{fy}$	$ff_{\alpha}{}^{\beta}\, fy^{\alpha}{}_{f}\, fy_{\beta}{}^{y} = yy^{y}{}_{f}$	$ff^{\alpha\beta}\, fy_{\alpha}{}^{f}\, fy_{\beta}{}^{y} = yy^{fy}$
$rr_{\alpha\beta}\, ry^{\alpha}{}_{r}\, ry^{\beta}{}_{y} = yy_{ry}$	$rr_{\alpha}{}^{\beta}\, ry^{\alpha}{}_{r}\, ry_{\beta}{}^{y} = yy^{y}{}_{r}$	$rr^{\alpha\beta}\, ry_{\alpha}{}^{r}\, ry_{\beta}{}^{y} = yy^{ry}$
$xx_{\alpha\beta}\, xy^{\alpha}{}_{x}\, xy^{\beta}{}_{y} = yy_{xy}$	$xx_{\alpha}{}^{\beta}\, xy^{\alpha}{}_{x}\, xy_{\beta}{}^{y} = yy^{y}{}_{x}$	$xx^{\alpha\beta}\, xy_{\alpha}{}^{x}\, xy_{\beta}{}^{y} = yy^{xy}$
$oo_{\alpha\beta}\, oy^{\alpha}{}_{o}\, oy^{\beta}{}_{y} = yy_{oy}$	$oo_{\alpha}{}^{\beta}\, oy^{\alpha}{}_{o}\, oy_{\beta}{}^{y} = yy^{y}{}_{o}$	$oo^{\alpha\beta}\, oy_{\alpha}{}^{o}\, oy_{\beta}{}^{y} = yy^{oy}$

3.3.4.5. Fixed System Metric Transformations:

Fixed: oN

$aa_{\alpha\beta}\, ao^{\alpha}{}_{a}\, ao^{\beta}{}_{o} = oo_{ao}$	$aa_{\alpha}{}^{\beta}\, ao^{\alpha}{}_{a}\, ao_{\beta}{}^{o} = oo^{o}{}_{a}$	$aa^{\alpha\beta}\, ao_{\alpha}{}^{a}\, ao_{\beta}{}^{o} = oo^{ao}$

$$ff_{\alpha\beta}\, fo^{\alpha}_{\;f}\, fo^{\beta}_{\;o} = oo_{fo} \qquad ff^{\beta}_{\;\alpha}\, fo^{\alpha}_{\;f}\, fo^{o}_{\;\beta} = oo^{o}_{\;f} \qquad ff^{\alpha\beta}\, fo^{f}_{\;\alpha}\, fo^{o}_{\;\beta} = oo^{fo}$$

$$rr_{\alpha\beta}\, ro^{\alpha}_{\;r}\, ro^{\beta}_{\;o} = oo_{ro} \qquad rr^{\beta}_{\;\alpha}\, ro^{\alpha}_{\;r}\, ro^{o}_{\;\beta} = oo^{o}_{\;r} \qquad rr^{\alpha\beta}\, ro^{r}_{\;\alpha}\, ro^{o}_{\;\beta} = oo^{ro}$$

$$xx_{\alpha\beta}\, xo^{\alpha}_{\;x}\, xo^{\beta}_{\;o} = oo_{xo} \qquad xx^{\beta}_{\;\alpha}\, xo^{\alpha}_{\;x}\, xo^{o}_{\;\beta} = oo^{o}_{\;x} \qquad xx^{\alpha\beta}\, xo^{x}_{\;\alpha}\, xo^{o}_{\;\beta} = oo^{xo}$$

$$oo_{\alpha\beta}\, oo^{\alpha}_{\;o}\, oo^{\beta}_{\;o} = oo_{oo} \qquad oo^{\beta}_{\;\alpha}\, oo^{\alpha}_{\;o}\, oo^{o}_{\;\beta} = oo^{o}_{\;o} \qquad oo^{\alpha\beta}\, oo^{o}_{\;\alpha}\, oo^{o}_{\;\beta} = oo^{oo}$$

3.4.
Norms of Metric Transformation and Bases to Bases Transformation:

Transformation of Metric of bases a to Metric of bases b:

$$aa_{MP}\,(ab^{M}_{\;N} = ba_{N}^{\;M})\,(ab^{P}_{\;Q} = ba_{Q}^{\;P}) = bb_{NQ} \qquad bb^{NQ} = (ab_{M}^{\;N} = ba^{N}_{\;M})\,(ab_{P}^{\;Q} = ba^{Q}_{\;P})\, aa^{MP}$$

$$aa^{MP}\,(ab_{M}^{\;N} = ba^{N}_{\;M})(ab_{P}^{\;Q} = ba^{Q}_{\;P}) = bb^{NQ} \qquad bb_{NQ} = (ab^{M}_{\;N} = ba_{N}^{\;M})\,(ab^{P}_{\;Q} = ba_{Q}^{\;P})\, aa_{MP}$$

Consider:

$$aa_{MP}\, ab^{M}_{\;N}\, ab^{P}_{\;Q} = bb_{NQ} \qquad bb^{NQ} = ba^{N}_{\;M}\, ba^{Q}_{\;P}\, aa^{MP}$$

$$aa^{MP}\, ab_{M}^{\;N}\, ab_{P}^{\;Q} = bb^{NQ} \qquad bb_{NQ} = ba_{N}^{\;M}\, ba_{Q}^{\;P}\, aa_{MP}$$

Then Norm of Metric Transformation or Determinant:

$$\|aa_{MP}\|\;\|ab^{M}_{\;N}\|\;\|ab^{P}_{\;Q}\| = \|bb_{NQ}\| \qquad \|bb^{NQ}\| = \|ba^{N}_{\;M}\|\;\|ba^{Q}_{\;P}\|\;\|aa^{MP}\|$$

$$\|aa^{MP}\|\;\|ab_{M}^{\;N}\|\;\|ab_{P}^{\;Q}\| = \|bb^{NQ}\| \qquad \|bb_{NQ}\| = \|ba_{N}^{\;M}\|\;\|ba_{Q}^{\;P}\|\;\|aa_{MP}\|$$

Or:

$$(\alpha\alpha)\;\|ab^{M}_{\;N}\|\;\|ab^{P}_{\;Q}\| = (\beta\beta) \qquad 1/(\beta\beta) = \|ba^{N}_{\;M}\|\;\|ba^{Q}_{\;P}\|\;1/(\alpha\alpha)$$

$$1/(\alpha\alpha)\;\|ab_{M}^{\;N}\|\;\|ab_{P}^{\;Q}\| = 1/(\beta\beta) \qquad (\beta\beta) = \|ba_{N}^{\;M}\|\;\|ba_{Q}^{\;P}\|\;(\alpha\alpha)$$

$$\underline{a}^{N}\, aa_{NM} = \underline{a}_{M}$$

$$\underline{a}_{M}\, aa^{MN} = \underline{a}^{N}$$

$$\|\underline{a}_{M}\| = \sqrt{(\underline{a}_{M}\cdot\underline{a}_{M})} = \sqrt{(\underline{aa}_{MM})} \text{ no sum}$$

$$\|\underline{a}^{N}\| = \sqrt{(\underline{a}^{N}\cdot\underline{a}^{N})} = \sqrt{(\underline{aa}^{NN})} \text{ no sum}$$

$$\|_{aa}\|\;\|^{a}_{\;b}\|\;\|^{a}_{\;b}\| = \|_{bb}\| \qquad \|^{bb}\| = \|^{b}_{\;a}\|\;\|^{b}_{\;a}\|\;\|^{aa}\|$$

$$\|^{aa}\|\;\|_{a}^{\;b}\|\;\|_{a}^{\;b}\| = \|^{bb}\| \qquad \|_{bb}\| = \|_{b}^{\;a}\|\;\|_{b}^{\;a}\|\;\|_{aa}\|$$

Or:

$$\|_{aa}\| * \|^{a}_{\;b}\|^{2} = \|_{bb}\| \qquad \|^{bb}\| = \|^{b}_{\;a}\|^{2} * \|^{aa}\|$$

$$\|^{aa}\| * \|_{a}^{\;b}\|^{2} = \|^{bb}\| \qquad \|_{bb}\| = \|_{b}^{\;a}\|^{2} * \|_{aa}\|$$

So:

Ratio of Metric Norms multiplied by Square of Norm of bases to bases Transformation is equal one

$$\|_{aa}\|*\|^{a}_{\;b}\|^{2}*\|^{bb}\| \equiv (\alpha\alpha)*\|^{a}_{\;b}\|^{2}*1/(\beta\beta) = 1 \qquad \|_{bb}\|*\|^{b}_{\;a}\|^{2}*\|^{aa}\| \equiv (\beta\beta)*\|^{b}_{\;a}\|^{2}*1/(\alpha\alpha) = 1$$

$$||^{aa}||*||_a{}^b||^2*||_{bb}|| \equiv 1/(\alpha\alpha)*||_a{}^b||^2*(\beta\beta) = 1 \qquad ||^{bb}||*||_b{}^a||^2*||_{aa}|| \equiv 1/(\beta\beta)*||^b{}_a||^2*(\alpha\alpha) = 1$$

And the Norm of bases a to bases b Transformation:

	$||^a{}_b|| = \beta/\alpha$	$||^b{}_a|| = \alpha/\beta$	
	$||_a{}^b|| = \alpha/\beta$	$||_b{}^a|| = \beta/\alpha$	

$$\alpha\,||^a{}_b||\,1/\beta = 1 = 1/\beta\,||_b{}^a||\,\alpha$$

$$1/\alpha\,||_a{}^b||\,\beta = 1 = \beta\,||^b{}_a||\,1/\alpha$$

$$||^a{}_b|| = ||_b{}^a|| = \beta/\alpha$$

$$||_a{}^b|| = ||^b{}_a|| = \alpha/\beta$$

Apply to different systems:

	General b_N	General Unitary g_N	Orthogonal s_N	Orthonormal y_N	Fixed o_N
a_M	$||^a{}_b|| = \beta/\alpha$	$||^a{}_g|| = \gamma/\alpha$	$||^a{}_s|| = \sigma/\alpha$	$||^a{}_y|| = 1/\alpha$	$||^a{}_o|| = 1/\alpha$
f_M	$||^f{}_b|| = \beta/1$	$||^f{}_g|| = \gamma/1$	$||^f{}_s|| = \sigma/1$	$||^f{}_y|| = 1/1$	$||^f{}_o|| = 1/1$
r_M	$||^r{}_b|| = \beta/\rho$	$||^r{}_g|| = \gamma/\rho$	$||^r{}_s|| = \sigma/\rho$	$||^r{}_y|| = 1/\rho$	$||^r{}_o|| = 1/\rho$
x_M	$||^x{}_b|| = \beta/1$	$||^x{}_g|| = \gamma/1$	$||^x{}_s|| = \sigma/1$	$||^x{}_y|| = 1/1$	$||^x{}_o|| = 1/1$
o_M	$||^o{}_b|| = \beta/1$	$||^o{}_g|| = \gamma/1$	$||^o{}_s|| = \sigma/1$	$||^o{}_y|| = 1/1$	$||_o|| = 1/1$

And:

	General b_N	General Unitary g_N	Orthogonal s_N	Orthonormal y_N	Fixed o_N
a_M	$||_a{}^b|| = \alpha/\beta$	$||_a{}^g|| = \alpha/1$	$||_a{}^s|| = \alpha/\sigma$	$||_a{}^y|| = \alpha/1$	$||_a{}^o|| = \alpha/1$
f_M	$||_f{}^b|| = 1/\beta$	$||_f{}^g|| = 1/1$	$||_f{}^s|| = 1/\sigma$	$||_f{}^y|| = 1/1$	$||_f{}^o|| = 1/1$
r_M	$||_r{}^b|| = \rho/\beta$	$||_r{}^g|| = \rho/1$	$||_r{}^s|| = \rho/\sigma$	$||_r{}^y|| = \rho/1$	$||_r{}^o|| = \rho/1$
x_M	$||_x{}^b|| = 1/\beta$	$||_x{}^g|| = 1/1$	$||_x{}^s|| = 1/\beta$	$||_x{}^y|| = 1/1$	$||_x{}^o|| = 1/1$
o_M	$||_o{}^b|| = 1/\beta$	$||_o{}^g|| = 1/1$	$||_o{}^s|| = 1/\beta$	$||_o{}^y|| = 1/1$	$||_o{}^o|| = 1/1$

3.5. *Isotropic3 Identity:*

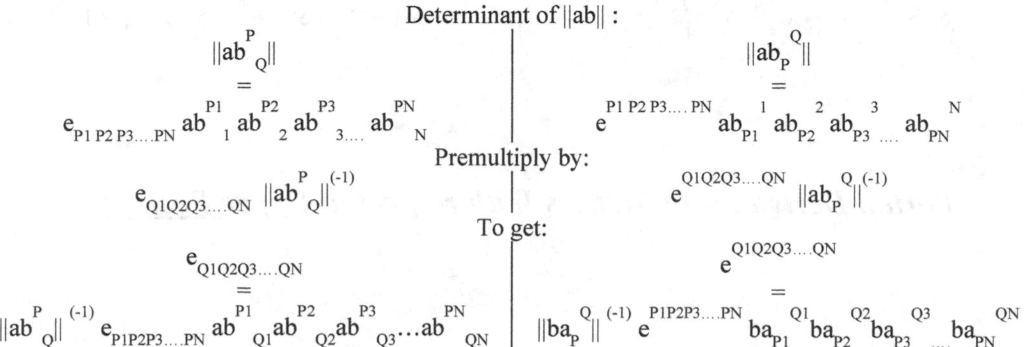

Determinant of $||ab||$:

$$||ab^P{}_Q|| = e_{P1\,P2\,P3....PN}\, ab^{P1}{}_1\, ab^{P2}{}_2\, ab^{P3}{}_{3....}\, ab^{PN}{}_N \qquad ||ab_P{}^Q|| = e^{P1\,P2\,P3....\,PN}\, ab_{P1}{}^1\, ab_{P2}{}^2\, ab_{P3}{}^3 .... ab_{PN}{}^N$$

Premultiply by:

$$e_{Q1Q2Q3....QN}\,||ab^P{}_Q||^{(-1)} \qquad e^{Q1Q2Q3....QN}\,||ab_P{}^Q||^{(-1)}$$

To get:

$$e_{Q1Q2Q3....QN} = ||ab^P{}_Q||^{(-1)}\, e_{P1P2P3....PN}\, ab^{P1}{}_{Q1}\, ab^{P2}{}_{Q2}\, ab^{P3}{}_{Q3} \cdots ab^{PN}{}_{QN}$$

$$e^{Q1Q2Q3....QN} = ||ba_P{}^Q||^{(-1)}\, e^{P1P2P3....PN}\, ba_{P1}{}^{Q1}\, ba_{P2}{}^{Q2}\, ba_{P3}{}^{Q3} .... ba_{PN}{}^{QN}$$

Isotropic3 Identity Determinant:

$$e_{abc}(a)\ \|ab^{P}{}_{Q}\| = e_{\alpha\beta\chi}(a)\ ab^{\alpha}{}_{a}\ ab^{\beta}{}_{b}\ ab^{\chi}{}_{c}$$

$$e_{\alpha\beta\chi}(b)\ \|ba^{P}{}_{Q}\| = e_{abc}(b)\ ba^{a}{}_{\alpha}\ ba^{b}{}_{\beta}\ ba^{c}{}_{\chi}$$

$$e^{abc}(a)\ \|ab_{P}{}^{Q}\| = e^{\alpha\beta\chi}(a)\ ab_{\alpha}{}^{a}\ ab_{\beta}{}^{b}\ ab_{\chi}{}^{c}$$

$$e^{\alpha\beta\chi}(b)\ \|ba_{P}{}^{Q}\| = e^{abc}(b)\ ba_{a}{}^{\alpha}\ ba_{b}{}^{\beta}\ ba_{c}{}^{\chi}$$

But Transformation of Isotropic3 Identity Determinant:

$$e_{\alpha\beta\chi}(a)\ ab^{\alpha}{}_{a}\ ab^{\beta}{}_{b}\ ab^{\chi}{}_{c} = e_{abc}(b)$$

$$e_{abc}(b)\ ba^{a}{}_{\alpha}\ ba^{b}{}_{\beta}\ ba^{c}{}_{\chi} = e_{\alpha\beta\chi}(a)$$

$$e^{\alpha\beta\chi}(a)\ ab_{\alpha}{}^{a}\ ab_{\beta}{}^{b}\ ab_{\chi}{}^{c} = e^{abc}(b)$$

$$e^{abc}(b)\ ba_{a}{}^{\alpha}\ ba_{b}{}^{\beta}\ ba_{c}{}^{\chi} = e^{\alpha\beta\chi}(a)$$

So:

Transformation of e:

$$e_{abc}(a)\ \|ab^{P}{}_{Q}\| = e_{abc}(b)$$

$$e_{\alpha\beta\chi}(b)\ \|ba^{P}{}_{Q}\| = e_{\alpha\beta\chi}(a)$$

$$e^{abc}(a)\ \|ab_{P}{}^{Q}\| = e^{abc}(b)$$

$$e^{\alpha\beta\chi}(b)\ \|ba_{P}{}^{Q}\| = e^{\alpha\beta\chi}(a)$$

But:

Transformation Norm of Degree One:

$$\|{}^{a}{}_{b}\| = \|{}^{a}{}_{b}\| = \beta/\alpha$$

$$\|{}_{a}{}^{b}\| = \|{}_{a}{}^{b}\| = \alpha/\beta$$

So Transformation of e:

$$e_{abc}(a)\ \|{}^{a}{}_{b}\| = e_{abc}(b)$$

$$e_{\alpha\beta\chi}(b)\ \|{}^{b}{}_{a}\| = e_{\alpha\beta\chi}(a)$$

$$e^{abc}(a)\ \|{}_{a}{}^{b}\| = e^{abc}(b)$$

$$e^{\alpha\beta\chi}(b)\ \|{}_{b}{}^{a}\| = e^{\alpha\beta\chi}(a)$$

$$e_{\alpha\beta\chi}(a) = aa_{\alpha\phi}\ aa_{\beta\gamma}\ aa_{\chi\eta}\ e^{\phi\gamma\eta}(a)$$

$$e^{\alpha\beta\chi}(a) = aa^{\alpha\phi}\ aa^{\beta\gamma}\ aa^{\alpha\eta}\ e_{\phi\gamma\eta}(a)$$

$$e^{abc} = bb^{af}\ bb^{bg}\ bb^{ch}\ e_{fgh}$$

$$e_{abc} = bb_{af}\ bb_{bg}\ bb_{ch}\ e^{fgh}$$

$$6 = e^{\alpha\beta\chi}\ e_{\alpha\beta\chi}$$

$$6 = e^{\alpha\beta\chi}\ aa_{\alpha\phi}\ aa_{\beta\gamma}\ aa_{\chi\eta}\ e^{\phi\gamma\eta}$$

$$6 = e_{\alpha\beta\chi}\ aa^{\alpha\phi}\ aa^{\beta\gamma}\ aa^{\alpha\eta}\ e_{\phi\gamma\eta}$$

$$6 = e_{abc}\ e^{abc}$$

$$6 = e_{abc}\ bb^{af}\ bb^{bg}\ bb^{ch}\ e_{fgh}$$

$$6 = e^{abc}\ bb_{af}\ bb_{bg}\ bb_{ch}\ e^{fgh}$$

$$e^{\alpha\beta\chi}\ aa_{\alpha\phi}\ aa_{\beta\gamma}\ e^{\phi\gamma\eta} = 2\ aa^{\alpha\eta}$$

$$e_{\alpha\beta\chi}\ aa^{\alpha\phi}\ aa^{\beta\gamma}\ e_{\phi\gamma\eta} = 2\ aa_{\chi\eta}$$

$$e_{abc}\ bb^{af}\ bb^{bg}\ e_{fgh} = 2\ bb_{ch}$$

$$e^{abc}\ bb_{af}\ bb_{bg}\ e^{fgh} = 2\ bb^{ch}$$

$$\|\delta\| = (1/6)\ e_{ijk}\ e_{pqr}\ \delta^{ip}\ \delta^{jq}\ \delta^{kr} = 1$$

$$\|\delta\| = (1/6)\ e_{ijk}\ e^{pqr}\ \delta^{i}{}_{p}\ \delta^{j}{}_{q}\ \delta^{k}{}_{r} = 1$$

$$\|\delta\| = (1/6)\ e_{ijk}\ e^{pqr}\ \delta_{p}{}^{i}\ \delta_{q}{}^{j}\ \delta_{r}{}^{k} = 1$$

$$\|\delta\| = (1/6)\ e^{ijk}\ e^{pqr}\ \delta_{pi}\ \delta_{qj}\ \delta_{rk} = 1$$

3.6.
Partial Derivative of Metrics With respect to Upper Position:

Partial Derivative of
Lower and *Upper* Metrics
With respect to *Upper Position*

The following

$$[aa_{MN}]_{,K} = + [a3a_{KMN} + a3a_{NKM}]$$

$$[aa^{MN}]_{,K} = - [a3a_{K}^{\ MN} + a3a_{N}^{\ KM}]$$

Will be discussed in what follows:

3.6.1. Trio Symbol:

Trio Symbol
and
Christoffel Two Symbols of
Upper and Lower Third Index

Trio Symbol with respect to first two indices $_{JK}$ Symmetry

Trio Symbol

$$a3a_{JKn} \equiv ½(aa_{nK,J} + aa_{Jn,K} - aa_{JK,n})$$

Upper and Lower Christoffel Symbols:

Upper Christoffel Symbol

$$a3a_{JK}^{\ \ Q} \equiv a3a_{JKn}\, aa^{nQ}$$

Lower Christoffel Symbol ≡ Trio Symbol

$$a3a_{JKn} \equiv ½ (aa_{nK,J} + aa_{Jn,K} - aa_{JK,n})$$

Symmetry of first two indices:

$$a3a_{JK}^{\ \ Q} = a3a_{KJ}^{\ \ Q} \qquad a3a_{JKQ} = a3a_{KJQ}$$

Raising and Lowering of Christoffel Symbols:

$$a3a_{JK}^{\ \ Q}\, aa_{QN} = a3a_{JKN} \qquad a3a_{JKQ}\, aa^{QN} = a3a_{JK}^{\ \ N}$$

Lower Christoffel Symbol Upper Christoffel Symbol

Raising of First Index in Christoffel Symbols:

$$aa^{IJ}\, a3a_{JN}^{\ \ Q} = a3a^{IQ}_{\ \ N} \qquad aa^{IJ}\, a3a_{JKQ} = a3a^{I}_{\ KQ}$$

Symmetry with respect to first two indices

$$a3a^{IQ}_{\ \ N} = a3a^{NQ}_{\ \ I} \qquad a3a^{I}_{\ KQ} = a3a^{K}_{\ IQ}$$

Raising of Second Index in Christoffel Symbols:

$$a3a_{JN}^{\ \ Q}\, aa^{NK} = a3a_{J}^{\ KQ} \qquad a3a_{JKQ}\, aa^{NK} = a3a_{J\ Q}^{\ K}$$

Symmetry with respect to first two indices

$a3a_{J}^{\ KQ} = a3a_{K}^{\ JQ}$ $\qquad$ $a3a_{J\ Q}^{\ K} = a3a_{K\ Q}^{\ J}$

Raising of First Index in Christoffel Symbols:

$aa^{IJ}\ a3a_{JN}^{\ \ Q}\ aa^{NK} = a3a^{IKQ}$ $\qquad$ $aa^{IJ}\ a3a_{JKQ}\ aa^{NK} = a3a^{IN}_{\ \ Q}$

Symmetry with respect to first two indices

$a3a^{IKQ} = a3a^{KIQ}$ $\qquad$ $a3a^{IN}_{\ \ Q} = a3a^{NI}_{\ \ Q}$

Symmetry of indices:

$a3a_{JK}^{\ \ Q} = a3a_{KJ}^{\ \ Q}$ $\qquad$ $a3a_{JKQ} = a3a_{KJQ}$

$a3a^{I\ Q}_{\ N} = a3a^{N\ Q}_{\ I}$ $\qquad$ $a3a^{I}_{\ KQ} = a3a^{K}_{\ IQ}$

$a3a_{J}^{\ KQ} = a3a_{K}^{\ JQ}$ $\qquad$ $a3a_{J\ Q}^{\ K} = a3a_{K\ Q}^{\ J}$

$a3a^{IKQ} = a3a^{KIQ}$ $\qquad$ $a3a^{IN}_{\ \ Q} = a3a^{NI}_{\ \ Q}$

Following articles are Discusseed In Terms of Trio Symbol (Christoffel)

3.6.2.
Partial Derivative of Metrics:

Partial Derivative of
Lower and *Upper* Metrics
With respect to *Upper Position*

The following

$$[aa_{MN}]_{,K} = +\ [\,a3a_{KMN} + a3a_{NKM}]$$

$$[aa^{MN}]_{,K} = -\ [\,a3a_{K}^{\ MN} + a3a_{N}^{\ KM}\,]$$

Will be discussed in what follows:

3.6.2.1.
Partial Derivative of *Lower* Metric With Respect to *Upper Position*

Add:

$$\tfrac{1}{2}\,(\,aa_{NJ,L} + aa_{NL,J} - aa_{JL,N}) = a3a_{JLN}$$

$$\tfrac{1}{2}\,(\,aa_{LJ,N} + aa_{LN,J} - aa_{JN,L}) = a3a_{JNL}$$

To get:

$$0 \ + \ aa_{NL,J} \ - 0 = a3a_{JLN} + \ a3a_{JNL}$$

:

$$\partial aa_{MN}/\partial a^{K} \equiv aa_{MN,K}$$

:

Partial Derivative of Lower Metric With respect to a^{K} *Upper Position*:

$$[aa_{MN}]_{,K} = + [a3a_{KMN} + a3a_{NKM}]$$

3.6.2.2.
Partial Derivative of *Upper* Metric With Respect to *Upper Position*:

Upper a^{q} P.D. of inners of upper and lower metrics:

$$aa^{jl} \; aa_{nl} = \delta^{j}_{n}$$

Partial Derivative is:

$$aa^{jl}_{,q} \; aa_{nl} + aa^{jl} \; aa_{nl,q} = 0$$

$$aa^{jl}_{,q} \; aa_{nl} = - \; aa^{jl} \; aa_{nl,q}$$

$$aa^{j}_{n,q} = - \; aa^{j}_{n,q}$$

Or Multiply

$$aa^{jl}_{,q} \; aa_{nl} = - \; aa^{jl} \; aa_{nl,q}$$

$$*$$

$$aa^{ns}$$

To get:

$$aa^{jl}_{,q} \; aa_{nl} \; aa^{ns} = - \; aa^{ns} \; aa^{jl} \; aa_{nl,q}$$

$$aa^{jl}_{,q} \; \delta_{l}^{s} = - \; aa^{ns} \; aa^{jl} \; aa_{nl,q}$$

But:

$$aa_{nl,q} = a3a_{qln} + a3a_{qnl} = aa_{nj} \; a3a_{ql}^{\;\;j} + aa_{lj} \; a3a_{qn}^{\;\;j}$$

So:

$$aa^{js}_{,q} \equiv aa^{js}_{,q} = - \; aa^{ns} \; aa^{jl} \; (a3a_{qln} + a3a_{qnl})$$

$$aa^{js}_{,q} \equiv aa^{js}_{,q} = - \; aa^{jl} \; a3a_{ql}^{\;\;s} - aa^{ns} \; a3a_{qn}^{\;\;j}$$

$$aa^{js}_{,q} \equiv aa^{js}_{,q} = - \; a3a_{q}^{\;js} - a3a_{q}^{\;sj}$$

So:

$$\partial aa^{MN} / \partial a^{K}$$

Partial Derivative of Upper Metric With respect to a^{K} *Upper Position*:

$$[aa^{MN}]_{,K} = - \; [a3a_{K}^{\;MN} + a3a_{N}^{\;KM}]$$

3.6.2.3.
Partial Derivative of *Lower* and *Upper* Metrics With respect to *Upper Position*:

Both Results:

Partial Derivative of
Lower and *Upper* Metrics
With respect to *Upper Position*

$$[aa_{MN}]_{,K} = + [a3a_{KMN} + a3a_{NKM}]$$

$$[aa^{MN}]_{,K} = - \; [a3a_{K}^{\;MN} + a3a_{N}^{\;KM}]$$

4.
Structural Del Operations

Define:
Del Operations

$$\underline{\text{Del}} \blacklozenge [\, T \,] \equiv \underline{\nabla} \blacklozenge [\, T \,]$$

:

$$\underline{\text{Del}} \blacklozenge [] \equiv \underline{\nabla} \blacklozenge [] \equiv \blacklozenge \partial[]/\partial\underline{a} = \blacklozenge \partial[]/\partial\underline{f} = \blacklozenge \partial[]/\partial\underline{r} = \blacklozenge \partial[]/\partial\underline{f} = \blacklozenge \partial[]/\partial\underline{x} = \blacklozenge \partial[]/\partial\underline{o}$$

:

$$\underline{\text{Del}} \blacklozenge [] \equiv \underline{\nabla} \blacklozenge [] \equiv \underline{a}_I \blacklozenge \partial[]/\partial a_I = \underline{f}_I \blacklozenge \partial[]/\partial f_I = \underline{r}_I \blacklozenge \partial[]/\partial r_I = \underline{x}_I \blacklozenge \partial[]/\partial x_I = \underline{o}_I \blacklozenge \partial[]/\partial o_I$$

$$\underline{\text{Del}} \blacklozenge [] \equiv \underline{\nabla} \blacklozenge [] \equiv \underline{a}^I \blacklozenge \partial[]/\partial a^I = \underline{f}^I \blacklozenge \partial[]/\partial f^I = \underline{r}^I \blacklozenge \partial[]/\partial r^I = \underline{x}^I \blacklozenge \partial[]/\partial x^I = \underline{o}^I \blacklozenge \partial[]/\partial o^I$$

Where:

$$\blacklozenge \partial[]/\partial a_I \neq \blacklozenge \partial[]/\partial f_I \neq \blacklozenge \partial[]/\partial r_I \neq \blacklozenge \partial[]/\partial x_I \neq \blacklozenge \partial[]/\partial o_I$$

$\underline{\text{Del}}$ of a Scalar T:

:

$$\underline{\text{Del}} \blacklozenge [T] \equiv \blacklozenge \partial[T]/\partial\underline{a} = \blacklozenge \partial[T]/\partial\underline{f} = \blacklozenge \partial[T]/\partial\underline{r} = \blacklozenge \partial[T]/\partial\underline{x} = \blacklozenge \partial[T]/\partial\underline{o}$$

:

$$\underline{\text{Del}} \blacklozenge [T] \equiv \underline{a}_I \blacklozenge \partial[T]/\partial a_I = \underline{f}_I \blacklozenge \partial[T]/\partial f_I = \underline{r}_I \blacklozenge \partial[T]/\partial r_I = \underline{x}_I \blacklozenge \partial[T]/\partial x_I = \underline{o}_I \blacklozenge \partial[T]/\partial o_I$$

$$\underline{\text{Del}} \blacklozenge [T] \equiv \underline{a}^I \blacklozenge \partial[T]/\partial a^I = \underline{f}^I \blacklozenge \partial[T]/\partial f^I = \underline{r}^I \blacklozenge \partial[T]/\partial r^I = \underline{x}^I \blacklozenge \partial[T]/\partial x^I = \underline{o}^I \blacklozenge \partial[T]/\partial o^I$$

Or:

$$\underline{\text{Del}} \blacklozenge [T] \equiv \underline{a}_I \blacklozenge Ta^I = \underline{f}_I \blacklozenge Tf^I = \underline{r}_I \blacklozenge Tr^I = \underline{x}_I \blacklozenge Tx^I = \underline{o}_I \blacklozenge To^I$$

$$\underline{\text{Del}} \blacklozenge [T] \equiv \underline{a}^I \blacklozenge Ta_I = \underline{f}^I \blacklozenge Tf_I = \underline{r}^I \blacklozenge Tr_I = \underline{x}^I \blacklozenge Tx_I = \underline{o}^I \blacklozenge To_I$$

$\underline{\text{Del}}$ of an Upper First Order T^m:

:

$$\underline{\text{Del}} \blacklozenge [T^m] \equiv \blacklozenge \partial[T^m]/\partial\underline{a} = \blacklozenge \partial[T^m]/\partial\underline{f} = \blacklozenge \partial[T^m]/\partial\underline{r} = \blacklozenge \partial[T^m]/\partial\underline{x} = \blacklozenge \partial[T^m]/\partial\underline{o}$$

:

$$\underline{\text{Del}} \blacklozenge [T^m] \equiv \underline{a}_I \blacklozenge \partial[\, T^m]/\partial a_I = \underline{f}_I \blacklozenge \partial[\, T^m]/\partial fa_I = \underline{r}_I \blacklozenge \partial[\, T^m]/\partial r_I = \underline{x}_I \blacklozenge \partial[\, T^m]/\partial x_I = \underline{o}_I \blacklozenge \partial[\, T^m]/\partial o_I$$

$$\underline{\text{Del}} \blacklozenge [T^m] \equiv \underline{a}^I \blacklozenge \partial[\, T^m]/\partial a^I = \underline{f}^I \blacklozenge \partial[\, T^m]/\partial f^I = \underline{r}^I \blacklozenge \partial[\, T^m]/\partial r^I = \underline{x}^I \blacklozenge \partial[\, T^m]/\partial x^I = \underline{o}^I \blacklozenge \partial[\, T^m]/\partial o^I$$

Or:

$$\underline{\text{Del}} \blacklozenge [T^m] \equiv \underline{a}_I \blacklozenge Ta^{m\,I} = \underline{f}_I \blacklozenge Tf^{m\,I} = \underline{r}_I \blacklozenge Tr^{mI} = \underline{x}_I \blacklozenge Tx^{m\,I} = \underline{o}_I \blacklozenge To^{m\,I}$$

$$\underline{\text{Del}} \blacklozenge [T^m] \equiv \underline{a}^I \blacklozenge Ta^m{}_I = \underline{f}^I \blacklozenge Tf^m{}_I = \underline{r}^I \blacklozenge Tr^m{}_I = \underline{x}^I \blacklozenge Tx^m{}_I = \underline{o}^I \blacklozenge To^m{}_I$$

:

$\underline{\text{Del}}$ of a Lower First Order T_m:

:

$$\underline{\text{Del}} \blacklozenge [T_m] \equiv \blacklozenge \partial[\, T_m]/\partial\underline{a} = \blacklozenge \partial[\, T_m]/\partial\underline{f} = \blacklozenge \partial[\, T_m]/\partial\underline{r} = \blacklozenge \partial[\, T_m]/\partial\underline{x} = \blacklozenge \partial[\, T_m]/\partial\underline{o}$$

:

$$\underline{Del} \blacklozenge [T_m] \equiv \underline{a}_I \blacklozenge \partial[T_m]/\partial a_I = \underline{f}_I \blacklozenge \partial[T_m]/\partial f_I = \underline{r}_I \blacklozenge \partial[T_m]/\partial r_I = \underline{x}_I \blacklozenge \partial[T_m]/\partial x_I = \underline{o}_I \blacklozenge \partial[T_m]/\partial o_I$$

$$\underline{Del} \blacklozenge [T_m] \equiv \underline{a}^I \blacklozenge \partial[T_m]/\partial a^I = \underline{f}^I \blacklozenge \partial[T_m]/\partial f^I = \underline{r}^I \blacklozenge \partial[T_m]/\partial r^I = \underline{x}^I \blacklozenge \partial[T_m]/\partial x^I = \underline{o}^I \blacklozenge \partial[T_m]/\partial o^I$$

Or:

$$\underline{Del} \blacklozenge [T_m] \equiv \underline{a}_I \blacklozenge Ta_m{}^I = \underline{f}_I \blacklozenge Tf_m{}^I = \underline{r}_I \blacklozenge Tr_m{}^I = \underline{x}_I \blacklozenge Tx_m{}^I = \underline{o}_I \blacklozenge To_m{}^I$$

$$\underline{Del} \blacklozenge [T_m] \equiv \underline{a}^I \blacklozenge Ta_{mI} = \underline{f}^I \blacklozenge Tf_{mI} = \underline{r}^I \blacklozenge Tr_{mI} = \underline{x}^I \blacklozenge Tx_{mI} = \underline{o}^I \blacklozenge To_{mI}$$

$\underline{Del}$ of an Upper First Order *Coordinate* a^m:

$$\underline{Del} \blacklozenge [a^m] \equiv \blacklozenge \partial[a^m]/\partial \underline{a} = \blacklozenge \partial[a^m]/\partial \underline{f} = \blacklozenge \partial[a^m]/\partial \underline{r} = \blacklozenge \partial[a^m]/\partial \underline{x} = \blacklozenge \partial[a^m]/\partial \underline{o}$$

:

$$\underline{Del} \blacklozenge [a^m] \equiv \underline{a}_I \blacklozenge \partial[a^m]/\partial a_I = \underline{f}_I \blacklozenge \partial[a^m]/\partial fa_I = \underline{r}_I \blacklozenge \partial[a^m]/\partial r_I = \underline{x}_I \blacklozenge \partial[a^m]/\partial x_I = \underline{o}_I \blacklozenge \partial[a^m]/\partial o_I$$

$$\underline{Del} \blacklozenge [a^m] \equiv \underline{a}^I \blacklozenge \partial[a^m]/\partial a^I = \underline{f}^I \blacklozenge \partial[a^m]/\partial f^I = \underline{r}^I \blacklozenge \partial[a^m]/\partial r^I = \underline{x}^I \blacklozenge \partial[a^m]/\partial x^I = \underline{o}^I \blacklozenge \partial[a^m]/\partial o^I$$

Or:

$$\underline{Del} \blacklozenge [a^m] \equiv \underline{a}_I \blacklozenge aa^{mI} = \underline{f}_I \blacklozenge af^{mI} = \underline{r}_I \blacklozenge ar^{mI} = \underline{x}_I \blacklozenge ax^{mI} = \underline{o}_I \blacklozenge ao^{mI}$$

$$\underline{Del} \blacklozenge [a^m] \equiv \underline{a}^I \blacklozenge aa^m{}_I = \underline{f}^I \blacklozenge af^m{}_I = \underline{r}^I \blacklozenge ar^m{}_I = \underline{x}^I \blacklozenge ax^m{}_I = \underline{o}^I \blacklozenge ao^m{}_I$$

$\underline{Del}$ of a Lower First Order *Coordinate* a_m:

:

$$\underline{Del} \blacklozenge [a_m] \equiv \blacklozenge \partial[a_m]/\partial \underline{a} = \blacklozenge \partial[a_m]/\partial \underline{f} = \blacklozenge \partial[a_m]/\partial \underline{r} = \blacklozenge \partial[a_m]/\partial \underline{x} = \blacklozenge \partial[a_m]/\partial \underline{o}$$

:

$$\underline{Del} \blacklozenge [a_m] \equiv \underline{a}_I \blacklozenge \partial[a_m]/\partial a_I = \underline{f}_I \blacklozenge \partial[a_m]/\partial f_I = \underline{r}_I \blacklozenge \partial[a_m]/\partial r_I = \underline{x}_I \blacklozenge \partial[a_m]/\partial x_I = \underline{o}_I \blacklozenge \partial[a_m]/\partial o_I$$

$$\underline{Del} \blacklozenge [a_m] \equiv \underline{a}^I \blacklozenge \partial[a_m]/\partial a^I = \underline{f}^I \blacklozenge \partial[a_m]/\partial f^I = \underline{r}^I \blacklozenge \partial[a_m]/\partial r^I = \underline{x}^I \blacklozenge \partial[a_m]/\partial x^I = \underline{o}^I \blacklozenge \partial[a_m]/\partial o^I$$

Or:

$$\underline{Del} \blacklozenge [a_m] \equiv \underline{a}_I \blacklozenge aa_m{}^I = \underline{f}_I \blacklozenge af_m{}^I = \underline{r}_I \blacklozenge ar_m{}^I = \underline{x}_I \blacklozenge ax_m{}^I = \underline{o}_I \blacklozenge ao_m{}^I$$

$$\underline{Del} \blacklozenge [a_m] \equiv \underline{a}^I \blacklozenge aa_{mI} = \underline{f}^I \blacklozenge af_{mI} = \underline{r}^I \blacklozenge ar_{mI} = \underline{x}^I \blacklozenge ax_{mI} = \underline{o}^I \blacklozenge ao_{mI}$$

$\underline{Del}$ of Coordinates a^m and a_m are Not Qqual:

$$a^m \qquad \neq \qquad a_m$$

:

$$\underline{Del} \blacklozenge a^m = \underline{o}^I \blacklozenge (ao)^m{}_{,I} = \underline{o}_I \blacklozenge (ao)^{m,I} \qquad \neq \qquad \underline{Del} \blacklozenge a_m = \underline{o}_I \blacklozenge (ao)_m{}^{,I} = \underline{o}^I \blacklozenge (ao)_{m,I}$$

And:

$$\blacklozenge (ao)^m{}_{,I} = \blacklozenge (ao)^{m,I} \qquad \neq \qquad \blacklozenge (ao)_m{}^I = \blacklozenge (ao)_{m,I}$$

:

$\underline{Del}$ of Fixed Coordinate O^m:

:

$$\underline{Del} \blacklozenge [O^m] \equiv \blacklozenge \partial[O^m]/\partial \underline{a} = \blacklozenge \partial[O^m]/\partial \underline{r} = \blacklozenge \partial[O^m]/\partial \underline{x} = \blacklozenge \partial[O^m]/\partial \underline{o}$$

:

$$\underline{Del} \blacklozenge [O^m] \equiv \underline{a}_I \blacklozenge \partial[O^m]/\partial a_I = \underline{f}_I \blacklozenge \partial[O^m]/\partial f_I = \underline{r}_I \blacklozenge \partial[O^m]/\partial r_I = \underline{x}_I \blacklozenge \partial[O^m]/\partial x_I = \underline{o}_I \blacklozenge \partial[O^m]/\partial o_I$$

$$\underline{Del} \blacklozenge [O^m] \equiv \underline{a}^I \blacklozenge \partial[O^m]/\partial a^I = \underline{f}^I \blacklozenge \partial[O^m]/\partial f^I = \underline{r}^I \blacklozenge \partial[O^m]/\partial r^I = \underline{x}^I \blacklozenge \partial[O^m]/\partial x^I = \underline{o}^I \blacklozenge \partial[O^m]/\partial o^I$$

Or:

$$\underline{Del} \blacklozenge [O^m] \equiv \underline{a}_I \blacklozenge Oa^{mI} = \underline{f}_I \blacklozenge Of^{mI} = \underline{r}_I \blacklozenge Or^{mI} = \underline{x}_I \blacklozenge Ox^{mI} = \underline{o}_I \blacklozenge Oo^{mI}$$

$$\underline{Del} \blacklozenge [O^m] \equiv \underline{a}^I \blacklozenge Oa^m{}_I = \underline{f}^I \blacklozenge Of^m{}_I = \underline{r}^I \blacklozenge Or^m{}_I = \underline{x}^I \blacklozenge Ox^m{}_I = \underline{o}^I \blacklozenge Oo^m{}_I$$

:

<u>Del</u> of Fixed Coordinate O_m:

:

$$\underline{Del} \blacklozenge [O_m] \equiv \blacklozenge \partial[O_m]/\partial\underline{a} = \blacklozenge \partial[O_m]/\partial\underline{f} = \blacklozenge \partial[O_m]/\partial\underline{r} = \blacklozenge \partial[O_m]/\partial\underline{x} = \blacklozenge \partial[O_m]/\partial\underline{o}$$

:

$$\underline{Del} \blacklozenge [O_m] \equiv \underline{a}_I \blacklozenge \partial[O_m]/\partial a_I = \underline{f}_I \blacklozenge \partial[O_m]/\partial f_I = \underline{r}_I \blacklozenge \partial[O_m]/\partial r_I = \underline{x}_I \blacklozenge \partial[O_m]/\partial x_I = \underline{o}_I \blacklozenge \partial[O_m]/\partial o_I$$

$$\underline{Del} \blacklozenge [O_m] \equiv \underline{a}^I \blacklozenge \partial[O_m]/\partial a^I = \underline{f}^I \blacklozenge \partial[O_m]/\partial f^I = \underline{r}^I \blacklozenge \partial[O_m]/\partial r^I = \underline{x}^I \blacklozenge \partial[O_m]/\partial x^I = \underline{o}^I \blacklozenge \partial[O_m]/\partial o^I$$

Or:

$$\underline{Del} \blacklozenge [O_m] \equiv \underline{a}_I \blacklozenge Oa_m^{\ I} = \underline{f}_I \blacklozenge Of_m^{\ I} = \underline{r}_I \blacklozenge Or_m^{\ I} = \underline{x}_I \blacklozenge Ox_m^{\ I} = \underline{o}_I \blacklozenge Oo_m^{\ I}$$

$$\underline{Del} \blacklozenge [O_m] \equiv \underline{a}^I \blacklozenge Oa_{mI} = \underline{f}^I \blacklozenge Of_{mI} = \underline{r}^I \blacklozenge Or_{mI} = \underline{x}^I \blacklozenge Ox_{mI} = \underline{o}^I \blacklozenge Oo_{mI}$$

:

<u>Del</u> of Fixed Coordinates O^m and O_m *are* Qqual:

$$O^m \quad = \quad O_m$$

$$\underline{Del} \blacklozenge O^m = \underline{a}^I \blacklozenge [Oa]^m_{\ ,I} = \underline{a}_I \blacklozenge [Oa]^{mI}_{\ ,} \quad = \quad \underline{Del} \blacklozenge O_m = \underline{a}_I \blacklozenge [Oa]_m^{\ I} = \underline{a}^I \blacklozenge [Oa]_{m,I}$$

Since : $O^m = O_m$:

$$\underline{Del} \blacklozenge O^m = \underline{a}^I \blacklozenge [Oa]_{m',I} = \underline{a}_I \blacklozenge [Oa]_{m'}^{\ I} \quad = \quad \underline{Del} \blacklozenge O_m = \underline{a}_I \blacklozenge [Oa]^{mI}_{\ ,} = \underline{a}^I \blacklozenge [Oa]^m_{\ ,I}$$

Whereas:

$$\blacklozenge [Oa]^m_{\ I} = \blacklozenge [Oa]_{mI} \quad \neq \quad \blacklozenge [Oa]_m^{\ I} = \blacklozenge [Oa]^{mI}_{\ ,}$$

$$\underline{Del} \blacklozenge [o^m] \equiv$$

$$\blacklozenge \partial[o^m]/\partial\underline{a} = \blacklozenge \partial[o^m]/\partial\underline{f} = \blacklozenge \partial[o^m]/\partial\underline{r} = \blacklozenge \partial[o^m]/\partial\underline{x} = \blacklozenge \partial[o^m]/\partial\underline{o}$$

Or:

$$\underline{a}_I \blacklozenge \partial[o^m]/\partial a_I = \underline{f}_I \blacklozenge \partial[o^m]/\partial f_I = \underline{r}_I \blacklozenge \partial[o^m]/\partial r_I = \underline{x}_I \blacklozenge \partial[o^m]/\partial x_I = \underline{o}_I \blacklozenge \partial[o^m]/\partial o_I$$

$$\underline{a}^I \blacklozenge \partial[o^m]/\partial a^I = \underline{f}^I \blacklozenge \partial[o^m]/\partial f^I = \underline{r}^I \blacklozenge \partial[o^m]/\partial r^I = \underline{x}^I \blacklozenge \partial[o^m]/\partial x^I = \underline{o}^I \blacklozenge \partial[o^m]/\partial o^I$$

Or:

$$\underline{a}_I \blacklozenge oa^{mI} = \underline{f}_I \blacklozenge of^{mI} = \underline{r}_I \blacklozenge or^{mI} = \underline{x}_I \blacklozenge ox^{mI} = \underline{o}_I \blacklozenge oo^{mI}$$

$$\underline{a}^I \blacklozenge oa^m_{\ I} = \underline{f}^I \blacklozenge of^m_{\ I} = \underline{r}^I \blacklozenge or^m_{\ I} = \underline{x}^I \blacklozenge ox^m_{\ I} = \underline{o}^I \blacklozenge oo^m_{\ I}$$

Del of Coordinate: $o^i = o_i$ and $a^i \neq a_i$ then :

$$\underline{Del} \blacklozenge o^m = \underline{a}^I [\blacklozenge oa]^m_{\ ,I} = \underline{a}_I [\blacklozenge oa]^{mI}_{\ ,} \quad = \quad \underline{Del} \blacklozenge o_m = \underline{a}_I [\blacklozenge oa]_m^{\ I} = \underline{a}^I [\blacklozenge oa]_{m,I}$$

$$\underline{Del} \blacklozenge a^I = \underline{o}^m [\blacklozenge ao]^I_{\ ,m} = \underline{o}_m [\blacklozenge ao]^{I,m} \quad \neq \quad \underline{Del} \blacklozenge a_I = \underline{o}_m [\blacklozenge ao]_{I,}^{\ m} = \underline{o}^m [\blacklozenge ao]_{I,m}$$

Since : $o^m = o_m$ then :

$$\underline{a}^I [\blacklozenge oa]^m_{\ I} = \underline{a}^I [\blacklozenge oa]_{m',I} \quad = \quad \underline{a}_I [\blacklozenge oa]^{mI}_{\ ,} = \underline{a}_I [\blacklozenge oa]_{m,}^{\ I}$$

Since: $a^m \neq a_m$ Then:

$$[\blacklozenge oa]^m_{\ I} = [\blacklozenge oa]_{mI} \quad \neq \quad [\blacklozenge oa]^{mI} = [\blacklozenge oa]_m^{\ I}$$

Since : $a^m \neq a_m$ then :

$$[\blacklozenge ao^I_{\ m}\ \underline{o}^m \neq \blacklozenge ao_{Im}\ \underline{o}^m] \quad \equiv \quad [\blacklozenge ao^{Im}\ \underline{o}_m \neq \blacklozenge ao_I^{\ m}\ \underline{o}_m]$$

So :

$$[\blacklozenge ao^I_{\ m} \neq \blacklozenge ao_{Im}] \quad \equiv \quad [\blacklozenge ao^{Im} \neq \blacklozenge ao_I^{\ m}]$$

While : $o^m = o_m$ then:

$$\blacklozenge ao^I_{\ m} = \blacklozenge ao^{Im} \quad \neq \quad \blacklozenge ao_{Im} = \blacklozenge ao_I^{\ m}$$

<u>Del</u> is commonly written::

Del ♦ [ ] ≡ ∇ ♦ [ ]

And in terms of Del Operations:

Gradient : ♦ = ( )

Divergence : ♦ = (.)   Curl : ♦ = (^)

Curl of Gradient is Zero

Div of Grad = Grad of Div   Gradient : ♦ = ( )

Divergence of Curl is Zero   Divergence : ♦ = (.)   → Curl : ♦ = (^) ↑

As follows:

∇ ^ ∇ U = 0

∇ . ∇ U = ∇ ∇ . V   ∇ U

∇ . ∇ ^ V = 0   ∇ . V   → *∇ ^ V* ↑

4.1. Gradient:

Gradient of U : ♦ = ( ) So: ∇ ♦ U = ∇ U

Curl of Gradient is Zero

Div of Grad = Grad of Div   Gradient : ♦ = ( )

Or:

∇ ^ ∇ U = 0

∇ . ∇ U = ∇ ∇ . V   ∇ U

Derivative With Respect To Vector a:

$$\text{Del } [\,] \equiv \nabla\, [] \equiv \partial[]/\partial \underline{a} = \partial[]/\partial \underline{f} = \partial[]/\partial \underline{r} = \partial[]/\partial \underline{x} = \partial[]/\partial \underline{o}$$

Or:

$$\text{Del } [\,] \equiv \nabla\, [] \equiv \underline{a}_I\ \partial[]/\partial a_I = \underline{f}_I\ \partial[]/\partial f_I = \underline{r}_I\ \partial[]/\partial r_I = \underline{x}_I\ \partial[]/\partial x_I = \underline{o}_I\ \partial[]/\partial o_I$$

$$\text{Del } [\,] \equiv \nabla\, [] \equiv \underline{a}^I\ \partial[]/\partial a^I = \underline{f}^I\ \partial[]/\partial f^I = \underline{r}^I\ \partial[]/\partial r^I = \underline{x}^I\ \partial[]/\partial x^I = \underline{o}^I\ \partial[]/\partial o^I$$

Where:

$$\partial[\,]/\partial a_I \neq \partial[\,]/\partial f_I \neq \partial[\,]/\partial r_I \neq \partial[\,]/\partial x_I \neq \partial[\,]/\partial o_I$$

:

$$\text{Del } [\,] \equiv \nabla\, [] \equiv \underline{a}_I\ []^{aI} = \underline{f}_I\ []^{fI} = \underline{r}_I\ []^{rI} = \underline{x}_I\ []^{xI} = \underline{o}_I\ []^{oI}$$

$$\underline{\text{Del}}\ [\,] \equiv \underline{\nabla}\ [\,] \equiv \underline{a}^{I}\ [\,]_{aI} = \underline{f}^{I}\ [\,]_{fI} = \underline{r}^{I}\ [\,]_{rI} = \underline{x}^{I}\ [\,]_{xI} = \underline{o}^{I}\ [\,]_{oI}$$

$$\underline{a}_{I}\,\partial[\,]/\partial a_{I} = \underline{a}_{I}\ [\,],{}^{a^{I}} = \underline{a}_{I}\ [\,]^{I}$$

$\underline{a}_{I}$	$\underline{a}_{1}$	$\underline{a}_{2}$	$\underline{a}_{3}$
$a[\,]^{I}$	$a[\,]^{1} = \partial[\,]/\partial a_{1}$	$a[\,]^{2} = \partial[\,]/\partial a_{2}$	$a[\,]^{3} = \partial[\,]/\partial a_{3}$

$$\underline{a}^{I}\,\partial[\,]/\partial a^{I} = \underline{a}^{I}\ [\,],{}_{a_{I}} = \underline{a}^{I}\ [\,]_{I}$$

$\underline{a}^{I}$	$\underline{a}^{1}$	$\underline{a}^{2}$	$\underline{a}^{3}$
$a[\,]_{I}$	$a[\,]_{1} = \partial[\,]/\partial a^{1}$	$a[\,]_{2} = \partial[\,]/\partial a^{2}$	$a[\,]_{3} = \partial[\,]/\partial a^{3}$

4.1.1.
Gradient of Coordinates:

Applied:

$$\underline{\text{Del}}\ [a_{k}] \equiv \underline{\nabla}\ [a_{k}] = [a_{k}]^{om}\ \underline{o}_{m} = ao_{k}^{\ m}\ \underline{o}_{m} = \underline{o}_{m}\ oa^{m}_{\ k}$$

$$\underline{\text{Del}}\ [a^{k}] \equiv \underline{\nabla}\ [a^{k}] = [a^{k}]_{om}\ \underline{o}^{m} = ao^{k}_{\ m}\ \underline{o}^{m} = \underline{o}^{m}\ oa_{m}^{\ k}$$

$\underline{a}_{k} = \underline{\nabla}\, a_{k} = \underline{o}_{m}\ oa^{m}_{\ k}$	$\underline{a}^{k} = \underline{\nabla}\, a^{k} = \underline{o}^{m}\ oa_{m}^{\ k}$
$\underline{f}_{k} = \underline{\nabla}\, f_{k} = \underline{o}_{m}\ of^{m}_{\ k}$	$\underline{f}^{k} = \underline{\nabla}\, f^{k} = \underline{o}^{m}\ of_{m}^{\ k}$
$\underline{r}_{k} = \underline{\nabla}\, r_{k} = \underline{o}_{m}\ or^{m}_{\ k}$	$\underline{r}^{k} = \underline{\nabla}\, r^{k} = \underline{o}^{m}\ or_{m}^{\ k}$
$\underline{x}_{k} = \underline{\nabla}\, x_{k} = \underline{o}_{m}\ ox^{m}_{\ k}$	$\underline{x}^{k} = \underline{\nabla}\, x^{k} = \underline{o}^{m}\ ox_{m}^{\ k}$
$\underline{o}_{k} = \underline{\nabla}\, o_{k} = \underline{o}_{m}\ oo^{m}_{\ k}$	$\underline{o}^{k} = \underline{\nabla}\, o^{k} = \underline{o}^{m}\ oo_{m}^{\ k}$

:

Bases Algorithm:

Product of Norms of Unit Conjugate Bases is equal One:

Ba[-k] * Ba[+k] = || Ba[-k] || * || Ba[+k] || = 1

:

:

Lower Case		Upper Case
	Norm of a Unit Basis is equal One:	
|| aBa(-k) || = 1		|| aBa[+k] || = 1
	Or:	
Ba[-k] = 1/ Ba[+k]		Ba[+k] = 1/ Ba[-k]
	Define:	
aBa(-k) ≡ 1/Ba[-<u>k</u>] * Ba(-k)		aBa(+k) ≡ 1/Ba[+<u>k</u>] * Ba(+k)
=		=
Ba[+<u>k</u>] * Ba(-k)		Ba[-<u>k</u>] * Ba(+k)
	:	
B(∇a(-k)) = Ba(-k) = Ba[-<u>k</u>] aBa(-k)		B(∇a(+k)) = Ba(+k) = Ba[+<u>k</u>] aBa(+k)
aBa(-k) ≡ 1/ Ba[-<u>k</u>] * Ba(-k)		aBa(+k) ≡ 1/ Ba[+<u>k</u>] * Ba(+k)
	:	
	But:	
Ba(-k)		Ba(+ k)
=		=
ao(- k,+m)*Bo(+m) = oa(+m,-k)* Bo(- m)		ao(+k,-m)*Bo(+m) = oa(+m,-k)*Bo(- m)
	Or:	
B(∇a(-k)) = Ba(-k) = Ba[-<u>k</u>] aBa(-k)		B(∇a(+k)) = Ba(+k) = Ba[+<u>k</u>] aBa(+k)
=		=
ao(- k,+m) Bo(+m) = oa(+m,- k) Bo(- m)		ao(+ k,-m) Bo(-m) = ao(+ k,-m) Bo(+ m)

Metrics:

Ba(- i) . Ba(- j) = aa(-i,-j)   Ba(+ i) . Ba(+ j) = aa(+i,+j)

| Ba(-k) | | Ba(-k) | = aa(- k , - k)

| Ba(-k) | = | (∇a(-k))B| = sqrt |aa(- k , - k)| |Ba(+k)| = |(∇a(+k))B| = sqrt |aa(+ k ,+ k)|

:

aBa(-k) ≡ 1/ Ba[-k] Ba(-k)   aBa(+k) ≡ 1/Ba[+k] Ba(+k)

=   =

Ba[+ k] Ba(-k)   Ba[-k] Ba(+k)

(∇a(-k))B = ao(- k,+m) Bo(- m) = Bo(- m) oa(+m,-k)

(∇a(+k))B = ao(+ k,-m) Bo(+ m) = Bo(+ m) oa(-m , +k)

Ba(-k) = ∇(a(-k))B = Bo(-m) oa(+m,-k)   Ba(+k) = ∇(a(+ k))B = Bo(+m) oa(-m,+k)

Bf(-k) = ∇(f(-k))B = Bo(-m) of(+m,-k)   Bf(+k) = ∇(f(+ k))B = Bo(+m) of(-m,+k)

Br(-k) = ∇(r(-k))B = Bo(-m) or(+m,-k)   Br(+k) = ∇(r(+k))B = Bo(+m) or(-m,+k)

Bx(-k) = ∇(x(-k))B = Bo(-m) ox(+m,-k)   Bx(+k) = ∇(x(+k))B = Bo(+m) ox(-m,+k)

Bo(-k) = ∇(o(-k))B = Bo(-m) oo(+m,-k)   Bo(+k) = ∇(o(+k))B = Bo(+m) oo(-m,+k)

Ba[-k] ≡ |Ba(-k)| Ba[- k] Ba[+ k] = 1   Ba[+k] ≡ |ba(+k)|

Bases (by Inverse partial Derivatives) Integrable

Bo(-k) = Ba(-m) ao(+m, -k)   Bo(+k) = Ba(+m) ao(-m, +k)

Bo(-k) = Ba(-m) ao(+m, -k)   Bo(+k) = Ba(+m) ao(-m, +k)

Bases (by partial Derivatives)

Ba(-k) = ∇(a(-k))B = Bo(-m) oa(+m,-k)   Ba(+k) = ∇(a(+ k))B = Bo(+m) oa(-m,+k)

Ba(-k) = ∇(a(-k))B = Bo(-m) oa(+m,-k)   Ba(+k) = ∇(a(+ k))B = Bo(+m) oa(-m,+k)

Ba[-k]≡|Ba(-k)| Ba[- k] Ba[+ k] = 1   Ba[+k]≡|Ba(+k)|

:

Del [T] ≡ ∇ [T] ≡

$$\partial[T]/\partial\underline{a} = \partial[T]/\partial\underline{f} = \partial[T]/\partial\underline{r} = \partial[T]/\partial\underline{x} = \partial[T]/\partial\underline{o}$$

Or:

$$\underline{a}_I\ \partial[T]/\partial a_I = \underline{f}_I\ \partial[T]/\partial f_I = \underline{r}_I\ \partial[T]/\partial r_I = \underline{x}_I\ \partial[T]/\partial x_I = \underline{o}_I\ \partial[T]/\partial o_I$$

$$\underline{a}^I\ \partial[T]/\partial a^I = \underline{f}^I\ \partial[T]/\partial f^I = \underline{r}^I\ \partial[T]/\partial r^I = \underline{x}^I\ \partial[T]/\partial x^I = \underline{o}^I\ \partial[T]/\partial o^I$$

Or:

$$\underline{a}_I\ Ta_,^{\ I} = \underline{f}_I\ Tf_,^{\ I} = \underline{r}_I\ Tr_,^{\ I} = \underline{x}_I\ Tx_,^{\ I} = \underline{o}_I\ To_,^{\ I}$$

$$\underline{a}^I\ Ta_{,I} = \underline{f}^I\ Tf_{,I} = \underline{r}^I\ Tr_{,I} = \underline{x}^I\ Tx_{,I} = \underline{o}^I\ To_{,I}$$

Or:

$$\underline{a}_I\ Ta^I = \underline{f}_I\ Tf^I = \underline{r}_I\ Tr^I = \underline{x}_I\ Tx^I = \underline{o}_I\ To^I$$

$$\underline{a}^I\ Ta_I = \underline{f}^I\ Tf_I = \underline{r}^I\ Tr_I = \underline{x}^I\ Tx_I = \underline{o}^I\ To_I$$

$$\underline{a}_I\ Ta_,^{\ I}$$

$$\underline{a}^I\ Ta_{,I}$$

4.1.2.
Gradient of Position Vector:

Position Vector:

$$\underline{a} = \underline{a}_i\ a^i = a_i\ \underline{a}^i = \underline{f}_i\ f^i = f_i\ \underline{f}^i = \underline{r}_i\ r^i = r_i\ \underline{r}^i = \underline{x}_i\ x^i = x_i\ \underline{x}^i = \underline{o}_i\ o^i = o_i\ \underline{o}^i$$

Del [aJ] = $\underline{aJ}$

$$\underline{a}_I\, \partial[a_J]/\partial a_I = \underline{f}_I\, \partial[a_J]/\partial f_I = \underline{r}_I\, \partial[a_J]/\partial r_I = \underline{x}_I\, \partial[a_J]/\partial x_I = \underline{o}_I\, \partial[a_J]/\partial o_I$$
$$\underline{a}^I\, \partial[a_J]/\partial a^I = \underline{f}^I\, \partial[a_J]/\partial f^I = \underline{r}^I\, \partial[a_J]/\partial r^I = \underline{x}^I\, \partial[a_J]/\partial x^I = \underline{o}^I\, \partial[a_J]/\partial o^I$$

Or:

$$\underline{a}_I\, aa_J^{\;I} = \underline{f}_I\, af_J^{\;I} = \underline{r}_I\, ar_J^{\;I} = \underline{x}_I\, ax_J^{\;I} = \underline{o}_I\, ao_J^{\;I}$$
$$\underline{a}^I\, aa_{JI} = \underline{f}^I\, af_{JI} = \underline{r}^I\, ar_{JI} = \underline{x}^I\, ax_{JI} = \underline{o}^I\, ao_{JI}$$

Del [xJ] = $\underline{xJ}$

$$\underline{a}_I\, \partial[x_J]/\partial a_I = \underline{f}_I\, \partial[x_J]/\partial f_I = \underline{r}_I\, \partial[x_J]/\partial r_I = \underline{x}_I\, \partial[x_J]/\partial x_I = \underline{o}_I\, \partial[x_J]/\partial o_I$$
$$\underline{a}^I\, \partial[x_J]/\partial a^I = \underline{f}^I\, \partial[x_J]/\partial f^I = \underline{r}^I\, \partial[x_J]/\partial r^I = \underline{x}^I\, \partial[x_J]/\partial x^I = \underline{o}^I\, \partial[x_J]/\partial o^I$$

Or:

$$\underline{a}_I\, xa_J^{\;I} = \underline{f}_I\, xf_J^{\;I} = \underline{r}_I\, xr_J^{\;I} = \underline{x}_I\, xx_J^{\;I} = \underline{o}_I\, xo_J^{\;I}$$
$$\underline{a}^I\, xa_{JI} = \underline{f}^I\, xf_{JI} = \underline{r}^I\, xr_{JI} = \underline{x}^I\, xx_{JI} = \underline{o}^I\, xo_{JI}$$

<u>Gradient</u> of Coordinate: $x^J = x_J$ but $a^J \neq a_J$ then :

$\underline{\nabla}\, x^J = \underline{a}^I\, xa^J_{\;I} = \underline{a}_I\, xa^{JI} = \underline{x}^J$ = $\underline{\nabla}\, x_J = \underline{a}_I\, xa_J^{\;I} = \underline{a}^I\, xa_{JI} = \underline{x}_J$

$\underline{\nabla}\, a^I = \underline{x}^J\, ax^I_{\;J} = \underline{x}_J\, ax^{IJ} = \underline{a}^I$ ≠ $\underline{\nabla}\, a_I = \underline{x}_J\, ax_I^{\;J} = \underline{x}^J\, ax_{IJ} = \underline{a}_I$

Since: $x^J = x_J$

$\underline{a}^I\, xa^J_{\;I} = \underline{a}^I\, xa_{JI}$ = $\underline{a}_I\, xa^{JI} = \underline{a}_I\, xa_J^{\;I}$

Since: $a^I \neq a_I$

$xa^J_{\;I}$ ≠ xa^{JI}

xa_{JI} ≠ $xa_J^{\;I}$

Since: $x^J = x_J$ and $a^I \neq a_I$

$xa^J_{\;I} = xa_{JI}$ ≠ $xa^{JI} = xa_J^{\;I}$

$ax^I_{\;J} = ax^{IJ}$ ≠ $ax_{IJ} = ax_I^{\;J}$

$\underline{x}^J\, ax^I_{\;J}$ = $\underline{x}_J\, ax^{IJ}$

$ax^I_{\;J}$ = ax^{IJ}

$\underline{x}^J\, ax_{IJ}$ = $\underline{x}_J\, ax_I^{\;J}$

ax_{IJ} = $ax_I^{\;J}$

$(\underline{x}^J\, ax^I_{\;J} \neq \underline{x}^J\, ax_{IJ})$ = $(\underline{x}_J\, ax^{IJ} \neq \underline{x}_J\, ax_I^{\;J})$

$(ax^I_{\;J} \neq ax_{IJ})$ = $(ax^{IJ} \neq ax_I^{\;J})$

Where:

$ax_I^{\;J} = xa^J_{\;I}$ ≠ $ax^I_{\;J} = xa_J^{\;I}$

a <u>Gradient</u> of o Conjugates : $\underline{r}[o] = \partial a \backslash \partial \underline{r} = \underline{\nabla}[\, o\,]$ so: $d\, o = \underline{\nabla}\,[o]\,.\, d\underline{r}$

$\underline{\nabla}[o^m] = \underline{a}^I\, \nabla_I[o^m] = \underline{a}\, oa^m_{\;I} = \underline{a}^I\, ao_I^{\;m} = \underline{o}^m$ $\underline{\nabla}[o_m] = \underline{a}_I\, \nabla^I[o_m] = \underline{a}_I\, oa_m^{\;I} = \underline{a}_I\, ao^I_{\;m} = \underline{o}_m$

$\| oa^m_{\;I}\| = \| ao_I^{\;m}\| = \underline{a}_1 \wedge \underline{a}_2\,.\,\underline{a}_3 = \|_a\|$ $\| oa_m^{\;I}\| = \|ao^I_{\;m}\| = \underline{a}^1 \wedge \underline{a}^2\,.\,\underline{a}^3 = \|^a\|$

o <u>Gradient</u> of Conjugates : $\underline{r}[a] = \partial o \backslash \partial \underline{r} = \underline{\nabla}[\, a\,]$ so: $d\, a = \underline{\nabla}\,[a]\,.\, d\underline{r}$

$\underline{\nabla}(a^I) = \underline{o}^m \nabla_m(a^I) = \underline{o}^m\, ao^I{}_m = \underline{o}^m\, oa_m{}^I = \underline{a}^I$ $\quad$ $\underline{\nabla}(a_I) = \underline{o}_m \nabla^m(a_I) = \underline{o}_m ao_I{}^m = \underline{o}_m oa^m{}_I = \underline{a}_I$

$\|oa_m{}^I\| = \|ao^I{}_m\| = \underline{a}^1 \wedge \underline{a}^2 \cdot \underline{a}^3 = \|^a\|$ $\quad$ $\|oa^m{}_I\| = \|ao_I{}^m\| = \underline{a}_1 \wedge \underline{a}_2 \cdot \underline{a}_3 = \|_a\|$

$\|oa^m{}_I\|\,\|ao^I{}_m\| = \|_a\|\ \|^a\| = 1$ $\quad$ $\|oa_m{}^I\|\,\|ao_I{}^m\| = \|^a\|\|_a\| = 1$

a Derivative of Vector = o Gradient of a Conjugate

$[\underline{r}]_a = \partial\underline{r}/\partial a = \underline{a}$	$d\underline{r}.d\underline{r} = da^I da_I$	$[a]_r = \partial a/\partial\underline{r} = \underline{\nabla}[a]$
$\underline{a}^I = \partial\underline{r}/\partial o_m\, \partial o_m/\partial a_I = \underline{o}^m oa_m{}^I$	$= \partial\underline{r}/\partial a_I = \partial a^I/\partial\underline{r} =$	$\underline{o}^m \nabla_m [a^I] = \underline{o}^m ao^I{}_m = \underline{a}^I$
$\underline{a}_I = \partial\underline{r}/\partial o^m\, \partial o^m/\partial a^I = \underline{o}_m oa^m{}_I$	$= \partial\underline{r}/\partial a^I = \partial a_I/\partial\underline{r} =$	$\underline{o}_m \nabla^m [a_I] = \underline{o}_m ao_I{}^m = \underline{a}_I$

o Derivative of Vector = a Gradient of o Conjugate

$o[\underline{r}] = \partial\underline{r}/\partial o = \underline{o}$	$d\underline{r}\,.\,d\underline{r} = do^m\ do_m$	$[o]_r = \partial o/\partial\underline{r} = \underline{\nabla}[o]$
$\underline{o}^m = \partial\underline{r}/\partial a_I \partial a_I/\partial o_m = \underline{a}^I ao_I{}^m$	$= \partial\underline{r}/\partial o_m = \partial o^m/\partial\underline{r} =$	$\underline{a}^I \nabla_I [o^m] = \underline{a}^I oa_I{}^m = \underline{o}^m$
$\underline{o}_m = \partial\underline{r}/\partial a^I \partial a^I/\partial o^m = \underline{a}_I ao^I{}_m$	$= \partial\underline{r}/\partial o^m = \partial o_m/\partial\underline{r} =$	$\underline{a}_I \nabla^I [o_m] = \underline{a}_I oa^I{}_m = \underline{o}_m$

4.1.3.
Gradient of a Scalar T:

Gradient of T: ♦ = () ≡ (blank) So: $\quad \underline{\nabla} \blacklozenge T = \underline{\nabla}\ T$

Del $[T] \equiv \underline{\nabla}\ [T] \equiv$

$\partial[T]/\partial\underline{a} = \partial[T]/\partial\underline{r} = \partial[T]/\partial\underline{x} = \partial[T]/\partial\underline{o}$

Or:

$\underline{a}_I\ \partial[T]/\partial a_I = \underline{r}_I\ \partial[T]/\partial r_I = \underline{x}_I\ \partial[T]/\partial x_I = \underline{o}_I\ \partial[T]/\partial o_I$

$\underline{a}^I\ \partial[T]/\partial a^I = \underline{r}^I\ \partial[T]/\partial r^I = \underline{x}^I\ \partial[T]/\partial x^I = \underline{o}^I\ \partial[T]/\partial o^I$

Or:

$\underline{a}_I\ Ta_,{}^I = \underline{r}_I\ Tr_,{}^I = \underline{x}_I\ Tx_,{}^I = \underline{o}_I\ To_,{}^I$

$\underline{a}^I\ Ta_{,I} = \underline{r}^I\ Tr_{,I} = \underline{x}^I\ Tx_{,I} = \underline{o}^I\ To_{,I}$

Or:

$\underline{a}_I\ Ta^I = \underline{r}_I\ Tr^I = \underline{x}_I\ Tx^I = \underline{o}_I\ To^I$

$\underline{a}^I\ Ta_I = \underline{r}^I\ Tr_I = \underline{x}^I\ Tx_I = \underline{o}^I\ To_I$

$\underline{a}_I\ Ta_,{}^I$

$\underline{a}^I\ Ta_{,I}$

Transfer Gradient of a Scalar T:

$(To_K\ \ oa^K{}_L) = (Ta_L)$	$\underline{\nabla}[T] = \underline{a}^I \partial[T]/\partial a^I = \underline{a}^I\ Ta_I$
$(Ta_L\ \ ao^L{}_K) = (To_K)$	$\underline{\nabla}[T] = \underline{o}^m \partial[T]/\partial o^m = \underline{o}^m\ To_m$
$(To^K\ \ oa_K{}^L) = (Ta^L)$	$\underline{\nabla}[T] = \underline{a}_I\ \partial[T]/\partial a_I = \underline{a}_I\ Ta^I$
$(Ta^L\ \ ao_L{}^K) = (To^K)$	$\underline{\nabla}[T] = \underline{o}_m\ \partial[T]/\partial o_m = \underline{o}_m\ To^m$

Gradient of T: $\quad \underline{\nabla}\,T = \underline{a}^I\ []_j\ T = \underline{a}^I\ [T\ ]_j = \underline{a}^I\ \partial T/\partial a^I$

$\underline{a}^I$	$\underline{a}^1 = \alpha^1\ \underline{A}^1$	$\underline{a}^2 = \alpha^2\ \underline{A}^2$	$\underline{a}^3 = \alpha^3\ \underline{A}^3$
$[T\ ]_I$	$[T]_1 = \partial T/\partial a^1$	$[T]_2 = \partial T/\partial a^2$	$[T]_3 = \partial T/\partial a^3$

Gradient of Coordinate: $o^1 o^2 o^3$: $oa^m{}_I = oa_{mI} \neq oa^{mI} = oa^I{}_m$

$\underline{a}^I$	$\underline{a}^1 = \alpha^1 \underline{A}^1$	$\underline{a}^2 = \alpha^2 \underline{A}^2$	$\underline{a}^3 = \alpha^3 \underline{A}^3$
$[o^1]_I$	$oa^1{}_1$	$oa^1{}_2$	$oa^1{}_3$
$[o^2]_I$	$oa^2{}_1$	$oa^2{}_2$	$oa^2{}_3$
$[o^3]_I$	$oa^3{}_1$	$oa^3{}_2$	$oa^3{}_3$

Gradient of T: $\underline{\nabla} T = [T]^I \underline{a}_I = \partial T / \partial a_I \; \underline{a}_I$

$\underline{a}_I$	$\underline{a}_1 = \alpha_1 \underline{A}_1$	$\underline{a}_2 = \alpha_2 \underline{A}_2$	$\underline{a}_3 = \alpha_3 \underline{A}_3$
$[T]^I$	$[T]^1 = \partial T / \partial a_1$	$[T]^2 = \partial T / \partial a_2$	$[T]^3 = \partial T / \partial a_3$

Gradient of Coordinate: $o_1 o_2 o_3$: $oa^m{}_I = oa_{mI} \neq oa^{mI} = oa^I{}_m$

$\underline{a}_I$	$\underline{a}_1 = \alpha_1 \underline{A}_1$	$\underline{a}_2 = \alpha_2 \underline{A}_2$	$\underline{a}_3 = \alpha_3 \underline{A}_3$
$[o_1]^I$	$oa_1{}^1$	$oa_1{}^2$	$oa_1{}^3$
$[o_2]^I$	$oa_2{}^1$	$oa_2{}^2$	$oa_2{}^3$
$[o_3]^I$	$oa_3{}^1$	$oa_3{}^2$	$oa_3{}^3$

Gradient of T $\equiv \underline{\nabla} T = o[]_m \underline{o}^m T = o[T]_m \underline{o}^m = \partial T / \partial o^m \; \underline{o}^m$

$\underline{o}^m$	$\underline{o}^1$	$\underline{o}^2$	$\underline{o}^3$
$o[T]_m$	$\partial T / \partial o^1$	$\partial T / \partial o^2$	$\partial T / \partial o^3$

Gradient of Coordinate: $a^1 a^2 a^3$: $ao^j{}_n = ao^{jn} \neq ao_{jn} = ao_j{}^n$

$\underline{o}^m$	$\underline{o}^1$	$\underline{o}^2$	$\underline{o}^3$
$[a^1]_I$	$ao^1{}_1$	$ao^1{}_2$	$ao^1{}_3$
$[a^2]_I$	$ao^2{}_1$	$ao^2{}_2$	$ao^2{}_3$
$[a^3]_I$	$ao^3{}_1$	$ao^3{}_2$	$ao^3{}_3$

Gradient of T $\equiv \underline{\nabla} T = []^n \underline{o}_n T = [T]^n \underline{o}_n = \partial T / \partial o_j \; \underline{o}_j$

$\underline{o}_m$	$\underline{o}_1$	$\underline{o}_2$	$\underline{o}_3$
$[T]^m$	$[T]^1 = \partial T / \partial o_1$	$[T]^1 = \partial T / \partial o_2$	$[T]^1 = \partial T / \partial o_3$

Gradient of Coordinate: $a_1 a_2 a_3$: $ao^j{}_n = ao^{jn} \neq ao_{jn} = ao_j{}^n$

$\underline{o}_m$	$\underline{o}_1$	$\underline{o}_2$	$\underline{o}_3$
$[a_1]^I$	$ao_1{}^1$	$ao_1{}^2$	$ao_1{}^3$
$[a_2]^I$	$ao_2{}^1$	$ao_2{}^2$	$ao_2{}^3$
$[a_3]^I$	$ao_3{}^1$	$ao_3{}^2$	$ao_3{}^3$

4.1.4. Order of Gradient of a Quantity:

$\underline{P1}(a) \equiv \underline{\nabla}\, p(a)$ Increase Order of P or Q by 1 to Become One $\underline{Q1}(b) \equiv \underline{\nabla}\, q(b)$

:

Composition: to Covariant :Resolution

From One Dimension To M Dimension

$$\underline{P1} = pa_1{}^M \; \underline{a}_M \qquad \underline{Q1} = qb^1{}_M \; \underline{b}^M$$

$$dP1 = pa^1{}_m \; da^m \qquad dQ1 = qb_1{}^m \; db_m$$

So:

$$\underline{P1}\, dP1 = \underline{a}_M \, pa_1{}^M \; pa^1{}_m \; da^m \qquad \underline{Q1}\, dQ1 = \underline{b}_M \, qb_1{}^M \; qb^1{}_m \; db^m$$

Compared With:

$$\underline{P1}\, dP1 = \underline{a}_M \, aP^M{}_1 \; pa^1{}_m \; da^m \qquad \underline{Q1}\, dQ1 = \underline{b}_M \, bQ^M{}_1 \; qb^1{}_m \; db^m$$

General Transformation:

$\underline{P1}\, Q1_1 dP1 = \underline{a}_M\, aP^M{}_1\, Q1_1\, pa^1{}_m\, da^m$ $\qquad$ $\underline{Q1}\, Q1 dQ1 = \underline{b}_M\, bQ^M{}_1\, Q1_1\, qb^1{}_m\, db^m$

And:

Composition: to Contravariant :Resolution

From One Dimension To M Dimension

$\underline{P1} = pa^1{}_M\, \underline{a}^M$ $\qquad$ $\underline{Q1} = qb^1{}_M\, \underline{b}^M$

$dP1 = pa_1{}^m\, da_m$ $\qquad$ $dQ1 = qb_1{}^m\, db_m$

So:

$\underline{P1}\, dP1 = \underline{a}^M\, pa^1{}_M\, pa_1{}^m\, da_m$ $\qquad$ $\underline{Q1}\, dQ1 = \underline{b}^M\, qb^1{}_M\, qb_1{}^m\, db_m$

Compared With:

$\underline{P1}\, dP1 = \underline{a}^M\, aP^1{}_M\, pa_1{}^m\, da_m$ $\qquad$ $\underline{Q1}\, dQ1 = \underline{b}^M\, bQ^1{}_M\, qb_1{}^m\, db_m$

General Transformation:

$\underline{P1}\, Q1^1 dP1 = \underline{a}^M\, aP^1{}_M\, Q1^1\, pa_1{}^m\, da_m$ $\qquad$ $\underline{Q1}\, Q1^1 dQ1 = \underline{b}^M\, bQ^1{}_M\, Q1^1\, qb_1{}^m\, db_m$

Covariant Bases:

Composition: :Resolution

From M Dimension To One Dimension

$\underline{b}_M = bQ^1{}_M\, \underline{Q1}$ $\qquad$ $\underline{a}_M = \nabla^I{}_M\, \underline{a}_I = aP^1{}_M\, pa^I{}_1\, \underline{a}_I = aP^1{}_M\, \underline{P1}$

$db^m = bQ^m{}_1\, dQ1$ $\qquad$ $da^m = aP^m{}_1\, dP1$

So:

$\underline{b}_M\, db^m = \underline{Q1}\, bQ^1{}_M\, bQ^m{}_1\, dQ1$ $\qquad$ $\underline{a}_M\, da^m = \underline{P1}\, aP^1{}_M\, aP^m{}_1\, dP1$

Compared With:

$\underline{b}_M\, db^m = \underline{Q1}\, Q\, b^1{}_M\, bQ^m{}_1\, dQ1$ $\qquad$ $\underline{a}_M\, da^m = \underline{P1}\, pa^1{}_M\, aP^m{}_1\, dP1$

General Transformation:

$\underline{b}_M\, Q^M{}_m\, db^m = \underline{Q1}\, Q\, b_1{}^M\, Q^m{}_M\, bQ^m{}_1\, dQ1$ $\qquad$ $\underline{a}_M\, Q^M{}_m\, da^m = \underline{P1}\, pa^1{}_M\, Q^M{}_m\, aP^m{}_1\, dP1$

Covariant:

Composition: One To M Dimension $\qquad$ M To One Dimension:Resolution

$\underline{P1} = pa_1{}^M\, \underline{a}_M$ $\qquad$ $\underline{a}_M = aP^1{}_M\, \underline{P1}$

$dP1 = pa^1{}_m\, da^m$ $\qquad$ $da^m = aP^m{}_1\, dP1$

So:

$\underline{P1}\, dP1 = \underline{a}_M\, pa_1{}^M\, pa^1{}_m\, da^m$ $\qquad$ $\underline{a}_M\, da^m = \underline{P1}\, aP^1{}_M\, aP^m{}_1\, dP1$

Compared With:

$\underline{P1}\, dP1 = \underline{a}_M\, aP^M{}_1\, pa^1{}_m\, da^m$ $\qquad$ $\underline{a}_M\, da^m = \underline{P1}\, pa^1{}_M\, aP^m{}_1\, dP1$

General Transformation:

$\underline{P1}\, Q1_1\, dP1 = \underline{a}_M\, aP^M{}_1\, Q1_1\, pa^1{}_m\, da^m$ $\qquad$ $\underline{a}_M\, Q^M{}_m\, da^m = \underline{P1}\, pa^1{}_M\, Q^m{}_M\, aP^m{}_1\, dP1$

Contravariant Bases:

Composition: :Resolution

From M Dimension To One Dimension

$\underline{b}^M = bQ^M{}_1\, \underline{Q1}$ $\qquad$ $\underline{a}^M = \nabla^M{}_I\, \underline{a}^I = aP^M{}_1\, pa^1{}_I\, \underline{a}^I = aP^M{}_1\, \underline{P1}$

$db_m = bQ^1{}_m\, dQ1$ $\qquad$ $da_m = aP^1{}_m\, dP1$

So:

$\underline{b}^M db_m = \underline{Q1}\, bQ^M{}_1\ bQ_m{}^1\ dQ1$ $\underline{a}^M da_m = \underline{P1}\, aP^M{}_1\ aP_m{}^1\ dP1$

Compared With:

$\underline{b}^M db_m = \underline{Q1}\, Q\, b_1{}^M\ bQ_m{}^1\ dQ1$ $\underline{a}^M da_m = \underline{P1}\, pa_1{}^M\ aP_m{}^1\ dP1$

General Transformation:

$\underline{b}^M Q_M{}^m db_m = \underline{Q1}\, Q\, b_1{}^M Q_M{}^m\ bQ_m{}^1\ dQ1$ $\underline{a}^M Q_M{}^m da_m = \underline{P1}\, pa_1{}^M Q_M{}^m\ aP_m{}^1\ dP1$

Contravariant:

Composition: One To M Dimension M To One Dimension:Resolution

$\underline{P1} = pa^1{}_M\ \underline{a}^M$ $\underline{a}^M = aP^M{}_1\ \underline{P1}$

$dP1 = pa_1{}^m\ da_m$ $da_m = aP_m{}^1\ dP1$

So:

$\underline{P1}\, dP1 = \underline{a}^M\ pa^1{}_M\ pa_1{}^m\ da_m$ $\underline{a}^M da_m = \underline{P1}\ aP^M{}_1\ aP_m{}^1\ dP1$

Compared With:

$\underline{P1}\, dP1 = \underline{a}^M\ aP_M{}^1\ pa_1{}^m\ da_m$ $\underline{a}^M da_m = \underline{P1}\, pa_1{}^M\ aP_m{}^1\ dP1$

General Transformation:

$\underline{P1}\, Q1^1\ dP1 = \underline{a}^M\ aP_M{}^1\ Q1^1\ pa_1{}^m\ da_m$ $\underline{a}^M Q_M{}^m da_m = \underline{P1}\, pa_1{}^M Q_M{}^m\ aP_m{}^1\ dP1$

Similarly:

Increase (N-1) the Order of $S^i{}_{jk\ldots}(a)$ or $T^i{}_{jk\ldots}(a)$ by 1 to Become of order N

Composition: :Resolution

$\underline{S}^1 \equiv \underline{S}^{1i}{}_{jk\ldots}(a) \equiv \underline{\nabla}\, S^i{}_{jk\ldots}(a)$ $\underline{T}^1 \equiv \underline{T}^{1i}{}_{jk\ldots}(b) \equiv \underline{\nabla} T^i{}_{jk\ldots}(b)$

<u>Gradient</u>:

Composition: :Resolution

From N Dimension To M Dimension

$\underline{T}^1 = Ta^1{}_M\ \underline{a}^M$ $\underline{S}^1 = Sb^1{}_M\ \underline{b}^M$

$dT_1 = Ta_1{}^m\ da_m$ $dS_1 = Sb_1{}^m\ db_m$

So:

$\underline{T}^1\, dT_1 = \underline{a}^M\ Ta^1{}_M\ Ta_1{}^m\ da_m$ $\underline{S}^1\, dS_1 = \underline{b}^M\ Sb^1{}_M\ Sb_1{}^m\ db_m$

Compared With:

$\underline{T}^1\, dT_1 = \underline{a}^M\ aT_M{}^1\ Ta_1{}^m\ da_m$ $\underline{S}^1\, dS_1 = \underline{b}^M\ bS_M{}^1\ Sb_1{}^m\ db_m$

General Transformation:

$\underline{T}^1\, Q1^1\ dT_1 = \underline{a}^M\ aT_M{}^1\ Q1^1\ Ta_1{}^m\ da_m$ $\underline{S}^1\, Q1^1\ dS_1 = \underline{b}^M\ bS_M{}^1\ Q1^1\ Sb_1{}^m\ db_m$

Composition: :Resolution

From M Dimension To N Dimension

$\underline{b}^M = bS^M{}_1\ \underline{S}^1$ $\underline{a}^M = \nabla^M{}_I\, \underline{a}^I = aT^M{}_1\, Ta^1{}_I\, \underline{a}^I = aT^M{}_1\ \underline{T}^1$

$db_m = bS_m{}^1\ dS_1$ $da_m = aT_m{}^1\ dT_1$

So:

$\underline{b}^M db_m = \underline{S}^1\ bS^M{}_1\ bS_m{}^1\ dS_1$ $\underline{a}^M da_m = \underline{T}^1\ aT^M{}_1\ aT_m{}^1\ dT_1$

Compared With:

$\underline{b}^M db_m = \underline{S}^1\ S\, b_1{}^M\ bS_m{}^1\ dS_1$ $\underline{a}^M da_m = \underline{T}^1\ Ta_1{}^M\ aT_m{}^1\ dT_1$

General Transformation:

$\underline{b}^M Q_M{}^m db_m = \underline{S}^1\ Sb_1{}^M Q_M{}^m\ bS_m{}^1\ dS_1$ $\underline{a}^M Q_M{}^m da_m = \underline{T}^1\ Ta_1{}^M Q_M{}^m\ aT_m{}^1\ dT_1$

Composition: N To M Dimension | M To N Dimension:Resolution

$\underline{T}^{1} = Ta^{1}_{M}\ \underline{a}^{M}$ | $\underline{a}^{M} = aT^{M}_{1}\ \underline{T}^{1}$

$dT_{1} = Ta_{1}^{m}\ da_{m}$ | $da_{m} = aT_{m}^{1}\ dT_{1}$

So:

$\underline{T}^{1}\ dT_{1} = \underline{a}^{M}\ Ta^{1}_{M}\ Ta_{1}^{m}\ da_{m}$ | $\underline{a}^{M}\ da_{m} = \underline{T}^{1}\ aT^{M}_{1}\ aT_{m}^{1}\ dT_{1}$

Compared With:

$\underline{T}^{1}\ dT_{1} = \underline{a}^{M}\ aT_{M}^{1}\ Ta_{1}^{m}\ da_{m}$ | $\underline{a}^{M}\ da_{m} = \underline{T}^{1}\ Ta_{1}^{M}\ aT_{m}^{1}\ dT_{1}$

General Transformation:

$\underline{T}^{1}\ Q1^{1}\ dT_{1} = \underline{a}^{M}\ aT_{M}^{1}\ Q1^{1}\ Ta_{1}^{m}\ da_{m}$ | $\underline{a}^{M}\ da_{m} = \underline{T}^{1}\ Ta_{1}^{M}\ Q_{M}^{m}\ aT_{m}^{1}\ dT_{1}$

Composition: M To N Dimension | N To M Dimension:Resolution

$\underline{b}^{M} = bS^{M}_{1}\ \underline{S}^{1}$ | $\underline{S}^{1} = Sb^{1}_{M}\ \underline{b}^{M}$

$db_{m} = bS_{m}^{1}\ dS_{1}$ | $dS_{1} = Sb_{1}^{m}\ db_{m}$

So:

$\underline{b}^{M}\ db_{m} = \underline{S}^{1}\ bS^{M}_{1}\ bS_{m}^{1}\ dS_{1}$ | $\underline{S}^{1}\ dS_{1} = \underline{b}^{M}\ Sb^{1}_{M}\ Sb_{1}^{m}\ db_{m}$

Compared With:

$\underline{b}^{M}\ db_{m} = \underline{S}^{1}\ Sb_{1}^{M}\ bS_{m}^{1}\ dS_{1}$ | $\underline{S}^{1}\ dS_{1} = \underline{b}^{M}\ bS_{M}^{1}\ Sb_{1}^{m}\ db_{m}$

General Transformation:

$\underline{b}^{M}\ Q_{M}^{m}\ db_{m} = \underline{S}^{1}\ Sb_{1}^{M}\ Q_{M}^{m}\ bS_{m}^{1}\ dS_{1}$ | $\underline{S}^{1}\ dS_{1} = \underline{b}^{M}\ bS_{M}^{1}\ Q1^{1}\ Sb_{1}^{m}\ db_{m}$

4.1.5.
Gradient and Direction Cosines:

Gradient and Orientation (Direction) Cosines:

Composition: | :Resolution

From One Dimension To One Dimension

π : positive angle between $\underline{P}$ and $\underline{Q}$ | $(-\pi)$: positive angle between $\underline{Q}$ and $\underline{P}$

$\Theta = \Pi \equiv \cos\pi \equiv \Pi = \Theta$ | $\Pi = \Theta \equiv \cos\theta \equiv \Theta = \Pi$

Where cosine is symmetric:

$\Pi^{M1} \equiv \cos(\underline{P}^{M}, \underline{Q1})$ $\quad \Pi^{M}{}_{1} \equiv \cos(\underline{P}^{M}, \underline{Q1})$ $\quad \cos(\underline{Q1}, \underline{P}^{M}) \equiv \Theta^{M}_{1}$ $\quad \cos(\underline{Q1}, \underline{P}^{M}) \equiv \Theta^{1M}$

$\Pi_{M}{}^{1} \equiv \cos(\underline{P}_{M}, \underline{Q1})$ $\quad \Pi_{M1} \equiv \cos(\underline{P}_{M}, \underline{Q1})$ $\quad \cos(\underline{Q1}, \underline{P}_{M}) \equiv \Theta_{1M}$ $\quad \cos(\underline{Q1}, \underline{P}_{M}) \equiv \Theta^{1}{}_{M}$

$T(\Pi 1M) = \Pi M1 \equiv \cos(\underline{PM}, \underline{Q1}) \quad = \quad \cos(\underline{Q1}, \underline{PM}) \equiv \Theta 1M = T(\Theta M1)$

Composition: Q into P | P into Q :Resolution:

$\Theta^{1M} = \Pi^{M1}$ $\quad \Theta_{1}^{M} = \Pi^{M}{}_{1}$ | $\Pi^{M}{}_{1} = \Theta_{1}^{M}$ $\quad \Pi^{M1} = \Theta^{1M}$

$\Theta^{1}{}_{M} = \Pi_{M}^{1}$ $\quad \Theta_{1M} = \Pi_{M1}$ | $\Pi_{M1} = \Theta_{1M}$ $\quad \Pi_{M}{}^{1} = \Theta^{1}{}_{M}$

$\Theta 1M = \Pi M1 \quad = \quad \Pi M1 = \Theta 1M$

α : positive angle between $\underline{a}$ and $\underline{Q}$ | $(-\alpha)$: positive angle between $\underline{Q}$ and $\underline{a}$

$\Theta = A$: $A = \Theta$

$A^{M1} \equiv \cos(\underline{a}^{M},\underline{Q1})$ $\quad A^{M}{}_{1} \equiv \cos(\underline{a}^{M},\underline{Q1})$ $\quad \cos(\underline{Q1},\underline{a}^{M}) \equiv \Theta_{1}{}^{M}$ $\quad \cos(\underline{Q1},\underline{a}^{M}) \equiv \Theta^{1M}$

$A_{M}{}^{1} \equiv \cos(\underline{a}_{M},\underline{Q1})$ $\quad A_{M1} \equiv \cos(\underline{a}_{M},\underline{Q1})$ $\quad \cos(\underline{Q1},\underline{a}_{M}) \equiv \Theta_{1M}$ $\quad \cos(\underline{Q1},\underline{a}_{M}) \equiv \Theta^{1}{}_{M}$

$T(A1M) = AM1 \equiv \cos(\underline{a}M,\underline{Q1})$ $=$ $\cos(\underline{Q1},\underline{a}M) \equiv \Theta 1M = T(\Theta M1)$

: Composition: Q into a — a into Q :Resolution:

$\Theta^{1M} = A^{M1}$ $\quad \Theta_{1}{}^{M} = A^{M}{}_{1}$ $\quad A^{M}{}_{1} = \Theta_{1}{}^{M}$ $\quad A^{M1} = \Theta^{1M}$

$\Theta^{1}{}_{M} = A_{M}{}^{1}$ $\quad \Theta_{1M} = A_{M1}$ $\quad A_{M1} = \Theta_{1M}$ $\quad A_{M}{}^{1} = \Theta^{1}{}_{M}$

$\Theta 1M = AM1$ $=$ $AM1 = \Theta 1M$

System Orientation Cosines:

Composition
ΘNM = AMN

	Grad Q1	General bN	Unitary gN	Orthogonal sN	Orthonormal yN	Fixed oN
P1	Θ11 = Π11	BN1 = Π1N	ΓN1 = Π1N	ΣN1 = Π1N	ΨN1 = Π1N	ON1 = Π1N
aM	Θ1M = AM1	B1M =AM1	Γ1M = AM1	Σ1M = AM1	Ψ1M =AM1	O1M = AM1
fM	Θ1M = ΓM1	B1M = ΓM1	Γ1M = ΓM1	Σ1M = ΓM1	Ψ1M = ΓM1	O1M = ΓM1
rM	Θ1M = PM1	B1M = PM1	Γ1M = PM1	Σ1M = PM1	Ψ1M = PM1	O1M = PM1
xM	Θ1M = Ξ11	B1M = Ξ11	Γ1M = Ξ11	Σ1M = Ξ11	Ψ1M = Ξ11	O1M = Ξ11
oM	Θ1M = OM1	B1M =OM1	Γ1M = OM1	Σ1M = OM1	Ψ1M =OM1	O1M = OM1

And:

Resolution
AMN = ΘNM

	Grad Q1	General bN	Unitary gN	Orthogonal sN	Orthonormal yN	Fixed oN
P1	Π 11=Θ 11	Π N1= B1N	Π N1= ΓN1	Π N1= ΣN1	Π N1=ΨN1	Π N1=ON1
aM	AM1=Θ1M	AM1=B1M	AM1=Γ1M	AM1=Σ1M	AM1=Ψ1M	AM1=O1M
fM	ΓM1=Θ1M	ΓM1=B1M	ΓM1=Γ1M	ΓM1=Σ1M	ΓM1=Ψ1M	ΓM1=O1M
rM	PM1=Θ1M	PM1=B1M	PM1=Γ1M	PM1=Σ1M	PM1=Ψ1M	PM1=O1M
xM	ΞM1=Θ1M	ΞM1=B1M	ΞM1=Γ1M	ΞM1=Σ1M	ΞM1=Ψ1M	ΞM1=O1M
oM	OM1=Θ1M	OM1=B1M	OM1=Γ1M	OM1=Σ1M	OM1=Ψ1M	OM1=O1M

	Grad Q1	General bN	Unitary gN	Orthogonal sN	Orthonormal yN	Fixed oN
P 1	$\pi^{1}\, PQ1^{1}\, \theta_{N}$ = $\Pi\Theta_{1}{}^{1} = \Theta\Pi^{1}{}_{1}$ = $\theta_{1}\, QP1_{1}\, \pi^{1}$	$\pi^{1}\, Pb_{1}{}^{N}\, \beta_{N}$ = $\Pi B_{1}{}^{N} = B\Pi^{N}{}_{1}$ = $\beta_{N}\, bP^{N}{}_{1}\, \pi^{1}$	$\pi^{1}\, Pg_{1}{}^{N}\, 1$ = $\Pi\Gamma_{1}{}^{N} = \Gamma\Pi^{N}{}_{1}$ = $\gamma_{N}\, gP^{N}{}_{1}\, \pi^{1}$	$\pi^{1}\, Ps_{1}{}^{N}\, \sigma_{N}$ = $\Pi\Sigma_{1}{}^{N} = \Sigma\Pi^{N}{}_{1}$ = $\sigma_{N}\, sP^{N}{}_{1}\, \pi^{1}$	$\pi^{1}\, Py_{1}{}^{N}\, 1$ = $\Pi\Psi_{1}{}^{N} = \Psi\Pi^{N}{}_{1}$ = $1\, yP^{N}{}_{1}\, \pi^{1}$	$\pi^{1}\, Po_{1}{}^{N}\, 1$ = $\Pi O_{1}{}^{N} = O\Pi^{N}{}_{1}$ = $1\, oP^{N}{}_{1}\, \pi^{1}$

a M	$\alpha^{M}\ aQ_{M}^{\ 1}\ \theta_{\underline{1}}$ = $A\Theta_{\underline{M}}^{\ 1}=\Theta A^{1}_{\ \underline{M}}$ = $\theta_{\underline{1}}\ Qa^{1}_{\ M}\ \alpha^{M}$	$\alpha^{M}\ ab_{M}^{\ 1}\ \beta_{\underline{1}}$ = $AB_{\underline{M}}^{\ 1}=BA^{1}_{\ \underline{M}}$ = $\beta_{\underline{1}}\ ba^{1}_{\ M}\ \alpha^{M}$	$\alpha^{M}\ ag_{M}^{\ 1}\ 1$ = $A\Gamma_{\underline{M}}^{\ 1}=\Gamma A^{1}_{\ \underline{M}}$ = $1\ ga^{1}_{\ M}\ \alpha^{M}$	$\alpha^{M}\ as_{M}^{\ 1}\ \sigma_{\underline{1}}$ = $A\Sigma_{\underline{M}}^{\ 1}=\Sigma A^{1}_{\ \underline{M}}$ = $\sigma_{\underline{1}}\ sa^{1}_{\ M}\ \alpha^{M}$	$\alpha^{M}\ ay_{M}^{\ 1}\ 1$ = $A\Psi_{\underline{M}}^{\ 1}=\Psi A^{1}_{\ \underline{M}}$ = $1\ ya^{1}_{\ M}\ \alpha^{M}$	$\alpha^{M}\ ao_{M}^{\ 1}\ 1$ = $AO_{\underline{M}}^{\ 1}=OA^{1}_{\ \underline{M}}$ = $1\ oa^{1}_{\ M}\ \alpha^{M}$
f M	$1\ fQ_{M}^{\ 1}\ \theta_{\underline{1}}$ = $\Phi\Theta_{\underline{M}}^{\ 1}=\Theta\Phi^{1}_{\ \underline{M}}$ = $\theta_{\underline{1}}\ Qf^{1}_{\ 1}\ 1$	$1\ fb_{M}^{\ 1}\ \beta_{\underline{1}}$ = $\Phi B_{\underline{M}}^{\ 1}=B\Phi^{1}_{\ \underline{M}}$ = $\beta_{\underline{1}}\ bf^{1}_{\ 1}\ 1$	$1\ fg_{M}^{\ 1}\ 1$ = $\Phi\Gamma_{\underline{M}}^{\ 1}=\Gamma\Phi^{1}_{\ \underline{M}}$ = $1\ gf^{1}_{\ 1}\ 1$	$1\ fs_{M}^{\ 1}\ \sigma_{\underline{1}}$ = $\Phi\Sigma_{\underline{M}}^{\ 1}=\Sigma\Phi^{1}_{\ \underline{M}}$ = $\sigma_{\underline{1}}\ sf^{1}_{\ 1}\ 1$	$1\ fy_{M}^{\ 1}\ 1$ = $\Phi\Psi_{\underline{M}}^{\ 1}=\Psi\Phi^{1}_{\ \underline{M}}$ = $1\ yf^{1}_{\ 1}\ 1$	$1\ fo_{M}^{\ 1}\ 1$ = $\Phi O_{\underline{M}}^{\ 1}=O\Phi^{1}_{\ \underline{M}}$ = $1\ of^{1}_{\ 1}\ 1$
r M	$\rho^{M}\ rQ_{M}^{\ 1}\ \theta_{\underline{1}}$ = $P\Theta_{\underline{M}}^{\ 1}=\Theta P^{1}_{\ \underline{M}}$ = $\theta_{\underline{1}}\ Qr^{1}_{\ M}\ \rho^{M}$	$\rho^{M}\ rb_{M}^{\ 1}\ \beta_{\underline{1}}$ = $PB_{\underline{M}}^{\ 1}=BP^{1}_{\ \underline{M}}$ = $\beta_{\underline{1}}\ br^{1}_{\ M}\ \rho^{M}$	$\rho^{M}\ rg_{M}^{\ 1}\ 1$ = $P\Gamma_{\underline{M}}^{\ 1}=\Gamma P^{1}_{\ \underline{M}}$ = $1\ gr^{1}_{\ M}\ \rho^{M}$	$\rho^{M}\ rs_{M}^{\ 1}\ \sigma_{\underline{1}}$ = $P\Sigma_{\underline{M}}^{\ 1}=\Sigma P^{1}_{\ \underline{M}}$ = $\sigma_{\underline{1}}\ sr^{1}_{\ M}\ \rho^{M}$	$\rho^{M}\ ry_{M}^{\ 1}\ 1$ = $P\Psi_{\underline{M}}^{\ 1}=\Psi P^{1}_{\ \underline{M}}$ = $1\ yr^{1}_{\ M}\ \rho^{M}$	$\rho^{M}\ ro_{M}^{\ 1}\ 1$ = $PO_{\underline{M}}^{\ 1}=OP^{1}_{\ \underline{M}}$ = $1\ or^{1}_{\ M}\ \rho^{M}$
x M	$1\ xQ_{M}^{\ 1}\ \theta_{\underline{1}}$ = $\Xi\Theta_{\underline{M}}^{\ 1}=\Theta\Xi^{1}_{\ \underline{M}}$ = $\theta_{\underline{1}}\ Qx^{1}_{\ M}\ 1$	$1\ xb_{M}^{\ 1}\ \beta_{\underline{1}}$ = $\Xi B_{\underline{M}}^{\ 1}=B\Xi^{1}_{\ \underline{M}}$ = $\beta_{\underline{1}}\ bx^{1}_{\ M}\ 1$	$1\ xg_{M}^{\ 1}\ 1$ = $\Xi\Gamma_{\underline{M}}^{\ 1}=\Gamma\Xi^{1}_{\ \underline{M}}$ = $1\ gx^{1}_{\ M}\ 1$	$1\ xs_{M}^{\ 1}\ \sigma_{\underline{1}}$ = $\Xi\Sigma_{\underline{M}}^{\ 1}=\Sigma\Xi^{1}_{\ \underline{M}}$ = $\sigma_{\underline{1}}\ sx^{1}_{\ M}\ 1$	$1\ xy_{M}^{\ 1}\ 1$ = $\Xi\Psi_{\underline{M}}^{\ 1}=\Psi\Xi^{1}_{\ \underline{M}}$ = $1\ yx^{1}_{\ M}\ 1$	$1\ xo_{M}^{\ 1}\ 1$ = $\Xi O_{\underline{M}}^{\ 1}=O\Xi^{1}_{\ \underline{M}}$ = $1\ ox^{1}_{\ M}\ 1$
o M	$1\ oQ_{M}^{\ 1}\ \theta_{\underline{1}}$ = $O\Theta_{\underline{M}}^{\ 1}=\Theta O^{1}_{\ \underline{M}}$ = $\theta_{\underline{1}}\ Qo^{1}_{\ M}\ 1$	$1\ ob_{M}^{\ 1}\ \beta_{\underline{1}}$ = $OB_{\underline{M}}^{\ 1}=BO^{1}_{\ \underline{M}}$ = $\beta_{\underline{1}}\ bo^{1}_{\ M}\ 1$	$o^{M}\ og_{M}^{\ 1}\ 1$ = $O\Gamma_{\underline{M}}^{\ 1}=\Gamma O^{1}_{\ \underline{M}}$ = $1\ go^{1}_{\ M}\ 1$	$1\ os_{M}^{\ 1}\ \sigma_{\underline{1}}$ = $O\Sigma_{\underline{M}}^{\ 1}=\Sigma O^{1}_{\ \underline{M}}$ = $\sigma_{\underline{1}}\ so^{1}_{\ M}\ 1$	$1\ oy_{M}^{\ 1}\ 1$ = $O\Psi_{\underline{M}}^{\ 1}=\Psi O^{1}_{\ \underline{M}}$ = $1\ yo^{1}_{\ M}\ 1$	$1\ oo_{M}^{\ 1}\ 1$ = $OO_{\underline{M}}^{\ 1}=OO^{1}_{\ \underline{M}}$ = $1\ oo^{1}_{\ M}\ 1$

4.1.6. Bases Transformation:

Bases Transformation in a System:

: Composition: b onto a

α : positive angle between a and b

$$\underline{a}^{M}\ ab_{M}^{\ N} = \underline{b}^{N}$$

a onto b :Resolution:

β : positive angle between b and a

$$\underline{a}^{M} = \underline{b}^{N}\ ba_{N}^{\ M}$$

: Composition: b onto a

$$\underline{a}^{M}\ ab_{M}^{\ N} = \underline{b}^{N}$$

	Grad Q1	General bN	Unitary gN	Orthogonal sN	Orthonormal yN	Fixed oN
P1	$\underline{P1}\ PQ1^{1} = \underline{Q1}$	$\underline{P1}\ Pb_{1}^{\ N} = \underline{b}^{N}$	Π N1= ΓN1	$\underline{P1}\ Ps_{1}^{\ N} = \underline{s}^{N}$	$\underline{P1}\ Py_{1}^{\ N} = \underline{y}^{N}$	$\underline{P1}\ Po_{1}^{\ N} = \underline{o}^{N}$
aM	$\underline{a}^{M}\ aQ_{M}^{\ N} = \underline{Q1}$	$\underline{a}^{M}\ ab_{M}^{\ N} = \underline{b}^{N}$	$\underline{a}^{M}\ ag_{M}^{\ N} = \underline{g}^{N}$	$\underline{a}^{M}\ as_{M}^{\ N} = \underline{s}^{N}$	$\underline{a}^{M}\ ay_{M}^{\ N} = \underline{y}^{N}$	$\underline{a}^{M}\ ao_{M}^{\ N} = \underline{o}^{N}$
fM	ΓM1=Θ1M	$\underline{f}^{M}\ fb_{M}^{\ N} = \underline{b}^{N}$	$\underline{f}^{M}\ fg_{M}^{\ N} = \underline{g}^{N}$	$\underline{f}^{M}\ fs_{M}^{\ N} = \underline{s}^{N}$	$\underline{f}^{M}\ fy_{M}^{\ N} = \underline{y}^{N}$	$\underline{f}^{M}\ fo_{M}^{\ N} = \underline{o}^{N}$
rM	PM1=Θ1M	$\underline{r}^{M}\ rb_{M}^{\ N} = \underline{b}^{N}$	$\underline{r}^{M}\ rg_{M}^{\ N} = \underline{g}^{N}$	$\underline{r}^{M}\ rs_{M}^{\ N} = \underline{s}^{N}$	$\underline{r}^{M}\ ry_{M}^{\ N} = \underline{y}^{N}$	$\underline{r}^{M}\ ro_{M}^{\ N} = \underline{o}^{N}$
xM	ΞM1=Θ1M	$\underline{x}^{M}\ xb_{M}^{\ N} = \underline{b}^{N}$	$\underline{x}^{M}\ xg_{M}^{\ N} = \underline{g}^{N}$	$\underline{x}^{M}\ xs_{M}^{\ N} = \underline{s}^{N}$	$\underline{x}^{M}\ xy_{M}^{\ N} = \underline{y}^{N}$	$\underline{x}^{M}\ xo_{M}^{\ N} = \underline{o}^{N}$

oM	OM1=Θ1M	$\underline{o}^M\, ob_M{}^N = \underline{b}^N$	$\underline{o}^M\, og_M{}^N = \underline{g}^N$	$\underline{o}^M\, os_M{}^N = \underline{s}^N$	$\underline{o}^M\, oy_M{}^N = \underline{y}^N$	$\underline{o}^M\, oo_M{}^N = \underline{o}^N$

And:

a onto b :Resolution:

$$\underline{a}^M = \underline{b}^N\, b_N{}^M$$

	Grad Q1	General bN	Unitary gN	Orthogonal sN	Orthonormal yN	Fixed oN
P1	$\underline{P1}=\underline{Q1}\,QP1^1$	$\underline{P1}=\underline{b}^N\,bP_N{}^1$	$\underline{P1}=\underline{g}^N\,gP_N{}^1$	$\underline{P1}=\underline{s}^N\,sP_N1$	$\underline{P1}=\underline{y}^N\,yP_N1$	$\underline{P1}=\underline{o}^N\,oP_N1$
aM	$\underline{a}^M=\underline{Q1}\,Qa_1{}^M$	$\underline{a}^M=\underline{b}^N\,ba_N{}^M$	$\underline{a}^M=\underline{g}^N\,ga_N{}^M$	$\underline{a}^M=\underline{s}^N\,sa_N{}^M$	$\underline{a}^M=\underline{y}^N\,ya_N{}^M$	$\underline{a}^M=\underline{o}^N\,oa_N{}^M$
fM	$\underline{f}^M=\underline{Q1}\,Qf_1{}^M$	$\underline{f}^M=\underline{b}^N\,bf_N{}^M$	$\underline{f}^M=\underline{g}^N\,gf_N{}^M$	$\underline{f}^M=\underline{s}^N\,sf_N{}^M$	$\underline{f}^M=\underline{y}^N\,yf_N{}^M$	$\underline{f}^M=\underline{o}^N\,of_N{}^M$
rM	$\underline{r}^M=\underline{Q1}\,Qr_1{}^M$	$\underline{r}^M=\underline{b}^N\,br_N{}^M$	$\underline{r}^M=\underline{g}^N\,gr_N{}^M$	$\underline{r}^M=\underline{s}^N\,sr_N{}^M$	$\underline{r}^M=\underline{y}^N\,yr_N{}^M$	$\underline{r}^M=\underline{o}^N\,or_N{}^M$
xM	$\underline{x}^M=\underline{Q1}\,Qx_1{}^M$	$\underline{x}^M=\underline{b}^N\,bx_N{}^M$	$\underline{x}^M=\underline{g}^N\,gx_N{}^M$	$\underline{x}^M=\underline{s}^N\,sx_N{}^M$	$\underline{x}^M=\underline{y}^N\,yx_N{}^M$	$\underline{x}^M=\underline{o}^N\,ox_N{}^M$
oM	$\underline{o}^M=\underline{Q1}\,Qo_1{}^M$	$\underline{o}^M=\underline{b}^N\,bo_N{}^M$	$\underline{o}^M=\underline{g}^N\,go_N{}^M$	$\underline{o}^M=\underline{s}^N\,so_N{}^M$	$\underline{o}^M=\underline{y}^N\,yo_N{}^M$	$\underline{o}^M=\underline{o}^N\,oo_N{}^M$

Apply Grad to General components:
Component Gradient Is Qqual Its Conjugate Basis:

Composition: $\underline{\nabla} aI \equiv \underline{aI}$ to $\underline{a}$ (m)
One dimension to m dimensions

$$\underline{a}^I \equiv \underline{\nabla} a^I = aa^I{}_M\,\underline{a}^M = ar^I{}_N\,\underline{r}^N = ax^I{}_N\,\underline{x}^N = ao^I{}_M\,\underline{o}^M$$

$$\underline{a}_I \equiv \underline{\nabla} a_I = aa_I{}^M\,\underline{a}_M = ar_I{}^N\,\underline{r}_N = ax_I{}^N\,\underline{x}_N = ao_I{}^M\,\underline{o}_M$$

$\underline{a}$ (m) to $\underline{\nabla}\, a^I \equiv \underline{a}^I$:Resolution
m dimensions to One dimension

$$aa_M{}^K\,\underline{a}_K = \nabla_M{}^I\,\underline{a}_I = \underline{a}_M$$
$$aa^M{}_K\,\underline{a}^K = \nabla^M{}_K\,\underline{a}^K = \underline{a}^M$$
$$ff_M{}^K\,\underline{f}_K = \nabla_M{}^I\,\underline{f}_I = \underline{f}_M$$
$$ff^M{}_K\,\underline{f}^K = \nabla^M{}_K\,\underline{f}^K = \underline{f}^M$$
$$rr_M{}^K\,\underline{r}_K = \nabla_M{}^K\,\underline{r}_K = \underline{r}_M$$
$$rr^M{}_K\,\underline{r}^K = \nabla^M{}_K\,\underline{r}^K = \underline{r}^M$$
$$xx_M{}^K\,\underline{x}_K = \nabla_M{}^K\,\underline{x}_K = \underline{x}_M$$
$$xx^M{}_K\,\underline{x}^K = \nabla^M{}_K\,\underline{x}^K = \underline{x}^M$$
$$oo_M{}^K\,\underline{o}_K = \nabla_M{}^K\,\underline{o}_K = \underline{o}_M$$
$$oo^M{}_K\,\underline{o}^K = \nabla^M{}_K\,\underline{o}^K = \underline{o}^M$$

(f coordinates is treated like a coordinates from now on)

4.1.7.
Differential, Derivative and Metric Distance:

: Composition: b onto a
α : positive angle between a and b
bNM daM = dbN

a onto b :Resolution:
β : positive angle between b and a
daM = aMN dbN

: Composition: b onto a
bNM daM = dbN

Grad	General	Orthogonal	Orthonormal	Fixed

	Q1	bN	sN	yN	oN
P1	Q11 dP1= dQ1	bN1 dP1 = dbN	sN1 dP1 = dsN	yN1 dP1 = dyN	oN1 dP1 = doN
aM	Q1M daM= dQ1				
rM	Q1M drM= dQ1				
xM	Q1M dxM= dQ1				
oM	Q1M doM= dQ1				

And:

a onto b :Resolution:
daM = aMN dbN

	Q1	bN	sN	yN	oN
P1	dP1=P11 dQ1	dP1 = P1N dbN	dP1 = P1N dsN	dP1 = P1N dyN	dP1 = P1N doN
aM	daM=aM1 dQ1				
rM	drM=rM1 dQ1				
xM	dxM=xM1 dQ1				
oM	doM=oM1 dQ1				

System Derivatives:

Composition
bNM = aMN

	Grad Q1	General bN	Orthogonal sN	Orthonormal yN	Fixed oN
P1	Q11 = P 11	bN1 = P1N	sN1 = P 1N	yN1 = P 1N	oN1 = P 1N
aM	Q1M = aM1	bNM = aMN			
rM	Q 1M = rM1				
xM	Q 1M = xM1				
oM	Q 1M = oM1				

Resolution
aMN = bNM

	Q1	bN	sN	yN	oN
P1	P 11 = Q11	P1N = bN1	P 1N = sN1	P 1N = yN1	P 1N = oN1
aM	aM1 = Q1M	aMN = bNM			
rM	rM1 = Q1M				
xM	xM1 = Q1M				
oM	oM1 = Q1M				

Metric Distance:

	Q1	bN	sN	yN	oN
P1	dP1dP1=dQ1dQ1	dP1dP1 = $db_N db^N$	dP1dP1 = $ds_N ds^N$	dP1dP1 = $dy_N dy^N$	dP1dP1 = $do_N do^N$
aM	$da_M da^M$ = Q1dQ1				
rM	$dr_M dr^M$ = Q1dQ1				
xM	$dx_M dx^M$ = Q1dQ1				
oM	$do_M do^M$ = Q1dQ1				

4.1.8.
General Transformation:

: Composition: b onto a a onto b :Resolution:

α : positive angle between a and b β : positive angle between b and a

Recall:

Differential Transformation:

bnm dam = dbn dam = amn dbn

And:

Basis Transformation: (Integral)

aM aMN = bN aM = bN b NM

To Get General Transformation:

aM aMN S(a)Nn bnm dam = bN S(b)Nn dbn | aM S(a)Mm dam = bN bNM S(b)Mm amn dbn

Referncered General Transformation In System:

	Grad $\underline{Q1}$	General $\underline{b}^{N}$	Orthogonal $\underline{s}^{N}$	Orthonormal $\underline{y}^{N}$	Fixed $\underline{o}^{N}$
$\underline{P1}$	$\underline{P1}P1^{1}QTQ1^{1}Q1^{1}dP1$ = $\underline{Q1}Q1^{1}PTP1^{1}P1^{1}dQ1$	$\underline{P1}P1^{N}bTb_{N}^{n}b_{n}^{1}dP1$ = $\underline{b}^{N}b_{N}^{1}PTP1^{1}P1^{n}db_{n}$	$\underline{P1}P1^{N}sTs_{N}^{n}s_{n}^{1}dP1$ = $\underline{s}^{N}s_{N}^{1}PTP1^{1}P1^{n}ds_{n}$	$\underline{P1}P1^{N}yTy_{N}^{n}y_{n}^{1}dP1$ = $\underline{y}^{N}y_{N}^{1}PTP1^{1}P1^{n}dy_{n}$	$\underline{P1}P1^{N}oTo_{N}^{n}o_{n}^{1}dP1$ = $\underline{o}^{N}o_{N}^{1}PTP1^{1}P1^{n}do_{n}$
$\underline{a}^{M}$	$\underline{a}^{M}a_{M}^{1}QTQ1^{1}Q1^{m}da_{m}$ = $\underline{Q1}Q1^{M}aTa_{M}^{m}a_{m}^{1}dQ1$				
$\underline{r}^{M}$	$\underline{r}^{M}r_{M}^{1}QTQ1^{1}Q1^{m}dr_{m}$ = $\underline{Q1}Q1^{M}rTr_{M}^{m}r_{m}^{1}dQ1$				
$\underline{x}^{M}$	$\underline{x}^{M}x_{M}^{1}QTQ1^{1}Q1^{m}dx_{m}$ = $\underline{Q1}Q1^{M}xTx_{M}^{m}x_{m}^{1}dQ1$				
$\underline{o}^{M}$	$\underline{o}^{M}o_{M}^{1}QTQ1^{1}Q1^{m}do_{m}$ = $\underline{Q1}Q1^{M}oTo_{M}^{m}o_{m}^{1}dQ1$				

And

General Transformation In System:

	Grad $\underline{Q1}$	General $\underline{b}^{N}$	Orthogonal $\underline{s}^{N}$	Orthonormal $\underline{y}^{N}$	Fixed $\underline{o}^{N}$
$\underline{P1}$	$P1^{1}QTQ1^{1}Q1^{1}$ = $Q1^{1}PTP1^{1}P1^{1}$	$P1^{N}bTb_{N}^{n}b_{n}^{1}$ = $b_{N}^{1}PTP1^{1}P1^{n}$	$P1^{N}sTs_{N}^{n}s_{n}^{1}$ = $s_{N}^{1}PTP1^{1}P1^{n}$	$P1^{N}yTy_{N}^{n}y_{n}^{1}$ = $y_{N}^{1}PTP1^{1}P1^{n}$	$P1^{N}oTo_{N}^{n}o_{n}^{1}$ = $o_{N}^{1}PTP1^{1}P1^{n}$
$\underline{a}^{M}$	$a_{M}^{1}QTQ1^{1}Q1^{m}$ = $Q1^{M}aTa_{M}^{m}a_{m}^{1}$	$a_{M}^{N}bTb_{N}^{n}b_{n}^{m}$ = $b_{N}^{M}aTa_{M}^{m}a_{m}^{n}$	$a_{M}^{N}sTs_{N}^{n}s_{n}^{m}$ = $s_{N}^{M}aTa_{M}^{m}a_{m}^{n}$	$a_{M}^{N}yTy_{N}^{n}y_{n}^{m}$ = $y_{N}^{M}aTa_{M}^{m}a_{m}^{n}$	$a_{M}^{N}oTo_{N}^{n}o_{n}^{m}$ = $o_{N}^{M}aTa_{M}^{m}a_{m}^{n}$
$\underline{r}^{M}$	$r_{M}^{1}QTQ1^{1}Q1^{m}$ = $Q1^{M}rTr_{M}^{m}r_{m}^{1}$	$r_{M}^{N}bTb_{N}^{n}b_{n}^{m}$ = $b_{N}^{M}rTr_{M}^{m}r_{m}^{n}$	$r_{M}^{N}sTs_{N}^{n}s_{n}^{m}$ = $s_{N}^{M}rTr_{M}^{m}r_{m}^{n}$	$r_{M}^{N}yTy_{N}^{n}y_{n}^{m}$ = $y_{N}^{M}rTr_{M}^{m}r_{m}^{n}$	$r_{M}^{N}oTo_{N}^{n}o_{n}^{m}$ = $o_{N}^{M}rTr_{M}^{m}r_{m}^{n}$
$\underline{x}^{M}$	$x_{M}^{1}QTQ1^{1}Q1^{m}$ = $Q1^{M}xTx_{M}^{m}x_{m}^{1}$	$x_{M}^{N}bTb_{N}^{n}b_{n}^{m}$ = $b_{N}^{M}xTx_{M}^{m}x_{m}^{n}$	$x_{M}^{N}sTs_{N}^{n}s_{n}^{m}$ = $s_{N}^{M}xTx_{M}^{m}x_{m}^{n}$	$x_{M}^{N}yTy_{N}^{n}y_{n}^{m}$ = $y_{N}^{M}xTx_{M}^{m}x_{m}^{n}$	$x_{M}^{N}oTo_{N}^{n}o_{n}^{m}$ = $o_{N}^{M}xTx_{M}^{m}x_{m}^{n}$

$\underline{o}^{M}$	$o_M{}^{1}QTQ1{}^{1}Q1^{m}$ = $Q1^{M}oTo_M{}^{m}o_m{}^{1}$	$o_M{}^{N}bTb_N{}^{n}b_n{}^{m}$ = $b_N{}^{M}oTo_M{}^{m}o_m{}^{n}$	$o_M{}^{N}sTs_N{}^{n}s_n{}^{m}$ = $s_N{}^{M}oTo_M{}^{m}o_m{}^{n}$	$o_M{}^{N}yTy_N{}^{n}y_n{}^{m}$ = $y_N{}^{M}oTo_M{}^{m}o_m{}^{n}$	$o_M{}^{N}oTo_N{}^{n}o_n{}^{m}$ = $o_N{}^{M}oTo_M{}^{m}o_m{}^{n}$

Gradient of a Quantity:

$$\underline{\nabla}T\ ,\underline{\nabla}T^{i}\ ,\underline{\nabla}T^{i}_{j}\ ,\quad \underline{\nabla}T^{i}_{jk}$$

$\underline{\nabla}T \equiv [\,T\,]_{\underline{p}} \equiv \underline{Tp} \equiv \partial T/\partial\underline{p}$ $\qquad$ $\underline{\nabla}T \equiv [\,T\,]_{\underline{q}} \equiv \underline{Tq} \equiv \partial T/\partial\underline{q}$

: Composition: x onto o $\qquad$ o onto x :Resolution:

α : positive angle between o and x $\qquad$ β : positive angle between x and o

Initial-Material Description $\qquad$ Current-Spatial Description

$T^{i}_{jk} = T^{i}_{jk}(p^{M}, t) = T^{i}_{jk}(o^{M}, t)$ $\qquad$ $T^{i}_{jk} = T^{i}_{jk}(q^{N}, t) = T^{i}_{jk}(x^{N}, t)$

$T^{i}_{jk} = T^{i}_{jk}(p_{M}, t) = T^{i}_{jk}(o_{M}, t)$ $\qquad$ $T^{i}_{jk} = T^{i}_{jk}(q_{N}, t) = T^{i}_{jk}(x_{N}, t)$

Gradient of a Quantity T^{i}_{jk}: $\underline{\nabla}T^{i}_{jk}$ =

$\partial T^{i}_{jk}/\partial p_M\ \underline{p}_M \equiv Tp^{i\ M}_{jk}\ \underline{p}_M$ $\qquad$ $\partial T^{i}_{jk}/\partial q_M\ \underline{q}_M \equiv Tq^{i\ M}_{jk}\ \underline{q}_M$

$\partial T^{i}_{jk}/\partial p^{M}\ \underline{p}^{M} \equiv Tp^{i}_{jkM}\ \underline{p}^{M}$ $\qquad$ $\partial T^{i}_{jk}/\partial q^{M}\ \underline{q}^{M} \equiv Tq^{i}_{jkM}\ \underline{q}^{M}$

Gradient of Coordinate:

$\partial q_N/\partial p_M\ \underline{p}_M = qp_N{}^{M}\ \underline{p}_M$ $\qquad$ $\partial p_N/\partial q_M\ \underline{q}_M = pq_N{}^{M}\ \underline{q}_M$

$\partial q^{N}/\partial p^{M}\ \underline{p}^{M} = qp^{N}{}_{M}\ \underline{p}^{M}$ $\qquad$ $\partial p^{N}/\partial q^{M}\ \underline{q}^{M} = pq^{N}{}_{M}\ \underline{q}^{M}$

Transfer of Grad f to basis and vice versa are log type

4.2. *Divergence:*

:

Divergence of $\underline{U}$: ♦ = (.) So: $\underline{\nabla} ♦ \underline{U} = \underline{\nabla}\,.\,\underline{U}$

$\underline{\nabla}\,.\,\underline{\nabla}\ U = \underline{\nabla}\ \underline{\nabla}\,.\,\underline{V}$

$\underline{V}\,.\,\underline{V} \wedge \underline{V} = 0$ $\qquad$ $\underline{\nabla}\,.\,\underline{V}$

Div of Grad = Grad of Div

Divergence of Curl is Zero $\qquad$ Divergence : ♦ = (.)

Del . [ ] ≡ $\underline{\nabla}$. [] ≡

$. \partial[]/\partial\underline{a} = . \partial[]/\partial\underline{r} = . \partial[]/\partial\underline{x} = . \partial[]/\partial\underline{o}$

Or:

$\underline{a}_I\,.\,\partial[]/\partial a_I = \underline{r}_I\,.\,\partial[]/\partial r_I = \underline{x}_I\,.\,\partial[]/\partial x_I = \underline{o}_I\,.\,\partial[]/\partial o_I$

$$\underline{a}^I . \partial[]/\partial a^I = \underline{r}^I . \partial[]/\partial r^I = \underline{x}^I . \partial[]/\partial x^I = \underline{o}^I . \partial[]/\partial o^I$$

Or:

$$\underline{a}_I . []a^I = \underline{r}_I . []r^I = \underline{x}_I . []x^I = \underline{o}_I . []o^I$$
$$\underline{a}^I . []a_I = \underline{r}^I . []r_I = \underline{x}^I . []x_I = \underline{o}^I . []o_I$$

Divergence of $\underline{U}$: ♦ = (.) So: $\underline{\nabla}$ ♦ $\underline{U} = \underline{\nabla} . \underline{U}$

$\underline{Del} . [U] \equiv \underline{\nabla} . [U] \equiv$

$$. \partial[U]/\partial\underline{a} = . \partial[U]/\partial\underline{r} = . \partial[U]/\partial\underline{x} = . \partial[U]/\partial\underline{o}$$

Or:

$$\underline{a}_I . \partial[U]/\partial a_I = \underline{r}_I . \partial[U]/\partial r_I = \underline{x}_I . \partial[U]/\partial x_I = \underline{o}_I . \partial[U]/\partial o_I$$
$$\underline{a}^I . \partial[U]/\partial a^I = \underline{r}^I . \partial[U]/\partial r^I = \underline{x}^I . \partial[U]/\partial x^I = \underline{o}^I . \partial[U]/\partial o^I$$

Or:

$$\underline{a}_I . Ua^I = \underline{r}_I . Ur^I = \underline{x}_I . Ux^I = \underline{o}_I . Uo^I$$
$$\underline{a}^I . Ua_I = \underline{r}^I . Ur_I = \underline{x}^I . Ux_I = \underline{o}^I . Uo_I$$

$\underline{Del} . [\,\underline{U}\,] \equiv \underline{\nabla} . [\,\underline{U}\,]$

$$. \partial[\,\underline{U}]/\partial\underline{a} = . \partial[\,\underline{U}]/\partial\underline{r} = . \partial[\,\underline{U}]/\partial\underline{x} = . \partial[\,\underline{U}]/\partial\underline{o}$$

Or:

$$\underline{a}_I . \partial[\,\underline{U}]/\partial a_I = \underline{r}_I . \partial[\,\underline{U}]/\partial r_I = \underline{x}_I . \partial[\,\underline{U}]/\partial x_I = \underline{o}_I . \partial[\,\underline{U}]/\partial o_I$$
$$\underline{a}^I . \partial[\,\underline{U}]/\partial a^I = \underline{r}^I . \partial[\,\underline{U}]/\partial r^I = \underline{x}^I . \partial[\,\underline{U}]/\partial x^I = \underline{o}^I . \partial[\,\underline{U}]/\partial o^I$$

Or:

$$\underline{a}_I . \underline{Ua}^I = \underline{r}_I . \underline{Ur}^I = \underline{x}_I . \underline{Ux}^I = \underline{o}_I . \underline{Uo}^I$$
$$\underline{a}^I . \underline{Ua}_I = \underline{r}^I . \underline{Ur}_I = \underline{x}^I . \underline{Ux}_I = \underline{o}^I . \underline{Uo}_I$$

Or:

$$\underline{a}_I . [\,\underline{a}_m \partial U^m /\partial a_I\,] = \underline{r}_I . [\,\underline{r}_m \partial U^m /\partial r_I\,] = \underline{x}_I . [\,\underline{x}_m \partial U^m /\partial x_I\,] = \underline{o}_I . [\,\underline{o}_m \partial U^m /\partial o_I\,]$$
$$\underline{a}^I . [\,\underline{a}^m \partial U_m /\partial a^I\,] = \underline{r}^I . [\,\underline{r}^m \partial U_m /\partial r^I\,] = \underline{x}^I . [\,\underline{x}^m \partial U_m /\partial x^I\,] = \underline{o}^I . [\,\underline{o}^m \partial U_m /\partial o^I\,]$$

Or:

$$\underline{a}_I . \underline{a}_m\, Ua^{m\,I}_{\ ,} = \underline{r}_I . \underline{r}_m\, Ur^{m\,I}_{\ ,} = \underline{x}_I . \underline{x}_m\, Ux^{m\,I}_{\ ,} = \underline{o}_I . \underline{o}_m\, Uo^{m\,I}_{\ ,}$$
$$\underline{a}^I . \underline{a}^m\, Ua_{m\,,\,I} = \underline{r}^I . \underline{r}^m\, Ur_{m\,,\,I} = \underline{x}^I . \underline{x}^m\, Ux_{m\,,\,I} = \underline{o}^I . \underline{o}^m\, Uo_{m\,,\,I}$$

Or:

$$aa_{IJ}\, Ua^{JI}_{\ ,} = rr_{IJ}\, Ur^{JI}_{\ ,} = xx_{IJ}\, Ux^{JI}_{\ ,} = oo_{IJ} Uo^{JI}_{\ ,}$$
$$aa^{IJ}\, Ua_{J,I} = rr^{IJ}\, Ur_{J,I} = xx^{IJ}\, Ux_{J,I} = oo^{IJ}\, Uo_{J,I}$$

Similarly:

$$aa_I^J\, Ua_{J,}^I = rr_I^J\, Ur_{J,}^I = xx_I^J\, Ux_{J,}^I = oo_I^J\, Uo_{J,}^I$$
$$aa_J^I\, Ua^J_{\ ,I} = rr_J^I\, Ur^J_{\ ,I} = xx_J^I\, Ux^J_{\ ,I} = oo_J^I\, Uo^J_{\ ,I}$$

So:

$$aa_{IJ}Ua^{JI}_{\ ,} = rr_{IJ}Ur^{JI}_{\ ,} = xx_{IJ}Ux^{JI}_{\ ,} = oo_{IJ}Uo^{JI}_{\ ,}$$
$$aa_J^I Ua^J_{\ ,I} = rr_J^I Ur^J_{\ ,I} = xx_J^I Ux^J_{\ ,I} = oo_J^I Uo^J_{\ ,I}$$

$$aa_I^J\, Ua_{J,}^I = rr_I^J\, Ur_{J,}^I = xx_I^J\, Ux_{J,}^I = oo_I^J\, Uo_{J,}^I$$
$$aa^{IJ}\, Ua_{J,I} = rr^{IJ}\, Ur_{J,I} = xx^{IJ}\, Ux_{J,I} = oo^{IJ}\, Uo_{J,I}$$

Where:

$$aa_{IJ}\, Ua^{JI}_{\ ,} = rr_{IJ}\, Ur^{JI}_{\ ,} = Ux^{II}_{\ ,}$$
$$Ua^I_{\ ,I}$$

$$Ua_{I,}^I$$
$$aa^{IJ}\, Ua_{J,I} = rr^{IJ}\, Ur_{J,I} = Ux_{I,I}$$

Divergence of a Position:

Position Vector:

$$\underline{U} = \underline{a}_n a^n = a_n \underline{a}^n = \underline{r}_n r^n = r_n \underline{r}^n = \underline{x}_n x^n = x_n \underline{x}^n = \underline{o}_n o^n = o_n \underline{o}^n$$

$$aa_{IJ}\, aa^{JI}_{,} = rr_{IJ}\, rr^{JI}_{,} = xx^{II}_{,} = 3 \qquad aa_{I}^{I} = 3$$

$$aa^{I}_{,I} = 3 \qquad aa^{IJ}\, aa_{J,I} = rr^{IJ}\, rr_{J,I} = xx_{I,I} = 3$$

$$\underline{a}_I . \partial[\underline{a}^J a_J]/\partial a_I = \underline{r}_I . \partial[\underline{a}^J a_J]/\partial r_I = \underline{x}_I . \partial[\underline{a}^J a_J]/\partial x_I = \underline{o}_I . \partial[\underline{a}^J a_J]/\partial o_I$$

$$\underline{a}^I . \partial[\underline{a}^J a_J]/\partial a^I = \underline{r}^I . \partial[\underline{a}^J a_J]/\partial r^I = \underline{x}^I . \partial[\underline{a}^J a_J]/\partial x^I = \underline{o}^I . \partial[\underline{a}^J a_J]/\partial o^I$$

Or:

$$\underline{a}_I . \underline{a}^J\, aa_J^{I} = \underline{r}_I . \underline{a}^J\, ar_J^{I} = \underline{x}_{II} . \underline{a}^J\, ax_J^{I} = \underline{o}_I . \underline{a}^J\, ao_J^{I}$$

$$\underline{a}^I . \underline{a}^J\, aa_{JI} = \underline{r}^I . \underline{a}^J\, ar_{JI} = \underline{x}^I . \underline{a}^J\, ax_{JI} = \underline{o}^I . \underline{a}^J\, ao_{JI}$$

$$\underline{a}_I . \partial[\underline{x}^J x_J]/\partial a_I = \underline{r}_I . \partial[\underline{x}^J x_J]/\partial r_I = \underline{x}_I . \partial[\underline{x}^J x_J]/\partial x_I = \underline{o}_I . \partial[\underline{x}^J x_J]/\partial o_I$$

$$\underline{a}^I . \partial[\underline{x}^J x_J]/\partial a^I = \underline{r}^I . \partial[\underline{x}^J x_J]/\partial r^I = \underline{x}^I . \partial[\underline{x}^J x_J]/\partial x^I = \underline{o}^I . \partial[\underline{x}^J x_J]/\partial o^I$$

Or:

$$\underline{a}_I . \underline{x}^J\, xa_J^{I} = \underline{r}_I . \underline{x}^J\, xr_J^{I} = \underline{x}_{II} . \underline{x}^J\, xx_J^{I} = \underline{o}_I . \underline{x}^J\, xo_J^{I}$$

$$\underline{a}^I . \underline{x}^J\, xa_{JI} = \underline{r}^I . \underline{x}^J\, xr_{JI} = \underline{x}^I . \underline{x}^J\, xx_{JI} = \underline{o}^I . \underline{x}^J\, xo_{JI}$$

$$\underline{\nabla} . \underline{U} = \underline{a}^J . \partial/\partial a^J\, \underline{U} = \underline{a}^J . \underline{U}_{,J} = U^J_{,J} \qquad \underline{\nabla} . \underline{U} = \underline{a}_J . \partial/\partial a_J\, \underline{U} = \underline{a}_J . \underline{U}^J_{,} = U_J{}^{,J}$$

$$\underline{\nabla} . \underline{U} = \underline{o}^J . \partial/\partial o^J\, \underline{U} = \underline{o}^J . \underline{U}_{,I} = U^J_{,I} \qquad \underline{\nabla} . \underline{U} = \underline{o}_I . \partial/\partial o_I\, \underline{U} = \underline{o}_I . \underline{U}^I = U_I{}^{,I}$$

Divergence of a Position: $o^i \underline{o}_i$: trace oa^J_J

$$[o^J]_J = oa^J_J = oa^1_1 + oa^2_2 + oa^3_3$$

Divergence of a Position: $o_i \underline{o}^i$: trace oa_J^J

$$[o_J]^J = oa_J^J = oa_1^1 + oa_2^2 + oa_3^3$$

Divergence of a Position: $a^i \underline{a}_i$: trace ao^J_J

$$[a^J]_J = ao^J_J = ao^1_1 + ao^2_2 + ao^3_3$$

Divergence of a Position: $o_i \underline{o}^i$: trace ao_J^J

$$[a_J]^J = ao_J^J = ao_1^1 + ao_2^2 + ao_3^3$$

$p(a)1MM \equiv \underline{\nabla} . \underline{P1}(a)$ Decrease Order by One $q(b)1NN \equiv \underline{\nabla} . \underline{Q1}(b)$

Composition: :Resolution

$$\underline{\nabla} . \underline{Q1} \equiv (\underline{a}_M \partial/\partial a_M).(Q1_N \underline{a}^N) = Qa^{1\,M}_{\;\;M} \qquad \underline{\nabla} . \underline{P1} \equiv (\underline{b}_M \partial/\partial b_M).(P1_N \underline{b}^N) = Pb^{1\,M}_{\;\;M}$$

$$\underline{\nabla} . \underline{Q1} \equiv (\underline{a}^M \partial/\partial a^M) . (Q1^N \underline{a}_N) = Qa_{1\;N}^{\;N} \qquad \underline{\nabla} . \underline{P1} \equiv (\underline{b}^M \partial/\partial b^M) . (P_1{}^N \underline{b}_N) = pa_{1\;N}^{\;N}$$

Where:

$$\underline{a}_M . \underline{a}^N = \underline{b}_N . \underline{b}^M = \delta^N_M$$

$$\underline{a}^M . \underline{a}_N = \underline{b}^M . \underline{b}_N = \delta^M_N$$

System Divergence:

: Composition: b onto a

	Grad	General	Orthogonal	Orthonormal	Fixed
	Q1	b^N	s^N	y^N	o^N
P1	Q11 = P 11	bNN = P11	sNN = P 11	yNN = P 11	oNN = P 11
a^M	Q11 = aMM	bNN = aMM	sNN = aMM	yNN = aMM	oNN = aMM
r^M	Q 11 = rMM	bNN = rMM	sNN = rMM	yNN = rMM	oNN = rMM
x^M	Q 11 = xMM	bNN = xMM	sNN = xMM	yNN = xMM	oNN = xMM
o^M	Q 11 = oMM	bNN = oMM	sNN = oMM	yNN = oMM	oNN = oMM

And:

a onto b :Resolution:

	Q1	b^N	s^N	y^N	o^N
P1	P 11 = Q11	P11 = bNN	P11 = sNN	P11 = yNN	P11 = oNN
a^M	aMM = Q11	aMM = bNN	aMM = sNN	aMM = yNN	aMM = oNN
r^M	rMM = Q11	rMM = bNN	rMM = sNN	rMM = yNN	rMM = oNN
x^M	xMM = Q11	xMM = bNN	xMM = sNN	xMM = yNN	xMM = oNN
o^M	oMM = Q11	oMM = bNN	oMM = sNN	oMM = yNN	oMM = oNN

:

4.3. Curl:

Curl of U : ♦ = (^) So: ∇ ♦ U = ∇ ^ U

∇ ^ ∇ U = 0

∇ . ∇ ^ V = 0 → ∇ ^ V ↑

Curl of Gradient is Zero

Divergence of Curl is Zero → Curl : ♦ = (^) ↑

Del * [ ] ≡ ∇ * [] ≡ * ∂[]/∂a = * ∂[]/∂r = * ∂[]/∂x = * ∂[]/∂o

Or:

a_I * ∂[]/∂a_I = r_I * ∂[]/∂r_I = x_I * ∂[]/∂x_I = o_I * ∂[]/∂o_I

a^I * ∂[]/∂a^I = r^I * ∂[]/∂r^I = x^I * ∂[]/∂x^I = o^I * ∂[]/∂o^I

Or:

$$\underline{a}_I * []a^I = \underline{r}_I * []r^I = \underline{x}_I * []x^I = \underline{o}_I * []o^I$$
$$\underline{a}^I * []a_I = \underline{r}^I * []r_I = \underline{x}^I * []x_I = \underline{o}^I * []o_I$$

$$\underline{Del} * [U] \equiv \underline{\nabla} * [U] \equiv * \partial[U]/\partial\underline{a} = * \partial[U]/\partial\underline{r} = * \partial[U]/\partial\underline{x} = * \partial[U]/\partial\underline{o}$$

Or:

$$\underline{a}_I * \partial[U]/\partial a_I = \underline{r}_I * \partial[U]/\partial r_I = \underline{x}_I * \partial[U]/\partial x_I = \underline{o}_I * \partial[U]/\partial o_I$$
$$\underline{a}^I * \partial[U]/\partial a^I = \underline{r}^I * \partial[U]/\partial r^I = \underline{x}^I * \partial[U]/\partial x^I = \underline{o}^I * \partial[U]/\partial o^I$$

Or:

$$\underline{a}_I * Ua^I = \underline{r}_I * Ur^I = \underline{x}_I * Ux^I = \underline{o}_I * Uo^I$$
$$\underline{a}^I * Ua_I = \underline{r}^I * Ur_I = \underline{x}^I * Ux_I = \underline{o}^I * Uo_I$$

$$\underline{Del} * [\underline{U}] \equiv \underline{\nabla} * [\underline{U}] \equiv * \partial[\underline{U}]/\partial\underline{a} = * \partial[\underline{U}]/\partial\underline{r} = * \partial[\underline{U}]/\partial\underline{x} = * \partial[\underline{U}]/\partial\underline{o}$$

Or:

$$\underline{a}_I * \partial[\underline{U}]/\partial a_I = \underline{r}_I * \partial[\underline{U}]/\partial r_I = \underline{x}_I * \partial[\underline{U}]/\partial x_I = \underline{o}_I * \partial[\underline{U}]/\partial o_I$$
$$\underline{a}^I * \partial[\underline{U}]/\partial a^I = \underline{r}^I * \partial[\underline{U}]/\partial r^I = \underline{x}^I * \partial[\underline{U}]/\partial x^I = \underline{o}^I * \partial[\underline{U}]/\partial o^I$$

Or:

$$\underline{a}_I * \underline{Ua}^I = \underline{r}_I * \underline{Ur}^I = \underline{x}_I * \underline{Ux}^I = \underline{o}_I * \underline{Uo}^I$$
$$\underline{a}^I * \underline{Ua}_I = \underline{r}^I * \underline{Ur}_I = \underline{x}^I * \underline{Ux}_I = \underline{o}^I * \underline{Uo}_I$$

Or:

$$\underline{a}_I *[\underline{a}_J \partial U^J/\partial a_I] = \underline{r}_I * [\underline{r}_J \partial U^J/\partial r_I] = \underline{x}_I * [\underline{x}_J \partial U^J/\partial x_I] = \underline{o}_I *[\underline{o}_J \partial U^J/\partial o_I]$$
$$\underline{a}^I *[\underline{a}^J \partial U_J/\partial a^I] = \underline{r}^I * [\underline{r}^J \partial U_J/\partial r^I] = \underline{x}^I *[\underline{x}^J \partial U_J/\partial x^I] = \underline{o}^I *[\underline{o}^J \partial U_J/\partial o^I]$$

Or:

$$\underline{a}^K \varepsilon_{KJI} Ua^{J,I} = \underline{r}^K \varepsilon_K{}^{JI} Ur_{J,I} = \underline{x}^K \varepsilon_K{}^{JI} Ux_{J,I} = \underline{o}^K \varepsilon_K{}^{JI} Uo_{J,I}$$
$$\underline{a}_K \varepsilon^{KJI} Ua_{J,I} = \underline{r}_K \varepsilon^K{}_{JI} Ur^{J,I} = \underline{x}_K \varepsilon^K{}_{JI} Ux^{J,I} = \underline{o}_K \varepsilon^K{}_{JI} Uo^{J,I}$$

Similarly:

$$\underline{a}^K \varepsilon_{KJ}{}^I Ua^{J,}{}_I = \underline{r}^K \varepsilon_{KJ}{}^I Ur^{J,}{}_I = \underline{x}^K \varepsilon_{KJ}{}^I Ux^{J,}{}_I = \underline{o}^K \varepsilon_{KJ}{}^I Uo^{J,}{}_I$$
$$\underline{a}_K \varepsilon^{KJ}{}_I Ua_{J,}{}^I = \underline{r}_K \varepsilon^{KJ}{}_I Ur_{J,}{}^I = \underline{x}_K \varepsilon^{KJ}{}_I Ux_{J,}{}^I = \underline{o}_K \varepsilon^{KJ}{}_I Uo_{J,}{}^I$$

So:

$$\underline{a}^K \varepsilon_{KJI} Ua^{J,I} \qquad\qquad \underline{a}^K \varepsilon_{KJ}{}^I Ua^{J,}{}_I$$
$$\underline{a}_K \varepsilon^{KJ}{}_I Ua_{J,}{}^I \qquad\qquad \underline{a}_K \varepsilon^{KJI} Ua_{J,I}$$

$\underline{p}(a)M \equiv \underline{\nabla} \wedge \underline{P1}(a)$ Same Order $\underline{q}(b)N \equiv \underline{\nabla} \wedge \underline{Q1}(b)$

Curl:

$$\underline{\nabla} \wedge \underline{P1} \equiv \partial/\partial a_i\, \underline{a}_i \wedge P_1{}^j \underline{a}_j \equiv 1/\|_a\|\ \varepsilon_{ijk}\, \partial^{,i}\, pa_1{}^j\, \underline{a}^k \equiv 1/\|_a\|\ \varepsilon_{ijk}\, pa_1{}^{j,i}\, \underline{a}^k$$
$$\underline{\nabla} \wedge \underline{P1} \equiv \partial/\partial a^i\, \underline{a}^i \wedge P1_j \underline{a}^j \equiv 1/\|^a\|\ \varepsilon^{ijk}\, \partial_{,i}\, pa^1{}_j\, \underline{a}_k \equiv 1/\|^a\|\ \varepsilon^{ijk}\, pa^1{}_{j,i}\, \underline{a}_k$$

$$\boxed{\underline{\nabla} \wedge \underline{P1}} = \boxed{1/\|_a\|} * \begin{vmatrix} \underline{a}^1 & \underline{a}^2 & \underline{a}^3 \\ \partial/\partial a_1 & \partial/\partial a_2 & \partial/\partial a_3 \\ P1^1 & P1^2 & P1^3 \end{vmatrix}$$

Curl of a Position:

Position Vector:

$$\underline{U} = \underline{a}_i a^i = a_i \underline{a}^i = \underline{r}_i r^i = r_i \underline{r}^i = \underline{x}_i x^i = x_i \underline{x}^i = \underline{o}_i o^i = o_i \underline{o}^i$$

$\underline{a}^K \varepsilon_{KJI}\, aa^{J,I} = 0$ $\qquad$ $\underline{a}^K \varepsilon_{KJ}{}^{I}\, aa^{J,}{}_{I} = 0$

$\underline{a}_K \varepsilon^{KJ}{}_{I}\, aa_{J,}{}^{I} = 0$ $\qquad$ $\underline{a}_K \varepsilon^{KJI}\, aa_{J,I} = 0$

Outer Product $\underline{U} \wedge \underline{V}$

$\underline{a}^K \varepsilon_{KJI}\, U^J V^I$ $\qquad$ $\underline{a}^K \varepsilon_{KJ}{}^{I}\, U^J V_I$

$\underline{a}_K \varepsilon^{KJ}{}_{I}\, U_J V^I$ $\qquad$ $\underline{a}_K \varepsilon^{KJI}\, U_J V_I$

Outer Product

$\underline{a}_I \wedge \underline{a}_J = \underline{a}^K \varepsilon_{KJI}\, 1^J 1^I$ $\qquad$ $\underline{a}^K \varepsilon_{KJ}{}^{I}\, U^J V_I$

$\underline{a}_K \varepsilon^{KJ}{}_{I}\, U_J V^I$ $\qquad$ $\underline{a}^I \wedge \underline{a}^J = \underline{a}_K \varepsilon^{KJI}\, 1_J 1_I$

$$\underline{\text{Del}} \wedge[\underline{U}] = \underline{\nabla}\wedge\underline{U} = [\wedge\underline{U}]_{\underline{a}} = \partial[\wedge\underline{U}]/\partial\underline{a} = \nabla[\underline{U}]\wedge\underline{a} = \nabla[\underline{U}]\wedge\underline{o} = xU\,\underline{a}\wedge\underline{a} = oU\,\underline{o}\wedge\underline{o}$$

Outer Product of o basis

$\underline{o}^I \wedge \underline{o}^J = \underline{o}_K \varepsilon^{KIJ}$ $\qquad$ $\underline{o}^I \wedge \underline{o}_J = \underline{o}^K \varepsilon_{K}{}^{I}{}_{J}$

$\underline{o}_I \wedge \underline{o}^J = \underline{o}_K \varepsilon^{K}{}_{I}{}^{J}$ $\qquad$ $\underline{o}_I \wedge \underline{o}_J = \underline{o}^K \varepsilon_{KIJ}$

Curl of o basis

$\underline{\nabla}\wedge\underline{o}^J = \underline{o}^I\, \partial/\partial o^I \wedge \underline{o}^J = \underline{o}_K \varepsilon^{KIJ}\, \partial/\partial o^I$ $\qquad$ $\underline{\nabla}\wedge\underline{o}_J = \underline{o}^I\, \partial/\partial o^I \wedge \underline{o}_J = \underline{o}^K \varepsilon_{K}{}^{I}{}_{J}\, \partial/\partial o^I$

$\underline{\nabla}\wedge\underline{o}^J = \underline{o}_I\, \partial/\partial o_I \wedge \underline{o}^J = \underline{o}_K \varepsilon^{KJ}{}_{I}\, \partial/\partial o_I$ $\qquad$ $\underline{\nabla}\wedge\underline{o}_J = \underline{o}_I\, \partial/\partial o_I \wedge \underline{o}_J = \underline{o}^K \varepsilon_{KIJ}\, \partial/\partial o_I$

Outer Product $\underline{u}\wedge\underline{v}$

$\underline{o}^I u_I \wedge \underline{o}^J v_J = \underline{o}_K \varepsilon^{KIJ}\, u_I v_J$ $\qquad$ $\underline{o}^I u_I \wedge \underline{o}_J v^J = \underline{o}^K \varepsilon_{K}{}^{I}{}_{J}\, u_I v^J$

$\underline{o}_I u^I \wedge \underline{o}^J v_J = \underline{o}_K \varepsilon^{KJ}{}_{I}\, u^I v_J$ $\qquad$ $\underline{o}_I u^I \wedge \underline{o}_J v^J = \underline{o}^K \varepsilon_{KIJ}\, u^I v^J$

$\underline{\text{Curl}}$ of $\underline{v} \equiv \underline{\nabla}\wedge\underline{v}$

$\underline{o}^I\, \partial/\partial o^I \wedge \underline{o}^J v_J = \underline{o}_K \varepsilon^{KIJ}\, \partial/\partial o^I v_J$ $\qquad$ $\underline{o}^I\, \partial/\partial o^I \wedge \underline{o}_J v^J = \underline{o}^K \varepsilon_{K}{}^{I}{}_{J}\, \partial/\partial o^I v^J$

$\underline{o}_I\, \partial/\partial o_I \wedge \underline{o}^J v_J = \underline{o}_K \varepsilon^{KJ}{}_{I}\, \partial/\partial o_I v_J$ $\qquad$ $\underline{o}_I\, \partial/\partial o_I \wedge \underline{o}_J v^J = \underline{o}^K \varepsilon_{KIJ}\, \partial/\partial o^I v^J$

So : $\underline{\nabla}\wedge\underline{v}$

$\underline{o}_K \varepsilon^{KIJ}\, v_{J;I} = \underline{o}_K \varepsilon^{KIJ}\, vo_{JI}$ $\qquad$ $\underline{o}^K \varepsilon_{K}{}^{I}{}_{J}\, v^{J}{}_{;I} = \underline{o}^K \varepsilon_{K}{}^{I}{}_{J}\, vo^{J}{}_{I}$

$\underline{o}_K \varepsilon^{KJ}{}_{I}\, v_{J;}{}^{I} = \underline{o}_K \varepsilon^{KJ}{}_{I}\, vo_{J}{}^{I}$ $\qquad$ $\underline{o}^K \varepsilon_{KIJ}\, v^{J;I} = \underline{o}^K \varepsilon_{KIJ}\, vo^{JI}$

Recall

$\|\underline{a}\| \equiv 1/6\,\varepsilon^{KIJ}\, \varepsilon_{pMn}\, ao^{p}{}_{K}\, ao^{M}{}_{I}\, ao^{n}{}_{J}$ $\qquad$ $\|\underline{a}\| \equiv 1/6\,\varepsilon_{pMn}\, \varepsilon^{KIJ}\, ao_{K}{}^{p}\, ao_{I}{}^{M}\, ao_{J}{}^{n}$

For $\qquad i \neq j \neq k \neq i$

$$\underline{o}^{I} \wedge \underline{o}^{J} . \underline{o}^{K} = 1 \qquad \underline{o}_{I} \wedge \underline{o}_{J} . \underline{o}_{K} = 1$$

$$e^{KIJ} \underline{o}_{K} \equiv \underline{o}^{I} \wedge \underline{o}^{J} / \|^{o}\| \qquad e_{KIJ} \underline{o}^{K} \equiv \underline{o}_{I} \wedge \underline{o}_{J} / \|_{o}\|$$

$$1 = \underline{o}_{\underline{K}} . \underline{o}^{\underline{K}} = \underline{o}^{I} \wedge \underline{o}^{J} . \underline{o}^{K} / \|^{o}\| = \|^{o}\| / \|^{o}\| = 1 \qquad 1 = \underline{o}^{\underline{K}} . \underline{o}_{\underline{K}} = \underline{o}_{I} \wedge \underline{o}_{J} . \underline{o}_{K} / \|_{o}\| = \|_{o}\| / \|_{o}\| = 1$$

$$\|^{o}\| \underline{o}_{\underline{K}} . \underline{o}^{\underline{K}} = \underline{o}^{I} \wedge \underline{o}^{J} . \underline{o}^{K} \qquad \|_{o}\| \underline{o}^{\underline{K}} . \underline{o}_{\underline{K}} = \underline{o}_{I} \wedge \underline{o}_{J} . \underline{o}_{K}$$

For $i \neq j \neq k \neq i$

$$\underline{a}^{i} \wedge \underline{a}^{j} . \underline{a}^{\underline{k}} \equiv \|^{a}\| \qquad \underline{a}_{i} \wedge \underline{a}_{j} . \underline{a}_{k} \equiv \|_{a}\|$$

$$e^{kij} \underline{a}_{k} \equiv \underline{a}^{i} \wedge \underline{a}^{j} / \|^{a}\| \qquad e_{kij} \underline{a}^{k} \equiv \underline{a}_{i} \wedge \underline{a}_{j} / \|_{a}\|$$

$$1 = \underline{a}_{\underline{k}} . \underline{a}^{\underline{k}} = \underline{a}^{i} \wedge \underline{a}^{j} . \underline{a}^{k} / \|^{a}\| = \|^{a}\| / \|^{a}\| = 1 \qquad 1 = \underline{a}^{\underline{k}} . \underline{a}_{\underline{k}} = \underline{a}_{i} \wedge \underline{a}_{j} . \underline{a}_{k} / \|_{a}\| = \|_{a}\| / \|_{a}\| = 1$$

$$\|^{a}\| \underline{a}_{\underline{k}} . \underline{a}^{\underline{k}} = \underline{a}^{i} \wedge \underline{a}^{j} . \underline{a}^{k} \qquad \|_{a}\| \underline{a}^{\underline{k}} . \underline{a}_{\underline{k}} = \underline{a}_{i} \wedge \underline{a}_{j} . \underline{a}_{k}$$

For $i \neq j \neq k \neq i$ Outer Product of a basis

$$\underline{a}^{i} \wedge \underline{a}^{j} \equiv \varepsilon^{kij} \underline{a}_{k} \|^{a}\|$$

$$\underline{a}_{i} \wedge \underline{a}_{j} \equiv \varepsilon_{kij} \underline{a}^{k} \|_{a}\|$$

or

$$\underline{a}^{i} \wedge \underline{a}^{j} = ao^{i}_{m} \underline{o}^{m} \wedge ao^{j}_{n} \underline{o}^{n} = \underline{o}_{k} \varepsilon^{kmn} ao^{i}_{m} ao^{j}_{n} = \underline{o}_{k} \varepsilon^{kmn} \alpha^{i} C(^{i}, _{m}) \alpha^{j} C(^{j}, _{n})$$

$$\underline{a}^{i} \wedge \underline{a}_{j} = ao^{i}_{m} \underline{o}^{m} \wedge ao_{j}^{n} \underline{o}_{n} = \underline{o}^{k} \varepsilon_{k}{}^{m}{}_{n} ao^{i}_{m} ao_{j}^{n} = \underline{o}_{k} \varepsilon_{k}{}^{m}{}_{n} \alpha^{i} C(^{i}, _{m}) \alpha_{j} C(_{j}, ^{n})$$

$$\underline{a}_{i} \wedge \underline{a}^{j} = ao_{i}^{m} \underline{o}_{m} \wedge ao^{j}_{n} \underline{o}^{n} = \underline{o}_{k} \varepsilon^{k}{}_{m}{}^{n} ao_{i}^{m} ao^{j}_{n} = \underline{o}_{k} \varepsilon^{k}{}_{m}{}^{n} \alpha_{i} C(_{i}, ^{m}) \alpha^{j} C(^{j}, _{n})$$

$$\underline{a}_{i} \wedge \underline{a}_{j} = ao_{i}^{m} \underline{o}_{m} \wedge ao_{j}^{n} \underline{o}_{n} = \underline{o}^{k} \varepsilon_{kmn} ao_{i}^{m} ao_{j}^{n} = \underline{o}^{k} \varepsilon_{kmn} \alpha_{i} C(_{i}, ^{m}) \alpha_{j} C(_{j}, ^{n})$$

Notice

$$\underline{a}^{i} \wedge \underline{a}^{j} . \underline{a}^{k} = \underline{o}_{p} \varepsilon^{pmn} ao^{i}_{m} ao^{j}_{n} . \underline{a}^{k} = \underline{o}_{p} \varepsilon^{pmn} ao^{i}_{m} ao^{j}_{n} . \underline{o}^{q} ao^{k}_{q} = e^{qmn} ao^{i}_{m} ao^{j}_{n} ao^{k}_{q}$$

multiply by ε_{kij} then

$$e_{kij} \underline{a}^{i} \wedge \underline{a}^{j} . \underline{a}^{k} = 6\|^{a}\| = \varepsilon_{kij} \varepsilon^{qmn} ao^{i}_{m} ao^{j}_{n} ao^{k}_{q}$$

$$\underline{a}_{i} \wedge \underline{a}_{j} . \underline{a}_{k} = \underline{o}^{p} \varepsilon_{pmn} ao_{i}^{m} ao_{j}^{n} . \underline{a}_{k} = \underline{o}^{p} \varepsilon_{pmn} ao_{i}^{m} ao_{j}^{n} . \underline{o}_{q} ao_{k}^{q} = \varepsilon_{qmn} ao_{i}^{m} ao_{j}^{n} ao_{k}^{q}$$

multiply by ε^{kij} then

$$\varepsilon^{kij} \underline{a}_{i} \wedge \underline{a}_{j} . \underline{a}_{k} = 6\|_{a}\| = \varepsilon^{kij} \varepsilon_{qmn} ao_{i}^{m} ao_{j}^{n} ao_{k}^{q}$$

Curl of a basis

$$\underline{\nabla} \wedge \underline{a}^{j} = \underline{a}^{i} \partial/\partial a^{i} \wedge \underline{a}^{j} = \underline{a}_{k} \|^{a}\| \varepsilon^{kij} \partial/\partial a^{i} \qquad \underline{\nabla} \wedge \underline{a}_{j} = \underline{a}^{i} \partial/\partial a^{i} \wedge \underline{a}_{j}$$

$$\underline{\nabla} \wedge \underline{a}^{j} = \underline{a}_{i} \partial/\partial a_{i} \wedge \underline{a}^{j} \qquad \underline{\nabla} \wedge \underline{a}_{j} = \underline{a}_{i} \partial/\partial a_{i} \wedge \underline{a}_{j} = \underline{a}^{k} \|_{a}\| \varepsilon_{kij} \partial/\partial a_{i}$$

Outer Product $\underline{u} \wedge \underline{v}$

$\underline{a}^i u_i \wedge \underline{a}^j v_j = \underline{a}_k \|a\| \varepsilon^{kij} u_i v_j =$ $ao^i_m \underline{o}^m \wedge ao^j_n \underline{o}^n u_i v_j =$ $\underline{o}_k \varepsilon^{kmn} ao^i_m ao^j_n u_i v_j =$ $\underline{o}_k \varepsilon^{kmn} \alpha^i C(,_m) \alpha^j C(,_n) u_i v_j$	$\underline{a}^i u_i \wedge \underline{a}_j v^j$
$\underline{a}_i u^i \wedge \underline{a}^j v_j$	$\underline{a}_i u^i \wedge \underline{a}_j v^j = \underline{a}^k \|a\| \varepsilon_{kij} u^i v^j =$ $ao_i^m \underline{o}_m \wedge ao_j^n \underline{o}_n u^i v^j =$ $\underline{o}^k \varepsilon_{kmn} ao_i^m ao_j^n u^i v^j =$ $\underline{o}^k \varepsilon_{kmn} \alpha_i C(_i,^m) \alpha_j C(_j,^n) u^i v^j$
Curl of $\underline{v}$:	$\underline{\nabla} \wedge \underline{v}$
$\underline{\nabla} \wedge \underline{v} = \underline{a}^i \partial/\partial a^i \wedge \underline{a}^j v_j = \underline{a}_k \varepsilon^{kij} \partial/\partial a^i v_j$ $=$ $\underline{a}_k \varepsilon^{kij} va_{ji}$	$\underline{\nabla} \wedge \underline{v}$ $=$ $\underline{a}^k \varepsilon_{kj}^{\ \ i} va^j_{\ i}$
$\underline{\nabla} \wedge \underline{v}$ $=$ $\underline{a}_k \varepsilon^{kj}_{\ \ i} va_j^{\ i}$	$\underline{\nabla} \wedge \underline{v} = \underline{a}_i \partial/\partial a_i \wedge \underline{a}_j v^j = \underline{a}^k \varepsilon_{kij} \partial/\partial a^i v^j$ $=$ $\underline{a}^k \varepsilon_{kij} va^{ji}$

4.4.
Two Sequential Operations:

$\underline{\nabla} \wedge \underline{\nabla}\, U = 0$

$\underline{\nabla} \cdot \underline{\nabla}\, U = \underline{\nabla}\ \underline{\nabla} \cdot \underline{V}$

$\underline{\nabla} \cdot \underline{\nabla} \wedge \underline{V} = 0$

Curl of Gradient is Zero

Div of Grad = Grad of Div

Divergence of Curl is Zero

4.4.1.
(Laplacian)
Divergence Of Gradient
Or:
Gradient Of Divergence :

$\underline{\nabla} \cdot \underline{\nabla}\, U = \underline{\nabla}\ \underline{\nabla} \cdot \underline{V}$

Div of Grad = Grad of Div

Divergence of Gradient or Laplacian of T:

$$\text{Div}\ \underline{\text{Grad}}\ T = \underline{\nabla} . \underline{\nabla} T = \{(\partial/\partial a^J\ \underline{a}^J).(\partial/\partial a_K\ \underline{a}_K)\}\ T = T_{,J}^{\ J}$$

$aa_{IJ}\ Ta^{JI}_{,} = rr_{IJ}\ Tr^{JI}_{,} = Tx^{II}_{,}$	$Ta^{I}_{,I}$
$Ta^{I}_{,I}$	$aa^{IJ}\ Ta_{,JI} = rr^{IJ}\ Tr_{,JI} = Tx_{,II}$

Gradient of Divergence of U :

$$\underline{\text{Grad}}\ \text{Div}\ \underline{U} = \underline{\text{Grad}}\ [(\partial/\partial a^J\ \underline{a}^J)(U^m\ \underline{a}_m)] = \underline{\text{Grad}}\ U^m_{,m}$$

$$\underline{\text{Grad}}\ U^m_{,m} = (\partial/\partial a_K\ \underline{a}_K)\ U^m_{,m} = (\partial/\partial a^K\ \underline{a}^K)\ U^m_{,m}$$

$$\underline{\text{Grad}}\ U^m_{,m} = U^m_{,m}{}^K\ \underline{a}_K = U^m_{,mK}\ \underline{a}^K$$

$aa_{IJ}\ Ua^{JI}_{,K}\ \underline{a}^K = rr_{IJ}\ Ur^{JI}_{,K}\ \underline{r}^K = Ux^{II}_{,K}\ \underline{x}^K$	$Ua^{I}_{,I\ K}\ \underline{a}^K$
$Ua^{I}_{,IK}\ \underline{a}^K$	$aa^{IJ}\ Ua_{J,IK}\ \underline{a}^K = rr^{IJ}\ Ur_{J,IK}\ \underline{r}^K = Ux_{I,IK}\ \underline{x}^K$

$aa_{IJ}\ Ua^{JIK}_{,}\ \underline{a}_K = rr_{IJ}\ Ur^{JIK}_{,}\ \underline{r}_K = Ux^{IIK}_{,}\ \underline{x}_K$	$Ua^{IK}_{I,}\ \underline{a}_K$
$Ua^{IK}_{,I}\ \underline{a}_K$	$aa^{IJ}\ Ua_{J,I}{}^K\ \underline{a}_K = rr^{IJ}\ Ur_{J,I}{}^K\ \underline{r}_K = Ux_{I,I}{}^K\ \underline{x}_K$

4.4.2.
Divergence of Curl is Zero:

$$\underline{\nabla} . \underline{\nabla} \wedge \underline{V} = 0$$

Divergence of Curl is Zero

Divergence of Curl = 0 :

$$\underline{\text{Del}} . [\underline{U}] = \underline{\nabla} . \underline{U} = \underline{a}[.\underline{U}] = \partial[.\underline{U}]/\partial\underline{a} = \nabla[\underline{U}].\underline{a} = aU\ \underline{a}\ .\ \underline{a}$$

$$\underline{\text{Del}} . [\underline{\nabla} \wedge \underline{V}] = \underline{\nabla} . \underline{\nabla} \wedge \underline{V} = \nabla[\underline{\nabla} \wedge \underline{V}].\underline{a} = a[\underline{\nabla} \wedge \underline{V}]\ \underline{a}\ .\ \underline{a} = \text{Div}\ \underline{\text{Curl}}\ V = 0$$

$\underline{a}^K\ \varepsilon_{KJI}\ Va^{J,I}$	$\underline{a}^K\ \varepsilon_{KJ}{}^I\ Va^{J,}{}_I$
$\underline{a}_K\ \varepsilon^{KJ}{}_I\ Va_{J,}{}^I$	$\underline{a}_K\ \varepsilon^{KJI}\ Va_{J,I}$

So:

$aa_{IJ}\ Ua^{JI}_{,} = rr_{IJ}\ Ur^{JI}_{,} = Ux^{II}_{,}$	$Ua^{I}_{\ I,}$
$Ua^{I}_{\ ,I}$	$aa^{IJ}\ Ua_{J,I} = rr^{IJ}\ Ur_{J,I} = Ux_{I,I}$

$J = I$

$\underline{a}^{K}\ \varepsilon_{KII}\ Va^{I,I} = 0$	$\underline{a}^{K}\ \varepsilon_{KI}^{\ \ I}\ Va^{I,}_{\ \ I} = 0$
$\underline{a}_{K}\ \varepsilon^{KI}_{\ \ \ I}\ Va^{I}_{\ I,} = 0$	$\underline{a}_{K}\ \varepsilon^{KII}\ Va_{I,I} = 0$

4.4.3.
Curl of Gradient is Zero:

$\underline{\nabla} \wedge \underline{\nabla}\ U = 0$

Curl of Gradient is Zero

Curl of Gradient = 0 :

$\underline{Del}\ \wedge[\underline{U}] = \underline{\nabla}\wedge\underline{U} = \underline{a}\ [\wedge\underline{U}] = \partial[\wedge\underline{U}]/\partial\underline{a} = \nabla[\underline{U}]\wedge\underline{a} = aU\ \underline{a}\wedge\underline{a}$
$\underline{Del}\ \wedge[\underline{\nabla}T] = \underline{\nabla}\wedge\underline{\nabla}T = \nabla[\underline{\nabla}T]\wedge\underline{a} = a\nabla T\ \underline{a}\wedge\underline{a}$
$\underline{Curl}\ \underline{Grad}\ T = \underline{Curl}\ (\underline{a}^{m}\ T_{,m}) = \underline{Curl}\ (\underline{a}_{m}\ T^{m}_{,}) = \underline{0}$

$\underline{a}_{I}\ Ta^{I}_{,}$	
	$\underline{a}^{I}\ Ta_{,I}$

$\underline{a}^{K}\ \varepsilon_{KII}\ Ua^{I,I} = 0$	$\underline{a}^{K}\ \varepsilon_{KI}^{\ \ I}\ Ua^{I,}_{\ \ I} = 0$
$\underline{a}_{K}\ \varepsilon^{KI}_{\ \ \ I}\ Ua^{I}_{\ I,} = 0$	$\underline{a}_{K}\ \varepsilon^{KII}\ Ua_{I,I} = 0$

$\underline{U}\wedge(\underline{A}\wedge\underline{B}) = (\underline{U}\cdot\underline{B})\underline{A} - (\underline{U}\cdot\underline{A})\underline{B}$

$\underline{\nabla}\wedge(u\ \ \underline{U}) = \underline{\nabla}u\wedge\underline{U} - u\ \ \underline{\nabla}\wedge\underline{U}$

Let:

$\underline{U} = \underline{\nabla}\, a^{J}$

Such That:

$\underline{\nabla}\wedge\underline{U} = \underline{\nabla}\wedge\underline{\nabla}\, a^{J} = 0$

Then:

$$\underline{\nabla} \wedge (u \quad \underline{U} \quad) = \underline{\nabla} \wedge [u \ \underline{\nabla} a^{J}] = \underline{\nabla} u \wedge [\underline{\nabla} a^{J}] \quad - 0$$

Let:

$$u = a^{I}$$

Then:

$$\underline{\nabla} \wedge (u \quad \underline{U} \quad) = \underline{\nabla} \wedge [a^{I} \ \underline{\nabla} a^{J}] = \underline{\nabla} a^{I} \wedge [\underline{\nabla} a^{J}] = \underline{a}^{I} \wedge \underline{a}^{J}$$

But:

$$\underline{a}^{I} \wedge \underline{a}^{J} = \underline{a}_{k} \ \|oa_{m}^{\ n}\| = \underline{a}_{K} \ / \|oa^{n}_{\ m}\|$$

Recall:

$$\underline{a}^{I} \wedge \underline{a}^{J} \ . \ \underline{a}^{K} = \|oa_{m}^{\ n}\| = 1 / \|oa^{n}_{\ m}\|$$

So:

$$\underline{\nabla} \wedge (u \quad \underline{U} \quad) = \underline{\nabla} \wedge [a^{I} \ \underline{\nabla} a^{J}] = \underline{a}_{K} \ \|oa_{m}^{\ n}\| = \underline{a}_{K} \ / \|oa^{n}_{\ m}\|$$

In General:

$$\underline{a}^{K} / \|oa_{n}^{\ m}\| = \underline{\nabla} a^{K} / \| oa_{n}^{\ m} \| = \nabla \wedge [a_{I} \ \underline{\nabla} a_{J}] \qquad \underline{a}_{K} \ / \|oa^{n}_{\ m}\| = \underline{\nabla} a_{K} / \| oa^{n}_{\ m} \| = \nabla \wedge [a^{I} \ \underline{\nabla} a^{J}]$$

$$\underline{U} \wedge (\underline{A} \wedge \underline{B}) = (\underline{U} \ . \ \underline{B}) \underline{A} \ - (\underline{U} \ . \ \underline{A}) \ \underline{B}$$

Curl of $\underline{U}$: $\quad \nabla \wedge \underline{U}$

$\underline{\nabla} \wedge \underline{U} = \underline{a}^{K} \partial/\partial a^{K} \wedge \underline{a}^{J} U_{J} = \underline{a}_{K} \varepsilon^{KIJ} \partial/\partial a^{I} U_{J}$ $=$ $\underline{a}_{K} \ \varepsilon^{KIJ} \ Ua_{J\,I}$	$\underline{\nabla} \wedge \underline{U}$ $=$ $\underline{a}^{K} \ \varepsilon_{K\,J}^{\ \ I} \ Ua^{J}_{\ I}$
$\underline{\nabla} \wedge \underline{U}$ $=$ $\underline{a}_{K} \ \varepsilon^{K\,J}_{\ \ \ I} \ Ua_{J}^{\ I}$	$\underline{\nabla} \wedge \underline{U} = \underline{a}_{I} \partial/\partial a_{I} \wedge \underline{a}_{J} U^{J} = \underline{a}^{K} \ \varepsilon_{K\,I\,J} \ \partial/\partial a^{I} U^{J}$ $=$ $\underline{a}^{K} \ \varepsilon_{K\,I\,J} \ Ua^{J\,I}$

Curl of a basis

$$\underline{\nabla} \wedge \underline{a}^{J} = \underline{a}^{K} \partial/\partial a^{K} \wedge \underline{a}^{J} = \varepsilon^{KIJ} \ \underline{a}_{K} \ \|_{a}\| \ \partial/\partial a^{I}$$

$$\underline{\nabla} \wedge \underline{a}^{J} = \underline{a}_{I} \ \partial/\partial a_{I} \wedge \underline{a}^{J}$$

$$\underline{\nabla} \wedge \underline{a}_{J} = \underline{a}^{I} \partial/\partial a^{I} \wedge \underline{a}_{J} = \underline{a}^{K} \ \|^{a}\| \ \varepsilon_{KIJ} \partial/\partial a_{I}$$

$$\underline{\nabla} \wedge \underline{a}_{J} = \underline{a}_{I} \ \partial/\partial a_{I} \wedge \underline{a}_{J}$$

These Bases can be found in Other Ways: where i,j,k ordered indices

$$\underline{a}^{k} = \underline{\nabla} a^{k} = \underline{\nabla} a_{i} \wedge \underline{\nabla} a_{j} / \|oa^{n}_{\ m}\| \qquad \underline{a}_{k} = \underline{\nabla} a_{k} = \underline{\nabla} a^{i} \wedge \underline{\nabla} a^{j} / \| oa_{n}^{\ m} \|$$

$$\underline{a}^{k} / \|oa_{n}^{\ m}\| = \underline{\nabla} a^{k} / \|oa_{n}^{\ m}\| = \nabla \wedge (a_{i} (\underline{\nabla} a_{j})) \qquad \underline{a}_{k} / \| oa^{n}_{\ m}\| = \underline{\nabla} a_{k} / \|oa^{n}_{\ m}\| = \nabla \wedge (a^{i} (\nabla a^{j}))$$

4.5.
Differential and Integral Operations:

$$d[\textstyle\int_{a}^{b} f(s) \, ds] / dt = db/dt \ f(s = b) - da/dt \ f(s = a) + \textstyle\int_{a}^{b} d f(s)/dt \ ds$$

$$d[\textstyle\int_{a}^{b} f(s) \, ds]/ds \ ds/dt = db/ds \ ds/dt \ f(s = b) - da/ds \ ds/dt \ f(s = a) + \textstyle\int_{a}^{b} df(s)/ds \ ds/dt \ ds$$

If t = s then:

$$d[\textstyle\int_{a}^{b} f(s) \, ds]/ds = db/ds \ f(s = b) - da/ds \ f(s = a) + \textstyle\int_{a}^{b} df(s)/ds \ ds$$

$$d[\textstyle\int_{a}^{b} f(s) \, ds]/ds = db/ds \ f(s = b) - da/ds \ f(s = a) + f(s) \,|_{0}^{t}$$

Integral operations

$$\iiint T,^{r} dx_r\, dx_s\, dx_t = \iint T\, n^{r}\, dx_s\, dx_t$$

$$\int_{Volume} T,^{i} dV = \int_{Surface} T\, n^{i}\, dS$$

$$\int_{Volume} T,_{j} dV = \int_{Surface} T\, n_{j}\, dS$$

Condition of existence of $T,_{j}$ everywhere inside S (Does not hold in shock wave)

Gauss:

If T is a scalar: $T = \varphi$ then:

$$\int_{Volume} \underline{Grad}\, \varphi\, dV = \int_{Volume} \underline{\nabla}\, \varphi\, dV = \int_{Surface} \varphi\, \underline{n}\, dS$$

$$\int_{Volume} \varphi,^{i}\, dV = \int_{Surface} \varphi\, n^{i}\, dS$$

$$\int_{Volume} \varphi,_{j}\, dV = \int_{Surface} \varphi\, n_{j}\, dS$$

In two dimensions:

$$\int_{Area} \underline{Grad}\, \varphi\, dA = \int_{Area} \underline{\nabla}\, \varphi\, dA = \int_{Curve} \varphi\, \underline{n}\, dC$$

$$\int_{Area} \varphi,^{i}\, dA = \int_{Curve} \varphi\, n^{i}\, dC$$

$$\int_{Area} \varphi,_{j}\, dA = \int_{Curve} \varphi\, n_{j}\, dC$$

In one dimension:

$$\int \underline{Grad}\, \varphi\, dx = \int \underline{\nabla}\, \varphi\, dx = \varphi(x+) - \varphi(x-)$$

$$\int_{Curve} \varphi,^{i}\, dx = \varphi(x^{+}) - \varphi(x^{-})$$

$$\int_{Curve} \varphi,_{j}\, dx = \varphi(x_{+}) - \varphi(x_{-})$$

Divergence Theorem:

If T is a vector: $T = \underline{U}$ then:

$$\int_{Volume} U^{m,j}\, dV = \int_{Surface} U^{m}\, n^{j}\, dS$$

$$\int_{Volume} U_{m},_{j}\, dV = \int_{Surface} U_{m}\, n_{j}\, dS$$

Contracting this to Get:

$$\int_{Volume} U^{m,m}\, dV = \int_{Surface} U^{m}\, n^{m}\, dS$$

$$\int_{Volume} U_{m},_{m}\, dV = \int_{Surface} U_{m}\, n_{m}\, dS$$

$$\int_{Volume} \text{divergence}\ \underline{U}\ dV = \int_{Surface} \underline{U} \cdot \underline{n}\ dS$$

Curl (Stoke) Theorem:

Since:

$$\int_{Volume} U_{m,j}\ dV = \int_{Surface} U_m\ n_j\ dS$$

Contracting this with ε_{IJK} to Get:

$$\int_{Volume} U_{I,J}\ \varepsilon_{IJK}\ dV = \int_{Surface} U_I\ n_J\ \varepsilon_{IJK}\ dS$$

$$\int_{Volume} \underline{\text{curl}\ U}\ dV = \int_{Surface} \underline{n} \wedge \underline{U}\ dS$$

Laplacian:

If T is : $T = \varphi,_j$ then:

$$\int_{Volume} \varphi,_{jj}\ dV = \int_{Surface} \varphi,_j\ n_j\ dS$$

$$\int_{Volume} \nabla^2\ \varphi\ dV = \int_{Surface} d\varphi / dn\ dS$$

And if $\nabla^2 \varphi = 0$ Get: $\int_{Surface} d\varphi / dn\ dS = 0$

Apply to Integration By parts:

Since

$$\int_{Volume} \psi\ \underline{\text{Grad}}\ \varphi\ dV = \int_{Volume} \text{Grad}(\psi\ \varphi) dV - \int_{Volume} \varphi\ \underline{\text{Grad}}\ \psi\ dV$$

Apply to the first term to the right

$$\int_{Volume} \underline{\text{Grad}}(\psi\ \varphi) dV = \int_{Surface} (\psi\ \varphi)\ \underline{n}\ dS$$

Then

$$\int_{Volume} \psi\ \underline{\text{Grad}}\ \varphi\ dV = \int_{Surface} (\psi\ \varphi)\ \underline{n}\ dS - \int_{Volume} \varphi\ \underline{\text{Grad}}\ \psi\ dV$$

Tensor Integral (Cauchy):

$$\int_{V(n+1)} dV(n+1)\ \ \underline{\nabla}\ \ T = \int_{S(n)} \underline{dS}(n)\ \ T$$

$$\int_{V(n+1)} dV(n+1)\ \ \underline{\nabla} \cdot T = \int_{S(n)} \underline{dS}(n) \cdot T$$

$$\int_{V(n+1)} dV(n+1)\ \ \underline{\nabla} \wedge T = \int_{S(n)} \underline{dS}(n) \wedge T$$

Surface(2) and Volume(2+1):

$\int_{Volume} T^m,_j\ dV = \int_{Surface} T^m\ n_j\ dS$	$\int_{Volume} T_m,^i\ dV = \int_{Surface} T_m\ n^i\ dS$
$\int_{Volume} T^{m,j}\ dV = \int_{Surface} T^m\ n^j\ dS$	$\int_{Volume} T_{m,i}\ dV = \int_{Surface} T_m\ n_i\ dS$

Condition: existence of $T,_j$ and $T,^i$ inside S (Does not hold in shock wave)

A scalar Tensor φ then (Gauss):

$$\int_{Volume} \underline{\text{Grad}}\ \varphi\ dV = \int_{Volume} \underline{\nabla}\ \varphi\ dV = \int_{Surface} \varphi\ \underline{n}\ dS$$

$\int_{Volume} \varphi,_j\ dV = \int_{Surface} \varphi\ n_j\ dS$	$\int_{Volume} \varphi,^i\ dV = \int_{Surface} \varphi\ n^i\ dS$

Curve(1) and Area(1+1):

$$\int_{Area} \underline{\text{Grad}}\,\varphi \; dA = \int_{Area} \underline{\nabla}\,\varphi \; dA = \int_{Curve} \varphi \; \underline{n} \; dC$$

$\int_{Area} \varphi,_{j} \; dA = \int_{Curve} \varphi \; n_{j} \; dC$	$\int_{Area} \varphi,^{i} \; dA = \int_{Curve} \varphi \; n^{i} \; dC$

Position(0) and Curve(0+1):

$$\int \underline{\text{Grad}}\,\varphi \; dx = \int \underline{\nabla}\,\varphi \; dx = \varphi\,(x+) - \varphi\,(x-)$$

$\int_{Curve} \varphi,_{j} \; dx = \varphi\,(x_{+}) - \varphi\,(x_{-})$	$\int_{Curve} \varphi,^{i} \; dx = \varphi\,(x^{+}) - \varphi\,(x^{-})$

Divergence Theorem:
Contracting:

$$\int_{Volume} \text{Pvergence}\; \underline{U} \;\; dV = \int_{Surface} \underline{U} \,.\, \underline{n} \; dS$$

$\int_{Volume} U^{m},_{m} \; dV = \int_{Surface} U^{m}\, n_{m} \; dS$	$\int_{Volume} U_{m}{}^{,m} \; dV = \int_{Surface} U_{m}\, n^{m} \; dS$
$\int_{Volume} U^{m,m} \; dV = \int_{Surface} U^{m}\, n^{m} \; dS$	$\int_{Volume} U_{m,m} \; dV = \int_{Surface} U_{m}\, n_{m} \; dS$

Contracting a <u>Gradient</u> then: (Laplacian):

$$\int_{Volume} \nabla^{2} \varphi \; dV = \int_{Surface} d\,\varphi \,/\, d\,n \;\; dS$$

$\int_{Volume} \varphi^{,j},_{j} \; dV = \int_{Surface} \varphi^{,j}\, n_{j} \; dS$	$\int_{Volume} \varphi,_{j}{}^{,j} \; dV = \int_{Surface} \varphi,_{j}\; n^{j} \; dS$
$\int_{Volume} \varphi^{,j,j} \; dV = \int_{Surface} \varphi^{,j}\, n^{j} \; dS$	$\int_{Volume} \varphi,_{j\,j} \; dV = \int_{Surface} \varphi,_{j}\; n_{j} \; dS$

$$\text{And if } \nabla^{2} \varphi = 0 \text{ Get: } \int_{Surface} d\,\varphi \,/\, d\,n \;\; dS \qquad = 0$$

Curl (Stoke) Theorem: Contracting:

$$\int_{Volume} \underline{\text{curl}}\; \underline{U} \;\; dV = \int_{Surface} \underline{n} \wedge \underline{U}\, dS$$

Contracting this with e_{IJK} to Get:

$\int_{V} \varepsilon_{i}{}^{j}{}_{k}\, T^{i},_{j}\, dV = \int_{S} \varepsilon_{i}{}^{j}{}_{k}\, T^{i}\, n_{j} \; dS$	$\int_{V} \varepsilon^{i}{}_{jk}\, T_{i},^{j}\, dV = \int_{S} \varepsilon^{i}{}_{jk}\, T_{i}\, n^{j} \; dS$
$\int_{V} \varepsilon_{i}{}^{jk}\, T^{i},_{j}\, dV = \int_{S} \varepsilon_{i}{}^{jk}\, T^{i}\, n_{j} \; dS$	$\int_{V} \varepsilon_{j}{}^{ik}\, T_{i},^{j}\, dV = \int_{S} \varepsilon_{j}{}^{ik}\, T_{i}\, n^{j} \; dS$

$\int_{V} \varepsilon_{ijk}\, T^{i,j}\, dV = \int_{S} \varepsilon_{i}{}^{j}{}_{k}\, T^{i}\, n^{j} \; dS$	$\int_{V} \varepsilon^{ij}{}_{k}\, T_{i},_{j}\, dV = \int_{S} \varepsilon^{ij}{}_{k}\, T_{i}\, n_{j} \; dS$
$\int_{V} \varepsilon_{ij}{}^{k}\, T^{i,j}\, dV = \int_{S} \varepsilon_{ij}{}^{k}\, T^{i}\, n^{j} \; dS$	$\int_{V} \varepsilon^{ijk}\, T_{i},_{j}\, dV = \int_{S} \varepsilon^{ijk}\, T_{i}\, n_{j} \; dS$

Integration by parts:
Since

$$\int_{Volume} \psi \; \underline{\text{Grad}}\,\varphi \; dV = \int_{Volume} \text{Grad}(\psi\,\varphi) dV - \int_{Volume} \varphi \; \underline{\text{Grad}}\,\psi \;\; dV$$

Apply to the first term to the right

$$\int_{Volume} \underline{\text{Grad}}(\psi\,\varphi) dV = \int_{Surface} (\,\psi\;\varphi\,)\; \underline{n} \; dS$$

Then

$$\int_{Volume} \psi \; \underline{\text{Grad}}\,\varphi \; dV = \int_{Surface} (\,\psi\;\varphi\,)\; \underline{n} \; dS - \int_{Volume} \varphi \; \underline{\text{Grad}}\,\psi \;\; dV$$

5.
Structural Variables Time Derivatives:

Naturally Time is progressive.

5.1.
Operations on Global and Local Systems:

5.1.1.
Operations on Coordinate and Fixed Systems:

A Tensor F is to be written in a Global System o and in a Local System a:

Fixed Global Frame of Reference o:	General Local Frame of Reference a:
o is Independent of t	a is Dependent on o and t
(o) System	(a) System
Basis $\underline{o}^{o}$ is equal $\underline{o}_{o}$	Basis $\underline{a}^{a}$ is *not* equal. $\underline{a}_{a}$
F_{o} is equal F^{o}	F_{a} is *not* equal. F^{a}

o_{o}	o^{o}	$a_{a} = a_{a}(o_{o}, t)$	$a^{a} = a^{a}(o^{o}, t)$

5.1.2.
Operations on Two Systems:

A Tensor F is to be written in a System p and in a System q:

p is Independent of t	q is Dependent on p and t
(p) System	(q) System
Basis $\underline{p}^{p}$ is *not* equal $\underline{p}_{p}$	Basis $\underline{q}^{q}$ is *not* equal. $\underline{q}_{q}$
F_{p} is *not* equal F_{p}^{i}	F_{q} is *not* equal. F^{q}

p_{p}	p^{P}	$q_{q} = q_{q}(p_{p}, t)$	$q^{q} = q^{q}(p^{P}, t)$

5.1.3.
Scalars in Two Systems:

Initial-Material Description	Current-Spatial Description
$\rho = \rho(\underline{p}, t) = \rho(p^{i}, t) = \rho(o^{i}, t)$	$\rho = \rho(\underline{q}, t) = \rho(q^{i}, t) = \rho(x^{i}, t)$
$\rho = \rho(\underline{p}, t) = \rho(p_{N}, t) = \rho(o_{N}, t)$	$\rho = \rho(\underline{q}, t) = \rho(q_{N}, t) = \rho(x_{N}, t)$

where

: Composition: q onto p	p onto q :Resolution:
α : positive angle between p and q	β : positive angle between q and p

$q^j = q^j(p^i, t)$	$p^j = p^j(q^i, t)$
$q_N = q_N(p_M, t)$	$p_N = p_N(q_M, t)$

5.2.
Velocity Operations:

5.2.1.
Gradient of Velocity:

:

Two coordinate System: Initial Reference and Current Time Dependent:

At t = 0 Initial Material Description: Lagrangian $p^i = p^i(q^j, t)$

At t Current Spatial Description: Eulerian $q^i = q^i(p^j, t)$

$\partial p^i / \partial p^m = pp^i{}_m = \delta^i{}_m$	$\partial q^i / \partial q^m = qq^i{}_m = \delta^i{}_m$
$\partial p_n / \partial p_j = pp_n{}^j = \delta_n{}^j$	$\partial q_n / \partial q_j = qq_n{}^j = \delta_n{}^j$

Gradient of quantity F

$\underline{Fp}$	$\underline{Fq}$

:

Fixed Coordinate System:
And:
Time Dependent Position Vector:

:

0 = t, Lagrangian	0 < t, Eulerian
Initial-Material Description	Current Spatial Description
p is Independent of t	q is Dependent on p and t

Initial (p) Before an Additional Deformation	Current (q) After an Additional Deformation
$\underline{p}(t) \equiv \underline{p}^N \; p_N[q(t),t]$	$\underline{q}(t) \equiv \underline{q}^N \; q_N[p(t),t]$
$\underline{p}(t) \equiv \underline{p}_N \; p^N[q(t),t]$	$\underline{q}(t) \equiv \underline{q}_N \; q^N[p(t),t]$

:

Assumed System Bases to be Fixed:

$(\underline{p}^N)$ = Fixed	$(\underline{q}^N)$ Not Fixed
$(\underline{p}_N)$ = Fixed	$(\underline{q}_N)$ Not Fixed

Bases derivatives are zeros:

$\partial \underline{p}^N / \partial t = (\underline{pt}^N) = \underline{u}^N = 0$	$\partial \underline{q}^N / \partial t = (\underline{qt}^N) = \underline{v}^N$ *Not Equal 0*
$\partial \underline{p}_N / \partial t = (\underline{pt}_N) = \underline{u}_N = 0$	$\partial \underline{q}_N / \partial t = (\underline{qt}_N) = \underline{v}_N$ *Not Equal 0*

Global Velocity Vector :

$\underline{p}_m \; pT^m \equiv \underline{p}_m \; pt^m + \underline{p}_m \; pq^m{}_n \; qt^n$	$\underline{q}_m \; qT^m \equiv \underline{q}_m \; qt^m + \underline{q}_m \; qp^m{}_n \; pt^n$
$\underline{p}^m \; pT_m \equiv \underline{p}^N \; pt_m + \underline{p}^N \; pq_m{}^n \; qt_n$	$\underline{q}^m \; qT_m \equiv \underline{q}^m \; qt_m + \underline{q}^m \; qp_m{}^n \; pt_n$

Velocity Components:

$pT^m \equiv pt^m + pq^m{}_n \; qt^n$	$qT^m \equiv qt^m + qp^m{}_n \; pt^n$

$pT_m \equiv pt_m + pq_m{}^n \; qt_n$	$qT_m \equiv qt_m + qp_m{}^n \; pt_n$

Or:

$pT^m \equiv u^m + pq^m{}_n \; v^n$	$qT^m \equiv v^m + qp^m{}_n \; u^n$
$pT_m \equiv u_m + pq_m{}^n \; v_n$	$qT_m \equiv v_m + qp_m{}^n \; u_n$

Where
Time rates:

$u \equiv pt$	$v \equiv qt$

And gradients:

pp	pq	qp	qq

Position Gradient:

Material Gradient of Global Position: Local Gradient of Global Position:	Material Gradient of Local Position: Local Gradient of Local Position:
$\underline{\text{Grad}}\ \underline{p}$	$\underline{\text{Grad}}\ \underline{q}$
$\underline{\text{grad}}\ \underline{p}$	$\underline{\text{grad}}\ \underline{q}$

$\partial p_n/\partial p_N \equiv \delta^N{}_n$	$\partial p^n/\partial p^N \equiv \delta^n{}_N$	$\partial q_n/\partial p_N \equiv Q^N{}_n$	$\partial q^n/\partial p^N \equiv Q^n{}_N$
$\partial p_n/\partial q_N \equiv P^N{}_n$	$\partial p^n/\partial q^N \equiv P^n{}_N$	$\partial q_n/\partial q_N \equiv \delta^N{}_n$	$\partial q^n/\partial q^N \equiv \delta^n{}_N$

$\underline{p}^n \delta^N{}_n = \underline{\text{Grad}}\ \underline{p} = \underline{p}_n \delta^n{}_N$	$\underline{p}^n qp^N{}_n = \underline{\text{Grad}}\ \underline{q} = \underline{p}_n qp^n{}_N$ $\underline{p}^n Q^N{}_n = \underline{\text{Grad}}\ \underline{q} = \underline{p}_n Q^n{}_N$
$\underline{q}^n pq^N{}_n = \underline{\text{grad}}\ \underline{p} = \underline{q}_n pq^n{}_N$ $\underline{q}^n P^N{}_n = \underline{\text{grad}}\ \underline{p} = \underline{q}_n P^n{}_N$	$\underline{q}^n \delta^N{}_n = \underline{\text{grad}}\ \underline{q} = \underline{q}_n \delta^n{}_N$

Gradients of Velocity
(Convection):

Material Gradient of Global Velocity:	Material Gradient of Local Velocity:
Local Gradient of Global Velocity:	Local Gradient of Local Velocity:
$I \equiv \underline{\nabla}_p\ u \equiv up = uq\ \ qp$	$K \equiv \underline{\nabla}_p\ v \equiv vp = vq\ \ qp$
$J \equiv \underline{\nabla}_q\ u \equiv uq = up\ \ pq$	$L \equiv \underline{\nabla}_q\ v \equiv vq = vp\ \ pq$
$I = J\ \ Q$	$K = L\ \ Q$
$J = I\ \ P$	$L = K\ \ P$
$\underline{\text{Grad}}\ \underline{u}$	$\underline{\text{Grad}}\ \underline{v}$
$\underline{p}^n\ up^N{}_n = \underline{p}_n\ up^n{}_N$	$\underline{p}^n\ vp^N{}_n = \underline{p}_n\ vp^n{}_N$

$\underline{p}^{n}\, uq_{n}^{K}\, qp_{K}^{N} = \underline{p}_{n}\, uq^{n}{}_{K}\, qp^{K}{}_{N}$	$\underline{p}^{n}\, vq_{n}^{K}\, qp_{K}^{N} \quad \underline{p}_{n}\, vq^{n}{}_{K}\, qp^{K}{}_{N}$

$\underline{\text{grad}}\ \underline{u}$	$\underline{\text{grad}}\ \underline{v}$
$\underline{q}^{n}\, uq_{n}^{N} = \underline{q}_{n}\, uq^{n}{}_{N}$	$\underline{q}^{n}\, vq_{n}^{N} = \underline{q}_{n}\, vq^{n}{}_{N}$
$\underline{q}^{n}\, up_{n}^{K}\, pq_{K}^{N} = \underline{q}_{n}\, up^{n}{}_{K}\, pq^{K}{}_{N}$	$\underline{q}^{n}\, vp_{n}^{K}\, qp_{K}^{N} = \underline{q}_{n}\, vp^{n}{}_{K}\, qp^{K}{}_{N}$

$\underline{\text{Grad}}\ \underline{u}$	$\underline{\text{Grad}}\ \underline{v}$
$\underline{p}^{n}\, I_{n}^{N} = \underline{p}_{n}\, I^{n}{}_{N}$	$\underline{p}^{n}\, K_{n}^{N} = \underline{p}_{n}\, K^{n}{}_{N}$
$\underline{p}^{n}\, J_{n}^{K}\, Q_{K}^{N} = \underline{p}_{n}\, J^{n}{}_{K}\, Q^{K}{}_{N}$	$\underline{p}^{n}\, L_{n}^{K}\, Q_{K}^{N} = \underline{p}_{n}\, L^{n}{}_{K}\, Q^{K}{}_{N}$

$\underline{\text{grad}}\ \underline{u}$	$\underline{\text{grad}}\ \underline{v}$
$\underline{q}^{n}\, J_{n}^{N} = \underline{q}_{n}\, J^{n}{}_{N}$	$\underline{q}^{n}\, L_{n}^{N} = \underline{q}_{n}\, L^{n}{}_{N}$
$\underline{q}^{n}\, I_{n}^{K}\, P_{K}^{N} = \underline{q}_{n}\, I^{n}{}_{K}\, P^{K}{}_{N}$	$\underline{q}^{n}\, K_{n}^{K}\, Q_{K}^{N} = \underline{q}_{n}\, K^{n}{}_{K}\, Q^{K}{}_{N}$

$I_{M}^{N} \equiv up_{M}^{N} \equiv ptp_{M}^{N}$	$J_{M}^{N} \equiv uq_{M}^{N} \equiv ptq_{M}^{N}$	$K_{M}^{N} \equiv vp_{M}^{N} \equiv qtp_{M}^{N}$	$L_{M}^{N} \equiv vq_{M}^{N} \equiv qtq_{M}^{N}$
$I^{N}{}_{M} \equiv up^{N}{}_{M} \equiv ptp^{N}$	$J^{N}{}_{M} \equiv uq^{N}{}_{M} \equiv ptq^{N}{}_{M}$	$K^{N}{}_{M} \equiv vp^{N}{}_{M} \equiv qtp^{N}$	$L^{N}{}_{M} \equiv vq^{N}{}_{M} \equiv qtq^{N}{}_{M}$

5.2.2. Divergence of Velocity:

Divergence of a Function, $\underline{F}$ is a Function of one Less Rank:

$\underline{\nabla}p.\underline{F} \equiv Fp = \partial F^{n}/\partial p^{n} \equiv Fp^{n}{}_{n} \equiv Fp_{n}^{n}$	$\underline{\nabla}q.\underline{F} \equiv Fq = \partial F^{n}/\partial q^{n} \equiv Fq^{n}{}_{n} \equiv Fq_{n}^{n}$
$\underline{\nabla}p.\underline{F} \equiv Fp = \partial F_{n}/\partial p_{n} \equiv Fp_{n}^{n} \equiv Fp^{n}{}_{n}$	$\underline{\nabla}q.\underline{F} \equiv Fq = \partial F^{n}/\partial q^{n} \equiv Fq^{n}{}_{n} \equiv Fq_{n}^{n}$

Fp_{N}^{N}	Fq_{N}^{N}
$Fp^{N}{}_{N}$	$Fq^{N}{}_{N}$

Fixed Coordinate Divergence of Velocity:

$(\underline{\nabla}.\,\underline{u}) \equiv uo \equiv uo_{N}^{N} = pto_{N}^{N}$	$(\underline{\nabla}.\,\underline{v}) \equiv vo \equiv vo_{N}^{N} \equiv qto_{N}^{N}$
$(\underline{\nabla}.\,\underline{u}) \equiv uo \equiv uo^{N}{}_{N} = pto^{N}{}_{N}$	$(\underline{\nabla}.\,\underline{v}) \equiv vo \equiv vo^{N}{}_{N} \equiv qto^{N}{}_{N}$

Divergence of Velocity (u) and (v) with respect to Initial Time Coordinates [Initial p]:

Initial (u)	Current (v)
$(up) \equiv up_{N}^{N} = ptp_{N}^{N}$	$(vp) \equiv vp_{N}^{N} = qtp_{N}^{N}$

$(up) \equiv up^{N}{}_{N} = ptp^{N}{}_{N}$	$(vp) \equiv vp^{N}{}_{N} = qtp^{N}{}_{N}$
$I_{N}{}^{N} \equiv up_{N}{}^{N} \equiv uq_{N}{}^{K}\, qp_{K}{}^{N} = J_{N}{}^{K}\, qp_{K}{}^{N}$	$K_{N}{}^{N} \equiv vp_{N}{}^{N} = vq_{N}{}^{K}\, qp_{K}{}^{N} = L_{N}{}^{K}\, qp_{K}{}^{N}$
$I^{N}{}_{N} \equiv up^{N}{}_{N} = uq^{N}{}_{K}\, qp^{K}{}_{N} = J^{N}{}_{K}\, qp^{K}{}_{N}$	$K^{N}{}_{N} \equiv vp^{N}{}_{N} = vq^{N}{}_{K}\, qp^{K}{}_{N} = L^{N}{}_{K}\, qp^{K}{}_{N}$

Divergence of Velocity (u) and (v) with respect to Current Time Coordinates [Current q]:

Initial (u) Coordinate	Current (v) Coordinate
$(uq) \equiv uq_{N}{}^{N} = ptq_{N}{}^{N}$	$(vq) \equiv vq_{N}{}^{N} \equiv qtq_{N}{}^{N}$
$(uq) \equiv uq^{N}{}_{N} = ptq^{N}{}_{N}$	$(vq) \equiv vq^{N}{}_{N} \equiv qtq^{N}{}_{N}$
$J_{N}{}^{N} \equiv uq_{N}{}^{N} = up_{N}{}^{K}\, pq_{K}{}^{N} = I_{N}{}^{K}\, pq_{K}{}^{N}$	$L_{N}{}^{N} \equiv vq_{N}{}^{N} = vp_{N}{}^{K}\, pq_{K}{}^{N} = K_{N}{}^{K}\, pq_{K}{}^{N}$
$J^{N}{}_{N} \equiv uq^{N}{}_{N} = up^{N}{}_{K}\, pq^{K}{}_{N} = I^{N}{}_{K}\, pq^{K}{}_{N}$	$L^{N}{}_{N} \equiv vq^{N}{}_{N} = vp^{N}{}_{K}\, pq^{K}{}_{N} = K^{N}{}_{K}\, pq^{K}{}_{N}$

So:

Divergence of Initial (u) Velocity:

Divergence of Initial (u) Coordinate Velocity with respect to Initial Time Coordinates [Initial p]	Divergence of Initial (u) Coordinate Velocity with respect to Current Time Coordinates [Current q]
$I_{N}{}^{N} \equiv up_{N}{}^{N} = J_{N}{}^{K}\, qp_{K}{}^{N}$	$J_{N}{}^{N} \equiv uq_{N}{}^{N} = I_{N}{}^{K}\, pq_{K}{}^{N}$
$I^{N}{}_{N} \equiv up^{N}{}_{N} = J^{N}{}_{K}\, qp^{K}{}_{N}$	$J^{N}{}_{N} \equiv uq^{N}{}_{N} = I^{N}{}_{K}\, pq^{K}{}_{N}$

Similarly:

Divergence of Current (v) Velocity:

Divergence of Current (v) Coordinate Velocity with respect to Initial Time Coordinates [Initial p]	Divergence of Current (v) Coordinate Velocity with respect to Current Time Coordinates [Current q]
$K^{N}{}_{N} \equiv vp^{N}{}_{N} = L^{N}{}_{K}\, qp^{K}{}_{N}$	$L_{N}{}^{N} \equiv vq_{N}{}^{N} = K_{N}{}^{K}\, pq_{K}{}^{N}$
$K^{N}{}_{N} \equiv vp^{N}{}_{N} = L^{N}{}_{K}\, qp^{K}{}_{N}$	$L^{N}{}_{N} \equiv vq^{N}{}_{N} = K^{N}{}_{K}\, pq^{K}{}_{N}$

5.3.
Divergence of a Factored Quantity:

Derivative of First multiply Second plus First multiply Derivative of Second

Divergence of a factored Quantity, f F is, fFp . It is a sum of two quantities:	Divergence of a factored Quantity, f F is, fFq . It is a sum of two quantities:

$$(fFp) \equiv (\underline{\nabla p}) \,.\, (f\,\underline{F}) \qquad\qquad (fFq) \equiv (\underline{\nabla q}) \,.\, (f\,\underline{F})$$
$$(fFp) = \underline{fp} \,.\, \underline{F} + f \,.\, Fp \qquad\qquad (fFq) = \underline{fq} \,.\, \underline{F} + f \,.\, Fq$$

$$(fFp) \equiv \partial f/\partial p^n \, F^n + f \;\; \partial F^n/\partial p^n \qquad\qquad fFq \equiv \partial f/\partial q^n \, F^n + f \;\; \partial F^n/\partial q^n$$
$$(fFp) \equiv \partial f/\partial p_n \, F_n + f \;\; \partial F_n/\partial p_n \qquad\qquad fFq \equiv \partial f/\partial q_n \, F_n + f \;\; \partial F_n/\partial q_n$$

$$(fFp) \equiv fp_n \, F^n + \;\; f \;\; Fp^n{}_n \qquad\qquad (fFq) \equiv fq^n \, F_n + \;\; f \;\; Fq_n{}^n$$
$$(fFp) \equiv fp^n \, F_n + \;\; f \;\; Fp_n{}^n \qquad\qquad (fFq) \equiv fq^n \, F_n + \;\; f \;\; Fq_n{}^n$$

5.3.1.
Divergence of Factored Initial Velocity:

Divergence of Factored Initial ($\underline{v}$) Velocity:

$$\underline{\nabla p}.(f\underline{u}) \equiv fup \equiv \partial(f\underline{u})/\partial\underline{p} \equiv \underline{fp} \,.\, \underline{u} + f\,up \qquad\qquad \underline{\nabla q}.(f\underline{u}) \equiv fuq \equiv \partial(f\underline{u})/\partial\underline{q} \equiv \underline{fq} \,.\, \underline{u} + f\,uq$$

$$fup \equiv \partial f/\partial p^n \, u^n + f \; \partial u^n/\partial p^n \qquad\qquad fuq \equiv \partial f/\partial q^n \, u^n + f \; \partial u^n/\partial q^n$$
$$fup \equiv \partial f/\partial p_n \, u_n + f \; \partial u_n/\partial p_n \qquad\qquad fuq \equiv \partial f/\partial q_n \, u_n + f \; \partial u_n/\partial q_n$$

Or:

$$fup \equiv fp_n \, u^n + f \; up^n{}_n \qquad\qquad fuq \equiv fq_n \, u^n + f \; uq^n{}_n$$
$$fup \equiv fp^n \, u_n + f \; up_n{}^n \qquad\qquad fuq \equiv fq^n \, u_n + f \; uq_n{}^n$$

Divergence of Initial (u) Velocity:

$$(f) = 1$$

$$(\underline{fp}) = \underline{1p} = \partial\, 1/\partial p^n = \underline{0} \qquad\qquad (\underline{fq}) = \underline{1q} = \partial\, 1/\partial q^n = \underline{0}$$

$$\underline{\nabla p}.(1\underline{u}) \equiv 1up \equiv \partial(1\underline{u})/\partial\underline{p} \equiv \underline{0} \,.\, \underline{u} + 1\,up \qquad\qquad \underline{\nabla q}.(1\,\underline{u}) \equiv (1\,u)q \equiv \partial(\underline{u})/\partial\underline{q} \equiv \underline{0} \,.\, \underline{u} + 1\,uq$$

$$\underline{\nabla p}.(\underline{u}) \equiv (up) \equiv \partial u^n/\partial p^n \qquad\qquad \underline{\nabla q}.(\underline{u}) \equiv (uq) \equiv \partial u^n/\partial q^n$$
$$\underline{\nabla p}.(\underline{u}) \equiv (up) \equiv \partial u_n/\partial p_n \qquad\qquad \underline{\nabla q}.(\underline{u}) \equiv (uq) \equiv \partial u_n/\partial q_n$$

5.3.2.
Divergence of Factored Current Velocity:

Divergence of Factored Current ($\underline{v}$) Velocity:

$$\underline{\nabla p} \,.\, (f\underline{v}) \equiv fvp \equiv \partial(f\underline{v})/\partial\underline{p} \equiv \underline{fp} \,.\, \underline{v} + f\;vp \qquad\qquad \underline{\nabla q} \,.\, (f\underline{v}) \equiv fvq \equiv \partial(f\underline{v})/\partial\underline{q} \equiv \underline{fq} \,.\, \underline{v} + f\;vq$$

$$fvp \equiv \partial f/\partial p^n \, v^n + \;\; f \;\; \partial v^n/\partial p^n \qquad\qquad fvq \equiv \partial f/\partial q^n \, v^n + \;\; f \;\; \partial v^n/\partial q^n$$

$fvp \equiv \partial f/\partial p_n\ v_n + f\ \partial v_n/\partial p_n$	$fvq \equiv \partial f/\partial q_n\ v_n + f\ \partial v_n/\partial q_n$

Or:

$fvp \equiv fp_n\ v^n + f\ vp^n{}_n$	$fvq \equiv fq_n\ v^n + f\ vq^n{}_n$
$fvp \equiv fp^n\ v_n + f\ vp_n{}^n$	$fvq \equiv fq^n\ v_n + f\ vq_n{}^n$

Divergence of Current (v) Velocity:

$$(f) = 1$$

$(\underline{fp}) = \underline{1p} = \partial 1/\partial p^n = \underline{0}$	$(\underline{fq}) = \underline{1q} = \partial 1/\partial q^n = \underline{0}$
$\underline{\nabla p}.(1\underline{v}) \equiv 1vp \equiv \partial(1\underline{v})/\partial\underline{p} \equiv \underline{0}\,.\,\underline{v} + 1\ vp$	$\underline{\nabla q}.(1\,\underline{v}) \equiv (1\,v)q \equiv \partial(\underline{v})/\partial\underline{q} \equiv \underline{0}\,.\,\underline{v} + 1\ vq$
$\underline{\nabla p}.(\underline{v}) \equiv (vp) \equiv \partial v^n/\partial p^n$	$\underline{\nabla q}.(\underline{v}) \equiv (vq) \equiv \partial v^n/\partial q^n$
$\underline{\nabla}p.(\underline{v}) \equiv (vp) \equiv \partial v_n/\partial p_n$	$\underline{\nabla}q.(\underline{v}) \equiv (vq) \equiv \partial v_n/\partial q_n$

5.4.
Intrinsic Time Derivative:

5.4.1.
Partial Derivative of *Lower* Metric With Respect to *__Upper Position__*

Add:

$$\tfrac{1}{2}(aa_{NJ,L} + aa_{NL,J} - aa_{JL,N}) = a3a_{JLN}$$
$$\tfrac{1}{2}(aa_{LJ,N} + aa_{LN,J} - aa_{JN,L}) = a3a_{JNL}$$

To get:

$$0 + aa_{NL,J} - 0 = a3a_{JLN} + a3a_{JNL}$$

:

$$\partial aa_{MN}/\partial a^K \equiv aa_{MN,K}$$

:

Partial Derivative of Lower Metric With respect to a^K *__Upper Position__*:

$$[aa_{MN}]_{,K} = +[a3a_{KMN} + a3a_{NKM}]$$

5.4.2.
Partial Derivative of *Upper* Metric With Respect to *__Upper Position__*:

Upper a^q P.D. of inners of upper and lower metrics:

$$aa^{j1}\ aa_{n1} = \delta^j{}_n$$

Partial Derivative is:

$$aa^{jl}{}_{,q}\ aa_{nl} + aa^{jl}\ aa_{nl,q} = 0$$

$$aa^{jl}{}_{,q}\ aa_{nl} = -\ aa^{jl}\ aa_{nl,q}$$

$$aa^{j}{}_{n,q} = -\ aa_{n}{}^{j}{}_{,q}$$

Or Multiply

$$aa^{jl}{}_{,q}\ aa_{nl} = -\ aa^{jl}\ aa_{nl,q}$$

$$*$$

$$aa^{ns}$$

To get:

$$aa^{jl}{}_{,q}\ aa_{nl}\ aa^{ns} = -\ aa^{ns}\ aa^{jl}\ aa_{nl,q}$$

$$aa^{jl}{}_{,q}\ \delta_{l}{}^{s} = -\ aa^{ns}\ aa^{jl}\ aa_{nl,q}$$

But:

$$aa_{nl,q} = a3a_{qln} + a3a_{qnl} = aa_{nj}\ a3a_{ql}{}^{j} + aa_{lj}\ a3a_{qn}{}^{j}$$

So:

$$aa^{js}{}_{,q} \equiv aa^{js}{}_{,q} = -\ aa^{ns}\ aa^{jl}\ (a3a_{qln} + a3a_{qnl})$$

$$aa^{js}{}_{,q} \equiv aa^{js}{}_{,q} = -\ aa^{jl}\ a3a_{ql}{}^{s} - aa^{ns}\ a3a_{qn}{}^{j}$$

$$aa^{js}{}_{,q} \equiv aa^{js}{}_{,q} = -\ a3a_{q}{}^{js} - a3a_{q}{}^{sj}$$

So:

$$\partial aa^{MN}/\partial a^{K}$$

Partial Derivative of Upper Metric With respect to a^{K} *Upper Position*:

$$[aa^{MN}]_{,K} = -\ [a3a_{K}{}^{MN} + a3a_{N}{}^{KM}]$$

5.4.3.
Partial Derivative of *Lower* and *Upper* Metrics With respect to *Upper Position*

Both Results:

Partial Derivative of
Lower and *Upper* Metrics
With respect to *Upper Position*

$$[aa_{MN}]_{,K} = +\ [a3a_{KMN} + a3a_{NKM}]$$
$$[aa^{MN}]_{,K} = -\ [a3a_{K}{}^{MN} + a3a_{N}{}^{KM}]$$

:

5.4.4.
Intrinsic Time Derivative:

Intrinsic Time Derivative of f:

$$\delta[f])/\delta q^{\delta}\ \ dq^{\delta}/dt \equiv [f]|_{\delta}\ dq^{\delta}/dt \equiv [f]_{,\delta}\ qT^{\delta}$$

:

$$[qT_\lambda]|_\delta \; dq^\delta/dt \equiv \{ [qT_\lambda]_{,\delta} - [qT_m]\; q3q_{\lambda\delta}{}^m \} \; qT^\delta$$

$$[qT^\upsilon]|_\delta \; dq^\delta/dt \equiv \{ [qT^\upsilon]_{,\delta} + [qT^p]\; q3q_{p\delta}{}^\upsilon \} \; qT^\delta$$

In General:

$$T(q)_\lambda{}^\upsilon|_\delta \; dq^\delta/dt \equiv [T(q)_\lambda{}^\upsilon{}_{,\delta} - T(q)_m{}^\upsilon \; q3q_{\lambda\delta}{}^m + T(q)_\lambda{}^p \; q3q_{p\delta}{}^\upsilon] \; qT^\delta$$

Where:

Upper Christoffel Symbol

$$a3a_{JL}{}^q \equiv aa^{qn} \; a3a_{JLn}$$

Lower Christoffel Symbol ≡ Trio Symbol

$$a3a_{JLn} \equiv \tfrac{1}{2}(aa_{nJ,L} + aa_{nL,J} - aa_{JL,n})$$

:

Define Intrinsic Derivative:

$$V^m|_k \equiv \delta V^m/\delta a^k \equiv V^m{}_{,k} + V^i \, aa^m{}_{ik}$$

$$\underline{V}_{,k} = \underline{a}_m V^m|_k \equiv \underline{a}_m \; \delta V^m/\delta a^k$$

$$V_m|_k \equiv \delta V_m/\delta a^k \equiv V_{m,k} - V_n \, aa^n{}_{mk}$$

$$\underline{V}_{,k} = \underline{a}^m V_m|_k \equiv \underline{a}^m \; \delta V_m/\delta a^k$$

:

5.5.
Conserved Time Derivative:

:

$$Cf \equiv ft + \quad fvq \quad \equiv ft + \underline{fq} \,.\, \underline{v} + f \; vq$$

Cf is a sum of two scalars:

Cf ≡ f rate + divergence of f factored velocity vector

$$Cf \equiv \partial f/\partial t \; + \; \underline{\nabla}.(f\underline{v}) \equiv ft \; + \; fvq^n{}_n \equiv ft \; + \; fvq_n{}^n$$

Cf is a sum of three scalars:

Cf ≡ f rate + <u>Gradient</u> of scalar . Velocity+ f factored of Velocity divergence

$$Cf \equiv \partial f/\partial t + \underline{\nabla} f \,.\, \underline{v} + f \; \underline{\nabla}\,.\,\underline{v} \equiv ft + fq^n v_n + f \; vq^n{}_n \equiv ft + fq_n v^n + f \; vq_n{}^n$$

Time derivative of scalar, fT:

(fT) ≡ f rate + <u>Gradient</u> of scalar f . Velocity

$$(fT) \equiv (ft + fq^n \; v_n) \equiv (ft + fq_n \; v^n)$$

Cf is a sum of two scalars;

Cf ≡ time derivative of scalar f + f factored of Velocity divergence

$$Cf \equiv (fT) + \{f \,.\, vq^n{}_n\} \equiv (fT) + \{f \,.\, vq_n{}^n\}$$

So:

$$Cf \equiv ft + [\quad \{fvq^n{}_n\}] \equiv ft + [\quad \{fvq_n{}^n\}]$$

$$Cf \equiv (ft + [fq_n \; v^n) + \{f \; vq^n{}_n\}] \equiv (\partial f/\partial t + [fq^n \; v_n) + \{f \; vq_n{}^n\}]$$

$$Cf \equiv (fT) + \{f \; vq^n{}_n\} \equiv (fT) + \{f \; vq_n{}^n\}$$

:

$$(fvq) \equiv fq^n \,.\, v_n + f \,.\, vq^n{}_n \equiv fq_n \,.\, v^n + f \,.\, vq_n{}^n$$

$$(fT) \equiv ft + fq^n \,.\, v_n \equiv ft + fq_n \,.\, v^n$$

Of:

$$Cf \equiv \partial f / \partial t + [\quad \partial \{f v^n\} / \partial q^n] \equiv \partial f / \partial t + [\quad \partial \{f v_n\} / \partial q_n]$$
$$Cf \equiv (\partial f/\partial t + [\partial f /\partial q^n \; v^n) + (f \; \partial v^n / \partial q^n)] \equiv (\partial f/\partial t + [\partial f /\partial q_n \; v_n) + (f \; \partial v_n / \partial q_n)]$$
$$Cf \equiv (df / dt) + f \; \partial v^n / \partial q^n \equiv (df / dt) + f \; \partial v_n / \partial q_n$$

$$Cf \equiv ft + \quad fvq \equiv ft + \underline{fq} . \underline{v} + f \; vq \equiv fT + f \; vq$$

Where:

$$(fvq) \equiv fq^n v_n + f \; vq^n{}_n \equiv fq_n v^n + f \; vq_n{}^n$$

$$(fT) \equiv ft + fq^n v_n \equiv ft + fq_n v^n$$

Apply To f^m and f_m

$$Cf^m \equiv ft^m + fq^{mn} v_n + f^m \; vq^n{}_n \equiv ft^m + fq^m{}_n v^n + f^m \; vq_n{}^n$$

$$Cf_m \equiv ft_m + fq_m{}^n v_n + f_m \; vq^n{}_n \equiv ft_m + fq_{mn} v^n + f_m \; vq_n{}^n$$

:

$$Cf^m \equiv ft^m + \quad fvq^{mn}{}_n$$

$$Cf_m \equiv ft_m + \quad fvq_{mn}{}^n$$

$$Cf^m \equiv ft^m + fq^m{}_n v^n + f^m \; vq^n{}_n$$

$$Cf_m \equiv ft_m + fq_m{}^n v_n + f_m \; vq^n{}_n$$

$$Cf^m \equiv fT^m + f^m \; vq^n{}_n$$

$$Cf_m \equiv fT_m + f_m \; vq^n{}_n$$

Where:

$$(fT^m) \equiv ft^m + fq^m{}_n v^n$$

$$(fT_m) \equiv ft_m + fq_m{}^n v_n$$

Apply To velocity: v^m and v_m

$$Cv^m \equiv vt^m + vq^m{}_n v^n + v^m \; vq^n{}_n$$
$$Cv_m \equiv vt_m + vq_m{}^n v_n + v_m \; vq^n{}_n$$

$$Cv^m \equiv vT^m + v^m \; vq^n{}_n$$
$$Cv_m \equiv vT_m + v_m \; vq^n{}_n$$

$$Cv^m \equiv vt^m + \quad vvq^{mn}{}_n$$
$$Cv_m \equiv vt_m + \quad vvq_{mn}{}^n$$

Where:

$$(vT^m) \equiv vt^m + vq^m{}_n v^n$$
$$(vT_m) \equiv vt_m + vq_m{}^n v_n$$

:

(vtm): position acceleration relative to moving frame

(vqnn vn): Acceleration of frame of reference

(vqnn): Angular velocity of frame of reference

Rate of momentum is force

$$P_m \equiv (mvT_m) \equiv (mvt_m) + (mvq_m^{\ n})\, v_n \qquad P^m \equiv (mvT^m) \equiv (mvt^m) + (mvq^m_{\ n}\, v^n)$$

:

Time Derivative of a scalar, f is a sum of two scalars; time rate of f plus inner product of Gradient of f by velocity:

$$(fT) \equiv df/dt \equiv \partial f/\partial t + \underline{\nabla} f \,.\, \underline{u} \qquad (fT) \equiv df/dt \equiv \partial f/\partial t + \underline{\nabla} f \,.\, \underline{v}$$

$$(fT) \equiv ft + \underline{fp} \,.\, \underline{u} \equiv ft + fp_n\, u^n \qquad (fT) \equiv ft + \underline{fq} \,.\, \underline{v} \equiv ft + fq_n\, v^n$$

$$(fT) \equiv ft + \underline{fp} \,.\, \underline{u} \equiv ft + fp^n\, u_n \qquad (fT) \equiv ft + \underline{fq} \,.\, \underline{v} \equiv ft + fq^n\, v_n$$

Define Time Derivatives of Scalar f:

fT:

$$ft + \underline{fp} \,.\, \underline{u} \equiv ft + fp_n\, u^n \equiv ft + fp^n\, u_n \qquad ft + \underline{fq} \,.\, \underline{v} \equiv ft + fq_n\, v^n \equiv ft + fq^n\, v_n$$

:

Define Conserved Time Derivatives of Tensor f:

In p coordinates S(f):

$$S(f) \equiv ft + \underline{\nabla}_p \,.\, (f\,\underline{u}) \equiv ft + \underline{\nabla}_p f \,.\, \underline{u} + f\ \underline{\nabla}_p \,.\, \underline{u} \equiv fT + f\ \underline{\nabla}_p \,.\, \underline{u}$$

$$S(f) \equiv ft + (fu)p \equiv ft + \underline{fp} \,.\, \underline{u} + f\ \ up \equiv fT + f\ \ up$$

$$S(f) = ft + 0 \equiv fT + f\ \ 0$$

In q coordinates T(f):

$$T(f) \equiv ft + \underline{\nabla}_q \,.\, (f\,\underline{v}) \equiv ft + \underline{\nabla}_p f \,.\, \underline{u} + f\ \underline{\nabla}_p \,.\, \underline{u} \equiv fT + f\ \underline{\nabla}_q \,.\, \underline{v}$$

$$T(f) \equiv ft + (fv)q \equiv ft + \underline{fq} \,.\, \underline{v} + f\ \ vq \equiv fT + f\ \ vq$$

Where:

$$fvq^n_{\ n} \equiv fq_n\ \ v^n + f\ \ vq^n_{\ n} \equiv fq^n\ \ v_n + f\ \ vq^n_{\ n} \equiv\ \ fvq_n^{\ n}$$

:

$$S(f) = ft + 0 \qquad T(f) = ft + \underline{fq} \,.\, \underline{v} + f\ \ vq$$

$$S(f) =\ \ fT + f\ \ 0 \qquad T(f) \equiv fT + f\ \ vq$$

$$T(f) = ft + \underline{fq} \,.\, \underline{v} + f\ \ L$$

$$T(f) \equiv fT + f\ \ L$$

:

So:

$$Tf \equiv ft + \qquad fvq^n_{\ n} \equiv ft + \qquad fvq_n^{\ n}$$

$$Tf \equiv ft + fq^n\ \ v_n + f\ \ vq^n_{\ n} \equiv ft + fq_n\ \ v^n + f\ \ vq_n^{\ n}$$

$$Tf \equiv \quad fT \quad + f\ \ vq^n_{\ n} \equiv \quad fT \quad + f\ \ vq_n^{\ n}$$

Where:

$$fvq \equiv fq^n\ \ v_n + f\ \ vq^n_{\ n} \equiv fq_n\ \ v^n + f\ \ vq_n^{\ n}$$

$$fT \equiv ft + fq^n\ \ v_n \equiv ft + fq_n\ \ v^n$$

Or:

$$Tf \equiv \partial f/\partial t + \qquad \partial(f\,v^n)/\partial q^n \equiv \partial f/\partial t + \qquad \partial(f\,v_n)/\partial q_n$$

$$Tf \equiv \partial f/\partial t + \partial f/\partial q^n\ \ v^n + f\ \ \partial v^n/\partial q^n \equiv \partial f/\partial t + \partial f/\partial q_n\ \ v_n + f\ \ \partial v_n/\partial q_n$$

$$Tf \equiv \qquad df/dt \qquad + f\ \ \partial v^n/\partial q^n \equiv \qquad df/dt \qquad + f\ \ \partial v_n/\partial q_n$$

5.5.1.
Time Derivatives of Product of Tensors:

:

Time Derivative and Conserved Time Derivative of a product of tensors, fg:

$$(fg)T \equiv (fg)t \;+\; (fg)q_n * v^n \equiv (fg)t \;+\; (fg)q^n * v_n$$

$$T(fg) \equiv (fg)t + (fg)q_n * v^n + fg * vq^n{}_n \equiv (fg)T + fg * L \equiv (fg)t + (fg)q^n * v_n + fg * vq^n{}_n$$

But:

$$fgT \equiv fT \;*\; g \;+ f \;*\; gT$$

Then:

$$T(fg) \equiv fT * g + f * gT + fg * vq^n{}_n \equiv fT * g + f * gT + fg * L \equiv fT * g + f * gT + fg * vq_n{}^n$$

Or:

$$T(fg) \equiv fT * \;g \;+ f \;(\,gT + g * L\,) \;\equiv fT \quad g \;+ f \;(\,gT + g * L\,)$$

So:

$$T(fg) \equiv \; fT * g \;+\; f * T(g) \qquad\qquad T(gf) \equiv \; gT * f \;+\; g * T(f)$$

Conserved Time Derivative Is Rearranged in A <u>Symmetric</u> Form:

$$T(fg) \equiv fT * g + (fg) * vq^n{}_n + f \; * gT \equiv fT * g + (fg) * L + f * gT \equiv fT * g + (fg) * vq_n{}^n + f * gT$$

So:

$$T(fg) \equiv (\,fT + f * \;L\,) * g + f * gT \quad \equiv (\,fT + f * L\,) * g + f * gT$$

Conserved Time Derivative of fg and of gf are Equal:

$$T(fg) \;\equiv T(f) \quad g + \;f \;\; gT \qquad = \qquad T(gf) \;\equiv\; T(g) \;\; f + \;\; g \;\; fT$$

Where:

$$gt + gq_n * v^n \equiv gT \equiv gt + gq^n * v_n \qquad\qquad ft \;+\; fq_n * v^n \equiv fT \equiv ft \;+\; fq^n * v_n$$

$$ft + fq_n * v^n \equiv T(f) \;- f * L \equiv T(f) * ft + fq^n * v_n \qquad\qquad gt \;+\; gq_n * v^n \equiv T(g) - \; g * L \equiv gt \;+\; gq^n * v_n$$

Similarly:

Conserved Time Derivative of fgh:

$$T(fgh) = T(f)\; gh + f \;\; ghT \;= T(ghf) \;= T(g) \;\; hf \;+ g \;\; hfT = T(hfg) = T(h) \;\; fg + h \;\; fgT$$

$$T(fgh) = T(fg)\; h + fg \;\; hT = T(ghf) \;= T(gh) \;\; f \;+ gh \;\; fT \;= T(hfg) \;= T(hf) \;\; g + hf \;\; gT$$

Now Generalize:

Time Derivative and Conserved Time Derivative, of f:

$$fT \equiv ft + fq_n \;* v^n \qquad\qquad \equiv ft \;+ fq^n \;\; v_n$$

$$T(f) \equiv ft + fq_n * v^n + f * vq^n{}_n \equiv fT \;+ f * L \;\equiv ft + fq^n * v_n + f * vq^n{}_n$$

And:

Time Derivative and Conserved Time Derivative of fg:

$$fgT \equiv fT * g \;+ f * gT$$

$$T(fg) \equiv T(f) * g \;+\; \; f * gT \equiv T(g) * f \;+ g * fT \equiv T(gf)$$

To:

Time Derivative and Conserved Time Derivative of fgh•••:

$$gh\ldots T \equiv gh\cdots t + \;\; gh\cdots q_n \;* v^n \equiv gh\cdots t + \;\; gh\cdots q^n \;\; * v_n$$

$$T(f\,gh\cdots) \equiv [T(f)]\,(gh\cdots) + (f)\,(gh\cdots T) \equiv [T(gh\cdots)]\,(f) + (gh\cdots)\,(f\,T) \equiv T(gh\cdots f)$$

Where:

gh…T ≡ gT * h••• + g * h•••T

5.5.2.
Timely Conserved Tensor:

Apply To f^m and f_m

:

$S(f_m)=ft_m=fT_m$	$S(f^m)=ft^m=fT^m$	$T(f_m) \equiv fT_m + f_m\ L$	$T(f^m) \equiv fT^m + f^m\ L$

In q coordinates:

$$T(f_m) \equiv ft_m + fq_m{}^n\, v_n + f_m\ vq^n{}_n\, ft_m + fq_{mn}\, v^n + f_m\ vq_n{}^n$$

$$T(f^m) \equiv ft^m + fq^{mn}\, v_n + f^n\ vq^m{}_n \equiv fT^m + f^m\ L \equiv ft^m + fq^m{}_n\, v^n + f^m\ vq_n{}^n$$

Of which Select:

$T(f_m) \equiv\ ft_m +\ fvq_m{}_n{}^n$	$T(f^m) \equiv\ ft^m +\ fvq^{mn}{}_n$
$T(f_m) \equiv\ ft_m + fq_m{}^n\, v_n + f_m\ vq^n{}_n$	$T(f^m) \equiv\ ft^m + fq^m{}_n\, v^n + f^m\ vq^n{}_n$
$T(f_m) \equiv\ fT_m + f_m\ L$	$T(f^m) \equiv\ fT^m + f^m\ L$

Where:

$fT_m \equiv ft_m + fq_m{}^n\, v_n$	$fT^m \equiv ft^m + fq^m{}_n\, v^n$

Apply To Position: fm = pm

In p coordinates:

$S(p^m) \equiv pt^m \equiv pT^m = 0$	$S(p_m) \equiv pt_m \equiv pT_m = 0$

In q coordinates:

$T(p_m) \equiv pt_m + pq_m{}^n\, v_n + p_m\ vq^n{}_n$	$T(p^m) \equiv pt^m + pp^m{}_n\, v^n + p^m\ vq^n{}_n$
$T(p_m) \equiv 0 + p_m\ L$	$T(p^m) \equiv 0 + p^m\ L$

Apply To Position: fm = qm

In p coordinates:

$S(q_m) \equiv qt_m \equiv qT_m$	$S(q^m) \equiv qt^m \equiv qT^m$

In q coordinates:

$T(q_m) \equiv qt_m + qq_m{}^n\, v_n + q_m\ vq^n{}_n$	$T(q^m) \equiv qt^m + qq^m{}_n\, v^n + q^m\ vq^n{}_n$
$T(q_m) \equiv qt_m + v_m + q_m\ vq^n{}_n$	$T(q^m) \equiv qt^m + v^m + q^m\ vq^n{}_n$
$T(q_m) \equiv qT_m + q_m\ vq^n{}_n$	$T(q^m) \equiv qT^m + q^m\ vq^n{}_n$

qt^m	**Position velocity relative to moving frame**
$qq^m{}_n\, v^n$	**Velocity of frame of reference**

Acceleration: Apply To Velocity: fm = vm

In p coordinates:

$S(v_m) \equiv vt_m \equiv vT_m$	$S(v^m) \equiv vt^m \equiv vT^m$

In p coordinates $S(v_m)$:

$$S(v_m) \equiv (v_m)t + \underline{\nabla}_p \cdot [(v_m)\,\underline{u}] \equiv (v_m)t + [\underline{\nabla}_p\,(v_m)] \cdot \underline{u} + (v_m)\,[\underline{\nabla}_p \cdot \underline{u}] \equiv (v_m)T + (v_m)\,\underline{\nabla}_p \cdot \underline{u}$$

$$S(v_m) \equiv (v_m)t + [(v_m)u]p \equiv (v_m)t + (v_m)\underline{p} \cdot \underline{u} + (v_m)\ up \equiv (v_m)T + (v_m)\ up$$

$$S(v_m) = (v_m)t + 0 \equiv (v_m)T + f\ 0$$

In p coordinates $S(v^m)$:

$$S(v^{m}) \equiv (v^{m})t + \underline{\nabla}_p \cdot [(v^{m})\,\underline{u}] \equiv (v^{m})t + [\underline{\nabla}_p\,(v^{m})] \cdot \underline{u} + (v^{m})\,[\underline{\nabla}_p \cdot \underline{u}] \equiv (v^{m})T + (v^{m})\,\underline{\nabla}_p \cdot \underline{u}$$

$$S(v^{m}) \equiv (v^{m})t + [(v^{m})u]p \equiv (v^{m})t + (v^{m})\underline{p} \cdot \underline{u} + (v^{m})\ \ up \equiv (v^{m})T + (v^{m})\ \ up$$

$$S(v^{m}) = (v^{m})t + 0 \equiv (v^{m})T + f\ \ 0$$

In q coordinates:

$T(v_m) \equiv vt_m + vvq_{m\,n}{}^{n}$	$T(v^{m}) \equiv vt^{m} + vvq^{m\,n}{}_{n}$
$T(v_m) \equiv vt_m + vq_m{}^{n}\,v_n + v_m\,vq^{n}{}_{n}$	$T(v^{m}) \equiv vt^{m} + vq^{m}{}_{n}\,v^{n} + v^{m}\,vq^{n}{}_{n}$
$T(v_m) \equiv vT_m + v_m\,vq^{n}{}_{n}$	$T(v^{m}) \equiv vT^{m} + v^{m}\,vq^{n}{}_{n}$

Where:

$vT_m \equiv vt_m + vq_m{}^{n}\,v_n$	$vT^{m} \equiv vt^{m} + vq^{m}{}_{n}\,v^{n}$

حيث:

vtm:	**position acceleration relative to moving frame**
vqnn vn:	**Acceleration of frame of reference**
vqnn:	**Angular velocity of frame of reference**

Rate of momentum is force:

$$F_m \equiv (mvT_m) \equiv (mvt_m) + (mvq_m{}^{n})\,v_n \qquad F^{m} \equiv (mvT^{m}) \equiv (mvt^{m}) + (mvq^{m}{}_{n}\,v^{n})$$

5.6.
Time Derivative of Spatial Integral Description:

Given:

$$I(t) \equiv \int_{volume} f * d\,volume$$

Then: Its differential is equal change within volume plus change on surface

change in I(t) = volumetric change + convection on surface

That is:

$$dI = \int_{volume} df * d\,volume + \int_{surface} f * (d\underline{q} \,.\, \underline{d\,surface})$$

But:

$$(d\underline{q} \,.\, \underline{d\,surface}) \equiv (\underline{v}\,dt \,.\, \underline{n}\,d\,surface) \equiv (\underline{v} \,.\, \underline{n})\,d\,surface\ \ dt$$

So:

$$dI = \int_{volume} df * d\,volume + \int_{surface} f * (\underline{v} \,.\, \underline{n})\ \ d\,surface\ \ dt$$

$$dI/dt = \int_{volume} \partial f/\partial t * d\,volume + \int_{surface} f * (\underline{v} \,.\, \underline{n})\ \ d\,surface$$

Where:

$$(\underline{v} \,.\, \underline{n}) = v^{n}\cos(\underline{v}, \underline{n}) = vn^{n}{}_{n}$$

$$(\underline{v} \,.\, \underline{n}) = v_n\cos(\underline{v}, \underline{n}) = vn_{n}{}^{n}$$

$$[f * (\underline{v} \,.\, \underline{n})] = (f * \underline{v} \,.\, \underline{n}) = fv^{n}\cos(\underline{v}, \underline{n}) = fvn^{n}{}_{n}$$

$$[f * (\underline{v} \,.\, \underline{n})] = (f * \underline{v} \,.\, \underline{n}) = fv_n\cos(\underline{v}, \underline{n}) = fvn_{n}{}^{n}$$

Then:

$$\int_{surface} f * (\underline{v} \,.\, \underline{n})\,d\,surface = \int_{surface} (f\,\underline{v}) \,.\, \underline{n}\,d\,surface = \int_{v} \underline{\nabla} \,.\, (f\,\underline{v})\,dv = \int_{v} fvq\,dv$$

And:

$$dI/dt \equiv \int_v Cf\, dv = \int_v [ft + \underline{\nabla}.(f\underline{v})]\, dv = \int_v [ft + fvq]\, dv = \int_v [ft + fvq^k{}_k]\, dv = \int_v [ft + fvq_k{}^k]\, dv$$

So Given:

$$I \equiv \int_{volume} f \quad d\,volume$$

Then:

$$IT = \int_{volume} Cf \quad d\,volume$$

Where:

$$Cf \equiv ft + fvq_n{}^n \equiv ft + fvq^n{}_n$$

$$Cf \equiv ft + fq_n \,.\, v^n + f \,.\, vq^n{}_n \equiv ft + fq^n \,.\, v_n + f \,.\, vq_n{}^n$$

Convection term:

Convection of (fv) ≡ Convection of first * second + first * Convection of second

Convection of (fv) ≡ Convection of f multiply v + f multiply Convection of v

$$fvq_n{}^n \equiv fq_n\, v^n + f \;\; vq^n{}_n \equiv fq_n\, v^n + \quad f \;\; L$$

$$fvq^n{}_n \equiv fq^n\, v_n + f \;\; vq_n{}^n \equiv fq^n\, v_n + \quad f \;\; L$$

5.7.
Generalized Coordinates Derivatives:

Gradient of a Function:

$$\underline{\nabla}T \;, \underline{\nabla}T^i \;, \underline{\nabla}T^i{}_j \;, \quad \underline{\nabla}T^i{}_{jk}$$

$$\underline{\nabla}T \equiv [\,T\,]_{\underline{p}} \equiv \underline{Tp} \equiv \partial T/\partial \underline{p} \qquad\qquad \underline{\nabla}T \equiv [\,T\,]_{\underline{q}} \equiv \underline{Tq} \equiv \partial T/\partial \underline{q}$$

: Composition: x onto o	o onto x :Resolution:
α : positive angle between o and x	β : positive angle between x and o
Initial-Material Description	Current-Spatial Description
$T^i_{jk} = T^i_{jk}(p^M, t) = T^i_{jk}(o^M, t)$	$T^i_{jk} = T^i_{jk}(q^N, t) = T^i_{jk}(x^N, t)$
$T^i_{jk} = T^i_{jk}(p_M, t) = T^i_{jk}(o_M, t)$	$T^i_{jk} = T^i_{jk}(q_N, t) = T^i_{jk}(x_N, t)$

Gradient of a Function T^i_{jk}: $\underline{\nabla}T^i_{jk}$ =

$\partial T^i_{jk}/\partial p_M \;\; \underline{p}_M \equiv Tp^{i\;M}_{jk} \;\; \underline{p}_M$	$\partial T^i_{jk}/\partial q_M \;\; \underline{q}_M \equiv Tq^{i\;M}_{jk} \;\; \underline{q}_M$
$\partial T^i_{jk}/\partial p^M \;\; \underline{p}^M \equiv Tp^i_{jkM} \;\; \underline{p}^M$	$\partial T^i_{jk}/\partial q^M \;\; \underline{q}^M \equiv Tq^i_{jkM} \;\; \underline{q}^M$

Gradient of a Coordinate:

$\partial q_N/\partial p_M \;\; \underline{p}_M = qp_N{}^M \;\; \underline{p}_M$	$\partial p_N/\partial q_M \;\; \underline{q}_M = pq_N{}^M \;\; \underline{q}_M$
$\partial q^N/\partial p^M \;\; \underline{p}^M = qp^N{}_M \;\; \underline{p}^M$	$\partial p^N/\partial q^M \;\; \underline{q}^M = pq^N{}_M \;\; \underline{q}^M$

Initial-Material Description	Current-Spatial Description
$T^i_{jk} = T^i_{jk}(p^M, t) = T^i_{jk}(o^M, t)$	$T^i_{jk} = T^i_{jk}(q^N, t) = T^i_{jk}(x^N, t)$
$T^i_{jk} = T^i_{jk}(p_M, t) = T^i_{jk}(o_M, t)$	$T^i_{jk} = T^i_{jk}(q_N, t) - T^i_{jk}(x_N, t)$

Transfer of Grad f to basis and vice versa are bases type

6.
Structural Mechanics Tensorials

6.1.
Tensor:

A linear transformation is extended to multi-dimensional manifolds.
Transformation is subjected to a well defined tensorial transformation rules.
Inversely, any quantity follows these transformation rules is a tensor.
Tensor of order 0 is a scalar
Tensor of order 1 is a vector
Tensor of order 2 is a matrix
And so on …
Order of a tensor is a non-negative integer.

6.1.1.
Order of a Tensor:

A Tensor is an alphanumerically named variable referenced by a number of indices.
The total number of indeces is called its order.
A Tensor index is either a *Lower* index or an *Upper* one.
A Tensor Transformation from a coordinate system does not change its order.
:

Examples:		Indices = Lower + Upper = Order
	A *Scalar* is A Tensor with Zero index (no index):	
Temperature	Temp	0
Mass	Mass	0
	:	
	A *Vector* is A Tensor with *One* index:	
Coordinates of a joint	$JtLoc^{M}$	0+1=1
Displacement of a joint	$JointDis^{M}$	0+1=1
Velocity	Vel^{M}	0+1=1
Force	F^{M}	0+1=1
	:	
	A *Matrix* is A Tensor with *Two* indices:	
Strain	E_{M}^{N}	1+1=2
Stress	$Stress_{M}^{N}$	1+1=2
A System Transformation	$ab^{m}{}_{N}$	1+1=2
Dirac Delta Tensor	$\delta^{m}{}_{n}$	1+1=2
Lower Metric Tensor	aa_{mn}	2+0=2
Upper Metric Tensor	aa^{mn}	0+2=2
	:	
	Tensors with *Three* indices:	
Lower Isotropic 3-Identity	e_{uvw}	3+0=3
Upper Isotropic 3-Identity	e^{uvw}	0+3=3
	:	
	Tensors with *Four* indices:	
Stress Strain Constitution	C_{fg}^{vw}	2+2=4

	$C_{fg}{}^{vw}$	2+2=4
	…	
	Tensor with *Six* indices:	
Isotropic 6-Identity	$E^{\pi\theta\rho}{}_{\iota\theta\rho} \equiv e^{\pi\theta\rho}\, e_{\iota\theta\rho}$	3+3=6
	…	
	Tensor with *Number* of indices:	
A tensor	$T^{AB\ldots K}{}_{LM\ldots N.}$	N
	Trio *Symbols*:	
Lower Christoffel Symbol	$a3a_{JKn}$	3+0=3
Upper Christoffel Symbol	$a3a_{JK}{}^{q}$	2+1=3
	Fourth *Symbols*: Not Tensors:	
Lower Fourth Symbol	$a4a_{ldep}$	4+0=4
Upper Fourth Symbol	$a4a_{lde}{}^{p}$	3+1=4
	Intrinsic Tensor Derivative increases its order by one:	
$T(b)_{AM}{}^{N}$ Intrinsic Derivative	$T(b)_{AM}{}^{N}\big\vert_{B}$	(2+1)+1=4
$T(b)_{A}{}^{CD}{}_{M}{}^{N}$ Intrinsic Derivative	$T(b)_{A}{}^{CD}{}_{M}{}^{N}\big\vert_{B}$	(2+3)+1=6
$T^{AB\ldots K}{}_{LM\ldots N}$ tensor Intrinsic Derivative	$T^{AB\ldots K}{}_{LM\ldots N.}\big\vert_{d}$	(… + …)+1= N + 1
	Forming Tensors:	
Ricci Tensor	$a2a^{m}{}_{n}$	1+1=2
Einstein Tensor	$a2a^{m}{}_{n} - 1/2\ a2a^{K}{}_{K}\ \delta^{m}{}_{n}$	1+1=2

6.1.2.
Symmetric and Anti-Symmetric Tensor:

Symmetric Tensor in i and k

i, k

$$T_{i}{}^{j}{}_{kl} = +T_{k}{}^{j}{}_{il}$$

Anti-Symmetric Tensor i and k

i, k

$$T_{i}{}^{j}{}_{kl} = -\,T_{k}{}^{j}{}_{il}$$

Any tensor can be split into a symmetric and anti-symmetric tensors

$$A_{i}{}^{j}{}_{kl} = S_{i}{}^{j}{}_{kl} + N_{i}{}^{j}{}_{kl}$$

$$A_{i}{}^{j}{}_{kl} = 1/2\ (A_{i}{}^{j}{}_{kl} + A_{k}{}^{j}{}_{il}) + 1/2\ (A_{i}{}^{j}{}_{kl} - A_{k}{}^{j}{}_{il})$$

Symmetric:

$$S_{i\ kl}^{\ j} = 1/2\ (A_{i\ kl}^{\ j} + A_{k\ il}^{\ j}) = + S_{k\ il}^{\ j}$$

Anti-symmetric:

$$N_{i\ kl}^{\ j} = 1/2\ (A_{i\ kl}^{\ j} - A_{k\ il}^{\ j}) = - N_{k\ il}^{\ j}$$

6.1.3. Differential-Preliminaries of Tensors:

If $Q(a)^{\alpha\beta}_{\ \ \gamma}$ components are zeros in a then $Q(b)^{ab}_{\ \ c}$ are zeros in b.

If a theorem is true in a- coordinates then it is valid in b- coordinates.

Equal tensors in a-coordinates stay equal in b- coordinates.

Linear combination of tensors in a-coordinates is a tensor.

Linear combination tensor of tensors in a, is a tensor in b.

6.1.4. Tensor Contracting:

Contracting *upper* and *lower* indices β and ε of a tensor $Q^{\alpha\beta\chi}_{\ \ \ \ \varepsilon\phi\gamma}$ is a tensor:

$$Q^{\alpha\beta\chi}_{\ \ \ \ \varepsilon\phi\gamma}\ \delta^{\varepsilon}_{\beta} = Q^{\alpha\varepsilon\chi}_{\ \ \ \ \varepsilon\phi\gamma} = Q^{\alpha\beta\chi}_{\ \ \ \ \beta\phi\gamma} = Q^{\alpha\chi}_{\ \ \ \phi\gamma}$$

6.1.5. Outer Product of Tensors:

Multiplication of Tensors is an Outer Product:

$$P^{\alpha\beta\chi}_{\ \ \ \ \varepsilon\phi\gamma}\ Q^{\eta\iota}_{\ \ \ \varphi\kappa} = R^{\alpha\beta\chi\ \ \ \ \eta\iota}_{\ \ \ \ \varepsilon\phi\gamma\ \ \ \ \varphi\kappa}$$

6.1.6. Inner Product of Tensors:

If an upper index in the first tensor and a lower index in the second tensor are contracted then it is an Inner Product:

$$P^{\alpha\beta\chi}_{\ \ \ \ \varepsilon\phi\gamma}\ Q^{\eta\iota}_{\ \ \ \varphi\kappa}\ \delta^{\kappa}_{\beta} = R^{\alpha\beta\chi\ \ \ \ \eta\iota}_{\ \ \ \ \varepsilon\phi\gamma\ \ \ \ \varphi\beta}$$

6.1.7. Tensor of Two Quantities Product:

If T Is a Tensor Results of a Product of Two Quantities Q and R:
T = Q R
Where:
Resulting T is a Tensor of an order
(LM,UN)
And One of these Two Quantities is a Tensor of a certain order
(LI,UJ)
Then Second Quantity is a Tensor of order
(LM-LI,UN-UJ)

6.2. *Tensor Transformation:*

Tensor Introduction from Coordinate Transformations:

$$(ab)_{L}/^{KK}\backslash_{L} = (ba)^{K}\backslash_{LL}/^{K}$$

Need to be evaluated:

$(ab)^{K}\backslash_{L}$ $\qquad$ $(ab)_{M}/^{N}$

And get:

$(ab)^{K}_{L}$ ◄≡ $(ab)^{K}_{L}$ $\qquad$ $(ab)_{M}^{N}$ =► $(ba)^{N}_{M}$

Then:

(a) da = (b) db $\qquad$ (da)/a = (db) / b

Bases-Post-Transformer (pre-tensor)	Differential-Pre-Transformer (post-tensor)
(a) (da/ db) = (b)	(a) (db / da) = (b)
(a) (ab) = (b)	(a) (ba) = (b)
Basis	Differential
$(\underline{a}_K)\,(ab)^{K}_{L} = (\underline{b})_{L}$	$(db)^{N} = (ba)^{N}_{M}\,(da)^{M}$
Weight	
$(a)_K\;[(ab)^{K}_{L}]^{**}W = (b_L)$	
(∇) (covariant)	(contravariant)

Bases-Post-Transformer (pre-tensor)	Differential-Pre-Transformer (post-tensor)
$T(b)^{b} = (ba)^{b}_{a}\,T(a)^{a}$.\	$T(a)_{a} = T(b)_{b}\,(ba)^{b}_{a}$

Such That:

$$(\underline{a}_K)\,(ab)^{K}_{L}\,[bTb]^{L}_{M}\,(db)^{M}$$
$$= (\underline{a}_K)\,[aTb]^{K}_{L}\,(da)^{L} = (\underline{b})_{k}\,[bTb]^{k}_{l}\,(db)^{l} =$$
$$(\underline{b})_{k}\,[bTb]^{k}_{l}\,(ba)^{l}_{L}\,(da)^{L}$$

So:

$$[aTa]^{K}_{L} = (ab)^{K}_{k}\,[bTb]^{k}_{l}\,(ba)^{l}_{L}$$

Mixed Order:

$$(\underline{aaa})_{ABD}\,(ab)^{A}_{a}\,(ab)^{B}_{b}\,(ab)^{D}_{d}\,[bbbTb]^{ab\,d}_{c}\,(db)^{D}$$
$$= (\underline{aaa}_{ABD})\,[aaaTb]^{AB\,D}_{C}\,(da)^{C} = (\underline{bbb})_{abd}\,[bbbTb]^{ab\,d}_{c}\,(db)^{c} =$$
$$(\underline{bbb})_{abd}\,[bbbTb]^{ab\,d}_{c}\,(ba)^{c}_{C}\,(da)^{C}$$

So:

$$[aaaTa]^{AB\,D}_{C} = (ab)^{A}_{a}\,(ab)^{B}_{b}\,(ab)^{D}_{d}\,[bbbTb]^{ab\,d}_{c}\,(ba)^{c}_{C}$$

:

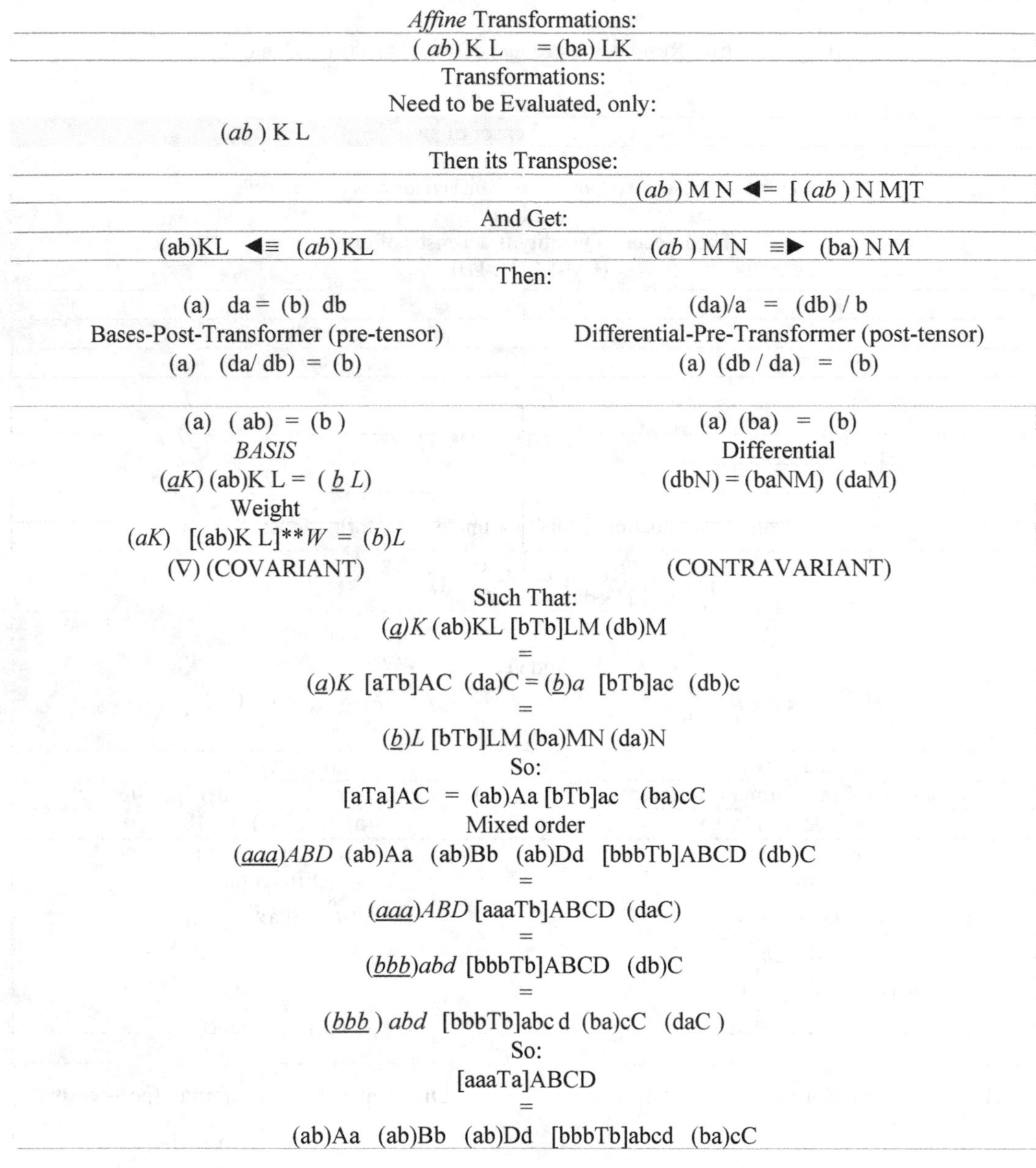

Affine Transformations:

(*ab*) K L = (ba) LK

Transformations:

Need to be Evaluated, only:

(*ab*) K L

Then its Transpose:

(*ab*) M N ◄= [(*ab*) N M]T

And Get:

(ab)KL ◄≡ (*ab*) KL (*ab*) M N ≡► (ba) N M

Then:

(a) da = (b) db	(da)/a = (db) / b
Bases-Post-Transformer (pre-tensor)	Differential-Pre-Transformer (post-tensor)
(a) (da/ db) = (b)	(a) (db / da) = (b)
(a) (ab) = (b)	(a) (ba) = (b)
BASIS	Differential
(*a̲K*) (ab)K L = (*b̲ L*)	(dbN) = (baNM) (daM)
Weight	
(*aK*) [(ab)K L]***W* = (*b*)*L*	
(∇) (COVARIANT)	(CONTRAVARIANT)

Such That:

(*a̲*)*K* (ab)KL [bTb]LM (db)M

=

(*a̲*)*K* [aTb]AC (da)C = (*b̲*)*a* [bTb]ac (db)c

=

(*b̲*)*L* [bTb]LM (ba)MN (da)N

So:

[aTa]AC = (ab)Aa [bTb]ac (ba)cC

Mixed order

(*a̲a̲a̲*)*ABD* (ab)Aa (ab)Bb (ab)Dd [bbbTb]ABCD (db)C

=

(*a̲a̲a̲*)*ABD* [aaaTb]ABCD (daC)

=

(*b̲b̲b̲*)*abd* [bbbTb]ABCD (db)C

=

(*b̲b̲b̲*) *abd* [bbbTb]abc d (ba)cC (daC)

So:

[aaaTa]ABCD

=

(ab)Aa (ab)Bb (ab)Dd [bbbTb]abcd (ba)cC

6.2.1.
First Order Tensor Transformation:

(pre-tensor)

$T(b)^{b} = (ba)^{b}{}_{a}\, T(a)^{a}$.\

Bases-Post-Transformer

Basis = $\underline{b}_{b}$ (da) Differential

$\underline{b}_{b}\, T(b)^{b} = \underline{b}_{b}\, (ba)^{b}{}_{a}\, T(a)^{a}$.\ \.

$\underline{b}_{b}\, T(b)^{b}\, (db) = \underline{b}_{b}\, (ba)^{b}{}_{a}\, T(a)^{a}\, (db)$.\ \.

(post-tensor)

\\. $T(a)_a = T(b)_b \ (ba)^b{}_a$

Differential-Pre-Transformer

Basis = $\underline{b}$ $(da)^a$

.\\ \\. $T(a)_a (da)^a = T(b)_b \ (ba)^b{}_a (da)^a$

.\\ \\. $\underline{b}\ T(a)_a (da)^a = \underline{b}\ T(b)_b \ (ba)^b{}_a (da)^a$

So:

Bases-Post-Transformer (pre-tensor) — Differential-Pre-Transformer (post-tensor)

$T(b)^b = (ba)^b{}_a T(a)^a$.\\ $T(a)_a = T(b)_b \ (ba)^b{}_a$

Obviously there is no mixed tensor of otder one.

6.2.2.
Second Order Tensor Transformation:

Transformation of Second Order *Lower* and *Upper* Tensors
Is A Special Case of The Symbol of
First Derivative of Transformation of First Order *Lower* and *Upper* Tensors
With Respect to <u>Upper Position</u>:

Bases-Post-Transformer (pre-tensor) .\\ \\. Differential-Pre-Transformer (post-tensor)

$\underline{\underline{aa}}_{\mu\theta}\ ab^{\mu}{}_M\ ab^{\theta}{}_Q = \underline{\underline{b}}_{MQ}$.\\ \\. $T(b)^{NP} = ba^N{}_\nu\ ba^P{}_\pi\ T(a)^{\nu\pi}$

6.2.2.1.
Second Order *Lower* Tensor Transformation:

Transformation of Second Order *Lower* Tensor
Is A Special Case of The Symbol of
First Derivative of Transformation of First Order *Lower* Tensor With Respect to <u>Upper Position</u>:

Transformation of First Order *Lower* Tensor w.r.to <u>Upper Position</u>:

$$T(b)_M = T(a)_\mu\ ab^{\mu}{}_M$$

Symbol of
First Derivative of Transformation of First Order *Lower* Tensor With Respect to <u>Upper Position</u>:

$$\partial\,[T(b)_M]/\partial b^N = \partial\,[T(a)_\mu]/\partial b^N * ab^{\mu}{}_M + T(a)_\mu * \partial\,[ab^{\mu}{}_M]/\partial b^N$$

Substituting:

$$\partial\,[T(a)_\mu]/\partial b^N = \partial\,[T(a)_\mu]/\partial a^k\ \ \partial a^k/\partial b^N \qquad \partial\,[ab^{\mu}{}_M]/\partial b^N$$

Or:

$$T(b)_{M,N} = T(a)_{\mu,k}\ \ ab^k{}_N \qquad [ab^{\mu}{}_M]_{,N}$$

It is just a Symbol

$$T(b)_{M,N} = T(a)_{\mu}\ [ab^{\mu}_{M}]_{,N} + T(a)_{\mu,k}\ ab^{\mu}_{M}\ ab^{k}_{N}$$

6.2.2.2.
Second Order *Upper* Tensors Transformation:

Transformation of Second Order *Upper* Tensor
Is A Special Case of The Symbol of
First Derivative of Transformation of First Order *Upper* Tensor With Respect to <u>Upper Position</u>:

Transformation of First Order *Upper* Tensor w.r.to <u>Upper Position</u>:

$$T(b)^{M} = ba^{M}_{\mu}\ T(a)^{\mu}$$

Symbol of First Derivative of Transformation of First Order *Upper* Tensor
With Respect to <u>Upper Position</u>:

$$\partial [T(b)^{M}]/\partial b^{N} = \partial [ba^{M}_{\mu}]/\partial b^{N}\ T(a)^{\mu} + ba^{M}_{\mu}\ \partial [T(a)^{\mu}]/\partial b^{N}$$

Substituting:

$$\partial [ba^{M}_{\mu}]/\partial b^{N} \qquad \partial [T(a)^{\mu}]/\partial b^{N} = \partial [T(a)^{\mu}]/\partial a^{K}\ ab^{k}_{N}$$

Or:

$$[ba^{M}_{\mu}]_{,N} \qquad T(a)^{\mu}_{,N} = T(a)^{\mu}_{,k}\ ab^{k}_{N}$$

It is just a Symbol

$$T(b)^{M}_{,N} = [ba^{M}_{\mu}]_{,N}\ T(a)^{\mu} + ba^{M}_{\mu}\ T(a)^{\mu}_{,k}\ ab^{k}_{N}$$

So both results:
Transformation of P.D. of Lower and Upper Tensor w.r.to <u>Upper Position</u>

Symbols

$$T(b)_{M,N} = T(a)_{\mu}\ [ab^{\mu}_{M}]_{,N} + T(a)_{\mu,k}\ ab^{\mu}_{M}\ ab^{k}_{N} = [ba_{M}^{\mu}]_{,N}\ T(a)_{\mu} + ba_{M}^{\mu}\ ba_{N}^{k}\ T(a)_{\mu,k}$$

$$T(b)^{M}_{,N} = [ba^{M}_{\mu}]_{,N}\ T(a)^{\mu} + ba^{M}_{\mu}\ ba_{N}^{k}\ T(a)^{\mu}_{,k} = T(a)^{\mu}\ [ab^{M}_{\mu}]_{,N} + T(a)^{\mu}_{,k}\ ab^{M}_{\mu}\ ab^{k}_{N}$$

P.D. of a vector is a Symbol but not a tensor.

However if:

$$[ba^{M}_{\mu}]_{,N} = 0 = [ab^{\mu}_{M}]_{,N}$$

Then these are tensors:

$$T(b)_{M,N} = T(a)_{\mu,k}\ ab^{\mu}_{M}\ ab^{k}_{N} = ba_{M}^{\mu}\ ba_{N}^{k}\ T(a)_{\mu,k}$$

$$T(b)^{M}_{,N} = ba^{M}_{\mu}\ ba_{N}^{k}\ T(a)^{\mu}_{,k} = T(a)^{\mu}_{,k}\ ab^{M}_{\mu}\ ab^{k}_{N}$$

$$\underline{aa}_{\alpha\beta}\ ab^{\alpha}{}_{a}\ ab^{\beta}{}_{b} = \underline{bb}_{ab} \qquad T(b)^{ab} = ba^{a}{}_{\alpha}\ ba_{\beta}\ T(a)^{\alpha\beta}$$

$$T(a)_{\alpha\beta}\ ab^{\alpha}{}_{a}\ ab^{\beta}{}_{b} = T(b)_{ab} \qquad T(b)^{ab} = ba^{a}{}_{\alpha}\ ba_{\beta}\ T(a)^{\alpha\beta}$$

:

$$\underline{aa}_{\alpha\beta}\ ab^{\alpha}{}_{a}\ ab^{\beta}{}_{b} = \underline{bb}_{ab} \qquad T(b)^{ab} = ba^{a}{}_{\alpha}\ ba^{b}{}_{\beta}\ T(a)^{\alpha\beta}$$

$$\underline{aa}^{\alpha\beta}\ ab^{a}{}_{\alpha}\ ab^{b}{}_{\beta} = \underline{bb}^{ab} \qquad T(b)_{ab} = T(a)_{\alpha\beta}\ ba^{a}{}_{\alpha}\ ba^{b}{}_{\beta}$$

$$\underline{aa}_{\alpha}{}^{\beta}\ ab^{\alpha}{}_{a}\ ab^{b}_{\beta} = \underline{bb}_{a}{}^{b} \qquad T(b)^{ab} = ba^{a}{}_{\alpha}\ ba^{b}{}_{\beta}\ T(a)^{\alpha\beta}$$

:

$$\underline{bb}_{ab}\ T(b)^{ab} = \underline{bb}_{ab}\ ba^{a}{}_{\alpha}\ ba_{\beta}\ T(a)^{\alpha\beta}$$

$$\underline{bb}^{ab}\ T(b)_{ab} = \underline{bb}^{ab}\ T(a)_{\alpha\beta}\ ab^{\alpha}{}_{a}\ ab^{\beta}{}_{b}$$

$$\underline{bb}_{a}{}^{b}\ T(b)^{a}{}_{b} = \underline{bb}_{a}{}^{b}\ ba^{a}{}_{\alpha}\ T(a)^{\alpha}{}_{\beta}\ ab^{\beta}{}_{b}$$

:

$$T(b)^{ab} = ba^{a}{}_{\alpha}\ ba_{\beta}\ T(a)^{\alpha\beta}$$

$$T(b)_{ab} = T(a)_{\alpha\beta}\ ab^{\alpha}{}_{a}\ ab^{\beta}{}_{b}$$

$$T(b)^{a}{}_{b} = ba^{a}{}_{\alpha}\ T(a)^{\alpha}{}_{\beta}\ ab^{\beta}{}_{b}$$

FORTRAN Program:

```
      EQUIVALENT (AT(+J,-I),A(+I,-J),AN(-J,+I),BN(+I,-J),B(-J,+I),BT(-I,+J))
      EQUIVALENT (BT(+J,-I),B(+I,-J),BN(-J,+I),AN(+I,-J),A(-J,+I),AT(-I,+J))
C     ... ... ... ... ... ... ... ... BSA(+i,-j) = B(+i,-m) SA(+m,-j)
      DO 1000   I=1   ,IMAX
      SBIJ = 0
      DO 100    J=1   ,JMAX
C     ... ... ... ... ... ... ... ... SA(+m,-j) = Sa(+m,-n) A(+n,-j)
      DO 10     M=1   ,MMAX
      SAMJ = 0
      DO 1      N=1   ,NMAX
      SAMJ = SAMJ+SA(+M,-N)*A(+N,-J)
1     CONTINUE
      SA(+M,-J)= SAMJ
10    CONTINUE
      SBIJ = SBIJ+B(+I,-M)*SA(+M,-J)
100   CONTINUE
      SB(+I,-J) = SBIJ
1000  CONTINUE
C     ... ... ... ... ... ... ... ... ... ... ... ... ... ... ... ... ... ... ... ... ... ... ... ...
```

Affine:

Sb(i,j)=B(i,m) Sa(m,n) BT(n,j) | Sb(i,j)=AT(i,m) Sa(m,n) A(n,j)

All Affine indices are positive.

:

6.2.3.
Third Order Tensor Transformation:

$$T(b)^{ab}{}_{c} = ba^{a}{}_{\alpha}\ ba_{\beta}\ T(a)^{\alpha\beta}{}_{\gamma}\ ab^{\gamma}{}_{c}$$

:

6.2.4.
Any Order Tensor Transformation:

$$T(b)^{ab\ \ f}_{\ \ \ cde} = ba^{a}_{\alpha}\ ba^{b}_{\beta}\ ba^{f}_{\phi}\ T(a)^{\alpha\beta\ \ \phi}_{\ \ \ \chi\delta\varepsilon}\ ab^{\chi}_{c}\ ab^{\delta}_{d}\ ab^{\varepsilon}_{e}$$

Apply:

a : Composition: b into a

$\underline{a}M$ $$T(b)^{ab\ \ f}_{\ \ \ cde} = ba^{ab}_{\alpha}\ ba_{\beta}\ ba^{f}_{\phi}\ T(a)^{\alpha\beta\ \ \phi}_{\ \ \ \chi\delta\varepsilon}\ ab^{\chi}_{c}\ ab^{\delta}_{d}\ ab^{\varepsilon}_{e}$$

To Different Systems:

General: bN

$\underline{a}M$ $$T(b)^{ab\ \ f}_{\ \ \ cde} = ba^{ab}_{\alpha}\ ba_{\beta}\ ba^{f}_{\phi}\ T(a)^{\alpha\beta\ \ \phi}_{\ \ \ \chi\delta\varepsilon}\ ab^{\chi}_{c}\ ab^{\delta}_{d}\ ab^{\varepsilon}_{e}$$

$\underline{r}M$ $$T(b)^{ab\ \ f}_{\ \ \ cde} = br^{rb}_{\alpha}\ br_{\beta}\ br^{f}_{\phi}\ T(r)^{\alpha\beta\ \ \phi}_{\ \ \ \chi\delta\varepsilon}\ rb^{\chi}_{c}\ rb^{\delta}_{d}\ rb^{\varepsilon}_{e}$$

$\underline{x}M$ $$T(b)^{xb\ \ f}_{\ \ \ cde} = bx^{xb}_{\alpha}\ bx_{\beta}\ bx^{f}_{\phi}\ T(x)^{\alpha\beta\ \ \phi}_{\ \ \ \chi\delta\varepsilon}\ xb^{\chi}_{c}\ xb^{\delta}_{d}\ xb^{\varepsilon}_{e}$$

$\underline{o}M$ $$T(b)^{ob\ \ f}_{\ \ \ cde} = bo^{ob}_{\alpha}\ bo_{\beta}\ bo^{f}_{\phi}\ T(o)^{\alpha\beta\ \ \phi}_{\ \ \ \chi\delta\varepsilon}\ ob^{\chi}_{c}\ ob^{\delta}_{d}\ ob^{\varepsilon}_{e}$$

Orthogonal: sN

$\underline{a}M$ $$T(s)^{as\ \ f}_{\ \ \ cde} = sa^{a}_{\alpha}\ sa_{\beta}\ sa^{f}_{\phi}\ T(a)^{\alpha\beta\ \ \phi}_{\ \ \ \chi\delta\varepsilon}\ as^{\chi}_{c}\ as^{\delta}_{d}\ as^{\varepsilon}_{e}$$

$\underline{r}M$ $$T(s)^{rs\ \ f}_{\ \ \ cde} = sr^{r}_{\alpha}\ sr_{\beta}\ sr^{f}_{\phi}\ T(r)^{\alpha\beta\ \ \phi}_{\ \ \ \chi\delta\varepsilon}\ rs^{\chi}_{c}\ rs^{\delta}_{d}\ rs^{\varepsilon}_{e}$$

$\underline{x}M$ $$T(s)^{xs\ \ f}_{\ \ \ cde} = sx^{x}_{\alpha}\ sx_{\beta}\ sx^{f}_{\phi}\ T(x)^{\alpha\beta\ \ \phi}_{\ \ \ \chi\delta\varepsilon}\ xs^{\chi}_{c}\ xs^{\delta}_{d}\ xs^{\varepsilon}_{e}$$

$\underline{o}M$ $$T(s)^{os\ \ f}_{\ \ \ cde} = so^{o}_{\alpha}\ so_{\beta}\ so^{f}_{\phi}\ T(o)^{\alpha\beta\ \ \phi}_{\ \ \ \chi\delta\varepsilon}\ os^{\chi}_{c}\ os^{\delta}_{d}\ os^{\varepsilon}_{e}$$

Orthonormal: yN

$\underline{a}M$ $$T(y)^{ay\ \ f}_{\ \ \ cde} = ya^{a}_{\alpha}\ ya_{\beta}\ ya^{f}_{\phi}\ T(a)^{\alpha\beta\ \ \phi}_{\ \ \ \chi\delta\varepsilon}\ ay^{\chi}_{c}\ ay^{\delta}_{d}\ ay^{\varepsilon}_{e}$$

$\underline{r}M$ $$T(y)^{ry\ \ f}_{\ \ \ cde} = yr^{r}_{\alpha}\ yr_{\beta}\ yr^{f}_{\phi}\ T(r)^{\alpha\beta\ \ \phi}_{\ \ \ \chi\delta\varepsilon}\ ry^{\chi}_{c}\ ry^{\delta}_{d}\ ry^{\varepsilon}_{e}$$

$\underline{x}M$ $$T(y)^{xy\ \ f}_{\ \ \ cde} = yx^{x}_{\alpha}\ yx_{\beta}\ yx^{f}_{\phi}\ T(x)^{\alpha\beta\ \ \phi}_{\ \ \ \chi\delta\varepsilon}\ xy^{\chi}_{c}\ xy^{\delta}_{d}\ xy^{\varepsilon}_{e}$$

$\underline{o}M$ $$T(y)^{oy\ \ f}_{\ \ \ cde} = yo^{o}_{\alpha}\ yo_{\beta}\ yo^{f}_{\phi}\ T(o)^{\alpha\beta\ \ \phi}_{\ \ \ \chi\delta\varepsilon}\ oy^{\chi}_{c}\ oy^{\delta}_{d}\ oy^{\varepsilon}_{e}$$

Fixed: oN

$\underline{a}M$ $$T(o)^{ao\ \ f}_{\ \ \ cde} = oa^{a}_{\alpha}\ oa_{\beta}\ oa^{f}_{\phi}\ T(a)^{\alpha\beta\ \ \phi}_{\ \ \ \chi\delta\varepsilon}\ ao^{\chi}_{c}\ ao^{\delta}_{d}\ ao^{\varepsilon}_{e}$$

$\underline{r}M$ $$T(o)^{ro\ \ f}_{\ \ \ cde} = or^{r}_{\alpha}\ or_{\beta}\ or^{f}_{\phi}\ T(r)^{\alpha\beta\ \ \phi}_{\ \ \ \chi\delta\varepsilon}\ ro^{\chi}_{c}\ ro^{\delta}_{d}\ ro^{\varepsilon}_{e}$$

$\underline{x}M$ $$T(o)^{xo\ \ f}_{\ \ \ cde} = ox^{x}_{\alpha}\ ox_{\beta}\ ox^{f}_{\phi}\ T(x)^{\alpha\beta\ \ \phi}_{\ \ \ \chi\delta\varepsilon}\ xo^{\chi}_{c}\ xo^{\delta}_{d}\ xo^{\varepsilon}_{e}$$

$\underline{o}M$ $$T(o)^{oo\ \ f}_{\ \ \ cde} = oo^{o}_{\alpha}\ oo_{\beta}\ oo^{f}_{\phi}\ T(o)^{\alpha\beta\ \ \phi}_{\ \ \ \chi\delta\varepsilon}\ oo^{\chi}_{c}\ oo^{\delta}_{d}\ oo^{\varepsilon}_{e}$$

:

T Weighted Tensor Integrable Transformation

:

6.2.5.
W Weighted Tensor

Integrable Transformation:

Weight is Bases-Post-Transformer (pre-tensor)
While
Derivative is Differential-Pre-Transformer (post-tensor)
fo(-j) = X[+s-r]**W fx(-m) X(+m -j)
fx(-j) = O[+s-r]**W fo(- m) O(+m -j)
Differential is Differential-Pre-Transformer (post-tensor)

$dg^n = (gf^n{}_m)^W \; df^m$ $\qquad df^m = (fg^m{}_n)^W \; dg^n$

$dg_n = (gf_n{}^m)^W \; df_m$ $\qquad df_m = (fg_m{}^n)^W \; dg_n$

Contravariant Transformation of W Weight:
(Differential-Pre-Transformation)
Weight Bases-Post-Transformer (pre-tensor)
qo(+k) = |xo(+s,-r)|**W ox(+k,-n) qx(+n)
Inverse Contravariant Transformation of Weight W:
(Differential-Pre-Transformation)
Weight Bases-Post-Transformer (pre-tensor)
qx(+k) = |ox(+s,-r)|**W xo(+k,-n) qo(+n)

Covariant Transformation of Weight W:
Bases-Post-Transformer (pre-tensor)
qo(-j) = xo[+s,-r]**W qx(-m) xo(+m , - j)
Covariant Inverse Transformation of Weight W:
Bases-Post-Transformer (pre-tensor)
qx(-j) = ox[+s,-r]**W qo(- m) ox(+m , -j)

6.3.
Raising and Lowering of Indices:

Recall:

$oo_{ik} = \underline{o}_i . \underline{o}_k = \delta_{ik}$ $\qquad oo_I{}^K \equiv \underline{o}_I . \underline{o}^K = \delta_I{}^K$

$oo^I{}_K \equiv \underline{o}^I . \underline{o}_K = \delta^I{}_K$ $\qquad oo^{ik} = \underline{o}^i . \underline{o}^k = \delta^{ik}$

$aa_{IK} \equiv \underline{a}_I . \underline{a}_K = \alpha_I \alpha_K \cos(_{IK})$ $\qquad aa_I{}^K \equiv \underline{a}_I . \underline{a}^K = \delta_I{}^K$

$aa^I{}_K \equiv \underline{a}^I . \underline{a}_K = \delta^I{}_K$ $\qquad aa^{IK} \equiv \underline{a}^I . \underline{a}^K = \alpha^I \alpha^K \cos(^{IK})$

Metric Tensor Reciprocals

$aa_{mi} \; aa^{in} = \delta_m{}^n$ $\qquad aa^{mi} \; aa_{in} = \delta^m{}_n$

Raising and Lowering of first order tensor by Metric Tensors:

$v_j = aa_{jl} \; v^l$ $\qquad v^m = aa^{mk} \; v_k$

This is multiplied as it follows:

$aa^{mj} v_j = aa^{mj} aa_{jl} v^l = \delta^m{}_l v^l = v^m$ $\qquad aa_{nm} v^m = aa_{nm} aa^{mk} v_k = \delta^k{}_n v_k = v_n$

In general:
Raising and Lowering of Indices:

$aa_{ql} \; S_{j}{}^{l}{}_{n} = S_{jqn}$ $\quad$ $aa^{qj} \; S_{j}{}^{l}{}_{n} = S^{ql}{}_{n}$

Bases Transformation:

$\underline{a}^{N} \; aa_{NM} = \underline{a}_{M}$ $\quad$ $\underline{a}_{M} \; aa^{MN} = \underline{a}^{N}$

6.4. *Isotropic 3-Identity:*

Knowing:

$$6 = e^{uvw}(a)\; e_{uvw}(a) = e^{uvw}(r)\; e_{uvw}(r) = e^{uvw}(x)\; e_{uvw}(x)$$

Determinant:

$(1/6)\, e_{uvw}(a)\, e^{pqr}(a)\, ab_{p}{}^{u}\, ab_{q}{}^{v}\, ab_{r}{}^{w} \equiv ||ab||$ $\quad$ $(1/6)\, e_{uvw}(a)\, e^{pqr}(a)\, ba^{u}{}_{p}\, ba^{v}{}_{q}\, ba^{w}{}_{r} \equiv ||ba||$

Multiplying by:

$$6 = e^{pqr}(a)\; e_{pqr}(a)$$

Gives:

$e_{uvw}(a)\, e^{pqr}(a)\; ab_{p}{}^{u}\; ab_{q}{}^{v}\; ab_{r}{}^{w} = ||ab||\, e^{pqr}(a)\, e_{pqr}(a)$ $\quad$ $e_{uvw}(a)\, e^{pqr}(a)\; ba^{u}{}_{p}\; ba^{v}{}_{q}\; ba^{w}{}_{r} = ||ba||\, e^{pqr}(a)\, e_{pqr}(a)$

Or:

$e_{uvw}(a)\, ab_{p}{}^{u}\, ab_{q}{}^{v}\, ab_{r}{}^{w} = ||ab_{p}{}^{u}||\, e_{pqr}(a)$ $\quad$ $e_{uvw}(a)\; ba^{u}{}_{p}\; ba^{v}{}_{q}\; ba^{w}{}_{r} = ||ba^{u}{}_{p}||\, e_{pqr}(a)$

:

Also, Multiplying by:

$$6 = e^{uvw}(a)\; e_{uvw}(a)$$

Gives:

$e_{uvw}(a)\; e^{pqr}(a)\; ab_{p}{}^{u}\; ab_{q}{}^{v}\; ab_{r}{}^{w} = ||ab||\; e^{uvw}(a)\; e_{uvw}(a)$ $\quad$ $e_{uvw}(a)\; e^{pqr}(a)\; ba^{u}{}_{p}\; ba^{v}{}_{q}\; ba^{w}{}_{r} = ||ba||\; e^{uvw}(a)\; e_{uvw}(a)$

Or:

$e^{pqr}(a)\, ab_{p}{}^{u}\, ab_{q}{}^{v}\, ab_{r}{}^{w} = ||ab_{p}{}^{u}||\, e^{uvw}(a)$ $\quad$ $e^{pqr}(a)\, ba^{u}{}_{p}\, ba^{v}{}_{q}\, ba^{w}{}_{r} = ||ba^{u}{}_{p}||\, e^{uvw}(a)$

Determinant of: ||ab|| and ||ba||.

Similarly: (o)

$(1/6)\, e_{ijk}(o)\; e^{pqr}(o)\; \delta_{p}{}^{i}\, \delta_{q}{}^{j}\, \delta_{r}{}^{k} = ||\delta||$ $\quad$ $(1/6)\, e_{ijk}(o)\, e^{pqr}(o)\, \delta^{i}{}_{p}\, \delta^{j}{}_{q}\, \delta^{k}{}_{r} = ||\delta||$

Isotropic Dual Identity Determinant: ||δ|| = 1

$e^{pqr}(o)\; \delta_{p}{}^{u}\; \delta_{q}{}^{v}\; \delta_{r}{}^{w} = ||\delta_{p}{}^{u}||\, e^{uvw}(o)$ $\quad$ $e_{uvw}(o)\; \delta^{u}{}_{p}\; \delta^{v}{}_{q}\; \delta^{w}{}_{r} = ||\delta^{u}{}_{p}||\, e_{pqr}(o)$

Determinant of: || δ ||=1

$e^{pqr}(o)\; \delta_{p}{}^{u}\; \delta_{q}{}^{v}\; \delta_{r}{}^{w} = 1\, e^{uvw}(o)$ $\quad$ $e^{pqr}(o)\; \delta^{u}{}_{p}\; \delta^{v}{}_{q}\; \delta^{w}{}_{r} = 1\, e^{uvw}(o)$

Transformatian of e:

$e_{abc}(b) = e_{\alpha\beta\chi}(a)\; ab^{\alpha}{}_{a}\; ab^{\beta}{}_{b}\; ab^{\chi}{}_{c}$ $\quad$ $e_{\alpha\beta\chi}(a) = e_{abc}(b)\; ba^{a}{}_{\alpha}\; ba^{b}{}_{\beta}\; ba^{c}{}_{\chi}$

$e^{abc}(b) = e^{\alpha\beta\chi}(a)\; ab_{\alpha}{}^{a}\; ab_{\beta}{}^{b}\; ab_{\chi}{}^{c}$ $\quad$ $e^{\alpha\beta\chi}(a) = e^{abc}(b)\; ba_{a}{}^{\alpha}\; ba_{b}{}^{\beta}\; ba_{c}{}^{\chi}$

But, Determinant of: || ab || and || ba||:

$e_{uvw}(a)\, ab^{u}_{p}\, ab^{v}_{q}\, ab^{w}_{r} = \|ab^{u}_{p}\|\, e_{pqr}(b)$	$e^{pqr}\, ba^{u}_{p}\, ba^{v}_{q}\, ba^{w}_{r} = \|ba^{u}_{p}\|\, e^{uvw}$
$e^{pqr}(a)\, ab_{p}^{u}\, ab_{q}^{v}\, ab_{r}^{w} = \|ab_{p}^{u}\|\, e^{uvw}$	$e_{uvw}\, ba^{u}_{p}\, ba^{v}_{q}\, ba^{w}_{r} = \|ba^{u}_{p}\|\, e_{pqr}$

So:

$$e_{abc}(b) = e_{\alpha\beta\chi}(a)\, ab^{\alpha}_{a}\, ab^{\beta}_{b}\, ab^{\chi}_{c} = e_{abc}(a)\, \|ab^{P}_{Q}\|$$

$$e_{\alpha\beta\chi}(a) = e_{abc}(b)\, ba^{a}_{\alpha}\, ba^{b}_{\beta}\, ba^{c}_{\chi} = e_{\alpha\beta\chi}(b)\, \|ba^{P}_{Q}\|$$

And:

$$e^{abc}(b) = e^{\alpha\beta\chi}(a)\, ab_{\alpha}^{a}\, ab_{\beta}^{b}\, ab_{\chi}^{c} = e^{abc}(a)\, \|ab_{P}^{Q}\|$$

$$e^{\alpha\beta\chi}(b)\, \|ba_{P}^{Q}\| = e^{\alpha\beta\chi}(a) = e^{abc}(b)\, ba_{a}^{\alpha}\, ba_{b}^{\beta}\, ba_{c}^{\chi}$$

But:

$\|ab^{\alpha}_{a}\| = \beta/\alpha$	$\alpha/\beta = \|ba^{a}_{\alpha}\|$
$\|ab_{\alpha}^{a}\| = \alpha/\beta$	$\beta/\alpha = \|ba_{a}^{\alpha}\|$

So:

$e_{abc}(b) = e_{\alpha\beta\chi}(a)\, ab^{\alpha}_{a}\, ab^{\beta}_{b}\, ab^{\chi}_{c} = e_{abc}(a)\, \|ab^{P}_{Q}\| = e_{abc}(a) * \beta/\alpha$	$e_{\alpha\beta\chi}(a) = e_{abc}(b)\, ba^{a}_{\alpha}\, ba^{b}_{\beta}\, ba^{c}_{\chi} = e_{\alpha\beta\chi}(b)\, \|ba^{P}_{Q}\| = e_{abc}(b) * \alpha/\beta$
$e^{abc}(b) = e^{\alpha\beta\chi}(a)\, ab_{\alpha}^{a}\, ab_{\beta}^{b}\, ab_{\chi}^{c} = e^{abc}(a)\, \|ab_{P}^{Q}\| = e^{abc}(a) * \alpha/\beta$	$e^{\alpha\beta\chi}(a) = e^{abc}(b)\, ba_{a}^{\alpha}\, ba_{b}^{\beta}\, ba_{c}^{\chi} = e^{\alpha\beta\chi}(b)\, \|ba_{P}^{Q}\| = e^{\alpha\beta\chi}(b) * \beta/\alpha$

$$e_{\alpha\beta\chi}(b)\ /\ \beta = e_{\alpha\beta\chi}(a)\ /\ \alpha$$

$$e^{abc}(b) * \beta = e^{abc}(a) * \alpha$$

Or

$$e_{\alpha\beta\chi}(b) * \alpha = e_{\alpha\beta\chi}(a) * \beta$$

$$e^{abc}(b)\ /\ \alpha = e^{abc}(a)\ /\ \beta$$

For example:

$$e_{abc}(s)\ /\ \sigma = e_{abc}(o)$$

$$e^{abc}(s) * \sigma = e^{abc}(o)$$

Or:

$$e_{abc}(s) = e_{abc}(o) * \sigma$$

$$e^{abc}(s) = e^{abc}(o)\ /\ \sigma$$

These Relationships of (e)

$e_{\alpha\beta\chi}(a) = aa_{\alpha\phi}\, aa_{\beta\gamma}\, aa_{\chi\eta}\, e^{\phi\gamma\eta}(a)$	$e^{abc} = bb^{af}\, bb^{bg}\, bb^{ch}\, e_{fgh}$
$e^{\alpha\beta\chi}(a) = aa^{\alpha\phi}\, aa^{\beta\gamma}\, aa^{\chi\eta}\, e_{\phi\gamma\eta}(a)$	$e_{abc} = bb_{af}\, bb_{bg}\, bb_{ch}\, e^{fgh}$

Are Multiplied by:

$e^{\alpha\beta\chi}$	e_{abc}
$e_{\alpha\beta\chi}$	e^{abc}

Knowing:

$6 = e_{\alpha\beta\chi}\, e^{\alpha\beta\chi}$	$6 = e^{abc}\, e_{abc}$
$6 = e^{\alpha\beta\chi}\, e_{\alpha\beta\chi}$	$6 = e_{abc}\, e^{abc}$

To get:

$6 = (aa_{\alpha\phi}\, aa_{\beta\gamma})\, aa_{\chi\eta}\, e^{\phi\gamma\eta}\, e^{\alpha\beta\chi}$	$6 = (bb^{af}\, bb^{bg})\, bb^{ch}\, e_{fgh}\, e_{abc}$
$6 = (aa^{\alpha\phi}\, aa^{\beta\gamma})\, aa^{\chi\eta}\, e_{\phi\gamma\eta}\, e_{\alpha\beta\chi}$	$6 = bb_{af}\, bb_{bg}\, bb_{ch}\, e^{fgh}\, e^{abc}$

If multiplied by:

$aa^{\chi\eta}$ $\quad bb_{ch}$

$aa_{\chi\eta}$ $\quad bb^{ch}$

Get:

$$2*3\;aa^{\chi\eta} = aa^{\chi\eta}(aa_{\alpha\phi}\,aa_{\beta\gamma})\,aa_{\chi\eta}\,e^{\phi\gamma\eta}\,e^{\alpha\beta\chi} \qquad 2*3\;bb_{ch} = bb_{ch}\,(bb^{af}\,bb^{bg})\,bb^{ch}\,e_{fgh}\,e_{abc}$$

$$2*3\;aa_{\chi\eta} = aa_{\chi\eta}\,(aa^{\alpha\phi}\,aa^{\beta\gamma})\,aa^{\chi\eta}\,e_{\phi\gamma\eta}\,e_{\alpha\beta\chi} \qquad 2*3\;bb^{ch} = bb^{ch}\,bb_{\varepsilon\kappa}\,bb_{af}\,bb_{bg}\,bb_{ch}\,e^{fgh}\,e^{abc}$$

But:

$$3 = aa^{\chi\eta}\;aa_{\chi\eta} \qquad 3 = bb_{ch}\;bb^{ch}$$

$$3 = aa_{\chi\eta}\;aa^{\chi\eta} \qquad 3 = bb^{ch}\;bb_{ch}$$

So:

$$2\;aa^{\chi\eta} = (aa_{\alpha\phi}\;aa_{\beta\gamma})\,e^{\phi\gamma\eta}\;e^{\alpha\beta\chi} \qquad 2\;bb_{ch} = bb^{af}\;bb^{bg}\;e_{fgh}\,e_{abc} =$$

$$2\;aa_{\chi\eta} = (aa^{\alpha\phi}\;aa^{\beta\gamma})\,e_{\phi\gamma\eta}\;e_{\alpha\beta\chi} \qquad 2\;bb^{ch} = bb_{af}\;bb_{bg}\,e^{abc}\;e^{fgh} =$$

Isotropic 6-Identity:

$$E^{\pi\theta\rho}{}_{\iota\theta\rho} \equiv e^{\pi\theta\rho}\;e_{\iota\theta\rho} = e^{\pi\theta\rho}\;aa_{\iota\phi}\;aa_{\theta\gamma}\;aa_{\rho\eta}\;e^{\phi\gamma\eta} = aa_{\iota\phi}\;2\;aa^{\pi\phi} = 2\,\delta^{\pi}_{\iota} = aa^{\pi\phi}\;aa^{\theta\gamma}\;aa^{\rho\eta}\;e_{\phi\gamma\eta}\;e_{\iota\theta\rho} = aa^{\pi\phi}\;2\;aa_{\iota} = 2\,\delta^{\pi}_{\iota}$$

$$E^{pqr}{}_{iqr} \equiv e^{pqr}\;e_{iqr} = bb^{pf}\;bb^{qg}\;bb^{rh}\;e_{fgh}\,e_{iqr} = bb^{pf}\;2\,bb_{if} = 2\,\delta^{p}_{i} = e^{pqr}\,bb_{if}\;bb_{qg}\;bb_{rh}\;e^{fgh} = (bb_{if})\,2\,bb^{pf} = 2\,\delta^{p}_{i}$$

6.5.
Surface and Volume:

Basis amd differential transformations

$\underline{o}_{\mu}\;oa^{\mu}_{M} = \underline{a}_{M}$ $\quad$ \=/ $\;$ /=\ $\quad$ $da^{N} = ao^{N}_{n}\;do^{n}$

$\underline{o}^{\mu}\;oa^{M}_{\mu} = \underline{a}^{M}$ $\quad$ /=\ $\;$ \=/ $\quad$ $da_{N} = ao^{n}_{N}\;do_{n}$

$\underline{a}_{m}\;ao^{m}_{M} = \underline{o}_{M}$ $\quad$ \=/ $\;$ /=\ $\quad$ $do^{\nu} = oa^{\nu}_{n}\;da^{n}$

$\underline{a}^{m}\;ao^{M}_{m} = \underline{o}^{M}$ $\quad$ /=\ $\;$ \=/ $\quad$ $do_{\nu} = oa^{n}_{\nu}\;da_{n}$

6.5.1.
Surface in 3-D Space:

2-D Surface in 3-D Space is Oriented in the Third Dimension:

:

$\underline{a}_P \wedge \underline{a}_Q$

Outer Product:

$$\underline{a}_p \wedge \underline{a}_q = e_{ijk}(o)\, a_p{}^{i}\, a_q{}^{j}\, \underline{o}^{k}$$

$$\underline{a}^p \wedge \underline{a}^q = e^{ijk}(o)\, a^p{}_{i}\, a^q{}_{j}\, \underline{o}_{k}$$

(e) transformation

$$e_{abc}(o) = e_{abc}(a) / \alpha$$

$$e^{abc}(o) = e^{abc}(a) * \alpha$$

So Surface or Outer Product:

$$\underline{a}_p \wedge \underline{a}_q = e_{ijk}(o)\, a_p{}^{i}\, a_q{}^{j}\, \underline{o}^{k} = e_{ijk}(a)/\alpha\ \ a_p{}^{i}\, a_q{}^{j}\, \underline{o}^{k}$$

$$\underline{a}^p \wedge \underline{a}^q = e^{ijk}(o)\, a^p{}_{i}\, a^q{}_{j}\, \underline{o}_{k} = e^{ijk}(a)*\alpha\ \ a^p{}_{i}\, a^q{}_{j}\, \underline{o}_{k}$$

(N-1)-D Surface in N-D Space is Oriented in the N-th Dimension:

$$e_{ijk}(0)\ a_{p1p2...pN-1}{}^{i}\ a_{q1q2...qN-1}{}^{j}\ \underline{o}^{k} = e_{ijk}(a)/\alpha\ \ a_{p1p2...pN-1}{}^{i}\ a_{q1q2...qN-1}{}^{j}\ \underline{o}^{k}$$

$$e^{ijk}(0)*\alpha\ \ a^{p1p2...pN-1}{}_{i}\ a^{q1q2...qN-1}{}_{j}\ \underline{o}_{k} = e^{ijk}(a)*\alpha\ \ a^{p1p2...pN-1}{}_{i}\ a^{q1q2...qN-1}{}_{j}\ \underline{o}_{k}$$

In Complex Space:
Dimension of Complex Space is Even 2N
(2N-1)-D Surface in 2N-D Space is Oriented in the 2N-th Dimension:

6.5.2.
Volume

Mixed Products:

$$0 < \|_{a}\| \equiv (\underline{a}_p \wedge \underline{a}_q . \underline{a}_r) = \|_{oa}\|\ e_{pqr}(o) \le \alpha_1\, \alpha_2\, \alpha_3$$

$$0 < \|^{a}\| \equiv (\underline{a}^p \wedge \underline{a}^q . \underline{a}^r) = \|^{ao}\|\ e^{pqr}(o) \le \alpha^1\, \alpha^2\, \alpha^3$$

:

$$e_{pqr}(a) = (\underline{a}_p \wedge \underline{a}_q . \underline{a}_r) = \|_{oa}\|\ e_{pqr}(a) = \alpha\ \ e_{pqr}(o)$$

$$e^{pqr}(a) = (\underline{a}^p \wedge \underline{a}^q . \underline{a}^r) = \|^{ao}\|\ e^{pqr}(a) = 1/\alpha\ \ e^{pqr}(o)$$

Incremental Volume or Outer Product:

$$0 < (\underline{a}_p da^p \wedge \underline{a}_q da^q . \underline{a}_r da^r) = \|_{a}\|\ e_{pqr}(o)\ da^p da^q da^r = \alpha\ e_{pqr}(o) da^p da^q da^r \le \alpha_1 \alpha_2 \alpha_3 da^p da^q da^r$$

$$0 < (\underline{a}^p da_p \wedge \underline{a}^q da_q . \underline{a}^r da_r) = \|^{a}\|\ e^{pqr}(o)\ da_p\, da_q da_r = 1/\alpha\ e^{pqr}(o) da_p da_q da_r \le \alpha^1 \alpha^2 \alpha^3 da_p da_q da_r$$

6.6.
Partial Derivative of Metrics

Partial Derivative of
Lower and *Upper* Metrics
With respect to Upper Position

The following

$$[aa_{MN}]_{,K} = + [\, a3a_{KMN} + a3a_{NKM}]$$

$$[aa^{MN}]_{,K} = - [\, a3a_{K}^{\;MN} + a3a_{N}^{\;KM}\,]$$

Will be discussed in what follows:

6.6.1. Trio Symbol:

Trio Symbol
and
Christoffel Two Symbols of
Upper and Lower Third Index

Trio Symbol with respect to first two indices $_{JK}$ Symmetry

Trio Symbol

$$a3a_{JKn} \equiv \tfrac{1}{2}(\, aa_{nK,J} + aa_{Jn,K} - aa_{JK,n})$$

Upper and Lower Christoffel Symbols:

Upper Christoffel Symbol

$$a3a_{JK}^{\;\;Q} \equiv a3a_{JKn}\, aa^{nQ}$$

Lower Christoffel Symbol ≡ Trio Symbol

$$a3a_{JKn} \equiv \tfrac{1}{2}\,(aa_{nK,J} + aa_{Jn,K} - aa_{JK,n})$$

Symmetry of first two indices:

$$a3a_{JK}^{\;\;Q} = a3a_{KJ}^{\;\;Q} \qquad a3a_{JK\,Q} = a3a_{KJ\,Q}$$

Raising and Lowering of Christoffel Symbols:

$$a3a_{JK}^{\;\;Q}\; aa_{QN} = a3a_{JKN} \qquad a3a_{JK\,Q}\; aa^{QN} = a3a_{JK}^{\;\;N}$$

Lower Christoffel Symbol — Upper Christoffel Symbol

Raising of First Index in Christoffel Symbols:

$$aa^{IJ}\; a3a_{JN}^{\;\;Q} = a3a^{I\;Q}_{\;N} \qquad aa^{IJ}\; a3a_{JK\,Q} = a3a^{I}_{\;KQ}$$

Symmetry with respect to first two indices

$$a3a^{I\;Q}_{\;N} = a3a^{N\;Q}_{\;I} \qquad a3a^{I}_{\;KQ} = a3a^{K}_{\;IQ}$$

Raising of Second Index in Christoffel Symbols:

$$a3a_{JN}^{\;\;Q}\; aa^{NK} = a3a_{J}^{\;KQ} \qquad a3a_{JK\,Q}\; aa^{NK} = a3a_{J\;Q}^{\;K}$$

Symmetry with respect to first two indices

$$a3a_{J}^{\;KQ} = a3a_{K}^{\;JQ} \qquad a3a_{J\;Q}^{\;K} = a3a_{K\;Q}^{\;J}$$

Raising of First Index in Christoffel Symbols:

$$aa^{IJ}\ a3a_{JN}{}^{Q}\ aa^{NK} = a3a^{IKQ} \qquad aa^{IJ}\ a3a_{JKQ}\ aa^{NK} = a3a^{IN}{}_{Q}$$

Symmetry with respect to first two indices

$$a3a^{IKQ} = a3a^{KIQ} \qquad a3a^{IN}{}_{Q} = a3a^{NI}{}_{Q}$$

Symmetry of indices:

$$a3a_{JK}{}^{Q} = a3a_{KJ}{}^{Q} \qquad a3a_{JKQ} = a3a_{KJQ}$$

$$a3a^{I}{}_{N}{}^{Q} = a3a^{N}{}_{I}{}^{Q} \qquad a3a^{I}{}_{KQ} = a3a^{K}{}_{IQ}$$

$$a3a_{J}{}^{KQ} = a3a_{K}{}^{JQ} \qquad a3a_{J}{}^{K}{}_{Q} = a3a_{K}{}^{J}{}_{Q}$$

$$a3a^{IKQ} = a3a^{KIQ} \qquad a3a^{IN}{}_{Q} = a3a^{NI}{}_{Q}$$

<u>Following articles</u> are Discusseed In Terms of
Trio Symbol (Christoffel)

6.6.2.
Partial Derivative of Metrics With Respect to *Upper Position*:

Partial Derivative of
Lower and *Upper* Metrics
With respect to <u>*Upper Position*</u>

The following

$$[aa_{MN}]_{,K} = +\ [a3a_{KMN} + a3a_{NKM}]$$

$$[aa^{MN}]_{,K} = -\ [a3a_{K}{}^{MN} + a3a_{N}{}^{KM}]$$

Will be discussed in what follows:

6.6.2.1.
Partial Derivative of *Lower* Metric With Respect to *Upper Position*

Add:

$$\tfrac{1}{2}(aa_{NJ,L} + aa_{NL,J} - aa_{JL,N}) = a3a_{JLN}$$

$$\tfrac{1}{2}(aa_{LJ,N} + aa_{LN,J} - aa_{JN,L}) = a3a_{JNL}$$

To get:

$$0 + aa_{NL,J} - 0 = a3a_{JLN} + a3a_{JNL}$$

:

$$\partial aa_{MN}/\partial a^{K} \equiv aa_{MN,K}$$

:

Partial Derivative of Lower Metric With respect to a^{K} <u>*Upper Position*</u>:

$$[aa_{MN}]_{,K} = +\ [a3a_{KMN} + a3a_{NKM}]$$

6.6.2.2.
Partial Derivative of *Upper* Metric With Respect to *Upper Position*:

Upper a^q P.D. of inners of upper and lower metrics:

$$aa^{jl}\ aa_{nl} = \delta^j_n$$

Partial Derivative is:

$$aa^{jl}_{,q}\ aa_{nl} + aa^{jl}\ aa_{nl,q} = 0$$

$$aa^{jl}_{,q}\ aa_{nl} = -\ aa^{jl}\ aa_{nl,q}$$

$$aa^{j}_{n,q} = -\ aa^{j}_{n,q}$$

Or Multiply

$$aa^{jl}_{,q}\ aa_{nl} = -\ aa^{jl}\ aa_{nl,q}$$

*

$$aa^{ns}$$

To get:

$$aa^{jl}_{,q}\ aa_{nl}\ \ aa^{ns} = -\ aa^{ns}\ \ aa^{jl}\ \ aa_{nl,q}$$

$$aa^{jl}_{,q}\ \delta_l^{\ s} = -\ \ aa^{ns}\ aa^{jl}\ \ aa_{nl,q}$$

But:

$$aa_{nl,q} = a3a_{qln} + \ \ a3a_{qnl} = aa_{nj}\ \ a3a_{ql}^{\ \ j} + aa_{lj}\ \ a3a_{qn}^{\ \ j}$$

So:

$$aa^{js}_{,q} \equiv aa^{js}_{,q} = -\ aa^{ns}\ aa^{jl}\ (a3a_{qln} + a3a_{qnl}\ )$$

$$aa^{js}_{,q} \equiv aa^{js}_{,q} = -\ \ aa^{jl}\ \ a3a_{ql}^{\ \ s} -\ aa^{ns}\ \ \ a3a_{qn}^{\ \ j}$$

$$aa^{js}_{,q} \equiv aa^{js}_{,q} = -\ \ \ a3a_{q}^{\ \ js} \ \ -\ \ \ a3a_{q}^{\ \ sj}$$

So:

$$\partial aa^{MN} / \partial a^K$$

Partial Derivative of Upper Metric With respect to a^K <u>Upper Position</u>:

$$[aa^{MN}]_{,K} = -\ [\ a3a_K^{\ \ MN} + a3a_N^{\ \ KM}\ ]$$

6.6.2.3.
Partial Derivative of *Lower* and *Upper* Metrics With respect to *Upper Position*

Both Results:

Partial Derivative of
Lower and *Upper* Metrics
With respect to <u>Upper Position</u>

$$[aa_{MN}]_{,K} = +\ [\ a3a_{KMN} + a3a_{NKM}]$$
$$[aa^{MN}]_{,K} = -\ [\ a3a_K^{\ \ MN} + a3a_N^{\ \ KM}\ ]$$

6.7.
Partial Derivative of Bases with respect to Upper Position:

Partial Derivative of
Lower Basis and *Upper* Basis w.r.t. to <u>Upper Position</u>:

$$\underline{a}_M \, aa_{NK}^{\;\;\;\;M} \equiv \underline{a}_{N,K} \equiv \underline{a}^M \, aa_{NKM}$$

$$\underline{a}_M \, aa^{N\;\;M}_{\;\;K} \equiv (-\underline{a}^N_{\;,K}) \equiv \underline{a}^M \, aa^N_{\;\;KM}$$

:

Partial Derivative

$$\underline{a}^{,n} = \partial(\underline{a}^i)/\partial a_n \, a_i + \underline{a}^i \, \partial a_i / \partial a_n \qquad \underline{a}_{,n} = \partial(\underline{a}_i)/\partial a^n \, a^i + \underline{a}_i \, \partial a^i / \partial a^n$$

$$\underline{a}^{,n} = \underline{a}^{i,n} \, a_i + \underline{a}^i \, \delta_i^{\;n} \qquad \underline{a}_{,n} = \underline{a}_{i,n} \, a^i + \underline{a}_i \, \delta^i_{\;n}$$

$$\underline{a}^{,n} = \underline{a}^{i}{}_{,}{}^{n} \, a_i + \underline{a}^n \qquad \underline{a}_{,n} = \underline{a}_{i,n} \, a^i + \underline{a}_n$$

Second Partial Derivative

$$\underline{a}^{,mn} = \underline{a}^{i,mn} a_i + \underline{a}^{i,m} a_i{}_{,}{}^{n} + \underline{a}^{m,n} \neq \underline{a}_{,mn} = \underline{a}_{i,mn} a^i + \underline{a}_{i,m} a^i_{\;,n} + \underline{a}_{m,n}$$

And:

$$\underline{a}^{m,n} \neq \underline{a}_{m,n}$$

P.D. of A Position Vector of Fixed Bases:
For Fixed Bases:

$$\underline{a}^{i,n} = 0 \qquad \underline{a}_{i,n} = 0$$

Then:

$$\underline{a}^{,n} = \underline{a}^{i}{}_{,}{}^{n} \, a_i + \underline{a}^n = \underline{a}^n \qquad \underline{a}_{,n} = \underline{a}_{i,n} \, a^i + \underline{a}_n = \underline{a}_n$$

Now:
Let:

$$\underline{o} = \underline{o}^i \, o_i \qquad \underline{o} = \underline{o}_i \, o^i$$

Then:

$$\underline{o}^{,n} = (\underline{o}^i)^{,n} \, o_i + (\underline{o}^i) \, o_i^{\;,n} \neq \underline{o}_{,n} = (\underline{o}_i)_{,n} \, o^i + \underline{o}_i \, o^i_{\;,n}$$

But:

$$(\underline{o}^i)^{,n} = 0 \qquad \partial(\underline{o}_i)/\partial o^n = 0$$

Then:

$$\underline{o}^{,n} = (\underline{o}^i) \, o_i^{\;,n} \neq \underline{o}_{,n} = (\underline{o}_i) \, o^i_{\;,n}$$

And:

$$\underline{o}^{,mn} = \underline{o}^i \, o_{i,}^{\;\;mn} \neq \underline{o}_{,mn} = \underline{o}_i \, o^{i,}_{\;\;m,n}$$

$$\underline{o}_\mu \, \partial^2 o^\mu / \partial a^m / \partial a^n = \underline{a}_{m,n}$$

$$\underline{o}^\mu \, \partial^2 o_\mu / \partial a_m / \partial a_n = \underline{a}^{m,n}$$

$$\underline{o}_\mu \, \partial(oa^\mu_{\;\;m}) / \partial a^n = \underline{a}_{m,n}$$

$$\underline{o}^\mu \, \partial(oa_\mu^{\;\;m}) / \partial a_n = \underline{a}^{m,n}$$

$$\underline{a}_p \, ao^p_{\;\;M} = \underline{a}_p \, oa^p_{\;\;M} = \underline{o}_M$$

$$\underline{a}^p \, ao^M_{\;\;p} = \underline{a}^p \, oa^M_{\;\;p} = \underline{o}^M$$

$$\underline{a}_p \, o^p_{\;\;\mu} \, \partial oa^\mu_{\;\;m} / \partial a^n = \underline{a}_{m,n}$$

$$\underline{a}^p \, oa^\mu_{\;\;p} \, \partial oa_\mu^{\;\;m} / \partial a_n = \underline{a}^{m,n}$$

Partial Derivative of *Lower* Basis and *Upper* Basis w.r.t. to <u>Upper Position</u>

6.7.1.
Partial Derivative of *Lower* Basis With Respect to *Upper Position*:

Partial Derivative of *Lower* Basis w.r. to *Upper Position*

Define aa_{NKM} :

$$\underline{a}^{M}\; aa_{NKM} \equiv \underline{a}_{N,K}$$

Differential-Previous Definition is inner Differential-Premultiplied by $\underline{a}^{n}$:

$$\underline{a}^{n} \cdot \underline{a}^{M}\; aa_{NKM} = \underline{a}^{n} \cdot \underline{a}_{N,K}$$

$$(aa^{nM}\; aa_{NKM}) = \underline{a}^{n} \cdot \underline{a}_{N,K}$$

$$(aa_{NK}^{\;\;n}) = \underline{a}^{n} \cdot \underline{a}_{N,K}$$

$$(aa_{NK}^{\;\;n}) = (\delta^{n}_{M}\; aa_{NK}^{\;\;M}) = (\underline{a}^{n} \cdot \underline{a}_{M}\; aa_{NK}^{\;\;M}) = \underline{a}^{n} \cdot \underline{a}_{N,K}$$

So Equal Factors of $\underline{a}^{n}$:

$$\underline{a}_{M}\; aa_{NK}^{\;\;M} \equiv \underline{a}_{N,K} \equiv \underline{a}^{M}\; aa_{NKM}$$

Which Define $aa_{NK}^{\;\;M}$ and aa_{NKM}

6.7.2.
Partial Derivative of *Upper* Basis With Respect to *Upper Position*:

Partial Derivative of *Upper* Basis w.r. to *Upper Position*:

Define $aa^{N\;\;M}_{\;\;K}$:

$$\underline{a}_{M}\; aa^{N\;\;M}_{\;\;K} = (-\,\underline{a}^{N}_{\;\;,K})$$

Differential-Previous Definition is inner Differential-Premultiplied by $\underline{a}_{n}$:

$$\underline{a}_{n} \cdot \underline{a}_{M}\; aa^{N\;\;M}_{\;\;K} = \underline{a}_{n} \cdot (-\,\underline{a}^{N}_{\;\;,K})$$

$$(aa_{nM}\; aa^{N\;\;M}_{\;\;K}) = \underline{a}_{n} \cdot (-\,\underline{a}^{N}_{\;\;,K})$$

$$(aa^{N}_{\;\;Kn}) = \underline{a}_{n} \cdot (-\,\underline{a}^{N}_{\;\;,K})$$

$$(aa^{N}_{\;\;Kn}) = (\delta^{M}_{n}\; aa^{N}_{\;\;KM}) = (\underline{a}_{n} \cdot \underline{a}^{M}\; aa^{N}_{\;\;KM}) = \underline{a}_{n} \cdot (-\,\underline{a}^{N}_{\;\;,K})$$

So Equal Factors of $\underline{a}^{n}$:

$$\underline{a}_{M}\; aa^{N\;\;M}_{\;\;K} \equiv (-\,\underline{a}^{N}_{\;\;,K}) \equiv \underline{a}^{M}\; aa^{N}_{\;\;KM}$$

Both past Results:
Partial Derivative of
Lower Basis and *Upper* Basis w.r. to *Upper Position*:

$$\underline{a}_{M}\; aa_{NK}^{\;\;M} \equiv \underline{a}_{N,K} \equiv \underline{a}^{M}\; aa_{NKM}$$

$$\underline{a}_{M}\; aa^{N\;\;M}_{\;\;K} \equiv (-\,\underline{a}^{N}_{\;\;,K}) \equiv \underline{a}^{M}\; aa^{N}_{\;\;KM}$$

Partial Derivative w.r. to a^{k} of:

$$\underline{a}^{M} \cdot \underline{a}_{N} = \delta^{M}_{N}$$

Is:

$$\underline{a}^{M}_{,K} \cdot \underline{a}_{N} + \underline{a}^{M} \cdot \underline{a}_{N,K} = 0$$

$$(-\,\underline{a}^{M}_{,K}) \cdot \underline{a}_{N} = \underline{a}^{M} \cdot \underline{a}_{N,K}$$

$$\underline{a}^{n}\, aa^{M}_{Kn} \cdot \underline{a}_{N} = \underline{a}^{M} \cdot \underline{a}_{m}\; aa^{m}_{NK}$$

$$\delta^{n}_{N}\, aa^{M}_{Kn} = \delta^{M}_{m}\; aa^{m}_{NK}$$

$$[aa^{M}_{KN}] = [aa^{M}_{NK}]$$

$[\,{}^{M}\backslash_{K\to N}]$ symmetry about equal sign $[\,{}_{N\leftarrow K}/^{M}]$

6.8.
Partial Derivative of Lower Metric Determinant:

Partial Derivative of
Lower Metric Determinant
With respect to Both:
Lower Metric and <u>*Upper Position*</u>:

6.8.1.
Partial Derivative of *Lower* Metric Determinant with respect to *Lower* Metric:

AA is Cofactor of aa

$$AA^{nI} \equiv \|aa_{..}\|\, aa^{nI} = (\alpha\alpha)\, aa^{nI} \qquad AA_{nI} \equiv \|aa^{..}\|\, aa_{nI} = 1/(\alpha\alpha)\, aa_{nI}$$

$$aa^{nI} = AA^{nI}/\|aa_{..}\| = AA^{nI}/(\alpha\alpha) \qquad aa_{nI} = AA_{nI}/\|aa^{..}\| = AA_{nI} * (\alpha\alpha)$$

$$\delta_{m}^{n} = aa_{mI}\, aa^{nI} = aa_{mI}\, AA^{nI}/(\alpha\alpha) \qquad \delta^{m}_{n} = aa^{mI}\, aa_{nI} = aa^{mI}\, AA_{nI} * (\alpha\alpha)$$

$$aa_{mI}\, AA^{nI} = \alpha\alpha\; \delta^{m}_{n} = \alpha^{2}\, \delta^{m}_{n} \qquad aa^{mI}\, AA_{nI} = 1/\alpha\alpha\; \delta^{m}_{n} = 1/\alpha^{2}\, \delta^{m}_{n}$$

Since: $\delta^{\underline{m}}_{\underline{m}} = 1$ then:

$$aa_{\underline{m}I} * AA^{\underline{m}I} = (\alpha\alpha) \qquad aa^{\underline{m}I} * AA_{\underline{m}I} = 1/(\alpha\alpha)$$

And:

$$\partial(\alpha\alpha)/\partial aa_{mJ} = \partial aa_{\underline{m}I}/\partial aa_{mJ} * AA^{\underline{m}I} + aa_{\underline{m}I} * \partial AA^{\underline{m}I}/\partial aa_{mJ}$$

$$\partial[1/(\alpha\alpha)]/\partial aa^{mJ} = \partial aa^{\underline{m}I}/\partial aa^{mJ} * AA_{\underline{m}I} + aa^{\underline{m}I} * \partial AA_{\underline{m}I}/\partial aa^{mJ}$$

$(AA^{\underline{m}I})$ contains no (aa_{mJ}) and $(AA_{\underline{m}I})$ contains no (aa^{mJ}) then: $\partial AA/\partial aa = 0$:

$\partial AA^{mI}/\partial aa_{mJ}=\partial[(\alpha\alpha)aa^{mI}]/\partial aa_{mJ}=0$ $\qquad$ $\partial AA_{mI}/\partial aa^{mJ}=\partial[1/(\alpha\alpha)aa_{mI}]/\partial aa^{mJ}=0$

And:

$\partial aa_{mI}/\partial aa_{mJ}*AA^{mI} = \delta_I^J*AA^{mI}$ $\qquad$ $\partial aa^{mI}/\partial aa^{mJ}*AA_{mI} = \delta^I_J*AA_{mI}$

▼

$\partial(\alpha\alpha)/\partial aa_{mJ} = AA^{mJ} = (\alpha\alpha)\, aa^{mJ}$ $\qquad$ $\partial[1/(\alpha\alpha)]/\partial aa^{mJ}=AA_{mJ}=1/(\alpha\alpha)aa_{mJ}$

$\partial\|aa_{..}\|/\partial aa_{mJ} = AA^{mJ} = \|aa_{..}\|\, aa^{mJ}$ $\qquad$ $\partial\|aa^{..}\|/\partial aa^{mJ} = AA_{mJ} = \|aa^{..}\|\, aa_{mJ}$

Or:

P.D. of *Lower* Metric Determinant w.r.to *Lower* Metric:

▼

$\partial(\alpha\alpha)/\partial aa_{mJ}/(\alpha\alpha) = aa^{mJ}$ $\qquad$ $\partial[1/(\alpha\alpha)]/\partial aa^{mJ}/[1/(\alpha\alpha)] =aa_{mJ}$

$\partial[Log(\alpha^2)]/\partial aa_{mJ} = aa^{mJ}$ $\qquad$ $\partial\{Log[1/(\alpha\alpha)]\}/\partial aa^{mJ} = aa_{mJ}$

6.8.2.
Partial Derivative of *Lower* Metric Determinant with respect to *Upper Position*:

Lower Metric Determinant P.D. with respect to Upper Position
In terms of: Symmetric
Lower Metric Determinant P.D. with respect to Lower Metric:

$$(\alpha\alpha)_{,N} \equiv \partial(\alpha\alpha)/\partial a^N = \partial(\alpha\alpha)/\partial aa_{JL}\ \partial aa_{JL}/\partial a^N = AA^{JL}\ aa_{JL,N}$$

But:

$$AA^{JL} = (\alpha\alpha)\ aa^{JL}$$

So:

$$(\alpha\alpha)_{,N} = (\alpha\alpha)\ aa^{JL}\ aa_{JL,N}$$

$$(\alpha\alpha)_{,N}=(\alpha\alpha)*aa^{JL}(a3a_{NL}^{\ \ Q}+a3a_{NJL})= (\alpha\alpha)*aa^{JL}(aa_{QJ}a3a_{NL}^{\ \ Q}+aa_{QL}a3a_{NJ}^{\ \ Q})$$

$$(\alpha\alpha)_{,N}=(\alpha\alpha)(\delta^L_Q a3a_{NL}^{\ \ Q}+\delta^J_Q a3a_{NJ}^{\ \ Q})=(\alpha\alpha)(a3a_{NQ}^{\ \ Q}+a3a_{NQ}^{\ \ Q})$$

So:

$$(\alpha\alpha)_{,N} \equiv \partial(\alpha\alpha)/\partial a^N = 2*\ (\alpha\alpha)\ * a3a_{NQ}^{\ \ Q}$$

P.D. of Lower Metric Determinant w.r. to Upper Position:

▼

$$(\alpha\alpha)_{,N}/[\ 2*(\alpha\alpha)\ ] \equiv [\partial(\alpha\alpha)/\partial a^N]/[\ 2*(\alpha\alpha)\ ] = a3a_{NQ}^{\ \ Q}$$

$$\partial[Log\ (\|aa_{nl}\|)^{1/2}]/\partial a^N \equiv \{Log\ (\alpha\alpha)^{1/2}\}_{,N} \equiv [Log(\alpha)]_{,N} = a3a_{NQ}^{\ \ Q} \equiv a1a_N$$

Results of Partial Derivatives:

With respect to	Partial Derivative of	
Upper Position a^K	Lower Metric aa_{MN}	$aa_{MN,K} =+a3a_{KNM}+a3a_{MKN}$
Upper Position a^K	Upper Metric aa^{MN}	$aa^{MN}_{\ \ ,K}=-a3a_K^{\ MN}-a3a_K^{\ NM}$
Upper Position a^K	Covariant Basis $\underline{a}_N$	$\underline{a}_{N,K} \equiv \underline{a}_M\, aa^M_{\ NK} \equiv \underline{a}^M\, aa_{MNK}$
Upper Position a^K	Contravariant Basis $\underline{a}^N$	$(-\underline{a}^N_{\ ,K}) \equiv \underline{a}_M\, aa^{NM}_{\ \ K} \equiv \underline{a}^M\, aa^N_{\ MK}$
Upper Position a^K	Lower Metric Determinant	► $[Log(\alpha)]_{,N} = a3a_{NQ}^{\ \ Q} \equiv a1a_N$

Lower Metric aa_{MN} Lower Metric Determinant ► $\partial[\,\mathrm{Log}(\alpha^2)]/\partial aa_{mJ} = aa^{mJ}$

6.9.
Intrinsic Derivative Is With Respect to Upper Position:

P.D. of *Lower* Basis and *Upper* Basis w.r. to Upper Position:

$$\underline{a}_M\, aa^M_{NK} \equiv \underline{a}_{N,K}$$

$$(-\underline{a}^N_{,K}) \equiv \underline{a}^M\, aa^N_{KM}$$

:

A Vector $\underline{V}$:

$$\underline{V} = V^M \underline{a}_M \qquad \underline{V} = V_M\, \underline{a}^{M,}$$

This Vector has a Partial Derivative:

$$\underline{V}_{,K} = V^M_{,K}\, \underline{a}_M + V^M \underline{a}_{M,K} \qquad \underline{V}_{,K} = V_{M,K}\, \underline{a}^M + V_M\, \underline{a}^M_{,K}$$

$$\underline{V}_{,K} = V^M_{,K}\, \underline{a}_M + V^M \underline{a}_N\, aa^N_{MK} \qquad \underline{V}_{,K} = V_{M,K}\, \underline{a}^M - V_M\, \underline{a}^N\, aa^M_{KN}$$

$$\underline{V}_{,K} = V^N_{,K}\, \underline{a}_N + V^M \underline{a}_N\, aa^N_{MK} \qquad \underline{V}_{,K} = V_{N,K}\, \underline{a}^N - V_M\, \underline{a}^N\, aa^M_{KN}$$

$$\underline{V}_{,K} = \underline{a}_N (V^N_{,K} + V^M aa^N_{MK}) \qquad \underline{V}_{,K} = \underline{a}^N (V_{N,K} - V_M\, aa^M_{KN})$$

Define Intrinsic Derivative:

$$V^N|_K \equiv \delta V^N/\delta a^K \equiv V^N_{,K} + V^M aa^N_{MK} \qquad V_N|_K \equiv \delta V_N/\delta a^K \equiv V_{N,K} - V_M\, aa^M_{KN}$$

Then:

$$\underline{V}_{,K} = \underline{a}_N V^N|_K \qquad \underline{V}_{,K} = \underline{a}^N V_N|_K$$

Define Tensor *Intrinsic* Derivative:

$$\delta[T(a)^\upsilon]/\delta a^\delta\, da^\delta/dt \equiv T(a)^\upsilon|_\delta\, aT^\delta \qquad \delta[T(a)_\lambda]/\delta a^\delta\, da^\delta/dt \equiv T(a)_\lambda|_\delta\, aT^\delta$$

$$T(a)^\upsilon|_\delta \equiv T(a)^\upsilon_{,\delta} + T(a)^p\, a3a^\upsilon_{p\delta} \qquad T(a)_\lambda|_\delta \equiv T(a)_{\lambda,\delta} - T(a)_m\, a3a^m_{\lambda\delta}$$

And Mixed Tensor Intrinsic Derivative:

$$\delta[T(a)_\lambda^\upsilon])/\delta a^\delta\, da^\delta/dt \equiv T(a)_\lambda^\upsilon|_\delta\, aT^\delta$$

$$T(a)_\lambda^\upsilon|_\delta \equiv T(a)_{\lambda,\delta}^\upsilon - T(a)_m^\upsilon\, a3a^m_{\lambda\delta} + T(a)_\lambda^p\, a3a^\upsilon_{p\delta}$$

With respect to Upper Position a^K Partial Derivative of a Vector, $\underline{V}$

$$\underline{V} = \underline{a}_m V^m \qquad \underline{V} = \underline{a}^m V_m$$

:

$$\underline{V}_{,k} = \underline{a}_m V^{m,k} + \underline{a}_{i,k} V^i \qquad \underline{V}_{,k} = \underline{a}^m\, \partial V_m/\partial a^k + \underline{a}^n_{,k} V_n$$

$$= \underline{a}_m V^m_{,k} + \underline{a}_m\, aa^m_{ik} V^i \qquad = \underline{a}^m V_{m,k} - \underline{a}^m\, aa^n_{mk} V_n$$

$$= \underline{a}_m (V^m_{,k} + aa^m_{ik} V^i) \qquad = \underline{a}^m (V_{m,k} - aa^n_{mk} V_n)$$

$$\equiv \underline{a}_m V^m|_k \equiv \underline{a}_m\, \delta V^m/\delta a^k \qquad \equiv \underline{a}^m V_m|_k \equiv \underline{a}^m\, \delta V_m/\delta a^k$$

With respect to *Upper Position* a^K Intrinsic Derivative of a Tensor, T :

$$T^m|_k \equiv T^m_{,k} + T^n\ a3a_{nk}^{\ \ m} \qquad\qquad T_m|_k \equiv T_{m,k} - T_n\ a3a_{mk}^{\ \ n}$$

With respect to *Upper Position*
Intrinsic Derivative of a Mixed Tensor, $T(a)_\lambda^{\ \upsilon}$:

$$T(a)_\lambda^{\ \upsilon}|_\delta \equiv T(a)_{\lambda\ ,\delta}^{\ \upsilon} - T(a)_m^{\ \upsilon}\ a3a_{\lambda\delta}^{\ \ m} + T(a)_m^{\ p}\ a3a_{p\delta}^{\ \ \upsilon}$$

Time,T Derivative of a Tensor, $T(a)_\lambda^{\ \upsilon}$ is the Product of
Intrinsic Derivative and Coordinate Time Derivative:

$$\delta[T(a)_\lambda^{\ \upsilon}] / \delta a^\delta \ * \ da^\delta/dt \equiv T(a)_\lambda^{\ \upsilon}|_\delta\ \ aT^\delta$$

6.10.
Transformation of Partial Derivative of Lower Metric With Resprct to Upper Position:

$$aa_{\alpha\beta}\ ab^\alpha_{\ a}\ ab^\beta_{\ b} = bb_{ab}$$

$$bb_{ab} = ba_a^{\ \alpha}\ ba_b^{\ \beta}\ aa_{\alpha\beta}$$

$$aa_{nl,j} = a3a_{jln} + a3a_{jnl} = aa_{nq}\ a3a_{jl}^{\ \ q} + aa_{lq}\ a3a_{jn}^{\ \ q}$$

$\partial/\partial y^j$ In terms of : $\partial/\partial x^n$:
P.D. of lower metric: yy_{nl} does not transform like tensor:

$$bb_{ab,j} = \partial bb_{ab}/\partial b^j$$

.Not Equal.

$$b3b_{jba} + b3b_{jab} = bb_{aq}\ b3b_{jb}^{\ \ q} + bb_{bq}\ b3b_{ja}^{\ \ q}$$

$$bb_{ab} = ab^\alpha_{\ a}\ ab^\beta_{\ b}\ aa_{\alpha\beta}$$

$$bb_{ab} = ba_a^{\ \alpha}\ ba_b^{\ \beta}\ aa_{\alpha\beta}$$

:

$$bb_{ab,j} = + \partial/\partial b^j [ab^\alpha_{\ a}]\ ab^\beta_{\ b}\ aa_{\alpha\beta} + ab^\alpha_{\ a}\ \partial/\partial b^j [ab^\beta_{\ b}])\ aa_{\alpha\beta} + ab^\alpha_{\ a}\ ab^\beta_{\ b}\ \partial/\partial b^j [aa_{\alpha\beta}]$$

$$bb_{ab,j} = + \partial/\partial b^j [ba_a^{\ \alpha}]\ ba_b^{\ \beta}\ aa_{\alpha\beta} + ba_a^{\ \alpha}\ \partial/\partial b^j [ba_b^{\ \beta}])\ aa_{\alpha\beta} + ba_a^{\ \alpha}\ ba_b^{\ \beta}\ \partial/\partial b^j [aa_{\alpha\beta}]$$

But:

$$aa_{\alpha\beta:j} \equiv \partial/\partial b^j [aa_{\alpha\beta}] = \partial/\partial a^k [aa_{\alpha\beta}]\ \partial a^k/\partial b^j = ab^k_{\ j}\ aa_{\alpha\beta k}$$

:

$$bb_{ab,j} = + ab^{\alpha}_{a,j}\, ab^{\beta}_{b}\, aa_{\alpha\beta} + ab^{\alpha}_{a}\, ab^{\beta}_{b,j}\, aa_{\alpha\beta} + ab^{\alpha}_{a}\, ab^{\beta}_{b}\, ab^{k}_{j}\, aa_{\alpha\beta,k}$$

$$bb_{ab,j} = + ba_{a}{}^{\alpha}{}_{:j}\, ba_{b}{}^{\beta}\, aa_{\alpha\beta} + ba_{a}{}^{\alpha}\, ba_{b}{}^{\beta}{}_{:j}\, aa_{\alpha\beta} + ba_{a}{}^{\alpha}\, ba_{b}{}^{\beta}\, ab^{k}_{j}\, aa_{\alpha\beta,k}$$

:

$$bb_{ab,j} = + ab^{\alpha}_{a,j}\, ab^{\beta}_{b}\, aa_{\alpha\beta} + ab^{\alpha}_{b,j}\, ab^{\beta}_{a}\, aa_{\alpha\beta} + ab^{\alpha}_{a}\, ab^{\beta}_{b}\, ab^{k}_{j}\, aa_{\alpha\beta,k}$$

$$bb_{ab,j} = + ba_{a}{}^{\alpha}{}_{:j}\, ba_{b}{}^{\beta}\, aa_{\alpha\beta} + ba_{b}{}^{\alpha}{}_{:j}\, ba_{a}{}^{\beta}\, aa_{\alpha\beta} + ba_{a}{}^{\alpha}\, ba_{b}{}^{\beta}\, ab^{k}_{j}\, aa_{\alpha\beta,k}$$

So :

$$aa_{nl,j} = a3a_{jln} + a3a_{jnl} = aa_{nq}\, a3a_{jl}{}^{q} + aa_{lq}\, a3a_{jn}{}^{q}$$

:

Transformation of P.D. of Lower Metric w.r. to <u>*Upper Position*</u>

$$bb_{ab,j} = ab^{\alpha}_{a,j}\, ab^{\beta}_{b}\, aa_{\alpha\beta} + ab^{\alpha}_{b,j}\, ab^{\beta}_{a}\, aa_{\alpha\beta} + ab^{\alpha}_{a}\, ab^{\beta}_{b}\, ab^{k}_{j}\, aa_{\alpha\beta,k}$$

6.11.
Transformation of Trio Symbol:

$$bb_{ab,j} = ab^{\alpha}_{a,j}\, ab^{\beta}_{b}\, aa_{\alpha\beta} + ab^{\alpha}_{b,j}\, ab^{\beta}_{a}\, aa_{\alpha\beta} + ab^{\alpha}_{a}\, ab^{\beta}_{b}\, ab^{k}_{j}\, aa_{\alpha\beta,k}$$

:

$$b3b_{abm} = 1/2\ (bb_{am,b} + bb_{mb,a} - bb_{ab,m})$$

:

$$bb_{am,b} = ab^{\alpha}_{a,b}\, ab^{\beta}_{m}\, aa_{\alpha\beta} + ab^{\alpha}_{m,b}\, ab^{\beta}_{a}\, aa_{\alpha\beta} + ab^{\alpha}_{a}\, ab^{\beta}_{m}\, ab^{k}_{b}\, aa_{\alpha\beta,k}$$

$$bb_{mb,a} = ab^{\alpha}_{m,a}\, ab^{\beta}_{b}\, aa_{\alpha\beta} + ab^{\alpha}_{b,a}\, ab^{\beta}_{m}\, aa_{\alpha\beta} + ab^{\alpha}_{m}\, ab^{\beta}_{b}\, ab^{k}_{a}\, aa_{\alpha\beta,k}$$

$$bb_{ab,m} = ab^{\alpha}_{a,m}\, ab^{\beta}_{b}\, aa_{\alpha\beta} + ab^{\alpha}_{b,m}\, ab^{\beta}_{a}\, aa_{\alpha\beta} + ab^{\alpha}_{a}\, ab^{\beta}_{b}\, ab^{k}_{m}\, aa_{\alpha\beta,k}$$

:

$$ab^{\alpha}_{a,b} = ab^{\alpha}_{b,a}$$

:

$$1/2(ab^{\alpha}_{a,b}\, ab^{\beta}_{m} + ab^{\alpha}_{m,b}\, ab^{\beta}_{a} + ab^{\alpha}_{m,a}\, ab^{\beta}_{b} + ab^{\alpha}_{b,a}\, ab^{\beta}_{m} - ab^{\alpha}_{a,m}\, ab^{\beta}_{b} - ab^{\alpha}_{b,m}\, ab^{\beta}_{a})$$

$$=$$

$$ab^{\alpha}_{a,b}\, ab^{\beta}_{m}$$

And:

$$(aa_{\alpha k,\beta} + aa_{\alpha k,\beta} - aa_{\alpha\beta,k}) = 2 * a3a_{\alpha\beta k}$$

And:

$$aa_{\alpha\beta,k}\,(ab^{\alpha}_{ab}\, ab^{\beta}_{m}\, ab^{k}_{b} + ab^{\alpha}_{m}\, ab^{\beta}_{b}\, ab^{k}_{ab} - ab^{\alpha}_{ab}\, ab^{\beta}_{b}\, ab^{k}_{m}) =$$

$$aa_{\alpha k,\beta}\, ab^{\alpha}_{ab}\, ab^{k}_{m}\, ab^{\beta}_{b} + aa_{k\beta,\alpha}\, ab^{k}_{m}\, ab^{\beta}_{b}\, ab^{\alpha}_{ab} - aa_{\alpha\beta,k}\, ab^{\alpha}_{ab}\, ab^{\beta}_{b}\, ab^{k}_{m} =$$

$$(aa_{\alpha k,\beta} + aa_{\alpha k,\beta} - aa_{\alpha\beta,k})ab^{\alpha}_{ab}\, ab^{\beta}_{b}\, ab^{k}_{m} = 2\ ab^{\alpha}_{ab}\, ab^{\beta}_{b}\, ab^{k}_{m}\ a3a_{\alpha\beta k}$$

Transformation of Christoffel Third Indexed, Lower:

$$b3b_{abm} = ab^{\alpha}_{a,b}\, ab^{\beta}_{m}\ aa_{\alpha\beta} + ab^{\alpha}_{a}\, ab^{\beta}_{b}\, ab^{k}_{m}\ a3a_{\alpha\beta k}$$

Only the following is a tensor:

$$ab^{\alpha}_{a}\, ab^{\beta}_{b}\, ab^{k}_{m} \quad a3a_{\alpha\beta k}$$

But:

$$ab^{\alpha}_{a,b}\, ab^{\beta}_{m} \quad aa_{\alpha\beta}$$

Is Not a tensor

This is zero when:

$$ab^{\beta}_{m} = 0$$

That is when:

$$ab^{\alpha}_{a,b} = 0$$

ab^{β}_{m} is affine, that is $ba^{n} = C^{n}_{\alpha}\; ab^{\alpha}$ with C^{n}_{α} are constants so: $ab^{\alpha}_{a,b} = 0$

Raising thied index in

$$b3b_{abm} = ab^{\alpha}_{a,b}\, ab^{\beta}_{m} \quad aa_{\alpha\beta} + ab^{\alpha}_{a}\, ab^{\beta}_{b}\, ab^{k}_{m} \quad a3a_{\alpha\beta k}$$

By multiplying by:

*

$$aa_{\alpha\beta}$$

=

$$b3b_{ab}^{\;\;n} = ab^{\alpha}_{a,b}\; ba^{n}_{\alpha} + ab^{\alpha}_{a}\, ab^{\beta}_{b} \quad ba^{n}_{k} \quad a3a^{k}_{\alpha\beta}$$

Transformation of Christoffel Third Indexed to Upper.

Transformation of Christoffel Symbol Third Indexed, Upper:

$$a3a_{ab}^{\;\;q} = aa^{qn} \quad a3a_{abn}$$

:

$$b3b_{ab}^{\;\;q} = bb^{qm} \quad b3b_{abm}$$

:

$$b3b_{abm} = ab^{\alpha}_{a,b}\, ab^{\beta}_{m} \; aa_{\alpha\beta} + ab^{\alpha}_{a}\, ab^{\beta}_{b}\, ab^{k}_{m} \; a3a_{\alpha\beta k}$$

:

$$b3b_{ab}^{\;\;q} = bb^{qm} \quad ab^{\alpha}_{a,b}\, ab^{\beta}_{m} \; aa_{\alpha\beta} + bb^{qm}\, ab^{\alpha}_{a}\, ab^{\beta}_{b}\, ab^{k}_{m} \; a3a_{\alpha\beta k}$$

Consider:

$$bb^{qm} = ba^{q}_{A}\, ba^{m}_{B}\, aa^{AB}$$

Substituted to get:

$$b3b_{ab}^{\;\;q} = ba^{q}_{A}\, ba^{m}_{B}\; aa^{AB}\; ab^{\alpha}_{a,b}\, ab^{\beta}_{m}\; aa_{\alpha\beta} + ba^{q}_{A}\, ba^{m}_{B}\; aa^{AB}\; ab^{\alpha}_{a}\, ab^{\beta}_{b}\, ab^{k}_{m}\; a3a_{\alpha\beta k}$$

$$b3b_{ab}^{\;\;q} = \delta^{\beta}_{B}\, ba^{q}_{A}\, ab^{\alpha}_{a,b}\; aa^{AB}\; aa_{\alpha\beta} + \delta^{k}_{B}\; ba^{q}_{A}\; ab^{\alpha}_{a}\, ab^{\beta}_{b}\; aa^{AB}\; a3a_{\alpha\beta k}$$

$$b3b_{ab}^{\;\;q} = ba^{q}_{A}\, ab^{\alpha}_{a,b}\; aa^{A\beta}\; aa_{\alpha\beta} + ba^{q}_{A}\; ab^{\alpha}_{a}\, ab^{\beta}_{b}\; aa^{Ak}\; a3a_{\alpha\beta k}$$

But:

$$aa^{mi} \quad aa_{in} = \delta^{m}_{n} \qquad\qquad aa_{mi} \quad aa^{in} = \delta^{m}_{n}$$

So:

$$b3b_{ab}^{\;\;q} = ba^{q}_{A}\, ab^{\alpha}_{a,b}\; \delta^{A}_{\alpha} + ba^{q}_{A}\; ab^{\alpha}_{a}\, ab^{\beta}_{b}\; a3a^{A}_{\alpha\beta}$$

Indices change:

$$b3b_{ab}^{\;\;n} = ab^{\alpha}_{a,b}\; ba^{n}_{k} \quad \delta^{k}_{\alpha} + ab^{\alpha}_{a}\, ab^{\beta}_{b} \quad ba^{n}_{k} \quad a3a^{k}_{\alpha\beta}$$

Transformation of Christoffel Symbol Third Indexed, Upper:

$$b3b_{ab}^{\;\;n} = ab^{\alpha}_{a,b}\; ba^{n}_{\alpha} + ab^{\alpha}_{a}\, ab^{\beta}_{b} \quad ba^{n}_{k} \quad a3a^{k}_{\alpha\beta}$$

$$b3b_{ab}^{\;\;n} = ab^{\alpha}_{a,b} \quad ab^{n}_{\alpha} + ab^{\alpha}_{a}\, ab^{\beta}_{b} \quad ba^{n}_{k} \quad a3a^{k}_{\alpha\beta}$$

Only the following is a tensor:

$$ab^{\alpha}_{a}\ ab^{\beta}_{b}\ ab^{k}_{m}\quad a3a_{\alpha\beta k}$$

But:

$$ab^{\alpha}_{a,b}\ ba^{n}_{k}\quad \delta^{k}_{\alpha} = ab^{\alpha}_{a,b}\ ba^{n}_{\alpha} = ab^{\alpha}_{a,b}\ ba^{n}_{\alpha}$$

Is Not a tensor

This is zero when:

$$ba^{n}_{\alpha} = 0$$

That is when:

$$ab^{\alpha}_{a,b} = 0$$

ab^{β}_{m} is affine, that is $ba^{n} = C^{n}_{\alpha}\ ab^{\alpha}$ with C^{n}_{α} are constants so: $ab^{\alpha}_{a,b} = 0$

Lowering third index of:

$$b3b_{ab}^{n} = ab^{\alpha}_{a,b}\ ab^{n}_{\alpha} + ab^{\alpha}_{a}\ ab^{\beta}_{b}\quad ba^{n}_{k}\quad a3a^{k}_{\alpha\beta}$$

By multiplying by:

*

$$aa_{\alpha\beta}$$

=

$$b3b_{abm} = ab^{\alpha}_{a,b}\ ab^{\beta}_{m}\quad aa_{\alpha\beta} + ab^{\alpha}_{a}\ ab^{\beta}_{b}\ ab^{k}_{m}\quad a3a_{\alpha\beta k}$$

Transformation of Christoffel Third Indexed, Lower:

6.12.
Second P.D. of Upper Position With Respect to Upper Position

:

Later on, the following is needed in Transformation:

Second P.D. of <u>Upper Position</u>

With Respect to <u>Upper Position</u> :: $[ab^{\mu}_{,M}]_{N} \equiv ab^{\mu}_{,MN}$

Transformation of symbols:

$$b3b_{abm} = ab^{\alpha}_{a,b}\quad ab^{\beta}_{m}\quad aa_{\alpha\beta} + ab^{\alpha}_{a}\ ab^{\beta}_{b}\quad ab^{k}_{m}\quad a3a_{\alpha\beta k}$$

$$b3b_{ab}^{n} = ab^{\alpha}_{a,b}\quad ba^{n}_{k}\quad \delta^{k}_{\alpha} + ab^{\alpha}_{a}\ ab^{\beta}_{b}\quad ba^{n}_{k}\quad a3a^{k}_{\alpha\beta}$$

Discussion of:

$$ab^{\alpha}_{a,b}$$

If

$$b3b_{ab}^{n} = ab^{\alpha}_{a,b}\quad ba^{n}_{\alpha} + ab^{\alpha}_{a}\ ab^{\beta}_{b}\quad ba^{n}_{k}\quad a3a^{k}_{\alpha\beta}$$

Is multiplied by: ab^{γ}_{n} to get:

$$b3b_{ab}^{n}\ ab^{\gamma}_{n} = ab^{\alpha}_{a,b}\quad ba^{n}_{\alpha}\quad a^{\gamma}_{n} + ab^{\alpha}_{a}\ ab^{\beta}_{b}\quad ba^{n}_{k}\quad a3a^{k}_{\alpha\beta}\quad ab^{\gamma}_{n}$$

$$b3b_{ab}^{n}\ ab^{\gamma}_{n} = ab^{\alpha}_{a,b}\quad \delta^{\gamma}_{\alpha} + ab^{\alpha}_{a}\ ab^{\beta}_{b}\quad \delta^{\gamma}_{k}\quad a3a^{k}_{\alpha\beta}$$

Or:

$$b3b_{ab}^{n}\ ab^{\gamma}_{n} = ab^{\gamma}_{a,b} + ab^{\alpha}_{a}\ ab^{\beta}_{b}\quad a3a^{\gamma}_{\alpha\beta}$$

Second P.D.

Of

$$ab^{\gamma} = a^{\gamma}(b)$$

with respect to b:

$$[ab^{\gamma}_{a}]_{,b} = b3b_{ab}{}^{n}\, ab^{\gamma}_{n} - ab^{\alpha}_{a}\, ab^{\beta}_{b}\, a3a_{\alpha\beta}{}^{\gamma} = [ba_{a}{}^{\gamma}]_{,b}$$

Also:

Second P.D.

Of

$$ba^{\gamma} = b^{\gamma}(a)$$

with respect to b:

$$[ba^{\gamma}_{a}]_{,b} = a3a_{ab}{}^{n}\, ba^{\gamma}_{n} - ba^{\alpha}_{a}\, ba^{\beta}_{b}\, b3b_{\alpha\beta}{}^{\gamma} = [ab_{a}{}^{\gamma}]_{,b}$$

6.12.1.
Transformation of Tensor Intrinsic Derivatives:

Transformation of Intrinsic Derivative of *Lower and Upper* Tensors w.r. to Upper Position

6.12.1.1.
Transformation of Intrinsic Derivative of *Lower* Tensor

From these:

$$T(b)_{M,N} = T(a)_{\mu}\, [ab^{\mu}_{M}]_{,N} + T(a)_{\mu,k}\, ab^{\mu}_{M}\, ab^{k}_{N}$$

$$T(b)_{M,N} = [ba_{M}{}^{\mu}]_{,N}\, T(a)_{\mu} + ba_{M}{}^{\mu}\, ba_{N}{}^{k}\, T(a)_{\mu,k}$$

$$T(b)^{M}{}_{,N} = T(a)^{\mu}\, [ab^{\mu}_{M}]_{,N} + T(a)^{\mu}{}_{,k}\, ab_{\mu}{}^{M}\, ab^{k}_{N}$$

$$T(b)^{M}{}_{,N} = [ba^{M}{}_{\mu}]_{,N}\, T(a)^{\mu} + ba^{M}{}_{\mu}\, ba_{N}^{k}\, T(a)^{\mu}{}_{,k}$$

Select:

$$T(b)_{A,B} = T(a)_{\mu}\, [ab^{\mu}_{A}]_{,B} + T(a)_{\mu,k}\, ab^{\mu}_{A}\, ab^{k}_{B}$$

Substitute:

$$[ab^{\mu}_{A}]_{,B} = ab^{\mu}_{n}\, b3b_{AB}{}^{n} - ab^{\alpha}_{A}\, ab^{\beta}_{B}\, a3a_{\alpha\beta}{}^{\mu}$$

To get:

$$T(b)_{A,B} = T(a)_{\mu}\, ab^{\mu}_{n}\, b3b_{AB}{}^{n} - T(a)_{\mu}\, ab^{\alpha}_{A}\, ab^{\beta}_{B}\, a3a_{\alpha\beta}{}^{\mu} + T(a)_{\alpha,\beta}\, ab^{\alpha}_{A}\, ab^{\beta}_{B}$$

So:

Transformation of Intrinsic Derivative of Lower Tensor w.r. to Upper Position

$$T(b)_{A}|_{B} \equiv T(b)_{A,B} - T(b)_{n}\, b3b_{AB}{}^{n} = [\, T(a)_{\alpha,\beta} - T(a)_{\mu}\, a3a_{\alpha\beta}{}^{\mu}\,]\, ab^{\alpha}_{A}\, ab^{\beta}_{B}$$

$$T(a)_{\lambda}|_{\delta} \equiv T(a)_{\lambda,\delta} - T(a)_{m}\, a3a_{\lambda\delta}{}^{m}$$

$$T(b)_{l}|_{d} \equiv T(b)_{l,d} - T(b)_{n}\, b3b_{ld}{}^{n}$$

Transformation:

$$T(b)_{l}|_{d} = T(a)_{\lambda}|_{\delta}\, ab^{\lambda}_{l}\, ab^{\delta}_{d}$$

For Zero order tensor:

$$T(a)\,|_{\delta} \equiv T(a)_{,\delta}$$

$$T(b)\,|_{d} \equiv T(b)_{,d}$$

Transformation:

$$T(b)|_{d} = T(a)|_{\delta}\, ab^{\delta}_{d}$$

6.12.1.2.
Transformation of Intrinsic Derivative of *Upper* Tensor

From these:

$$T(b)_{M,N} = T(a)_{\mu}\,[ab^{\mu}_{M}]_{,N} + T(a)_{\mu,k}\,ab^{\mu}_{M}\,ab^{k}_{N}$$

$$T(b)_{M,N} = [ba_{M}^{\mu}]_{,N}\,T(a)_{\mu} + ba_{M}^{\mu}\,ba_{N}^{k}\,T(a)_{\mu,k}$$

$$T(b)^{M}_{,N} = T(a)^{\mu}\,[ab^{M}_{\mu}]_{,N} + T(a)^{\mu}_{,k}\,ab^{M}_{\mu}\,ab^{k}_{N}$$

$$T(b)^{M}_{,N} = [ba^{M}_{\mu}]_{,N}\,T(a)^{\mu} + ba^{M}_{\mu}\,ba_{N}^{k}\,T(a)^{\mu}_{,k}$$

Select:

$$T(b)^{A}_{,B} = T(a)^{\mu}\,[ab^{A}_{\mu}]_{,B} + T(a)^{\mu}_{,k}\,ab^{A}_{\mu}\,ab^{k}_{B}$$

Substitute:

$$a3a^{n}_{\mu B}\,ba^{A}_{n} - ba^{\alpha}_{\mu}\,ba^{\beta}_{B}\,b3b^{A}_{\alpha\beta} = [ab^{A}_{\mu}]_{,N}$$

To get:

$$T(b)^{A}_{,B} = T(a)^{\mu}\,a3a^{n}_{\mu B}\,ba^{A}_{n} - T(a)^{\mu}\,ba^{\alpha}_{\mu}\,ba^{\beta}_{B}\,b3b^{A}_{\alpha\beta} + T(a)^{\mu}_{,k}\,ab^{A}_{\mu}\,ab^{k}_{B}$$

So:

Transformation of Intrinsic Derivative of Upper Tensor w.r. to <u>Upper Position</u>

$$T(b)^{A}|_{B} \equiv T(b)^{A}_{,B} + T(b)^{\mu}\,b3b^{A}_{\mu B} = [T(a)^{\alpha}_{,\beta} + T(a)^{\mu}\,a3a^{\alpha}_{\mu\beta}]\,ba^{A}_{\alpha}\,ab^{\beta}_{B}$$

$$T(a)^{\upsilon}|_{\delta} \equiv T(a)^{\upsilon}_{,\delta} + T(a)^{p}\,a3a^{\upsilon}_{p\delta}$$

$$T(b)^{u}|_{d} \equiv T(b)^{u}_{,d} + T(b)^{q}\,b3b^{u}_{q\delta}$$

Transformation

$$T(b)^{u}|_{d} = T(a)^{\upsilon}|_{\delta}\,ba^{u}_{\upsilon}\,ab^{\delta}_{d}$$

$$T(a)|_{\delta} \equiv T(a)_{,\delta}$$

$$T(b)|_{d} \equiv T(b)_{,d}$$

$$T(b)|_{d} = T(a)|_{\delta}\,ab^{\delta}_{d}$$

Intrinsic Derivative of Mixed Tensor w.r. to <u>Upper Position</u>:

$$T(a)^{\upsilon}_{\lambda}|_{\delta} \equiv T(a)^{\upsilon}_{\lambda,\delta} + T(a)^{p}_{\lambda}\,a3a^{\upsilon}_{p\delta} - T(a)^{\upsilon}_{m}\,a3a^{m}_{\lambda\delta}$$

$$T(b)^{u}_{l}|_{d} \equiv T(b)^{u}_{l,d} + T(b)^{q}_{l}\,b3b^{u}_{q\delta} - T(b)^{u}_{n}\,b3b^{n}_{ld}$$

Mixed tensor transformation:

$$T(b)^{u}_{l}|_{d} = T(a)^{\upsilon}_{\lambda}|_{\delta}\,ba^{u}_{\upsilon}\,ab^{\lambda}_{l}\,ab^{\delta}_{d}$$

Intrinsic derivative of Mixed tensor transformation:

$$T(a)^{\upsilon}_{\lambda\mu}|_{\delta} \equiv T(a)^{\upsilon}_{\lambda\mu,\delta} + T(a)^{p}_{\lambda\mu}\,a3a^{\upsilon}_{p\delta} - T(a)^{\upsilon}_{p\mu}\,a3a^{p}_{\lambda\delta} - T(a)^{\upsilon}_{\lambda p}\,a3a^{p}_{\mu\delta}$$

$$T(b)^{u}_{lm}|_{d} \equiv T(b)^{u}_{lm,d} + T(b)^{q}_{lm}\,b3b^{u}_{q\delta} - T(b)^{u}_{qm}\,b3b^{q}_{ld} - T(b)^{u}_{lq}\,b3b^{q}_{md}$$

And:

$$T(b)^{u}_{lm}|_{d} = T(a)^{\upsilon}_{\lambda\mu}|_{\delta}\,ba^{u}_{\upsilon}\,ab^{\lambda}_{l}\,ab^{\mu}_{m}\,ab^{\delta}_{d}$$

6.12.1.3. Transformation of Intrinsic Derivative of Tensor of W Weight with respect to *Upper Position*:

:

Order 0:

Transformation of Order 0:

$$T(b) = T(a) \; \| ab \|^{W}$$

:

$$T(b)_{,d} = T(a)_{,\delta} \; ab^{\delta}_{d} \; \| ab \|^{W} + T(a) \; W \; \| ab \|^{W-1} \; \| ab \|_{,d}$$

:

$$T(a)_{,\delta} \; ab^{\delta}_{d} \; \| ab \|^{W} + T(a) \; W \; \| ab \|^{W-1} \; \| ab \|_{,d}$$

$$=$$

$$T(a) \, W \, \| ab \|^{W} \; b3b_{kd}^{\;k} + T(a)_{,\delta} \; ab^{\delta}_{d} - T(a) \; ab^{\delta}_{d} \; W \; a3a_{k\delta}^{\;k}$$

$$T(a)_{,\delta} \; ab^{\delta}_{d} \; \| ab \|^{W} = T(a)_{,\delta} \; ab^{\delta}_{d}$$

$$\| ab \|^{W-1} \; \| ab \|_{,d} = \| ab \|^{W} \; b3b_{kd}^{\;k} - a3a_{k\delta}^{\;k}$$

To get:

$$T(b)_{,d} = T(a) \, W \, \| ab \|^{W} \; b3b_{kd}^{\;k} + T(a)_{,\delta} \; ab^{\delta}_{d} - T(a) \; ab^{\delta}_{d} \; W \; a3a_{k\delta}^{\;k}$$

$$T(b)_{,d} - T(a) \, W \, \| ab \|^{W} \; b3b_{kd}^{\;k} = T(a)_{,\delta} \; ab^{\delta}_{d} - T(a) \; ab^{\delta}_{d} \; W \; a3a_{k\delta}^{\;k}$$

Or:

$$T(b)_{,d} - T(b) \; W \; b3b_{kd}^{\;k} = \{ \, T(a)_{,\delta} - T(a) \; W \; a3a_{k\delta}^{\;k} \, \} \; ab^{\delta}_{d}$$

Transformation of Intrinsic Derivative of Order 0:

$$T(b) \,|_{d} = T(a) \,|_{\delta} \; ab^{\delta}_{d}$$

:

$$T(a)|_{\delta} \equiv T(a)_{,\delta} - T(a) \; W \; a3a_{k\delta}^{\;k}$$

$$T(b)|_{d} \equiv T(b)_{,d} - T(b) \; W \; b3b_{kd}^{\;k}$$

:

Any Order:

Transformation of Mixed Tensor of Weight W:

$$T(b)^{u}_{\;l} = T(a)^{\upsilon}_{\;\lambda} \; \| ab \|^{W}$$

Transformation of Intrinsic Derivative of Mixed Tensor:

$$T(b)^{u}_{\;l}|_{d} = T(a)^{\upsilon}_{\;\lambda}|_{\delta} \; ba^{u}_{\upsilon} \; ab^{\lambda}_{l} \; ab^{\delta}_{d}$$

:

$$T(a)^{\upsilon}_{\lambda}|_{\delta} \equiv T(a)^{\upsilon}_{\lambda,\delta} - T(a)^{\upsilon}_{\lambda} \, W \; a3a_{k\delta}^{\;k} + T(a)^{p}_{\lambda} \; a3a_{p\delta}^{\;\upsilon} - T(a)^{\upsilon}_{m} \; a3a_{\lambda\delta}^{\;m}$$

$$T(b)^{u}_{l}|_{d} \equiv T(b)^{u}_{l,d} - T(b)^{u}_{l} \, W \; b3b_{kd}^{\;k} + T(b)^{q}_{l} \; b3b_{qd}^{\;u} - T(b)^{u}_{n} \; b3b_{ld}^{\;n}$$

Weight is Bases-Post-Transformer (pre-tensor)

While

Derivative is Differential-Pre-Transformer (post-tensor)

fo(-j) = X[+s-r]**t fx(-m) X(+m - j)

fx(-j) = O[+s-r]**t fo(- m) O(+m -j)

Differential is Differential-Pre-Transformer (post-tensor)

$$dg^{n} = (gf^{n}_{\;m})^{t} \; df^{m} \qquad\qquad df^{m} = (fg^{m}_{\;n})^{t} \; dg^{n}$$

$$dg_n = (gf_n{}^m)^t \; df_m \qquad\qquad df_m = (fg_m{}^n)^t \; dg_n$$

6.12.2.
Metric Intrinsic Derivative is Null:
(Ricci's Theorem):

$$aa^{\upsilon\varpi}|_\delta \equiv aa^{\upsilon\varpi}{}_{,\delta} + aa^{p\varpi} \; a3a_{p\delta}{}^{\upsilon} + aa^{\upsilon p} \; a3a_{p\delta}{}^{\varpi}$$

$$bb^{uv}|_d \equiv bb^{uv}{}_{,d} + bb^{qv} \; b3b_{q\delta}{}^{u} + bb^{uq} \; b3b_{q\delta}{}^{v}$$

$$bb^{uv}|_d = aa^{\upsilon\varpi}|_\delta \; ba^u{}_\upsilon \; ba^v{}_\varpi \; ab^\delta{}_d$$

Intrinsic derivative:

$$aa_{\lambda\mu}|_\delta \equiv aa_{\lambda\mu,\delta} - aa_{r\mu} \; a3a_{\lambda\delta}{}^{r} - aa_{\lambda s} \; a3a_{\mu\delta}{}^{s}$$

$$bb_{lm}|_d \equiv bb_{lm,d} - bb_{rm} \; b3b_{ld}{}^{r} - bb_{ls} \; b3b_{md}{}^{s}$$

$$bb_{lm}|_d = aa_{\lambda\mu}|_\delta \; ab^\lambda{}_l \; ab^\mu{}_m \; ab^\delta{}_d$$

But:

$$aa_{nl,q} = a3a_{qln} + a3a_{qnl} = aa_{nj} \; a3a_{ql}{}^{j} + aa_{lj} \; a3a_{qn}{}^{j}$$

$$aa^{js}{}_{,q} = -a3a_q{}^{sj} - a3a_q{}^{js} = -aa^{sp} \; a3a_{qp}{}^{j} - aa^{jp} \; a3a_{qp}{}^{s}$$

So:

$$aa_{nl,q} - aa_{nj} \; a3a_{ql}{}^{j} - aa_{lj} \; a3a_{qn}{}^{j} = 0$$

$$aa^{js}{}_{,q} + aa^{sp} \; a3a_{qp}{}^{j} + aa^{jp} \; a3a_{qp}{}^{s} = 0$$

(Ricci's Theorem):

$$aa_{nl}|_q = aa_{nl,q} - aa_{nj} \; a3a_{ql}{}^{j} - aa_{lj} \; a3a_{qn}{}^{j} = 0$$

$$aa^{js}|_q = aa^{js}{}_{,q} + aa^{sp} \; a3a_{qp}{}^{j} + aa^{jp} \; a3a_{qp}{}^{s} = 0$$

Where:

$$aa^{js} \; aa_{is} = \delta^j{}_i$$

:

$$aa^{js}|_n \; aa_{is} + aa^{js} \; aa_{is}|_n = \delta^j{}_{i|n} = 0$$

Corollary:

Given:

$$T(a)^u{}_l$$

Then:

$$(aa_{jm} \; T_i{}^j{}_k)|_n = aa_{jm}|_n \; T_i{}^j{}_k + aa_{jm} \, (T_i{}^j{}_k)|_n = 0 + aa_{jm} \, (T_i{}^j{}_k)|_n$$

$$(aa_{jm} \; T_i{}^j{}_k)|_n = aa_{jm} (T_i{}^j{}_k)|_n$$

6.12.3.
Second Intrinsic Derivative of a Tensor:

$$(T_l|_d)|_e \quad .NE. \quad (T_l|_e)|_d$$

$$T(a)_l|_d \equiv T(a)_{l,d} - T(a)_n \; a3a_{ld}^{\;n}$$
$$T_l|_d \equiv T_{l,d} - T_p \; a3a_{ld}^{\;p}$$

$$(T_l|_d)|_e = (T_l|_d)_{,e} - (T_l|_m)\; a3a_{de}^{\;m} - (T_n|_d)\; a3a_{le}^{\;n}$$
$$(T_l|_d)|_e = (T_{l,d} - T_p\; a3a_{ld}^{\;p})_{,e} - (T_{l,m} - T_p\; a3a_{lm}^{\;p})\; a3a_{de}^{\;m} - (T_{n,d} - T_p\, a3a_{nd}^{\;p})\, a3a_{le}^{\;n}$$

$$(T_l|_e)|_d = (T_l|_e)_{,d} - (T_l|_m)\; a3a_{ed}^{\;m} - (T_n|_e)\; a3a_{ld}^{\;n}$$
$$(T_l|_e)|_d = (T_{l,e} - T_p\; a3a_{le}^{\;p})_{,d} - (T_{l,m} - T_p\; a3a_{lm}^{\;p})\; a3a_{ed}^{\;m} - (T_{n,e} - T_p\, a3a_{ne}^{\;p})\; a3a_{ld}^{\;n}$$

6.12.4.
Difference of Second Intrinsic Derivatives:

Subtract:

$$(T_l|_d)|_e \equiv (T_{l,d} - T_p\; a3a_{ld}^{\;p})_{,e} - (T_{l,m} - T_p\; a3a_{lm}^{\;p})\, a3a_{de}^{\;m} - (T_{n,d} - T_p\, a3a_{nd}^{\;p})\, a3a_{le}^{\;n}$$
$$(T_l|_e)|_d \equiv (T_{l,e} - T_p\; a3a_{le}^{\;p})_{,d} - (T_{l,m} - T_p\; a3a_{lm}^{\;p})\, a3a_{ed}^{\;m} - (T_{n,e} - T_p\, a3a_{ne}^{\;p})\, a3a_{ld}^{\;n}$$

To get:

Upper Fourth Symbol

$$(T_l|_d)|_e - (T_l|_e)|_d \equiv T_p \quad a4a_{lde}^{\;p}$$

6.12.5.
Upper and Lower Fourth Symbols

6.12.5.1.
Upper Fourth Symbol

Fourth order symbol:

$$a4a_{lde}^{\;p} \equiv - a3a_{ld\;,e}^{\;p} + a3a_{le\;,d}^{\;p} + a3a_{nd}^{\;p}\, a3a_{le}^{\;n} - a3a_{ne}^{\;p}\, a3a_{ld}^{\;n}$$

Written:

$$a4a_{lde}^{\;p}$$
$$\equiv$$

$\partial/\partial a^d \equiv ,d$	$\partial/\partial a^e \equiv ,e$	+	$a3a_{nd}^{\;p}$	$a3a_{ne}^{\;p}$
$a3a_{ld}^{\;p}$	$a3a_{le}^{\;p}$		$a3a_{ld}^{\;n}$	$a3a_{le}^{\;n}$

6.12.5..2.

Lower Fourth Symbol

$$a4a_{ldeq} \equiv aa_{pq} \quad a4a_{lde}^{\ \ \ p}$$

Non-zero values:

N=(n+1)*n*n*(n–1)/12=(n*n*n*n–n*n)/12

From (n*n*n*n) onlb non-zeros: N=(n+1)*n*n*(n–1)/12=(n*n*n*n–n*n)/12

a_{ijij} : two distinct	a_{ijkj} : three distinct	a_{ijkl} : four distinct
a_{ijij} : 1212,1313,2323	a_{ijkj} :1213,2123,3132	none in 3D
a_{ijij} : 1212	none in 2D	none in 2D
C_n^2 =n(n-1)/2	$3C_n^2$ = n(n-1)(n-2)/2	$2C_n^4$=n(n-1)(n-2)(n-3)/12

:

Properties:

two slides sym	midis anti-sym	edges anti-sym
$_{ijkl} =+ _{klij}$	$_{ijkl} = - _{ikjl}$	$_{ijkl} = - _{ljki}$

No change of fourth

$$a4a_{ijkl} + a4a_{jkil} + a4a_{kijl} = 0$$

6.12.6.
Two Indexed Ricci and Einstein Tensors:

6.12.6.1.
Two Indexed Ricci Tensor:

Contracting Fourth and Third Indices in Fourth Indexed Symbol is

$$(T_l|_d)|_e - (T_l|_e)|_d$$
$$=$$
$$T_p \quad a4a_{lde}^{\ \ \ p}$$
$$=$$
$$-T_p\, a3a_{ld}^{\ \ p}{}_{,e} + T_p a3a_{nd}^{\ \ p}\, a3a_{le}^{\ \ n}$$
$$+T_p\, a3a_{le}^{\ \ p}{}_{,d} - T_p a3a_{ne}^{\ \ p}\, a3a_{ld}^{\ \ n}$$

Fourth Indexed Tensor

$$a4a_{lde}^{\ \ \ p} \equiv - a3a_{ld}^{\ \ p}{}_{,e} + a3a_{le}^{\ \ p}{}_{,d} + a3a_{nd}^{\ \ p}\, a3a_{le}^{\ \ n} - a3a_{ne}^{\ \ p}\, a3a_{ld}^{\ \ n}$$

Contracted to get Lower Second Indexed Lower Ricci

$$a2a_{l\,d} \equiv a4a_{ldq}{}^{q} = -\ a3a_{l\,d}{}^{q}{}_{,q} + a3a_{l\,q}{}^{q}{}_{,d} + a3a_{n\,d}{}^{q}\, a3a_{l\,q}{}^{n} - a3a_{n\,q}{}^{q}\, a3a_{l\,d}{}^{n}$$

But:

$$\partial(\alpha\alpha)/\partial a^{n} \equiv (\alpha\alpha)_{,n} = 2\ (\alpha\alpha)\ \ a3a_{n\,q}{}^{q}$$

That is:

$$[\mathrm{Log}(\alpha)]_{,p} \equiv \partial[\mathrm{Log}(\alpha)]\,/\partial a^{p} \equiv a3a_{p\,q}{}^{q}$$

Contracted to get Lower Second Indexed Lower Ricci

$$(R2T_{l\,d}) \equiv (a2a_{l\,d}) \equiv a4a_{ldq}{}^{q} =$$

$$-a3a_{ld\,,q}{}^{q} + [\mathrm{Log}\sqrt{(\alpha\alpha)}]_{,l\,,d} + a3a_{n\,d}{}^{q}\, a\,3a_{l\,q}{}^{n} - [\mathrm{Log}\sqrt{(\alpha\alpha)}]_{,n}\ a3a_{l\,d}{}^{n}$$

$$(R2T_{l\,d}) \equiv (a2a_{l\,d}) \equiv a4a_{ldq}{}^{q} =$$

$$-a3a_{ld\,,q}{}^{q} + [\mathrm{Log}(\alpha)]_{,l\,,d} + a3a_{n\,d}{}^{q}\, a\,3a_{l\,q}{}^{n} - [\mathrm{Log}(\alpha)]_{,n}\ a3a_{l\,d}{}^{n}$$

$a_{l\,i} = a_{i\,l}$ symmetric $\quad$ n(n+1)/2 distinct components.

(Upper Ricci Tensor)

Upper Ricci Tensor

$$(R2T^{m}{}_{n}) \equiv (a2a^{m}{}_{n}) \equiv aa^{L\,m}\ \ a2a_{L\,n}$$

6.12.6.2. Two Indexed Einstein Tensor:

Upper Ricci Tensor

$$(R2T^{m}{}_{n}) \equiv (a2a^{m}{}_{n}) \equiv aa^{L\,m}\ \ a2a_{L\,n}$$

Contracted to get:

The Scalar

$$a0a \equiv a2a^{M}{}_{M}$$

(Einstein Tensor)

$$E2T^{m}{}_{n} \equiv a2a^{m}{}_{n} - \tfrac{1}{2}\ a0a\ \delta^{m}{}_{n}$$

6.12.6.3. Contracting Derivative of Einstein Tensor is Null:

Partial Derivative of Einstein Tensor

$$=$$

$$E2T^{m}{}_{n,\,m} \equiv a2a^{m}{}_{n\,,m} - \tfrac{1}{2}\ a0a_{,m}\ \delta^{m}{}_{n} - \tfrac{1}{2}\ a0a\ \delta^{m}{}_{n\,,m}$$

$$=$$

$$a2a^{m}{}_{n\,,m} - \tfrac{1}{2}\ a0a_{,m}\ \delta^{m}{}_{n} - 0$$

:

Partial Derivative of No Change of Fourth:

$$a4a_{ijkl} + a4a_{jkil} + a4a_{kijl} = 0$$

$$a4a_{ijk\ l,m} + a4a_{jkil,m} + a4a_{kij\ l,m} = 0$$

Multiply by:

$$aa^{li}\ aa^{jk}$$

Get:

$$aa^{li}\ aa^{jk}\ a4a_{ijkl,m} + aa^{li}\ aa^{jk}\ a4a_{jkil,m} + aa^{li}\ aa^{jk}\ a4a_{kijl,m} = 0$$

Contracting Derivative of Einstein Tensor is Null:

$$2\ a2a^{m}_{n,m} - a0a_{,n} = 0$$

$$E1T_{n} \equiv E2T^{m}_{n,m} \equiv a2a^{m}_{n,m} - \tfrac{1}{2}\ a0a_{,n} = 0$$

$$E1T_{n} \equiv E2T^{m}_{n,m} = a2a^{m}_{n,m} - \tfrac{1}{2}\ a0a_{,m}\ \delta^{m}_{n} = 0$$

7. General Bases Structural Applications:

Composition: a into o:

$\underline{o}_n \; oa^n{}_j \; T^j{}_k(a) \; ao^k{}_m \; do^m$

$a(+k) = a(+k)[o(+m)]$

$\partial a(+k)/\partial o(+m) \equiv ao(+k,-m)$

$oo(-m,-n) \equiv ao(+k,-m) \; ao(+k,-n)$

Resolution: o into a :

$\underline{a}_n \; ao^n{}_j \; T^j{}_k(o) \; oa^k{}_m \; da^m$

$o(+k) = o(+k)[a(+m)]$

$\partial o(+k)/\partial a(+m) \equiv oa(+k,-m)$

$aa(-m,-n) \equiv oa(+k,-m) \; oa(+k,-n)$

Bases Inner Products are Metrics of Coordinate Bases:

$$0 \leq |aa_{IK}| = \alpha_I \alpha_K \, |AA_{IK}| \leq \alpha_I \alpha_K$$

If: $I = K$

$$aa_{\underline{I}\underline{I}} = (\alpha_I)^2$$

If: $I \neq K$

$$0 \leq |aa_{IK}| \leq \alpha_I \alpha_K$$

Ordered Mixed Product: (Determinant): (Bases Volume):

$\underline{a}_M$

$$0 < |_a| \equiv (\underline{a}_1 \wedge \underline{a}_2 . \underline{a}_3) = |oa(+n,-m)| \leq \equiv \alpha_1 \alpha_2 \alpha_3$$

……… (not published due to page number requirement)

8. Orthogonal Bases Structural Applications:

Composition: r into o:

$\underline{o}_n \; or^n_j \; T^j_k \; ro^k_m \; do^m$

r(+k) = r(+k)[o(+m)]

∂r(+k)/∂o(+m) ≡ ro(+k,-m)

oo(-m,-n) ≡ ro(+k,-m) ro(+k,-n)

Resolution: o into r :

$\underline{r}_n \; ro^n_j \; T^j_k \; or^k_m \; dr^m$

o(+k) = o(+k)[r(+m)]

∂o(+k)/∂r(+m) ≡ or(+k,-m)

rr(-m,-n) ≡ or(+k,-m) or(+k,-n)

Bases Inner Products are Metrics of Coordinate Bases:

$$rr_{IK} = \rho_I \rho_K PP_{IK} = \rho_I \rho_K \delta_{IK}$$

If: I = K

$$rr_{\underline{I}\underline{I}} = (\rho_I)^2$$

If: I ≠ K

$$rr_{IK} = 0$$

Ordered Mixed Product: (Determinant): (Bases Volume):

$\underline{r}_M$ $$0 < |_r| \equiv (\underline{r}_1 \wedge \underline{r}_2 \cdot \underline{r}_3) = |or(+n,-m)| \le \rho_1 \rho_2 \rho_3 \equiv \rho_{123}$$

……… (not published due to page number requirement)

9. Orthonormal Bases Structural Applications:

Composition: x into o:

$ox^{n}_{j}\ T^{j}_{k}\ xo^{k}_{m}$

x(+k) = x(+k) [o(+m)]

∂x(+k) / ∂o(+m) ≡ xo(+k,-m)

oo(-m,-n) ≡ xo(+k,-m) xo(+k,-n)

Resolution: o into x :

$xo^{n}_{j}\ T^{j}_{k}\ ox^{k}_{m}$

o(+k) = o(+k) [x(+m)]

∂o(+k) / ∂x(+m) ≡ ox(+k,-m)

xx(-m,-n) ≡ ox(+k,-m) ox(+k,-n)

Bases Inner Products are Metrics of Coordinate Bases:

$$xx_{IK} = \xi_I \xi_K\ \Xi\Xi_{IK} = \delta_{IK}$$

If: I = K

$$xx_{\underline{I}\underline{I}} = 1$$

If: I ≠ K

$$xx_{IK} = 0$$

Ordered Mixed Product: (Determinant) : (Bases Volume) :

$\underline{x}_M$ $$0 < \|_x\| \equiv (\underline{x}_1 \wedge \underline{x}_2 \cdot \underline{x}_3) = \|ox(+n,-m)\| = 1$$

… … … (not published due to page number requirement)

10.
Exact Kinematics of Structures:

10.1.
Initial and Current
Descriptions of Structural Contiuum:

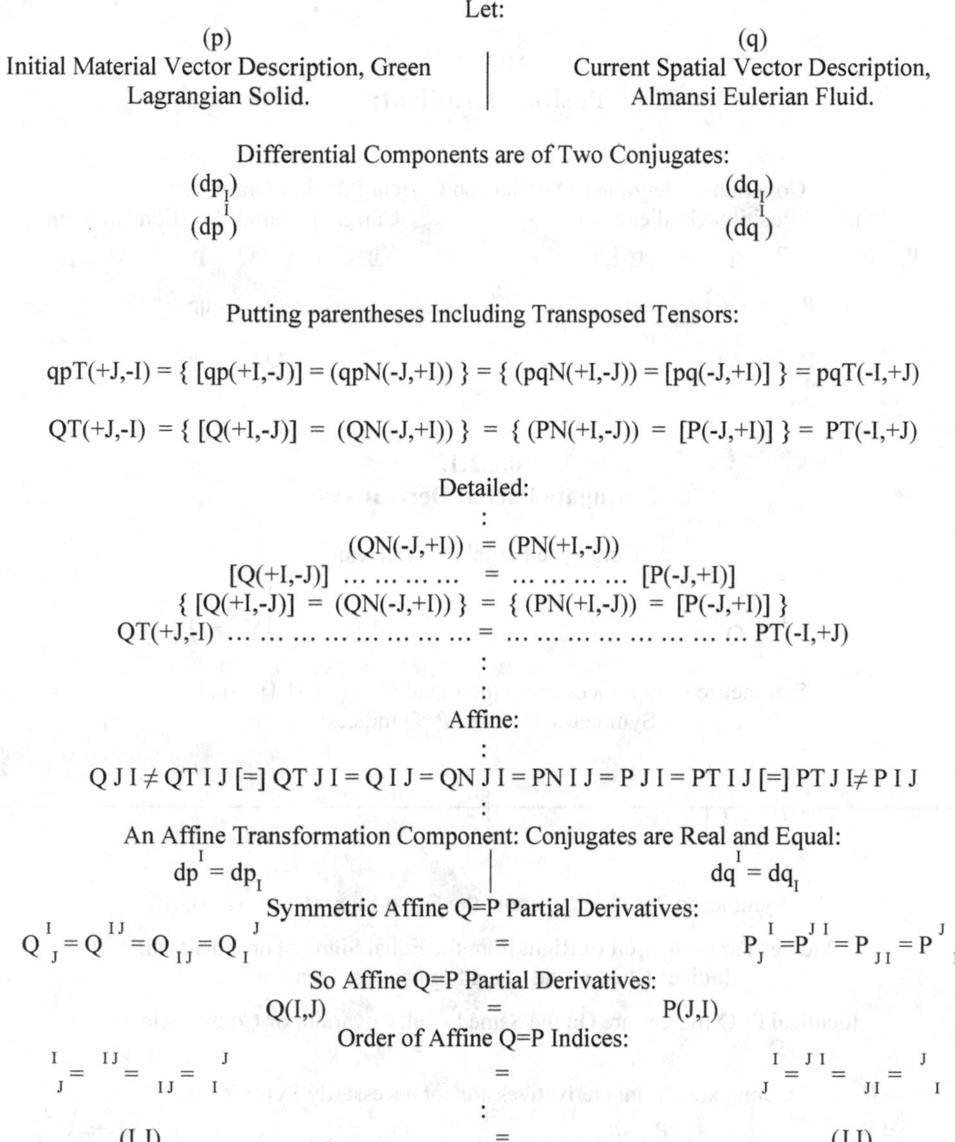

Let:

(p) Initial Material Vector Description, Green Lagrangian Solid.	(q) Current Spatial Vector Description, Almansi Eulerian Fluid.

Differential Components are of Two Conjugates:

(dp_I)	(dq_I)
(dp^I)	(dq^I)

Putting parentheses Including Transposed Tensors:

qpT(+J,-I) = { [qp(+I,-J)] = (qpN(-J,+I)) } = { (pqN(+I,-J)) = [pq(-J,+I)] } = pqT(-I,+J)

QT(+J,-I) = { [Q(+I,-J)] = (QN(-J,+I)) } = { (PN(+I,-J)) = [P(-J,+I)] } = PT(-I,+J)

Detailed:

:

(QN(-J,+I)) = (PN(+I,-J))

[Q(+I,-J)] … … … … = … … … … [P(-J,+I)]

{ [Q(+I,-J)] = (QN(-J,+I)) } = { (PN(+I,-J)) = [P(-J,+I)] }

QT(+J,-I) … … … … … … … … = … … … … … … … … PT(-I,+J)

:

:

Affine:

:

Q J I ≠ QT I J [=] QT J I = Q I J = QN J I = PN I J = P J I = PT I J [=] PT J I≠ P I J

:

An Affine Transformation Component: Conjugates are Real and Equal:

$dp^I = dp_I$	$dq^I = dq_I$

Symmetric Affine Q=P Partial Derivatives:

$$Q^I{}_J = Q^{IJ} = Q_{IJ} = Q_I{}^J \quad = \quad P_J{}^I = P^{JI} = P_{JI} = P^J{}_I$$

So Affine Q=P Partial Derivatives:

$$Q(I,J) \quad = \quad P(J,I)$$

Order of Affine Q=P Indices:

$${}^I{}_J = {}^{IJ} = {}_{IJ} = {}_I{}^J \quad = \quad {}_J{}^I = {}^{JI} = {}_{JI} = {}^J{}_I$$

:

$$(I,J) \quad = \quad (J,I)$$

10.2.
Srain Kinematics of Structural Contiuum:

10.2.1. Equal Metrics:

The length of $\underline{dp}$ is equal dp, which is the square root of metric.
The length of $\underline{dq}$ is equal dq, which is the square root of metric.
(Metric: Derivative, Logarithmic and Contravariant)
Undeformable Equal Metric is Sum of Products of Components Conjugates:

$$(dp)^2 \equiv dp^I dp_I = dq^J dq_J \equiv (dq)^2$$

If there is No Deformation, then it is Coordinates Transformation:

10.2.2. Position Gradient:

Covariant: Integrable: Material and Current Position Gradients:

Material Position Gradient

$$P_N{}^n\, \underline{q}_n \equiv P^N{}_n\, \underline{q}^n \equiv \underline{grad}\ \underline{p}$$

$$P^N{}_n \equiv pq^K{}_n$$

$$P_N{}^n \equiv pq_K{}^n$$

Current (Spatial) Position Gradient

$$\underline{Grad}\ \underline{q} \equiv Q^N{}_n\, \underline{p}^n \equiv Q_N{}^n\, \underline{p}_n$$

$$Q^N{}_n \equiv qp^K{}_n$$

$$Q_N{}^n \equiv qp_K{}^n$$

10.2.2.1. Conjugate Partial Derivatives:

Conjugate Partial Derivatives:

$$P^I{}_J = Q_J{}^I \qquad P_I{}^J = Q^J{}_I$$

$$P^{IJ} = Q^{JI} \qquad P_{IJ} = Q_{JI}$$

Symmetric P=Q Indices about the Equal Sign (=): (I J) (=) (J I):
Symmetric Order of P=Q Indices:

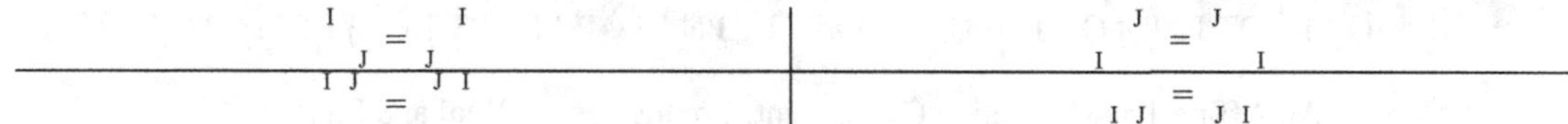

Symmetric P=Q Indices about the Equal Sign: (I J) (=) (J I):

Indices Far to the Left or Right from the Equal Sign (=) are Identical.
Indices Close to the Equal Sign (=) are Identical.

Identical P=Q Indices are On the Same Level, Covariant or Contravariant.

Conjugate Partial Derivatives are not necessarily Symmetric:

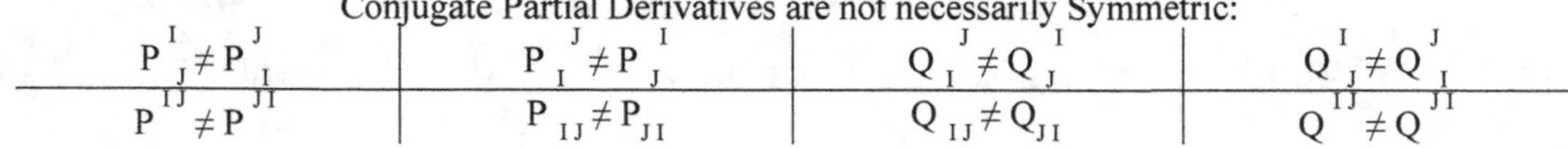

$P^I{}_J \neq P_I{}^J$	$P_I{}^J \neq P^I{}_J$	$Q_I{}^J \neq Q^I{}_J$	$Q^I{}_J \neq Q_I{}^J$
$P^{IJ} \neq P^{JI}$	$P_{IJ} \neq P_{JI}$	$Q_{IJ} \neq Q_{JI}$	$Q^{IJ} \neq Q^{JI}$

10.2.2.2. Affine Partial Derivatives:

An Affine Metric Component: Conjugates are Real, and hence are Equal:

$dp^I = dp_I$ | $dq^I = dq_I$

A Metric Component:

A Product of Affine Two Reals is Real:

$$dp^I dp_I = dp^I dp^I = dp^I dp_I = dp_I dp_I = dq^J dq_J = dq^J dq_J = dq^J dq^J = dq_J dq_J$$

Symmetric Affine P=Q Partial Derivatives:

$$P^I_{\ J} = P^{IJ} = P_{IJ} = P_I^{\ J} \quad = \quad Q_J^{\ I} = Q^{JI} = Q_{JI} = Q^J_{\ I}$$

Affine Partial Derivatives:

P(I,J) = Q(J,I)

But Affine Conjugate Partial Derivatives are not necessarily Symmetric:

P(I,J) $\neq$ P(J,I) | Q(I,J) $\neq$ Q(J,I)

So Affine P=Q Partial Derivatives:

P(I,J) = Q(J,I)

Order of Affine P=Q Indices:

$${}^I_J = {}^{IJ} = {}_{IJ} = {}_I^J \quad = \quad {}_J^I = {}^{JI} = {}_{JI} = {}^J_I$$

So Order of P=Q Indices:

(I,J) = (J,I)

Detailed:

Symmetric Affine P=Q Partial Derivatives:

$P^I_{\ J} = Q^I_{\ J}$	$P^I_{\ J} = Q^{JI}$	$P^I_{\ J} = Q_{JI}$	$P_I^{\ J} = Q^J_{\ I}$
$P^{IJ} = Q^I_{\ J}$	$P^{IJ} = Q^{JI}$	$P^{IJ} = Q_{JI}$	$P_{IJ} = Q^J_{\ I}$
$P_{IJ} = Q^I_{\ J}$	$P_{IJ} = Q^{JI}$	$P_{IJ} = Q_{JI}$	$P_I^{\ J} = Q^J_{\ I}$
$P_I^{\ J} = Q^I_{\ J}$	$P_I^{\ J} = Q^{JI}$	$P_I^{\ J} = Q_{JI}$	$P_{IJ} = Q^J_{\ I}$

Detailed: Symmetry of Order of Affine P=Q Indices:

${}^I_J = {}^I_J$	${}^I_J = {}^{JI}$	${}^I_J = {}_{JI}$	${}_I^J = {}^J_I$
${}^{IJ} = {}^I_J$	${}^{IJ} = {}^{JI}$	${}^{IJ} = {}_{JI}$	${}_{IJ} = {}^J_I$
${}_{IJ} = {}^I_J$	${}_{IJ} = {}^{JI}$	${}_{IJ} = {}_{JI}$	${}_I^J = {}^J_I$
${}_I^J = {}^I_J$	${}_I^J = {}^{JI}$	${}_I^J = {}_{JI}$	${}_{IJ} = {}^J_I$

Affine P=Q Partial Derivatives As in Partial Derivatives:

(I): Indices Far to the Left or Right from the Equal Sign (=) are Identical.

(J): Indices Close to the Equal Sign (=) are Identical.

But: Affine P=Q Index Levels are Indifferent, not As in Partial Derivatives:

Affine Identical P=Q Indices are <u>not</u> necessarily On the Same Level.

Affine Conjugate Partial Derivatives As in Partial Derivatives:

Affine Conjugate Partial Derivatives are not necessarily Symmetric:

P(I,J) $\neq$ P(J,I) | Q(I,J) $\neq$ Q(J,I)

10.2.3.
Orthonormal Partial Derivatives:

Partial Derivatives Orthonormality Identity:

$$Q^M_{\ I} P^I_{\ N} = \delta^M_{\ N} = \delta_N^{\ M} = P_N^{\ J} Q_J^{\ M} \quad | \quad P^M_{\ I} Q^I_{\ N} = \delta^M_{\ N} = \delta_N^{\ M} = Q_N^{\ J} P_J^{\ M}$$

Orthonormality Identity Transformation:

$$Q^M_{\ m}\delta^m_{\ n}P^n_{\ N} = \delta^M_{\ N} = \delta_N^{\ M} = P_N^{\ n}\delta_n^{\ m}Q_m^{\ M} \quad | \quad P^M_{\ m}\delta^m_{\ n}Q^n_{\ N} = \delta^M_{\ N} = \delta_N^{\ M} = Q_N^{\ n}\delta_n^{\ m}P_m^{\ M}$$

Since Partial Derivatives are Symmetric:

$P^I_{\ J} = Q^I_{\ J}$ | $P_I^{\ J} = Q^J_{\ I}$

$P^{IJ} = Q^{JI}$ | $P_{IJ} = Q_{JI}$

Then Orthonormality Identity:

$Q^{M}{}_{I} Q_{N}{}^{I} = \delta^{M}{}_{N} = \delta_{N}{}^{M} = P_{N}{}^{J} P^{M}{}_{J}$ $\qquad$ $P^{M}{}_{I} P_{N}{}^{I} = \delta^{M}{}_{N} = \delta_{N}{}^{M} = Q_{N}{}^{J} Q^{M}{}_{J}$

And Orthonormality Identity Transformation:

$Q^{M}{}_{m} \delta^{m}{}_{n} Q_{N}{}^{n} = \delta^{M}{}_{N} = \delta_{N}{}^{M} = P_{N}{}^{n} \delta_{n}{}^{m} P^{M}{}_{m}$ $\qquad$ $P^{M}{}_{m} \delta^{m}{}_{n} P_{N}{}^{n} = \delta^{M}{}_{N} = \delta_{N}{}^{M} = Q_{N}{}^{n} \delta_{n}{}^{m} Q^{M}{}_{m}$

Affine Orthonormality Identity:

Q(M,I) P(I,N) = δ(M,N) = δ(N,M) = P(N,J) Q(J,M)

Affine Orthonormality Identity:

Q(M,I) P(I,N) = δ(M,N) = δ(N,M) = P(N,J) Q(J,M)

Affine Orthonormality Identity Transformation:

Q(M,m) δ(m,n) P(n,N) = δ(M,N) = δ(N,M) = P(N,n) δ(n,m) Q(m,M)

Since Partial Derivatives are Symmetric:

P(I,J) = Q(J,I)

Then Affine Orthonormality Identity:

Q(M,I) Q(N,I) = δ(M,N) = δ(N,M) = P(N,J) P(M,J)

And Affine Orthonormality Identity Transformation:

Q(M,m) δ(m,n) Q(N,n) = δ(M,N) = δ(N,M) = P(N,n) δ(n,m) P(M,m)

10.2.4. Partial Derivatives Symmetric Directors:

Define:

$q^{M} \equiv \| \underline{q}^{M} \|$ $\qquad$ $p^{M} \equiv \| \underline{p}^{M} \|$

$q_{N} \equiv \| \underline{q}_{N} \|$ $\qquad$ $p_{N} \equiv \| \underline{p}_{N} \|$

Symmetric Direction cosines:

$\cos(\underline{q}^{M}, \underline{p}_{I}) = \Theta^{M}{}_{I} = \Pi_{I}{}^{M}$ $\qquad$ $\cos(\underline{p}^{M}, \underline{q}_{I}) = \Pi^{M}{}_{I} = \Theta_{I}{}^{M}$

$\cos(\underline{q}_{N}, \underline{p}^{J}) = \Theta_{N}{}^{J} = \Pi^{J}{}_{N}$ $\qquad$ $\cos(\underline{p}_{N}, \underline{q}^{J}) = \Pi_{N}{}^{J} = \Theta^{J}{}_{N}$

First Vector Norm * Direction Cosine * Second Vector Norm:

$Q^{M}{}_{N} = q^{\underline{M}} \Theta^{M}{}_{N} p_{\underline{N}}$ $\qquad$ $P^{M}{}_{N} = p^{\underline{M}} \Pi^{M}{}_{N} q_{\underline{N}}$

$P_{N}{}^{M} = p_{\underline{N}} \Pi_{N}{}^{M} q^{\underline{M}}$ $\qquad$ $Q_{N}{}^{M} = q_{\underline{N}} \Theta_{N}{}^{M} p^{\underline{M}}$

Symmetry Due to Symmetric Direction cosines:

$Q^{M}{}_{N} = q^{\underline{M}} \Theta^{M}{}_{N} p_{\underline{N}} = P_{N}{}^{M} = p_{\underline{N}} \Pi_{N}{}^{M} q^{\underline{M}}$ $\qquad$ $P^{M}{}_{N} = p^{\underline{M}} \Pi^{M}{}_{N} q_{\underline{N}} = Q_{N}{}^{M} = q_{\underline{N}} \Theta_{N}{}^{M} p^{\underline{M}}$

Affine Direction cosines:

ΘMN = ΠNM

Affine Symmetry Due to Symmetric Direction cosines:

QMN = qM ΘMN pN = pN ΠNM qM = PNM

10.2.5. Differential:

Differential:

$dq^{I} = Q^{I}{}_{J}\ dp^{J}$ $\qquad$ $dp^{I} = P^{I}{}_{J}\ dq^{J}$

$dq_{I} = Q_{I}{}^{J}\ dp_{J}$ $\qquad$ $dp_{I} = P_{I}{}^{J}\ dq_{J}$

Since P=Q Partial Derivatives are Symmetric then:

$dq^{I} = P_{J}{}^{I}\ dp^{J} = P^{JI}\ dp_{J}$ $\qquad$ $dp^{I} = Q_{J}{}^{I}\ dq^{J} = Q^{JI} dq_{J}$

$dq_I = P^J{}_I\, dp_J = P_{JI}\, dp^J$ $\qquad$ $dp_I = Q^J{}_I\, dq_J = Q_{JI}\, dq^J$

Differential and Partial Derivatives:

$dq^I = Q^I{}_J\, dp^J = P_J{}^I dp^J = P^{JI} dp_J$ $\qquad$ $dp^I = P^I{}_J\, dq^J = Q_J{}^I dq^J = Q^{JI} dq_J$

$dq_I = Q_I{}^J\, dp_J = P^J{}_I dp_J = P_{JI} dp^J$ $\qquad$ $dp_I = P_I{}^J\, dq_J = Q^J{}_I dq_J = Q_{JI} dq^J$

10.2.6.
Affine Differential:

Affine Index Levels are Indifferent, so Differential:

$dq^I = Q_I{}^J dp^J = Q^{IJ} dp^J = Q_{IJ} dp^J = Q^I{}_J dp^J$ $\qquad$ $dp^I = P_I{}^J dq^J = P^{IJ} dq^J = P_{IJ} dq^J = P^I{}_J dq^J$

=

$dq_I = Q_I{}^J dp_J = Q^{IJ} dp_J = Q_{IJ} dp_J = Q^I{}_J dp_J$ $\qquad$ $dp_I = P_I{}^J dq_J = P^{IJ} dq_J = P_{IJ} dq_J = P^I{}_J dq_J$

So:

dq(I) =Q(I,J) dp(J) $\qquad$ dp(I) = P(I,J) dq(J)

Since Affine P=Q Partial Derivatives are Symmetric then:

$dq^I = P^J{}_I dp^J = P^{JI} dp^J = P_{JI} dp^J = P_J{}^I dp^J$ $\qquad$ $dp^I = Q^J{}_I dq^J = Q^{JI} dq^J = Q_{JI} dq^J = Q_J{}^I dq^J$

=

$dq_I = P_I{}^J dp_J = P^{IJ} dp_J = P_{IJ} dp_J = P^I{}_J dp_J$ $\qquad$ $dp_I = Q^J{}_I dq_J = Q^{JI} dq_J = Q_{JI} dq_J = Q_J{}^I dq_J$

So:

dq(I) = P(J,I) dp(J) $\qquad$ dp(I) = Q(J,I) dq(J)

Affine Differential and Partial Derivatives:

$dq^I = Q_I{}^J dp^J = Q^{IJ} dp^J = Q_{IJ} dp^J = Q^I{}_J dp^J$

$dq^I = P^J{}_I dp^J = P^{JI} dp^J = P_{JI} dp^J = P_J{}^I dp^J$

=

$dq_I = Q_I{}^J dp_J = Q^{IJ} dp_J = Q_{IJ} dp_J = Q^I{}_J dp_J$

$dq_I = P_I{}^J dp_J = P^{IJ} dp_J = P_{IJ} dp_J = P^I{}_J dp_J$

$dp^I = P_I{}^J dq^J = P^{IJ} dq^J = P_{IJ} dq^J = P^I{}_J dq^J$

$dp^I = Q^J{}_I dq^J = Q^{JI} dq^J = Q_{JI} dq^J = Q_J{}^I dq^J$

=

$dp_I = P_I{}^J dq_J = P^{IJ} dq_J = P_{IJ} dq_J = P^I{}_J dq_J =$

$dp_I = Q^J{}_I dq_J = Q^{JI} dq_J = Q_{JI} dq_J = Q_J{}^I dq_J$

So:

Affine Differential:

dq(I) =Q(I,J) dp(J) =P(J,I) dp(J) $\qquad$ dp(I) = P(I,J) dq(J) = Q(J,I) dq(J)

10.2.7.
Metric and Relative Metric Difference:

Metric:

$(dp)^2 = P^J{}_K\, dq_J\, P^K{}_M\, dq^M$ $\qquad$ $(dq)^2 = Q_K{}^J\, dp_J\, Q^K{}_M\, dp^M$

$(dp)^2 = P^K{}_M\, dq^M\, P_K{}^J\, dq_J$ $\qquad$ $(dq)^2 = Q^K{}_M\, dp^M\, Q_K{}^J\, dp_J$

Metric Difference:

$2\, e^J{}_M\, dq_J\, dq^M \equiv$

$(\delta^J{}_M - P_K{}^J P^K{}_M)\, dq_J\, dq^M \equiv$

$2\, IPP^J{}_M\, dq_J\, dq^M \equiv$

$\equiv (dq)^2 - (dp)^2 \equiv$

$\equiv 2\, E^J{}_M\, dp_J dp^M$

$\equiv (Q_K{}^J Q^K{}_M - \delta^J{}_M)\, dp_J\, dp^M$

$\equiv 2\, QQI^J{}_M\, dp_J dp^M$

Relative Metric Difference:

$2\, e \equiv 2\, IPP \equiv (1 - PP)$

Almansi's Strain: Eulerian Strain

$2\, e^J{}_M \equiv 2\, IPP^J{}_M \equiv \delta^J{}_M - P_K{}^J P^K{}_M$

$2\, E \equiv 2\, QQI \equiv (QQ - 1)$

Green's Strain: Lagrangian Strain

$2\, E^J{}_M \equiv 2\, QQI^J{}_M \equiv Q_K{}^J Q^K{}_M - \delta^J{}_M$

$$2\,e_M{}^J \equiv 2\,IPP_M{}^J \equiv \delta_M{}^J - P^K{}_M P_K{}^J \qquad\Big|\qquad 2\,E_M{}^J \equiv 2\,QQI_M{}^J \equiv Q^K{}_M Q_K{}^J - \delta_M{}^J$$

Where:

$$\underline{p} = \underline{q} - \underline{r} \qquad\Big|\qquad \underline{q} = \underline{p} + \underline{r}$$

Metric Difference:

$$2\,e^J{}_M\, dq_J\, dq^M \equiv \qquad \equiv 2\,E^J{}_M\, dp_J\, dp^M$$

$$(\delta^J{}_M - P_K{}^J P^K{}_M)\, dq_J\, dq^M \equiv \qquad \equiv (dq)^2 - (dp)^2 \equiv \qquad \equiv (Q_K{}^J Q^K{}_M - \delta^J{}_M)\, dp_J\, dp^M$$

$$2\,IPP^J{}_M\, dq_J\, dq^M \equiv \qquad \equiv 2\,QQI^J{}_M\, dp_J\, dp^M$$

$$2\,e_M{}^J\, dq^M\, dq_J \equiv \qquad \equiv 2\,E^J{}_M\, dp_J\, dp^M$$

$$(\delta^J{}_M - P^K{}_M P_K{}^J)\, dq^M\, dq_J \equiv \qquad \equiv (dq)^2 - (dp)^2 \equiv \qquad \equiv (Q^K{}_M Q_K{}^J - \delta^J{}_M)\, dp^M\, dp_J$$

$$2\,IPP^J{}_M\, dq^M\, dq_J \equiv \qquad \equiv 2\,QQI^J{}_M\, dp^M\, dp_J$$

Relative Metric Difference:

Almansi's Strain: Eulerian Strain	Green's Strain: Lagrangian Strain
$2\,e^J{}_M \equiv 2\,IPP^J{}_M \equiv \delta^J{}_M - P_K{}^J P^K{}_M$	$2\,E^J{}_M \equiv 2\,QQI^J{}_M \equiv Q_K{}^J Q^K{}_M - \delta^J{}_M$
$2\,e_M{}^J \equiv 2\,IPP_M{}^J \equiv \delta_M{}^J - P^K{}_M P_K{}^J$	$2\,E_M{}^J \equiv 2\,QQI_M{}^J \equiv Q^K{}_M Q_K{}^J - \delta_M{}^J$

So:

$$e^J{}_M \equiv IPP^J{}_M \equiv IPP_M{}^J \equiv e_M{}^J \qquad\Big|\qquad E^J{}_M \equiv QQI^J{}_M \equiv QQI_M{}^J \equiv E_M{}^J$$

But:

$$e^J{}_M \neq e^M{}_J \equiv e_J{}^M \neq e_M{}^J \qquad\Big|\qquad E^J{}_M \neq E^M{}_J \equiv E_J{}^M \neq E_M{}^J$$

Differential:

$$dp^I = (\delta^I{}_J - rq^I{}_J)\, dq^J \qquad\Big|\qquad dq^I = (\delta^I{}_J + rp^I{}_J)\, dp^J$$

$$dp_I = (\delta_I{}^J - rq_I{}^J)\, dq_J \qquad\Big|\qquad dq_I = (\delta_I{}^J + rp_I{}^J)\, dp_J$$

$$2e^J{}_M \equiv 2e_M{}^J$$

$$= \delta^J{}_M - (\delta_K{}^J - rq_K{}^J)(\delta^K{}_M - rq^K{}_M)$$

$$= rq_M{}^J + rq^J{}_M - rq_K{}^J\, rq^K{}_M$$

$$2E^J{}_M \equiv 2E_M{}^J$$

$$(\delta_K{}^J + rp_K{}^J)(\delta^K{}_M + rp^K{}_M) - \delta^J{}_M = rp_M{}^J + rp^J{}_M + rp_K{}^J\, rp^K{}_M$$

But:

$$e^J{}_M \neq e^M{}_J \equiv e_J{}^M \neq e_M{}^J \qquad\Big|\qquad E^J{}_M \neq E^M{}_J \equiv E_J{}^M \neq E_M{}^J$$

Affine Metric Difference: $(dq)^2 - (dp)^2$:

2 e(J,M) dq(J) dq(M) ≡ [δ(J,M)-P(K,J) P(K,M)]dq(J) dq (M)

2 E(J,M) dp(J) dp(M) ≡ [Q(K,J) Q(K,M)-δ(J,M)] dp(J) dp(M)

Relative Metric Difference::

2 e(J,M) ≡2 IPP (J,M) ≡ δ(J,M) - P(K,J) P(K,M)

2 E(J,M) ≡ 2 QQI (J,M) ≡ Q(K,J) Q(K,M) - δ(J,M)

Symmetric:

e(J,M) = e(M,J)

E(J,M) = E(M,J)

10.2.10. Relative Incremental Metric Difference:

Orthonormality Identity:

$$Q^{M}{}_{i}P^{i}{}_{N} = Q^{M}{}_{n}\delta^{n}{}_{m}P^{m}{}_{N} = \delta^{M}{}_{N} \qquad\qquad P^{M}{}_{i}Q^{i}{}_{N} = P^{M}{}_{n}\delta^{n}{}_{m}Q^{m}{}_{N} = \delta^{M}{}_{N}$$

$$P_{N}{}^{j}Q_{j}{}^{M} = P_{N}{}^{m}\delta_{m}{}^{n}Q_{n}{}^{M} = \delta_{N}{}^{M} \qquad\qquad Q_{N}{}^{j}P_{j}{}^{M} = Q_{N}{}^{m}\delta_{m}{}^{n}P_{n}{}^{M} = \delta_{N}{}^{M}$$

[Example: If A rigid body rotates ($-\theta$): that is axis p rotates (resolves) ($+\theta$) to q.]

Affine Transformation Partial Derivatives:

P(K,J) = Q(J, K)

And: Affine Orthonormality Identity:

Q(M,I) P(I,N) = δ(M,N) P(M,I) Q(I,N) = δ(M,N)

Then Metric Difference:

$$(\delta^{J}{}_{M} - P_{K}{}^{J}P^{K}{}_{M})\,dq_{J}\,dq^{M} \equiv \qquad\qquad \equiv (Q_{K}{}^{J}Q^{K}{}_{M} - \delta^{J}{}_{M})\,dp_{J}\,dp^{M}$$

$$(\delta^{J}{}_{M} - Q_{J}{}^{K}P^{K}{}_{M})\,dq_{J}\,dq^{M} \equiv \quad \equiv (dq)^{2} - (dp)^{2} \equiv \quad \equiv (P_{J}{}^{K}Q^{K}{}_{M} - \delta^{J}{}_{M})\,dp_{J}\,dp^{M}$$

$$0 = [\delta^{J}{}_{M} - \delta^{J}{}_{M}]\,dq_{J}\,dq^{M} \equiv \qquad\qquad \equiv [\delta^{J}{}_{M} - \delta^{J}{}_{M}]\,dp_{J}\,dp^{M} = 0$$

Initial Material Description: Lagrangian Solid: p

Current Spatial Description: Eulerian Fluid: q

Let:

$$\underline{p} = \underline{q} - \underline{r} \qquad\qquad \underline{q} = \underline{p} + \underline{r}$$

Conjugates:

$$p^{I} = q^{I} - r^{I} \qquad\qquad q^{I} = p^{I} + r^{I}$$

$$p_{I} = q_{I} - r_{I} \qquad\qquad q_{I} = p_{I} + r_{I}$$

Then:

$$\partial p^{I}/\partial q^{J} = \delta^{I}{}_{J} - \partial r^{I}/\partial q^{J} \qquad\qquad \partial q^{I}/\partial p^{J} = \delta^{I}{}_{J} + \partial r^{I}/\partial p^{J}$$

$$\partial p_{I}/\partial q_{J} = \delta_{I}{}^{J} - \partial r_{I}/\partial q_{J} \qquad\qquad \partial q_{I}/\partial p_{J} = \delta_{I}{}^{J} + \partial r_{I}/\partial p_{J}$$

Identically:

$$P^{I}{}_{J} = \delta^{I}{}_{J} - rq^{I}{}_{J} \qquad\qquad Q^{I}{}_{J} = \delta^{I}{}_{J} + rp^{I}{}_{J}$$

$$P_{I}{}^{J} = \delta_{I}{}^{J} - rq_{I}{}^{J} \qquad\qquad Q_{I}{}^{J} = \delta_{I}{}^{J} + rp_{I}{}^{J}$$

Differential:

$$dp^{I} = (\delta^{I}{}_{J} - rq^{I}{}_{J})\,dq^{J} \qquad\qquad dq^{I} = (\delta^{I}{}_{J} + rp^{I}{}_{J})\,dp^{J}$$

$$dp_{I} = (\delta_{I}{}^{J} - rq_{I}{}^{J})\,dq_{J} \qquad\qquad dq_{I} = (\delta_{I}{}^{J} + rp_{I}{}^{J})\,dp_{J}$$

$$dp^{I}dp_{I} = (\delta^{I}{}_{J} - rq^{I}{}_{J})\,dq^{J}\,(\delta_{I}{}^{K} - rq_{I}{}^{K})\,dq_{K} \qquad\qquad dq^{I}dq_{I} = (\delta^{I}{}_{J} + rp^{I}{}_{J})\,dp^{J}\,(\delta_{I}{}^{K} + rp_{I}{}^{K})\,dp_{K}$$

Or:

$$dq^{I}dq_{I} - dp^{I}dp_{I} = rq^{K}{}_{J}dq^{J}dq_{K} + rq_{J}{}^{K}dq^{J}dq_{K} - rq^{I}{}_{J}rq_{I}{}^{K}dq^{J}dq_{K}$$

$$dq^{I}dq_{I} - dp^{I}dp_{I} = rp^{K}{}_{J}dp^{J}dp_{K} + rp_{J}{}^{K}dp^{J}dp_{K} + rp^{I}{}_{J}rp_{I}{}^{K}dp^{J}dp_{K}$$

Added:

$$2\,(dq^{I}dq_{I} - dp^{I}dp_{I}) =$$

$$rp^{K}{}_{J}dp^{J}dp_{K} + rq^{K}{}_{J}dq^{J}dq_{K} + rp_{J}{}^{K}dp^{J}dp_{K} + rq_{J}{}^{K}dq^{J}dq_{K} + rp^{I}{}_{J}rp_{I}{}^{K}dp^{J}dp_{K} - rq^{I}{}_{J}rq_{I}{}^{K}dq^{J}dq_{K}$$

And:

$2\,e^{J}{}_{M} = \delta^{J}{}_{M} - (\delta_{K}{}^{J} - rq_{K}{}^{J})(\delta^{K}{}_{M} - rq^{K}{}_{M})$	$2\,E^{J}{}_{M} = (\delta_{K}{}^{J} + rp_{K}{}^{J})(\delta^{K}{}_{M} + rp^{K}{}_{M}) - \delta^{J}{}_{M}$
$= rq^{J}{}_{M} + rq^{J}{}_{M} - rq_{K}{}^{J}\,rq^{K}{}_{M}$	$= rp^{J}{}_{M} + rp^{J}{}_{M} + rp_{K}{}^{J}\,rp^{K}{}_{M}$
$p^{I} = q^{I} - r^{I}$	$q^{I} = p^{I} + r^{I}$
$\partial p^{I}/\partial q^{J} = \delta^{I}{}_{J} - \partial r^{I}/\partial q^{J}$	$\partial q^{I}/\partial p^{J} = \delta^{I}{}_{J} + \partial r^{I}/\partial p^{J}$

$$IPP^I_K = 1/2\,(\delta^I_K - \partial p^J/\partial q^I\ \partial p^J/\partial q^K)$$
$$=1/2[\delta^I_K - (\delta^J_I - \partial r^J/\partial q^I)(\delta^J_K - \partial r^J/\partial q^K)]$$
$$=1/2[\delta^I_K - (\delta^I_K - \partial r^I/\partial q^K - \partial r^K/\partial q^I + \partial r^J/\partial q^I\ \partial r^J/\partial q^K]$$
$$=1/2(\partial r^I/\partial q^K + \partial r^K/\partial q^I + \partial r^J/\partial q^I\ \partial r^J/\partial q^K)$$

$$QQI^I_K = 1/2\,[\partial q^J/\partial p^I\ \partial q^J/\partial p^K - \delta^I_K]$$
$$=1/2\,[(\delta^J_I + \partial r^J/\partial p^I)(\delta^J_K + \partial r^J/\partial p^K) - \delta^I_K]$$
$$=1/2[(\delta^I_K + \partial r^I/\partial p^K + \partial r^K/\partial p^I + \partial r^J/\partial p^I\ \partial r^J/\partial p^K) - \delta^I_K]$$
$$=1/2(\partial r^I/\partial p^K + \partial r^K/\partial p^I + \partial r^J/\partial p^I\ \partial r^J/\partial p^K)$$

$$IPP^I_J \cong 1/2\,(\partial r^I/\partial q^J + \partial r^J/\partial q^I)$$

$$QQI^I_K \cong 1/2\,(\partial r^I/\partial p^J + \partial r^J/\partial p^I)$$

$$2\,e^J_M \equiv 2\,IPP^J_M \equiv \delta^J_M - P^J_K P^K_M$$
$$2\,e^J_M = \delta^J_M - (\delta^J_K - rq^J_K)(\delta^K_M - rq^K_M)$$
$$2\,e^J_M = rq^J_M + rq^J_M - rq^J_K\ rq^K_M$$

$$2\,E^J_M \equiv 2\,QQI^J_M \equiv Q^J_K Q^K_M - \delta^J_M$$
$$2\,E^J_M = (\delta^J_K + rp^J_K)(\delta^K_M + rp^K_M) - \delta^J_M$$
$$2\,E^J_M = rp^J_M + rp^J_M + rp^J_K\ rp^K_M$$

Metric Difference Approximation:

$$dq^2 - dp^2 \cong 2*1/2(\partial r^I/\partial q^J + \partial r^J/\partial q^I)dq_I\ dq^J$$
$$dq^2 - dp^2 \cong (\partial r^I/\partial q^J + \partial r^J/\partial q^I)dq_I\ dq^J$$

$$dq^2 - dp^2 \cong 2*1/2(\partial r^I/\partial p^J + \partial r^J/\partial p^I)\,dp_I\,dp^J$$
$$dq^2 - dp^2 \cong (\partial r^I/\partial p^J + \partial r^J/\partial p^I)\,dp_I\,dp^J$$

Infinitesimal Strain Tensor:

$$\varepsilon^I_J = 1/2\,(\partial r^I/\partial q^J + \partial r^J/\partial q^I)$$

So

$$\|dq\|^2 - \|dp\|^2 \cong 2 * \varepsilon^I_J\, dq_I\, dq^J = 2 * 1/2\,(rq^I_J + rq^J_I)\,dq_I\,dq^J$$

10.2.9.
Affine Metric or Rigid Body:

Rigid Body Metric Difference:

$$(\delta^J_M - P^J_K P^K_M)\,dq_J\,dq^M \equiv$$
$$(\delta^J_M - Q^K_J P^K_M)\,dq_J\,dq^M \equiv$$
$$0 = [\delta^J_M - \delta^J_M]\,dq_J\,dq^M \equiv$$

$$\equiv (dq)^2 - (dp)^2 \equiv$$

$$\equiv (Q^J_K Q^K_M - \delta^J_M)\,dp_J\,dp^M$$
$$\equiv (P^K_J Q^K_M - \delta^J_M)\,dp_J\,dp^M$$
$$\equiv [\delta^J_M - \delta^J_M]dp_J\,dp^M = 0$$

Fixed Metric Kinematics:

$$(dq^I dq_I - dp^I dp_I) = 0$$

Or:

$$(rp^K_J + rp^K_J)dp^J dp_K + (rq^K_J + rq^K_J)dq^J dq_K + rp^I_J rp^K_I\ dp^J dp_K - rq^I_J rq^K_I\ dq^J dq_K = 0$$

Metric Difference:

$$(\delta^J_M - P^J_K P^K_M)dq_J\,dq^M \equiv (dq)^2 - (dp)^2 \equiv (Q^J_K Q^K_M - \delta^J_M)dp_J\,dp^M$$

$$2\,e^J_M \equiv 2\,IPP^J_M \equiv \delta^J_M - P^J_K P^K_M$$
$$2\,e^J_M = \delta^J_M - (\delta^J_K - rq^J_K)(\delta^K_M - rq^K_M)$$
$$2\,e^J_M = rq^J_M + rq^J_M - rq^J_K\ rq^K_M$$

$$2\,E^J_M \equiv 2\,QQI^J_M \equiv Q^J_K Q^K_M - \delta^J_M$$
$$2\,E^J_M = (\delta^J_K + rp^J_K)(\delta^K_M + rp^K_M) - \delta^J_M$$
$$2\,E^J_M = rp^J_M + rp^J_M + rp^J_K\ rp^K_M$$

10.2.10..

Approximations:

10.2.10.1.
Metric Difference Ratio Approximation:

Metric Difference Ratio:

$(dq+dp)(dq-dp)/dq^2 \equiv (dq^2-dp^2)/dq^2 \equiv 2e$ | $2E \equiv (dq^2-dp^2)/dp^2 \equiv (dq+dp)(dq-dp)/dp^2$

Approximate:

$(dq + dp)/dq \cong 2$ | $(dq + dp)/dp \cong 2$

Then:

$2\ (1- P) \cong 2\ e$ | $2\ E \cong 2\ (Q -1)$

Or:

$(1- P) \cong e$ | $E \cong (Q -1)$

10.2.10.2.
Metric Difference Approximations:

Metric Difference:

$$2\ e^J_{\ M}\ dq_J\ dq^M \equiv (\delta^J_{\ M} - P_K^{\ J}\ P^K_{\ M})\ dq_J\ dq^M \equiv (dq+dp)\ (dq-dp)$$

$$\equiv (dq)^2-(dp)^2 \equiv$$

$$2\ E^J_{\ M}\ dp_J dp^M \equiv (Q_K^{\ J}\ Q^K_{\ M} - \delta^J_{\ M})\ dp_J\ dp^M \equiv (dq+dp)\ (dq-dp)$$

Approximated:

$(dq + dp) \cong 2\ dq$ | $(dq + dp) \cong 2\ dp$

Then:

$$2\ dq\ (dq–dp) \equiv 2\ e\ (dq)^2 \quad \cong (dq)^2-(dp)^2 \cong \quad 2\ E\ (dp)^2 \equiv 2\ dp\ (dq–dp)$$

10.2.10.3.
The Engineering Strain is a Length Ratio Approximation:

Since

$P^2 \equiv dp^2/dq^2 = 1\text{-}2e = 1 - (1- dp^2/ dq^2)$ | $Q^2 \equiv dq^2/dp^2 = 1+2E = 1 + (dq^2/dp^2-1)$

Or:

$P^2 \equiv dp^2/dq^2 = 1 - (dq^2 - dp^2) / dq^2$ | $Q^2 \equiv dq^2/dp^2 = 1+ (dq^2 - dp^2) / dp^2$

Then:

The Length Ratio:

$P \equiv dp/dq = [1 - (dq^2 - dp^2) / dq^2]^{1/2}$ | $Q \equiv dq/dp = [1+ (dq^2 - dp^2) / dp^2]^{1/2}$

Approximated:

$P \cong 1 - 1/2\ (dq^2 - dp^2) / dq^2 + \ldots$ | $Q \cong 1 + 1/2\ (dq^2 - dp^2)/ dp^2 + \ldots$

Merge these by:

Defining the Length Elongation Ratio, ε by the Engineering Strain:

$$1/[dp/dq]-1 \equiv dq/dp -1 \equiv Q – 1 \equiv (dq – dp) / dp \equiv \varepsilon$$

Then Engineering Strain, ε is Approximated:

$$1/[1-1/2(dq^2-dp^2)/dq^2+\ldots]–1 \cong \varepsilon \cong 1+1/2(dq^2-dp^2)/dp^2+\ldots -1$$

Or:

$$1 + 1/2(dq^2 - dp^2)/dq^2 - 1 \cong \varepsilon \cong 1/2 \ (dq^2 - dp^2)/dp^2$$
$$1/2\,(2\,e) \cong \varepsilon \cong 1/2\,(2\,E)$$
$$1/2\ (dq^2 - dp^2)/dq^2 \cong \varepsilon \cong 1/2\,(dq^2 - dp^2)/\,dp^2$$
$$1 - P = (dq - dp)/dq \cong \varepsilon \equiv (dq - dp)/\,dp = Q - 1$$
$$1 - P \cong \varepsilon \equiv Q - 1$$

Sum of Left and Right Sides:

$$2\,\varepsilon \cong 1 - P + Q - 1 \equiv Q - P \equiv dq/dp - dp/dq$$
$$\varepsilon \cong 1/2\,(Q - P)$$

The Engineering Strain:

$$\varepsilon \equiv Q - 1 \equiv dq/dp - 1 \equiv (dq - dp)\,/\,dp$$

Infinitesimal Engineering strain ≡ material Engineering strain

$\varepsilon = Q - 1 = (1\text{-}P)\,Q \cong (1\text{-}P)$ $\qquad$ $\varepsilon = Q - 1$

Engineering strain equals half its metric difference ratio:

spatial Engineering strain | material Engineering strain

$$1 - dp/dq \cong 1/\,2\,(1 - dp^2/dq^2) \qquad dq/\,dp - 1 \cong 1/\,2\,(dq^2/\,dp^2 - 1)$$

$$1 - dp/dq \cong IPP^I{}_M\, dq_I\, dq^M /\ dq_K\, dq^K \qquad dq/dp - 1 \cong QQI^I{}_M\, dp_I\, dp^M \,/\, dp_K\, dp^K$$

10.2.10.4. Metric difference ratio in terms of Engineering Strain:

$\varepsilon = Q - 1 = (1\text{-}P)\,Q \cong (1\text{-}P)$ $\qquad$ $\varepsilon = Q - 1$

$$2\,dq\,(dq - dp) \qquad \cong (dq)^2 - (dp)^2 \cong \qquad 2\,dp\,(dq - dp)$$

So:

Metric Difference:

$$2\,dq\,(dq - dp) \cong 2\,e\ dq^2 \equiv dq^2 - dp^2 \equiv 2\,E\ dp^2 \cong 2\,dp\,(dq - dp)$$

Metric Difference Ratio:

$$2(1\text{-}P) \equiv 2(dq - dp)/dq \cong 2e \equiv 1 - dp^2/dq^2 \cong dq^2/dp^2 - 1 \equiv 2E \cong 2(dq - dp)/dp \equiv 2(Q-1) \equiv 2\varepsilon$$

$2\,(1 - P) \cong 2\,e \cong 2\,\varepsilon$ $\qquad$ $2\,E \cong 2\,(Q - 1) \equiv 2\,\varepsilon$

$e \cong \varepsilon$ $\qquad$ $E \cong \varepsilon$

Length Ratio Approximation:

$P \cong 1 - \varepsilon \cong 1 - e$ $\qquad$ $Q \equiv 1 + \varepsilon \cong 1 + E$

Engineering strain equals half its metric difference ratio:

Spatial Engineering strain | Material Engineering strain

$1 - P \cong \varepsilon \cong e$ $\qquad$ $Q - 1 \cong \varepsilon \cong E$

Approximations:

Approximate Twice Relative Length:

$$2 \approx 1 + P \equiv (dq + dp)\,/\,dq \approx 2 \approx (dq + dp)\,/\,dp \equiv Q + 1 \approx 2$$

Then:

$2\,e \equiv 2\,IPP \equiv (1 - PP) \approx 2\ (1 - P)$ | $2\,E \equiv 2\,QQI \equiv (QQ - 1) \approx 2\,(Q - 1)$

Introduce an Approximation:

Length Ratio Approximation:

$P \approx 1 - e$ | $Q \approx 1 + E$

Introduce another Approximation:

Engineering Strain:

Spatial Engineering strain | Material Engineering strain

$1 - P$ | $Q - 1$

Infinitesimal Engineering strain ≡ material Engineering strain: $\varepsilon = Q - 1$

$\varepsilon = Q - 1 = (1-P)\,Q \approx (1-P) \approx e$ | $\varepsilon = Q - 1 \approx E$

Introduce another Approximation:

Metric Difference Approximation:

$$(rq^{I}{}_{J}+rq^{J}{}_{I})\, dq_I dq^J \approx 2e\,(dq)^2 \approx 2\, dq\,(dq-dp) \equiv 2\,(dq)^2\,[\,1-P]$$

$$\approx (dq)^2-(dp)^2 \approx$$

$$(rp^{I}{}_{J}+rp^{J}{}_{I})\, dp_I dp^J \approx 2E(dp)^2 \approx 2dp(dq-dp) \equiv 2\,(dp)^2\,[\,Q-1]$$

Or:

$$(dq+dp)(dq-dp)/dq^2 \equiv (dq^2-dp^2)/dq^2 \equiv 2e \approx 2E \equiv (dq^2-dp^2)/dp^2 \equiv (dq+dp)(dq-dp)/dp^2$$

$$(dq)^2\,[\,1-PP] \equiv 2e \approx 2E \equiv (dp)^2\,[\,QQ-1]$$

Engineering strain equals approximately half its metric difference ratio:

$1 - dp/dq \approx 1/2\,(1- dp^2/dq^2)$ | $dq/dp - 1 \approx 1/2\,(dq^2/dp^2 - 1)$

$1 - dp/dq \approx IPP^{I}{}_{M}\, dq_I\, dq^M /\, dq_K\, dq^K$ | $dq/dp - 1 \approx QQI^{I}{}_{M}\, dp_I\, dp^M /\, dp_K\, dp^K$

Engineering strain is approximately half metric difference ratio:

$$2e \approx 2\,\varepsilon \approx 2(\,rq^{I}{}_{J} + rq^{J}{}_{I}\,) \approx 2E$$

Since

$P^2 \equiv dp^2/dq^2 = 1\text{-}2e = 1 - (1- dp^2/dq^2)$ | $Q^2 \equiv dq^2/dp^2 = 1+2E = 1 + (dq^2/dp^2-1)$

Or:

$P^2 \equiv dp^2/dq^2 = 1 - (dq^2 - dp^2)/dq^2$ | $Q^2 \equiv dq^2/dp^2 = 1+ (dq^2 - dp^2)/dp^2$

Then:

The Length Ratio:

$P \equiv dp/dq = [1 - (dq^2 - dp^2)/dq^2]^{1/2}$ | $Q \equiv dq/dp=[1+ (dq^2 - dp^2)/dp^2]^{1/2}$

Approximated:

$P \approx 1 - 1/2\ (dq^2 - dp^2)/dq^2 + \ldots$ | $Q \approx 1 + 1/2\ (dq^2 - dp^2)/dp^2 + \ldots$

Merge these by:

Defining the Length Elongation Ratio, ε by the Engineering Strain:

$$1/[dp/dq]-1 \equiv dq/dp -1 \equiv Q-1 \equiv (dq-dp)/dp \equiv \varepsilon$$

Then Engineering Strain, ε is Approximated:

$$1/[1-1/2(dq^2-dp^2)/dq^2+\ldots]-1 \approx \varepsilon \approx 1+1/2(dq^2-dp^2)/dp^2+\ldots\ -1$$

Or:

$$1 + 1/2(dq^2-dp^2)/dq^2 - 1 \approx \varepsilon \approx 1/2\ (dq^2 - dp^2)/dp^2$$

$$1/2\,(2\,e) \approx \varepsilon \approx 1/2\,(2\,E)$$

$$1/2\ (dq^2-dp^2)/dq^2 \approx \varepsilon \approx 1/2\,(dq^2 - dp^2)/dp^2$$

$$1 - P = (dq-dp)/dq \approx \varepsilon \equiv (dq - dp)/dp = Q - 1$$

$$1 - P \approx \varepsilon \equiv Q - 1$$

Sum of Left and Right Sides:

$$2\,\varepsilon \approx 1 - P + Q - 1 \equiv Q - P \equiv dq/dp - dp/dq$$

$$\varepsilon \approx 1/2\,(Q-P)$$

10.2.10.5.
Different Approximations:

Not Affine:

Conjugate Partial Derivatives:

$P^{I}{}_{J} = Q_{J}{}^{I}$ | $P_{I}{}^{J} = Q^{J}{}_{I}$

$P^{IJ} = Q^{JI}$ | $P_{IJ} = Q_{JI}$

Symmetric P=Q Indices about the Equal Sign (=): (I J) (=) (J I):

But:

Conjugate Partial Derivatives are not necessarily Symmetric:

$P^{I}_{J} \neq P^{J}_{I}$	$P^{J}_{I} \neq P^{I}_{J}$	$Q^{J}_{I} \neq Q^{I}_{J}$	$Q^{I}_{J} \neq Q^{J}_{I}$
$P^{IJ} \neq P^{JI}$	$P_{IJ} \neq P_{JI}$	$Q_{IJ} \neq Q_{JI}$	$Q^{IJ} \neq Q^{JI}$

Partial Derivatives Orthonormality Identity:

$$Q^{M}_{I}\,P^{I}_{N} = \delta^{M}_{N} = \delta^{M}_{N} = P_{N}^{J}\,Q_{J}^{M} \quad\Big|\quad P^{M}_{I}\,Q^{I}_{N} = \delta^{M}_{N} = \delta^{M}_{N} = Q_{N}^{J}\,P_{J}^{M}$$

Affine:

Affine P=Q Partial Derivatives:

P(I,J)	=	Q(J,I)

But:

Affine Conjugate Partial Derivatives are not necessarily Symmetric:

$P(I,J) \neq P(J,I)$	$Q(I,J) \neq Q(J,I)$

Affine Orthonormality Identity:

$$Q(M,I)\,P(I,N) = \delta(M,N) = \delta(N,M) = P(N,J)\,Q(J,M)$$

Metric Difference:

$$2\,e^{J}_{M}\,dq_J\,dq^M \equiv \qquad\qquad \equiv 2\,E^{J}_{M}\,dp_J dp^M$$

$$(\delta^{J}_{M} - P_{K}^{J}\,P^{K}_{M})\,dq_J\,dq^M \equiv \quad \equiv (dq)^2 - (dp)^2 \equiv \quad \equiv (Q_{K}^{J}\,Q^{K}_{M} - \delta^{J}_{M})\,dp_J\,dp^M$$

$$2\,IPP^{J}_{M}\,dq_J\,dq^M \equiv \qquad\qquad \equiv 2\,QQI^{J}_{M}\,dp_J dp^M$$

Relative Metric Difference:

2 e ≡ 2 IPP ≡ (1- PP)	2 E ≡ 2 QQI ≡ (QQ -1)
Almansi's Strain: Eulerian Strain	Green's Strain: Lagrangian Strain
$2\,e^{J}_{M} \equiv 2\,IPP^{J}_{M} \equiv \delta^{J}_{M} - P_{K}^{J}\,P^{K}_{M}$	$2\,E^{J}_{M} \equiv 2\,QQI^{J}_{M} \equiv Q_{K}^{J}\,Q^{K}_{M} - \delta^{J}_{M}$
$2\,e_{M}^{J} \equiv 2\,IPP_{M}^{J} \equiv \delta_{M}^{J} - P^{K}_{M}\,P_{K}^{J}$	$2\,E_{M}^{J} \equiv 2\,QQI_{M}^{J} \equiv Q^{K}_{M}\,Q_{K}^{J} - \delta_{M}^{J}$

Where:

$$\underline{p} = \underline{q} - \underline{r} \qquad\qquad \underline{q} = \underline{p} + \underline{r}$$

Differential:

$$dp^I = (\delta^{I}_{J} - rq^{I}_{J})\,dq^J \qquad\qquad dq^I = (\delta^{I}_{J} + rp^{I}_{J})\,dp^J$$

$$dp_I = (\delta_{I}^{J} - rq_{I}^{J})\,dq_J \qquad\qquad dq_I = (\delta_{I}^{J} + rp_{I}^{J})\,dp_J$$

$$2e^{J}_{M} \equiv 2e_{M}^{J}$$
$$= \delta^{J}_{M} - (\delta_{K}^{J} - rq_{K}^{J})(\delta^{K}_{M} - rq^{K}_{M}) =$$
$$rq_{M}^{J} + rq^{J}_{M} - rq_{K}^{J}\,rq^{K}_{M}$$

$$2E^{J}_{M} \equiv 2E_{M}^{J}$$
$$= (\delta_{K}^{J} + rp_{K}^{J})(\delta^{K}_{M} + rp^{K}_{M}) - \delta^{J}_{M} =$$
$$rp_{M}^{J} + rp^{J}_{M} + rp_{K}^{J}\,rp^{K}_{M}$$

But:

$$e^{J}_{M} \neq e^{M}_{J} \equiv e_{J}^{M} \neq e_{M}^{J} \qquad\qquad E^{J}_{M} \neq E^{M}_{J} \equiv E_{J}^{M} \neq E_{M}^{J}$$

Approximations:

Approximate Twice Relative Length:

$$2 \approx 1 + P \equiv (dq + dp) / dq \approx 2 \approx (dq + dp) / dp \equiv Q + 1 \approx 2$$

Then:

$$2\,e \equiv 2\,IPP \equiv (1 - PP) \approx 2\,(1 - P) \quad\Big|\quad 2\,E \equiv 2\,QQI \equiv (QQ - 1) \approx 2\,(Q - 1)$$

Introduce an Approximation:

Length Ratio Approximation:

$$P \approx 1 - e \qquad\qquad Q \approx 1 + E$$

Introduce another Approximation:

Engineering Strain:

spatial Engineering strain: $1 - P$ material Engineering strain: $Q - 1$

Infinitesimal Engineering strain ≡ material Engineering strain: $\varepsilon = Q - 1$

$$\varepsilon = Q - 1 = (1-P)\,Q \approx (1-P) \approx e \qquad\qquad \varepsilon = Q - 1 \approx E$$

Introduce another Approximation:

Metric Difference Approximation:

$$(rq^I_{\ J}+rq^J_{\ I})\ dq_I dq^J \approx 2e\ (dq)^2 \approx 2\ dq\ (dq-dp) \equiv 2\ (dq)^2\ [\ 1-P]$$

$$\approx (dq)^2-(dp)^2 \approx$$

$$(rp^I_{\ J}+rp^J_{\ I})\ dp_I dp^J \approx 2E(dp)^2 \approx 2dp(dq-dp) \equiv 2\ (dp)^2\ [\ Q-1]$$

Or:

Metric Difference Ratio Approximation:

$$(dq+dp)(dq-dp)/dq^2 \equiv (dq^2-dp^2)/dq^2 \equiv 2e \approx 2E \equiv (dq^2-dp^2)/dp^2 \equiv (dq+dp)(dq-dp)/dp^2$$

$$(dq)^2\ [\ 1-PP]\ \equiv 2e \approx 2E \equiv (dp)^2\ [\ QQ-1]$$

Engineering strain equals approximately half its metric difference ratio:

$$1\ -\ dp/dq\ \approx 1/2\ (1-\ dp^2/dq^2) \qquad dq/\ dp\ -\ 1 \approx 1/2\ (dq^2/\ dp^2\ -\ 1)$$

$$1\ -\ dp/dq\ \approx IPP^I_{\ M}\ dq_I\ dq^M\ /\ dq_K\ dq^K \qquad dq/dp\ -\ 1 \approx QQI^I_{\ M}\ dp_I\ dp^M\ /\ dp_K\ dp^K$$

Engineering strain is approximately half metric difference ratio:

$$2e \approx 2\ \varepsilon \approx 2(\ rq^I_{\ J}\ +rq^J_{\ I}\ ) \approx 2E$$

Since

$$P^2 \equiv dp^2/dq^2 = 1\text{-}2e = 1\ -\ (1-\ dp^2/\ dq^2) \qquad Q^2 \equiv dq^2/dp^2 = 1+2E = 1 + (dq^2/dp^2-1)$$

Or:

$$P^2 \equiv dp^2/dq^2 = 1\ -\ (dq^2\ -\ dp^2)\ /\ dq^2 \qquad Q^2 \equiv dq^2/dp^2\ = 1+ (dq^2\ -\ dp^2)\ /\ dp^2$$

Then:

The Length Ratio:

$$P \equiv dp/dq = [1\ -\ (dq^2\ -\ dp^2)\ /\ dq^2]^{1/2} \qquad Q \equiv dq/dp = [1+ (dq^2\ -\ dp^2)\ /\ dp^2]^{1/2}$$

Approximated:

$$P \approx 1\ -\ 1/2\ \ (dq^2\ -\ dp^2)\ /\ dq^2 + \ \ldots \qquad Q \approx 1 + 1/2\ \ (dq^2\ -\ dp^2)/\ dp^2\ +\ \ldots$$

Merge these by:

Defining the Length Elongation Ratio, ε by the Engineering Strain:

$$1/[dp/dq]\text{-}1 \equiv dq/dp\ \text{-}1 \equiv Q-1 \equiv (dq-dp)\ /\ dp \equiv \varepsilon$$

Then Engineering Strain, ε is Approximated:

$$1/[1\text{-}1/2(dq^2\text{-}dp^2)/dq^2+\ldots]\text{-}1\ \approx\ \varepsilon\ \approx\ 1+1/2(dq^2\text{-}dp^2)/dp^2+\ldots\ \ \text{-}1$$

Or:

$$1 + 1/2(dq^2\text{-}dp^2)/dq^2\text{-}\ 1\ \approx\ \varepsilon \approx\ \ 1/2\ \ (dq^2\ -\ dp^2)/dp^2$$

$$1/2\ (2\ e)\ \approx\ \varepsilon \approx\ 1/2\ (2\ E)$$

$$1/2\ \ (dq^2\text{-}dp^2)/dq^2\ \approx\ \varepsilon \approx 1/2\ (dq^2\ -\ dp^2)/\ dp^2$$

$$1\ -\ P\ \ = (dq-dp\ )/dq\ \approx\ \varepsilon \equiv (dq\ -\ dp)/\ dp\ =\ Q\ -\ 1$$

$$1\ -\ P\ \approx\ \varepsilon\ \equiv\ Q\ -\ 1$$

Sum of Left and Right Sides:

$$2\ \varepsilon\ \approx 1\ -\ P\ + Q\ -\ 1\ \equiv\ Q-P \equiv dq/dp\ -\ dp/dq$$

$$\varepsilon\ \approx 1/2\ (Q-P)$$

Compatability 1: $\varepsilon_{\underline{K}\,\underline{K}\,,I\,I} + \varepsilon_{I\,I,\underline{K}\,\underline{K}} = 2\,\varepsilon_{\underline{K}\,I\,,\underline{K}\,I}$

10.3.
Infinitesimal Strain Tensor:

Infinitesimal Strain Tensor:

$$\varepsilon^I_{\ J} \quad = \quad 1/2\ \ (\ \ \partial\, r^I / \partial q^J\ \ + \partial\, r^J/\partial q^I\ \ )$$

Engineering strain is approximately half metric difference ratio:

$$||dq||^2 - ||dp||^2 \cong 2 * \varepsilon^I{}_J \, dq_I \, dq^J = 2 * \; 1/2 \; (rq^I{}_J + rq^J{}_I) \, dq_I \, dq^J$$

Global Coordinates o and General a are Two coordinate systems:

$a^j = a^j(o^i, t)$ $\quad a_N = a_N(o_M, t)$ $\quad o^j = o^j(a^i, t)$ $\quad o_N = o_N(a_M, t)$

Two Timed Descriptions: At Initial Time t = 0 and At Time t:

At t= 0 Initial Material Description: Lagrangian $p = p(q \;, t)$

At t Current Spatial Description: Eulerian $q = q(p \;, t)$

$q^i = q^i(p^j, t)$ $\quad q_i = q_i(p^j, t)$ $\quad p^i = p^i(q^j, t)$ $\quad p_i = p_i(q_j, t)$

A Domain Description:

Current (Spatial) Description Initial-Material Description

$q^i = q^i(p^j, t) = a^i(o^j, t)$ $\quad q^i = q^i(p_j, t) = a^i(o_j, t)$ $\quad p^i = p^i(q^j, t) = o^i(a^j, t)$ $\quad p^i = p^i(q_j, t) = o^i(a_j, t)$

$q_i = q_i(p^j, t) = a_i(o^j, t)$ $\quad q_i = q_i(p_j, t) = a_i(o_j, t)$ $\quad p_i = p_i(q^j, t) = o_i(a^j, t)$ $\quad p_i = p_i(q_j, t) = o_i(a_j, t)$

Eample Material Density ρ:

$$\rho(q^i, t) = \rho[a^i(o^j, t), t] = \rho[\, a^i(o_j, t), t] = \rho[\, a_i(o^j, t), t] = \rho[a_i(o_j, t), t] = \rho(q_i, t)$$

$$\rho(p^i, t) = \rho[o^i(a^i, t), t\,] = \rho[o^i(a_j, t), t] = \rho[o_i(a^i, t), t\,] = \rho[o_i(a_j, t), t] = \rho(p_i, t)$$

Position Gradients Components:

Material Gradient with respect to Material

$\partial p^N/\partial p^n = \delta^N{}_n$ $\quad \partial p_N/\partial p_n = \delta^n{}_N$

Material Gradient with respect to Current

$\partial p^N/\partial q^n = pq^N{}_n \equiv P^N{}_n$ $\quad \partial p_N/\partial q_n = pq_N{}^n \equiv P^n{}_N$

Current Gradient with respect to Material

$\partial q^N/\partial p^n = qp^N{}_n \equiv Q^N{}_n$ $\quad \partial q_N/\partial p_n = qp_N{}^n \equiv Q^n{}_N$

Current Gradient with respect to Current

$\partial q^N/\partial q^n = \delta^N{}_n$ $\quad \partial q_N/\partial q_n = \delta^n{}_N$

Position Gradient:

Material Gradient with respect to Material

$\underline{p}^n \, \delta^N{}_n = \underline{\text{Grad}} \; \underline{p} = \underline{p}_n \, \delta^n{}_N$

Material Gradient with respect to Current

$pq^N{}_n \, \underline{q}^n = \underline{\text{grad}} \; \underline{p} = pq_N{}^n \, \underline{q}_n$

Current Gradient with respect to Material

$\underline{p}^n \, qp^N{}_n = \underline{\text{Grad}} \; \underline{q} = \underline{p}_n \, qp^n{}_N$

Current Gradient with respect to Current

$\underline{q}^n \, \delta^N{}_n = \underline{\text{grad}} \; \underline{q} = \underline{q}_n \, \delta^n{}_N$

Velocity:

Knowing Fixed Global Coordinates (Local Material) $\underline{p}$ Then:

0 : Fixed Global Material Particle Velocity $\underline{u} = 0$:

$$\partial \underline{p}/\partial t = \underline{p}t = \underline{u} = 0$$

$$d\underline{p}/dt \equiv \underline{p}T = \partial \underline{p}/\partial t + \partial \underline{p}/\partial \underline{p} \,.\, \partial \underline{p}/\partial t = 0$$

$$\underline{p}_n \, u^n \equiv \underline{p}_n \, pT^n \equiv \underline{p}_n \, pt^n = 0 = \underline{p}^n \, pt_n \equiv \underline{p}^n \, pT_n \equiv \underline{p}^n \, u_n$$

Knowing Local Coordinates (Spatial position) $\underline{q}$ is function of $\underline{p}$ and t Then:

$\underline{v}$: Local Spatial Particle Velocity $\underline{v} \neq 0$:

$$\partial \underline{q}/\partial t = \underline{q}t = \underline{v}$$

$$d\underline{q}/dt \equiv \underline{q}T = \partial \underline{q}/\partial t + \partial \underline{q}/\partial \underline{p} \,.\, \partial \underline{p}/\partial t = \underline{q}t + 0 = \underline{q}t \equiv \underline{v}$$

$$\underline{q}_n \, v^n \equiv \underline{q}_n \, qT^n \equiv \underline{q}_n \, qt^n \neq 0 \neq \underline{q}^n \, qt_n \equiv \underline{q}^n \, qT_n \equiv \underline{q}^n \, v_n$$

Velocity Gradients: Convections:

Material Gradient of Global Velocity:

$\partial u_n/\partial p_N \equiv up_n{}^N$ $\quad \partial u^n/\partial p^N \equiv up^n{}_N$

Local Gradient of Global Velocity:

$\partial u_n/\partial q_N \equiv uq_n{}^N$ $\quad \partial u^n/\partial q^N \equiv uq^n{}_N$

Material Gradient of Local Velocity

Local Gradient of Local Velocity:

$K_n^{\;N} \equiv vp_n^{\;N} \equiv \partial v_n / \partial p_N$ $\quad K^n_{\;N} \equiv vp^n_{\;N} \equiv \partial v^n / \partial p^N$ $\quad L_n^{\;N} \equiv vq_n^{\;N} \equiv \partial v_n / \partial q_N$ $\quad L^n_{\;N} \equiv vq^n_{\;N} \equiv \partial v^n / \partial q^N$

$vp_n^{\;N} = vq_n^{\;K}\, qp_K^{\;N}$ $\quad vp^n_{\;N} = vq^n_{\;K}\, qp^K_{\;N}$ $\quad vq_n^{\;N} = vp_n^{\;K}\, pq_K^{\;N}$ $\quad vq^n_{\;N} = vp^n_{\;K}\, pq^K_{\;N}$

$K_n^{\;N} = L_n^{\;K}\, Q_K^{\;N}$ $\quad K^n_{\;N} = L^n_{\;K}\, Q^K_{\;N}$ $\quad L_n^{\;N} = K_n^{\;K}\, P_K^{\;N}$ $\quad L^n_{\;N} = K^n_{\;K}\, P^K_{\;N}$

Velocity Gradients Trace: Convections Trace:

$L^n_{\;K}\, Q^K_{\;n} = (K \equiv vp) = L_n^{\;K}\, Q_K^{\;n}$ $\qquad K^n_{\;K}\, P^K_{\;n} = (L \equiv vq) = K_n^{\;K}\, P_K^{\;n}$

Velocity Gradient: Convection:

Material Gradient of Global Velocity: $\qquad$ Local Gradient of Global Velocity

$up_n^{\;N}\, \underline{p}^n = \underline{\text{Grad}}\ \underline{u} = 0 = up_N^{\;n}\, \underline{p}_n$ $\qquad uq_n^{\;N}\, \underline{q}^n = \underline{\text{grad}}\ \underline{u} = uq_N^{\;n}\, \underline{q}_n$

Material Gradient of Local Velocity: $\qquad$ Local Gradient of Local Velocity:

$\underline{p}^n\, vp_n^{\;N} = \underline{\text{Grad}}\ \underline{v} = \underline{p}_n\, vp^n_{\;N}$ $\qquad \underline{q}^n\, vq_n^{\;N} = \underline{\text{grad}}\ \underline{v} = \underline{q}_n\, vq^n_{\;N}$

$\underline{p}^n\, K_n^{\;N} = \underline{\text{Grad}}\ \underline{v} = \underline{p}_n\, K^n_{\;N}$ $\qquad \underline{q}^n\, L_n^{\;N} = \underline{\text{grad}}\ \underline{v} = \underline{q}_n\, L^n_{\;N}$

$\underline{p}^n\, L_n^{\;K}\, Q_K^{\;N} = \underline{\text{Grad}}\ \underline{v} = \underline{p}_n\, L^n_{\;K}\, Q^K_{\;N}$ $\qquad \underline{q}^n\, K_n^{\;K}\, P_K^{\;N} = \underline{\text{grad}}\ \underline{v} = \underline{q}_n\, K^n_{\;K}\, P^K_{\;N}$

Two coordinate systems: o and x

$x^j = x^j(o^i, t)$ $\qquad x_N = x_N(o_M, t)$ $\qquad o^j = o^j(x^i, t)$ $\qquad o_N = o_N(x_M, t)$

Two Timed Descriptions: At Initial Time t = 0 and At Time t:

At t= 0 Initial Material Description: Lagrangian $\quad p = p(q\ \ , t)$

At t $\quad$ Current Spatial Description: Eulerian $\quad q = q(p\ \ , t)$

$q^i = q^i(p^j, t) = x^{ji}(o^i, t)$ $\qquad q_i = q_i(p^j, t) = x_{ij}(o_j, t)$ $\qquad p^i = p^i(q^j, t) = o^{ji}(x^i, t)$ $\qquad p_i = p_i(q_j, t) = o_{ij}(x_j, t)$

So:

A Domain Description:

A quantity, (say material density): ρ

Current (Spatial) Description $\rho(\underline{q}, t)$ $\qquad$ Initial-Material Description $\rho(\underline{p}, t)$

$\rho(q^i, t) = \rho[x^{ji}(o^i, t), t]\,] = \rho[\,x_{NM}(o_M, t), t] = \rho(q_N, t)$ $\qquad \rho(p^i, t) = \rho[\,o^{ji}(x^i, t), t]\,] = \rho[\,o_{NM}(x_M, t), t] = \rho(p_N, t)$

Position Gradients:

Material Gradient with respect to Material $\qquad$ Material Gradient with respect to Current

$\partial p^N / \partial p^n = \delta^N_{\;n}$ $\qquad \partial p_N / \partial p_n = \delta_N^{\;n}$ $\qquad \partial p^N / \partial q^n = pq^N_{\;n} \equiv P^N_{\;n}$ $\qquad \partial p_N / \partial q_n = pq_N^{\;n} \equiv P_N^{\;n}$

Current Gradient with respect to Material $\qquad$ Current Gradient with respect to Current

$\partial q^N / \partial p^n = qp^N_{\;n} \equiv Q^N_{\;n}$ $\qquad \partial q_N / \partial p_n = qp_N^{\;n} \equiv Q_N^{\;n}$ $\qquad \partial q^N / \partial q^n = \delta^N_{\;n}$ $\qquad \partial q_N / \partial q_n = \delta_N^{\;n}$

No Delta Position Gradient Components:

Material Gradient with respect to Material $\qquad$ Material Gradient with respect to Current

$\delta^N_{\;n}\, \underline{p}^n = \underline{\text{Grad}}\ \underline{p} = \delta_N^{\;n}\, \underline{p}_n$ $\qquad pq^N_{\;n}\, \underline{q}^n = \underline{\text{grad}}\ \underline{p} = pq_N^{\;n}\, \underline{q}_n$

Current Gradient with respect to Material $\qquad$ Current Gradient with respect to Current

$qp^N_{\;n}\, \underline{p}^n = \underline{\text{Grad}}\ \underline{q} = qp_N^{\;n}\, \underline{p}_n$ $\qquad \delta^N_{\;n}\, \underline{q}^n = \underline{\text{grad}}\ \underline{q} = \delta_N^{\;n}\, \underline{q}_n$

Velocity:

Knowing Fixed Global Coordinates (Local Material) $\underline{p}$ Then:

0 : Fixed Global Material Particle Velocity $\underline{u} = 0$:

$$\partial \underline{p}/\partial t = \underline{p}t = \underline{u} = 0$$

$$d\underline{p}/dt \equiv \underline{p}T = \partial \underline{p}/\partial t + \partial \underline{p}/\partial \underline{p} \,.\, \partial \underline{p}/\partial t = 0$$

$$\underline{p}_n\, u^n \equiv \underline{p}_n\, pT^n \equiv \underline{p}_n\, pt^n = 0 = \underline{p}^n\, pt_n \equiv \underline{p}^n\, pT_n \equiv \underline{p}^n\, u_n$$

Knowing Local Coordinates (Spatial position) $\underline{q}$ is function of $\underline{p}$ and t Then:

$\underline{v}$: Local Spatial Particle Velocity $\underline{v} \neq 0$:

$$\partial \underline{q}/\partial t = \underline{q}t = \underline{v}$$

$$d\underline{q}/dt \equiv \underline{q}T = \partial \underline{q}/\partial t + \partial \underline{q}/\partial \underline{p} \,.\, \partial \underline{p}/\partial t = \underline{q}t + 0 = \underline{q}t \equiv \underline{v}$$

$$\underline{q}_n\, v^n \equiv \underline{q}_n\, qT^n \equiv \underline{q}_n\, qt^n \neq 0 \neq \underline{q}^n\, qt_n \equiv \underline{q}^n\, qT_n \equiv \underline{q}^n\, v_n$$

Velocity Gradients: Convection:

Global Gradient of Global Velocity

$$up^{N}_{\ n}\,\underline{p}^{n} = \underline{\mathrm{Grad}}\ \underline{u} = up_{N}^{\ n}\,\underline{p}_{n}$$

Local Gradient of Global Velocity

$$uq^{N}_{\ n}\,\underline{q}^{n} = \underline{\mathrm{grad}}\ \underline{u} = uq_{N}^{\ n}\,\underline{q}_{n}$$

Global Gradient of Local Velocity

$$vp^{N}_{\ n}\,\underline{p}^{n} = \underline{\mathrm{Grad}}\ \underline{v} = vp_{N}^{\ n}\,\underline{p}_{n}$$

Local Gradient of Local Velocity

$$vp^{N}_{\ n}\,\underline{q}^{n} = \underline{\mathrm{grad}}\ \underline{v} = vp_{N}^{\ n}\,\underline{q}_{n}$$

Or:

Material Gradient with respect to Material

$$\partial u^{N}/\partial p^{n} \equiv up^{N}_{\ n} \qquad \partial u_{N}/\partial p_{n} \equiv up_{N}^{\ n}$$

Material Gradient with respect to Current

$$\partial u^{N}/\partial q^{n} \equiv uq^{N}_{\ n} \qquad \partial u_{N}/\partial q_{n} \equiv uq_{N}^{\ n}$$

Current Gradient with respect to Material

$$K^{m}_{\ n} \equiv vp^{m}_{\ n} \equiv \partial v^{m}/\partial p^{n} \qquad \partial v_{n}/\partial p_{m} \equiv vp_{n}^{\ m} \equiv K_{n}^{\ m}$$

$$vp^{m}_{\ n} = vq^{m}_{\ K}\,qp^{K}_{\ n} \qquad vp_{n}^{\ m} = vq_{n}^{\ K}\,qp_{K}^{\ m}$$

$$K^{m}_{\ n} = L^{m}_{\ K}\,qp^{K}_{\ n} \qquad K_{n}^{\ m} = L_{n}^{\ K}\,qp_{K}^{\ m}$$

Current Gradient with respect to Current

$$L^{m}_{\ n} \equiv vq^{m}_{\ n} \equiv \partial v^{m}/\partial q^{n} \qquad \partial v_{n}/\partial q_{m} \equiv vq_{n}^{\ m} \equiv L_{n}^{\ m}$$

$$vq^{m}_{\ n} = vp^{m}_{\ K}\,pq^{K}_{\ n} \qquad vq_{n}^{\ m} = vp_{n}^{\ K}\,pq_{K}^{\ m}$$

$$L^{m}_{\ n} = K^{m}_{\ K}\,pq^{K}_{\ n} \qquad L_{n}^{\ m} = K_{n}^{\ K}\,pq_{K}^{\ m}$$

Velocity Gradients Trace: Convection Trace:

Current Gradient with respect to Material

$$L^{n}_{\ K}\,qp^{K}_{\ n} = (K \equiv vp) = L_{n}^{\ K}\,qp_{K}^{\ n}$$

Current Gradient with respect to Current

$$K^{n}_{\ K}\,pq^{K}_{\ n} = (L \equiv vq) = K_{n}^{\ K}\,pq_{K}^{\ n}$$

And:

Material Gradient with respect to Material

$$\underline{p}_{N}\,up^{N}_{\ ;n} \equiv \underline{p}_{N}\,pTp^{N}_{\ ;n} \equiv \underline{p}_{N}\,ptp^{N}_{\ ;n}$$

$$= 0 =$$

$$\underline{p}^{N}\,ptp_{N}^{\ ;n} \equiv \underline{p}^{N}\,pTp_{N}^{\ ;n} \equiv \underline{p}^{N}\,up_{N}^{\ ;n}$$

Material Gradient with respect to Current

Current Gradient with respect to Material

Current Gradient with respect to Current

$$\underline{q}_{N}\,vq^{N}_{\ ,n} \equiv \underline{q}_{N}\,qTq^{N}_{\ ,n} \equiv \underline{q}_{N}\,qtq^{N}_{\ ,n}$$

$$\neq 0 \neq$$

$$\underline{q}^{N}\,qtq_{N}^{\ ,n} \equiv \underline{q}^{N}\,qTq_{N}^{\ ,n} \equiv \underline{q}^{N}\,vq_{N}^{\ ,n}$$

or:

$$\underline{q}_{N}\,vp^{N}_{\ ;n}\,pq^{n}_{\ ,m} \equiv \underline{q}_{N}\,qTp^{N}_{\ ;n}\,pq^{n}_{\ ,m} \equiv \underline{q}_{N}\,qtp^{N}_{\ ;n}\,pq^{n}_{\ ,m}$$

$$\neq 0 \neq$$

$$\underline{q}^{N}\,qtp_{N}^{\ ;n}\,pq_{n}^{\ ,m} \equiv \underline{q}^{N}\,qTq_{N}^{\ ;n}\,pq_{n}^{\ ,m} \equiv \underline{q}^{N}\,vp_{N}^{\ ;n}\,pq_{n}^{\ ,m}$$

or:

$$\underline{q}_{N}L^{N}_{\ ,n} = \underline{q}_{N}K^{N}_{\ ;n}\,pq^{n}_{\ ,m} \neq 0 \neq \underline{q}^{N}K_{N}^{\ ;n}\,pq_{n}^{\ ,m} = \underline{q}^{N}L_{N}^{\ ,n}$$

Deformation, Strain and Rate:

Domain Description:

Two coordinate systems: o and x

$$x^{j} = x^{j}(o^{i}, t) \qquad o^{j} = o^{j}(x^{i}, t)$$

$$x_{N} = x_{N}(o_{M}, t) \qquad o_{N} = o_{N}(x_{M}, t)$$

At t= 0 Initial Material Description: Lagrangian $p^{i} = p^{i}(q^{j}, t)$

At t Current Spatial Description: Eulerian $q^{i} = q^{i}(p^{j}, t)$

$$q^{i} = q^{i}(p^{j}, t) \qquad p^{i} = p^{i}(q^{j}, t)$$

$$q_{N} = q_{N}(p^{j}, t) \qquad p_{N} = p_{N}(q_{M}, t)$$

A quantity, (say material density): ρ

Current-Spatial Description | Initial-Material Description

$\rho = \rho(\underline{q}, t) = \rho(q^i, t) = \rho(x^i, t)$ $\quad\quad$ $\rho = \rho(\underline{p}, t) = \rho(p^i, t) = \rho(o^i, t)$

$\rho = \rho(\underline{q}, t) = \rho(q_N, t) = \rho(x_N, t)$ $\quad\quad$ $\rho = \rho(\underline{p}, t) = \rho(p_N, t) = \rho(o_N, t)$

Where:

$q^j = q^j(p^i, t) = x^{ji}(o^i, t)$ $\quad\quad$ $p^j = p^j(q^i, t) = o^{ji}(x^i, t)$

$q_N = q_N(p_M, t) = x_{NM}(o_M, t)$ $\quad\quad$ $p_N = p_N(q_M, t) = o_{NM}(x_M, t)$

Derivative with respect to Vector $\underline{r}$:

Gradient of a Tensor R: $\underline{\text{Del}}\ R \equiv \underline{\nabla}\ R$

$\partial R/\partial \underline{q}$ $\quad\quad$ $\partial R/\partial \underline{p}$

Gradient of a Tensor R:

$\partial R/\partial q_M\ \underline{q}_M = Rq^M\ \underline{q}_M$ $\quad\quad$ $\partial R/\partial p_M\ \underline{p}_M = Rp^M\ \underline{p}_M$

$\partial R/\partial q^M\ \underline{q}^M = Rq_M\ \underline{q}^M$ $\quad\quad$ $\partial R/\partial p^M\ \underline{p}^M = Rp_M\ \underline{p}^M$

Gradient of a Coordinate:

$\partial p_N/\partial q_M\ \underline{q}_M = P_N^M\ \underline{q}_M$ $\quad\quad$ $\partial q_N/\partial p_M\ \underline{p}_M = Q_N^M\ \underline{p}_M$

$\partial p^N/\partial q^M\ \underline{q}^M = P^N_M\ \underline{q}^M$ $\quad\quad$ $\partial q^N/\partial p^M\ \underline{p}^M = Q^N_M\ \underline{p}^M$

Time Derivative:

Coordinate Per Time is Velocity

Global Velocity

$u_N \equiv pt_N \equiv \partial p_N/\partial t$

$u^N \equiv pt^N \equiv \partial p^N/\partial t$

Local Velocity

$v_N \equiv qt_N \equiv \partial q_N/\partial t$

$v^N \equiv qt^N \equiv \partial q^N/\partial t$

Fixed Global Coordinates, $\underline{p}$:

Material position: $\underline{p}$ is fixed: $\partial \underline{p}/\partial t \equiv \underline{p}t = 0$:

Material Particle Velocity:

$D\underline{p}/Dt \equiv d\underline{p}/dt \equiv \underline{p}T = \partial \underline{p}/\partial t + \partial \underline{p}/\partial \underline{p}\ .\ \partial \underline{p}/\partial t = 0$

Spatial position: $\underline{q}$ is function of $\underline{p}$ and of t:

Spatial Particle Velocity:

$D\underline{q}/Dt \equiv \underline{q}T = \partial \underline{q}/\partial t + \partial \underline{q}/\partial \underline{p}\ .\ \partial \underline{p}/\partial t = \underline{q}t + 0 = \underline{q}t \equiv \underline{v}$

$u_N \equiv pt_N \equiv \partial p_N/\partial t = 0$ $\quad\quad$ $v_N \equiv qt_N \equiv \partial q_N/\partial t \neq 0$

$u^N \equiv pt^N \equiv \partial p^N/\partial t = 0$ $\quad\quad$ $v^N \equiv qt^N \equiv \partial q^N/\partial t \neq 0$

Gradient of a Local Velocity ≡ Gradient of a Coordinate Per Time ≡ Convection:

Since $\partial u_N/\partial p_K \equiv 0$ it is of no use later $\quad\quad$ $\partial v_N/\partial p_n\ \underline{p}_n = \partial v_N/\partial q_K\ \partial q_K/\partial p_n\ \underline{p}_n$

Since $\partial u^N/\partial p^K \equiv 0$ it is of no use later $\quad\quad$ $\partial v^N/\partial p^n\ \underline{p}^n = \partial v^N/\partial q^K\ \partial q^K/\partial p^n\ \underline{p}^n$

Convection:

$$\partial v_N/\partial p_n\ \underline{p}_n = \partial\, qt_N/\partial q_K\ \partial q_K/\partial p_n\ \underline{p}_n$$

$$\partial v^N/\partial p^n\ \underline{p}^n = \partial\, qt^N/\partial q^K\ \partial q^K/\partial p^n\ \underline{p}^n$$

Time Rate of Deformation Material Gradient and Its Conjugate $\underline{K}$

Time Rate of Deformation Current Gradient and Its Conjugate: $\underline{L}$

Position Gradient Vector and Components:

Material Gradient

$$\underline{grad}\ \underline{p} \equiv {}^{K}\underline{P}$$
$$P^{N}{}_{n} \equiv pq^{K}{}_{n}$$
$$P_{N}{}^{n} \equiv pq_{K}{}^{n}$$

Current (Spatial) Gradient

$$\underline{Q}_{N} \equiv \underline{Grad}^{K}\ \underline{q}$$
$$Q^{n}{}_{n} \equiv qp^{n}$$
$$Q_{N}{}^{n} \equiv qp_{K}{}^{n}$$

Velocity Vector:

$$\underline{u} \equiv \underline{p}_{n}\, u^{n} \equiv \underline{p}_{n}\, pT^{n} \equiv \underline{p}_{n}\, pt^{n} \qquad \underline{v} \equiv \underline{q}_{n}\, v^{n} \equiv \underline{q}_{n}\, qT^{n} \equiv \underline{q}_{n}\, qt^{n}$$
$$\underline{u} \equiv \underline{p}^{n}\, u_{n} \equiv \underline{p}^{n}\, pT_{n} \equiv \underline{p}^{n}\, pt_{n} \qquad \underline{v} \equiv \underline{q}^{n}\, v_{n} \equiv \underline{q}^{n}\, qT_{n} \equiv \underline{q}^{n}\, qt_{n}$$

Components of Velocity:

Global Velocity Constant

$$u_{N} \equiv pt_{N} = pT_{N}$$
$$u^{N} \equiv pt^{N} = pT^{N}$$

Local Velocity

$$v_{N} \equiv qt_{N} = qT_{N} \neq 0$$
$$v^{N} \equiv qt^{N} = qT^{N} \neq 0$$

Global Gradient:

Global Gradient Vector of Velocity:

Global Gradient of Global Velocity≡0

Global Coordinate Reference

$$up_{N}{}^{;K}\, \underline{p}_{K} \equiv up_{N}{}^{K}\, \underline{p}_{K} \equiv 0$$
$$up^{N}{}_{;K}\, \underline{p}^{K} \equiv up^{N}{}_{K}\, \underline{p}^{K} \equiv 0$$

Global Gradient of Local Velocity

Global Convection

$$vp_{N}{}^{;K}\, \underline{p}_{K} \equiv vp_{N}{}^{n}\, \underline{p}_{n} = vq_{N}{}^{K}\, Q_{K}{}^{n}\, \underline{p}_{n}$$
$$vp^{N}{}_{;K}\, \underline{p}^{K} \equiv vp^{N}{}_{n}\, \underline{p}^{n} = vq^{N}{}_{K}\, Q^{K}{}_{n}\, \underline{p}^{n}$$

And:

Global Convection: Global Gradient Components of Local Velocity

$$qTp_{N}{}^{;n} \equiv QT_{N}{}^{;n} \equiv v_{N}{}^{;n} \equiv vp_{N}{}^{;n} \equiv vp_{N}{}^{n} = vq_{N}{}^{K}\, Q_{K}{}^{n} \equiv v_{N}{}^{K}\, Q_{K}{}^{n} \equiv L_{N}{}^{K}\, Q_{K}{}^{n}$$
$$qTp^{N}{}_{;n} \equiv QT^{N}{}_{;n} \equiv v^{N}{}_{;n} \equiv vp^{N}{}_{;n} \equiv vp^{N}{}_{n} = vq^{N}{}_{K}\, Q^{K}{}_{n} \equiv v^{N}{}_{K}\, Q^{K}{}_{n} \equiv L^{N}{}_{K}\, Q^{K}{}_{n}$$

Local Gradient:

Local Gradient Vector of Velocity:

Local Gradient of Local Velocity

Local Convection

$$vq_{N}{}^{,n}\, \underline{q}_{n} \equiv vq_{N}{}^{n}\, \underline{q}_{n} \equiv v_{N}{}^{,n}\, \underline{q}_{n} \equiv v_{N}{}^{n}\, \underline{q}_{n}$$
$$vq^{N}{}_{,n}\, \underline{q}^{n} \equiv vq^{N}{}_{n}\, \underline{q}^{n} \equiv v^{N}{}_{,n}\, \underline{q}^{n} \equiv v^{N}{}_{n}\, \underline{q}^{n}$$

And:

Local Convection: Local Gradient Components of Local Velocity

$$L_{N}{}^{K} \equiv v_{N}{}^{K} \equiv vq_{N}{}^{K} \equiv \partial(dqT_{N})/\partial q_{K} \neq 0$$
$$L^{N}{}_{K} \equiv v^{N}{}_{K} \equiv vq^{N}{}_{K} \equiv \partial(dqT^{N})/\partial q^{K} \neq 0$$

Global Gradient and Local Gradient are Defining:

Global Convection:

$$K^{N}{}_{n} \equiv vp^{N}{}_{n} \equiv vq^{N}{}_{K}\, Q^{K}{}_{n} \equiv v^{N}{}_{K}\, Q^{K}{}_{n}$$
$$K_{N}{}^{n} \equiv vp_{N}{}^{n} \equiv vq_{N}{}^{K}\, Q_{K}{}^{n} \equiv v_{N}{}^{K}\, Q_{K}{}^{n}$$

Local Convection:

$$L^{N}{}_{n} \equiv v^{N}{}_{n} \equiv vq^{N}{}_{n} \equiv vp^{N}{}_{K}\, P^{K}{}_{n}$$
$$L_{N}{}^{n} \equiv v_{N}{}^{n} \equiv vq_{N}{}^{n} \equiv vp_{N}{}^{K}\, P_{K}{}^{n}$$

Or:

$$K^{n}{}_{N} = L^{n}{}_{K}\, Q^{K}{}_{N}$$
$$vp^{n}{}_{N} = v^{n}{}_{K}\, Q^{K}{}_{N}$$
$$K_{N}{}^{n} = L_{N}{}^{K}\, Q_{K}{}^{n}$$
$$vp_{N}{}^{n} = v_{N}{}^{K}\, Q_{K}{}^{n}$$

$$L^{n}{}_{N} = K^{n}{}_{K}\, P^{K}{}_{N}$$
$$v^{n}{}_{N} = vp^{n}{}_{K}\, P^{K}{}_{N}$$
$$L_{N}{}^{n} = K_{N}{}^{K}\, P_{K}{}^{n}$$
$$v_{N}{}^{n} = vp_{N}{}^{K}\, P_{K}{}^{n}$$

And:

Trace of Global Convection:

$$K \equiv K^{n}_{n} \equiv vp^{n}_{n} = K_{n}^{n} \equiv vp_{n}^{n}$$

$$vp \equiv vp^{n}_{n} = vp_{n}^{n}$$

Trace of Local Convection:

$$L \equiv L^{n}_{n} \equiv vq^{n}_{n} \equiv v^{n}_{n} = L_{n}^{n} \equiv vq_{n}^{n} \equiv v_{n}^{n}$$

$$v \equiv v^{n}_{n} = v_{n}^{n}$$

Metric Difference:

$$2E^{J}_{M}\,dp_{J}dp^{M} = (Q_{K}^{J}Q^{K}_{M} - \delta^{J}_{M})\,dp_{J}dp^{M}$$

$$= dq^{2} - dp^{2} =$$

$$2e^{J}_{M}\,dq_{J}\,dq^{M} = (\delta^{J}_{M} - P_{K}^{J}P^{K}_{M})\,dq_{J}\,dq^{M}$$

Derivative of Metric Difference:

$$d(2E^{J}_{M}\,dp_{J}dp^{M})/dt \qquad d(2e^{J}_{M}\,dq_{J}\,dq^{M})/dt$$

Since $d(dp_{J})/dt = 0$, Left Side isFound Directly, Derivative of Metric Difference:

$$2ET^{J}_{M}\,dp_{J}\,dp^{M} = (QT_{K}^{J}Q^{K}_{M} + Q_{K}^{J}QT^{K}_{M})\,dp_{J}dp^{M} \quad \ldots \ldots \quad \ldots \ldots$$

If Right Side Derivative of Metric Difference is Treated:

$$\ldots \ldots \quad 2eT^{J}_{M}[dq_{J}dq^{M} + d(dq_{J}dq^{M})/dt] = 2eT^{J}_{M}[dq_{J}dq^{M} + dqT_{J}dq^{M} + dq_{J}dqT^{M}]$$

$$= (-PT_{K}^{J}P^{K}_{M} - P_{K}^{J}PT^{K}_{M})\,dq_{J}\,dq^{M} + (\delta^{J}_{M} - P_{K}^{J}P^{K}_{M})\,d(dq_{J}\,dq^{M})/dt$$

However:

Directly: $d(dq^{2} - dp^{2})/dt$:

Time Derivative of Metric Difference:

$$\ldots \ldots \quad = d(dq^{2} - dp^{2})/dt = \partial(dq)^{2}/\partial t = \partial(dq_{J}\,dq^{J})/\partial t = \quad \ldots \ldots$$

$$d(dq_{J}\,dq^{J})/dt = 2\,dq_{J}\,dqT^{J} \equiv dqT_{J}\,dq^{J} + dq_{K}\,dqT^{K}$$

$$= \partial(dqT_{J})/\partial q_{K}\,dq_{K}\,dq^{J} + dq_{K}\,\partial(dqT^{K})/\partial q^{J}\,dq^{J}$$

$$\equiv L_{J}^{K}\,dq_{K}\,dq^{J} + dq_{K}\,L^{K}_{J}\,dq^{J}$$

$$\equiv (L_{J}^{K} + L^{K}_{J})\,dq_{K}\,dq^{J} = 2L^{K}_{J}\,dq_{K}\,dq^{J} = 2L_{J}^{K}\,dq_{K}\,dq^{J}$$

So:

$$2ET^{N}_{n} = 2ET_{n}^{N}$$

$$K_{I}^{N}Q^{I}_{n} + Q_{I}^{N}K^{I}_{n} = K_{n}^{I}Q_{I}^{N} + Q_{n}^{I}K_{I}^{N}$$

$$vp_{I}^{N}Q^{I}_{n} + Q_{I}^{N}vp^{I}_{n} = vp_{n}^{I}Q_{I}^{N} + Q_{n}^{I}vp_{I}^{N}$$

$$L^{N}_{n} \equiv v^{N}_{n} = v_{n}^{N} \equiv L_{n}^{N}$$

$$L^{N}_{n} = K^{N}_{I}P^{I}_{n} = K_{n}^{I}P_{I}^{N} = L_{n}^{N}$$

$$v^{N}_{n} = vp^{N}_{I}P^{I}_{n} = vp_{n}^{I}P_{I}^{N} = v_{n}^{N}$$

Where:

$$2ET^{J}_{M}\,dp_{J}dp^{M} = (QT_{K}^{J}Q^{K}_{M} + Q_{K}^{J}QT^{K}_{M})\,dp_{J}dp^{M} = (vp_{K}^{J}Q^{K}_{M} + Q_{K}^{J}vp^{K}_{M})\,dp_{J}dp^{M}$$

$$= [(dq)^{2}]T = 2L_{J}^{K}\,dq_{K}\,dq^{J} = 2L^{K}_{J}\,dq_{K}\,dq^{J} = 2v_{J}^{K}\,dq_{K}\,dq^{J} = 2v^{K}_{J}\,dq_{K}\,dq^{J}$$

$$K^{n}_{N} = L^{n}_{K}\,Q^{K}_{N}$$

$$vp^{n}_{N} = v^{n}_{K}\,Q^{K}_{N}$$

$$K_{N}^{n} = L_{N}^{K}\,Q_{K}^{n}$$

$$vp_{N}^{n} = v_{N}^{K}\,Q_{K}^{n}$$

$$L^{n}_{N} = K^{n}_{K}\,P^{K}_{N}$$

$$v^{n}_{N} = vp^{n}_{K}\,P^{K}_{N}$$

$$L_{N}^{n} = K_{N}^{K}\,P_{K}^{n}$$

$$v_{N}^{n} = vp_{N}^{K}\,P_{K}^{n}$$

Divide $(L_{J}^{K} \equiv v_{J}^{K})$ and $(L^{K}_{J} \equiv v^{K}_{J})$ into two parts D_{JK} and W^{JK}:

Define:

Symmetric $$D_{JK} \equiv 1/2\,(L_{J}^{K} + L^{J}_{K}) \equiv +D_{KJ}$$

Antisymmetric $$W^{JK} \equiv 1/2\,(L_{J}^{K} - L^{J}_{K}) \equiv -W^{KJ}$$

Such That Sum:

$$D_{JK} + W^{JK} = L_J^{\ K} \equiv v_J^{\ K}$$

And Subtract:

$$D_{JK} - W^{JK} = L^J_{\ K} \equiv v^J_{\ K}$$

So:

$$(D_{JK} + W^{JK})\, dq_K\, dq^J \equiv L_J^{\ K}\, dq_K\, dq^J$$

$$(D_{JK} - W^{JK})\, dq_K\, dq^J \equiv L^J_{\ K}\, dq_J\, dq^K$$

Time Rate of Metric Difference

$$[(dq)^2]T = 2L_J^{\ K}\, dq_K\, dq^J = 2L^K_{\ J}\, dq_K\, dq^J$$

$$=$$

$$2(D_{JK}+W^{JK})dq_K dq^J = 2(D_{JK}-W^{JK})dq_K dq^J$$

Fixed Global Coordinates, $\underline{p}$:

Material position: $\underline{p}$ is fixed: $\partial\underline{p}/\partial t \equiv \underline{p}t = 0$:

Material Particle Velocity:

$$D\underline{p}/Dt \equiv d\underline{p}/dt \equiv \underline{p}T = \partial\underline{p}/\partial t + \partial\underline{p}/\partial\underline{p} \,.\, \partial\underline{p}/\partial t = 0$$

Spatial position: $\underline{q}$ is function of $\underline{p}$ and of t: $\underline{q} = \underline{q}(\underline{p}, t)$

Spatial Particle Velocity:

$$D\underline{q}/Dt \equiv \underline{q}T = \partial\underline{q}/\partial t + \partial\underline{q}/\partial\underline{p} \,.\, \partial\underline{p}/\partial t = \underline{q}t + 0 = \underline{q}t \equiv \underline{v}$$

$$\underline{q}(\underline{p}, t)$$

$$\underline{v} \equiv \underline{v}[\underline{q}(\underline{p}, t), t]$$

So Particle Velocity $\underline{v}$ is function of $\underline{q}$ and t , and in turn $\underline{q}$ is function of $\underline{p}$ and t.

$$v^N \equiv v^N [q^n (p^m, t), t]$$

$$v_N \equiv v_N [q_n (p_m, t), t]$$

Material Particle Function Time Derivative:

Any Material Function:

$$f = f[\underline{q}(\underline{p}, t), t]$$

$$f^{\cdot} \equiv Df/Dt \equiv fT = \partial f/\partial t + \partial f/\partial\underline{q} \,.\, \partial\underline{q}/\partial t$$

$f^{\cdot} \equiv$ Local rate + Convected rate

$$f^{\cdot} \equiv Df/Dt \equiv fT = \partial f/\partial t + \partial f/\partial\underline{q} \,.\, \underline{v}$$

Initial Material Description	Flow Spatial Description
At p^i = constant p_i = constant	At q^i = constant q_i = constant
Material Derivative: $d(f)/dt = (fT)$	Spatial Derivative: $\partial(f)/\partial t = (f)_{,t}$ Local
Global Material Particle Rate	Spatial Current Rate

$$f = f[\underline{q}(\underline{p}, t), t]$$

local function rate + Convected rate	local function rate
$fT \equiv df/dt = \partial f/\partial t + \partial f/\partial q^n \,.\, v^n$	$ft \equiv \partial f/\partial t$
$fT \equiv df/dt = \partial f/\partial t + \partial f/\partial q_n \,.\, v_n$	$ft \equiv \partial f/\partial t$

$$\rho = \rho[\underline{q}(\underline{p}, t), t]$$

local function rate + Convected rate

$\rho T \equiv d\rho/dt = \partial\rho/\partial t + \partial\rho/\partial q^{n} . v^{n}$

$\rho T \equiv d\rho/dt = \partial\rho/\partial t + \partial\rho/\partial q_{n} . v_{n}$

local function rate

$\rho t \equiv \partial\rho/\partial t$

$\rho t \equiv \partial\rho/\partial t$

Material Particle Acceleration and Convection:

Apply to Material Velocity $\underline{v} = \underline{v}[\underline{q}(\underline{p}, t), t]$

Material Acceleration:

$\partial\underline{v}/\partial t$

local acceleration=local velocity rate

$vt^{N} \equiv qtt^{N} \equiv \partial v^{N}/\partial t \equiv \partial\partial q_{N}/\partial t/\partial t$

$vt_{N} \equiv qtt_{N} \equiv \partial v_{N}/\partial t \equiv \partial\partial q^{N}/\partial t/\partial t$

$\underline{v}' = \partial\underline{v}/\partial t + \partial\underline{v}/\partial\underline{q} . \underline{v}$

local acceleration+Convected rate

$vT^{N} \equiv v^{N'} = vt^{N} + vq^{N}_{n} . v^{n}$

$vT_{N} \equiv v_{N}' = vt_{N} + vq_{N}^{n} . v_{n}$

Almansi's and Green's Strain Rate Tensors:

Recall Material Acceleration:

$\partial\underline{v}/\partial t$

local acceleration=local velocity rate

$vt_{N} \equiv qtt_{N} \equiv \partial v^{N}/\partial t \equiv \partial\partial q_{N}/\partial t/\partial t$

$vt^{N} \equiv qtt^{N} \equiv \partial v_{N}/\partial t \equiv \partial\partial q^{N}/\partial t/\partial t$

$\underline{v}' = \partial\underline{v}/\partial t + \partial\underline{v}/\partial\underline{q} . \underline{v}$

local acceleration+Convected rate

$v^{N'} = \partial v^{N}/\partial t + vq^{N}_{n} . v^{n}$

$v_{N}' = \partial v_{N}/\partial t + vq_{N}^{n} . v_{n}$

Recall: Convection $\equiv \partial/\partial t$

Almansi Eulerian Fluid in terms of q^{i}

Green Lagrangian Solid in terms of p^{i}

Local Convection:

$qtq^{N}_{n} \equiv vq^{N}_{n} = vp^{N}_{K} P^{K}_{n}$

$qtq_{N}^{n} \equiv vq_{N}^{n} = vp_{N}^{K} P_{K}^{n}$

Global Convection:

$qtp^{N}_{n} \equiv vp^{N}_{n} = vq^{N}_{K} Q^{K}_{n}$

$qtp_{N}^{n} \equiv vp_{N}^{n} = vq_{N}^{K} Q_{K}^{n}$

Define: d/dt

$Vq^{N}_{n} \equiv \partial V^{N}/\partial q^{n} = Vp^{N}_{K} P^{K}_{n}$

$Vp_{N}^{n} \equiv \partial V_{N}/\partial q_{n} = Vp_{N}^{K} P_{K}^{n}$

$Vp^{N}_{n} \equiv \partial V^{N}/\partial p^{n} \equiv dQ^{N}_{n}/dt \equiv d(Q^{N}_{n})/dt$

$Vp_{N}^{n} \equiv \partial V_{N}/\partial p_{n} \equiv dQ_{N}^{n}/dt \equiv d(Q_{N}^{n})/dt$

$$QQI^{I}_{M} = 12\,(Q_{K}^{I}\, Q^{K}_{M} - \delta^{I}_{M})$$

$$QQ^{I}_{M} = Q_{K}^{I}\, Q^{K}_{M}$$

$$VppI^{I}_{M} \equiv 12\; Vpp^{I}_{M} \equiv d\,(QQI^{I}_{M})/dt \equiv 12\; d\,(QQ^{I}_{M})/dt$$

$$Vpp^{I}_{M} = Vp_{K}^{I}\, Q^{K}_{M} + Q_{K}^{I}\; Vp^{K}_{M}$$

$$= \partial V_{K}/\partial p_{I}\; \partial q^{K}/\partial p^{M} + \partial q_{K}/\partial p_{I}\; \partial V^{K}/\partial p^{M}$$

And since:

$$(dq^{2} - dp^{2}) = 2\,(QQI^{I}_{M}\, dp_{I}\, dp^{M})$$

So

$$d(dq^2 - dp^2)/dt \quad = 2\, d(QQI^{I}_{M})/dt \;\; dp_I\, dp^M$$

$$= 2\, VppI^{/I}_{\backslash M}\, dp_I\, dp^M \qquad = Vpp^{/I}_{\backslash M}\, dp_I\, dp^M$$

$$Vpp^{/I}_{\backslash M}\, dp_I\, dp^M = (Vp_K{}^{I}\, Q^{K}_{M} + Q_K{}^{I}\;\; Vp^{K}_{M}) \qquad dp_I\, dp^M$$

$$= \qquad (\partial V_K/\partial p_I\;\; \partial q^K/\partial p^M \;+\; \partial q_K/\partial p_I\;\; \partial V^K/\partial p^M)\;\; dp_I\, dp^M$$

Previous QQI^{I}_{M} approach can not be used with IPP^{I}_{M} because:

$$d(dq^2 - dp^2)/dt \quad = 2\, d(IPP^{I}_{M}\, dq_I\, dq^M)/dt$$

$$\neq 2\, d[1/2\, (\delta^{I}_{M} - P_K{}^{I}\, P^{K}_{M}\;)]/dt\; dq_I\, dq^M$$

But substituting:

$$Vp^{N}_{\;n} \;=\; Vq^{N}_{\;L}\;\; Q^{L}_{\;n} = \;Vq^{N}_{\;L}\;\; Q^{L}_{\;n}$$

$$Vp_N{}^{n} \;=\; Vq_N{}^{L}\;\; Q_L{}^{n} = \;Vq_N{}^{L}\;\; Q_L{}^{n}$$

in

$$Vpp^{/I}_{\backslash M} \;=\; Vp_K{}^{I}\, Q^{K}_{M} + Q_K{}^{I}\;\; Vp^{K}_{M}$$

$$= \;\; Vq_K{}^{L}\;\; Q_L{}^{I}\;\; Q^{K}_{M} + Q_K{}^{I}\;\; Vq^{K}_{L}\; Q^{L}_{M}$$

$$= \;\; Vq_L{}^{K}\;\; Q_K{}^{I}\;\; Q^{L}_{M} + Q_K{}^{I}\;\; Vq^{K}_{L}\; Q^{L}_{M}$$

$$= \;\; (Vq_L{}^{K} \;+ Vq^{K}_{\;L}\;)\, Q_K{}^{I}\;\; Q^{L}_{M}$$

$$= \qquad (\,\partial V_L/\partial q_K + \partial V^K/\partial q^L)\;\;\; \partial q_K/\partial p_I \;\;\; \partial q^L/\partial p^M$$

Define:

$$V^{/K\backslash}_{\backslash L/} \;\equiv\; 1/2\, (Vq^{K}_{\;L} + Vq_L{}^{K}\;\;)$$

$$W^{/K\backslash}_{\backslash L/} \;\equiv\; 1/2\, (Vq^{K}_{\;L} - Vq_L{}^{K}\;\;)$$

Metric Difference Time Derivative

$$d(dq^2 - dp^2)/dt = 2\, VppI^{/I}_{\backslash M}\;\; dp_I\, dp^M$$

And

$$VppI^{/I}_{\backslash M} \;\equiv\; 1/2\;\; Vpp^{/I}_{\backslash M}$$

Where $Vpp^{/I}_{\backslash M} =$

$$Vpp^{/I}_{\backslash M} = (Vq_L{}^{K} + Vq^{K}_{\;L}\,)\, Q_K{}^{I}\;\; Q^{L}_{M} \;=\; (\partial V_L/\partial q_K + \partial V^K/\partial q^L)\partial q_K/\partial p_I\, \partial q^L/\partial p^M$$

$$Vpp^{/I}_{\backslash M} = Vp_K{}^{I}\, Q^{K}_{M} + Q_K{}^{I}\;\; Vp^{K}_{M} \;=\; \partial V_K/\partial p_I\, \partial q^K/\partial p^M + \partial q_K/\partial p_I\, \partial V^K/\partial p^M$$

A different approach:

$$d(dq^2 - dp^2)/dt \quad = \quad [d\, (dq_I)/dt]\, dq^I + dq_I\;\; [d(dq^I)/dt]$$

$$= [d(\partial q_I/\partial p_K\, dp_K)/dt]\, dq^I + dq_I\, [d(\partial q^I/\partial p^K\, dp^K)/dt]$$

$$= d(Q_I^K\, dp_K)/dt\, dq^I + dq_I d(Q^I_K dp^K)/dt$$

$$= (\partial V_I/\partial p_K\, dp_K)\, dq^I + dq_I\, (\partial V^I/\partial p^K\, dp^K)$$

$$= (Vp_I^K\, dp_K)\, dq^I + dq_I\, (Vp^I_K\, dp^K)$$

$$= (Vq_I^K\, dq_K)\, dq^I + dq_I\, (Vq^I_K\, dq^K)$$

$$= (Vq_K^I + Vq^I_K)\, (dq_I)\, (dq^K)$$

So

$$d(dq^2\text{-}dp^2)/dt = (Vq_K^I + Vq^I_K)\, dq_I\, dq^K$$

$$d(dq^2\text{-}dp^2)/dt = (Vp_K^I Q^K_M + Q^I_K Vp^K_M)\, dp_I\, dp^M$$

$$d(dq^2\text{-}dp^2)/dt = (Vq_K^I + Vq^I_K)\, (dq_I)\, (dq^K)$$

$$= (Vq_K^I + Vq^I_K)\, (Q_I^L\, dp_L)\, (Q^K_M\, dp^M)$$

$$= (Vq_K^I + Vq^I_K)\, Q_I^L\, Q^K_M\, dp_L\, dp^M$$

Which is the same result found earlier:

$$Vpp^{/I}_{\backslash M} = (Vq_L^K + Vq^K_L)\, Q_K^I\, Q^L_M = (\partial V_L/\partial q_K + \partial V^K/\partial q^L)\partial q_K/\partial p_I\, \partial q^L/\partial p^M$$

$$Vpp^{/I}_{\backslash M} = Vp_K^I\, Q^K_M + Q_K^I\, Vp^K_M = \partial V_K/\partial p_I\, \partial q^K/\partial p^M + \partial q_K/\partial p_I\, \partial V^K/\partial p^M$$

$$d(dq^2\text{-}dp^2)/dt = dV_J\, dq^J + dq_I\, dV^I$$

$$= 2\, dV_J\, dq^J \qquad = 2\, dV^I\, dq_I$$

$$= 2\, Vq_J^I\, dq_I\, dq^J \qquad = 2\, Vq^I_J\, dq_I\, dq^J$$

If affine:

$$d(dq^2\text{-}dp^2)/dt = (Vq_K^I + Vq^I_K)\, Q_I^L\, Q^K_M\, dp_L\, dp^M$$

$$d(dq^2\text{-}dp^2)/dt = (Vq_I^K\, dq_K)\, dq^I + (Vq^I_K\, dq^K)$$

$$= dq_I\, (Vq^K_I + Vq^I_K)\, dq^K = dq_I\, (2\, V^{/K\backslash}_{\backslash L/})\, dq^K$$

Some Results:

$$IPP^I_M = 1/2\, (\delta^I_M - \partial p_K/\partial q_I\, \partial p^K/\partial q^M)$$

$$QQI^I_M = 1/2\, (\partial q_K/\partial p_I\, \partial q^K/\partial p^M - \delta^I_M)$$

$$dq^2\text{-}dp^2$$

$$dq^2\text{-}dp^2 = 2\, IPP^I_M\, dq_I\, dq^M \qquad dq^2\text{-}dp^2 = 2\, QQI^I_M\, dp_I\, dp^M$$

$$d(dq^2-dp^2)/dt = 2\, VppI^{/I}_{\backslash M}\, dp_I\, dp^M$$

$$VppI^{/I}_{\backslash M} \equiv 1/2\ Vpp^{/I}_{\backslash M}$$

Where $Vpp^{/I}_{\backslash M}=$

$$Vpp^{/I}_{\backslash M} = (Vq_L{}^K + Vq^K{}_L)\, Q_K{}^I\, Q^L{}_M \qquad Vpp^{/I}_{\backslash M} = Vp_K{}^I\, Q^K{}_M + Q_K{}^I\, Vp^K{}_M$$

$$= (\partial V_L/\partial q_K + \partial V^K/\partial q^L)\partial q_K/\partial p_I\, \partial q^L/\partial p^M \qquad = \partial V_K/\partial p_I\, \partial q^K/\partial p^M + \partial q_K/\partial p_I\, \partial V^K/\partial p^M$$

$$d(dq^2)/dt = d(dq^2-dp^2)/dt$$

$$(Vq_K{}^I + Vq^I{}_K)\, dq_I\, dq^K \qquad (Vp_K{}^I Q^K{}_M + Q_K{}^I Vp^K{}_M)\, dp_I\, dp^M$$

Example:

$$dq = dq^1 \qquad dp = dp^1$$

Metric difference:

$$dp^2-dq^2 = (dp^1+dq^1)(dp^1-dq^1) \qquad dq^2-dp^2 = (dq^1+dp^1)(dq^1-dp^1)$$

Approximately

$$\cong 2dq^1\,(dp^1-dq^1) \qquad \cong 2dp^1\,(dq^1-dp^1)$$

Example:

$$|dq^2|\,|dq^3| \cos\theta = |dq^2|\,|dq^3| \sin(\pi/2-\theta)$$

$$|dq^2|\,|dq^3|\, 2\, QQI^2{}_3$$

$$dq^2/dp^2 = (1 + 2\ QQI^2{}_2)^{1/2}$$

$$dq^3/dp^3 = (1 + 2\ QQI^3{}_3)^{1/2}$$

$$\cos\theta = 2QQI^2{}_3/[(1+2QQI^2{}_2)(1+2QQI^3{}_3)]^{1/2}$$

Linear Almansi and Green Strain Tensors:

Displacement Tensor: r

$$r^I = q^I - p^I$$

Almansi

$$p^I = q^I - r^I$$

$$\partial p^I/\partial q^J = \delta^I{}_J - \partial r^I/\partial q^J$$

Green

$$q^I = p^I + r^I$$

$$\partial q^I/\partial p^J = \delta^I{}_J + \partial r^I/\partial p^J$$

Almansi and Green Strain Tensors:

$$IPP^I{}_K = 1/2\,(\delta^I{}_K - \partial p^J/\partial q^I\, \partial p^J/\partial q^K)$$

$$= 1/2[\delta^I{}_K - (\delta^J{}_I - \partial r^J/\partial q^I)(\delta^J{}_K - \partial r^J/\partial q^K)]$$

$$= 1/2[\delta^I{}_K - (\delta^I{}_K - \partial r^I/\partial q^K - \partial r^K/\partial q^I + \partial r^J/\partial q^I\, \partial r^J/\partial q^K]$$

$$= 1/2(\partial r^I/\partial q^K + \partial r^K/\partial q^I - \partial r^J/\partial q^I\, \partial r^J/\partial q^K)$$

$$QQI^I{}_K = 1/2\,[\partial q^J/\partial p^I\, \partial q^J/\partial p^K - \delta^I{}_K]$$

$$= 1/2\,[(\delta^J{}_I + \partial r^J/\partial p^I)(\delta^J{}_K + \partial r^J/\partial p^K) - \delta^I{}_K]$$

$$= 1/2[(\delta^I{}_K + \partial r^I/\partial p^K + \partial r^K/\partial p^I + \partial r^J/\partial p^I\, \partial r^J/\partial p^K) - \delta^I{}_K]$$

$$= 1/2(\partial r^I/\partial p^K + \partial r^K/\partial p^I + \partial r^J/\partial p^I\, \partial r^J/\partial p^K)$$

Approximately
Linear Almansi and Green Strain Tensors:

$$IPP^I{}_K \cong 1/2\ (\partial r^I/\partial q^J + \partial r^J/\partial q^I) \qquad QQI^I{}_J \cong 1/2\ (\partial r^I/\partial p^J + \partial r^J/\partial p^I)$$
$$\cong 1/2\ (U^I{}_J + U^J{}_I) \qquad \cong 1/2\ (U^I{}_J + U^J{}_I)$$

Metric difference

$$dq^2 - dp^2 = 2\ IPP^I{}_M\ dq_I\ dq^M \qquad dq^2 - dp^2 = 2\ QQI^I{}_M\ dp_I\ dp^M$$
$$\cong (U^I{}_J + U^J{}_I) dq_I\ dq^J \qquad \cong (U^I{}_J + U^J{}_I)\ dp_I\ dp^J$$

Metric difference ratio

$$(dq^2 - dp^2)/dq^2 = 1 - dp^2/dq^2 \qquad (dq^2 - dp^2)/dp^2 = dq^2/dp^2 - 1$$
$$\cong \qquad \cong$$
$$2(1 - dp/dq) \cong 2\ IPP^I{}_M\ dq_I\ dq^M /\ dq_K\ dq^K \qquad 2(dq/dp - 1) \cong 2QQI^I{}_M\ dp_I\ dp^M / dp_K\ dp^K$$
$$\cong \qquad \cong$$
$$(U^I{}_J + U^J{}_I) \qquad (U^I{}_J + U^J{}_I)$$

Engineering strain approximately equals half its metric difference ratio:

spatial Engineering strain — material Engineering strain

$$1 - dp/dq \cong 1/2\ (1 - dp^2/dq^2) \qquad dq/dp - 1 \cong 1/2\ (dq^2/dp^2 - 1)$$
$$\cong 1/2\ (U^I{}_J + U^J{}_I) \qquad \cong 1/2\ (U^I{}_J + U^J{}_I)$$

Length Ratio

$$dp/dq \cong 1 - (1 - dp/dq) \qquad dq/dp \cong 1 + (dq/dp - 1)$$
$$\cong 1 - 1/2\ (U^I{}_J + U^J{}_I) \qquad \cong 1 + 1/2\ (U^I{}_J + U^J{}_I)$$

Infinitesimal Engineering strain ≡ material Engineering strain

$$\varepsilon \cong IPP^I{}_M\ dq_I\ dq^M /\ dq_K\ dq^K\ \ dq/dp \qquad \varepsilon \cong QQI^I{}_M\ dp_I\ dp^M / dp_K\ dp^K$$
$$\varepsilon = dq/dp - 1 = (1 - dp/dq) dq/dp \qquad \varepsilon = dq/dp - 1$$
$$\cong 1/2\ (U^I{}_J + U^J{}_I)\ dq/dp \qquad \cong 1/2\ (U^I{}_J + U^J{}_I)$$
$$\varepsilon^I{}_J = 1/2\ (U^I{}_J + U^J{}_I)$$

Infinitesimal Strain Tensor:

$$||dq||^2 - ||dp||^2 \cong 2\ dq_I\ \ 1/2\ (\partial r^I/\partial q^J + \partial r^J/\partial q^I)\ dq^J$$
$$\varepsilon^I{}_J = 1/2\ (\partial r^I/\partial q^J + \partial r^J/\partial q^I) = 1/2\ (U^I{}_J + U^J{}_I)$$

Example:

$$dq = dq^1 \qquad dp = dp^1$$

Metric difference

$$dq^2 - dp^2 = 2\ IPP^1{}_1\ dq_1\ dq^1 \qquad dq^2 - dp^2 = 2\ QQI^1{}_1\ dp_1\ dp^1$$

Metric difference ratio

$$2(1 - dp/dq) \cong 2\ IPP^1{}_1 \qquad 2(dq/dp - 1) \cong 2\ QQI^1{}_1$$

Length Ratio

$$dp/dq = (1 - 2\ IPP^1{}_1)^{1/2} \qquad dq/dp = (1 + 2\ QQI^1{}_1)^{1/2}$$

Engineering strain approximately equals half its metric difference ratio:

spatial Engineering strain — material Engineering strain

$$1 - dp/dq \cong 1/2\ (1 - dp^2/dq^2) \qquad dq/dp - 1 \cong 1/2\ (dq^2/dp^2 - 1)$$

$1 - dp/dq \cong IPP^{1}_{1}$ $\quad dq/dp - 1 \cong QQI^{1}_{1}$

Length Ratio approximately

$dp/dq \cong 1 - IPP^{1}_{1}$ $\quad dq/dp \cong 1 + QQI^{1}_{1}$

Infinitesimal Engineering strain ≡ material Engineering strain

$\varepsilon = dq/dp - 1 = (1-dp/dq)\ dq/dp$ $\quad \varepsilon = dq/dp - 1$

$\varepsilon \cong IPP^{1}_{1}\ dq/dp \cong IPP^{1}_{1}(1 - IPP^{1}_{1})$ $\quad \varepsilon \cong QQI^{1}_{1}$

Rigid Body: or any affine transformation

Almansi's Strain: Eulerian Strain

$IPP^{I}_{M} = 1/2\ (\delta^{I}_{M} - P_{K}^{I}\ P^{K}_{M})$

Green's Strain: Lagrangian Strain

$QQI^{I}_{M} = 1/2\ (Q_{K}^{I}\ Q^{K}_{M} - \delta^{I}_{M})$

Metric difference

$dq^{2} - dp^{2} = 2\ IPP^{I}_{M}\ dq_{I}\ dq^{M}$ $\quad dq^{2} - dp^{2} = 2\ QQI^{I}_{M}\ dp_{I}\ dp^{M}$

There is No Deformation, that is Only Transformation of Coordinates:

Orthonormality of Metric 10:

$P_{K}^{m}\ P^{K}_{n} = \delta^{m}_{n}$ $\quad Q_{K}^{m}\ Q^{K}_{n} = \delta^{m}_{n}$

Orthonormality of Affine 10:

$P^{K}_{m}\ P^{K}_{n} = \delta^{m}_{n}$ $\quad Q^{K}_{m}\ Q^{K}_{n} = \delta^{m}_{n}$

So

$IPP^{I}_{M} = 1/2\ (\delta^{I}_{M} - P_{K}^{I}\ P^{K}_{M}) = 0$ $\quad QQI^{I}_{M} = 1/2\ (Q_{K}^{I}\ Q^{K}_{M} - \delta^{I}_{M}) = 0$

Displacement Tensor: r

$r^{I} = q^{I} - p^{I}$

Almansi

$p^{I} = q^{I} - r^{I}$

$P^{I}_{J} = \delta^{I}_{J} - \partial r^{I} / \partial q^{J}$

Green

$q^{I} = p^{I} + r^{I}$

$Q^{I}_{J} = \delta^{I}_{J} + \partial r^{I} / \partial p^{J}$

Gradient of Displacement Tensor:

$U^{I}_{J} = \partial r^{I} / \partial q^{J} = \delta^{I}_{J} - P^{I}_{J}$ $\quad U^{I}_{J} = \partial r^{I} / \partial p^{J} = Q^{I}_{J} - \delta^{I}_{J}$

Infinitesimal Engineering strain ≡ material Engineering strain

$$\varepsilon^{I}_{J} = 1/2\,(U^{I}_{J} + U^{J}_{I})$$

$$\varepsilon^{I}_{J} = 1/2(\delta^{I}_{J} - P^{I}_{J} + \delta^{J}_{I} - P^{J}_{I}) = \delta^{I}_{J} - 1/2\,(P^{I}_{J} + P^{J}_{I}) \neq 0$$

However If $(\delta^{I}_{J} - P^{I}_{J})$ is Anti symmetric then

$$\varepsilon^{I}_{J} = 1/2(\delta^{I}_{J} - P^{I}_{J} + \delta^{J}_{I} - P^{J}_{I}) = 0$$

Rotation of Rigid Body: R^{M}_{N}

Rotation of Rigid Body So:

$IPP^{I}_{M} = 1/2\ (\delta^{I}_{M} - P_{K}^{I}\ P^{K}_{M}) = 0$ $\quad QQI^{I}_{M} = 1/2\ (Q_{K}^{I}\ Q^{K}_{M} - \delta^{I}_{M}) = 0$

\

$\underline{q}_{m}\, Q^{m}_{M} = \underline{p}_{M}$ $\quad dp^{N} = P^{N}_{n}\ dq^{n}$

q_m $\quad Q^m_{\ M} = P^M_{\ m} = \delta^M_{\ m} + R^M_{\ m}$ $\qquad Q^n_{\ N} = P^N_{\ n} = \delta^N_{\ n} + R^N_{\ n}$ $\quad dq^n$

Where

$$R^M_{\ N} = -R^N_{\ M}$$

So

$$\varepsilon^I_{\ J} = \delta^I_{\ J} - 1/2\,(P^I_{\ J} + P^J_{\ I}) = \delta^I_{\ J} - 1/2\,(\delta^I_{\ J} + R^I_{\ J} + \delta^J_{\ I} + R^J_{\ I}) = -1/2\,(R^I_{\ J} + R^I_{\ I}) = 0$$

10.4.
Forming a Thin Plate into a Circle (Pipeline):

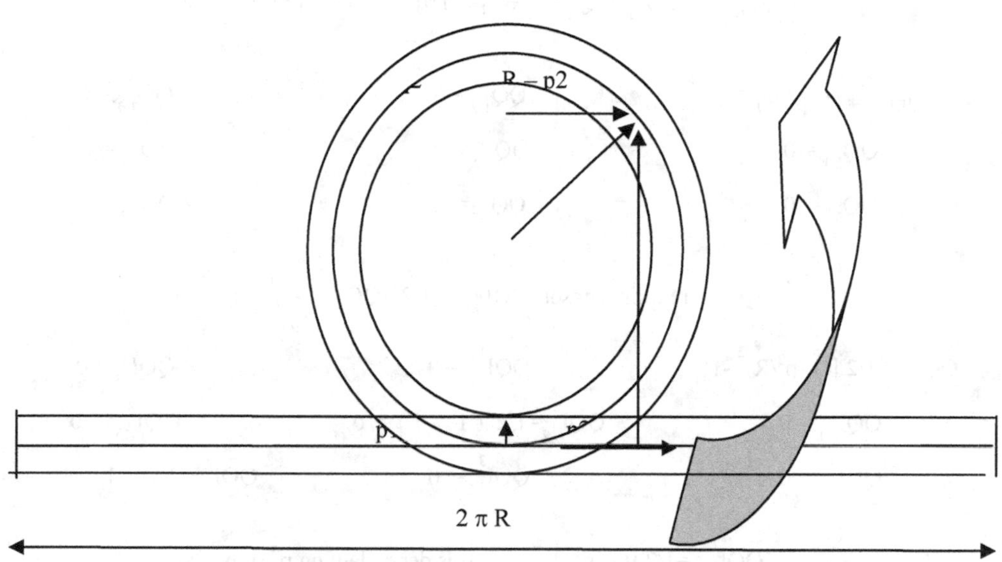

$$\theta = p^1/R$$

$$q^1 = (R - p^2)\sin(p^1/R)$$

$$q^2 = R - (R - p^2)\cos(p^1/R)$$

q^I
q^1 $\qquad \theta = p^1/R$
q^2 $\qquad q^1 = (R - p^2)\sin(p^1/R)$
q^3 $\qquad q^2 = R - (R - p^2)\cos(p^1/R)$
$q^3 = \qquad p^3$

From a Metric Component: Symmetric P=Q Partial Derivative:

Not Applicable. $\qquad (P_I^{\ J}) = Q^J_{\ I}$

Not Applicable. $\qquad$ Not Applicable.

$$(P_1^{\ 1}) = Q^1_{\ 1} = \partial q^1/\partial p^1 = \partial[(R-p^2)\sin(p^1/R)]/\partial p^1 = (R-p^2)\cos\theta\ /R = (1-p^2/R)\cos\theta$$

$$(P_1^{\ 2}) = Q^2_{\ 1} = \partial q^2/\partial p^1 = \partial[R-(R-p^2)\cos(p^1/R)]/\partial p^1 = (R-p^2)\sin\theta\ /R = (1-p^2/R)\sin\theta$$

$$(P_2^{\ 1}) = Q^1_{\ 2} = \partial q^1/\partial p^2 = \partial[(R-p^2)\sin(p^1/R)]/\partial p^2 = -\sin(p^1/R) = -\sin\theta$$

$$(P_2^{\ 2}) = Q^2_{\ 2} = \partial q^2/\partial p^1 = \partial[R-(R-p^2)\cos(p^1/R)]/\partial p^2 = \cos(p^1/R) = \cos\theta$$

$P_1^{\ 1} = Q_{\ 1}^{1} = (1-p^2/R)\cos\theta$ $\quad P_2^{\ 1} = Q_{\ 2}^{1} = -\sin\theta$ $\quad P_3^{\ 1} = Q_{\ 3}^{1} = 0$

$P_1^{\ 2} = Q_{\ 1}^{2} = (1-p^2/R)\sin\theta$ $\quad P_2^{\ 2} = Q_{\ 2}^{2} = \cos\theta$ $\quad P_3^{\ 2} = Q_{\ 3}^{2} = 0$

$P_1^{\ 3} = Q_{\ 1}^{3} = 0$ $\quad P_3^{\ 2} = Q_{\ 3}^{2} = 0$ $\quad P_3^{\ 3} = Q_{\ 3}^{3} = 1$

$$QQ_{IM} = Q^K_{\ I} Q^K_{\ M} = \text{sum of (column I *column M)}$$

$$QQ_{11} = Q^1_{\ 1}Q^1_{\ 1} + Q^2_{\ 1}Q^2_{\ 1} = [(1-p^2/R)\cos\theta]^2 + [(1-p^2/R)\sin\theta]^2$$

$$QQ_{21} = Q^1_{\ 2}Q^1_{\ 1} + Q^2_{\ 2}Q^2_{\ 1} = (-\sin\theta)(1-p^2/R)\cos\theta + \cos\theta\ (1-p^2/R)\sin\theta = 0$$

$$QQ_{12} = Q^1_{\ 1}Q^1_{\ 2} + Q^2_{\ 1}Q^2_{\ 2} = (1-p^2/R)\cos\theta\,(-\sin\theta) + (1-p^2/R)\sin\theta\cos\theta = 0$$

$$QQ_{22} = Q^1_{\ 2}Q^1_{\ 2} + Q^2_{\ 2}Q^2_{\ 2} = [-\sin\theta]^2 + [\cos\theta]^2 = 1$$

$QQ_{11} = (1-p^2/R)^2$ $\quad QQ_{12} = 0$ $\quad QQ_{13} = 0$

$QQ_{21} = 0$ $\quad QQ_{22} = 1$ $\quad QQ_{23} = 0$

$QQ_{31} = 0$ $\quad QQ_{32} = 0$ $\quad QQ_{33} = 1$

Green Strain Tensor: $QQI^I_{\ M} = 1/2\ [QQ_{IM} - \delta^I_{\ M}]$:

$QQI^1_{\ 1} = 1/2\ [(1-p^2/R)^2 - 1]$ $\quad QQI^1_{\ 2} = 0$ $\quad QQI^1_{\ 3} = 0$

$QQI^2_{\ 1} = 0$ $\quad QQI^2_{\ 2} = 1/2\ (1 - 1) = 0$ $\quad QQI^2_{\ 3} = 0$

$QQI^3_{\ 1} = 0$ $\quad QQI^3_{\ 2} = 0$ $\quad QQI^3_{\ 3} = 1/2\ (1 - 1) = 0$

$QQI^1_{\ 1} = 1/2\ [(1-p^2/R)^2 - 1\ ]$: It is dependant on p^2 only:

$p^2 = -h$	$p^2 = h$
$1/2[(1+h/R)^2-1] = 1/2[+2(h/R)+(h/R)^2]$	$1/2[(1-h/R)^2-1] = 1/2[-2(h/R)+(h/R)^2]$

Approximated:

$1/2[+2(h/R)+(h/R)^2] \cong h/R$	$1/2[-2(h/R)+(h/R)^2] \cong -h/R$

Selected Values of Strains:

	Outside Circle	Center Circle	Inside Circle
p^2	-h	0	h
$1/2[(1-p^2/R)^2-1]$	$1/2[+2(h/R)+(h/R)^2]$	0	$1/2[-2(h/R)+(h/R)^2]$
Circumference	$2\pi(R+h)$	$2\pi R$	$2\pi(R-h)$
Elongation	$2\pi\ h$	0	$-2\pi\ h$
Strain	h/R	0	- h/R

Selected Values of Green Strain Tensor Approximations:

	Outside Circle	Center Circle	Inside Circle
p^2	-h	0	h
$1/2[(1-p^2/R)^2-1]$	$\cong h/R$	0	$\cong -h/R$

Range of Values of Green Strain Tensor:

$-h/R < QQI^1_{\ 1} < h/R \qquad QQI^1_{\ 2} = 0$

$QQI^2_{\ 1} = 0 \qquad QQI^2_{\ 2} = 0$

<u>Linear</u> Green Strain Tensor, $\varepsilon^I_{\ J}$ is an Approximation to Green Strain Tensor:

$$QQI^I_{\ J} \cong \varepsilon^1_{\ 1} = 1/2\,(\partial r^I / \partial p^J + \partial r^J/\partial p^I\,)$$

$$q^I = p^I + r^I$$

$$\partial q^I / \partial p^J = \delta^I_{\ J} + \partial r^I / \partial p^J$$

Then:

$$\partial r^I/\partial p^J = -\,\delta^I_{\ J} + Q^I_{\ J}$$

$$\partial r^1/\partial p^1 = -1 + Q^1_{\ 1} = -1 + (1-p^2/R)\cos\theta$$

$$\partial r^2/\partial p^1 = Q^2_{\ 1} = (1-p^2/R)\sin\theta$$

$$\partial r^1/\partial p^2 = Q^1_{\ 2} = -\sin\theta$$

$$\partial r^2/\partial p^2 = -1 + Q^2_{\ 2} = -1 + \cos\theta$$

$$QQI^I_{\ J} \cong \varepsilon^1_{\ 1} = 1/2\,(\partial r^I / \partial p^J + \partial r^J/\partial p^I\,)$$

$$QQI^1_{\ 1} \cong 1/2\,(\partial r^1 / \partial p^1 + \partial r^1/\partial p^1\,) = -1 + (1-p^2/R)\cos\theta$$

$$QQI^2_{\ 1} \cong 1/2\,(\partial r^2 / \partial p^1 + \partial r^1/\partial p^2\,) = 1/2\,[(1-p^2/R)\sin\theta + (-\sin\theta\,)] = 1/2(p^2/R)\sin\theta$$

$$QQI^1_{\ 2} \cong 1/2\,(\partial r^1 / \partial p^2 + \partial r^2/\partial p^1\,) = 1/2(p^2/R)\sin\theta$$

$$QQI^2_{\ 2} \cong 1/2\,(\partial r^2 / \partial p^2 + \partial r^2/\partial p^2\,) = -1 + \cos\theta$$

Linear Green Strain Tensor: Linear $QQI^1_{\ 1} \cong \varepsilon^I_{\ J}$ is dependant on p^2 and θ :

$QQI^1_{\ 1} \cong \varepsilon^1_{\ 1} = -1 + (1-p^2/R)\cos\theta \qquad QQI^1_{\ 2} \cong \varepsilon^1_{\ 2} = 1/2\ \ p^2/R\ \ \sin\theta \qquad 0$

$QQI^2_{\ 1} \cong \varepsilon^2_{\ 1} = 1/2\ \ p^2/R\ \ \sin\theta \qquad QQI^2_{\ 2} \cong \varepsilon^2_{\ 2} = -1 + \cos\theta \qquad 0$

$0 \qquad 0 \qquad 0$

Whereas:

Green Strain Tensor: $QQI^I_{\ M} = 1/2\ [QQ_{IM} - \delta^I_{\ M}]$:

$QQI^1_{\ 1} = 1/2\,[(1-p^2/R)^2 - 1] \qquad QQI^1_{\ 2} = 0 \qquad QQI^1_{\ 3} = 0$

$QQI^2_{\ 1} = 0 \qquad QQI^2_{\ 2} = 1/2\ (1\ -1) = 0 \qquad QQI^2_{\ 3} = 0$

$QQI^3_{\ 1} = 0 \qquad QQI^3_{\ 2} = 0 \qquad QQI^3_{\ 3} = 1/2\ (1\ -1) = 0$

Range of Values of Linear $QQI^1_{\ 1} \cong \varepsilon^I_{\ J}$:

$-2 - h/R < QQI^1_{\ 1} \cong \varepsilon^1_{\ 1} < h/R \qquad -1/2\ h/R < QQI^1_{\ 2} \cong \varepsilon^1_{\ 2} < 1/2\ h/R$

$-1/2\ h/R < QQI^2_{\ 1} \cong \varepsilon^2_{\ 1} < 1/2\ h/R \qquad -2 < QQI^2_{\ 2} \cong \varepsilon^2_{\ 2} < 0$

Whereas:

Range of Values of Green Strain Tensor:

$-h/R < QQI^1_{\ 1} < h/R \qquad QQI^1_{\ 2} = 0$

$QQI^2_{\ 1} = 0 \qquad QQI^2_{\ 2} = 0$

10.5.
Plate in Pure Shear:

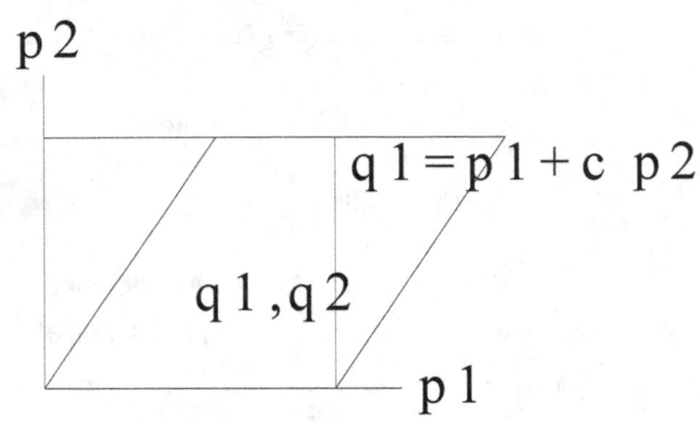

(C) ≡ tan γ

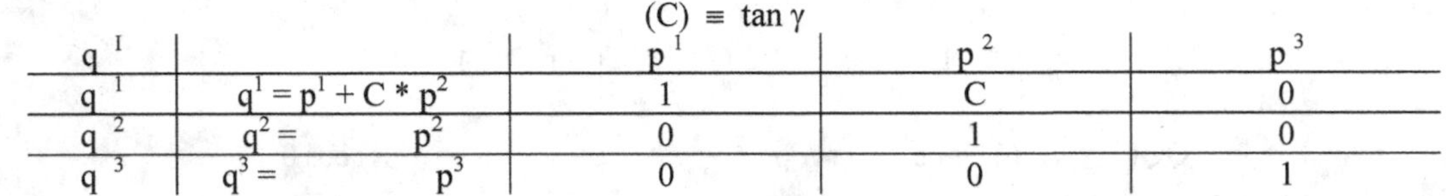

q^I		p^1	p^2	p^3
q^1	$q^1 = p^1 + C * p^2$	1	C	0
q^2	$q^2 = p^2$	0	1	0
q^3	$q^3 = p^3$	0	0	1

In order to find Green Strain Tensor $QQI^I_M = 1/2(Q^K_I\, Q^K_M - \delta^I_M)$

$$Q^K_M$$

$Q^1_1 = \partial q^1/\partial p^1 = 1$ $\qquad$ $Q^1_2 = \partial q^1/\partial p^2 = C$ $\qquad$ $Q^1_3 = \partial q^1/\partial p^3 = 0$

$Q^2_1 = \partial q^2/\partial p^1 = 0$ $\qquad$ $Q^2_2 = \partial q^2/\partial p^2 = 1$ $\qquad$ $Q^2_3 = \partial q^2/\partial p^3 = 0$

$Q^3_1 = \partial q^3/\partial p^1 = 0$ $\qquad$ $Q^3_2 = \partial q^2/\partial p^2 = 0$ $\qquad$ $Q^3_3 = \partial q^3/\partial p^3 = 1$

$QQ_{IM} = Q^K_I\, Q^K_M =$ sum of column I *column M

$QQ_{11} = 1$ $\qquad$ $QQ_{12} = C$ $\qquad$ $QQ_{13} = 0$

$QQ_{21} = C$ $\qquad$ $QQ_{22} = C * C + 1$ $\qquad$ $QQ_{23} = 0$

$QQ_{31} = 0$ $\qquad$ $QQ_{32} = 0$ $\qquad$ $QQ_{33} = 1$

Green Strain Tensor

$$QQI^I_M = 1/2\ [QQ_{IM} - \delta^I_M]$$

$QQI^1_1 = 0$ $\qquad$ $QQI^1_2 = 1/2\ C$ $\qquad$ $QQI^1_3 = 0$

$QQI^2_1 = 1/2\ C$ $\qquad$ $QQI^2_2 = 1/2\ C * C$ $\qquad$ $QQI^2_3 = 0$

$QQI^3_1 = 0$ $\qquad$ $QQI^3_2 = 0$ $\qquad$ $QQI^3_3 = 0$

<u>Linear</u> Green Strain Tensor, ε^I_J is an Approximation to Green Strain Tensor:

$$QQI^I_J \cong \varepsilon^I_J = 1/2\ (\partial r^I / \partial p^J + \partial r^J / \partial p^I\ )$$

$$q^I = p^I + r^I$$

$$\partial q^I / \partial p^J = \delta^I{}_J + \partial r^I / \partial p^J$$

Then:

$$\partial r^I / \partial p^J = -\delta^I{}_J + Q^I{}_J$$

$$\partial r^1/\partial p^1 = -1 + Q^1{}_1 = -1 + 1 = 0$$
$$\partial r^2/\partial p^1 = Q^2{}_1 = 0$$
$$\partial r^1/\partial p^2 = Q^1{}_2 = C$$
$$\partial r^2/\partial p^2 = -1 + Q^2{}_2 = -1 + 1 = 0$$
$$\partial r^3/\partial p^3 = -1 + Q^3{}_3 = -1 + 1 = 0$$

$\partial r^1/\partial p^1 = 0$	$\partial r^1/\partial p^2 = C$	$\partial r^1/\partial p^3 = 0$
$\partial r^2/\partial p^1 = 0$	$\partial r^2/\partial p^2 = 0$	$\partial r^2/\partial p^3 = 0$
$\partial r^3/\partial p^1 = 0$	$\partial r^3/\partial p^2 = 0$	$\partial r^3/\partial p^3 = 0$

$$QQI^I{}_J \cong \varepsilon^1{}_1 = 1/2\,(\partial r^I / \partial p^J + \partial r^J/\partial p^I)$$

$$QQI^1{}_1 \cong 1/2\,(\partial r^1 / \partial p^1 + \partial r^1/\partial p^1) = 0$$
$$QQI^2{}_1 \cong 1/2\,(\partial r^2 / \partial p^1 + \partial r^1/\partial p^2) = 1/2\,[0 + C] = 1/2\,C$$
$$QQI^1{}_2 \cong 1/2\,(\partial r^1 / \partial p^2 + \partial r^2/\partial p^1) = 1/2\,C$$
$$QQI^2{}_2 \cong 1/2\,(\partial r^2 / \partial p^2 + \partial r^2/\partial p^2) = 0$$

Linear Green Strain Tensor: Linear $QQI^1{}_1 \cong \varepsilon^I{}_J = 1/2\,(\partial r^I / \partial p^J + \partial r^J/\partial p^I)$:

$QQI^1{}_1 \cong \varepsilon^1{}_1 = 0$	$QQI^1{}_2 \cong \varepsilon^1{}_2 = 1/2\ C = 1/2\ C$	0
$QQI^2{}_1 \cong \varepsilon^2{}_1 = 1/2\ C = 1/2\ C$	$QQI^2{}_2 \cong \varepsilon^2{}_2 = 0$	0
0	0	0

Whereas:
Values of Green Strain Tensor:

$QQI^1{}_1 = 0$	$QQI^1{}_2 = 1/2\ C \equiv 1/2\ \tan\gamma$
$QQI^2{}_1 = 1/2\ C \equiv 1/2\ \tan\gamma$	$QQI^2{}_2 = 1/2\ C * C \equiv 1/2\ (\tan\gamma)^2$

10.6.
Plate Orthotropic Deformation:

q^I		p^1	p^2	p^3
q^1	$q^1 = A * p^1$	A	0	0
q^2	$q^2 = B * p^2$	0	B	0
q^3	$q^3 = C * p^3$	0	0	C

$$Q^K{}_M$$

$Q^1{}_1 = \partial q^1/\partial p^1 = A$	$Q^1{}_2 = \partial q^1/\partial p^2 = 0$	$Q^1{}_3 = \partial q^1/\partial p^3 = 0$
$Q^2{}_1 = \partial q^2/\partial p^1 = 0$	$Q^2{}_2 = \partial q^2/\partial p^2 = B$	$Q^2{}_3 = \partial q^2/\partial p^3 = 0$

$Q^3_{\ 1} = \partial q^3/\partial p^1 = 0$ $\quad Q^3_{\ 2} = \partial q^2/\partial p^2 = 0$ $\quad Q^3_{\ 3} = \partial q^3/\partial p^3 = C$

$QQ_{\backslash IM} = Q^K_{\ I} Q^K_{\ M} =$ sum of column I *column M

$QQ_{\backslash 11} = A * A$ $\quad QQ_{\backslash 12} = 0$ $\quad QQ_{\backslash 13} = 0$

$QQ_{\backslash 21} = 0$ $\quad QQ_{\backslash 22} = B * B$ $\quad QQ_{\backslash 23} = 0$

$QQ_{\backslash 31} = 0$ $\quad QQ_{\backslash 32} = 0$ $\quad QQ_{\backslash 33} = C * C$

Green Strain Tensor

$$QQI^I_{\ M} = 1/2\,[QQ_{\backslash IM} - \delta^I_{\ M}]$$

$QQI^1_{\ 1} = 1/2\,[A * A - 1]$ $\quad QQI^1_{\ 2} = 0$ $\quad QQI^1_{\ 3} = 0$

$QQI^2_{\ 1} = 0$ $\quad QQI^2_{\ 2} = 1/2\,[B * B - 1]$ $\quad QQI^2_{\ 3} = 0$

$QQI^3_{\ 1} = 0$ $\quad QQI^3_{\ 2} = 0$ $\quad QQI^3_{\ 3} = 1/2\,[C * C - 1]$

<u>Linear</u> Green Strain Tensor, $\varepsilon^I_{\ J}$ is an Approximation to Green Strain Tensor:

$$QQI^I_{\ J} \cong \varepsilon^I_{\ 1} = 1/2\,(\partial r^I / \partial p^J + \partial r^J/\partial p^I)$$

$$q^I = p^I + r^I$$

$$\partial q^I / \partial p^J = \delta^I_{\ J} + \partial r^I / \partial p^J$$

Then:

$$\partial r^I/\partial p^J = -\delta^I_{\ J} + Q^I_{\ J}$$

$$\partial r^1/\partial p^1 = -1 + Q^1_{\ 1} = -1 + A$$
$$\partial r^2/\partial p^1 = Q^2_{\ 1} = 0$$
$$\partial r^1/\partial p^2 = Q^1_{\ 2} = 0$$
$$\partial r^2/\partial p^2 = -1 + Q^2_{\ 2} = -1 + B$$
$$\partial r^3/\partial p^3 = -1 + Q^3_{\ 3} = -1 + C$$

$\partial r^1/\partial p^1 = -1 + A$ $\quad \partial r^1/\partial p^2 = 0$ $\quad \partial r^1/\partial p^3 = 0$

$\partial r^2/\partial p^1 = 0$ $\quad \partial r^2/\partial p^2 = -1 + B$ $\quad \partial r^2/\partial p^3 = 0$

$\partial r^3/\partial p^1 = 0$ $\quad \partial r^3/\partial p^2 = 0$ $\quad \partial r^3/\partial p^3 = -1 + C$

$$QQI^I_{\ J} \cong \varepsilon^1_{\ 1} = 1/2\,(\partial r^I / \partial p^J + \partial r^J/\partial p^I)$$

$$QQI^1_{\ 1} \cong 1/2\,(\partial r^1 / \partial p^1 + \partial r^1/\partial p^1) = -1 + A$$
$$QQI^2_{\ 2} \cong 1/2\,(\partial r^2 / \partial p^2 + \partial r^2/\partial p^2) = -1 + B$$
$$QQI^3_{\ 3} \cong 1/2\,(\partial r^3 / \partial p^3 + \partial r^3/\partial p^3) = -1 + C$$

Linear Green Strain Tensor: Linear $QQI^1_{\ 1} \cong \varepsilon^I_{\ J} = 1/2\,(\partial r^I / \partial p^J + \partial r^J/\partial p^I)$:

$QQI^1_{\ 1} \cong \varepsilon^1_{\ 1} = A - 1$	0	0
0	$QQI^2_{\ 2} \cong \varepsilon^2_{\ 2} = B - 1$	0
0	0	$QQI^3_{\ 3} \cong \varepsilon^3_{\ 3} = C - 1$

Whereas:
Values of Green Strain Tensor:

$QQI^1_{\ 1} = 1/2\,[A * A - 1]$ $\quad QQI^1_{\ 2} = 0$ $\quad QQI^1_{\ 3} = 0$

$QQI^2_{\ 1} = 0$ $\quad QQI^2_{\ 2} = 1/2\,[B * B - 1]$ $\quad QQI^2_{\ 3} = 0$

$QQI^3_{\ 1} = 0$ $\quad QQI^3_{\ 2} = 0$ $\quad QQI^3_{\ 3} = 1/2\,[C * C - 1]$

10.7. Plate Extension Deformation:

q^I		p^1	p^2	p^3
q^1	$q^1 = A\ p^1$	A	0	0
q^2	$q^2 = p^2$	0	1	0
q^3	$q^3 = p^3$	0	0	1

$Q^K_{\ M}$

$Q^1_{\ 1} = \partial q^1/\partial p^1 = A$ $\quad Q^1_{\ 2} = \partial q^1/\partial p^2 = 0$ $\quad Q^1_{\ 3} = \partial q^1/\partial p^3 = 0$

$Q^2_{\ 1} = \partial q^2/\partial p^1 = 0$ $\quad Q^2_{\ 2} = \partial q^2/\partial p^2 = 1$ $\quad Q^2_{\ 3} = \partial q^2/\partial p^3 = 0$

$Q^3_{\ 1} = \partial q^3/\partial p^1 = 0$ $\quad Q^3_{\ 2} = \partial q^2/\partial p^2 = 0$ $\quad Q^3_{\ 3} = \partial q^3/\partial p^3 = 1$

$QQ_{\backslash IM} = Q^K_{\ I}\, Q^K_{\ M} =$ sum of column I *column M

$QQ_{\backslash 11} = A * A$ $\quad QQ_{\backslash 12} = 0$ $\quad QQ_{\backslash 13} = 0$

$QQ_{\backslash 21} = 0$ $\quad QQ_{\backslash 22} = 1$ $\quad QQ_{\backslash 23} = 0$

$QQ_{\backslash 31} = 0$ $\quad QQ_{\backslash 32} = 0$ $\quad QQ_{\backslash 33} = 1$

Green Strain Tensor

$$QQI^I_{\ M} = 1/2\ [QQ_{\backslash IM} - \delta^I_{\ M}]$$

$QQI^1_{\ 1} = 1/2\,[A * A - 1]$ $\quad QQI^1_{\ 2} = 0$ $\quad QQI^1_{\ 3} = 0$

$QQI^2_{\ 1} = 0$ $\quad QQI^2_{\ 2} = 0$ $\quad QQI^2_{\ 3} = 0$

$QQI^3_{\ 1} = 0$ $\quad QQI^3_{\ 2} = 0$ $\quad QQI^3_{\ 3} = 0$

Linear Green Strain Tensor, $\varepsilon^I_{\ J}$ is an Approximation to Green Strain Tensor:

$$QQI^I_{\ J} \cong \varepsilon^I_{\ 1} = 1/2\,(\partial r^I / \partial p^J + \partial r^J/\partial p^I\,)$$

$$q^I = p^I + r^I$$

$$\partial q^I / \partial p^J = \delta^I_{\ J} + \partial r^I / \partial p^J$$

Then:

$$\partial r^I/\partial p^J = -\,\delta^I_{\ J} + Q^I_{\ J}$$

$$\partial r^1/\partial p^1 = -1 + Q^1_{\ 1} = -1 + A$$

$$\partial r^2/\partial p^1 = Q^2_{\ 1} = 0$$

$$\partial r^1/\partial p^2 = Q^1_{\ 2} = 0$$

$$\partial r^2/\partial p^2 = -\,1 + Q^2_{\ 2} = -1 + 1 = 0$$

$$\partial r^3/\partial p^3 = -1 + Q^3_3 = -1 + 1 = 0$$

$\partial r^1/\partial p^1 = -1 + A$	$\partial r^1/\partial p^2 = 0$	$\partial r^1/\partial p^3 = 0$
$\partial r^2/\partial p^1 = 0$	$\partial r^2/\partial p^2 = 0$	$\partial r^2/\partial p^3 = 0$
$\partial r^3/\partial p^1 = 0$	$\partial r^3/\partial p^2 = 0$	$\partial r^3/\partial p^3 = 0$

$$QQI^I{}_J \cong \varepsilon^1{}_1 = 1/2\,(\partial r^I / \partial p^J + \partial r^J/\partial p^I\,)$$

$$QQI^1{}_1 \cong 1/2\,(\partial r^1 / \partial p^1 + \partial r^1/\partial p^1\,) = -1 + A$$
$$QQI^2{}_2 \cong 1/2\,(\partial r^2 / \partial p^2 + \partial r^2/\partial p^2\,) = -1 + 1 = 0$$
$$QQI^3{}_3 \cong 1/2\,(\partial r^3 / \partial p^3 + \partial r^3/\partial p^3\,) = -1 + 1 = 0$$

Linear Green Strain Tensor: Linear $QQI^1{}_1 \cong \varepsilon^I{}_J = 1/2\,(\partial r^I / \partial p^J + \partial r^J/\partial p^I\,)$:

$QQI^1{}_1 \cong \varepsilon^1{}_1 = A - 1$	0	0
0	$QQI^2{}_2 \cong \varepsilon^2{}_2 = 0$	0
0	0	$QQI^3{}_3 \cong \varepsilon^3{}_3 = 0$

Whereas:
Values of Green Strain Tensor:

$QQI^1{}_1 = 1/2\,[A * A - 1]$	$QQI^1{}_2 = 0$	$QQI^1{}_3 = 0$
$QQI^2{}_1 = 0$	$QQI^2{}_2 = 0$	$QQI^2{}_3 = 0$
$QQI^3{}_1 = 0$	$QQI^3{}_2 = 0$	$QQI^3{}_3 = 0$

10.8.

Interface Conditions:

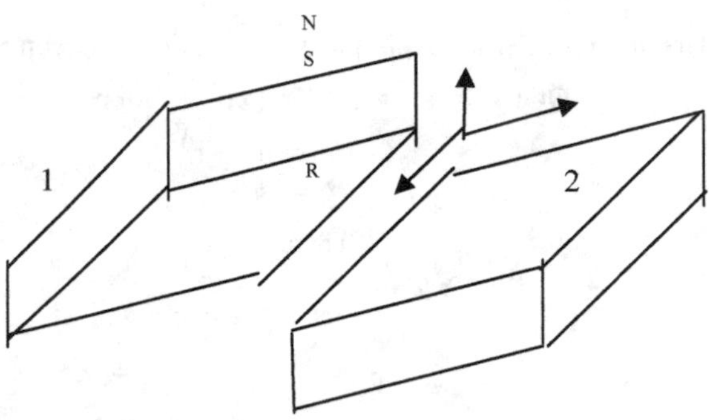

Displacements:
:

C^0 Continuous:

$$^1u^R = {}^2u^R$$
$$^1u^S = {}^2u^S$$
$$^1u^N = {}^2u^N$$

Strains:

:

C^1 Continuous:

$$^1\varepsilon^{RR} = {}^2\varepsilon^{RR} \qquad {}^1\varepsilon^{RS} = {}^2\varepsilon^{RS} \qquad {}^1\varepsilon^{RN} \neq {}^2\varepsilon^{RN}$$
$$^1\varepsilon^{SR} = {}^2\varepsilon^{SR} \qquad {}^1\varepsilon^{SS} = {}^2\varepsilon^{SS} \qquad {}^1\varepsilon^{SN} \neq {}^2\varepsilon^{SN}$$
$$^1\varepsilon^{NR} \neq {}^2\varepsilon^{NR} \qquad {}^1\varepsilon^{NS} \neq {}^2\varepsilon^{NS} \qquad {}^1\varepsilon^{NN} \neq {}^2\varepsilon^{NN}$$

Stresses:

$$^1\tau^{RR} \neq {}^2\tau^{RR} \qquad {}^1\tau^{RS} \neq {}^2\tau^{RS} \qquad {}^1\tau^{RN} = {}^2\tau^{RN}$$
$$^1\tau^{SR} \neq {}^2\tau^{SR} \qquad {}^1\tau^{SS} \neq {}^2\tau^{SS} \qquad {}^1\tau^{SN} = {}^2\tau^{SN}$$
$$^1\tau^{NR} = {}^2\tau^{NR} \qquad {}^1\tau^{NS} = {}^2\tau^{NS} \qquad {}^1\tau^{NN} = {}^2\tau^{NN}$$

11
Stress Based Solid Forces

11.1.
Stresses, Forces and Moments:

All Related Positive Quarter Coordinates: RST

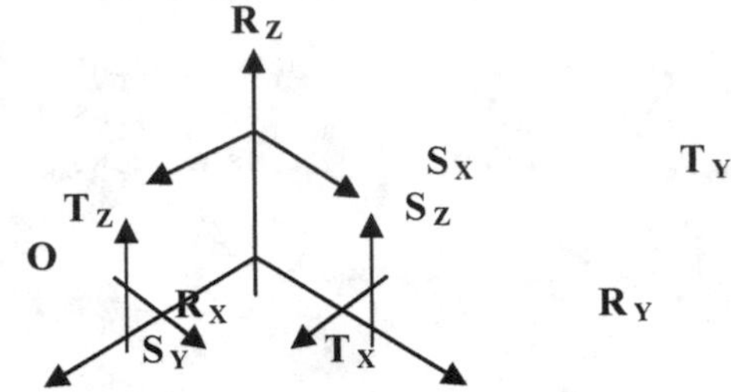

Cross Section Plane is R:

[1]

$\int\int Q\ldots RD_{,S,T} * dS * dT$ $\equiv$ $Q\ldots RD = \int Q\ldots RD_{,T} * dT$

$\int\int Q\ldots RD_{,S,T} * dS * dT$ $\equiv$ $Q\ldots RD = \int Q\ldots RD_{,S} * dS$

Its Derivative is:

[1/L]

$Q\ldots RD_{,T} = \int Q\ldots RD_{,T,S} * dS$ $\quad$ $Q\ldots RD_{,S} = \int Q\ldots RD_{,S,T} * dT$

Its Derivative is:

$[1/L^2]$

$Q\ldots RD_{,T,S} = \int Q\ldots RD_{,T,S,T}\ dT$ $\quad$ $Q\ldots RD_{,S,T} = \int Q\ldots RD_{S,T,S}\ dS$

Where D takes the 0, 1, 2, 3...Derivatives of:

R	S	T
	:	
$(d/dT)\int Q\ldots RD_{,T}*dT$	$=$T Derivative $= Q\ldots RD_{,T}$ $=$T Flow $=$	$\int Q\ldots RD_{,T}*dS$
$(d/dS)\int Q\ldots RD_{,S}*dS$	$=$S Derivative $= Q\ldots RD_{,S}$ $=$S Flow $=$	$\int Q\ldots RD_{,S}*dT$

11.2.
Cross Section General States:

11.2.1.
Cross Section Stresses:

3 D R-Plane Stresses:

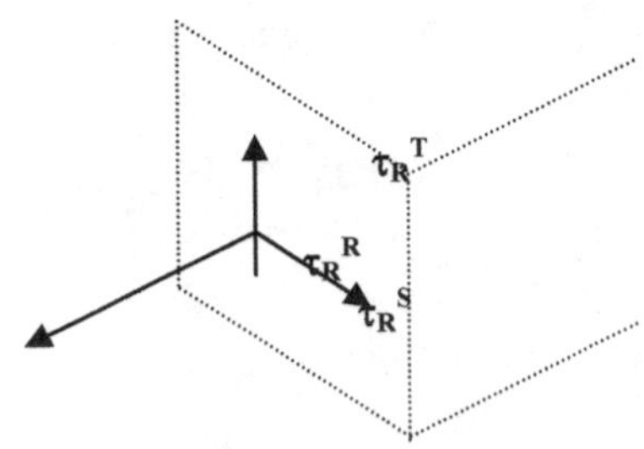

$\tau_R{}^R$ $\tau_R{}^S$ $\tau_R{}^T$

2 D

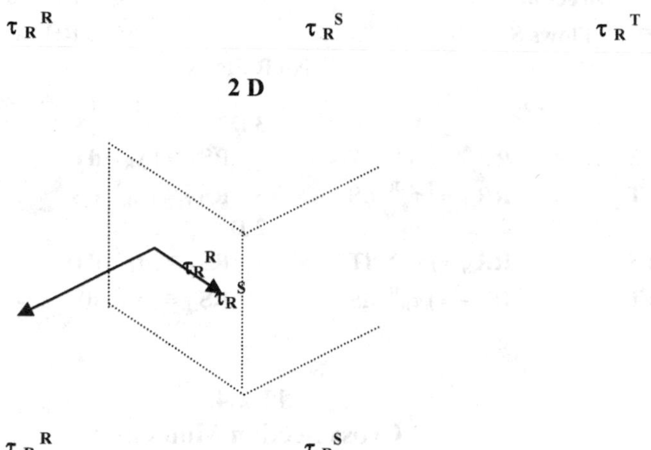

$\tau_R{}^R$ $\tau_R{}^S$

11.2.2. Cross Section Forces:

:

3 D

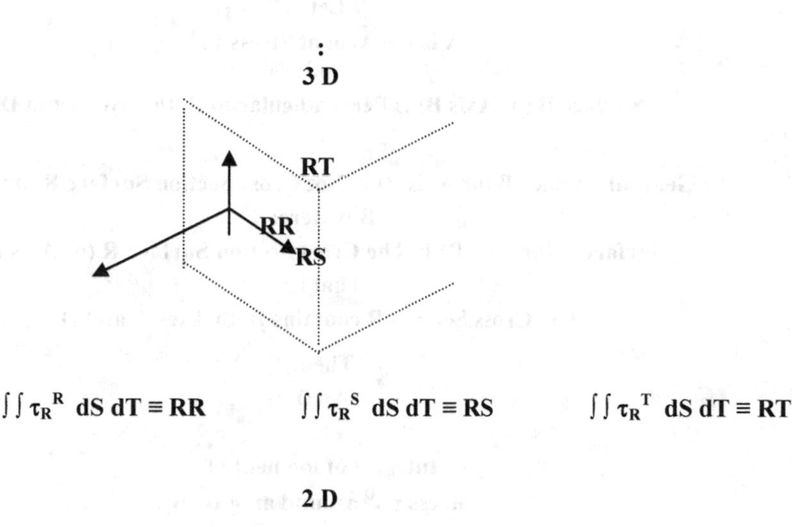

$\int\int \tau_R{}^R \; dS\, dT \equiv RR$ $\int\int \tau_R{}^S \; dS\, dT \equiv RS$ $\int\int \tau_R{}^T \; dS\, dT \equiv RT$

2 D

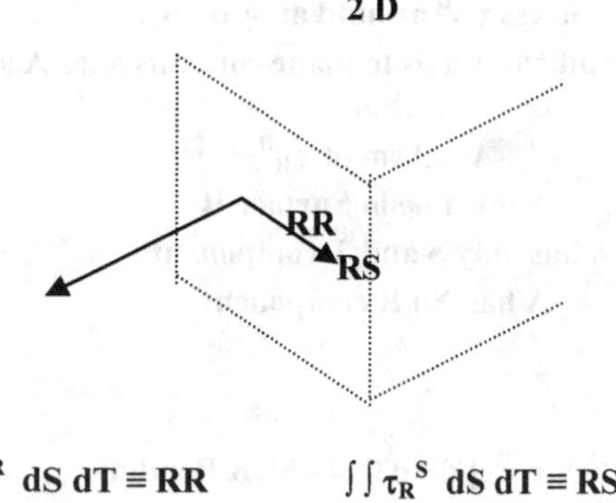

$\int\int \tau_R{}^R \; dS\, dT \equiv RR$ $\int\int \tau_R{}^S \; dS\, dT \equiv RS$

11.2.3. Cross Section Force Flows:

:

Direction D	**Q…RD≡ Q…RD**
Flows S	**SQ…RD≡ Q…RD$_{,S}$**

No R Flows

3 D

Flows S	$RR_{,S} \equiv \int \tau_R^{\ R}\ dT$	$RS_{,S} \equiv \int \tau_R^{\ S}\ dT$	$RT_{,S} \equiv \int \tau_R^{\ T}\ dT$
Flows T	$RR_{,T} \equiv \int \tau_R^{\ R}\ dS$	$RS_{,T} \equiv \int \tau_R^{\ S}\ dS$	$RT_{,T} \equiv \int \tau_R^{\ T}\ dS$

2 D

Flows S	$RR_{,S} \equiv \int \tau_R^{\ R}\ dT$	$RS_{,S} \equiv \int \tau_R^{\ S}\ dT$
Flows T	$RR_{,T} \equiv \int \tau_R^{\ R}\ dS$	$RS_{,T} \equiv \int \tau_R^{\ S}\ dS$

11.2.4.
Cross Section Moments:

:

$\tau_R^{\ D}$	RD
≡	≡
Stress on Cross Section Surface R Drecton D	**Force on Cross Section Surface R Drecton D**

Let:

A is the Arm of Stress $\tau_R^{\ D}$

And:

Surface B (or Axis B) is Perpendicular to Both Arm A and D

:

In General Surface B (or Axis B) Is Not Cross Section Surface R (or Axis R)

But Here:

Surface B (or Axis B) Is The Cross Section Surface R (or Axis R)

That is:

The Cross Section R contains Both Axes A and D

Then:

ARD

≡

Integral of moment of

Stress $\tau_R^{\ D}$ around an Axis B

Axis B is perpendicular axis to plane contains axes A and D

So:

A ≡ Arm of $\tau_R^{\ D}$

A must be in Surface R

A has only S and T components

A has No R component

3 D

$$\int\int [\, S * \tau_R^{\ R}\,]\, dS\, dT \equiv -\, SRR \text{ Bending}$$

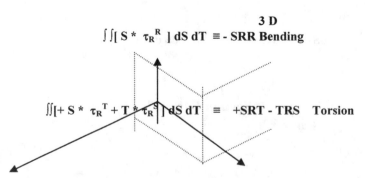

$$\int\int [+ S * \tau_R^{\ T} + T * \tau_R^{\ S}\,]\, dS\, dT \equiv +SRT - TRS \quad \text{Torsion}$$

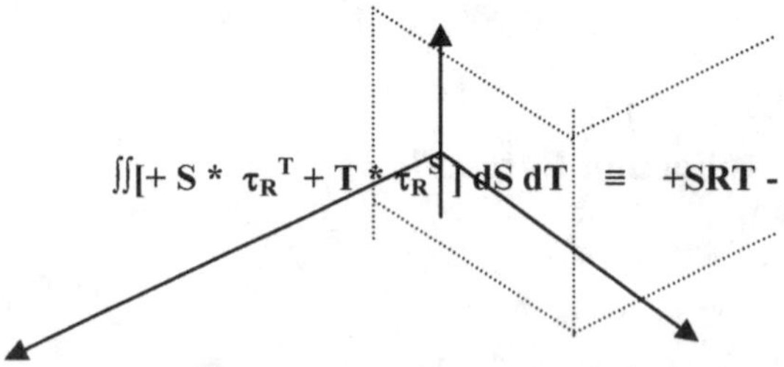

$\int\int$[T * $\tau_R{}^R$] dS dT ≡ + TRR Bending

R Twist Axis S Arm Torsion + T Arm Torsion	T Axis Rotation S Arm Bending	S Axis Rotation T Arm Bending
+SRT -TRS	-SRR	+TRR

:

2 D

$\int\int$[S * $\tau_R{}^R$] dS dT ≡ - SRR Bending

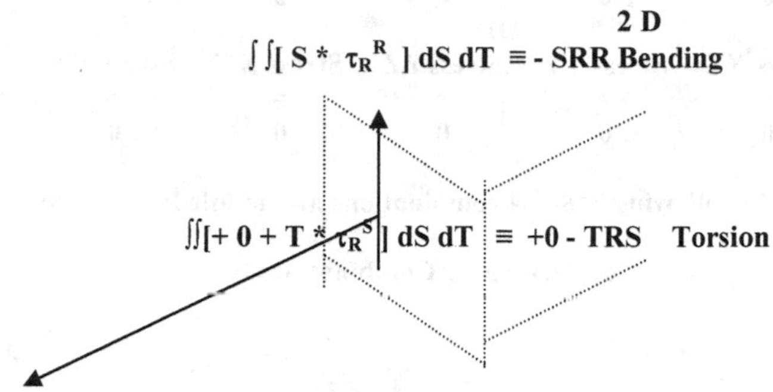

$\int\int$[+ 0 + T * $\tau_R{}^S$] dS dT ≡ +0 - TRS Torsion

R Twist Axis S Arm Torsion + T Arm Torsion	T Axis Rotation S Arm Bending
-TRS	-SRR

11.2.5.
Cross Section Moment Flows:

:

No R Flows

3 D
S and T Flows

Flow	R Twist Axis S Arm Torsion + T Arm Torsion	T Axis Rotation S Arm Bending	S Axis Rotation T Arm Bending
S	$+SRT_{,S} - TRS_{,S}$	$-SRR_{,S}$	$+TRR_{,S}$
T	$+SRT_{,T} - TRS_{,T}$	$-SRR_{,T}$	$+TRR_{,T}$

2 D
S and T Flows

Flow	R Twist Axis S Arm Torsion + T Arm Torsion	T Axis Rotation S Arm Bending
S	$-TRS_{,S}$	$-SRR_{,S}$
T	$-TRS_{,T}$	$-SRR_{,T}$

11.3.
Costumed Strain and Costumed Stress:

:

11.3.1.
Costumed Combinations:

Setting Zero Stress 64 Combinations:

:

	Strain XX	Strain YY	Strain ZZ	Strain XY	Strain YZ	Strain ZX
Strains	=	=	=	=	=	=
	0	0	0	0	0	0

Or: :

	Stress XX	Stress YY	Stress ZZ	Stress XY	Stress YZ	Stress ZX
Stresss	=	=	=	=	=	=
	0	0	0	0	0	0

In the following 8*8= 64 combinations are detailed:

8 Cases of Zero Shear Combinations

OO000						
OA011						ZX = 0
OA011					YZ = 0	
OA011				XY = 0		
OB112					YZ = 0	ZX = 0
OB112				XY = 0		ZX = 0
OB112				XY = 0	YZ = 0	
OC222				XY = 0	YZ = 0	ZX = 0

:

Plane States:

Plane State: Z is at least Less One Dimensional Order than X and Y:
8 Cases of Zero Single Axial ZZ = 0 and Shear Combinations

UO002			ZZ = 0			
UA013			ZZ = 0			ZX = 0
UA013			ZZ = 0		YZ = 0	
UA112			ZZ = 0	XY = 0		
UB114			ZZ = 0		YZ = 0	ZX = 0
UB123			ZZ = 0	XY = 0		ZX = 0
UB123			ZZ = 0	XY = 0	YZ = 0	
UC224			ZZ = 0	XY = 0	YZ = 0	ZX = 0

:

Plane State: Y is at least Less One Dimensional Order than Z and X:
8 Cases of Zero Single Axial YY = 0 and Shear Combinations

UO002		YY = 0				
UA112		YY = 0				ZX = 0
UA013		YY = 0			YZ = 0	
UA013		YY = 0		XY = 0		
UB123		YY = 0			YZ = 0	ZX = 0
UB123		YY = 0		XY = 0		ZX = 0
UB114		YY = 0		XY = 0	YZ = 0	
UC224		YY = 0		XY = 0	YZ = 0	ZX = 0

:

Plane State: X is at least Less One Dimensional Order than Y and Z:
8 Cases of Single One Axial XX = 0 and Shear Combinations

UO002	XX = 0					

UA013	XX = 0					ZX = 0
UA112	XX = 0				YZ = 0	
UA013	XX = 0			XY = 0		
UB123	XX = 0				YZ = 0	ZX = 0
UB114	XX = 0			XY = 0		ZX = 0
UB123	XX = 0			XY = 0	YZ = 0	
UC224	XX = 0			XY = 0	YZ = 0	ZX = 0

Lineal States:

Lineal State: Y and Z are at least Less One Dimensional Order than X:
8 Cases of Zero Two Axials YY = 0 and ZZ = 0, and Shear Combinations

VO0220		YY = 0	ZZ = 0			
VA123		YY = 0	ZZ = 0			ZX = 0
VA0330		YY = 0	ZZ = 0		YZ = 0	
VA123		YY = 0	ZZ = 0	XY = 0		
VB134		YY = 0	ZZ = 0		YZ = 0	ZX = 0
VB233		YY = 0	ZZ = 0	XY = 0		ZX = 0
VB134		YY = 0	ZZ = 0	XY = 0	YZ = 0	
VC244		YY = 0	ZZ = 0	XY = 0	YZ = 0	ZX = 0

Lineal State: Z and X are at least Less One Dimensional Order than Y:
8 Cases of Zero Two Axials ZZ = 0and XX = 0, and Shear Combinations

VO022	XX = 0		ZZ = 0			
VA033	XX = 0		ZZ = 0			ZX = 0
VA123	XX = 0		ZZ = 0		YZ = 0	
VA1203	XX = 0		ZZ = 0	XY = 0		
VB134	XX = 0		ZZ = 0		YZ = 0	ZX = 0
VB134	XX = 0		ZZ = 0	XY = 0		ZX = 0
VB233	XX = 0		ZZ = 0	XY = 0	YZ = 0	
VC244	XX = 0		ZZ = 0	XY = 0	YZ = 0	ZX = 0

Lineal State: X and Y are at least Less One Dimensional Order than Z:
8 Cases of Zero Two Axials XX = 0 and YY = 0, and Shear Combinations

VO022	XX = 0	YY = 0				
VA123	XX = 0	YY = 0				ZX = 0
VA123	XX = 0	YY = 0			YZ = 0	
VA033	XX = 0	YY = 0		XY = 0		
VB233	XX = 0	YY = 0			YZ = 0	ZX = 0
VB134	XX = 0	YY = 0		XY = 0		ZX = 0
VB134	XX = 0	YY = 0		XY = 0	YZ = 0	
VC244	XX = 0	YY = 0		XY = 0	YZ = 0	ZX = 0

Zero Axials:
8 Cases of Zero All Axials and Shear Combinations

WO222	XX = 0	YY = 0	ZZ = 0			
WA233	XX = 0	YY = 0	ZZ = 0			ZX = 0
WA233	XX = 0	YY = 0	ZZ = 0		YZ = 0	
WA233	XX = 0	YY = 0	ZZ = 0	XY = 0		

WB334	**XX = 0**	**YY = 0**	**ZZ = 0**		**YZ = 0**	**ZX = 0**
WB334	**XX = 0**	**YY = 0**	**ZZ = 0**	**XY = 0**		**ZX = 0**
WB334	**XX = 0**	**YY = 0**	**ZZ = 0**	**XY = 0**	**YZ = 0**	
WC444	**XX = 0**	**YY = 0**	**ZZ = 0**	**XY = 0**	**YZ = 0**	**ZX = 0**

11.3.2.
Costumed Combination Groups:

In the following 4+6+6+4= 20 No redundancy combinations groups are detailed:

O4 No Redundancy 4 Cases of Zero Shear Combinations

OO000	**1**						
OA011	**3**						**ZX = 0**
OB112	**3**					**YZ = 0**	**ZX = 0**
OC222	**1**				**XY = 0**	**YZ = 0**	**ZX = 0**

Plane State:
U6 No Redundancy 6 Cases of Zeoro Single Axial (llustrated for ZZ = 0) and Shear Combinations

UO002	**3**			**ZZ = 0**			
UA013	**6**			**ZZ = 0**			**ZX = 0**
UA112	**3**			**ZZ = 0**	**XY = 0**		
UB114	**3**			**ZZ = 0**		**YZ = 0**	**ZX = 0**
UB123	**6**			**ZZ = 0**	**XY = 0**		**ZX = 0**
UC22	**3**			**ZZ = 0**	**XY = 0**	**YZ = 0**	**ZX = 0**

Lineal State:
V6 No Redundancy 6 Cases of Zero Two Axial Stresses (llustrated for YY = 0 and ZZ = 0) and Shear Combinations

VO022	**3**		**YY = 0**	**ZZ = 0**			
VA123	**6**		**YY = 0**	**ZZ = 0**			**ZX = 0**
VA033	**3**		**YY = 0**	**ZZ = 0**		**YZ = 0**	
VB134	**6**		**YY = 0**	**ZZ = 0**		**YZ = 0**	**ZX = 0**
VB233	**3**		**YY = 0**	**ZZ = 0**	**XY = 0**		**ZX = 0**
VC244	**3**		**YY = 0**	**ZZ = 0**	**XY = 0**	**YZ = 0**	**ZX = 0**

:
W4
No Redundancy 4 Cases of Zeros of All Axial and Shear Combinations

WO222	1	**XX = 0**	**YY = 0**	**ZZ = 0**			
WA233	3	**XX = 0**	**YY = 0**	**ZZ = 0**			**ZX = 0**
WB334	3	**XX = 0**	**YY = 0**	**ZZ = 0**		**YZ = 0**	**ZX = 0**
WC444	1	**XX = 0**	**YY = 0**	**ZZ = 0**	**XY = 0**	**YZ = 0**	**ZX = 0**

11.4.
Costumed General States:

O Cases

8 Cases of Zero Shear Combinations
4 No redundancy combinations groups

OO000 **1**

OA011 **3** **ZX = 0**

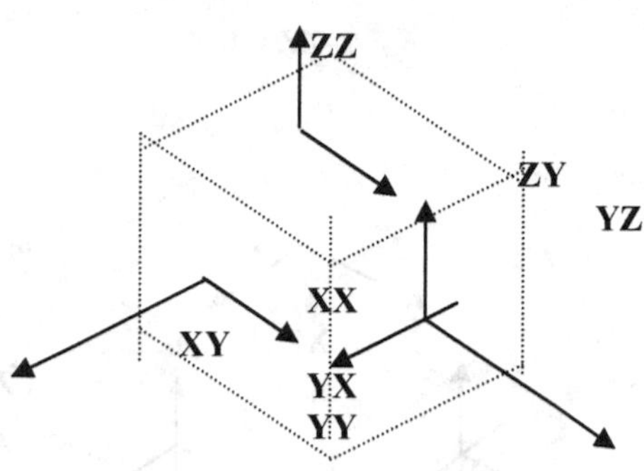

OB112 **3** **YZ = 0** **ZX = 0**

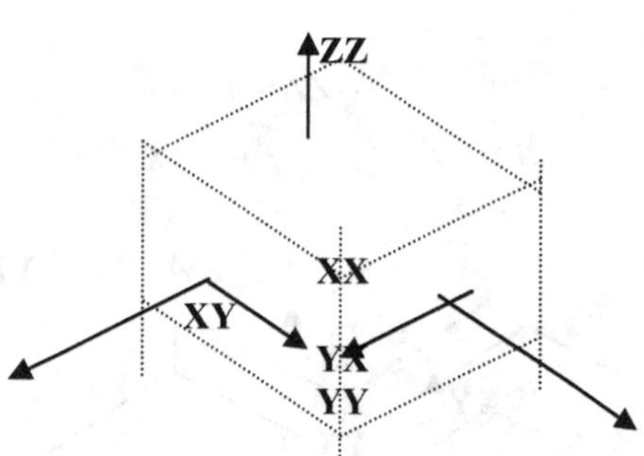

OC222 **1** **XY = 0** **YZ = 0** **ZX = 0**

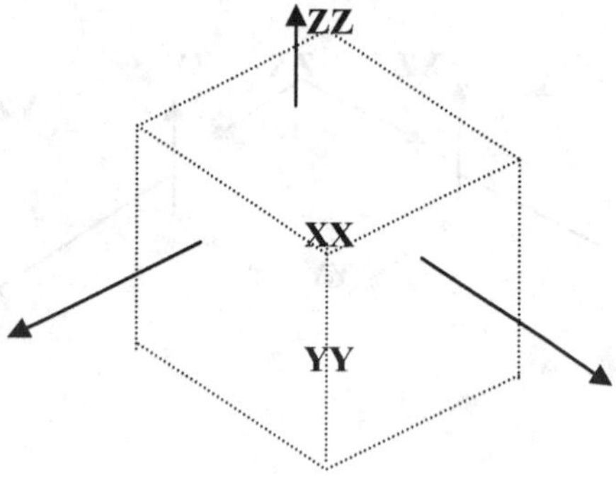

11.4.1.
Plane States:

U Cases
Plane States of 24 Combinations
6 No redundancy combinations groups

U6 of 6 Cases of Zeoro Single Axial (llustrated for 33) and Shear Combinations

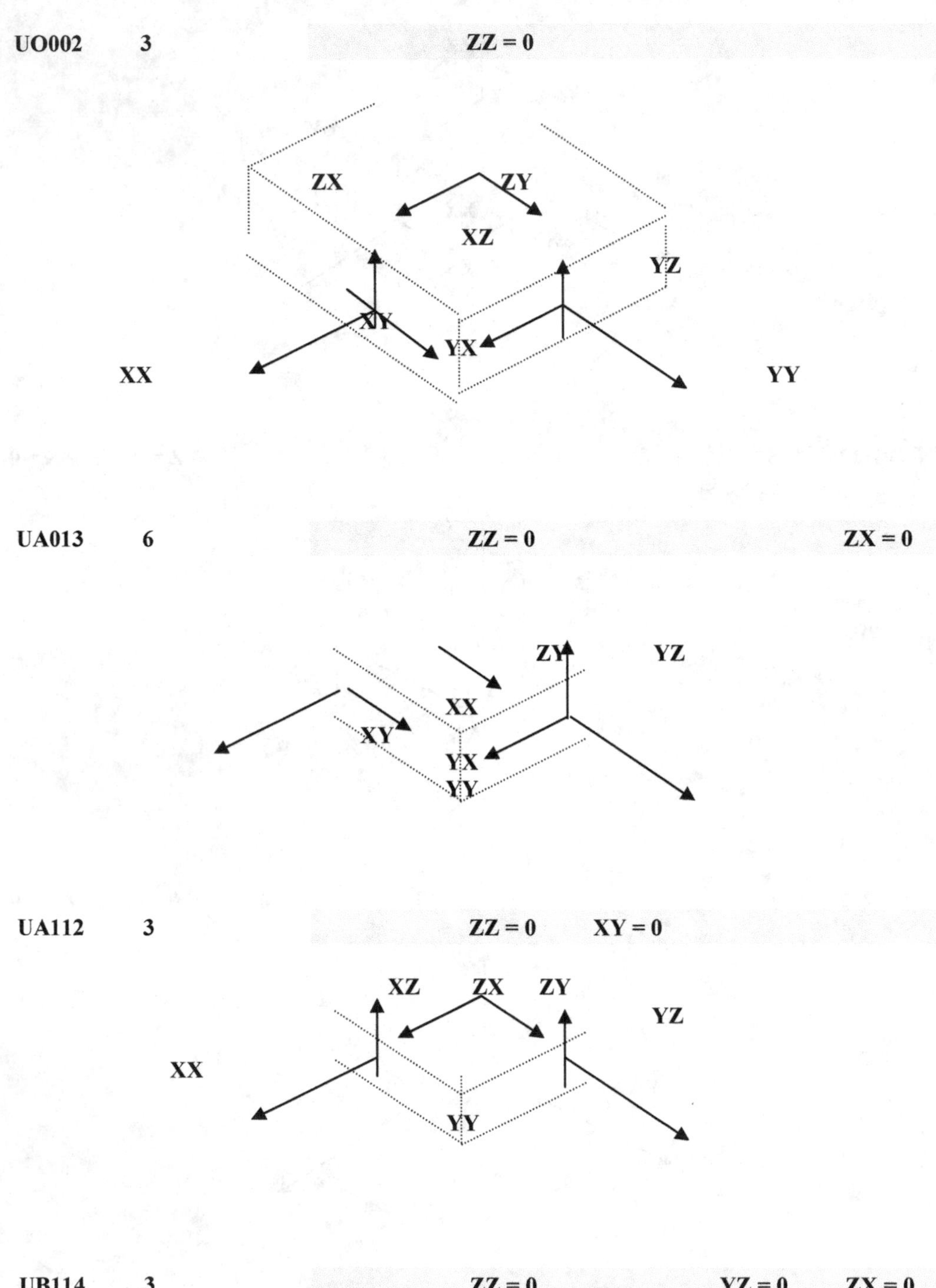

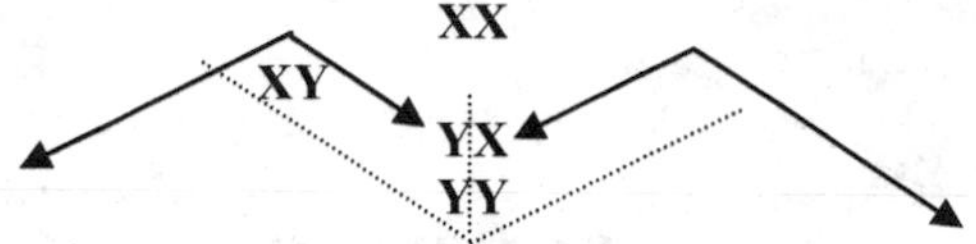

UB123	6	ZZ = 0	XY = 0		ZX = 0

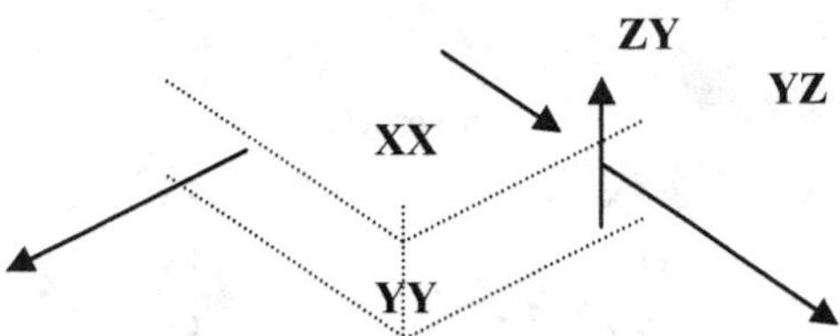

Notic that this is equivalent to:

UA013		ZZ = 0		YZ = 0	

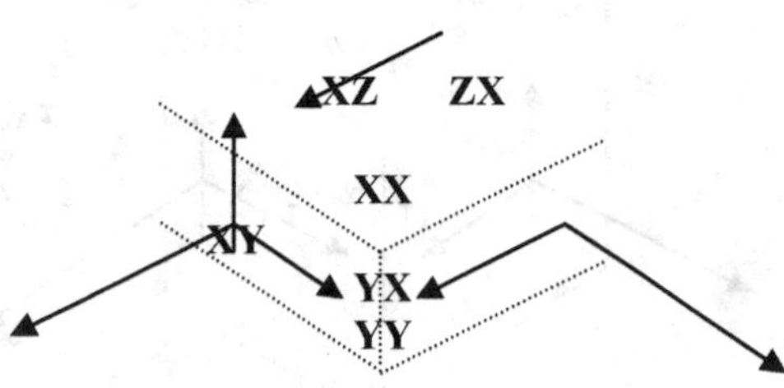

UC224	3	ZZ = 0	XY = 0	YZ = 0	ZX = 0

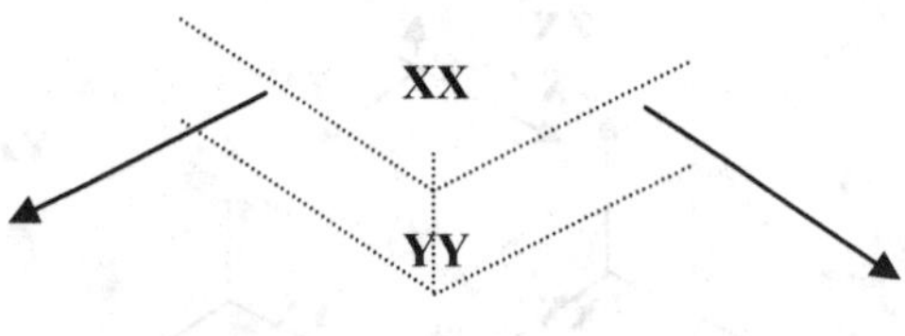

11.4.1.1. Deep Slab:

Z Deep Slab: Z is Less One Dimensional Order than X and Y:

:

			Stress 33	Stress 12	Stress 23	Stress 31
UO002 1/4	3	X-Y-Slab Z-Deep	ZZ = 0			

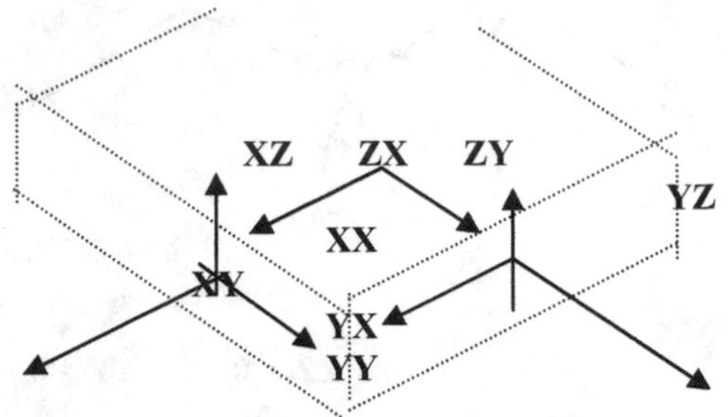

Stresses:

Surface X	$\tau_X^{\ X}$	$\tau_X^{\ Y}$	$\tau_X^{\ Z}$
Surface Y	$\tau_Y^{\ X}$	$\tau_Y^{\ Y}$	$\tau_Y^{\ Z}$
Surface Z	$\tau_Z^{\ X}$	$\tau_Z^{\ Y}$	$\tau_Z^{\ Z} = 0$

Stresses:

0

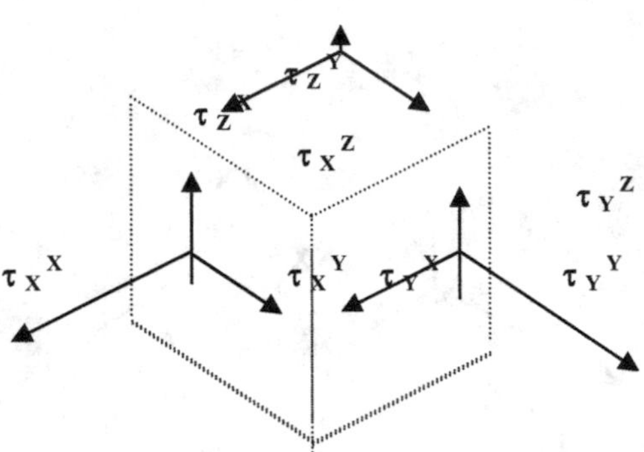

Forces:

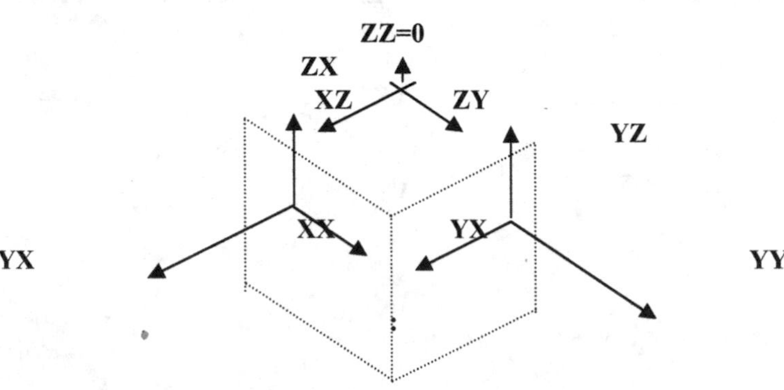

Z- Deep Solid: Z-Plane-Stress:X-Y Solid:

Forces:

Forces on X Face:	**XX**	**XY**	**XZ**
Forces on Y Face	**YX**	**YY**	**YZ**
Forces on Z Face	**ZX**	**ZY**	**ZZ = 0**

Flows:

X Flows on X Face:	$XX_{,X}$	$XY_{,X}$	$XZ_{,X}$
X Flows on Y Face	$YX_{,X}$	$YY_{,X}$	$YZ_{,X}$
X Flows on Z Face	$ZX_{,X}$	$ZY_{,X}$	$ZZ_{,X} = 0$

:

Y Flows on X Face:	$XX_{,Y}$	$XY_{,Y}$	$XZ_{,Y}$
Y Flows on Y Face	$YX_{,Y}$	$YY_{,Y}$	$YZ_{,Y}$
Y Flows on Z Face	$ZX_{,Y}$	$ZY_{,Y}$	$ZZ_{,Y} = 0$

:

Z Flows on X Face:	$XX_{,Z}$	$XY_{,Z}$	$XZ_{,Z}$
Z Flows on Y Face	$YX_{,Z}$	$YY_{,Z}$	$YZ_{,Z}$
Z Flows on Z Face	$ZX_{,Z}$	$ZY_{,Z}$	**NA**

:

X Plane Force Flows: :

Force Flows	**NA**	**Y Flows**	**Z Flows**
XX	**NA**	$XX_{,Y}$	$XX_{,Z}$
XY	**NA**	$XY_{,Y}$	$XY_{,Z}$
XZ	**NA**	$XZ_{,Y}$	$XZ_{,Z}$

Y Plane Force Flows:

Force Flows	**X Flows**	**NA**	**Z Flows**
YY	$YY_{,X}$	**NA**	$YY_{,Z}$
YZ	$YZ_{,X}$	**NA**	$YZ_{,Z}$
YX	$YX_{,X}$	**NA**	$YX_{,Z}$

Z Plane Force Flows:

Force Flows	**X Flows**	**Y Flows**	**NA**
ZZ	$ZZ_{,X} = 0$	$ZZ_{,Y} = 0$	**NA**
ZX	$ZX_{,X}$	$ZX_{,Y}$	**NA**
ZY	$ZY_{,X}$	$ZY_{,Y}$	**NA**

Moments:

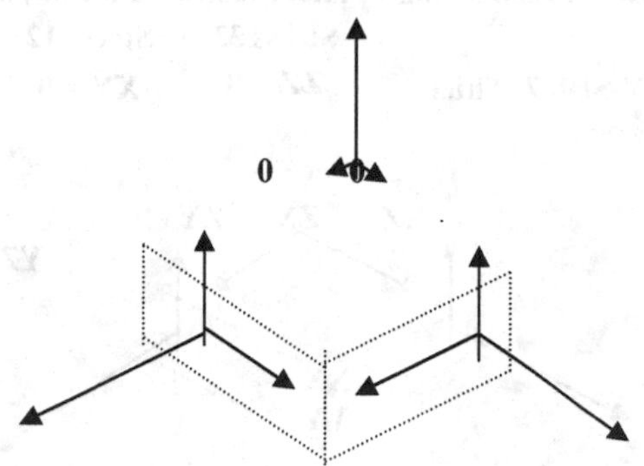

Moment Axis in Plane X		**Bending**	**Around X Torsion Axis**		**Torsion**
Moment of Y Arm and Z Axis	≡	**-YXX**	**Moment of Y Arm**	≡	**+YXZ**
Moment of Z Arm and Y Axis	≡	**+ZXX**	**Moment of Z Arm**	≡	**-ZXY**

Moment Axis in Plane Y		Bending	Around Y Xorsion Axis		Xorsion
Moment of Z Arm and X Axis	≡	-ZYY	Moment of Z Arm	≡	+ZYX
Moment of X Arm and Z Axis	≡	+XYY	Moment of X Arm	≡	-XYZ

Moment Axis in Plane Z		Bending	Around Z Yorsion Axis		Yorsion
Moment of X Arm and Y Axis	≡	-XZZ=0	Moment of X Arm	≡	+XZY
Moment of Y Arm and X Axis	≡	+YZZ=0	Moment of Y Arm	≡	-YZX

X Plane Moment Flows:

Moment Flows	Y Flows	Z Flows
ZXX Bending	$+ZXX_{,Y}$	$+ZXX_{,Z}$
YXX Bending	$-YXX_{,Y}$	$-YXX_{,Z}$
YXZ Torsion	$+YXZ_{,Y}$	$+YXZ_{,Z}$
ZXY Torsion	$-ZXY_{,Y}$	$-ZXY_{,Z}$

Y Plane Moment Flows:

Moment Flows	Z Flows	X Flows
XYY Bending	$+XYY_{,Z}$	$+XYY_{,X}$
ZYY Bending	$-ZYY_{,Z}$	$-ZYY_{,X}$
ZYX Torsion	$+ZYX_{,Z}$	$+ZYX_{,X}$
XYZ Torsion	$-XYZ_{,Z}$	$-XYZ_{,X}$

Z Plane Moment Flows:

Moment Flows	X Flows	Y Flows
YZZ Bending	$+YZZ_{,X} = 0$	$+YZZ_{,Y} = 0$
XZZ Bending	$-XZZ_{,X} = 0$	$-XZZ_{,Y} = 0$
XZY Torsion	$+XZY_{,X}$	$+XZY_{,Y}$
YZX Torsion	$-YZX_{,X}$	$-YZX_{,Y}$

11.4.1.2. Thick Slab:

Z Thick Slab: Z is Less Two **Dimensional Orders than X and Y:**

Plane State:

U6 No Redundancy 6 Cases of Zeoro Single Axial (llustrated for 33) and Shear Combinations

			Stress 33	Stress 12	Stress 23	Stress 31
UA112 1/16	3	X-Y-Slab Z-Thick	ZZ = 0	XY = 0		

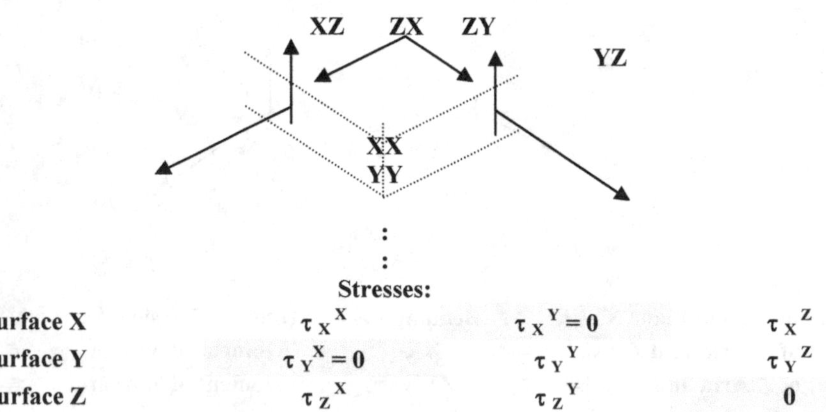

:

:

Stresses:

Surface X	$\tau_X^{\ X}$	$\tau_X^{\ Y} = 0$	$\tau_X^{\ Z}$
Surface Y	$\tau_Y^{\ X} = 0$	$\tau_Y^{\ Y}$	$\tau_Y^{\ Z}$
Surface Z	$\tau_Z^{\ X}$	$\tau_Z^{\ Y}$	0

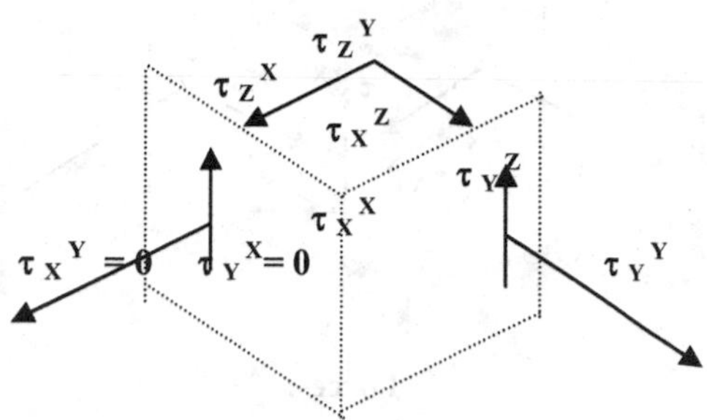

Forces:

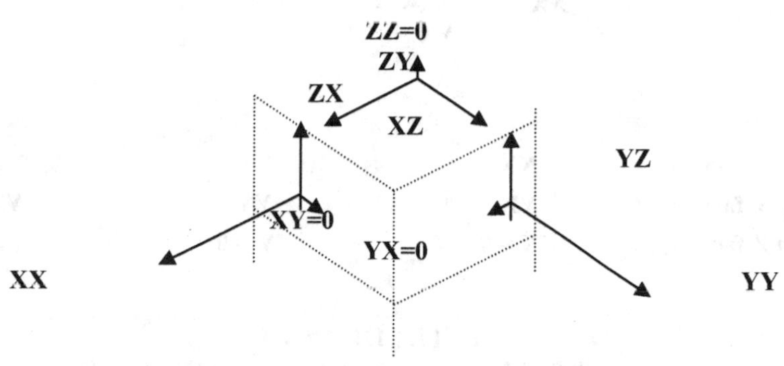

XX

Forces:

Forces on X face	**XX**	**XY = 0**	**XZ**
Forces on Y face	**YX = 0**	**YY**	**YZ**
Forces on Z face	**ZX**	**ZY**	**0**

11.4.1.3.
Thin Plate or Shell:

Z Thin X-Y Plate or Shell: Z is Less Three Dimensional Orders than X and Y:
: Plane State:
U6 No Redundancy 6 Cases of Zeoro Single Axial (Illustrated for 33) and Shear Combinations

			Stress 33	Stress 12	Stress 23	Stress 31
UB114 1/64	**3**	**X-Y Membrane Z-Thin:**	**ZZ = 0**		**YZ = 0**	**ZX = 0**

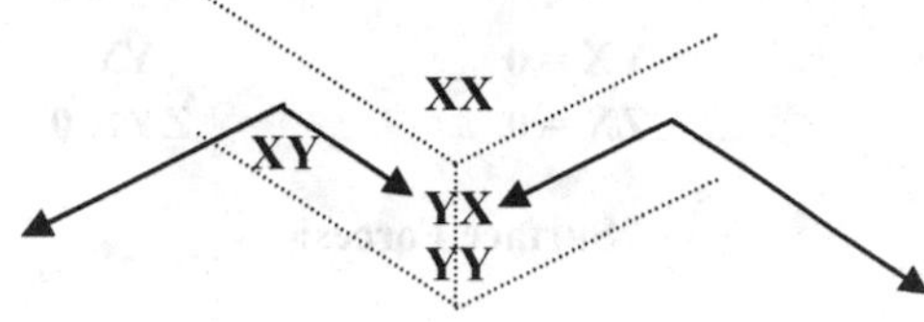

Stresses:

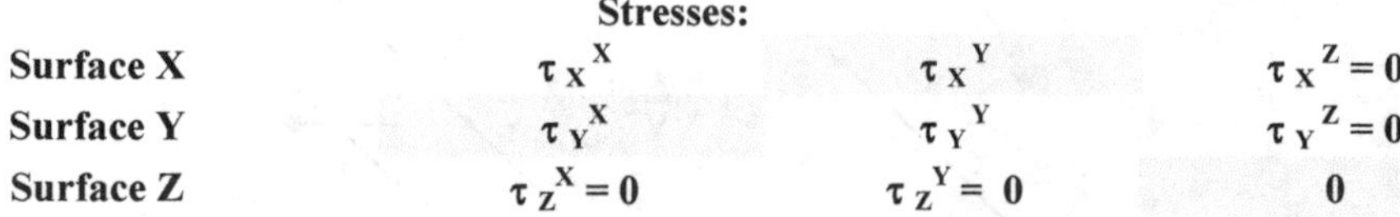

Surface X	τ_X^X	τ_X^Y	$\tau_X^Z = 0$
Surface Y	τ_Y^X	τ_Y^Y	$\tau_Y^Z = 0$
Surface Z	$\tau_Z^X = 0$	$\tau_Z^Y = 0$	**0**

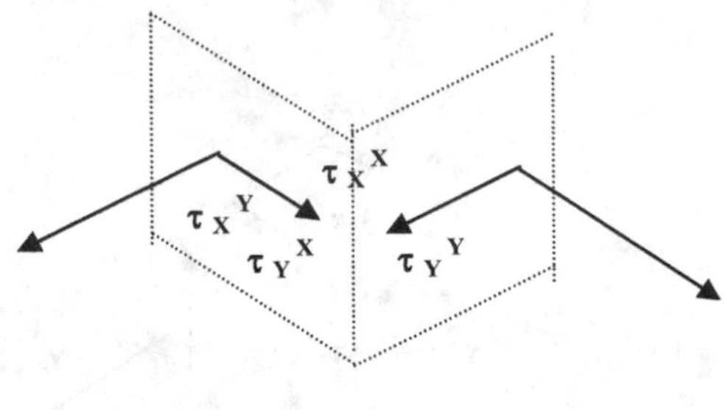

Forces:

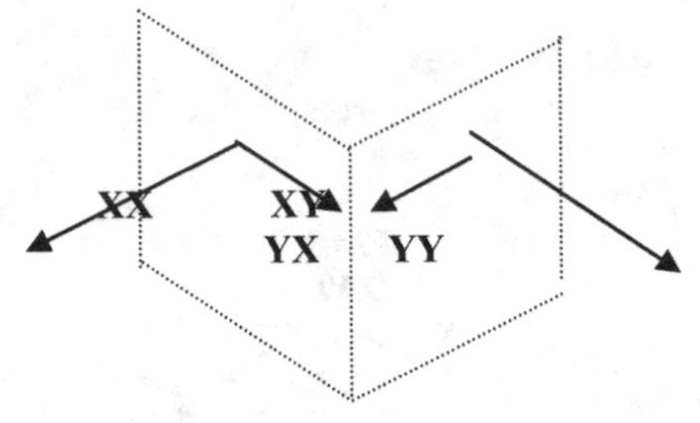

Forces on X face	XX	XY	XZ = 0
Forces on Y face	YX	YY	YZ = 0
Forces on Z face	ZX = 0	ZY = 0	ZZ = 0

11.4.1.4.
Axisymmetric Membrane and Axisymmetric Load:

Z-Axisymmetric Membrane and Axisymmetric Load

Plane State:

U6 No Redundancy 6 Cases of Zeoro Single Axial (llustrated for 33) and Shear Combinations

			Stress 33	Stress 12	Stress 23	Stress 31
UC224 1/64	3	Axisym Load Z-Thin:	ZZ = 0	XY = 0	YZ = 0	ZX = 0

11.4.1.5.
Unstressed Surface:

Surface Z has No Stress:
And X and Y Planes Have No Stress in Z Direction:
And All Surfaces have No moment:
And No Shear Stress:

Stresses on X face	XX	XY = 0	XZ = 0
Stresses on Y face	YX = 0	YY	YZ = 0
Stresses on Z face	ZX = 0	ZY = 0	ZZ = 0

Surface Forces:

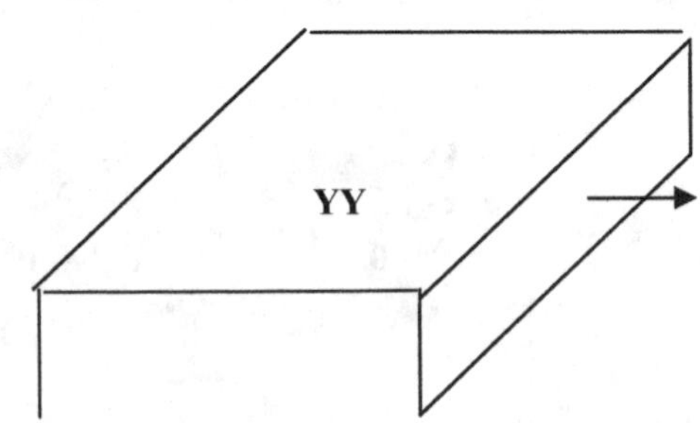

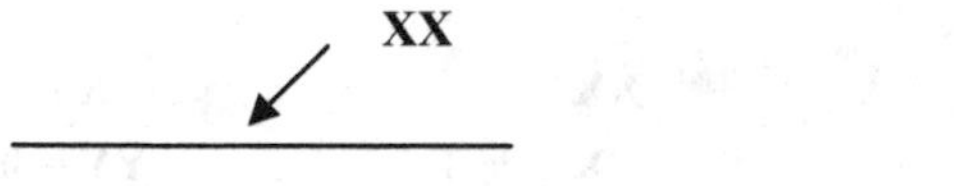

11.4.2.
Lineal States:

V Cases
Lineal States of 24 Combinations
6 No redundancy combinations groups

11.4.2.1.
No Shear:

V6 of 6 Cases of Zero Two Axial Stresses (llustrated for 22 and 33) and Shear Combinations

No sheaer VO022 $Z \approx Y < X$ YY = 0 ZZ = 0

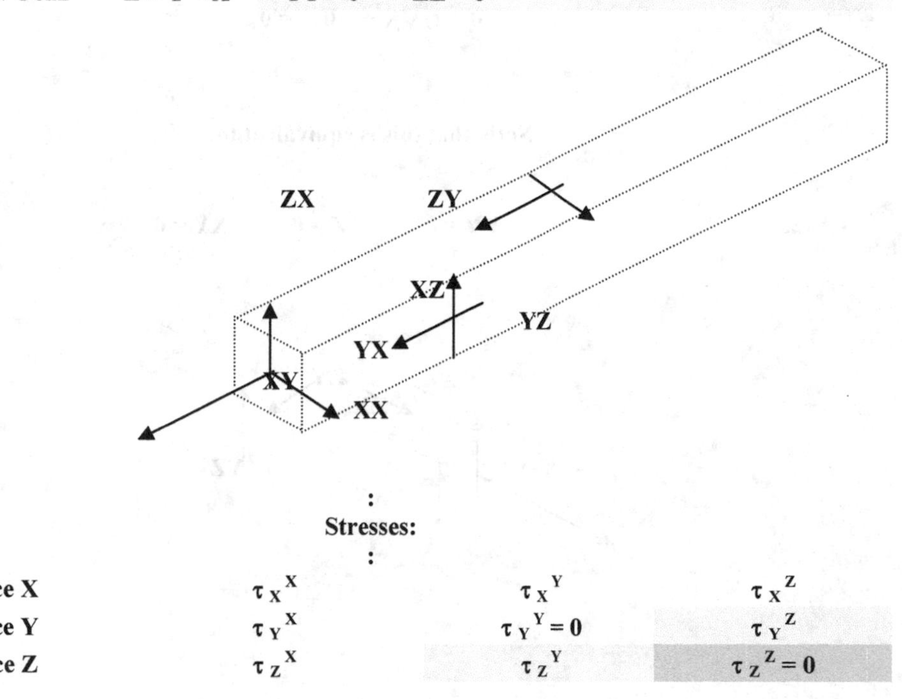

:
Stresses:
:

Surface X	$\tau_X{}^X$	$\tau_X{}^Y$	$\tau_X{}^Z$
Surface Y	$\tau_Y{}^X$	$\tau_Y{}^Y = 0$	$\tau_Y{}^Z$
Surface Z	$\tau_Z{}^X$	$\tau_Z{}^Y$	$\tau_Z{}^Z = 0$

$$\tau_{XX,X} + \tau_{XY,Y} + \tau_{XZ,Z} = 0$$
$$\tau_{YX,X} + 0 + \tau_{YZ,Z} = 0$$
$$\tau_{ZX,X} + \tau_{ZY,Y} + 0 = 0$$

Forces:

:

Forces on X face	**XX**	**XY**	**XZ**
Forces on Y face	**YX**	**YY = 0**	**YZ**
Forces on Z face	**ZX**	**ZY**	**ZZ = 0**

11.4.2.2. Y-Deep and Z-Thick:

Y-Deep and Z-Thick Member:

Y deep Z thick	**VA123**	**Z < Y < X**	**YY = 0**	**ZZ = 0**	**ZX = 0**

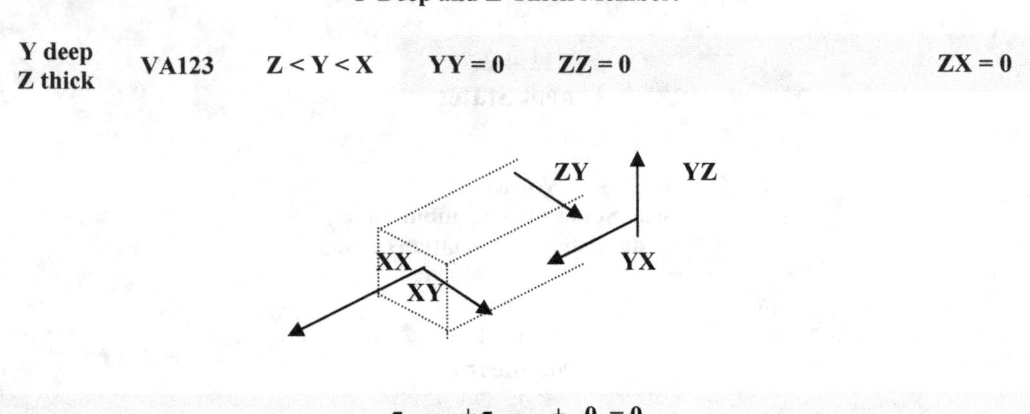

$$\tau_{XX,X} + \tau_{XY,Y} + 0 = 0$$
$$\tau_{YX,X} + 0 + \tau_{YZ,Z} = 0$$
$$0 + \tau_{ZY,Y} + 0 = 0$$

:

Notic that this is equivalent to:

VA123	**YY = 0**	**ZZ = 0**	**XY = 0**

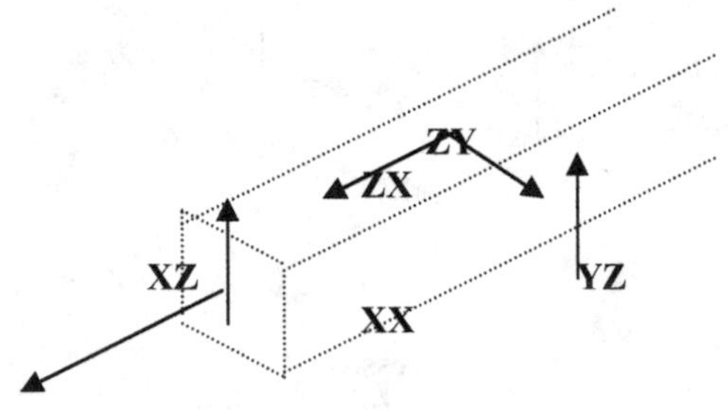

11.4.2.3. Torque:

Torque	**VA033**	**Z ≈ Y << X**	**YY = 0**	**ZZ = 0**	**YZ = 0**

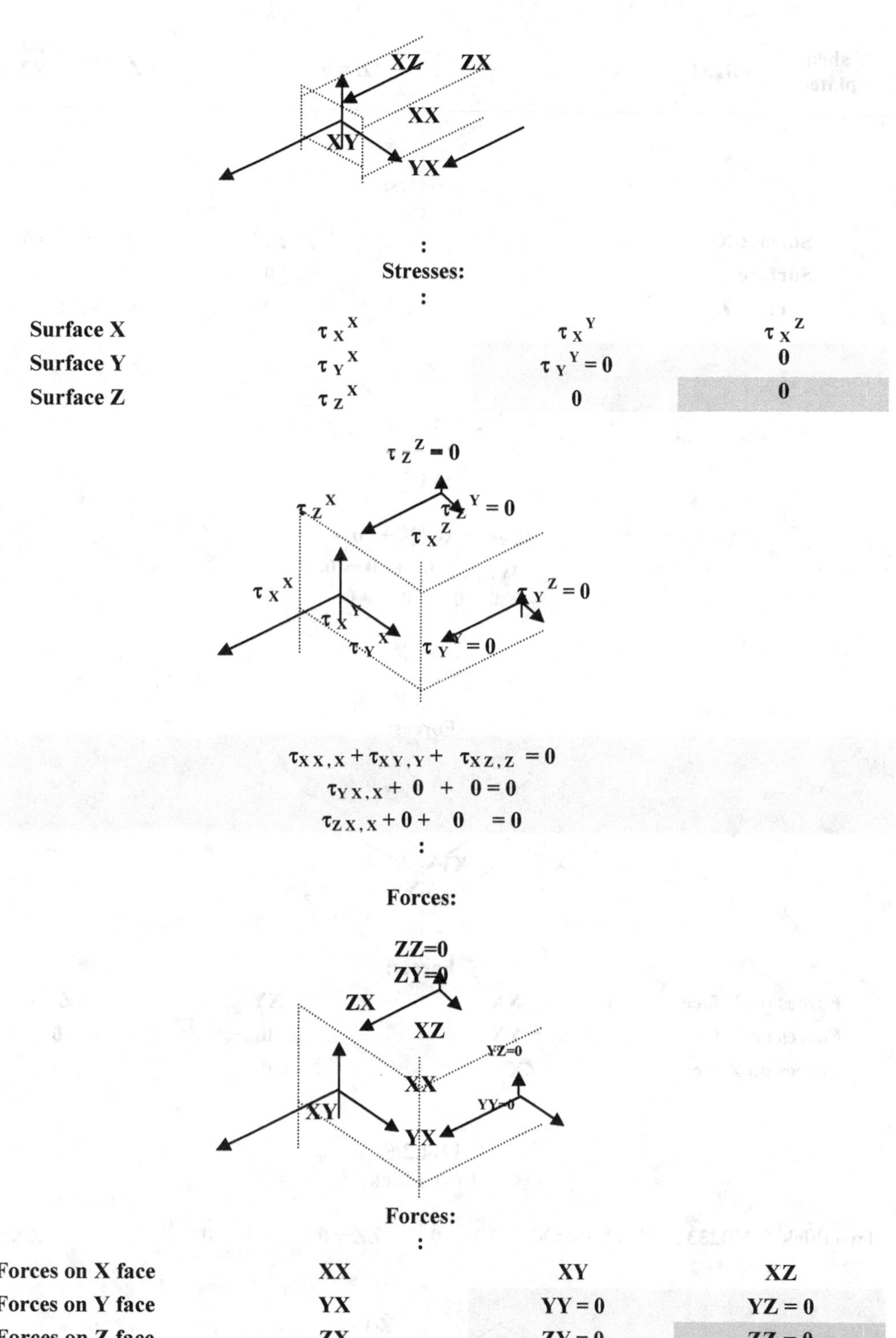

:

Stresses:

:

Surface X	$\tau_X{}^X$	$\tau_X{}^Y$	$\tau_X{}^Z$
Surface Y	$\tau_Y{}^X$	$\tau_Y{}^Y = 0$	**0**
Surface Z	$\tau_Z{}^X$	**0**	**0**

$$\tau_{XX,X} + \tau_{XY,Y} + \tau_{XZ,Z} = 0$$
$$\tau_{YX,X} + 0 + 0 = 0$$
$$\tau_{ZX,X} + 0 + 0 = 0$$

:

Forces:

Forces:

:

Forces on X face	**XX**	**XY**	**XZ**
Forces on Y face	**YX**	**YY = 0**	**YZ = 0**
Forces on Z face	**ZX**	**ZY = 0**	**ZZ = 0**

:

11.4.2.4.
Y-Shear Plate:

Y shear plate	VB134	Z < Y <<X	YY = 0	ZZ = 0		YZ = 0	ZX = 0

:

Stresses:

:

Surface X	$\tau_X{}^X$	$\tau_X{}^Y$	$\tau_X{}^Z = 0$
Surface Y	$\tau_Y{}^X$	0	0
Surface Z	$\tau_Z{}^X = 0$	0	0

$$\tau_{XX,X} + \tau_{XY,Y} + 0 = 0$$

$$\tau_{YX,X} + 0 + 0 = 0$$

$$0 + 0 + 0 = 0$$

:

Forces:

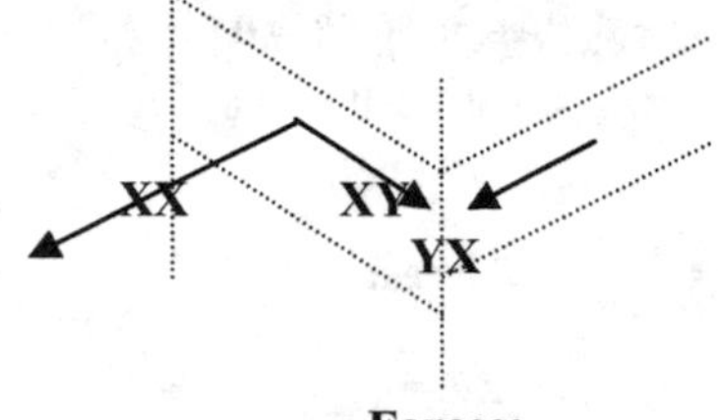

Forces:

Forces on X face	XX	XY	XZ = 0
Forces on Y face	YX	0	0
Forces on Z face	ZX = 0	0	0

11.4.2.5. Two Thick:

Two thick	VB233	Z ≈ Y <<< X	YY = 0	ZZ = 0	XY = 0		ZX = 0

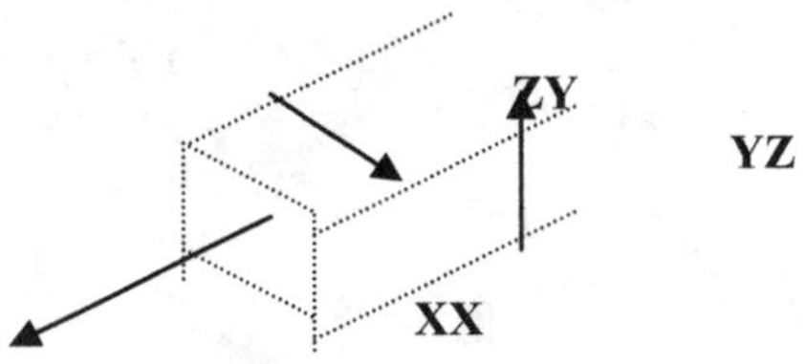

$$\tau_{XX,X} + \tau_{XY,Y} + 0 = 0$$

$$\tau_{YX,X} + 0 + 0 = 0$$

$$0 + 0 + 0 = 0$$

:

11.4.2.6.
Two Thin:

Two thin **VC244** **$Z \approx Y <<< X$** **YY = 0** **ZZ = 0** **XY = 0** **YZ = 0** **ZX = 0**

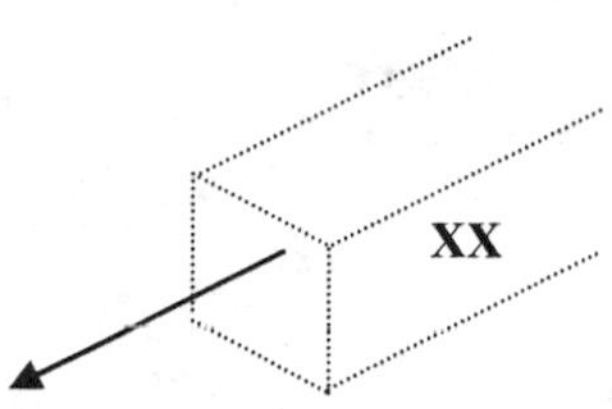

$$\tau_{XX,X} + 0 + 0 + = 0$$
$$0 + 0 + \tau_{YZ,Z} = 0$$
$$0 + \tau_{ZY,Y} + 0 = 0$$

11.4.3.
Zeros of All Axials:

W Cases
Zeros of All Axials States of 8 Combinations
4 No redundancy combinations groups

W4 No Redundancy 4 Cases of Zeros of All Axials and Shear Combinations

WO222 **1** **11** **22** **33**

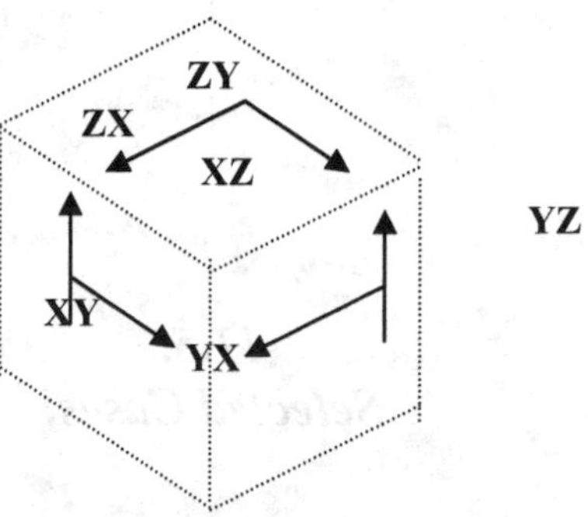

YZ

WA233 **3** **11** **22** **33** **31**

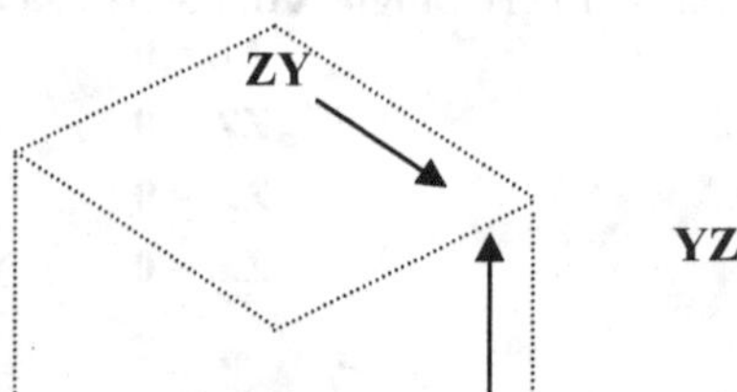

YZ

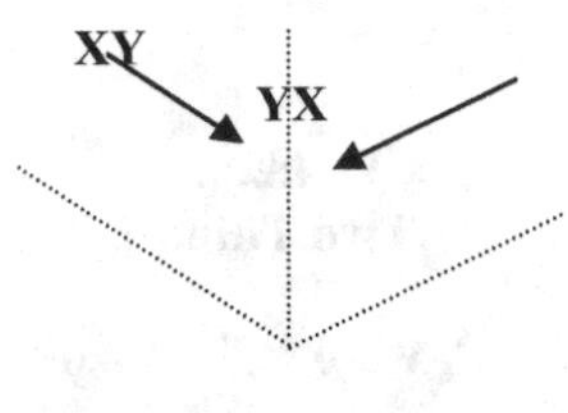

WB334	3	11	22	33		23	31

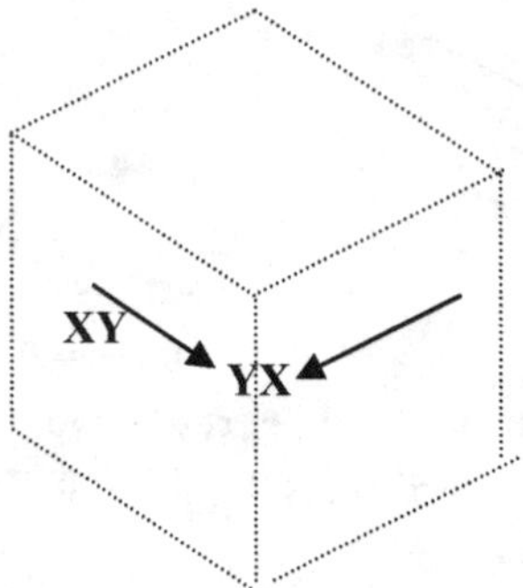

WC444	1	11	22	33	12	23	31

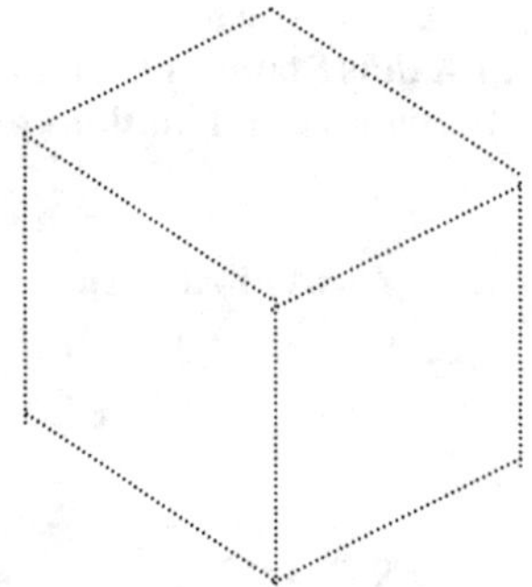

11.5.
Selected Cases:

Costumed XY Plane Stress States:

8 cases of Zero Single Axial Stress 33 and Shears

UO002 1/4	**ZZ = 0**			
UA013 1/16	**ZZ = 0**			**ZX = 0**
UA013 1/16	**ZZ = 0**		**YZ = 0**	
UA112 1/16	**ZZ = 0**	**XY = 0**		

UB114 1/64		**ZZ = 0**		**YZ = 0**	**ZX = 0**
UB123 1/64		**ZZ = 0**	**XY = 0**		**ZX = 0**
UB123 1/64		**ZZ = 0**	**XY = 0**	**YZ = 0**	
UC224 1/64		**ZZ = 0**	**XY = 0**	**YZ = 0**	**ZX = 0**

Selected Cases of Zero Single Axial Stress 33 and Shears

X-Y-Slab Z-Deep	**Z < Y ≈ X**	**ZZ = 0**			
Y-One Way Slab Z-Deep	**Z < Y < X**	**ZZ = 0**			**ZX = 0**
X- One Way Slab Z-Deep	**Z < X < Y**	**ZZ = 0**		**YZ = 0**	
X-Y-Slab Z-Thick	**Z << Y ≈ X**	**ZZ = 0**	**XY = 0**		
X-Y Membrane Z-Thin	**Z<<< Y ≈ X**	**ZZ = 0**		**YZ = 0**	**ZX = 0**
Y- One Way Slab Z-Thick	**Z << Y < X**	**ZZ = 0**	**XY = 0**		**ZX = 0**
X- One Way Slab Z-Thick	**Z << X < Y**	**ZZ = 0**	**XY = 0**	**YZ = 0**	
Axisym Load Z-Thin:	**Z<<< Y ≈ X**	**ZZ = 0**	**XY = 0**	**YZ = 0**	**ZX = 0**

:

2D Strsss $2: \tau_{XX} = E/(1-\nu^2)\,(+\varepsilon_{XX} + \nu\,\varepsilon_{YY}) - E/(1-\nu)\,\alpha h$

$2: \tau_{YY} = E/(1-\nu^2)\,(+\nu\,\varepsilon_{XX} + \varepsilon_{YY}) - E/(1-\nu)\,\alpha h$

$2: \tau_{ZZ} = 0$

11.6.
Torsional Analysis:

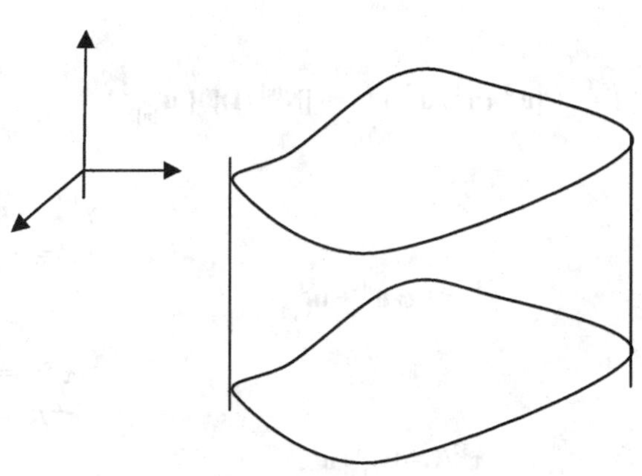

Strains:

:

C^1 Continuous:

$\varepsilon^{11} = 0$	$\varepsilon^{12} = 0$	$\varepsilon^{13} \neq 0$
$\varepsilon^{21} = 0$	$\varepsilon^{22} = 0$	$\varepsilon^{23} \neq 0$
$\varepsilon^{31} \neq 0$	$\varepsilon^{32} \neq 0$	$\varepsilon^{33} = 0$

Displacements:

:

C^0 Continuous:

$$[u^1] = -\varphi * u^3 * u^2 + u^1_0$$
$$[u^2] = +\varphi * u^3 * u^1 + u^2_0$$
$$[u^3] = +[u^3 (u^1, u^2)]$$

Strains:

:

C^1 Continuous:

$$\varepsilon^{11} = 0 \qquad \gamma^{12} = -\varphi u^3 + \varphi u^3 = 0 \qquad \gamma^{13} = -\varphi u^2 + u^3_{,1} \neq 0$$
$$\gamma^{21} = -\varphi u^3 + \varphi u^3 = 0 \qquad \varepsilon^{22} = 0 \qquad \gamma^{23} = +\varphi u^1 + u^3_{,2} \neq 0$$
$$\gamma^{31} = -\varphi u^2 + u^3_{,1} \neq 0 \qquad \gamma^{32} = +\varphi u^1 + u^3_{,2} \neq 0 \qquad \varepsilon^{33} = 0$$

:

Stresses:

$$\tau^{11}/G = 0 \qquad \tau^{12}/G = 0 \qquad \tau^{13}/G = -\varphi u^2 + u^3_{,1}$$
$$\tau^{21}/G = 0 \qquad \tau^{22}/G = 0 \qquad \tau^{23}/G = +\varphi u^1 + u^3_{,2}$$
$$\tau^{31}/G = -\varphi u^2 + u^3_{,1} \qquad \tau^{32}/G = +\varphi u^1 + u^3_{,2} \qquad \tau^{33}/G = 0$$

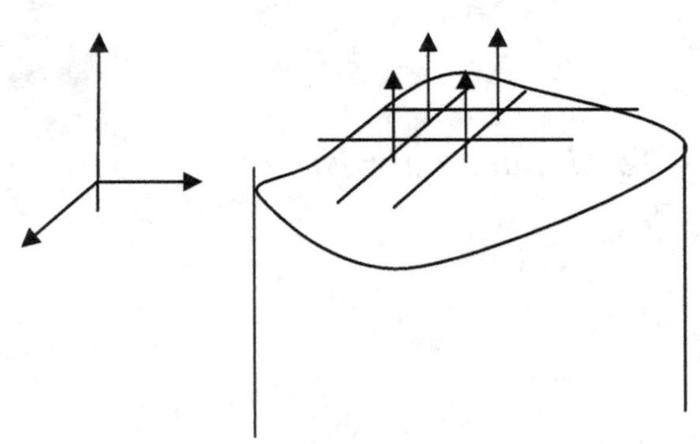

$$[u^3] = +[u^3 (u^1, u^2)] = [N^{[n]}(I)]\ [u_{[n]}^3]$$

$$\gamma^{13} = -\varphi u^2 + u^3_{,1}$$
$$\gamma^{23} = +\varphi u^1 + u^3_{,2}$$

$$\gamma^{31} = -\varphi u^2 + u^3_{,1} \qquad \gamma^{32} = +\varphi u^1 + u^3_{,2}$$

$$\tau^{13}/G = -\upsilon_0^2 + u^3_{,1}$$
$$\tau^{23}/G = +\upsilon_0^1 + u^3_{,2}$$

$$\tau^{31}/G = -\upsilon_0^2 + u^3_{,1} \qquad \tau^{32}/G = +\upsilon_0^1 + u^3_{,2}$$

Torsion:

$$T = \int [\ \tau^{31} * u^2 - \tau^{32} * u^1\ ]\ dArea$$

12. Analytical Mechanics:

12.1. Force and Displacement:

12.1.1. Newtonian Laws :

Let Ba, Ba + dBa position vectors at t, t+dt respectively then:

Newtonian Laws:

$Ba^{\bullet} \equiv d(Ba)/dt \equiv Bv$ $\qquad$ $Ba^{\bullet\bullet} \equiv d(Bv)/dt \equiv dd(Ba)/dt/dt \equiv Bacc$

Assumptions:

I. Body stays in the same frame of uniform momentum except when forced.
II. Rate of change of momentum is proportional to force
III. To every action there is an equal and opposite reaction

If (m) is inertial mass:

BForce = d(mass * Bv)/dt

If (mass) is invariant then

BForce = mass * d(Bv)/dt = mass * Bacc

Additionally If BForce = 0 then:

mass * Bv = constant

Bv = constant.

Law of Gravitation

$BF = k\, M_1\, M_2\, Br / |Br|^3$

12.1.2. Force and Displacement Differential:

$$d(R_m\, r^m) = d(r_n\, R^n)$$

$d(R_m r^m) = R_m dr^m + dR_m r^m$ $\qquad$ $d(r_m R^m) = r_m dR^m + dr_m R^m$

$R_m dr^m + dR_m r^m + dR_m dr^m$ $\qquad$ $r_m dR^m + dr_m R^m + dr_m dR^m$

$$dR_m\, dr^m = dr_n\, dR^n$$

Force Displacement Product:

R(r):

$Rr_m^{\ n} \equiv \partial R_m / \partial r_n$ $\qquad$ $Rr^m_{\ n} \equiv \partial R^m / \partial r^n$

$$\partial r_j \,/\, \partial R_i \equiv rR_j^{\;i} \qquad\qquad \partial r^j / \partial R^i \equiv rR_i^{\;j}$$

$$(r^j + dr^j)\,(R_j + dR_j) - r^j\, R_j = r^j\, dR_j + dr^j\, R_j + dr^j\, dR_j$$

$$dV = r^j\, dR(r)_j = r^j\, Rr(r)_j^{\;i}\, dr_i$$

$$dU = dr^j\, R(r)_j = dr^j\, Rr(r)_j^{\;i}\, r_i$$

$$dW = dr^j\, dR(r)_j = dr^j\, Rr(r)_j^{\;i}\, dr_i$$

$$V = \int r^j\, dR(r)_j = \int r^j\, Rr(r)_j^{\;i}\, dr_i$$

$$U = \int dr^j\, R(r)_j = \int dr^j\, Rr(r)_j^{\;i}\, r_i$$

$$U = \int R^j\, dr(R)_j = \int R^j\, rR(R)_j^{\;i}\, dR_i$$

$$V = \int dR^j\, r(R)_j = \int dR^j\, rR(R)_j^{\;i}\, R_i$$

$$U = \int R^j\, dr(R)_j \neq V = \int dR^j\, r_j(R)$$

$$(R^j + dR^j)\,(r_j + dr_j) - R^j\, r_j = R^j\, dr_j + dR^j\, r_j + dR^j\, dr_j$$

$$dU = R^j\, dr(R)_j = R^j\, rR(R)_j^{\;i}\, dR_i$$

$$dV = dR^j\, r(R)_j = dR^j\, rR(R)_j^{\;i}\, R_i$$

$$dW = dR^j\, dr_j = dR^j\, rR_j^{\;i}\, dR_i$$

$$U = \int R^j\, dr(R)_j = \int R^j\, rR(R)_j^{\;i}\, dR_i$$

$$V = \int dR^j\, r(R)_j = \int dR^j\, rR(R)_j^{\;i}\, R_i$$

$$V = \int r^j\, dR(r)_j = \int r^j\, Rr(r)_j^{\;i}\, dr_i$$

$$U = \int dr^j\, R(r)_j = \int dr^j\, Rr(r)_j^{\;i}\, r_i$$

$$V = \int r^j\, dR(r)_j \neq U = \int dr^j\, R(r)_j$$

In summary:

$$dU = dr^j\, R(r)_j = dr^j\, Rr(r)_j^{\;i}\, r_i = R^j\, dr(R)_j = R^j\, rR(R)_j^{\;i}\, dR_i$$
$$dV = r^j\, dR(r)_j = r^j\, Rr(r)_j^{\;i}\, dr_i = dR^j\, r(R)_j = dR^j\, rR(R)_j^{\;i}\, R_i$$

$$\int dU = \int dr^j\, R(r)_j = \int dr^j\, Rr(r)_j^{\;i}\, r_i = \int R^j\, dr(R)_j = \int R^j\, rR(R)_j^{\;i}\, dR_i$$
$$\int dV = r^j\, dR(r)_j = \int r^j\, Rr(r)_j^{\;i}\, dr_i = \int dR^j\, r(R)_j = \int dR^j\, rR(R)_j^{\;i}\, R_i$$

12.1.3. Energies Positive Definiteness:

$$0 \leq dU = dr^j \ R(r)_j = dr^j \ Rr(r)_j^{\ i} \ r_i = R^j \ dr(R)_j = R^j \ rR(R)_j^{\ i} \ dR_i$$
$$0 \leq dV = r^j \ dR(r)_j = r^j \ Rr(r)_j^{\ i} \ dr_i = dR^j \ r(R)_j = dR^j \ rR(R)_j^{\ i} \ R_i$$

$$0 \leq (r^j + dr^j)(R_j + dR_j) - r^j R_j \approx dU + dV$$

:

$$0 \leq dU + dV = dr^j \ R(r)_j + r^j \ dR(r)_j = R^j \ dr(R)_j + dR^j \ r(R)_j$$

$$0 \leq \int dU = \int dr^j \ R(r)_j = \int R^j \ dr(R)_j$$
$$0 \leq \int dV = \int r^j \ dR(r)_j = \int dR^j \ r(R)_j$$

$$0 \leq U + V = \int dU + \int dV = \int dr^j R(r)_j + \int r^j dR(r)_j = \int R^j \ dr(R)_j + \int dR^j \ r(R)_j$$

$$0 \leq U + V = \int dr^j Rr(r)_j^{\ i} r_i + \int r^j Rr(r)_j^{\ i} dr_i = \int R^j rR(R)_j^{\ i} dR_i + \int dR^j rR(R)_j^{\ i} R_i$$

$$0 \leq r^j R(r)_j = r^j Rr(r)_j^{\ i} r_i = R^j r(R)_j = R^j rR(R)_j^{\ i} R_i \approx U + V$$

$$0 \leq r^j R(r)_j = r^j Rr(r)_j^{\ i} r_i \approx U + V$$

$$0 \leq R^j r(R)_j = R^j rR(R)_j^{\ i} R_i \approx U + V$$

12.1.4.
Derivative of Energy With Respect To Displacement: Principle of Virtual Displacement Work: Vrtual Work:

$$dU = dr^j \ R(r)_j = dr^j \ Rr(r)_j^{\ i} \ r_i$$

$$\partial U / \partial r^j = R(r)_j = Rr(r)_j^{\ i} \ r_i$$

$$U + V = r^j \ R(r)_j = r^j \ Rr(r)_j^{\ i} \ r_i = R^j \ r(R)_j = R^j \ rR(R)_j^{\ i} \ R_i$$

12.1.5.
Derivative of Complementary Energy With Respect To Force: Principle of Virtual Force Work: Castigliano:

$$dV = dR^j \ r(R)_j = dR^j \ rR(R)_j^{\ i} \ R_i$$

$$\partial V / \partial R^j = r(R)_j = rR(R)_j^{\ i} \ R_i$$

$$U + V = r^j \ R(r)_j = r^j \ Rr(r)_j^{\ i} \ r_i = R^j \ r(R)_j = R^j \ rR(R)_j^{\ i} \ R_i$$

12.1.6.

Minimum Complementary Energy:

$$0 \leq R^j r(R)_j = R^j rR(R)_j{}^i R_i \approx U + V$$

$$V \approx R^j r(R)_j - U = R^j rR(R)_j{}^i R_i - U$$

$$0 = \partial V / \partial R^j \approx \partial(R^j r(R)_j)/\partial R^j - \partial U/\partial R^j = \partial(R^j rR(R)_j{}^i R_i) / \partial R^j - \partial U/\partial R^j$$

12.1.7. Minimum Potential Energy:

$$0 \leq r^j R(r)_j = r^j Rr(r)_j{}^i r_i \approx U + V$$

$$U \approx r^j R(r)_j - V = r^j Rr(r)_j{}^i r_i - V$$

$$0 = \partial U / \partial r^j \approx \partial(r^j R(r)_j)/\partial r^j - \partial V/\partial r^j = \partial(r^j Rr(r)_j{}^i r_i)/\partial r^j - \partial V/\partial r^j$$

12.1.8. Principle of Superposition:

$dr_j = rR_j{}^1\, dR_1$ $\quad$ $dr_j = rR_j{}^2\, dR_2$ $\quad$ $dr_j = rR_j{}^3\, dR_3$ $\quad$ $dr_j = rR_j{}^i\, dR_i$

$$dr_j = rR_j{}^i \, dR_i$$

$$\int dr_j = \int rR_j{}^i \, dR_i = rR_j{}^i \int dR_i$$

$$r_j = rR_j{}^i \, R_i$$

12.1.9. Reciprocal Theorem:

$Rr_i{}^j = Rr_j{}^i$ $\quad$ $Rr^j{}_i = Rr^i{}_j$

$rR_i{}^j = rR_j{}^i$ $\quad$ $rR^j{}_i = rR^i{}_j$

$Pp_i{}^j = Pp_j{}^i$ $\quad$ $Pp^j{}_i = Pp^i{}_j$

$pP_i{}^j = pP_j{}^i$ $\quad$ $pP^j{}_i = pP^i{}_j$

$$\int p^j\ dR_j = \int p^j\ Rp_j^i\ dp_i = Rp_j^i \int p^j\ dp_i = Rp_j^i\ (p^j\ p_{i\neq j} + 1/2\ p^j\ p_j)$$

$$\int dp^j\ R_j = \int dp^j\ Rp_j^i\ p_i = Rp_j^i \int dp^j\ p_i = Rp_j^i\ (p^j\ p_{i\neq j} + 1/2\ p^j\ p_j)$$

$$\int R^j\ dp_j = \int R^j\ pR_j^i\ dR_i = pR_j^i \int R^j\ dR_i = pR_j^i\ (R^j\ R_{i\neq j} + 1/2\ R^j\ R_j)$$

$$\int dR^j\ p_j = \int dR^j\ pR_j^i\ R_i = pR_j^i \int dR^j\ R_i = pR_j^i\ (R^j\ R_{i\neq j} + 1/2\ R^j\ R_j)$$

$$\int P^j\ dr_j = \int P^j\ rP_j^i\ dP_i = rP_j^i \int P^j\ dP_i = rP_j^i\ (P^j\ P_{i\neq j} + 1/2\ P^j\ P_j)$$

$$\int dP^j\ r_j = \int dP^j\ rP_j^i\ P_i = rP_j^i \int dP^j\ P_i = rP_j^i\ (P^j\ P_{i\neq j} + 1/2\ P^j\ P_j)$$

$$\int r^j\ dP_j = \int r^j\ Pr_j^i\ dr_i = Pr_j^i \int r^j\ dr_i = Pr_j^i\ (r^j\ r_{i\neq j} + 1/2\ r^j\ r_j)$$

$$\int dr^j\ P_j = \int dr^j\ Pr_j^i\ r_i = Pr_j^i \int dr^j\ r_i = Pr_j^i\ (r^j\ r_{i\neq j} + 1/2\ r^j\ r_j)$$

$$Rr_j^i = Pp_j^i = Rp_j^i = Pr_j^i$$
$$rR_j^i = pP_j^i = pR_j^i = rP_j^i$$

$$p^i\ R_i = R^j\ p_j = r_j\ P^j = r^j\ P_j$$

12.1.10.
Prototypes of Viscoelasticity:

Time Dependence:

$$Rr(t)_m^{\ n} \equiv \partial R(t)_m / \partial r(t)_n \qquad Rr(t)^m_{\ n} \equiv \partial R(t)^m / \partial r(t)^n$$

$$dR(t)_m = Rr(t-\tau)_m^{\ n}\ \ dr(t=\tau)_n \qquad dR(t)^m_{\ n} = Rr(t-\tau)^m_{\ n}\ \ dr(t-\tau)^n$$
$$dR(t)_m = Rr(t-\tau)_m^{\ n}\ (dr(t=\tau)_n/dt)\ d\tau \qquad dR(t)^m_{\ n} = Rr(t-\tau)^m_{\ n}\ (dr(t=\tau)^n/dt)\ d\tau$$

$$\partial r(t)_j / \partial R(t)_i \equiv rR(t)_j^i \qquad \partial r(t)^j / \partial R(t)^i \equiv rR(t)^j_i$$

$$dr(t)_j = rR(t-\tau)_j^i\ \ dR(t=\tau)_i \qquad dr(t)^j = rR(t-\tau)^j_i\ \ dR(t=\tau)^i$$
$$dr(t)_j = rR(t-\tau)_j^i\ (dR(t=\tau)_i / dt)\ d\tau \qquad dr(t)^j = rR(t-\tau)^j_i\ (dR(t=\tau)^i / dt)\ d\tau$$

$$Rr(t)_m^{\ n} \equiv \partial R(t)_m / \partial r(t)_n \qquad Rr(t)^m_{\ n} \equiv \partial R(t)^m / \partial r(t)^n$$

$$dR(t)_m = Rr(t-\tau)_m^{\ n}\ \ dr(t=\tau)_n \qquad dR(t)^m_{\ n} = Rr(t-\tau)^m_{\ n}\ \ dr(t-\tau)^n$$
$$dR(t)_m = Rr(t-\tau)_m^{\ n}\ (dr(t=\tau)_n/dt)\ d\tau \qquad dR(t)^m_{\ n} = Rr(t-\tau)^m_{\ n}\ (dr(t=\tau)^n/dt)\ d\tau$$

$$\partial r(t)_j / \partial R(t)_i \equiv rR(t)_j^i \qquad \partial r(t)^j / \partial R(t)^i \equiv rR(t)^j_i$$

$dr(t)_j = rR(t-\tau)_j^{\ i}\ dR(t=\tau)_i$

$dr(t)_j = rR(t-\tau)_j^{\ i}\ (dR(t=\tau)_i / dt)\ d\tau$

$dr(t)^j = rR(t-\tau)^j_{\ i}\ dR(t=\tau)^i$

$dr(t)^j = rR(t-\tau)^j_{\ i}\ (dR(t=\tau)^i / dt)\ d\tau$

12.2.
Particle Motion

$rb[t] = rx^i[t]\ xb_i[t]$ = rx(+i)[t] xb(-i)[t]

Curve C:

rx(+i) = rx(+i)[t] is the trajectory of a particle

:

v(+i) = d(rx(+i))/d(t)

:

Δ(Ax(+j,- i))/Δt = Ax(+j,-i |-k) dx(+k)/dt =

={∂[Ax(+j,-i)]/∂x(+k) + Ax(+n,-i) xbs[+j,-n,-k] - Ax(+j,-m) xbs[+m,-i ,-k]}dx(+k)/dt

=d[Ax(+j,-i)]/ dt + {Ax(+n,-i) xbs[+j,-n,-k] - Ax(+j,-m) xbs[+m,-i ,-k]} dx(+k)/dt

:

∂(ub)/ ∂x(+k) ={∂(ux(+ m))/ ∂x(+k) + ux(+n) xbs[+m,-n ,-k] } xb(-m)

= { Δ(ux(+ m))/Δx(+k)≡ ux(+m | -k) } xb(-m)

∂(ub)/ ∂x(+k) ={∂(x(+ m))/ ∂x(+k) + x(+n) xbs[+m,-n ,-k] } xb(-m)

= { Δ(ux(+ m))/Δx(+k)≡ ux(+m | -k) } xb(-m)

:

Δ(rx(+j))/Δ(t) = rx(+j |-k) d(x(+k))/d(t) =

Δ(x(+j))/Δ(t) = x(+j |-k) d(x(+k))/d(t) =

= (∂(x(+j))/∂x(+k) + x(+n) xbs[+j,-n,-k]) dx(+k)/d(t)

= ∂(x(+j))/ ∂x(+k) dx(+k)/d(t) + x(+n) xbs[+j,-n,-k]dx(+k)/d(t)

= Δ(+j,-k) v(+k) + 0 v(+k)

= v(+j)

:

Δ(vx(+j))/Δ(t) =vx(+j |-k) d(x(+k))/d(t) =

(∂(vx(+j))/∂x(+k)+vx(+n)xbs[+j,-n,-k])dx(+k)/d(t)

= ∂(vx(+j))/∂x(+k) dx(+k)/d(t) + vx(+n) xbs[+j,-n,-k] dx(+k)/d(t)

= d(vx(+j))/d(t) + vx(+n) xbs[+j,-n,-k] dx(+k)/d(t)

= d(d(x(+j))/d(t))/d(t) + d(x(+n))/d(t) xbs[+j,-n,-k] dx(+k)/d(t)

= $x^{\cdot\cdot}$ (+j)) + xbs[+j,-n,-k] $x^{\cdot}$ (+n) $x^{\cdot}$(+k)

= ax(+j)

:

fx(+j) = m Δ(vx(+j))/Δ(t) = m ax(+j)

fx(-i) = xbxb(-i,-j) fx(+j)

12.2.1.

Free Moving Particle

Application:

fo(-p) = 0 in

fx(-p) = d(∂(T[t])/ ∂($x^{\cdot}$(+p)) ∂($x^{\cdot}$(+p))/d(t) - ∂(T[t])/∂(x(+p))

fx(-p) = d(∂(T[t])/∂($x^{\cdot}$(+p)))/d(t) - ∂(T[t])/∂(x(+p)) = 0 Free

T[t] ≡ mass /2 [xbxb(-m,-n) $x^{\cdot}$ (+m) $x^{\cdot}$(+n)]

If global rectangular then

:

fo(-p) = d(∂(T[t])/∂(o˙(+p)))/d(t) - ∂(T[t])/∂(o(+p)) = 0 Free
T[t] = mass /2 Δ(-m,-n) o˙(+m) o˙(+n) = mass /2 o˙(+m) o˙(+m)
∂(T[t])/∂(o˙(+p)) = ∂(mass /2 o˙(+m) o˙(+m))/∂(o˙(+p)) = mass o˙(+p)
d(∂(T[t])/∂(o˙(+p)))/d(t) = d(mass o˙(+p))/d(t) = mass o¨(+p)

:

∂(T[t])/∂(o(+p)) = ∂(mass /2 o˙(+m) o˙(+m))/∂(o(+p)) = 0
d(∂(T[t])/∂(o˙(+p)))/d(t) - ∂(T[t])/∂(o(+p)) = mass o¨(+p) - 0 = 0
mass o¨(+p) = 0

12.2.2.

Constant Gravitational Field

Application:
V = mass * gravity * o(+3)
fx(-p) = d(∂(T[t])/ ∂(x˙(+p)) ∂(x˙(+p))/d(t) - ∂(T[t])/∂(x(+p))
fo(-p) = d(∂(T[t])/∂(o˙(+p)))/d(t) - ∂(T[t])/∂(o(+p))

T[t] ≡ mass /2 [xbxb(-m,-n) x˙ (+m) x˙(+n)]
T[t] = mass /2 Δ(-m,-n) o˙(+m) o˙(+n) = mass /2 o˙(+m) o˙(+m)
d(∂(T[t])/∂(o˙(+p)))/d(t) = d(mass o˙(+p))/d(t) = mass o¨(+p)
∂(T[t])/∂(o(+p)) = ∂(mass /2 o˙(+m) o˙(+m))/∂(o(+p)) = 0
∂(V)/∂(o(+p)) = ∂(mass * gravity * o(+3))/∂(o(+p)) = mass * gravity * o(+3) Δ(+3,-p)

d(∂(T[t])/∂(o˙(+p)))/d(t) - ∂(T[t])/∂(o(+p)) = - ∂(V)/∂(o(+p))
mass o¨(+p) - 0 = -mass * gravity * o(+3) Δ(+3,-p)
o¨(+p) = - gravity * o(+3) Δ(+3,-p)

o¨(+α) = 0 ; o¨(+3) = - gravity * o(+3)

Trajectory

o (+α) = Aα (t) + Bα ; o(+3) = - 1/2 gravity * o(+3) * o(+3) + A3 (t) + B3

12.2.3.

Simple Pendulum in Spherical Coordinates

Application: (r , γ , φ)
Spherical Coordinates Metric Tensor: xbxb(-m,-n)

1	0	0
0	x(1)**2	0

0 0 (x(1)sin(x(+2)))**2

Or:

1 0 0

0 r **2 0

0 0 (r sin(γ))**2

T[t] ≡ mass /2 [xbxb(-m,-n) x˙ (+m) x˙(+n)]

= mass /2[xbxb(-1,-1)x˙(+1)x˙(+1)+xbxb(-2,-2)x˙(+2)x˙(+2)+xbxb(-3,-3)x˙(+3)x˙(+3)]

= mass /2[1 x˙(+1)x˙(+1) + x(+1)**2 x˙(+2)x˙(+2)+ (x(+1)sin(x(+2))**2x˙(+3)x˙(+3)]

= mass /2[1 r˙ r˙ + r **2 γ˙ γ˙ + (r sin(γ))**2) φ˙ φ˙]

T[t] = mass/2 [r˙ **2 + r **2 γ˙ **2 + r **2 (sin(γ))**2 φ˙ **2]

T[t] = mass/2 [r˙ **2 + (r γ˙) **2 + (r sin(γ) φ˙)**2]

If x(+3) is downward

V = - mass * gravity * x(+1) cos(x(+2)) = - mass * gravity * r cos(γ)

Conservative Force fx(-i) = - ∂(V)/∂(x(+i))

fx(-1)=-∂ (-mass*gravity*x(+1)cos(x(+2)))/∂(x(+1)) = mass*gravity* cos(x(+2))

fx(-2)=-∂ (-mass*gravity*x(+1)cos(x(+2)))/∂(x(+2))

= mass*gravity* x(+1) (-sin(x(+2)))

Or

fx(-2)=-∂ (-mass*gravity*x(+1)cos(x(+2)))/∂(x(+2)) = 0

Or:

fx(-1) = mass*gravity* cos(γ)

fx(-2)= mass*gravity* r (-sin(γ))

fx(-2)= 0

T[t] = mass/2 [r˙ **2 + r**2 γ˙ **2 + r**2 sin(γ) **2 φ˙ **2] ∂(T[t])/∂(o˙(+p))

∂(T[t])/∂(r˙) = mass [r˙]

∂(T[t])/∂(γ˙) = mass [r**2 γ˙]

∂(T[t])/∂(φ˙) = mass [r**2 sin(γ) **2 φ˙]

d(∂(T[t])/∂(o˙(+p)))/d(t) :

d(∂(T[t])/∂(r˙))/d(t) = d(mass (r˙))/d(t) = mass (r¨)

d(∂(T[t])/∂(γ˙))/d(t)= d(mass (r**2 γ˙))/d(t) = mass (2 r r˙ γ˙+ r**2 γ¨)

d(∂(T[t])/∂(φ˙)))/d(t)= d(mass (r**2 sin(γ) **2 φ˙)))/d(t) =

= mass(2 r r˙ sin(γ) **2 φ˙ + r**2 2 sin(γ) cos(γ) γ˙ φ˙ + r**2 sin(γ) **2 φ¨)

T[t] = mass/2 [r˙ **2 + r**2 γ˙ **2 + r**2 sin(γ) **2 φ˙ **2]

∂(T[t])/∂(o(+p)) :

∂(T[t])/∂(r) = mass [r γ˙ **2 + r sin(γ) **2 φ˙ **2]

∂(T[t])/∂(γ) = mass [r**2 sin(γ) cos(γ) φ˙ **2]

∂(T[t])/∂(φ) = 0

V = - mass * gravity * r cos(γ)
∂(V)/∂(o(+p)) :
∂(V)/∂(r) = - mass * gravity cos(γ)

∂(V)/∂(γ) = - mass * gravity * r (-sin(γ))
∂(V)/∂(φ) = 0

d(∂(T[t])/∂(o˙(+p)))/d(t) - ∂(T[t])/∂(o(+p)) = - ∂(V)/∂(o(+p))
r˙˙ - [r γ˙ **2 + r sin(γ) **2 φ˙ **2] = gravity cos(γ)

2 r r˙ γ˙+ r**2 γ˙˙ - r**2 sin(γ) cos(γ) φ˙ **2 = gravity * r (-sin(γ))
d(r**2 sin(γ)**2 φ˙))/d(t)- = 0

In plane motion ⇔ φ˙ = 0
Then:
r˙˙ - r γ˙ **2 = gravity cos(γ)
2 r r˙ γ˙+ r**2 γ˙˙ = gravity * r (-sin(γ))
0 = 0

12.2.4.

Motion of a Particle on a Curve

Application: Motion of a Particle on a Curve C≡ x(+i) = x(+i)[cc]
dx(+m)/d(C) ≡λ(+m) 1 = λ(+m) λ(+n) xbxb(-m,-n)

$$\Delta/\Delta(C) \quad * \quad \begin{bmatrix} \lambda(+m) \\ \mu(+m) \\ \nu(+m) \end{bmatrix} = \begin{bmatrix} 0 & \kappa & 0 \\ -\kappa & 0 & \tau \\ 0 & -\tau & 0 \end{bmatrix} * \begin{bmatrix} \lambda(+m) \\ \mu(+m) \\ \nu(+m) \end{bmatrix}$$

ddx(+m)/(dC)/d(C) + xbs[+m,-j,-k] dx(+j)/dc dx(+k)/dc = κ μ(+m)

dλ(+m)/dC + xbs[+m,-j,-k] λ(+j) dx(+k)/dC = κ μ(+m)

dμ(+m)/dC + xbs[+m,-j,-k] μ(+j) dx(+k)/dC = - κ λ(+m) + τ ν(+m)

dν(+m)/dC + xbs[+m,-j,-k] ν(+j) dx(+k)/dC = - τ ν(+m)

vx(+m) = d(x(+m))/d(t) = d(x(+m))/d(C) d(C)/d(t) = λ(+m) (vs)
(vs) is a scalar value the magnitude of velocity so
Δ(vs)/ Δ(t) = ∂(vs)/∂(t)
ax(+m)=Δ(vx(+m))/Δ(t)=Δ(λ(+m)(vs))/Δ(t)=Δ(λ(+m))/Δ(t)(vs)+λ(+m)Δ(vs)/Δ(t)
= Δ (λ(+m))/ Δ (t) (vs) + λ(+m) d(vs)/d(t)

But

$$\Delta(\lambda(+m))/\Delta(t) = \Delta(\lambda(+m))/\Delta(x(+p))\ d(x(+p)/d(t)$$

$$= \Delta(\lambda(+m))/\Delta(x(+p))\ d(x(+p)/d(C) =$$

$$= \Delta(\lambda(+m))/\Delta(x(+p))\ d(x(+p)/d(C)\ d(C)/d(t)$$

$$= \Delta(\lambda(+m)\ \Delta(x(+p))\ d(x(+p)/d(C)\quad (vs)$$

$$= \qquad \Delta(\lambda(+m))/d(C) \qquad d(C)/d(t)$$

First equation

$$d\lambda(+m)/d(C) \quad + xbs[+m,-j,-k]\ \lambda(+j)\ \ dx(+k)/d(C) = \qquad \kappa\ \mu(+m)$$

multiplied by $d(C) / d(t) = vs$

$$d\lambda(+m)/d(C)\ \ d(C)/d(t) + xbs[+m,-j,-k]\ \lambda(+j)\ dx(+k)/d(C)\ d(C)/d(t) = \kappa\ \mu(+m)\ d(C)/d(t)$$

$$d\lambda(+m)/d(C) \quad (vs) \quad + xbs[+m,-j,-k]\ \lambda(+j)\ dx(+k)/d(C) \quad (vs) \quad = \kappa\ \mu(+m)\ (vs)$$

$$\Delta(\lambda(+m))/\Delta(t) \qquad\qquad = \kappa\ \mu(+m)\ (vs)$$

Then

$$ax(+m) = \Delta(\lambda(+m))/\Delta(t)\ (vs) + \lambda(+m)\ d(vs)/d(t)$$

$$= \kappa\ \mu(+m)\ (vs)\ (vs) + \lambda(+m)\ d(vs)/d(t)$$

$$ax(+m) = d(vs)/d(t)\ \ \lambda(+m)\ +\ \kappa\ \mu(+m)\ (vs)\ (vs)$$

i n oscillating plane $\kappa = 1/\rho$

$$ax(+m) = d(vs)/d(t)\ \ \lambda(+m)\ +\ (vs)\ (vs)\ 1/\rho\ \ \mu(+m)$$

The force

$$fx(+m) = mass * ax(+m) = mass\ (\ d(vs)/d(t)\ \ \lambda(+m)\ + \qquad (vs)(vs)\ 1/\rho\ \ \mu(+m)\ )$$

$$fx(+m) = mass * ax(+m) = mass * d(vs)/d(t)\ \ \lambda(+m)\ + mass * (vs)(vs)\ 1/\rho\ \ \mu(+m)$$

But kinetic energy

$$T[t] \equiv mass/2 \quad [xbxb(-i,-j)\ vx(+i)\ vx(+j)] = mass/2\ \ vb.vb = mass/2\ \ vs\ \ vs$$

$$d(T[t])/d(C) = mass * (vs)\ \ d(vs)/d(C))$$

$$= mass * d(C)/d(t)\ d(vs)/d(C) = mass * d(vs)/d(t$$

So

$$fx(+m) = d(T[t])/d(C) \quad \lambda(+m)\ +\ 2\ T[t]\ \ 1/\rho\ \ \mu(+m)$$

The tangent component of force $fx(+m)\ \lambda(-m) = d(T[t])/d(C)$

I f conservative $fx(+m)\ \lambda(-m) = fx(-n)\ \lambda(+n) = -\ \partial(V)/\partial(x(+n))\ \lambda(+n) = d(T[t])/d(C)$

$$fx(+m)\ \lambda(-m) = fx(-n)\ \lambda(+n) = -\ \partial(V)/\partial(x(+n))\ \lambda(+n) = d(T[t])/d(C)$$

$$= -\ \partial(V)/\partial(x(+n))\ d(x(+n))/d(C) = d(T[t])/d(C)$$

$$= \qquad -\ d(V)/d(C) = d(T[t])/d(C)$$

Or

$$d(T[t] + V)/d(C) = 0 \quad \text{null i.e.} \quad (T + V) = \text{constant along the curve } C$$

If Uniform Gravitational Field fx(+m) then it is Parallel Field $\Delta(fx(+m)) /\Delta(C) = 0$

If Uniform Gravitational Field fx(+m) then it is Parallel Field $\Delta(fx(+m)) /\Delta(C) = 0$

$$fx(+m) = mass * d(vs)/d(t) \ \ \lambda(+m) + mass * (vs)(vs) \kappa \ \mu(+m)$$

$$fx(+m) = d(T[t])/d(C) \ \ \lambda(+m) + 2 T[t] \ \kappa \ \mu(+m)$$

$$\Delta(fx(+m)) /\Delta(C) = \Delta(d(T[t])/d(C) \ \ \lambda(+m) + 2 T[t] \ \kappa \ \mu(+m)) /\Delta(C) = 0$$

$$= \Delta(d(T[t])/d(C))/\Delta(C) \ \lambda(+m) + d(T[t])/d(C) \ \Delta(\lambda(+m))/ \Delta(C) +$$
$$+2 \ \Delta(T[t])/\Delta(C) \ \kappa \ \mu(+m) + 2 T[t] \ \Delta(\kappa) /\Delta(C) \ \mu(+m) + 2 T[t] \ \kappa \ \Delta(\mu(+m)) /\Delta(C) = 0$$

$$= d(d(T[t]))/d(C))/d(C) \ \lambda(+m) + d(T[t])/d(C) \quad \kappa \ \mu(+m) \quad +$$
$$+2 \ d(T[t])/d(C) \ \kappa \ \mu(+m) +2 T[t] \ d(\kappa) /d(C) \ \mu(+m) + 2T[t]\kappa \ (- \kappa \ \lambda(+m)+\tau \ \nu(+m)) = 0$$

$$\Delta(fx(+m))/\Delta(C) = 0 :$$
$$2T[t]\kappa (\quad \tau \ \nu(+m)) +$$
$$+(d(d(T))/d(C))/d(C)+2T\kappa(- \kappa))\lambda(+m) + (3 \ d(T)/d(C) \ \kappa+2 T \ d(\kappa) /d(C))\mu(+m) = 0$$
$$\text{since } fx(+m) \ \nu(-m) = d(T[t])/d(C) \ \ \lambda(+m) \ \nu(-m) + 2 T[t] \ \kappa \ \mu(+m) \ \nu(-m) = 0+0=0$$
$$\text{amd } \Delta(fx(+m) \ \nu(-m)) /\Delta(C) = \Delta(fx(+m) /\Delta(C)) \ \nu(-m) + fx(+m) \ \Delta(\nu(-m))/\Delta(C) = 0$$

$$2T[t]\kappa (\quad \tau \ \nu(+m)) +$$
$$+(d(d(T))/d(C))/d(C)+2T\kappa(- \kappa))\lambda(+m) + (3 \ d(T)/d(C) \ \kappa+2 T \ d(\kappa) /d(C))\mu(+m) = 0$$
$$\text{since } fx(+m) \ \nu(-m) = d(T[t])/d(C) \ \ \lambda(+m) \ \nu(-m) + 2 T[t] \ \kappa \ \mu(+m) \ \nu(-m) = 0+0=0$$
$$\text{amd } \Delta(fx(+m) \ \nu(-m)) /\Delta(C) = \Delta(fx(+m) /\Delta(C)) \ \nu(-m) + fx(+m) \ \Delta(\nu(-m))/\Delta(C) = 0$$

$$\text{since } fx(+m) \ \nu(-m) = d(T[t])/d(C) \ \ \lambda(+m) \ \nu(-m) + 2 T[t] \ \kappa \ \mu(+m) \ \nu(-m) = 0+0=0$$
$$\text{amd } \Delta(fx(+m) \ \nu(-m)) /\Delta(C) = \Delta(fx(+m) /\Delta(C)) \ \nu(-m) + fx(+m) \ \Delta(\nu(-m))/\Delta(C) = 0$$

$$\text{since } fx(+m) \ \nu(-m) = d(T[t])/d(C) \ \ \lambda(+m) \ \nu(-m) + 2 T[t] \ \kappa \ \mu(+m) \ \nu(-m) = 0+0=0$$
$$\text{amd } \Delta(fx(+m) \ \nu(-m)) /\Delta(C) = \Delta(fx(+m) /\Delta(C)) \ \nu(-m) + fx(+m) \ \Delta(\nu(-m))/\Delta(C) = 0$$

$$\text{since } fx(+m) \ \nu(-m) = d(T[t])/d(C) \ \ \lambda(+m) \ \nu(-m) + 2 T[t] \ \kappa \ \mu(+m) \ \nu(-m) = 0+0=0$$
$$\text{amd } \Delta(fx(+m) \ \nu(-m)) /\Delta(C) = \Delta(fx(+m) /\Delta(C)) \ \nu(-m) + fx(+m) \ \Delta(\nu(-m))/\Delta(C) = 0$$

$$\text{parallel field} \quad = \quad 0 \quad + fx(+m) \ \Delta(\nu(-m))/\Delta(C) = 0$$
$$\text{then} \quad \Delta(\nu(-m))/\Delta(C) = 0 , \text{ but } \Delta(\nu(-m))/\Delta(C) = - \tau \ \mu(-m), \text{ so } \tau \ \mu(-m) = 0 \Leftrightarrow \tau=0.$$

$$\text{then} \quad \Delta(\nu(-m))/\Delta(C) = 0 , \text{ but } \Delta(\nu(-m))/\Delta(C) = - \tau \ \mu(-m), \text{ so } \tau \ \mu(-m) = 0 \Leftrightarrow \tau=0.$$

Hence

$$(d(d(T))/d(C))/d(C)+2T\kappa(- \kappa))\lambda(+m) + (3 \ d(T)/d(C) \ \kappa+2 T \ d(\kappa) /d(C))\mu(+m) = 0$$

or

$$d(d(T))/d(C))/d(C) + 2T \ \kappa(- \kappa) = 0 \qquad 3\kappa \ d(T)/d(C) + 2 T \ d(\kappa) /d(C) = 0$$

$d(d(T))/d(C))/d(C) = 2T \kappa \kappa$ $d(T) / T = - 2/3 \; d(\kappa)/\kappa$

$T = A \; \kappa**(-2/3)$

$d(d(\kappa**(-2/3)))/d(C))/d(C) = 2\kappa**(-2/3) \; \kappa \; \kappa = 2 \; \kappa**(4/3)$

;

Or:

$d(d(\kappa**(-2/3)))/d(C))/d(C) = 2 \; \kappa**(4/3)$

$d(d(\rho**(+2/3)))/d(C))/d(C) = 2 \; \rho**(-4/3)$

Let $R(C) \equiv \rho**(+2/3)$ then $R'' = 2 \; R**(-2)$ or $R''-2R**(-2) = R'' - 2 / R**2 = 0$

Let $R(C) \equiv \rho**(+2/3)$ then $R'' = 2 \; R**(-2)$ or $R''-2R**(-2) = R'' - 2 / R**2 = 0$

$R' = 2/3 \; \rho**(-1/3) \; \rho'$

$R'' = 2/3(-1/3) \; \rho**(-4/3) \; \rho' + 2/3 \; \rho**(-1/3) \; \rho''$

$2/3(-1/3) \; \rho**(-4/3) \; \rho' + 2/3 \; \rho**(-1/3) \; \rho'' = 2 \; \rho**(-4/3)$

$(-1/3) \; \rho**(-1) \; \rho' + \qquad \rho'' = 3 \; \rho**(-1)$

$\rho'' - \; 1/3 \; / \rho \; \rho' - 3/\rho = 0, \qquad \rho'' - \; \rho' /3/\rho \; - 3/\rho = 0$

:

$R'' = 2/3(-1/3) \; \rho**(-4/3) \; \rho' + 2/3 \; \rho**(-1/3) \; \rho''$

$d(d(\rho**(+2/3)))/d(C))/d(C) = 2 \; \rho**(-4/3)$ solution $\rho(C)$ is an intrinsic parabola

$2/3(-1/3) \; \rho**(-4/3) \; \rho' + 2/3 \; \rho**(-1/3) \; \rho'' = 2 \; \rho**(-4/3)$

$(-1/3) \; \rho**(-1) \; \rho' + \qquad \rho'' = 3 \; \rho**(-1)$

$\rho'' - \; 1/3 \; / \rho \; \rho' - 3/\rho = 0, \qquad \rho'' - \; \rho' /3/\rho \; - 3/\rho = 0$

$d(d(\rho**(+2/3)))/d(C))/d(C) = 2 \; \rho**(-4/3)$ solution $\rho(C)$ is an intrinsic parabola

12.2.5.

Motion of a Particle on a Surface:

A surface S:

$x(+m) = x(+m) \; [s(+\alpha)]$

:

Velocity:

$ux(+m) = \partial[x(+m)] / \partial[s(+\alpha)] \quad d[s(+\alpha)] / d(t)$

$ux(+m) = \; x(+m,-\alpha) \quad s'(+\alpha)$

$ux(+m) = \; x(+m,-\alpha) \quad us(+\alpha)$

:

Velocity:

ux(+m) = d [x(+m,-α)] / d (t) s (+α) + x(+m,-α) us(+α)

ux(+m) = ∂[x(+m,-α)]/ ∂[s(+β)] d[s(+β)]/d(t) s (+α) + x(+m,-α) s'(+α)

ux(+m) = ∂[x(+m,-α)]/ ∂[s(+β)] us(+β) s (+α) + x(+m,-α) s'(+α)

ux(+m) = x(+m,-α) s'(+α)

ux(+m) = x(+m,-α) us (+α)

:

Acceleration:

ax(+m) = d [x(+m,-α)] / d (t) us (+α) + x(+m,-α) as(+α)

ax(+m) = ∂(x(+m,-α)) / ∂(s(+β)) d[s(+β)]/d(t) us (+α) + x(+m,-α) as(+α)

ax(+m) = ∂[x(+m,-α)] / ∂[s(+β)] us(+β) us (+α) + x(+m,-α) as(+α)

:

Velocity:

ux(+m) = d [x(+m,-α)] / d (t) s (+α) + x(+m,-α) us(+α)

ux(+m) = x(+m,-α) s'(+α)

ux(+m) = x(+m,-α) us (+α)

Contravariant Coordinate of a Particle x^m: Programed [x(+m)], m=1,2,3.

Contravariant Coordinate of A Surface s^α : Programed [s(+α)], α =1,2.

Parameter (+α) is function of time, α (t).

Coordinate of a Particle [x(+m)] is a function of Surface [s(+α)]

A surface S:

x(+m) = x(+m)[s(+α)]

ux(+m) = ∂(x(+m) / ∂(s(+α)) d(s(+α)) / d(t)

= x(+m,-α) s'(+α) ≡ x(+m,-α) us (+α)

x(+m) * {s[+α (t)]}

Contravariant (+m):

Velocity:

[ux(+m)] = x(+m,-α) s'(+α)

≡ x(+m,-α) us (+α)

ax(+m) = Δ(ux(+m))/Δ(t) = Δ(x(+m,-α) us (+α)) / Δ(t)

[ux(+m)] = ∂(x(+m) / ∂(s(+α)) d(s(+α)) / d(t) = x(+m,-α) s'(+α) ≡ x(+m,-α) us (+α)

ax(+m) = Δ(ux(+m))/Δ(t) = Δ(x(+m,-α) us (+α)) / Δ(t)

[ux(+m)] = Δ(x(+m,-α)) / Δ(t) us (+α) + x(+m,-α) Δ(us (+α)) / Δ(t)

[ux(+m)] = ∂(x(+m,-α)) / ∂(s(+β)) d(s(+β)/d(t) us (+α) + x(+m,-α) as(+α)

[ux(+m)] = ∂(x(+m,-α)) / ∂(s(+β)) us(+β) us (+α) + x(+m,-α) as(+α)

[ux(+m)] = ∂ x(+m, s)/ ∂s d(s(+α)) / d(t) =

ux(+m) = x(+m,-α) s'(+α) ≡ x(+m,-α) us (+α)

[ux(+m)] = [x(+m,-α) us (+α)]

ax(+m) = Δ[ux(+m)]/Δ(t) = Δ[x(+m,-α) us (+α)] / Δ(t)

ax(+m) = Δ[x(+m,-α)] /Δ(t) us (+α) + x(+m,-α) Δ(us (+α)) / Δ(t)

ax(+m) = ∂[x(+m,-α)] / ∂[s(+β)] d[s(+β)]/d(t) us (+α) + x(+m,-α) as(+α)
ax(+m) = ∂(x(+m,-α)) / ∂(s(+β)) us(+β) us (+α) + x(+m,-α) as(+α)

A surface S:
x(+m) = x(+m)[s(+α)]

ux(+m) = ∂(x(+m) / ∂(s(+α)) d(s(+α)) / d(t) = x(+m,-α) s'(+α) ≡ x(+m,-α) us (+α)
ax(+m) = Δ(ux(+m))/Δ(t) = Δ(x(+m,-α) us (+α)) / Δ(t)

ux(+m) = ∂(x(+m) / ∂(s(+α)) d(s(+α)) / d(t) = x(+m,-α) s'(+α) ≡ x(+m,-α) us (+α)
ax(+m) = Δ(ux(+m))/Δ(t) = Δ(x(+m,-α) us (+α)) / Δ(t)

= Δ(x(+m,-α)) / Δ(t) us (+α) + x(+m,-α) Δ(us (+α)) / Δ(t)

= ∂(x(+m,-α)) / ∂(s(+β)) d(s(+β)/d(t) us (+α) + x(+m,-α) as(+α)

= ∂(x(+m,-α)) / ∂(s(+β)) us(+β) us (+α) + x(+m,-α) as(+α)

Motion of a Particle on a Surface
A surface S: x(+m) = x(+m)[s(+α)]
ux(+m) = ∂(x(+m) / ∂(s(+α)) d(s(+α)) / d(t)
= x(+m,-α) s'(+α) ≡ x(+m,-α) us (+α)

ax(+m) = Δ(ux(+m))/Δ(t) = Δ(x(+m,-α) us (+α)) / Δ(t)
= Δ(x(+m,-α)) / Δ(t) us (+α) + x(+m,-α) Δ(us (+α)) / Δ(t)
= ∂(x(+m,-α)) / ∂(s(+β)) d(s(+β)/d(t) us (+α) + x(+m,-α) as(+α)
= ∂(x(+m,-α)) / ∂(s(+β)) us(+β) us (+α) + x(+m,-α) as(+α)

12.3.
Energy :

12.3.1.
Work:

Work Increment:
d(W) = fx(-i) d(x(+i))
:
Work :
$\int_{t1}^{t2}$ fx(-i) d(x(+i))
= $\int_{t1}^{t2}$ xbxb(-i,-j) fx(+j) d(x(+i)) = ∫ xbxb(-i,-j) m Δ(vx(+j))/Δ(t) d(x(+i))
= $\int_{t1}^{t2}$ xbxb(-i,-j) m Δ(vx(+j))/Δ(t) vx(+i) d(t)
= $\int_{t1}^{t2}$ xbxb(-i,-j) m Δ(vx(+i))/Δ(t) vx(+j) d(t)
But:
Δ(xbxb(-i,-j) vx(+i) vx(+j)) / Δ(t) = 2 xbxb(-i,-j) Δ(vx(+i))/Δ(t) vx(+j) =
Since it is invariant:
Δ(xbxb(-i,-j) vx(+i) vx(+j)) / Δ(t) = d(xbxb(-i,-j) vx(+i) vx(+j)) / d(t)
:
xbxb(-i,-j) Δ(vx(+i))/Δ(t) vx(+j) = 1 / 2 d(xbxb(-i,-j) vx(+i) vx(+j)) / d(t)
:

Work:

$\int_{t1}^{t2}$ m /2 d(xbxb(-i,-j) vx(+i) vx(+j)) / d(t) d(t)

:

= m /2 [xbxb(-i,-j) vx(+i) vx(+j) $]_{[1]}^{[2]}$ = T[2] - T[1]

Where kinetic energy:

T[t]≡ mass /2 [xbxb(-i,-j) vx(+i) vx(+j)] = mass/2 vb.vb

:

Work done = Change in kinetic energy

If:

Force, fx(-i) is Independent of the Path

Then:

Work Increment is Exact Differential

:

d(W)= fx(-i) d(x(+i)) d(W) = - d(V) Work Increment is Negative Potential Increment

:

If

fx(-i) = - ∂(V)/∂(x(+i)) then ∂(fx(-i))/ ∂(x(+j)) = ∂(- ∂(V)/∂(x(+i)))/ ∂(x(+j))

:

And:

∂(fx(-j))/ ∂(x(+i)) = ∂(- ∂(V)/∂(x(+j)))/ ∂(x(+i))

So:

∂(fx(-i))/ ∂(x(+j)) = ∂(fx(-j))/ ∂(x(+i))

:

fx(-i) = - ∂(V)/∂(x(+i)) Conservative Force ⇔ ∂(fx(-i))/ ∂(x(+j)) = ∂(fx(-j))/ ∂(x(+i))

:

∂(fx(-i))/ ∂(x(+j)) = ∂(fx(-j))/ ∂(x(+i)) ⇔ Δ(fx(-i))/ Δ(x(+j)) = Δ(fx(-j))/ Δ(x(+i))

:

If:

∂(fx(-i))/ ∂(x(+j)) = ∂(fx(-j))/ ∂(x(+i))

Then:

Δ(fx(-i))/Δ(x(+j)) = ∂(fx(-i))/∂(x(+j)) - xbs[+k,-i,-j] fx(-k)

= ∂(fx(-j))/∂(x(+i)) - xbs[+k,-j,-i] fx(-k)

So :

Δ(fx(-i))/Δ(x(+j)) = Δ(fx(-j))/Δ(x(+i))

If:

Δ(fx(-i))/ Δ(x(+j)) = Δ(fx(-j))/ Δ(x(+i))

Then:

Δ(fx(-i))/Δ(x(+j)) = ∂(fx(-i))/∂(x(+j)) - xbs[+k,-i,-j] fx(-k)

Due Symm:

= ∂(fx(-i))/∂(x(+j)) - xbs[+k,-j,-i] fx(-k)

:

Δ(fx(-i))/Δ(x(+j)) = ∂(fx(-j))/∂(x(+i)) - xbs[+k,-j,-i] fx(-k)

So:

∂(fx(-i))/∂(x(+j)) = ∂(fx(-j))/∂(x(+i))

:

Corollary: Parallel field Δ(fx(-i))/Δ(x(+j)) = 0 is conservative.

12.3.2. Lagrangian Energy:

Energies:

Lagrange	Kinetic	Potential	Strain	2*(External)
L ≡ T – V	T	V = U – 2W	U	2W
L(uT_I , u_I)	T(uT_I)	V(u_I)	U(u_I)	W(u_I)

Energies Parameters:

Energies are Given in terms of Parameters:

(u^J), I = 1,…M

In Finite Elements Parameters are Displacements.

12.3.2.1.
Kinetic Energy:

Energies:

Lagrange	Kinetic	
$L \equiv T - V$	T	
$L(uT_I, u_I)$	$T(uT_I)$	

Kinetic Energy T is function of (uT_I) Only

$$T(uT_I) = \int_{volume} dT = \int_{volume} 1/2\, uT_I \;\; \rho \;\; uT^I \;\; (dVolume)$$

(uT^K) is written in terms of Nodal Degrees of Freedom

$$(uT^K) = B^K{}_J \;\; UT^J$$

Then:

$$T(uT_I) = \; UT_I \int_{volume} 1/2\, B^I{}_K(x,y,z) \;\; \rho \;\; B^K{}_J(x,y,z) \;\; (dVolume) \,] \;\; UT^J$$

Define Mass of of Nodal Degrees of Freedom

$$M^I{}_J = \int_{volume} B^I{}_K(x,y,z) \;\; \rho \;\; B^K{}_J(x,y,z) \;\; (dVolume)$$

Such That:

$$T(uT_I) \;=\; 1/2\, UT_I \; M^I{}_J \;\; UT^J$$

Its Partial Derivative:

$$[\, \partial T / \partial uT_I \,] \;=\; M^I{}_J \;\; UT^J$$

Time Derivative:

$$\{d\, [\, \partial T / \partial uT_I \,] / dt \,\} \;\;=\; M^I{}_J \;\; UTT^J$$

:

12.3.2.2.
Strain Energy:

Energies:

		Strain
		U
		$U(u_I)$

Strain Energy

$$U = \int_{volume} \int_0^{\varepsilon} \sigma_J \;\; d\varepsilon^J \;\; (dVolume)$$

Linear Strain Energy

$$U = \int_{volume} \{\, \varepsilon_I \;\; [1/2 * C^I{}_J \;\; \varepsilon^J{}_q \,]\} \; (dVolume)$$

12.3.2.3.
Distributed and Concentrated External Energy:

Energies:

		2*(External)

2W

$W(u_I)$

2*(External Energy)

$$2W = +\int f_J\, u^J\, (dVolume) + P_J\, u^J$$

12.3.2.4.
Potential Energy:

Energies:

Potential

$V = U - 2W$

$V(u_I)$

Potential Energy V is function of (u_I) Only

$$V(u_I) = U - 2W = \int_{volume} \int_0^{\varepsilon} \sigma_J\, d\varepsilon^J\, (dVolume) - [+\int f_J\, u^J\, (dVolume) + P_J\, u^J]$$

Linear Potential Energy V is function of (u_I) Only

$$V(u_I) = \int_{volume} \{ \varepsilon_I\, [1/2 * C^I{}_J\, \varepsilon^J{}_q]\}\, (dVolume) - [+\int f_J\, u^J\, (dVolume) + P_J\, u^J]$$

$$V(u_I) = 1/2\ u_I\, K^I{}_J\, u^J - u_I R^I$$

$$V(u_I) = 1/2\ u_I\, K^I{}_J\, u^J - u_I R^I$$

Its Partial Derivative:

$$\partial V/\partial u^J = K^I{}_J\, u^J - R^I$$

12.3.2.5.
Lagrangian Energy:

Lagrange	Kinetic	Potential	Strain	2*External
			U	2W
		Strain - 2External $V \equiv U - 2W \neq V(uT_K)$		
	$T \neq T(u_K)$			
Kinetic-Potential=Kinetic-Strain+2External $L \equiv T - V \equiv T - U + 2W$ $\{d[\partial L/\partial uT_K]/dt\} - \partial[L]/\partial u_K = 0$ $\{d[\partial(T-V)/\partial uT_K]/dt\} - \partial[T-V]/\partial u_K = 0$ $[d(\partial T/\partial uT_K)/dt - 0] - (0 - \partial V/\partial u_K) = 0$ $d(\partial T/\partial uT_K)/dt + \partial V/\partial u_K = 0$	$\partial T/\partial u_K = 0$	$\partial V/\partial uT_K = 0$		
Static: $0 + \partial V/\partial u_K = 0$ $\partial V/\partial u^J = 0$	Static: $\partial T/\partial uT_K = 0$	Static:		

Potential	Strain	2*External

V = U – 2 W	U	2W
	$\int_{volume} \int_0^{\varepsilon} \sigma_J \, d\varepsilon^J \, dVol$	$\int f_J u^J (dVol) + P_J u^J$
$1/2\, u_I K^I_J u^J - u_I B^I - u_I P^I$	$\int_{volume} \varepsilon_I [1/2 * C^I_J \varepsilon^J_q] \, dVol$	
$\partial V/\partial u^J = K^I_J u^J - R^I$		

Kinetic

$$T = \int_{volume} dT = \int_{volume} 1/2\, uT_I(x,y,z) \ \rho \ uT^I(x,y,z) \ (dVolume)$$

$$T = UT_I \int_{volume} 1/2\, B^I_K(x,y,z) \ \rho \ B^K_J(x,y,z) \ (dVolume)] \ UT^J$$

$$M^I_J = \int_{volume} B^I_K(x,y,z) \ \rho \ B^K_J(x,y,z) \ (dVolume)$$

$$T = 1/2\, UT_I \ M^I_J \ UT^J$$

$$[\partial T / \partial uT_I] = M^I_J \ UT^J$$

$$\{d [\partial T / \partial uT_I] / dt\} = M^I_J \ UTT^J$$

:

For Static Structural Systems:

$$\partial V/\partial u^J = K^I_J u^J - R^I = 0$$

For Dynamic Structural Systems:

$$\{d [\partial L / \partial uT_I] / dt\} - \partial[L] / \partial u_I = \{d [\partial T / \partial uT_I] / dt\} + \partial[V] / \partial u_I = 0$$

$$M^I_J \ UTT^J + [K^I_J u^J - R^I] = 0$$

$$M^I_J \ UTT^J + K^I_J u^J = R^I$$

Energies are given in terms of Parameters:

(u^J), I = 1,…M

In Finite Elements Parameters are Displacements.

Kinetic Energy T is function of (uT_I) Only Potential Energy V is function of (u_I) Only

Such That:

For Static Structural Systems:

$$\partial V/\partial u^J = 0$$

For Dynamic Structural Systems:

$$\{d [\partial L / \partial uT_K] / dt\} - \partial[L] / \partial u_K = 0$$

$$\{d [\partial (T-V) / \partial uT_K] / dt\} - \partial[T - V] / \partial u_K = 0$$

$$\{d [\partial T / \partial uT_K] / dt\} + \partial[V] / \partial u_K = 0$$

:

Where:

2*(External Energy)

$$2W = +\int f_J \ u^J \ (dVolume) + P_J \ u^J$$

:

Strain Energy

$$U = \int_{volume} \int_0^{\varepsilon} \sigma_J \ d\varepsilon^J \ (dVolume)$$

:

Linear Strain Energy

$$U = \int_{volume} \{\varepsilon_I \ [1/2 * C^I_J \ \varepsilon^J_q]\} \ (dVolume)$$

:

Potential Energy V is function of (u_I) Only

$$V(u_I) = U - 2W = \int_{volume} \int_0^{\varepsilon} \sigma_J \ d\varepsilon^J \ (dVolume) - [+\int f_J \ u^J \ (dVolume) + P_J \ u^J]$$

:

Linear Potential Energy

$$V(u_I) = \int_{volume} \{\varepsilon_I \ [1/2 * C^I_J \ \varepsilon^J_q]\} (dVolume) - [+\int f_J \ u^J \ (dVolume) + P_J \ u^J]$$

$$V(u_I) = 1/2 \ u_I \ K^I{}_J \ u^J \ - \ u_I B^I \ - u_I P^I$$

$$V(u_I) = 1/2 \ u_I \ K^I{}_J \ u^J \ - \ u_I R^I$$

And:

$$\partial V / \partial u^J = K^I{}_J \ u^J \ - \ R^I$$

:

Kinetic Energy T is function of (uT_I) Only

$$T(uT_I) = \int_{volume} dT = \int_{volume} 1/2 \ uT_I(x,y,z) \ \rho \ uT^I(x,y,z) \ \ (dVolume)$$

$$uT^K(x,y,z) = B^K{}_J(x,y,z) \ \ UT^J \quad \text{where } K=x,y,z$$

$$T(uT_I) = UT_I \int_{volume} 1/2 \ B^I{}_K(x,y,z) \ \rho \ B^K{}_J(x,y,z) \ \ (dVolume) \ ] \ \ UT^J$$

Define Mass:

$$M^I{}_J = \int_{volume} B^I{}_K(x,y,z) \ \rho \ B^K{}_J(x,y,z) \ \ (dVolume)$$

So That:

$$T = 1/2 \ UT_I \ M^I{}_J \ \ UT^J$$

And:

$$[\ \partial T / \partial uT_I\ ] = M^I{}_J \ \ UT^J$$

$$\{ d \ [\ \partial T / \partial uT_I\ ] / dt \ \} \ = M^I{}_J \ \ UTT^J$$

12.3.2.6. Static Lagrange Equation:

For Static Structural Systems:

Lagrange		Energies: Potential	Strain	2*(External)
$L = - V$		$V = U - 2W$	U	$2W$
$V(u_I)$		$V(u_I)$	$U(u_I)$	$W(u_I)$

For Static Structural Systems:

$$(\partial V / \ \partial u_I\ ) = \ \partial \{ Volume^* \ \tau^I{}_J \} / \ \partial u_I = \ 0$$

:

For Static Structural Systems:

$$\partial V / \partial u^J = \ K^I{}_J \ u^J \ - \ R^I = 0$$

12.3.2.7. Dynamic Lagrange Equation:

Lagrange	Kinetic	Energies: Potential	Strain	2*(External)
$L \equiv T - V$	T	$V = U - 2W$	U	$2W$
$L(uT_I , u_I)$	$T(uT_I)$	$V(u_I)$	$U(u_I)$	$W(u_I)$

For Dynamic Structural Systems:

$$\{ d \ (\ \partial L / \partial uT_I\ ) / dt \ \} \ - \ \partial L / \partial u_I \ = 0$$

Or:

$$\{ d \ (\ \partial L / \partial u\Omega_I\ ) / d\Omega \} \ - \ \partial L / \partial u_I \ = 0$$

So:

$$\{d\,[\,\partial L / \partial uT_I\,] / dt\,\} - \partial[L] / \partial u_I = \{d\,[\,\partial T / \partial uT_I\,] / dt\,\} + \partial[V] / \partial u_I = 0$$

$$M^I_J\ UTT^J + [\,K^I_J\ u^J - R^I] = 0$$

$$M^I_J\ UTT^J + K^I_J\ u^J = R^I$$

$(\partial L/ \partial uT_I)$	$(½\partial\{M(uT^j)(uT_J)\}- \partial\{Volume^*\ \tau^I_J\}+2\partial\{Force^I*(uT_J*T\})/\partial uT_I$
$(\partial L/\partial u\Omega_I)$	$(½\partial\{M(u\Omega^j)(u\,\Omega_J)\}- \partial\{Volume^*\tau^I_J\}+2\partial\{Force^I*(u\Omega_J*\Omega\})/\partial u\Omega_I$

(d) $(\partial L/ \partial uT_I)$ / (dt)	(d) $(½\partial\{M(uT^j)(uT_J)\}- \partial\{Volume^*\ \tau^I_J\}+2\partial\{Force^I*(uT_J*T\})/\partial uT_I$ / (dt)
(d) $(\partial L/\partial u\Omega_I)$ / $(d\Omega)$	(d) $(½\partial\{M(u\Omega^j)(u\,\Omega_J)\}- \partial\{Volume^*\tau^I_J\}+2\partial\{Force^I*(u\Omega_J*\Omega\})/\partial u\Omega_I$ / $(d\Omega)$

$(\partial L/ \partial u_I)$	$(½\partial\{M(uT^j)(uT_J)\}- \partial\{Volume^*\ \tau^I_J\}+2\partial\{Force^I*(uT_J*T\})/\partial u_I$
$(\partial L/\partial u_I)$	$(½\partial\{M(u\Omega^j)(u\,\Omega_J)\}- \partial\{Volume^*\tau^I_J\}+2\partial\{Force^I*(u\Omega_J*\Omega\})/\partial u_I$

(d) $(\partial L/ \partial uT_I)$ / (dt) - - $(\partial L/ \partial u_I)$ = 0	(d) $(½\partial\{M(uT^j)(uT_J)\}- \partial\{Volume^*\ \tau^I_J\}+2\partial\{Force^I*(uT_J*T\})/\partial uT_I$ / (dt) - $(½\partial\{M(uT^j)(uT_J)\}- \partial\{Volume^*\ \tau^I_J\}+2\partial\{Force^I*(uT_J*T\})/\partial u_I$ = 0
(d) $(\partial L/\partial u\Omega_I)$ / $(d\Omega)$ - $(\partial L/\partial u_I)$ = 0	(d) $(½\partial\{M(u\Omega^j)(u\,\Omega_J)\}- \partial\{Volume^*\tau^I_J\}+2\partial\{Force^I*(u\Omega_J*\Omega\})/\partial u\Omega_I$ / $(d\Omega)$ - $(½\partial\{M(u\Omega^j)(u\,\Omega_J)\}- \partial\{Volume^*\tau^I_J\}+2\partial\{Force^I*(u\Omega_J*\Omega\})/\partial u_I$ = 0

To Get:

$$(d(\partial T/ \partial uT_I)/dt) + \partial V / \partial u_I = (d\,(M\,(uT^I))\,/\,dt) + \partial\{Volume^*\ \tau^I_J\} / \partial u_I = 0$$

$$(d\,(\partial T/\partial u\Omega_I) / d\Omega) + \partial V / \partial u_I = (d\,(M\,(u\Omega^I))\,/\,d\Omega) + \partial\{Volume^*\ \tau^I_J\} / \partial u_I = 0$$

Directly:

Dynamic Lagrange Equation:

$$\{d\,(\,\partial L / \partial uT_I\,) / dt\,\} - \partial L / \partial u_I = 0$$

$$\{d\,(\,\partial L / \partial u\Omega_I\,) / d\Omega\} - \partial L / \partial u_I = 0$$

Or:

$$\{d\,(\,\partial\,(T-V\,)\,/ \partial uT_I\,) / dt\,\} - \partial(T-V\,) / \partial u_I = 0$$

$$\{d\,(\,\partial(T-V\,) / \partial u\Omega_I\,) / d\Omega\} - \partial(T-V\,) / \partial u_I = 0$$

Where:

V Is Not Function of (uT_I) T Is Not Function of (u_I)

V Is Not Function of $(u\Omega_I)$

To Get:

$$\{d(\partial T / \partial uT_I)/dt\} + \partial V / \partial u_I = 0$$

$$\{d(\partial T / \partial u\Omega_I)/dt\} + \partial V / \partial u_I = 0$$

:

Partial Derivative of T with respect to u Rate:

$$(\partial T/ \partial uT_I) = (\tfrac{1}{2}\partial\{M(uT^j)(uT_J)\})/\partial uT_I = M(uT^I)$$

$$(\partial T/\partial u\Omega_I) = (\tfrac{1}{2}\partial\{M(u\Omega^j)(u\Omega_J)\})/\partial u\Omega_I = M(u\Omega^I)$$

So Derivative of Partial Derivative of T:

$$(d(\partial T/ \partial uT_I)/dt) = (d(M(uT^I))/dt)$$

$$(d(\partial T/\partial u\Omega_I)/d\Omega) = (d(M(u\Omega^I))/d\Omega)$$

Added to:

$$\partial V / \partial u_I = \partial\{Volume^* \tau^I_J\}/\partial u_I$$

To Get:

$$(d(\partial T/ \partial uT_I)/dt) + \partial V / \partial u_I = (d(M(uT^I))/dt) + \partial\{Volume^* \tau^I_J\}/\partial u_I = 0$$

$$(d(\partial T/\partial u\Omega_I)/d\Omega) + \partial V / \partial u_I = (d(M(u\Omega^I))/d\Omega) + \partial\{Volume^* \tau^I_J\}/\partial u_I = 0$$

Kinetic Energy T is function of (uT_I) Only

$$T = \int_{volume} dT = \int_{volume} 1/2\, uT_I\ \rho\ uT^I\ (dVolume)$$

(uT^K) is written in terms of Nodal Degrees of Freedom

$$(uT^K) = B^K{}_J\ UT^J$$

Then:

$$T = UT_I \int_{volume} 1/2\, B^I{}_K(x,y,z)\ \rho\ B^K{}_J(x,y,z)\ (dVolume)]\ UT^J$$

Define Mass of of Nodal Degrees of Freedom

$$M^I{}_J = \int_{volume} B^I{}_K(x,y,z)\ \rho\ B^K{}_J(x,y,z)\ (dVolume)$$

Such That:

$$T = 1/2\, UT_I\ M^I{}_J\ UT^J$$

Its Partial Derivative:

$$[\partial T / \partial uT_I] = M^I{}_J\ UT^J$$

Time Derivative:

$$\{d[\partial T / \partial uT_I]/dt\} = M^I{}_J\ UTT^J$$

:

For Static Structural Systems:

$$\partial V/\partial u^J = K^I{}_J\ u^J - R^I = 0$$

:

:

For Dynamic Structural Systems:

$$\{d[\partial L / \partial uT_I]/dt\} - \partial[L]/\partial u_I = \{d[\partial T / \partial uT_I]/dt\} + \partial[V]/\partial u_I = 0$$

$$M^I{}_J\ UTT^J + [K^I{}_J\ u^J - R^I] = 0$$

$$R^I = M^I{}_J\ UTT^J + K^I{}_J\ u^J$$

Energies:

Lagrange	Kinetic	Potential	Strain	2*(External)
$L \equiv T - V$	T	$V = U - 2W$	U	$2W$

Energies are Given in terms of Parameters:

$(u^J), I = 1,\dots M$

In Finite Elements Parameters are Displacements.

Kinetic Energy T is function of (uT_I) Only   Potential Energy V is function of (u_I) Only

Such That:

For Static Structural Systems:

$$\partial V/\partial u^J = 0$$

And:

For Dynamic Structural Systems:

$$\{d[\partial L/\partial uT_I]/dt\} - \partial[L]/\partial u_I = 0$$

$$\{d[\partial(T-V)/\partial uT_I]/dt\} - \partial[T-V]/\partial u_I = 0$$

$$\{d[\partial T/\partial uT_I]/dt\} + \partial[V]/\partial u_I = 0$$

:

Where:

2*(External Energy)

$$2W = +\int f_J\, u^J\,(dVolume) + P_J\, u^J$$

:

Strain Energy

$$U = \int_{volume}\int_0^{\varepsilon} \sigma_J\, d\varepsilon^J\,(dVolume)$$

:

Linear Strain Energy

$$U = \int_{volume}\{\varepsilon_I\,[1/2 * C^I{}_J\, \varepsilon^J{}_q]\}\,(dVolume)$$

:

Potential Energy V is function of (u_I) Only

$$V(u_I) = U - 2W = \int_{volume}\int_0^{\varepsilon} \sigma_J\, d\varepsilon^J\,(dVolume) - [+\int f_J\, u^J\,(dVolume) + P_J\, u^J]$$

:

Linear Potential Energy

$$V(u_I) = \int_{volume}\{\varepsilon_I\,[1/2 * C^I{}_J\, \varepsilon^J{}_q]\}\,(dVolume) - [+\int f_J\, u^J\,(dVolume) + P_J\, u^J]$$

$$V(u_I) = 1/2\ u_I\, K^I{}_J\, u^J - u_I B^I - u_I P^I$$

$$V(u_I) = 1/2\ u_I\, K^I{}_J\, u^J - u_I R^I$$

And:

$$\partial V/\partial u^J = K^I{}_J\, u^J - R^I$$

:

Kinetic Energy T is function of (uT_I) Only

$$T(uT_I) = \int_{volume} dT = \int_{volume} 1/2\ uT_I(x,y,z)\ \rho\ uT^I(x,y,z)\ \ (dVolume)$$

$$uT^K(x,y,z) = B^K{}_J(x,y,z)\ \ UT^J \quad \text{where } K=x,y,z$$

$$T(uT_I) = UT_I \int_{volume} 1/2\, B^I{}_K(x,y,z)\ \rho\ B^K{}_J(x,y,z)\ \ (dVolume)]\ \ UT^J$$

Define Mass:

$$M^I{}_J = \int_{volume} B^I{}_K(x,y,z)\ \rho\ B^K{}_J(x,y,z)\ \ (dVolume)$$

So That:

$$T = 1/2\, UT_I\, M^I{}_J\ UT^J$$

And:

$$[\partial T/\partial uT_I] = M^I{}_J\ UT^J$$

$$\{d[\partial T/\partial uT_I]/dt\} = M^I{}_J\ UTT^J$$

:

For Static Structural Systems:

$\partial V/\partial u^J = K^I{}_J \ u^J \ - \ R^I = 0$

For Dynamic Structural Systems:

$\{d [\partial L / \partial uT_I] / dt \} \ - \partial[L] / \partial u_I \ =\{d [\partial T / \partial uT_I] / dt \} \ + \ \partial[V] / \partial u_I \ = \ 0$

$M^I{}_J \ UTT^J + [K^I{}_J \ u^J \ - \ R^I] \ = 0$

$M^I{}_J \ UTT^J + K^I{}_J \ u^J = R^I$

Incremental

:

fx(+j) = m Δ(vx(+j))/Δ(t) = m ax(+j)

:

fx(-i) = xbxb(-i,-j) fx(+j)

:

Work = $\int_{t1}^{t2}$ xbxb(-i,-j) fx(+j) d(x(+i)) = ∫ xbxb(-i,-j) m Δ(vx(+j))/Δ(t) d(x(+i))

:

Work = $\int_{t1}^{t2}$ mass /2 d(xbxb(-i,-j) vx(+i) vx(+j)) / d(t) d(t)

:

= mass /2 [xbxb(-i,-j) vx(+i) vx(+j) $]_{[1]}^{[2]}$ = T[2] - T[1]

work done = change in kinetic energy

Where kinetic energy

T[t] ≡ mass /2 [xbxb(-m,-n) vx(+m) vx(+n)] = mass/2 vb.vb

T[t] ≡ mass /2 [xbxb(-m,-n) $x^\bullet$ (+m) $x^\bullet$(+n)]

Derivative of T[t] w.r.to $x^\bullet$(+p)

∂(T[t])/∂($x^\bullet$ (+p))

= mass /2 [xbxb(-m,-n) Δ(+m,-p) $x^\bullet$(+n)]+ mass /2 [xbxb(-m,-n) $x^\bullet$ (+m)Δ(+n,-p)]

= mass /2 [xbxb(-p,-n) $x^\bullet$(+n)] + mass /2 [xbxb(-m,-p) $x^\bullet$ (+m)]

= mass [xbxb(-p,-q) $x^\bullet$(+q)]

:

d(∂(T[t])/∂(x•(+p)))/d(t) =

mass * [d(xbxb(-p,-q))/d(t) x'(+q)] +

+ mass * [xbxb(-p,-q) d($x^\bullet$(+q))/d(t)] =

= mass /2 * [d(xbxb(-p,-n))/d(t) $x^\bullet$(+n)] + mass /2 [d(xbxb(-m,-p))/d(t) $x^\bullet$ (+m)]+

+ mass [xbxb(-p,-q) $x^{\bullet\bullet}$(+q)] =

d(∂(T[t])/∂($x^\bullet$(+p)))/d(t) =

= mass *

{[xbxb(-p,-q) $x^{\bullet\bullet}$(+q)] +

+1/2[∂(xbxb(-p,-n))/∂(x(+m)) $x^\bullet$(+m) $x^\bullet$(+n)+

∂(xbxb(-m,-p))/∂(x(+n)) $x^\bullet$(+n)$x^\bullet$(+m)]}

d(∂(T[t])/∂($x^\bullet$(+p)))/d(t) = mass{[xbxb(-p,-q) $x^{\bullet\bullet}$(+q)] +

+ 1/2 [∂(xbxb(-p,-n))/∂(x(+m)) + ∂(xbxb(-m,-p))/∂(x(+n))] $x^\bullet$(+n)$x^\bullet$(+m)}

= mass *

{[xbxb(-p,-q)$x^{\bullet\bullet}$(+q)] + [xbs[-p,-m,-n]+1/2∂(xbxb(-m,-n))/∂(x(+p))] $x^\bullet$(+m)$x^\bullet$(+n)}

Derivative of T[t] w.r.to: x(+p)

∂(T[t])/∂(x(+p)) = mass/2 ∂(xbxb(-m,-n) $x^\bullet$ (+m) $x^\bullet$ (+n))/∂(x(+p)) =

∂(T[t])/∂(x(+p)) = mass/2 ∂(xbxb(-m,-n))/∂(x(+p)) $x^\bullet$ (+m) $x^\bullet$ (+n)

Subtract last derivatives

d(∂(T[t])/∂($x^\bullet$(+p)))/d(t) - ∂(T[t])/∂(x(+p))

= mass{[xbxb(-p,-q)$x^{\bullet\bullet}$(+q)] +

+ 1/2 [∂(xbxb(-p,-n))/∂(x(+m)) + ∂(xbxb(-m,-p))/∂(x(+n))] $x^\bullet$(+n)$x^\bullet$(+m)}

- mass/2 ∂(xbxb(-m,-n))/∂(x(+p)) x˙ (+m) x˙ (+n)

=

mass{[xbxb(-p,-q)x˙˙(+q)]

+

1/2 [∂(xbxb(-p,-n))/∂(x(+m)) +∂(xbxb(-m,-p))/∂(x(+n)) -∂(xbxb(-m,-n))/∂(x(+p))]*
x˙(+m)x˙(+n)

=

mass { [xbxb(-p,-q) x˙˙(+q)]

+

xbs[-p,-m,-n] x˙(+m)x˙(+n)}

= mass { [xbxb(-p,-q) x˙˙(+q)] + Δ(-p,+r) xbs[-r,-m,-n] x˙(+m)x˙(+n) }
= mass { [xbxb(-p,-q) x˙˙(+q)] + xbxb(-p,-q) xbxb(+q,+r) xbs[-r,-m,-n] x˙(+m)x˙(+n) }
= mass * xbxb(-p,-q) * {x˙˙(+q) + xbxb(+q,+r) xbs[-r,-m,-n] x˙(+m)x˙(+n)}
mass * xbxb(-p,-q) * {x˙˙(+q) + xbs[+q,-m,-n] x˙(+m)x˙(+n)} =
= mass * xbxb(-p,-q) * {∂x˙(+q)/∂(x(+m)) x˙(+m) + xbs[+q,-m,-n] x˙(+m)x˙(+n)}
= mass * xbxb(-p,-q) * {∂x˙(+q)/∂(x(+m)) + xbs[+q,-m,-n] x˙(+n) } x˙(+m)
= mass * xbxb(-p,-q) * Δ(x˙ (+q)) / Δ(x(+m)) x˙(+m)
= mass * xbxb(-p,-q) * Δ(x˙ (+q)) / Δ(t) = mass * xbxb(-p,-q) * Δ(vx(+q)) / Δ(t)
d(∂(T[t])/∂(x˙(+p)))/d(t) - ∂(T[t])/∂(x(+p))= mass * xbxb(-p,-q) * Δ(x˙ (+q)) / Δ(t)

=

= mass * xbxb(-p,-q) * Δ(vx(+q)) / Δ(t) =
= mass * xbxb(-p,-q) * ax(+q) =
= xbxb(-p,-q) * fx(+q)
d(∂(T[t])/∂(x˙(+p)))/d(t) - ∂(T[t])/∂(x(+p)) = fx(-p)

Lagrange

fx(-p) = d(∂(T[t])/∂(x˙(+p)))/d(t) - ∂(T[t])/∂(x(+p))

If rectangular then

fo(-p) = d(∂(T[t])/∂(o˙(+p)))/d(t) - ∂(T[t])/∂(o(+p))
T[t] = mass /2 Δ(-m,-n) o˙(+m) o˙(+n) = mass /2 o˙(+m) o˙(+m)

:

fo(-p) = d(∂(T[t])/∂(o˙(+p)))/d(t) - ∂(T[t])/∂(o(+p))

:

Physical components :

fx(-p) = fxs[p] xb[-p] = fxs[p] sqrt(xbxb[-p , -p])

:

fxs[p] sqrt(xbxb[-p , -p]) = d(∂(T[t])/∂(x˙(+p)))/d(t) - ∂(T[t])/∂(x(+p))

If conservative:

fx(-p) = - ∂(V)/∂(x(+p))

:

d(∂(T[t])/∂(x˙(+p)))/d(t) - ∂(T[t])/∂(x(+p)) = - ∂(V)/∂(x(+p))

:

d(∂(T[t])/∂(x˙(+p)))/d(t) - ∂(T[t] - V)/∂(x(+p)) = 0

Lagrange conservative

Define Lagrange Function L [t] ≡ T[t] - V

d (∂(L[t])/∂(x˙(+p))) /d(t) - ∂(L[t])/∂(x(+p)) = 0 Lagrange conservative

If global rectangular then

d (∂(L[t])/∂(o˙(+p))) /d(t) - ∂(L[t])/∂(o(+p)) = 0 Lagrange conservative

12.4.
Assembled Energies and Powers:

	R(τ)	S(Ω)
Power or Derivative of Energies:	R(τ) = (dEnergy)/dτ	S(Ω) = (dEnergy) /dΩ
Power or Rates of Energies:	$1/T*\int_{-T/2}^{T/2} R(\tau)*d\tau$ =	$1/\Omega*\int_{-\Omega/2}^{\Omega/2} S(\Omega)*d\Omega$ =

	$1/T*\int_{-T/2}^{T/2}$ (dEnergy)	$1/\Omega*\int_{-\Omega/2}^{\Omega/2}$ (dEnergy)
Incremental Energies:	$R(\tau)*d\tau$ = (dEnergy)	$S(\Omega)*d\Omega$= (dEnergy)
Limit of Power Or Limit of Rates of Energies:	$\text{limit}_{T\to\infty}$ $1/T*\int_{-T/2}^{T/2} R(\tau)*d\tau$ = $\text{limit}_{T\to\infty}$ $1/T*\int_{-T/2}^{T/2}$ (dEnergy)	$\text{limit}_{T\to\infty}$ $1/\Omega*\int_{-\Omega/2}^{\Omega/2} S(\Omega)*d\Omega$ == $\text{limit}_{T\to\infty}$ $1/\Omega*\int_{-\Omega/2}^{\Omega/2}$ (dEnergy)
Energies:	$\int R(\tau) * d\tau$ = $2\pi\ S(\omega=0)$ = $\int$(dEnergy) = 2π(dEnergy(ω=0))/dΩ	$\int S(\Omega) * d\Omega$ = $R(\tau=0)$ = $\int$ (dEnergy) = 2π (dEnergy(τ=0))/dτ
0 < Bounded Energies < ∞	$0<\int R(\tau)*d\tau<\infty$ = $0<\int$ (dEnergy) $<\infty$	$0<\int S(\Omega)*d\Omega<\infty$ = $0<\int$ (dEnergy) $<\infty$

Energies:

Energies:	Kinetic	-Strain	2*(External)
	T	-U	2W
	$T(uT_I)$	$-U(u_I)$	$W(u_I)$

Power
or
Derivative of Energies:

(dEnergy/dτ)	$½d\{M(uT^j)(uT_J)\}/d\tau$	$-d\{Volume*\ \tau^I_J\}/d\tau$	$2d\{Force^I*uT_J*T\}/d\tau$
(dEnergy/dΩ)	$½d\{M(u\Omega^j)(u\ \Omega_J)\}/d\Omega)$	$-d\{Volume*\tau^I_J\}/d\Omega)$	$2d\{Force^I*u\Omega_J*\Omega\}/d\Omega)$

Power
or
Rates of Energies:

$1/T*\int_{-T/2}^{T/2}$ dEnergy	$1/T*$ $½\{M(uT^j)(uT_J)\}\,\vert_{-T/2}^{T/2}$	$1/T*$ $\{-Volume*\tau^I_J\}\,\vert_{-T/2}^{T/2}$	$1/T*$ $2Force^I*d(uT_J*T)\,\vert_{-T/2}^{T/2}$
$1/\Omega*\int_{-\Omega/2}^{\Omega/2}$ dEnergy	$1/\Omega*$ $½M(u\ \Omega^j)(u\ \Omega_J)\,\vert_{-\Omega/2}^{\Omega/2}$	$1/\Omega*$ $\{-Volume*\tau^I_J\}\,\vert_{-\Omega/2}^{\Omega/2}$	$1/\Omega*$ $2Force^I*d(u\Omega_J*\Omega)\,\vert_{-\Omega/2}^{\Omega/2}$

Incremental Energies:

(dEnergy)	$½d\{M(uT^j)(uT_J)\}/$	$-d\{Volume*\ \tau^I_J\}$	$2d\{Force^I*uT_J*T\}$
(dEnergy)	$½d\{M(u\Omega^j)(u\ \Omega_J)\}$	$-d\{Volume*\tau^I_J\}$	$2d\{Force^I*u\Omega_J*\Omega\}$

Limit of Power
or
Limit of Rates of Energies:

$\text{limit}_{T\to\infty}$ $1/T*\int_{-T/2}^{T/2}$ dEnergy	$\text{limit}_{T\to\infty}$ $1/T*$ $½\{M(uT^j)(uT_J)\}\,\vert_{-T/2}^{T/2}$	$\text{limit}_{T\to\infty}$ $1/T*$ $\{-Volume*\tau^I_J\}\,\vert_{-T/2}^{T/2}$	$\text{limit}_{T\to\infty}$ $1/T*$ $2Force^I*d(uT_J*T)\,\vert_{-T/2}^{T/2}$
$\text{limit}_{\Omega\to\infty}$ $1/\Omega\int_{-\Omega/2}^{\Omega/2}$ dEnergy	$\text{limit}_{\Omega\to\infty}$ $1/\Omega*$ $½M(u\ \Omega^j)(u\ \Omega_J)\,\vert_{-\Omega/2}^{\Omega/2}$	$\text{limit}_{\Omega\to\infty}$ $1/\Omega*$ $\{-Volume*\tau^I_J\}\,\vert_{-\Omega/2}^{\Omega/2}$	$\text{limit}_{\Omega\to\infty}$ $1/\Omega*$ $2Force^I*d(u\Omega_J*\Omega)\,\vert_{-\Omega/2}^{\Omega/2}$

Energies:

$\int$dEnergy =	$\int(½d\{M(uT^j)(uT_J)\})$	$\int(-d\{Volume*\ \tau^I_J\})$	$\int(2d\{Force^I*(uT_J*T\})$

Energy	$= \frac{1}{2}\{M(uT^{j})(uT_{J})\}$	$= -Volume*\tau^{I}_{J}$	$= 2Force^{I}*d(uT_{J}*T)$
$\int dEnergy$ = Energy	$\int(\frac{1}{2}d\{M(u\Omega^{j})(u\,\Omega_{J})\})$ = $\frac{1}{2}\{M(u\,\Omega^{j})(u\,\Omega_{J})\}$	$\int(-d\{Volume*\tau^{I}_{J}\})$ = $-Volume*\tau^{I}_{J}$	$\int(2d\{Force^{I}*(u\Omega_{J}*\Omega\})$ = $2Force^{I}*d(u\Omega_{J}*\Omega)$

:

$Energy \mid_{-\infty}^{+\infty}$	$\frac{1}{2}\{M(uT^{j})(uT_{J})\} \mid_{-\infty}^{+\infty}$	$\{-Volume*\tau^{I}_{J}\} \mid_{-\infty}^{+\infty}$	$\int_{-\infty}^{+\infty}$ $2Force^{I}*d(uT_{J}*T) \mid_{-\infty}^{+\infty}$
$Energy \mid_{-\infty}^{+\infty}$	$\frac{1}{2}M(u\,\Omega^{j})(u\,\Omega_{J}) \mid_{-\infty}^{+\infty}$	$\{-Volume*\tau^{I}_{J}\} \mid_{-\infty}^{+\infty}$	$2Force^{I}*d(u\Omega_{J}*\Omega) \mid_{-\infty}^{+\infty}$

$0 <$ Bounded Energies $< \infty$:

$0<Energy<\infty$	$0<\frac{1}{2}\{M(uT^{j})(uT_{J})\}<\infty$	$0<Volume*\tau^{I}_{J}<\infty$	$0<2Force^{I}*d(uT_{J}*T)<\infty$
$0<Energy<\infty$	$0<\frac{1}{2}\{M(u\,\Omega^{j})(u\,\Omega_{J})\}<\infty$	$0<Volume*\tau^{I}_{J}<\infty$	$0<2Force^{I}*d(u\Omega_{J}*\Omega)<\infty$

Lagrange = Kinetic - Strain + 2*(External)

$L \equiv T - V = T - U + 2W$

Power

or

Derivative of Lagrange Energy

$(dL/d\tau)$	$(\frac{1}{2}d\{M(uT^{j})(uT_{J})\}-d\{Volume*\,\tau^{I}_{J}\}+2d\{Force^{I}*(uT_{J}*T\})/d\tau$
$(dL/d\Omega)$	$(\frac{1}{2}d\{M(u\Omega^{j})(u\,\Omega_{J})\}-d\{Volume*\tau^{I}_{J}\}+2d\{Force^{I}*(u\Omega_{J}*\Omega\})/d\Omega$

Power

or

Rates of Lagrange Energy

$1/T\int_{-T/2}^{T/2} dL$	$1/T(\frac{1}{2}\{M(uT^{j})(uT_{J})\}-Volume*\tau^{I}_{J}+2Force^{I}*d(uT_{J}*T\}) \mid_{-T/2}^{T/2}$
$1/\Omega\int_{-\Omega/2}^{\Omega/2} dL$	$1/\Omega\,(\frac{1}{2}\{M(u\,\Omega^{j})(u\,\Omega_{J})\}-Volume*\tau^{I}_{J}+2Force^{I}*d(u\Omega_{J}*\Omega\}) \mid_{-\Omega/2}^{\Omega/2}$

Incremental Lagrange Energy

(dL)	$\frac{1}{2}d\{M\,(uT^{j})(uT_{J})\} - d\{Volume*\,\tau^{I}_{J}\} + 2\,d\{Force^{I}*\,(uT_{J}*T\}$
(dL)	$\frac{1}{2}d\{M\,(u\Omega^{j})(u\,\Omega_{J})\} - d\{Volume*\,\tau^{I}_{J}\} + 2\,d\{Force^{I}*(u\Omega_{J}*\Omega\}$

Limit of Power

or

Limit of Rates of Lagrange Energy

$$Limit_{T\to\infty}\ 1/T\ (\frac{1}{2}\{M(uT^{j})(uT_{J})\}\ -\ Volume*\tau^{I}_{J}\ +\ 2Force^{I}*d(uT_{J}*T\}) \mid_{-T/2}^{T/2}$$

$$Limit_{\Omega\to\infty}\ 1/\Omega\ (\frac{1}{2}\{M(u\,\Omega^{j})(u\,\Omega_{J})\}\ -\ Volume*\tau^{I}_{J}\ +\ 2Force^{I}*d(u\Omega_{J}*\Omega\}) \mid_{-\Omega/2}^{\Omega/2}$$

Lagrange Energy

$\int dL$ = L	$\int(\frac{1}{2}d\{M\,(uT^{j})(uT_{J})\} - d\{Volume*\,\tau^{I}_{J}\} + 2\,d\{Force^{I}*\,(uT_{J}*T\})$ = $\frac{1}{2}\{M\,(uT^{j})(uT_{J})\} - Volume*\tau^{I}_{J} + 2\,Force^{I}*d(uT_{J}*T\}$
$\int dL$ = L	$\int(\frac{1}{2}d\{M\,(u\Omega^{j})(u\,\Omega_{J})\} - d\{Volume*\,\tau^{I}_{J}\} + 2\,d\{Force^{I}*(u\Omega_{J}*\Omega\})$ = $\frac{1}{2}\{M\,(u\,\Omega^{j})(u\,\Omega_{J})\} - Volume*\tau^{I}_{J} + 2\,Force^{I}*d(u\Omega_{J}*\Omega\}$

$L \mid_{-\infty}^{+\infty}$	$\frac{1}{2}\{M\,(uT^{j})(uT_{J})\} - Volume*\tau^{I}_{J} + 2\,Force^{I}*d(uT_{J}*T\} \mid_{-\infty}^{+\infty}$
$L \mid_{-\infty}^{+\infty}$	$\frac{1}{2}\{M\,(u\,\Omega^{j})(u\,\Omega_{J})\} - Volume*\tau^{I}_{J} + 2\,Force^{I}*d(u\Omega_{J}*\Omega\} \mid_{-\infty}^{+\infty}$

$0 <$ Bounded Lagrange Energy $< \infty$

$\int_{-\infty}^{+\infty} dL$	$0 < (½\{M(uT^j)(uT_j)\}\text{-Volume}*\tau^I_J+2\text{Force}^I*d(uT_J*T\}) \mid_{-\infty}^{+\infty} <\infty$
$\int_{-\infty}^{+\infty} dL$	$0 < (½\{M(u\,\Omega^j)(u\,\Omega_j)\}\text{-Volume}*\tau^I_J+2\text{Force}^I*d(u\Omega_J*\Omega\})_{-\infty}^{+\infty} <\infty$

12.5.
Structural Systems Potential Energy Stability:

Energies:		
Potential	Strain	2*(External)
$V = U - 2W$	U	$2W$
$V=V(u_I)$ Only	$U=U(u_I)$ Only	$W=W(u_I)$ Only

Define Internal Strain Energy:

$$W_q = \int_{volume} \{ \varepsilon_I\ [1/2 * C^I_J\ \varepsilon^J_q + \tau^I_q(0)]\}\ (dVolume) + Order(\ [u_M]^2\ )$$

$$W_q = \int_{volume} [\ \varepsilon_I * 1/2\ C^I_J\ \varepsilon^J_q]\ (dVolume)_q + \int_{volume} [\ \varepsilon_I\ \tau^I_q(0)\ \}\ (dVolume)_q$$

Writing:

$$\varepsilon_I = u_M\ a^M_I$$

Then:

$$W_q = 1/2 u_M\ [\int_{volumeq} a^M_I\ C^I_J\ a^J_{Nq}\ (dVolume)_q]\ u^N + u_M\ [\int_{volumeq} a^M_I\ \tau^I_q(0)\ (dVolume)_q]$$

Or:

$$W_q = 1/2\ u_M\ [\ K_{Elastic}]^M_{N\,q}\ u^N + u_M\ [\ B]^M_q$$

If Higher Order Values are Included:

$$W_q = 1/2\ u_M\ [\ K_{Elastic}]^M_{N\,q}\ u^N + u_M\ [\ B]^M_q + Order(\ [u_M]^3\ ) + Order(\ [u_M]^4\ )$$

Where:

$$[\ K_{Elastic}]^M_{N\,q} = [\ \int_{volumeq} a^M_I\ C^I_J\ a^J_{Nq}\ (dVolume)_q]$$

And:

$$[\ B]^M_q = [\int_{volumeq} a^M_I\ \tau^I_q(0)\ (dVolume)_q]$$

Adding Large Deflection Geometric Energy:

$$W_q = 1/2\ u_M\ [\ K_{Elastic}]^M_{N\,q}\ u^N + 1/2\ u_M\ [K_{Geometric}]^M_{N\,q}\ u^N + u_M\ [B]^M_{Nq}$$

$$[\ K_{Elastic}]^M_N = \Sigma_q\ [\ K_{Elastic}]^M_{N\,q} = \Sigma_q\ [\ \int_{volumeq} a^M_I\ C^I_J\ a^J_{Nq}\ (dVolume)_q]$$

$$[\ K_{Geometric}]^M_N = \Sigma_q\ [\ K_{Geometric}]^M_{N\,q} = \Sigma_q\ [\ \int_{volumeq} a^M_I\ C^I_J\ a^J_{Nq}\ (dVolume)_q]$$

$$[\ B]^M_N = \Sigma_q\ [B]^M_{Nq} = \Sigma_q\ [\int_{volumeq} a^M_I\ \tau^I_q(0)\ (dVolume)_q]$$

$$V = \Sigma_q\ V_q = \Sigma_q\ W_q - u_M\ [\ R]^M$$

$$V = 1/2\ u_M\ [\ K_{Elastic}]^M_N\ u^N + u_M\ [\ B]^M + 1/2\ u_M\ [\ K_{Geometric}]^M_N\ u^N - u_M\ [\ R]^M$$

Let:

$$\partial V / \partial u^I = 0$$

Then:

$$\partial V / \partial u^I = [\ K_{Elastic}]^M_N\ u^N + [\ B]^M + [\ K_{Geometric}]^M_N\ u^N - [\ R]^M = 0$$

Stability:

$$[\ K_{Elastic} + K_{Geometric}]^M_N\ u^N = 0$$

12.6.

Buckling of Plate

$$\tau^{critical}(x^I,t) = \lambda * \tau^1$$

$$\tau^{11}$$
$$\tau^{22}$$
$$\tau^{12}$$

$$\varepsilon^{11}(x^I,t) = \partial u^1(x^I,t)/\partial x^1 + 1/2 * [\partial u^3(x^I,t)/\partial x^1]^2$$
$$\varepsilon^{22}(x^I,t) = \partial u^2(x^I,t)/\partial x^2 + 1/2 * [\partial u^3(x^I,t)/\partial x^2]^2$$
$$2 * \varepsilon^{12}(x^I,t) = \partial u^1(x^I,t)/\partial x^2 + \partial u^2(x^I,t)/\partial x^1 + [\partial u^3(x^I,t)/\partial x^1] * [\partial u^3(x^I,t)/\partial x^2]$$

Or:

$$\varepsilon^{IJ}(x^I,t) = \varepsilon^{IJ}_{Small}(x^I,t) + \varepsilon^{IJ}_{Large}(x^I,t)$$

Where:

$$\varepsilon^{11}_{Large}(x^I,t) = 1/2 * [\partial u^3(x^I,t)/\partial x^1]^2 = 1/2\ [u^{3,1}]^2$$
$$\varepsilon^{22}_{Large}(x^I,t) = 1/2 * [\partial u^3(x^I,t)/\partial x^2]^2 = 1/2\ [u^{3,2}]^2$$
$$2 * \varepsilon^{12}_{Large}(x^I,t) = [\partial u^3(x^I,t)/\partial x^1] * [\partial u^3(x^I,t)/\partial x^2] = 2 * 1/2 * [u^{3,1}] * [u^{3,2}]$$

Due to Large Deformation:

$$\varepsilon^{11}_{Large}\ \tau_{11} = 1/2 * [u^{3,1}]^2\ \tau_{11}$$
$$\varepsilon^{22}_{Large}\ \tau_{22} = 1/2 * [u^{3,2}]^2\ \tau_{22}$$
$$2 * \varepsilon^{12}_{Large}(x^I,t)\ \tau_{12} = 2 * 1/2 * [u^{3,1}] * [u^{3,2}]\ \tau_{12}$$

Work Per Unit Volume Due to Large Deformation:

$$dW_{Large}/dVolume \equiv \varepsilon^{11}_{Large}\ \tau_{11} + \varepsilon^{22}_{Large}\ \tau_{22} + 2 * \varepsilon^{12}_{Large}(x^I,t)\ \tau_{12} =$$
$$1/2\ [u^{3,1}]^2\ \tau^{11} + 1/2\ [u^{3,2}]^2\ \tau^{11} + 2 * 1/2 * [u^{3,1}] * [u^{3,2}]\ \tau_{12}$$

$$W_{Large} \equiv \int_{Volume} (dW_{Large}/dVolume)\ dVolume \equiv \int_{Volume} \varepsilon^{IJ}_{Large}\ \tau_{IJ}\ dVolume$$
$$= \int_{Volume} 1/2[u^{3,I}]\ \tau_{IJ} [u^{3,J}]\ dVolume$$
$$W_{Large} = 1/2[u^3]\ \{ \int_{Volume} [\ ^{,I}]\ \tau_{IJ}\ [\ ^{,J}]\ dVolume \}\ [u^3]$$

Define Large Deformation Stiffness (Resilience):

$$K^3_{Large} \equiv \{ \int_{Volume} [\ ^{,I}]\ \tau_{IJ}\ [\ ^{,J}]\ dVolume \}$$

So that:

$$d[\text{Buckling Axial Force}] = [K^3_{Small} + K^3_{Large}]\ d[u^3]$$

12.7.
Geometric Stiffness (Power Resilience):

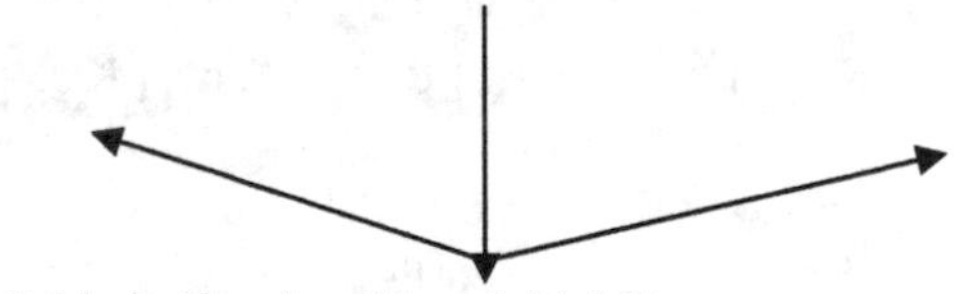

Cable in Tension Gives Initial Stresses.

$$\tau^{IJ}(x^I,t) = -E/(1-2\nu)\ \alpha T(x^I,t)\ \delta^{IJ} + E/(1+\nu)\ \varepsilon^{IJ} + E/(1+\nu)\ \nu/(1-2\nu)\ \varepsilon^I\ \delta^{IJ}$$

In General:

$$\tau^{IJ} = \tau(0)\ \delta^{IJ} + C^{IJ}_{KL}\ \varepsilon^{KL}$$

Initial Stress:

$$\tau(0)\ \delta^{IJ}$$

Strain:
Strain is Equal:
Linear in Displacement + Quadratic in Displacement

$$W = 1/2[u^m]\{\int_{Volume}[A^I]\tau_{IJ}[A^J]\,dVolume\}[u^m] + [u^m][BodyForce_m]$$

$$W = 1/2[u^m]\{K_{Small} + K_{Large}\}[u^m] + B$$

Potential:

$$V = -[u^m]R_m$$

$$+[u^m]\{\int_{Volume}[A^I]\tau(0)\,dVolume + 1/2[u^m]\{\int_{Volume}[A^I]\tau_{IJ}[A^J]\,dVolume\}[u^m]$$

Work of External Forces R:

$$[u^m]R_m =$$

$$[u^m]\{\int_{Volume}[A^I]\tau(0)\,dVolume + 1/2[u^m]\{\int_{Volume}[A^I]\tau_{IJ}[A^J]\,dVolume\}[u^m]$$

$$\partial V/\partial u^m = 0$$

$$[K^3{}_{Small} + K^3{}_{Large}]\,[u^3] + B - R = 0$$

Stability:

$$[K^3{}_{Small} + \lambda\, K^3{}_{Large}]\,[u^3] = 0$$

13. Conservation Principles:

13.1. Conservation of Mass:

Substitute: $f = \rho$ in:

$$fT \equiv ft + fq_n\, v^n \qquad \equiv ft + fq^n\, v_n$$

$$Cf \equiv ft + fq_n\, v^n + f\, vq^n{}_n \equiv fT + f\, L \equiv ft + fq^n\, v_n + f\, vq^n{}_n$$

To Get:

$$\rho T \equiv \rho t + \rho q_n\, v^n \qquad \equiv \rho t + \rho q^n\, v_n$$

$$C\rho \equiv \rho t + \rho q_n\, v^n + \rho\, vq^n{}_n \equiv \rho T + \rho\, L \equiv \rho t + \rho q^n\, v_n + \rho\, vq_n{}^n$$

No matter is added or subtracted, $C\rho \equiv 0$ Matter is Constant:

$$C\rho \equiv \rho t + \rho q_n\, v^n + \rho\, L \equiv \rho T + \rho\, L \equiv \rho t + \rho q^n\, v_n + \rho\, L = 0$$

If Incompressible Material (say Incompressible Fluid): $d\rho/dt = 0$

Trace of Gradient of a Local Velocity ≡ Trace of Convection≡ Convection = 0

Incompressible Material: $\rho T = d\rho/dt = 0$ Convection: $L = vq^k{}_k = vq_k{}^k = 0$

$$e^{ijk}\; dp_i\, dp_j\, dp_k = \text{Volume} \neq \text{volume} = e^{pqr}\; dq_p\, dq_q\, dq_r$$

But:

$$M(t) = \int_v \rho\; d\text{vol} = \text{constant}$$

$$d\,M/dt = \int_V [\rho t + \rho vq^k{}_k]\; dVol = 0$$

$$\rho_0\; d\,\text{Volume} = \rho\; d\,\text{volume}$$

$$\rho_0\; e^{ijk}\; dp_i\, dp_j\, dp_k = \rho\; e^{pqr}\; dq_p\, dq_q\, dq_r$$

$$\rho_0\; e^{ijk}\; dp_i\, dp_j\, dp_k = \rho\; e^{pqr}\; q_p{}^{;u}\, dp_u\; q_q{}^{;v}\, dp_v\; q_r{}^{;w}\, dp_w$$

$$(\rho_0\; e^{uvw} - \rho\; e^{pqr}\; q_p{}^{;u}\, q_q{}^{;v}\, q_r{}^{;w})\; dp_u\, dp_v\, dp_w = 0$$

$$\rho_0\; e^{uvw} = \rho\; e^{pqr}\; q_p{}^{;u}\, q_q{}^{;v}\, q_r{}^{;w}$$

$$\rho_0\; e^{uvw}\, e_{uvw} = \rho\; e^{pqr}\; e_{uvw}\, q_p{}^{;u}\, q_q{}^{;v}\, q_r{}^{;w}$$

$$\rho_0 = \rho\; e^{pqr}\; e_{uvw}\, q_p{}^{;u}\, q_q{}^{;v}\, q_r{}^{;w} \;/\; 6$$

$$\rho_0 = \rho\; \text{determinant}(\, q_m{}^{;n}\,)$$

Infinitesimal:

$$q_m{}^{;n} = (\, p_m + r_m)^{;n} = \delta_m{}^n + r_m{}^{;n}$$

$$\rho_0 = \rho\; \text{determinant}(\, \delta_m{}^n + r_m{}^{;n}\,)$$

$$\rho_0 = \rho\; (\, 1 + r_m{}^{;m}\,)$$

13.2. Conservation of Momentum:

Conserved Time Derivative of Product of Tensors, $fg = \rho\, v^m$: {kg/m^3• [m/s]}:

$$Cfg \equiv Cf\;\; g + f\;\; gT = fT\; g + f\;\; Cg$$

So: Mass/Volume•Acceleration : Force/Volume: {kg/m^3 • m/s^2}≡{kg/m^2/s^2}:

$$C(\rho v^m) \equiv C\rho\;\; v^m + \rho\;\; vT^m = \rho T\;\; v^m + \rho\;\; C v^m$$

But $C\rho \equiv 0$, So:

$$C(\rho v^m) \equiv 0 + \rho\;\; vT^m = \rho T\;\; v^m + \rho\;\; C v^m$$

Let $f = (\rho\, v^m)$ To get:

$$C(\rho v^{m}) \equiv (\rho v^{m})t + (\rho v^{m})q_k\ v^{k} + (\rho v^{m})\ L = \rho t\, v^{m} + \rho\, vt^{m} + (\rho v^{m})q_k\ v^{k} + \rho v^{m}\ L$$

$$C(\rho v^{m}) \equiv \rho t\ \ v^{m} + \rho\ \ vt^{m} + \rho q_k\ v^{m}\ \ v^{k} + \rho\ \ vq^{m}{}_{k}\ v^{k} + \ \rho v^{m}\ \ L$$

But by conservation of mass:

$$\rho t\ +\ \rho q_k\ \ v^{k} + \rho\ \ L\ = 0$$

So:

$$C(\rho v^{m}) \equiv 0 + \rho\, vt^{m} + 0 + \rho\ \ vq^{m}{}_{k}\ \ v^{k} + 0 = \rho\,(vt^{m}\ +\ vq^{m}{}_{k}\ \ v^{k})\ \equiv\ \rho\ (\ vT^{m}\ )$$

$\underline{F}$ Surface traction: (Force/Area): Not necessarily normal to Surface:

$$\underline{F}\ \ =\ \underline{g}_{m}\ (\ F^{m}{}_{n})\ n^{n}$$

$$F^{m}{}_{n}\ \ \ (3*3=9)$$

$$\int_s\ \underline{F}\ \ dS = \underline{g}_{m}\int_s (F^{m}{}_{n})\ n^{n}\ dS = \ \underline{g}_{m}\int_V (F^{m}{}_{n})^{,n}\ dV$$

$$\int_s (F^{m}{}_{n})\ \ n^{n}\ \ dS = \int_V (F^{m}{}_{n})^{,n}\ dV$$

So:

$$C(\ \rho v^{m}) = \rho\,(vt^{m}\ +\ vq^{m}{}_{k}\ \ v^{k})\ \equiv\ \rho(\ vT^{m}\ ) \equiv \rho\ \ d^{2}q^{m}/\,dt^{2}\ \ = \rho\ b^{m}\ +\ F^{m}{}_{n}{}^{,n}$$

$$C(\ \rho v_m) = \rho\,(vt_m\ +\ vq_{m}{}^{k}\ \ \ v_k)\ \equiv\ \rho(\ vT_m\ ) \equiv \rho\ \ d^{2}q_m/\,dt^{2}\ \ = \rho\ b_m\ +\ F_{m}{}^{n}{}_{,n}$$

Force/Volume = Body Force/Volume + Divergence of Surface Traction

Also by using integral form:

$$\underline{f}(t)\ \ =\ \int_V (\ \rho\ \underline{v}\ )\ d\,V$$

Rate of momentum is equal all exerted body force plus all surface traction force

$$d\,\underline{f}(t)/dt\ =\int_V [(\ \rho\ \ \underline{v}\ \ )t\ +\ (\ (\ \rho\ \ \underline{v}\ )vq)^{k}{}_{k}\ ]\ \ dV\ =\ \int_V\ \rho\ \ \underline{b}\ \ dV\ +\ \ \int_s\ \ \underline{F}\ \ dS$$

Or:

$$\underline{g}_{m}\ df^{m}/dt = \underline{g}_{m}\int_V [(\ \rho v^{m}\ )t + (\ (\ \rho\ \ v\ )vq)^{mk}{}_{k}]\ dV = \underline{g}_{m}\int_V\ \rho\, b^{m}\ \ dV\ + \underline{g}_{m}\int_s\ F^{m}\ \ \ dS$$

So:

$$df^{m}/dt = \int_V\ [(\rho v^{m})t + \rho v vq^{mk}{}_{k}]\ dV = \int_V\ \rho\, b^{m}\ \ dV\ +\int_V\ \ \partial(F^{m}{}_{n})\,/\,\partial\, q_n\ dV$$

But by conservation of mass

$$(\ \partial\rho/\partial t +\ \partial\rho v^{k}/\partial q^{k}\ ) = 0$$

So

$$\rho\, d^{2}q^{m}/dt^{2} \equiv \rho\ dv^{m}/dt =\ \ \rho\ \ b^{m}\ \ +\ \ \partial(\tau^{m}{}_{n})\,/\,\partial\, q_n \equiv\ \ \rho\ \ b^{m}\ \ +\ \ \ F^{m}{}_{n}{}^{,n}$$

$$\rho\, d^{2}q_m/dt^{2} \equiv \rho\ dv_m/dt =\ \ \rho\ \ b_m\ \ +\ \ \ \partial(\tau_{m}{}^{n})\,/\,\partial\, q_n \equiv\ \ \rho\ \ b_m\ \ +\ \ F_{m}{}^{n}{}_{,n}$$

$$\rho\ \ vT^{m}\ \ \ \ =\ \ \ \rho\ \ b^{m}\ \ +\ \ F^{m}{}_{n}{}^{,n}$$

Force/Volume=Body Force/Volume+Divergence of Surface Traction

If constant velocity and no body force:

$$\underline{v}T\ \ \ = 0\ \text{and}\ \ \ \underline{b}\ = 0\ \text{then}\ \ \ \ F^{m}{}_{n}{}^{,n} = \ 0\ =\ F_{m}{}^{n}{}_{,n}$$

13.3.
Conservation of Moment of Momentum:

Conserved Time Derivative of, fg = $\underline{q}\ \wedge(\rho\,\underline{v}\ )$:

$$Cfg \equiv\ \ fT\ \ g\ +\ f\ \ \ Cg$$

So:

$$C(\ \underline{q}_{l}\ \ \wedge(\ \rho\ \underline{v}\ )\ ) \equiv \underline{q}T_{l}\ \wedge(\ \rho\ \underline{v}\ ) + \underline{q}_{l}\wedge C(\ \rho\ \underline{v}\ )$$

By Conservation of Momentum:

$$\underline{g}_{m}\ C(\ \rho v^{m}) = \underline{g}_{m}\ \rho(vt^{m} + vq^{m}{}_{k}\, v^{k}) \equiv \underline{g}_{m}\ \ \rho(vT^{m}) \equiv \underline{g}_{m}\ \rho d^{2}q^{m}/dt^{2} = \underline{g}_{m}\,(\ \rho\ \ b^{m}\ +\ F^{m}{}_{n}{}^{,n})$$

So:

$$C(\ \underline{q}_{l}\ \ \wedge(\ \rho\ \underline{v}\ )_{m}\ ) \equiv \underline{q}_{l}\,T\wedge\underline{g}_{m}\ \ \ \rho v^{m}\ \ \ +\ \underline{q}_{l}\wedge\underline{g}_{m}\ \ \rho(vT^{m})$$

But $v^l v^m$ is symmetric so:

$$Cf = e_{klm}\ \rho\ [v^l v^m + v^l v^m + q^l\ vT^m] = \rho\ [0+0+ e_{klm}\ q^l\ vT^m] = e_{klm}\ \rho\ q^l\ vT^m$$

In fact:

Substitute: $f = [\underline{q} \wedge (\rho\,\underline{v})]_k = e_{klm}\,[q^l (\rho v^m)]$ **in:**

$$Cf \equiv ft + fq_n\ v^n + f\ L$$

To get:

$$Cf = e_{klm}\,[\,(q^l (\rho v^m))t + (q^l (\rho v^m))q_n\ v^n + (q^l (\rho v^m))\ L\,]$$

$$= e_{klm}\ [qt^l\ \rho v^m + q^l\ (\rho v^m)t + (q^l)q_n\ \rho v^m\ v^n + q^l\,((\rho v^m))q_n\, v^n + q^l\,\rho\, v^m\ L\,]$$

$$= e_{klm}[qt^l\ \rho v^m + q^l\ \rho t\, v^m + q^l\ \rho\, vt^m + (q^l)q_n\ \rho v^m\ v^n + q^l\ \rho q_n\, v^m\ v^n + q^l\ \rho\, vq^m{}_n\, v^n + q^l\ \rho\, v^m\ L]$$

But by Conservation of Mass:

$$\rho t + \rho q_n\ v^n + \rho\ L = 0$$

So:

$$Cf = e_{klm}\ [qt^l\ \rho v^m + 0 + q^l\ \rho\, vt^m + (q^l)q_n\ \rho v^m\ v^n + 0 + q^l\ \rho\, vq^m{}_n\, v^n + 0\,]$$

Also:

$$q^l\ \ \rho\,(vt^m + vq^m{}_n\ v^n) \equiv q^l\ \ \rho\ (\ vT^m\ )$$

And:

$$(q^l)q_n\ \ \rho\ v^m\ \ v^n = \delta^l_n\ \rho\, v^m\ v^n\ = \rho\, v^l\, v^m$$

And since: $(v^l\ v^m)$ is symmetric then:

$$e_{klm}\, v^l\ v^m\ = 0$$

So:

$$Cf = e_{klm}\ \rho\ [v^l v^m + v^l v^m + q^l\ \ vT^m] = e_{klm}\ \rho\ q^l\ \ vT^m$$

But $\underline{\tau}$ Surace traction:

$$\int_s (F^m{}_n)\ \ n^n\ dS = \int_V (F^m{}_n)^{,n}\ dV$$

And moment of $\underline{F}$ Surface traction:

$$\int_s (e_{klm}\ q^l\, \tau^m{}_n)\ \ n^n\ \ dS = \int_V (e_{klm}\ q^l\ F^m{}_n)^{,n}\ dV =$$

$$\int_V e_{klm}\ \ q^{l,n}\ \ \ F^m{}_n\ \ \ dV + \int_V e_{klm}\ \ q^l\ \ \ \ F^{m\ ,n}{}_n\ \ \ \ dV =$$

$$\int_V e_{klm}\ \ \delta^{l\,n}\ \ \ F^m{}_n\ \ \ dV + \int_V e_{klm}\ \ q^l\ \ \ \ F^{m\ ,n}{}_n\ \ \ \ dV\ \ =$$

$$\int_V e_k{}^n{}_m\ \ \ F^m{}_n\ \ \ dV + \int_V e_{klm}\ \ q^l\ \ \ \ F^{m\ ,n}{}_n\ \ \ \ dV$$

So:

$$e_{klm}\ \rho\ q^l\ \ vT^m = e_{klm}\, q^l\ b^m\ + e_k{}^n{}_m\ \ \ F^m{}_n\ \ + e_{klm}\ \ q^l\ \ \ \ F^{m\ ,n}{}_n$$

Or:

$$e_{klm}\, q^l\,(\,\rho\ \ \ vT^m\ -\ \ b^m\ -\ \ \ F^{m\ ,n}{}_n) = e_k{}^n{}_m\ \ \ \ F^m{}_n$$

But by Conservation of Momentum:

$$\rho\ \ \ vT^m\ -\ b^m\ -\ \ \ F^{m\ ,n}{}_n = 0$$

So:

$$0\ =\ e_k{}^n{}_m\ \ \ \ F^m{}_n$$

This means

Symmetric $F^m{}_n$ (3+1)3/2=((3+1) (3+1) - (3+1))/2=6

Also by using integral form:

$$\underline{G}(t) = \int_V\ \underline{q}\ \wedge\ (\,\rho\,\underline{v}\,)\ dV$$

Moment of Body force plus Moment of Surface Traction:

$$\int_V \underline{q}\ \wedge\ \rho\ \underline{b}\ dV + \int_s \underline{q} \wedge\ \ \underline{F}\ dS$$

Rate of Moment of Momentum is equal moment of exerted body force plus moment of Surface Traction

$$d\underline{G}/dt = \int_V\ \underline{q}\ \wedge\ [\rho\underline{v}t\ +\ \rho\underline{v}vq^k{}_k\,]\ dV\ = \int_V \underline{q} \wedge \rho\ \underline{b}\ \ dV\ + \int_S\ \underline{q}\ \wedge\ \underline{F}\ \ dS$$

$$\underline{q}_m\ dG^m/dt = \underline{q}_m \int_V \underline{q}\ \wedge\ [\rho\underline{v}t\ +\ \rho\underline{v}vq^k{}_k\,]\ dV = \underline{q}_m (\int_V \underline{q} \wedge \rho\ \underline{b}\ \ dV\ + \int_S\ \underline{q}\ \wedge\ \underline{F}\ \ dS)$$

But

$$\int_S\ \underline{q} \wedge\ F^m\ \ \ dS = \int_V \partial\,(\,\underline{q} \wedge F^m{}_n) / \partial\, q_n\ \ dV$$

So:

$$dG^m/dt = \int_V \underline{g} \wedge [\rho v t^m + \rho v v q^{mk}{}_k]\, dV =$$
$$\int_V \underline{g} \wedge \rho\, b^m\, dV + \int_V (\partial \underline{g}/\partial q_n) \wedge F^m{}_n\, dV + \int_V \underline{g} \wedge \partial(F^m{}_n)/\partial q_n\, dV$$

Or:

$$\int_V [\, \underline{g} \wedge \rho\, (\partial v^m/\partial t + v^k\, \partial v^m/\partial q^k) + \underline{g} \wedge v^m\, (\partial\rho/\partial t + \partial \rho v^k/\partial q^k)\,]\, dV =$$
$$\int_V [\, \underline{g} \wedge \rho\, b^m + (\partial \underline{g}/\partial q_n) \wedge F^m{}_n + \underline{g} \wedge \partial(F^m{}_n)/\partial q_n\,]\, dV$$

But by Conservation of Mass: $(\partial\rho/\partial t + \partial \rho v^k/\partial q^k) = 0$

$$\rho T + \rho\, L = \partial\rho/\partial t + \partial\rho/\partial q^k\, v^k + \rho\, \partial v^k/\partial q^k = 0$$

And by Conservation of Momentum:

$$\rho\, dv^m/dt = \rho\, b^m + \partial(F^m{}_n)/\partial q_n$$
$$\int_V e_k{}^n{}_m\, F^m{}_n\, dV = 0$$

So:

$\underline{F}$ Surface Traction (Force per Area) Is Symmetric:

$$F^m{}_n = F^n{}_m \qquad\qquad F_m{}^n = F_n{}^m$$

Define:

$$T_{pq} \equiv P_p{}^i\, F^i{}_j\, P^j{}_q \qquad\qquad T^{pq} \equiv P^p{}_i\, F_i{}^j\, P^q{}_j$$

Then:

$$Q_m{}^p\, T_{pq}\, Q^q{}_n = Q_m{}^p\, P_p{}^i\, F^i{}_j\, P^j{}_q\, Q^q{}_n = \delta_m{}^i\, F^i{}_j\, \delta^j{}_n = F^m{}_n \qquad Q^m{}_p\, T^{pq}\, Q_q{}^n = Q^m{}_p\, P^p{}_i\, F_i{}^j\, P^q{}_j\, Q_q{}^n = \delta^m{}_i\, F_i{}^j\, \delta_j{}^n = F_m{}^n$$

13.4.
Conservation of Energy:

A Unit Mass Kinetic Energy is Proportional to Velocity^2: {(m/s)^2}:
A Unit Mass Kinetic Power is Proportional to Velocity^2/Time: {(m/s)^2/s}:
Similarly:
e and eT: {(m/s)^2} and {(m/s)^2/s}:
A Unit Mass Has Internal Energy and Power:

Energy/Mass: e:	**{ J/kg }**	**Power/ Mass: eT**	**{W/kg}**
Force•Length/Mass:	**{ N•m/kg}**	**Force•Length/Mass/Time:**	**{N•m/kg/s }**
Lineal Force•Area/Mass	**{n•m^2/kg }**	**Lineal Force•Area/Mass/Time:**	**{n•m^2/kg /s}**
Stress•Volume/Mass:	**{Pa•m^3/kg}**	**Stress•Volume/Mass/Time:**	**{Pa•m^3/kg/s}**
Velocity^2: e:	**{(m/s)^2}**	**Velocity^2/Time: eT:**	**{(m/s)^2/s }**

Multiplied by {ρ : Mass/Volume}: To Get: ρ e and : ρ eT: {Pa} and {Pa/s}:
Energy and Power Per Unit Volume: Stress and Stress/Time:

Energy/Volume: ρ e:	**{ J/m^3}**	**Power/Volume: ρ eT:**	**{W/m^3}**
Force/Area:	**{ N/m^2}**	**Force/Area/Time:**	**{N/m^2/s}**
Line Force/Length:	**{n/m}**	**Lineal Force/Length/Time:**	**{n/m/s}**
Stress: ρ e:	**{Pa}**	**Stress/Time: ρ eT:**	**{Pa/s}**
Mass/length/time^2:	**{ kg/m/s^2 }**	**Mass/length/time^3:**	**{kg/m/s^3}**

Multiplied by Length: To Get: {n} and {n/s}:
Energy and Power Per Unit Area: Line Force and Lineal Force/Time:

Energy/Area:	**{ J/m^2}**	**Power/ Area:**	**{W/m^2}**
Force/Length:	**{ N/m}**	**Force/Area/Time:**	**{N/m/s}**
Line Force:	**{n}**	**Lineal Force/Time:**	**{n/s}**
Stress•Length:	**{Pa•m}**	**Stress•Length /Time:**	**{Pa•m/s}**
Mass/time^2:	**{ kg/s^2 }**	**Mass/time^3:**	**{kg/s^3}**

Multiplied by Length: To Get: {N} and {N/s}:
Hypothetical Energy and Power Per Unit Length: Force and Force/Time:

Energy/Length:	**{J/m}**	**Power/Length:**	**{W/m }**
Force:	**{N}**	**Force/Time:**	**{N/s}**
Lineal Force•Length:	**{n•m}**	**Lineal Force•Velocity:**	**{n•(m/s)}**
Stress•Area:	**{Pa•m^2}**	**Stress•Area/Time:**	**{Pa•m^2/s}**
Mass•Acceleration:	**{kg•m/s^2}**	**Mass•Acceleration /Time:**	**{kg •m/s^3}**

This Is Multiplied by Length: To Get: {J≡N•m≡cal/4•18} and {W≡J/s}:

Energy and Power:

Energy:	{J}	Power ≡ Energy/Time:	{W }
Force•Length:	{N•m}	Force•Velocity:	{N•(m/s)}
Lineal Force•Area:	$\{n \cdot m^2\}$	Lineal Force•Area/Time:	$\{n \cdot m^2/s\}$
Stress•Volume:	$\{Pa \cdot m^3\}$	Stress•Volume/time:	$\{Pa \cdot m^3/s\}$
Mass•Velocity²:	$\{kg \cdot (m/s)^2\}$	Mass•Velocity²/time:	$\{kg \cdot (m/s)^2/s\}$

Energy:

Energy/Mass:	Velocity²:	e:	{J/kg}:	$\{(m/s)^2\}$:
Energy/Volume:	Stress:	ρ e:	$\{J/m^3\}$:	{Pa}:
Energy/Area:	Line Force:	ρ e dX:	$\{J/m^2\}$:	{n}:
Energy/Length:	Force:	ρ e dS:	{J/m}:	{N}:
Energy:		ρ e dV:	{J}:	{J}:

Power:

Power/Mass:	Velocity²/Time:	eT:	{W/kg}:	$\{(m/s)^2/s\}$:
Power/Volume:	Stress/Time:	ρ eT:	$\{W/m^3\}$:	{Pa/s }:
Power/Area:	Line Force/Time:	ρ eT dX:	$\{W/m^2\}$:	{n/s }:
Power/Length:	Force/Time:	ρ eT dS:	{W/m }:	{N/s }:
Power:		ρ eT dV:	{W}:	{W}:

System Energies:

Internal:

$\int \rho\ e\, dV$ $\quad kg/m^3\ (m/s)^2\ m^3$

Kinetic:

$\int \frac{1}{2} \rho\, v^i v_i\, dV$ $\quad kg/m^3\ (m/s)^2\ m^3$

Body:

$\int \rho b^i v_i\, dV\, dt$ $\quad kg/m^3\ m/s^2\ m/s\ m^3\ s$

Volume Traction: $\int (F^i_n v_i)^{,n}\, dV\, dt =$ $\quad (Pa\ m/s)/m\ m^3\ s$

Surface Traction: $= \int F^i_n v_i n^n\, dS\, dt$ $\quad (Pa\ m/s)\ m^2\ s$

Volume Heat Supply: $\int h_n{}^{,n}\, dV\ dt$ $\quad (J/m^2/s)/m\ m^3\ s$

Surface Heat Supply: $= \int h_n n^n\, dS\, dt$ $\quad (J/m^2/s)\ m^3\ s$

System Powers:

Internal:

$\int \rho\ eT\, dV$ $\quad kg/m^3\ (m/s)^2\ m^3/s$

Kinetic:

$\int (\frac{1}{2} \rho\, v^i v_i) T\, dV$ $\quad kg/m^3\ (m/s)^2\ m^3/s$

Body:

$\int \rho b^i v_i\, dV$ $\quad kg/m^3\ m/s^2\ m/s\ m^3$

Volume Traction: $\int (F^i_n v_i)^{,n}\, dV =$ $\quad (Pa\ m/s)/m\ m^3$

Surface Traction: $= \int F^i_n v_i n^n\, dS$ $\quad (Pa\ m/s)\ m^2$

Volume Heat Supply: $\int h_n{}^{,n}\, dV =$ $\quad (J/m^2/s)/m\ m^3$

Surface Heat Supply: $= \int h_n n^n\, dS$ $\quad (J/m^2/s)\ m^3$

Define:

Divergence of Heat Gradient or Specific Heat Supply :

$$h_n{}^{,n} \equiv H \equiv h^n{}_{,n}$$

Power System:

Power **Intenal + Kinetic = Mechanical Supply+ Heat Supply**

Power Derivative:

$$\int [\rho\ \ eT + (\tfrac{1}{2} \rho\ v^i v_i) T\,]dV = (\int \rho b^i v_i\, dV + \int F^i_n v_i n^n\, dS + \int h_n n^n\, dS)$$

But:

$$\int (F^i_n v_i n^n - h_n)\, dS = \int (F^i_n v_i - h_n)^{,n}\, dV = \int (F^i_n{}^{,n}\ v_i + F_n\ \ v_i{}^{,n} + h_n{}^{,n})\, dV$$

So:

$$\int [\rho\ \ eT + (\tfrac{1}{2} \rho\ v^i v_i) T\,]dV = (\int [\rho b^i v_i + (F^i_n v_i + h_n)^{,n}\ ]\, dV)$$

But by Conservation of Momentum: $kg/m^3\ m/s^2 = Pa/m$

$$\rho \ vT^{m} = \rho \ b^{m} + F^{m}{}_{n}{}^{,n}$$

Force/Volume = Body Force/Volume + Divergence of Surface Traction

So:

Powers:

$$\int \rho \ \ eT \ \ dV = \int (\ F^{i}_{n} \ \ v_{i}{}^{;n} + H \ ) \ dV$$

Integrand is Power Per Unit of Volume, (ρ e)T:

$$\rho \ \ eT \ = \ F^{i}{}_{n} \ v_{i}{}^{;n} \ + H$$

Similarly Get:

$$\rho \ \ eT \ = \ F_{i}{}^{n} \ v^{i}{}_{,n} \ + H$$

But:

$$v_{N}{}^{;m} \equiv QT_{N}{}^{;m} \equiv L_{N}{}^{K} \ Q_{K}{}^{m} \qquad v^{N}{}_{;m} \equiv QT^{N}{}_{;m} \equiv L^{N}{}_{K} \ Q^{K}{}_{m}$$

And:

$$v_{N}{}^{n} \equiv v_{N}{}^{;m} P_{m}{}^{n} \equiv QT_{N}{}^{;m} P_{m}{}^{n} \equiv L_{N}{}^{K} Q_{K}{}^{m} P_{m}{}^{n} \qquad v^{N}{}_{n} \equiv v^{N}{}_{;m} P^{m}{}_{n} \equiv QT^{N}{}_{;m} P^{m}{}_{n} \equiv L^{N}{}_{K} Q^{K}{}_{m} P^{m}{}_{n}$$

So:

Specific Power Minus Specific Heat Supply : ΞT-H:

Power Per Unit of Volume-Divergence of Heat Supply Gradient =

Surface traction • Local Velocity Gradient

$$F^{i}{}_{n} \ v_{i}{}^{;n} \ = \ \rho \ eT \ - \ H \ = \ F_{i}{}^{n} \ v^{i}{}_{,n}$$

$$F^{i}_{n} \ v_{i}{}^{;m} \ P_{m}{}^{n} \equiv F^{i}_{n} \ QT_{i}{}^{m} \ P_{m}{}^{n} = \ \rho \ eT \ - \ H \ = F_{i}{}^{n} \ QT^{i}{}_{m} P^{m}{}_{n} \equiv F_{i}{}^{n} \ v^{i}{}_{;m} P^{m}{}_{n}$$

13.5. Conservation of Power:

Knowing: Specific Heat Supply :

$$h_{n}{}^{;n} \equiv H \equiv h^{n}{}_{,n}$$

$$vq^{n}{}_{n} \equiv L \equiv vq_{n}{}^{n}$$

$$fT \equiv ft + fq_{n} \ v^{n} + \quad 0 \qquad \equiv ft + fq^{n} \ v_{n} + \quad 0$$

$$Cf \equiv ft + fq_{n} \ v^{n} + \ f \ L \equiv \ fT \quad + \ f \ L \equiv ft + fq^{n} \ v_{n} + \ f \ L$$

Time Derivative and Conserved Time Derivative of fg:

$$fgT \equiv fT \ \ g + f \ \ gT$$

$$Cfg \equiv Cf \quad g + \ f \ \ gT \equiv Cg \quad f + \ g \ \ fT \equiv Cgf$$

Time Derivative and Conserved Time Derivative, of e:

eT:Velocity²/time:

$$eT \equiv et + eq_{n} \ v^{n} + \quad 0 \qquad \equiv et + eq^{n} \ v_{n} + \quad 0$$

$$Ce \equiv et + eq_{n} \ v^{n} + \ e \ vq^{n}{}_{n} \equiv \ eT \quad + \ e \ L \equiv et + eq^{n} \ v_{n} + \ e \ vq_{n}{}^{n}$$

Time Derivative and Conserved Time Derivative of fg =ρ e :

$$\rho eT \equiv \rho T \ \ e + \rho \ \ eT$$

$$C\rho e \equiv C\rho \quad e + \ \rho \ \ eT \equiv Ce \quad \rho + \ e \ \ \rho T \equiv Ce\rho$$

But Mass Conservation:

$$C\rho = (\rho T + \rho \ L) \ = 0$$

Then:

$$C\rho e = 0 \ + \ \rho \quad eT \ \equiv \ \rho / \rho 0 \ \ \Xi 0T \ \equiv \ \Xi T \ = Ce \quad \rho \ + \ e \quad \rho T \equiv Ce\rho$$

Notice:

$$C\rho e \equiv \rho \ \ eT \equiv Ce \ \ \rho + e \ \ \rho T = (eT + e \ L) \ \rho + e \ \ \rho T = \rho \ \ eT + e(\rho T + \rho \ L) = \rho \ \ eT + 0 \equiv Ce\rho$$

If Incompressible Material:

Incompressible Material: ρT = 0

$$C\rho e = \ \rho \quad eT \equiv \ \Xi T \ = \ Ce \quad \rho \ = Ce\rho$$

Introduce Internal Power per Volume or Specific Power : ΞT and Ξ0T:

$$\Xi 0T \equiv \rho 0\ eT \equiv \rho 0/\rho\ \Xi T \qquad \Xi T / \rho \equiv eT \equiv \Xi 0T / \rho 0 \qquad \Xi T \equiv \rho\ eT \equiv \rho / \rho 0\ \Xi 0T$$

Then:

Specific Power Minus Specific Heat Supply : ΞT-H:
Power Per Unit of Volume - Divergence of Heat Supply Gradient = Surface Traction * Local Velocity Gradient
Specific Power Minus Specific Heat Supply : ΞT-H: Is:

$$F^{i}_{\ n}\ v_{i}^{\ ,n} = \Xi T - H \equiv \rho\ \ eT - H = F_{i}^{\ n}\ v^{i}_{\ ,n}$$

$$F^{i}_{\ n}\ QT_{i}^{\ m}\ P_{m}^{\ n} = \Xi T - H \equiv \rho\ \ eT - H = F_{i}^{\ n}\ QT^{i}_{\ m} P^{m}_{\ n}$$

13.6.
Symmetric Finite Deformation Power:

Since F Is Symmetric:

$$F^{I}_{\ J}\, v_{I}^{\ ,J} = ½\,(F^{\ J}_{I}\, v_{J}^{\ ,I} + F^{I}_{\ J}\, v_{I}^{\ ,J})$$

Where:

$$v_{J}^{\ ,I} = v_{J}^{\ ;p}\, P_{p}^{\ I} = [v_{k}^{\ ;p}\ \delta_{k}^{\ J}]\ P_{p}^{\ I}$$

$$v_{I}^{\ ,J} = v_{I}^{\ ;q}\, P_{q}^{\ J} = [v_{k}^{\ ;q}\ \delta_{k}^{\ I}]\ P_{q}^{\ J}$$

Rewriting $\delta_{k}^{\ j}$ and $\delta_{k}^{\ i}$ In Two Reciprocal Forms:

$$\delta_{k}^{\ j} = (Q_{k}^{\ q}\ P_{q}^{\ j})$$

$$\delta_{k}^{\ i} = (P_{p}^{\ i}\ Q_{k}^{\ p})$$

Then:

$$v_{J}^{\ ,I} = [v_{k}^{\ ;p}\ (Q_{k}^{\ q}\ P_{q}^{\ J})]\ P_{p}^{\ I} \equiv [\,QT_{k}^{\ p}\ (Q_{k}^{\ q}\ P_{q}^{\ J})]\ P_{p}^{\ I}$$

$$v_{I}^{\ ,J} = [v_{k}^{\ ;q}\ (P_{p}^{\ I}\ Q_{k}^{\ p})]\ P_{q}^{\ J} \equiv [\,QT_{k}^{\ q}\ (P_{p}^{\ I}\ Q_{k}^{\ p})]\ P_{q}^{\ J}$$

To get:

$$F^{I}_{\ J}\, v_{I}^{\ ,J} \equiv ½\,(F^{\ J}_{I}\, v_{J}^{\ ,I} + F^{I}_{\ J}\, v_{I}^{\ ,J}) =$$

$$½[F^{\ J}_{I}\, v_{k}^{\ ;p}\ (Q_{k}^{\ q}\ P_{q}^{\ J})\ P_{p}^{\ I} + F^{I}_{\ J}\, v_{k}^{\ ;q} (P_{p}^{\ I}\ Q_{k}^{\ p})\ P_{q}^{\ J}]$$

= Rearranged =

$$½[v_{k}^{\ ;p} Q_{k}^{\ q}\ (P_{p}^{\ I}\ F^{I}_{\ J}\ P_{q}^{\ J}) + Q_{k}^{\ p}\, v_{k}^{\ ;q}\ (P_{p}^{\ I}\ F^{I}_{\ J}\ P_{q}^{\ J})] =$$

$$½\,(v_{k}^{\ ;p}\ Q_{k}^{\ q} + Q_{k}^{\ p}\ v_{k}^{\ ;q})\ P_{p}^{\ I}\ F^{I}_{\ J}\ P_{q}^{\ J} =$$

$$½\,(QT_{k}^{\ p}\ Q_{k}^{\ q} + Q_{k}^{\ p}\ QT_{k}^{\ q})\ (P_{p}^{\ I}\ F^{I}_{\ J}\ P_{q}^{\ J})$$

Introduce:
Stress T and Strain E:

$$T_{p\,q} \equiv\ P_{p}^{\ i}\ F^{i}_{\ j}\ \ P_{q}^{\ j} \qquad T^{p\,q} \equiv\ \ P^{p}_{\ i}\ F_{i}^{\ j}\ \ P^{q}_{\ j}$$

$$E^{pq} \equiv ½\,(Q_{k}^{\ p}\, Q_{k}^{\ q}) \qquad E_{pq} \equiv ½\,(Q^{k}_{\ p}\, Q^{k}_{\ q})$$

And ET:

$$ET^{pq} \equiv ½\,(QT_{k}^{\ p}\ Q_{k}^{\ q} + Q_{k}^{\ p}\ QT_{k}^{\ q}) \qquad ET_{pq} \equiv ½\,(QT^{k}_{\ p}\ Q^{k}_{\ q} + Q^{k}_{\ p}\ QT^{k}_{\ q})$$

Also Introduce: T0:

$$T0 \equiv\ \rho 0/\rho\ T \qquad T / \rho \equiv\ T0 / \rho 0 \qquad T \equiv \rho / \rho 0\ T0$$

So:

$$F^{I}_{\ J}\, v_{I}^{\ ,J} \equiv ½\,(F^{\ J}_{I}\, v_{J}^{\ ,I} + F^{I}_{\ J}\, v_{I}^{\ ,J}) = \qquad F_{I}^{\ J}\, v^{I}_{\ J} \equiv ½\,(F_{I}^{\ J}\, v^{I}_{\ J} + F_{I}^{\ ,J}\, v^{I}_{\ J}) =$$

$$ET^{pq}\ \ T_{p\,q} \equiv (ET^{pq})\ (\rho/\rho 0\ \ T0_{p\,q}) \qquad ET_{pq} T^{pq} \equiv (ET_{pq})\ (\rho/\rho 0\ \ T0^{p\,q})$$

This Is Substituted In: Specific Power Minus Specific Heat Supply : ΞT-H:

$$F^{i}_{\ n}\ v_{i}^{\ ,n} = \Xi T - H \equiv \rho\ eT - H = F_{i}^{\ n}\ v^{i}_{\ ,n}$$

To Get:

Specific Power Minus Specific Heat Supply : ΞT-H:

$$F^{I}_{J} v_{I}{}^{,J} \equiv ½ (F^{i}_{j} v_{i}{}^{,j} + F^{i}_{j} v_{i}{}^{,j}) \equiv ET^{pq} \; T_{pq} \equiv (ET^{pq}) (\rho/\rho0 \;\; T0_{pq})$$

$$= ΞT - H \equiv \rho \; eT - H =$$

$$F_{I}{}^{J} v^{I}{}_{,J} \equiv ½ (F_{i}{}^{j} v^{i}{}_{,j} + F_{i}{}^{,j} v^{i}{}_{j}) \equiv ET_{pq} \; T^{pq} \equiv (ET_{pq}) (\rho/\rho0 \;\; T0^{pq})$$

Now Introduce: E and TT:

$$TT_{pq} \equiv (P_{p}{}^{i} \; F^{i}_{j} \;\; P_{q}{}^{j})T \qquad\qquad TT^{pq} \equiv (P^{p}{}_{i} \; F_{i}{}^{j} \;\; P^{q}{}_{j})T$$

$$E^{pq} \equiv ½ (Q_{k}{}^{p} \, Q_{k}{}^{q} - \delta^{pq}) \qquad\qquad E_{pq} \equiv ½ (Q^{k}{}_{p} \, Q^{k}{}_{q} - \delta_{pq})$$

Also Introduce: T0T:

$$T0T \equiv \rho0/\rho \; TT \qquad\qquad TT / \rho \equiv T0T /\rho0 \qquad\qquad TT \equiv \rho /\rho0 \; T0T$$

Specific Power Minus Specific Heat Supply : ΞT-H: Power / Volume:

$$ET^{pq} \;\; T_{pq} = ΞT - H = \rho \;\; eT - H = ET_{pq} \;\; T^{pq}$$

Plus:

Complementary Specific Power Plus Specific Heat Supply : ΨT+H:

Complementary Power / Volume:

$$E^{pq} \;\; TT_{pq} = ΨT + H = \rho T \;\; e + H = E_{pq} \;\; TT^{pq}$$

Equal:

Specific Power Plus Complementary Specific Power:

Power and Complementary Power / Volume:

ΞT+ΨT:

$$(E^{pq} \;\; T_{pq})T = ΞT + ΨT = \rho eT = (E_{pq} \;\; T^{pq})T$$

Divergence of Heat Gradient: Convection of Heat Supply: $h_{n}{}^{,n} \equiv H \equiv h^{n}{}_{,n}$

Divergence of Velocity Gradient: Convection of Velocity: $vq^{n}{}_{n} \equiv L \equiv vq_{n}{}^{n}$

Energy / Volume: Differentials:

$$dE^{pq} \;\; T_{pq} = dΞ - H{*}dt = \rho \;\; de - H{*}dt = dE_{pq} \;\; T^{pq}$$

$$E^{pq} \;\; dT_{pq} = dΨ + H{*}dt = d\rho \;\; e + H{*}dt = E_{pq} \;\; dT^{pq}$$

$$d(E^{pq} \;\; T_{pq}) = dΞ + dΨ = d(\rho e) = d(E_{pq} \;\; T^{pq})$$

So:

Stress and Strain are derived from (Ξ + Ψ) = ρ (e + c)

$$T^{kl} = \rho \;\; \partial e / \partial E_{kl} = \partial Ξ / \partial E_{kl} \qquad\qquad T_{kl} = \rho \;\; \partial e / \partial E^{kl} = \partial Ξ / \partial T^{kl}$$

$$E^{kl} = \rho \;\; \partial c / \partial T_{kl} = \partial Ψ / \partial T_{kl} \qquad\qquad E_{kl} = \rho \;\; \partial c / \partial T^{kl} = \partial Ψ / \partial T^{kl}$$

13.7.
Strain and Stress are Reciprocals:

Stress T And Strain E:

$$T_{pq} \equiv P_{p}{}^{i} \; F^{i}_{j} \;\; P_{q}{}^{j} \qquad\qquad T^{pq} \equiv P^{p}{}_{i} \; F_{i}{}^{j} \;\; P^{q}{}_{j}$$

$$E^{pq} \equiv ½ (Q_{k}{}^{p} \, Q_{k}{}^{q}) \qquad\qquad E_{pq} \equiv ½ (Q^{k}{}_{p} \, Q^{k}{}_{q})$$

And

Stress Rate TT And Strain Rate ET:

$$TT_{pq} \equiv (P_{p}{}^{i} \; F^{i}_{j} \;\; P_{q}{}^{j})T \qquad\qquad TT^{pq} \equiv (P^{p}{}_{i} \; F_{i}{}^{j} \;\; P^{q}{}_{j})T$$

$$ET^{pq} \equiv ½ (QT_{k}{}^{p} \, Q_{k}{}^{q} + Q_{k}{}^{p} \, QT_{k}{}^{q}) \qquad\qquad ET_{pq} \equiv ½ (QT^{k}{}_{p} \, Q^{k}{}_{q} + Q^{k}{}_{p} \, QT^{k}{}_{q})$$

Where:

$$QT_{i}{}^{m} \;\; P_{m}{}^{n} \equiv v_{i}{}^{;m} \;\; P_{m}{}^{n} = v_{i}{}^{,n} \qquad\qquad v^{i}{}_{,n} = v^{i}{}_{;m} P^{m}{}_{n} \equiv QT^{i}{}_{m} P^{m}{}_{n}$$

T And ET:

$$T_{pq} \equiv P_p^{\;i} F^i_{\;j} P^j_{\;q} \qquad\qquad T^{pq} \equiv P^p_{\;i} F_i^{\;j} P^q_{\;j}$$

$$ET^{pq} \equiv ½\,(QT_k^{\;p} Q_k^{\;q} + Q_k^{\;p} QT_k^{\;q}) \qquad\qquad ET_{pq} \equiv ½\,(QT^k_{\;p} Q^k_{\;q} + Q^k_{\;p} QT^k_{\;q})$$

TT And E:

$$TT_{pq} \equiv (P_p^{\;i} F^i_{\;j} P^j_{\;q})T \qquad\qquad TT^{pq} \equiv (P^p_{\;i} F_i^{\;j} P^q_{\;j})T$$

$$E^{pq} \equiv ½\,(Q_k^{\;p} Q_k^{\;q} - \delta^{pq}) \qquad\qquad E_{pq} \equiv ½\,(Q^k_{\;p} Q^k_{\;q} - \delta_{pq})$$

Power / Volume: Stress/time:

$$ET^{pq}\; T_{pq} = \Xi T - H = \rho \quad eT - H = ET_{pq}\; T^{pq}$$

$$E^{pq}\; TT_{pq} = \Psi T + H = \rho T \quad e + H = E_{pq}\; TT^{pq}$$

$$(E^{pq}\; T_{pq})T = \Xi T + \Psi T = \rho eT = (E_{pq}\; T^{pq})T$$

Energy / Volume: Differentials: Stress:

$$dE^{pq}\; T_{pq} = d\Xi - H^*dt = \rho \quad de - H^*dt = dE_{pq}\; T^{pq}$$

$$E^{pq}\; dT_{pq} = d\Psi + H^*dt = d\rho \quad e + H^*dt = E_{pq}\; dT^{pq}$$

$$d(E^{pq}\; T_{pq}) = d\Xi + d\Psi = d(\rho e) = d(E_{pq}\; T^{pq})$$

Convection of Heat Supply: Stress/time: $h_n{}'^{\,n} \equiv h^n{}_{,n} \equiv \underline{\nabla} \bullet \underline{h} \equiv H \equiv \rho$ **heat**

Divergence of Velocity Gradient: Convection of Velocity:1/time: $vq^n{}_n \equiv L \equiv vq_n^{\;n}$

13.8.
Associated First Order Tensors With Second Order Tensors:

Assume:

(First Order Tensor, A Is Associated With Second Order Tensor, T)

And

(First Order Tensor, B Is Associated With Second Order Tensor, E)

Stress Stiffness (Resilience):

Stress = Stress * 1

$$T^{kl} = T^{kl}(E_{kl}, B_p) \qquad\qquad T_{kl} = T_{kl}(E^{kl}, B^p)$$

Strain Flexibility (Resilience):

1 = 1 /Stress * Stress

$$E^{kl} = E^{kl}(T_{kl}, A_p) \qquad\qquad E_{kl} = E_{kl}(T^{kl}, A^p)$$

Stress and Strain are Reciprocal functions of each other•

Time Rate of Stress Elastic Stiffness (Resilience):

Stress/ Time = Stress * 1/ Time

$$TT^{kl} = TT^{kl}(E_{kl}, B_p) \qquad\qquad TT_{kl} = TT_{kl}(E^{kl}, B^p)$$

Time Rate of Strain Elastic Flexibility (Resilience):

1/ Time = 1 /Stress * Stress/ Time

$$ET^{kl} = ET^{kl}(T_{kl}, A_p) \qquad\qquad ET_{kl} = ET_{kl}(T^{kl}, A^p)$$

Differential:

$$dT^{kl} = \partial T^{kl}/\partial E_{mn}\; dE_{mn} + \partial T^{kl}/\partial B_r\; dB_r \qquad\qquad dT_{kl} = \partial T_{kl}/\partial E^{mn}\; dE^{mn} + \partial T_{kl}/\partial B^r\; dB^r$$

$$dE^{kl} = \partial E^{kl}/\partial T_{mn}\; dT_{mn} + \partial E^{kl}/\partial A_c\; dA_c \qquad\qquad dE_{kl} = \partial E_{kl}/\partial T^{mn}\; dT^{mn} + \partial E_{kl}/\partial A^c\; dA^c$$

Divide by dt:

$$TT^{kl} = \partial T^{kl}/\partial E_{mn}\; ET_{mn} + \partial T^{kl}/\partial B_r\; BT_r \qquad\qquad TT_{kl} = \partial T_{kl}/\partial E^{mn}\; ET^{mn} + \partial T_{kl}/\partial B^r\; BT^r$$

$$ET^{kl} = \partial E^{kl}/\partial T_{mn}\; TT_{mn} + \partial E^{kl}/\partial A_c\; AT_c \qquad\qquad ET_{kl} = \partial E_{kl}/\partial T^{mn}\; TT^{mn} + \partial E_{kl}/\partial A^c\; AT^c$$

Or:

$$TT^{kl} = TE^{klmn}\; ET_{mn} + TE^{klr}\; BT_r \qquad\qquad TT_{kl} = TE_{klmn}\; ET^{mn} + TE_{klr}\; BT^r$$

$$ET^{kl} = ET^{klmn}\ TT_{mn} + ET^{klc}\ AT_c \qquad ET_{kl} = ET_{klmn}\ TT^{mn} + ET_{klc}\ AT^c$$

Or:

$$TT^{kl} = K^{klmn}\ ET_{mn} + \Pi^{klr}\ BT_r \qquad TT_{kl} = K_{klmn}\ ET^{mn} + \Pi_{klr}\ BT^r$$

$$ET^{kl} = \Gamma^{klmn}\ TT_{mn} + \Lambda^{klc}\ AT_c \qquad ET_{kl} = \Gamma_{klmn}\ TT^{mn} + \Lambda_{klc}\ AT^c$$

Or:

Stress/ Time = Stress * 1/ Time

$$TT^{kl} = TT^{kl}(E_{kl}, B_p) \qquad TT_{kl} = TT_{kl}(E^{kl}, B^p)\ TT_{kl}$$

$$= TE^{klmn}\ ET_{mn} + TE^{klr}\ BT_r \qquad = TE_{klmn}\ ET^{mn} + TE_{klr}\ BT^r$$

$$= K^{klmn}\ ET_{mn} + \Pi^{klr}\ BT_r \qquad = K_{klmn}\ ET^{mn} + \Pi_{klr}\ BT^r$$

And:

1/ Time = 1 /Stress * Stress/ Time

$$ET^{kl} = ET^{kl}(T_{kl}, A_p) \qquad ET_{kl} = ET_{kl}(T^{kl}, A^p)$$

$$= ET^{klmn}\ TT_{mn} + ET^{klc^p}\ AT_c \qquad = ET_{klmn}\ TT^{mn} + ET_{klc}\ AT^c$$

$$= \Gamma^{klmn}\ TT_{mn} + \Lambda^{klc}\ AT_c \qquad = \Gamma_{klmn}\ TT^{mn} + \Lambda_{klc}\ AT^c$$

13.9.
Stress:

Differentials:

$$dE^{pq}\ T_{pq} = d\Xi - H^*dt = \rho\ de - H^*dt = dE_{pq}\ T^{pq}$$

$$E^{pq}\ dT_{pq} = d\Psi + H^*dt = d\rho\ e + H^*dt = E_{pq}\ dT^{pq}$$

$$d(E^{pq}\ T_{pq}) = d\Xi + d\Psi = d(\rho e) = d(E_{pq}\ T^{pq})$$

Stress/time:

Split Internal Power/Volume Into Reversible And Irreversible Parts:

$\Xi \equiv \rho\ e = \rho$ e reversible (elastic) + ρ e irreversible (inelastic)

$$d\Xi \equiv \rho\ de$$

Split Complementary Power/Volume Into Reversible And Irreversible Parts:

$\Psi \equiv \rho\ c = \rho$ c reversible (elastic) + ρ c irreversible (inelastic)

$$d\Psi \equiv \rho\ dc$$

Assume:

Irreversible Or Inelastic Dissipation of Power / Mass Is: D: Velocity^2/Time:

So:

Irreversible Or Inelastic Dissipation of Power / Volume Is: ρ D: Stress/Time:

Then:

Power/Mass	**Velocity^2/Time**	**eT**	**heat**	**D**	**W/kg**	**(m/s)^2/s**
Power/Volume	**Stress/Time**	**ΞT=ρ eT**	**H=ρ heat**	**ρ D**	**W/m^3**	**Pa/s**

And:

Energy/Mass	**Velocity^2**	**de**	**heat dt**	**D dt**	**J/kg**	**(m/s)^2**
Energy/Volume	**Stress**	**dΞ=ρ de**	**H*dt=ρ heat dt**	**ρ D dt**	**J/m^3**	**Pa**

Differential: Stress:

$$dE^{kl}\ T_{kl} + \rho\ heat\ dt - \rho\ D\ dt = d\Xi \equiv \rho\ de = dE_{kl}\ T^{kl} + \rho\ heat\ dt - \rho\ D\ dt$$

$$E^{kl}\ dT_{kl} - \rho\ heat\ dt + \rho\ D\ dt = d\Psi \equiv d\rho\ e = E_{kl}\ dT^{kl} - \rho\ heat\ dt + \rho\ D\ dt$$

$$d(E^{kl}\ T_{kl}) = d(\Xi + \Psi) = d(E_{kl}\ T^{kl})$$

Divide by dt: Internal and Complementary Time Derivative:

$$ET^{kl}\ T_{kl} + \rho\ heat - \rho\ D = \Xi T = ET_{kl}\ T^{kl} + \rho\ heat - \rho\ D$$

$$E^{kl}\,TT_{kl} - \rho\, heat + \rho\, D \;=\; \Psi T \;=\; E_{kl}\,TT^{kl} - \rho\, heat + \rho\, D$$
$$(E^{kl}\,T_{kl})T = (\Xi + \Psi)T = \rho = (e + c)T = (E_{kl}\,T^{kl})T$$

If heat=0:

$$\Xi T \equiv \rho\; eT \;= ET^{kl}\,T_{kl} - \rho\; D = ET_{kl}\,T^{kl} - \rho\; D$$
$$\rho\; D = ET^{kl}\;T_{kl} - \rho\; eT = E_{kl}\,TT^{kl} - \rho\; eT \geq 0$$
$$= ET^{kl}\;T_{kl} - (\rho\,\partial e/\partial E_{kl}\,ET_{kl} + \rho\;\partial e/\partial B_r\;BT_r\;)$$
$$= \;ET^{kl}\;T_{kl} - ET^{kl}\;T_{kl} - \rho\;\partial e\,/\,\partial B_r\;BT_r = -\;\rho\;\partial e/\partial B_r\;BT_r \geq 0$$

And:

$$\Psi T \equiv \rho\;\; cT = E^{kl}\,TT_{kl} \;+\; \rho\; D = E_{kl}\,TT^{kl} \;+\; \rho\; D$$
$$\rho\;\; D = \rho\;\; cT - E^{kl}\;TT_{kl} = \rho\;\; cT - E_{kl}\,TT^{kl} \geq 0$$
$$= \rho\,\partial c/\partial T_{kl}\;\;TT_{kl} + \rho\;\;\partial c/\partial A_r\;\;AT_r - E^{kl}\;\;TT_{kl}$$
$$= \;E^{kl}\;\;TT_{kl} + \rho\;\partial c/\partial A_r\;\;AT_r - E^{kl}\;\;TT_{kl} = \rho\;\partial c/\partial A_r\;\;AT_r \geq 0$$

Then:

$$-\,\partial\Xi/\partial B_r\;\;BT_r = -\,\rho\;\;\partial e/\partial B_r\;\;BT_r \geq 0 \qquad -\,\partial\Xi/\partial B^r\;\;BT^r = -\,\rho\;\;\partial e/\partial B^r\;\;BT^r \geq 0$$
$$+\,\partial\Psi/\partial A_r\;\;AT_r = +\,\rho\;\;\partial c/\partial A_r\;\;AT_r \geq 0 \qquad +\,\partial\Psi/\partial A^r\;\;AT^r = +\,\rho\;\;\partial c/\partial A^r\;\;AT^r \geq 0$$

Identically:

Stress/ Time = Stress * 1/ Time

$$+\,\partial\Xi/\partial B_r\;\;BT_r = +\,\rho\;\;\partial e/\partial B_r\;\;BT_r \leq 0 \qquad +\,\partial\Xi/\partial B^r\;\;BT^r = +\,\rho\;\;\partial e/\partial B^r\;\;BT^r \leq 0$$
$$+\,\partial\Psi/\partial A_r\;\;AT_r = +\,\rho\;\;\partial c/\partial A_r\;\;AT_r \geq 0 \qquad +\,\partial\Psi/\partial A^r\;\;AT^r = +\,\rho\;\;\partial c/\partial A^r\;\;AT^r \geq 0$$

Where:

$$\dot{\Xi}\,(E^{kr}, B^r) = \Xi\,(E^{ii}, E^{ij}E^{ji}, B^r) = \Xi = \Xi\,(E_{ii}, E_{ij}E_{ji}, B_r) = \Xi\,(E_{kl}, B_r)$$
$$\Psi\,(T_{kl}, A_r) = \Psi\,(T_{ii}, T_{ij}T_{ji}, A_r) = \Psi = \Psi\,(T^{ii}, T^{ij}T^{ji}, A^r) = \Psi\,(T^{kr}, A^r)$$

13.10.
Strain and Stress Time derivatives:

Strain and Stress Time derivatives are Time Derivatives of Derived Forms

Since:

Strain and Stress are derived from (X + Ψ)

$$T^{kl}\;(E_{kl}, B_p) = \partial X\,/\,\partial E_{kl} \qquad T_{kl}\;(E^{kl}, B^p) = \partial X\,/\,\partial E^{kl}$$
$$E^{kl}\;(T_{kl}, A_p) = \partial\Psi\,/\,\partial T_{kl} \qquad E_{kl}\;(T^{kl}, A^p) = \partial\Psi\,/\,\partial T^{kl}$$

And:

$$dT^{kl} = \partial T^{kl}/\partial E_{mn}\;dE_{mn} + \partial T^{kl}/\partial B_r\;dB_r \qquad dT_{kl} = \partial T_{kl}/\partial E^{mn}\;dE^{mn} + \partial T_{kl}/\partial B^r\;dB^r$$
$$dE^{kl} = \partial E^{kl}/\partial T_{mn}\;dT_{mn} + \partial E^{kl}/\partial A_c\;dA_c \qquad dE_{kl} = \partial E_{kl}/\partial T^{mn}\;dT^{mn} + \partial E_{kl}/\partial A^c\;dA^c$$

To get Differential:

$$dT^{kl} = \rho\;\;\partial\,\partial e\,/\,\partial E_{kl}/\partial E_{mn}\;dE_{mn} + \qquad dT_{kl} = \rho\;\;\partial\,\partial e\,/\,\partial E^{kl}/\partial E^{mn}\;dE^{mn} +$$
$$+\;\rho\;\;\partial\,\partial e\,/\,\partial E_{kl}/\partial B_r\;\;dB_r \qquad +\,\rho\;\;\partial\,\partial e\,/\,\partial E^{kl}/\partial B^r\;\;dB^r$$
$$dE^{kl} = \rho\;\;\partial\;\partial c\,/\,\partial T_{kl}/\partial T_{mn}\;\;dT_{mn} + \qquad dE_{kl} = \rho\;\;\partial\,\partial c\,/\,\partial T^{kl}/\partial T^{mn}\;\;dT^{mn} +$$
$$+\;\rho\;\;\partial\;\partial c\,/\,\partial T_{kl}/\partial A_c\;\;dA_c \qquad +\,\rho\;\;\partial\,\partial c\,/\,\partial T^{kl}/\partial A^c\;\;dA^c$$

$$dT^{kl} = \partial\,\partial\,\Xi\,/\,\partial E_{kl}/\partial E_{mn}\;dE_{mn} + \qquad dT_{kl} = \partial\,\partial\,\Xi\,/\,\partial E^{kl}/\partial E^{mn}\;dE^{mn} +$$
$$+\;\partial\,\partial\,\Xi\,/\,\partial E_{kl}/\partial B_r\;\;dB_r \qquad +\,\partial\,\partial\,\Xi\,/\,\partial E^{kl}/\partial B^r\;\;dB^r$$
$$dE^{kl} = \partial\;\partial\Psi\,/\,\partial T_{kl}/\partial T_{mn}\;\;dT_{mn} + \qquad dE_{kl} = \partial\,\partial\Psi\,/\,\partial T^{kl}/\partial T^{mn}\;\;dT^{mn} +$$
$$+\;\;\partial\;\partial\Psi\,/\,\partial T_{kl}/\partial A_c\;\;dA_c \qquad +\;\;\partial\,\partial\Psi\,/\,\partial T^{kl}/\partial A^c\;\;dA^c$$

Divide Differentials by dt:
Time derivative of Stress and Strain are Time derivative of derived forms

$$TT^{kl} = \rho\ \partial\, \partial e / \partial E_{kl} / \partial E_{mn}\ ET_{mn} + \rho\ \partial\, \partial e / \partial E_{kl} / \partial B_r\ BT_r \qquad TT_{kl} = \rho\ \partial\, \partial e / \partial E^{kl} / \partial E^{mn}\ ET^{mn} + \rho\ \partial\, \partial e / \partial E^{kl} / \partial B^r\ BT^r$$

$$ET^{kl} = \rho\, \partial\, \partial c / \partial T_{kl} / \partial T_{mn}\ TT_{mn} + \rho\ \partial\, \partial c / \partial T_{kl} / \partial A_c\ AT_c \qquad ET_{kl} = \rho\ \partial\, \partial c / \partial T^{kl} / \partial T^{mn}\ TT^{mn} + \rho\ \partial\, \partial c / \partial T^{kl} / \partial A^c\ AT^c$$

Or:

$$TT^{kl} = \rho\ eEE^{klmn}\ ET_{mn} + \rho\ eEB^{klr}\ BT_r \qquad TT_{kl} = \rho\ eEE_{klmn}\ ET^{mn} + \rho\ eEB_{klr}\ BT^r$$

$$ET^{kl} = \rho\ cTT^{klmn}\ TT_{mn} + \rho\ cTA^{klc}\ AT_c \qquad ET_{kl} = \rho\ cTT_{klmn}\ TT^{mn} + \rho\ cTA_{klc}\ AT^c$$

Or:

$$TT^{kl} = \Xi EE^{klmn}\ ET_{mn} + \Xi EB^{klr}\ BT_r \qquad TT_{kl} = \Xi EE_{klmn}\ ET^{mn} + \Xi EB_{klr}\ BT^r$$

$$ET^{kl} = \Psi TT^{klmn}\ TT_{mn} + \Psi TA^{klc}\ AT_c \qquad ET_{kl} = \Psi TT_{klmn}\ TT^{mn} + \Psi TA_{klc}\ AT^c$$

Or:

$$TT^{kl} = K^{klmn}\ ET_{mn} + \Pi^{klr}\ BT_r \qquad TT_{kl} = K_{klmn}\ ET^{mn} + \Pi_{klr}\ BT^r$$

$$ET^{kl} = \Gamma^{klmn}\ TT_{mn} + \Lambda^{klc}\ AT_c \qquad ET_{kl} = \Gamma_{klmn}\ TT^{mn} + \Lambda_{klc}\ AT^c$$

So:

$$T^{kl} \equiv T^{kl}(E_{kl}, B_p) = \partial X / \partial E_{kl} \qquad T_{kl} \equiv T_{kl}(E^{kl}, B^p) = \partial X / \partial E^{kl}$$

$$E^{kl} \equiv E^{kl}(T_{kl}, A_p) = \partial \Psi / \partial T_{kl} \qquad E_{kl} \equiv E_{kl}(T^{kl}, A^p) = \partial \Psi / \partial T^{kl}$$

Time Derivatives:

$$TT^{kl} \equiv TT^{kl}(E_{kl}, B_p) = (\partial X / \partial E_{kl})T \qquad TT_{kl} \equiv TT_{kl}(E^{kl}, B^p) = (\partial X / \partial E^{kl})T$$

$$ET^{kl} \equiv ET^{kl}(T_{kl}, A_p) = (\partial \Psi / \partial T_{kl})T \qquad ET_{kl} \equiv ET_{kl}(T^{kl}, A^p) = (\partial \Psi / \partial T^{kl})T$$

Or:

Stress/ Time = Stress * 1/ Time

$$TT^{kl} = TT^{kl}(E_{kl}, B_p) \qquad TT_{kl} = TT_{kl}(E^{kl}, B^p)\ TT_{kl}$$

$$= \rho\ eEE^{klmn}\ ET_{mn} + \rho\ eEB^{klr}\ BT_r \qquad = \rho\ eEE_{klmn}\ ET^{mn} + \rho\ eEB_{klr}\ BT^r$$

$$= \Xi EE^{klmn}\ ET_{mn} + \Xi EB^{klr}\ BT_r \qquad = \Xi EE_{klmn}\ ET^{mn} + \Xi EB_{klr}\ BT^r$$

$$= K^{klmn}\ ET_{mn} + \Pi^{klr}\ BT_r \qquad = K_{klmn}\ ET^{mn} + \Pi_{klr}\ BT^r$$

And:

1/ Time = 1 /Stress * Stress/ Time

$$ET^{kl} = ET^{kl}(T_{kl}, A_p) \qquad ET_{kl} = ET_{kl}(T^{kl}, A^p)$$

$$= \rho\ cTT^{klmn}\ TT_{mn} + \rho\ cTA^{klc}\ AT_c \qquad = \rho\ cTT_{klmn}\ TT^{mn} + \rho\ cTA_{klc}\ AT^c$$

$$= \Psi TT^{klmn}\ TT_{mn} + \Psi TA^{klc}\ AT_c \qquad = \Psi TT_{klmn}\ TT^{mn} + \Psi TA_{klc}\ AT^c$$

$$= \Gamma^{klmn}\ TT_{mn} + \Lambda^{klc}\ AT_c \qquad = \Gamma_{klmn}\ TT^{mn} + \Lambda_{klc}\ AT^c$$

14.
Elastic and Inelastic Constitution:

Frame of Reference May Be Written:

$p_P \qquad p^P \qquad q_A = q_A(p_P, t) \qquad q^A = q^A(p^P, t)$

:

Example:
Material Density ρ:

$$\rho(p_i,t) = \rho[q_i(p_j,t),t] = \rho[q_i(p^j,t),t]$$
$$=$$
$$\rho(p^i,t) = \rho[q^i(p_j,t),t] = \rho[q^i(p^j,t),t]$$
$$:$$

Gradient of a Coordinate:

$$\partial p_N/\partial q_M\ \underline{q}_M = P_N^{\ M}\ \underline{q}_M \qquad \partial q_N/\partial p_M\ \underline{p}_M = Q_N^{\ M}\ \underline{p}_M$$
$$\partial p^N/\partial q^M\ \underline{q}^M = P^N_{\ M}\ \underline{q}^M \qquad \partial q^N/\partial p^M\ \underline{p}^M = Q^N_{\ M}\ \underline{p}^M$$

Rewriting $\delta_k^{\ j}$ and $\delta_k^{\ i}$ In Two Reciprocal Forms:

$$\delta_k^{\ j} = Q_k^{\ q} * P_q^{\ j}$$
$$\delta_k^{\ i} = P_p^{\ i} * Q_k^{\ p}$$

Time Derivative:

Coordinate Per Time is Velocity

Global Velocity

$$u_N \equiv pt_N \equiv \partial p_N/\partial t$$
$$u^N \equiv pt^N \equiv \partial p^N/\partial t$$

Local Velocity

$$v_N \equiv qt_N \equiv \partial q_N/\partial t$$
$$v^N \equiv qt^N \equiv \partial q^N/\partial t$$

Fixed Global Coordinates, $\underline{p}$:

Material position: $\underline{p}$ **is fixed:** $\partial\underline{p}/\partial t \equiv \underline{p}t = 0$:

Material Particle Velocity:

$$D\underline{p}/Dt \equiv d\underline{p}/dt \equiv \underline{p}T = \partial\underline{p}/\partial t + \partial\underline{p}/\partial\underline{p} \bullet \partial\underline{p}/\partial t = 0$$

Spatial position: $\underline{q}$ **is function of** $\underline{p}$ **and of t:**

Spatial Particle Velocity:

$$D\underline{q}/Dt \equiv \underline{q}T = \partial\underline{q}/\partial t + \partial\underline{q}/\partial\underline{p} \bullet \partial\underline{p}/\partial t = \underline{q}t + 0 = \underline{q}t \equiv \underline{v}$$

$$u_N \equiv pt_N \equiv \partial p_N/\partial t = 0 \qquad v_N \equiv qt_N \equiv \partial q_N/\partial t \neq 0$$
$$u^N \equiv pt^N \equiv \partial p^N/\partial t = 0 \qquad v^N \equiv qt^N \equiv \partial q^N/\partial t \neq 0$$

Gradient of a Local Velocity ≡ Gradient of a Coordinate Per Time ≡ Convection:

Since $\partial u_N/\partial p_K \equiv 0$ it is of no use later

Since $\partial u^N/\partial p^K \equiv 0$ it is of no use later

$$\partial v_N/\partial p_n\ \underline{p}_n = \partial v_N/\partial q_K\ \partial q_K/\partial p_n\ \underline{p}_n$$
$$\partial v^N/\partial p^n\ \underline{p}^n = \partial v^N/\partial q^K\ \partial q^K/\partial p^n\ \underline{p}^n$$

Convection:

$$\partial v_N/\partial p_n\ \underline{p}_n = \partial\, qt_N/\partial q_K\ \partial q_K/\partial p_n\ \underline{p}_n$$
$$\partial v^N/\partial p^n\ \underline{p}^n = \partial\, qt^N/\partial q^K\ \partial q^K/\partial p^n\ \underline{p}^n$$

Time Rate of Deformation Material Gradient and Its Conjugate $\underline{K}$

Time Rate of Deformation Current Gradient and Its Conjugate: $\underline{L}$

Position Gradient Vector and Components:

Material Gradient

$$\underline{grad}\ \underline{p} \equiv \underline{P}$$

Current (Spatial) Gradient

$$\underline{Q} \equiv \underline{Grad}\ \underline{q}$$

$$P^{N}{}_{n} \equiv pq^{K}{}_{n} \qquad Q^{N}{}_{n} \equiv qp^{K}{}_{n}$$
$$P_{N}{}^{n} \equiv pq_{K}{}^{n} \qquad Q_{N}{}^{n} \equiv qp_{K}{}^{n}$$

Velocity Vector:

$$\underline{u} \equiv \underline{p}_{n}\, u^{n} \equiv \underline{p}_{n}\, pT^{n} \equiv \underline{p}_{n}\, pt^{n} \qquad \underline{v} \equiv \underline{q}_{n}\, v^{n} \equiv \underline{q}_{n}\, qT^{n} \equiv \underline{q}_{n}\, qt^{n}$$
$$\underline{u} \equiv \underline{p}^{n}\, u_{n} \equiv \underline{p}^{n}\, pT_{n} \equiv \underline{p}^{n}\, pt_{n} \qquad \underline{v} \equiv \underline{q}^{n}\, v_{n} \equiv \underline{q}^{n}\, qT_{n} \equiv \underline{q}^{n}\, qt_{n}$$

Components of Velocity:

Global Velocity Constant / **Local Velocity**

$$u_{N} \equiv pt_{N} = pT_{N} \qquad v_{N} \equiv qt_{N} = qT_{N} \neq 0$$
$$u^{N} \equiv pt^{N} = pT^{N} \qquad v^{N} \equiv qt^{N} = qT^{N} \neq 0$$

Global Gradient:

Global Gradient Vector of Velocity:

Global Gradient of Global Velocity≡0 / **Global Gradient of Local Velocity**

Global Coordinate Reference / **Global Convection**

$$up_{N}{}^{;K}\, \underline{p}_{K} \equiv up_{N}{}^{K}\, \underline{p}_{K} \equiv 0 \qquad vp_{N}{}^{;K}\, \underline{p}_{K} \equiv vp_{N}{}^{n}\, \underline{p}_{n} = vq_{N}{}^{K}\, Q_{K}{}^{n}\, \underline{p}_{n}$$
$$up^{N}{}_{;K}\, \underline{p}^{K} \equiv up^{N}{}_{K}\, \underline{p}^{K} \equiv 0 \qquad vp^{N}{}_{;K}\, \underline{p}^{K} \equiv vp^{N}{}_{n}\, \underline{p}^{n} = vq^{N}{}_{K}\, Q^{K}{}_{n}\, \underline{p}^{n}$$

And:

Global Convection: Global Gradient Components of Local Velocity

$$qTp_{N}{}^{;n} \equiv QT_{N}{}^{;n} \equiv v_{N}{}^{;n} \equiv vp_{N}{}^{;n} \equiv vp_{N}{}^{n} = vq_{N}{}^{K}\, Q_{K}{}^{n} \equiv v_{N}{}^{K}\, Q_{K}{}^{n} \equiv L_{N}{}^{K}\, Q_{K}{}^{n}$$
$$qTp^{N}{}_{;n} \equiv QT^{N}{}_{;n} \equiv v^{N}{}_{;n} \equiv vp^{N}{}_{;n} \equiv vp^{N}{}_{n} = vq^{N}{}_{K}\, Q^{K}{}_{n} \equiv v^{N}{}_{K}\, Q^{K}{}_{n} \equiv L^{N}{}_{K}\, Q^{K}{}_{n}$$

Local Gradient:

Local Gradient Vector of Velocity:

Local Gradient of Local Velocity

Local Convection

$$vq_{N}{}^{,n}\, \underline{q}_{n} \equiv vq_{N}{}^{n}\, \underline{q}_{n} \equiv v_{N}{}^{,n}\, \underline{q}_{n} \equiv v_{N}{}^{n}\, \underline{q}_{n}$$
$$vq^{N}{}_{,n}\, \underline{q}^{n} \equiv vq^{N}{}_{n}\, \underline{q}^{n} \equiv v^{N}{}_{,n}\, \underline{q}^{n} \equiv v^{N}{}_{n}\, \underline{q}^{n}$$

And:

Local Convection: Local Gradient Components of Local Velocity

$$L_{N}{}^{K} \equiv v_{N}{}^{K} \equiv vq_{N}{}^{K} \equiv \partial(dqT_{N})/\partial q_{K} \neq 0$$
$$L^{N}{}_{K} \equiv v^{N}{}_{K} \equiv vq^{N}{}_{K} \equiv \partial(dqT^{N})/\partial q^{K} \neq 0$$

Global Gradient and Local Gradient are Defining:

Global Convection: / **Local Convection:**

$$K^{N}{}_{n} \equiv vp^{N}{}_{n} \equiv vq^{N}{}_{K}\, Q^{K}{}_{n} \equiv v^{N}{}_{K}\, Q^{K}{}_{n} \qquad L^{N}{}_{n} \equiv v^{N}{}_{n} \equiv vq^{N}{}_{n} \equiv vp^{N}{}_{K}\, P^{K}{}_{n}$$
$$K_{N}{}^{n} \equiv vp_{N}{}^{n} \equiv vq_{N}{}^{K}\, Q_{K}{}^{n} \equiv v_{N}{}^{K}\, Q_{K}{}^{n} \qquad L_{N}{}^{n} \equiv v_{N}{}^{n} \equiv vq_{N}{}^{n} \equiv vp_{N}{}^{K}\, P_{K}{}^{n}$$

Or:

$$K^{n}{}_{N} = L^{n}{}_{K}\, Q^{K}{}_{N} \qquad L^{n}{}_{N} = K^{n}{}_{K}\, P^{K}{}_{N}$$
$$vp^{n}{}_{N} = v^{n}{}_{K}\, Q^{K}{}_{N} \qquad v^{n}{}_{N} = vp^{n}{}_{K}\, P^{K}{}_{N}$$
$$K_{N}{}^{n} = L_{N}{}^{K}\, Q_{K}{}^{n} \qquad L_{N}{}^{n} = K_{N}{}^{K}\, P_{K}{}^{n}$$
$$vp_{N}{}^{n} = v_{N}{}^{K}\, Q_{K}{}^{n} \qquad v_{N}{}^{n} = vp_{N}{}^{K}\, P_{K}{}^{n}$$

And:

Trace of Global Convection: / **Trace of Local Convection:**

$$K \equiv K^{n}_{n} \equiv vp^{n}_{n} = K^{n}_{n} \equiv vp^{n}_{n} \qquad L \equiv L^{n}_{n} \equiv vq^{n}_{n} \equiv v^{n}_{n} = L^{n}_{n} \equiv vq^{n}_{n} \equiv v^{n}_{n}$$

$$vp \equiv vp^{n}_{n} = vp^{n}_{n} \qquad v \equiv v^{n}_{n} = v^{n}_{n}$$

$\underline{F}$ Surface Traction (Force per Area) Is Symmetric:

$$F^{m}_{n} = F^{n}_{m} \qquad F_{m}^{n} = F_{n}^{m}$$

Define:

$$\tau_{pq} \equiv P^{i}_{p} F^{i}_{j} P^{j}_{q} \qquad \tau^{pq} \equiv P^{p}_{i} F_{i}^{j} P^{q}_{j}$$

Then:

$$Q_{m}^{p} \tau_{pq} Q_{n}^{q} = Q_{m}^{p} P^{i}_{p} F^{i}_{j} P^{j}_{q} Q^{q}_{n} = \delta^{i}_{m} F^{i}_{j} \delta^{j}_{n} = F^{m}_{n} \qquad Q^{m}_{p} \tau_{pq} Q^{n}_{q} = Q^{m}_{p} P^{p}_{i} F_{i}^{j} P^{q}_{j} Q^{n}_{q} = \delta^{m}_{i} F_{i}^{j} \delta^{n}_{j} = F^{n}_{m}$$

14.1. Symmetric Finite Deformation Power:

Since F Is Symmetric:

$$F^{I}_{J} v_{I}^{,J} = ½ (F^{J}_{I} v_{J}^{,I} + F^{I}_{J} v_{I}^{,J})$$

Where:

$$v_{J}^{,I} = v_{J}^{;p} P^{I}_{p} = [v_{k}^{;p} \delta^{J}_{k}] P^{I}_{p}$$

$$v_{I}^{,J} = v_{I}^{;q} P^{J}_{q} = [v_{k}^{;q} \delta^{I}_{k}] P^{J}_{q}$$

Rewriting δ_{k}^{j} and δ_{k}^{i} In Two Reciprocal Forms:

$$\delta_{k}^{j} = (Q_{k}^{q} P^{j}_{q})$$

$$\delta_{k}^{i} = (P^{i}_{p} Q_{k}^{p})$$

Then:

$$v_{J}^{,I} = [v_{k}^{;p} (Q_{k}^{q} P^{J}_{q})] P^{I}_{p} \equiv [QT_{k}^{p} (Q_{k}^{q} P^{J}_{q})] P^{I}_{p}$$

$$v_{I}^{,J} = [v_{k}^{;q} (P^{I}_{p} Q_{k}^{p})] P^{J}_{q} \equiv [QT_{k}^{q} (P^{I}_{p} Q_{k}^{p})] P^{J}_{q}$$

To get:

$$F^{I}_{J} v_{I}^{,J} \equiv ½ (F^{J}_{I} v_{J}^{,I} + F^{I}_{J} v_{I}^{,J}) =$$

$$½[F^{J}_{I} v_{k}^{;p} (Q_{k}^{q} P^{J}_{q}) P^{I}_{p} + F^{I}_{J} v_{k}^{;q} (P^{I}_{p} Q_{k}^{p}) P^{J}_{q}]$$

$$= \text{Rearranged} =$$

$$½[v_{k}^{;p} Q_{k}^{q} (P^{I}_{p} F^{I}_{J} P^{J}_{q}) + Q_{k}^{p} v_{k}^{;q} (P^{I}_{p} F^{I}_{J} P^{J}_{q})] =$$

$$½ (v_{k}^{;p} Q_{k}^{q} + Q_{k}^{p} v_{k}^{;q}) P^{I}_{p} F^{I}_{J} P^{J}_{q} =$$

$$½ (QT_{k}^{p} Q_{k}^{q} + Q_{k}^{p} QT_{k}^{q}) (P^{I}_{p} F^{I}_{J} P^{J}_{q})$$

Introduce:

Stress τ and Strain E:

$$\tau_{pq} \equiv P^{i}_{p} F^{i}_{j} P^{j}_{q} \qquad \tau^{pq} \equiv P^{p}_{i} F_{i}^{j} P^{q}_{j}$$

$$E^{pq} \equiv ½ (Q_{k}^{p} Q_{k}^{q}) \qquad E_{pq} \equiv ½ (Q^{k}_{p} Q^{k}_{q})$$

And ET:

$$ET^{pq} \equiv ½ (QT_{k}^{p} Q_{k}^{q} + Q_{k}^{p} QT_{k}^{q}) \qquad ET_{pq} \equiv ½ (QT^{k}_{p} Q^{k}_{q} + Q^{k}_{p} QT^{k}_{q})$$

Also Introduce: τ0:

$$\tau0 \equiv \rho0/\rho\ \tau \qquad \tau/\rho \equiv \tau0/\rho0 \qquad \tau \equiv \rho/\rho0\ \tau0$$

So:

$F^I{}_J v_I{}^{,J} \equiv ½ (F^J{}_I v_J{}^{,I} + F^I{}_J v_I{}^{,J})$ $\quad$ $F_I{}^J v^I{}_J \equiv ½ (F_I{}^J v^I{}_J + F_I{}^{,J} v^I{}_J)$

$=$ $\quad$ $=$

$ET^{pq} \quad \tau_{pq} \equiv (ET^{pq})(\rho/\rho 0 \quad \tau 0_{pq})$ $\quad$ $ET_{pq} \quad \tau^{pq} \equiv (ET_{pq})(\rho/\rho 0 \quad \tau 0^{pq})$

This Is Substituted In: Specific Power Minus Specific Heat Supply : ΞT-H:

$F^i{}_n \; v_i{}^{,n} = \Xi T - H \equiv \rho \; eT - H = F_i{}^n \; v^i{}_{,n}$

To Get:

Specific Power Minus Specific Heat Supply : ΞT-H:

$F^I{}_J v_I{}^{,J} \equiv ½ (F^i{}_j v_i{}^{,j} + F^i{}_j v_i{}^{,j}) \equiv ET^{pq} \quad \tau_{pq} \equiv (ET^{pq})(\rho/\rho 0 \quad \tau 0_{pq})$

$= \Xi T - H \equiv \rho \; eT - H =$

$F_I{}^J v^I{}_J \equiv ½ (F_i{}^j v^i{}_j + F_i{}^{,j} v^i{}_j) \equiv ET_{pq} \quad \tau^{pq} \equiv (ET_{pq})(\rho/\rho 0 \quad \tau 0^{pq})$

Now Introduce: E and τT:

$\tau T_{pq} \equiv (P_p{}^i \; F^i{}_j \quad P_q{}^j)T$ $\quad$ $\tau T^{pq} \equiv (P^p{}_i \; F_i{}^j \quad P^q{}_j)T$

$E^{pq} \equiv ½ (Q_k{}^p Q_k{}^q - \delta^{pq})$ $\quad$ $E_{pq} \equiv ½ (Q^k{}_p Q^k{}_q - \delta_{pq})$

Also Introduce: τ0T:

$\tau 0T \equiv \rho 0/\rho \; \tau T$ $\quad$ $\tau T/\rho \equiv \tau 0T/\rho 0$ $\quad$ $\tau T \equiv \rho/\rho 0 \; \tau 0T$

Specific Power Minus Specific Heat Supply : ΞT-H: Power / Volume:

$ET^{pq} \quad \tau_{pq} = \Xi T - H = \rho \quad eT - H = ET_{pq} \; \tau^{pq}$

Plus:

Complementary Specific Power Plus Specific Heat Supply : ΨT+H:

Complementary Power / Volume:

$E^{pq} \quad \tau T_{pq} = \Psi T + H = \rho T \quad e + H = E_{pq} \quad \tau T^{pq}$

Equal:

Specific Power Plus Complementary Specific Power:

Power and Complementary Power / Volume:

ΞT+ΨT:

$(E^{pq} \quad \tau_{pq})T = \Xi T + \Psi T = \rho eT = (E_{pq} \; \tau^{pq})T$

Divergence of Heat Gradient: Convection of Heat Supply: $h_n{}^{,n}{}_n \equiv H \equiv h^n{}_{,n}$

Divergence of Velocity Gradient: Convection of Velocity: $vq^n{}_n \equiv L \equiv vq_n{}^n$

Energy / Volume: Differentials:

$dE^{pq} \quad \tau_{pq} = d\Xi - H^*dt = \rho \quad de - H^*dt = dE_{pq} \; \tau^{pq}$

$E^{pq} \quad d\tau_{pq} = d\Psi + H^*dt = d\rho \quad e + H^*dt = E_{pq} \quad d\tau^{pq}$

$d(E^{pq} \quad \tau_{pq}) = d\Xi + d\Psi = d(\rho e) = d(E_{pq} \; \tau^{pq})$

So:

Stress and Strain are derived from (Ξ + Ψ) = ρ (e + c)

$\tau^{kl} = \rho \; \partial e/\partial E_{kl} = \partial \Xi/\partial E_{kl}$ $\quad$ $\tau_{kl} = \rho \; \partial e/\partial E^{kl} = \partial \Xi/\partial \tau^{kl}$

$E^{kl} = \rho \; \partial c/\partial \tau_{kl} = \partial \Psi/\partial \tau_{kl}$ $\quad$ $E_{kl} = \rho \; \partial c/\partial \tau^{kl} = \partial \Psi/\partial \tau^{kl}$

14.2.
Strain and Stress are Reciprocals:

Stress τ And Strain E:

$\tau_{pq} \equiv P_p{}^i \; F^i{}_j \quad P_q{}^j$ $\quad$ $\tau^{pq} \equiv P^p{}_i \; F_i{}^j \quad P^q{}_j$

$E^{pq} \equiv ½ (Q_k{}^p Q_k{}^q)$ $\quad$ $E_{pq} \equiv ½ (Q^k{}_p Q^k{}_q)$

And

Stress Rate τT And Strain Rate ET:

$$\tau T_{pq} \equiv (P^i_p F^i_j P^j_q)T \qquad \tau T^{pq} \equiv (P^p_i F^j_i P^q_j)T$$

$$ET^{pq} \equiv ½(QT_k^p Q_k^q + Q_k^p QT_k^q) \qquad ET_{pq} \equiv ½(QT^k_p Q^k_q + Q^k_p QT^k_q)$$

Where:

$$QT_i^m P_m^n \equiv v_i^{;m} P_m^n = v_i^{,n} \qquad v^i_{,n} = v^i_{;m} P^m_n \equiv QT^i_m P^m_n$$

τ And ET:

$$\tau_{pq} \equiv P^i_p F^i_j P^j_q \qquad \tau^{pq} \equiv P^p_i F^j_i P^q_j$$

$$ET^{pq} \equiv ½(QT_k^p Q_k^q + Q_k^p QT_k^q) \qquad ET_{pq} \equiv ½(QT^k_p Q^k_q + Q^k_p QT^k_q)$$

τT And E:

$$\tau T_{pq} \equiv (P^i_p F^i_j P^j_q)T \qquad \tau T^{pq} \equiv (P^p_i F^j_i P^q_j)T$$

$$E^{pq} \equiv ½(Q_k^p Q_k^q - \delta^{pq}) \qquad E_{pq} \equiv ½(Q^k_p Q^k_q - \delta_{pq})$$

Power / Volume: Stress/time:

$$ET^{pq} \tau_{pq} = \Xi T - H = \rho \quad eT - H = ET_{pq} \tau^{pq}$$

$$E^{pq} \tau T_{pq} = \Psi T + H = \rho T \quad e + H = E_{pq} \tau T^{pq}$$

$$(E^{pq} \tau_{pq})T = \Xi T + \Psi T = \rho eT = (E_{pq} \tau^{pq})T$$

Energy / Volume: Differentials: Stress:

$$dE^{pq} \tau_{pq} = d\Xi - H^* dt = \rho \quad de - H^* dt = dE_{pq} \tau^{pq}$$

$$E^{pq} d\tau_{pq} = d\Psi + H^* dt = d\rho \quad e + H^* dt = E_{pq} d\tau^{pq}$$

$$d(E^{pq} \tau_{pq}) = d\Xi + d\Psi = d(\rho e) = d(E_{pq} \tau^{pq})$$

Convection of Heat Supply: Stress/time: $h_n{}^{,n} \equiv h^n{}_{,n} \equiv \underline{\nabla} \bullet \underline{h} \equiv H \equiv \rho$ **heat**

Divergence of Velocity Gradient: Convection of Velocity:1/time: $vq^n{}_n \equiv L \equiv vq_n{}^n$

14.3.
Associated First Order Tensors With Second Order Tensors:

Assume:

(First Order Tensor, A Is Associated With Second Order Tensor, τ)

And

(First Order Tensor, B Is Associated With Second Order Tensor, E)

:

Stress Stiffness (Resilience):

Stress = Stress * 1

$$\tau^{kl} = \tau^{kl}(E_{kl}, B_p) \qquad \tau_{kl} = \tau_{kl}(E^{kl}, B^p)$$

Strain Flexibility (Resilience):

1 = 1 /Stress * Stress

$$E^{kl} = E^{kl}(\tau_{kl}, A_p) \qquad E_{kl} = E_{kl}(\tau^{kl}, A^p)$$

Stress and Strain are Reciprocal functions of each other•

Time Rate of Stress Elastic Stiffness (Resilience):

Stress/ Time = Stress * 1/ Time

$$\tau T^{kl} = \tau T^{kl}(E_{kl}, B_p) \qquad \tau T_{kl} = \tau T_{kl}(E^{kl}, B^p)$$

Time Rate of Strain Elastic Flexibility (Resilience):

1/ Time = 1 /Stress * Stress/ Time

$$ET^{kl} = ET^{kl}(\tau_{kl}, A_p) \qquad ET_{kl} = ET_{kl}(\tau^{kl}, A^p)$$

Differential:

$$d\tau^{kl} = \partial\tau^{kl}/\partial E_{mn}\, dE_{mn} + \partial\tau^{kl}/\partial B_r\, dB_r \qquad d\tau_{kl} = \partial\tau_{kl}/\partial E^{mn}\, dE^{mn} + \partial\tau_{kl}/\partial B^r\, dB^r$$

$dE^{kl} = \partial E^{kl}/\partial\tau_{mn}\, d\tau_{mn} + \partial E^{kl}/\partial A_c\, dA_c$ $\qquad$ $dE_{kl} = \partial E_{kl}/\partial\tau^{mn}\, d\tau^{mn} + \partial E_{kl}/\partial A^c\, dA^c$

Divide by dt:

$\tau T^{kl} = \partial\tau^{kl}/\partial E_{mn}\, ET_{mn} + \partial\tau^{kl}/\partial B_r\, BT_r$ $\qquad$ $\tau T_{kl} = \partial\tau_{kl}/\partial E^{mn}\, ET^{mn} + \partial\tau_{kl}/\partial B^r\, BT^r$

$ET^{kl} = \partial E^{kl}/\partial\tau_{mn}\, \tau T_{mn} + \partial E^{kl}/\partial A_c\, AT_c$ $\qquad$ $ET_{kl} = \partial E_{kl}/\partial\tau^{mn}\, \tau T^{mn} + \partial E_{kl}/\partial A^c\, AT^c$

Or:

$\tau T^{kl} = \tau E^{klmn}\, ET_{mn} + \tau E^{klr}\, BT_r$ $\qquad$ $\tau T_{kl} = \tau E_{klmn}\, ET^{mn} + \tau E_{klr}\, BT^r$

$ET^{kl} = ET^{klmn}\, \tau T_{mn} + ET^{klc}\, AT_c$ $\qquad$ $ET_{kl} = ET_{klmn}\, \tau T^{mn} + ET_{klc}\, AT^c$

Or:

$\tau T^{kl} = K^{klmn}\, ET_{mn} + \Pi^{klr}\, BT_r$ $\qquad$ $\tau T_{kl} = K_{klmn}\, ET^{mn} + \Pi_{klr}\, BT^r$

$ET^{kl} = \Gamma^{klmn}\, \tau T_{mn} + \Lambda^{klc}\, AT_c$ $\qquad$ $ET_{kl} = \Gamma_{klmn}\, \tau T^{mn} + \Lambda_{klc}\, AT^c$

Or:

Stress/ Time = Stress * 1/ Time

$\tau T^{kl} = \tau T^{kl}(E_{kl}, B_p)$ $\qquad$ $\tau T_{kl} = \tau T_{kl}(E^{kl}, B^p)\, \tau T_{kl}$

$= \tau E^{klmn}\, ET_{mn} + \tau E^{klr}\, BT_r$ $\qquad$ $= \tau E_{klmn}\, ET^{mn} + \tau E_{klr}\, BT^r$

$= K^{klmn}\, ET_{mn} + \Pi^{klr}\, BT_r$ $\qquad$ $= K_{klmn}\, ET^{mn} + \Pi_{klr}\, BT^r$

And:

1/ Time = 1 /Stress * Stress/ Time

$ET^{kl} = ET^{kl}(\tau_{kl}, A_p)$ $\qquad$ $ET_{kl} = ET_{kl}(\tau^{kl}, A^p)$

$= ET^{klmn}\, \tau T_{mn} + ET^{klc}\, AT_c$ $\qquad$ $= ET_{klmn}\, \tau T^{mn} + ET_{klc}\, AT^c$

$= \Gamma^{klmn}\, \tau T_{mn} + \Lambda^{klc}\, AT_c$ $\qquad$ $= \Gamma_{klmn}\, \tau T^{mn} + \Lambda_{klc}\, AT^c$

14.4.
Stress:

Differentials:

$$dE^{pq}\ \tau_{pq} = d\Xi - H^*dt = \rho\ de - H^*dt = dE_{pq}\ \tau^{pq}$$

$$E^{pq}\ d\tau_{pq} = d\Psi + H^*dt = d\rho\ e + H^*dt = E_{pq}\ d\tau^{pq}$$

$$d(E^{pq}\ \tau_{pq}) = d\Xi + d\Psi = d(\rho e) = d(E_{pq}\ \tau^{pq})$$

Stress/time:

Split Internal Power/Volume Into Reversible And Irreversible Parts:

$\Xi \equiv \rho\ e = \rho$ e reversible (elastic) + ρ e irreversible (inelastic)

$d\Xi \equiv \rho\ de$

Split Complementary Power/Volume Into Reversible And Irreversible Parts:

$\Psi \equiv \rho\ c = \rho$ c reversible (elastic) + ρ c irreversible (inelastic)

$d\Psi \equiv \rho\ dc$

Assume:

Irreversible Or Inelastic Dissipation of Power / Mass Is: D: Velocity^2/Time:

So:

Irreversible Or Inelastic Dissipation of Power / Volume Is: ρ D: Stress/Time:

Then:

Power/Mass	**Velocity^2/Time**	**eT**	**heat**	**D**	**W/kg**	**(m/s)^2/s**
Power/Volume	**Stress/Time**	**$\Xi T=\rho$ eT**	**H=ρ heat**	**ρ D**	**W/m^3**	**Pa/s**

And:

Energy/Mass	**Velocity^2**	**de**	**heat dt**	**D dt**	**J/kg**	**(m/s)^2**
Energy/Volume	**Stress**	**$d\Xi=\rho$ de**	**H*dt=ρ heat dt**	**ρ D dt**	**J/m^3**	**Pa**

Differential: Stress:

$$dE^{kl}\ \tau_{kl} + \rho\ \text{heat}\ dt - \rho\ D\ dt = d\Xi \equiv \rho\ de = dE_{kl}\ \tau^{kl} + \rho\ \text{heat}\ dt - \rho\ D\ dt$$

$$E^{kl}\, d\tau_{kl} - \rho\, heat\, dt + \rho\ D\, dt \ = d\Psi \equiv d\rho\ e = E_{kl}\ d\tau^{kl} - \rho\, heat\ dt + \rho\ \ D\, dt$$
$$d(E^{kl}\tau_{kl}) \ = d(\Xi + \Psi) \ = \ d(E_{kl}\tau^{kl})$$

Divide by dt: Internal and Complementary Time Derivative:

$$ET^{kl}\tau_{kl} + \rho\, heat - \rho\, D \ = \Xi T = ET_{kl}\tau^{kl} + \rho\, heat\ - \rho\, D$$
$$E^{kl}\tau T_{kl} - \rho\, heat + \rho\, D \ = \Psi T = E_{kl}\ \tau T^{kl} - \rho\, heat + \rho\, D$$
$$(E^{kl}\tau_{kl})T = (\Xi + \Psi)T = \ \rho = (e + c)T = (E_{kl}\tau^{kl})T$$

If heat=0:

$$\Xi T \equiv \rho\ eT \ = ET^{kl}\tau_{kl} - \rho\ D = ET_{kl}\tau^{kl} - \rho\ D$$
$$\rho\ D = ET^{kl}\ \tau_{kl} - \rho\ eT = E_{kl}\tau T^{kl} - \rho\ eT \geq 0$$
$$= ET^{kl}\ \tau_{kl} - (\rho\ \partial e/\partial E_{kl}\ ET_{kl} + \rho\ \partial e/\partial B_r\ BT_r)$$
$$= \ ET^{kl}\ \tau_{kl} - ET^{kl}\ \tau_{kl} - \rho\ \partial e/\partial B_r\ BT_r = -\rho\ \partial e/\partial B_r\ BT_r \geq 0$$

And:

$$\Psi T \equiv \rho\ cT = E^{kl}\tau T_{kl} \ + \ \rho\ D = E_{kl}\tau T^{kl} \ + \ \rho\ D$$
$$\rho\ D = \rho\ cT - E^{kl}\ \tau T_{kl} = \rho\ cT - E_{kl}\tau T^{kl} \geq 0$$
$$= \rho\ \partial c/\partial \tau_{kl}\ \ \tau T_{kl} + \rho\ \ \partial c/\partial A_r\ \ AT_r - E^{kl}\ \ \tau T_{kl}$$
$$= \ E^{kl}\ \ \tau T_{kl} + \rho\ \partial c/\partial A_r\ \ AT_r - E^{kl}\ \ \tau T_{kl} = \rho\ \ \partial c/\partial A_r\ \ AT_r \geq 0$$

Then:

$$-\partial \Xi/\partial B_r\ \ BT_r = -\rho\ \ \partial e/\partial B_r\ \ BT_r \geq 0 \qquad -\partial \Xi/\partial B^r\ \ BT^r = -\rho\ \ \partial e/\partial B^r\ \ BT^r \geq 0$$
$$+\partial \Psi/\partial A_r\ \ AT_r = +\rho\ \ \partial c/\partial A_r\ \ AT_r \geq 0 \qquad +\partial \Psi/\partial A^r\ \ AT^r = +\rho\ \ \partial c/\partial A^r\ \ AT^r \geq 0$$

Identically:

Stress/ Time = Stress * 1/ Time

$$+\partial \Xi/\partial B_r\ \ BT_r = +\rho\ \ \partial e/\partial B_r\ \ BT_r \leq 0 \qquad +\partial \Xi/\partial B^r\ \ BT^r = +\rho\ \ \partial e/\partial B^r\ \ BT^r \leq 0$$
$$+\partial \Psi/\partial A_r\ \ AT_r = +\rho\ \ \partial c/\partial A_r\ \ AT_r \geq 0 \qquad +\partial \Psi/\partial A^r\ \ AT^r = +\rho\ \ \partial c/\partial A^r\ \ AT^r \geq 0$$

Where:

$$\Xi(E^{kr}, B^r) = \Xi(E^{ii}, E^{ij}E^{ji}, B^r) = \Xi = \Xi(E_{ii}, E_{ij}E_{ji}, B_r) = \Xi(E_{kl}, B_r)$$
$$\Psi(\tau_{kl}, A_r) = \Psi(\tau_{ii}, \tau_{ij}\tau_{ji}, A_r) = \Psi = \Psi(\tau^{ii}, \tau^{ij}\tau^{ji}, A^r) = \Psi(\tau^{kr}, A^r)$$

14.5.

Strain and Stress Time Derivatives

Strain and Stress Time derivatives are Time Derivatives of Derived Forms

Since:

Strain and Stress are derived from (X + Ψ)

$$\tau^{kl}(E_{kl}, B_p) = \partial X/\partial E_{kl} \qquad \tau_{kl}(E^{kl}, B^p) = \partial X/\partial E^{kl}$$
$$E^{kl}(\tau_{kl}, A_p) = \partial \Psi/\partial \tau_{kl} \qquad E_{kl}(\tau^{kl}, A^p) = \partial \Psi/\partial \tau^{kl}$$

And:

$$d\tau^{kl} = \partial \tau^{kl}/\partial E_{mn}\ dE_{mn} + \partial \tau^{kl}/\partial B_r\ dB_r \qquad d\tau_{kl} = \partial \tau_{kl}/\partial E^{mn}\ dE^{mn} + \partial \tau_{kl}/\partial B^r\ dB^r$$
$$dE^{kl} = \partial E^{kl}/\partial \tau_{mn}\ d\tau_{mn} + \partial E^{kl}/\partial A_c\ dA_c \qquad dE_{kl} = \partial E_{kl}/\partial \tau^{mn}\ d\tau^{mn} + \partial E_{kl}/\partial A^c\ dA^c$$

To get Differential:

$$d\tau^{kl} = \rho\ \partial\partial e / \partial E_{kl} / \partial E_{mn}\ dE_{mn} + \qquad d\tau_{kl} = \rho\ \partial\partial e / \partial E^{kl} / \partial E^{mn}\ dE^{mn} +$$
$$+ \rho\ \partial\partial e / \partial E_{kl} / \partial B_r\ dB_r \qquad + \rho\ \partial\partial e / \partial E^{kl} / \partial B^r\ dB^r$$
$$dE^{kl} = \rho\ \partial\ \partial c / \partial\tau_{kl} / \partial\tau_{mn}\ d\tau_{mn} + \qquad dE_{kl} = \rho\ \partial\partial c / \partial\tau^{kl} / \partial\tau^{mn}\ d\tau^{mn} +$$
$$+ \rho\ \partial\ \partial c / \partial\tau_{kl} / \partial A_c\ dA_c \qquad + \rho\ \partial\partial c / \partial\tau^{kl} / \partial A^c\ dA^c$$

$$d\tau^{kl} = \partial\partial\Xi / \partial E_{kl} / \partial E_{mn}\ dE_{mn} + \qquad d\tau_{kl} = \partial\partial\Xi / \partial E^{kl} / \partial E^{mn}\ dE^{mn} +$$
$$+ \partial\partial\Xi / \partial E_{kl} / \partial B_r\ dB_r \qquad + \partial\partial\Xi / \partial E^{kl} / \partial B^r\ dB^r$$
$$dE^{kl} = \partial\ \partial\Psi / \partial\tau_{kl} / \partial\tau_{mn}\ d\tau_{mn} + \qquad dE_{kl} = \partial\partial\Psi / \partial\tau^{kl} / \partial\tau^{mn}\ d\tau^{mn} +$$
$$+ \partial\ \partial\Psi / \partial\tau_{kl} / \partial A_c\ dA_c \qquad + \partial\partial\Psi / \partial\tau^{kl} / \partial A^c\ dA^c$$

Divide Differentials by dt:

Time derivative of Stress and Strain are Time derivative of derived forms

$$\tau T^{kl} = \rho\ \partial\partial e / \partial E_{kl} / \partial E_{mn}\ ET_{mn} + \qquad \tau T_{kl} = \rho\ \partial\partial e / \partial E^{kl} / \partial E^{mn}\ ET^{mn} +$$
$$+ \rho\ \partial\partial e / \partial E_{kl} / \partial B_r\ BT_r \qquad + \rho\ \partial\partial e / \partial E^{kl} / \partial B^r\ BT^r$$
$$ET^{kl} = \rho\ \partial\ \partial c / \partial\tau_{kl} / \partial\tau_{mn}\ \tau T_{mn} + \qquad ET_{kl} = \rho\ \partial\ \partial c / \partial\tau^{kl} / \partial\tau^{mn}\ \tau T^{mn} +$$
$$+ \rho\ \partial\ \partial c / \partial\tau_{kl} / \partial A_c\ AT_c \qquad + \rho\ \partial\partial c / \partial\tau^{kl} / \partial A^c\ AT^c$$

Or:

$$\tau T^{kl} = \rho\ eEE^{klmn}\ ET_{mn} + \rho\ eEB^{klr}\ BT_r \qquad \tau T_{kl} = \rho\ eEE_{klmn}\ ET^{mn} + \rho\ eEB_{klr}\ BT^r$$
$$ET^{kl} = \rho\ c\tau\tau^{klmn}\ \tau T_{mn} + \rho\ c\tau A^{klc}\ AT_c \qquad ET_{kl} = \rho\ c\tau\tau_{klmn}\ \tau T^{mn} + \rho\ c\tau A_{klc}\ AT^c$$

Or:

$$\tau T^{kl} = \Xi EE^{klmn}\ ET_{mn} + \Xi EB^{klr}\ BT_r \qquad \tau T_{kl} = \Xi EE_{klmn}\ ET^{mn} + \Xi EB_{klr}\ BT^r$$
$$ET^{kl} = \Psi\tau\tau^{klmn}\ \tau T_{mn} + \Psi\tau A^{klc}\ AT_c \qquad ET_{kl} = \Psi\tau\tau_{klmn}\ \tau T^{mn} + \Psi\tau A_{klc}\ AT^c$$

Or:

$$\tau T^{kl} = K^{klmn}\ ET_{mn} + \Pi^{klr}\ BT_r \qquad \tau T_{kl} = K_{klmn}\ ET^{mn} + \Pi_{klr}\ BT^r$$
$$ET^{kl} = \Gamma^{klmn}\ \tau T_{mn} + \Lambda^{klc}\ AT_c \qquad ET_{kl} = \Gamma_{klmn}\ \tau T^{mn} + \Lambda_{klc}\ AT^c$$

So:

$$\tau^{kl} \equiv \tau^{kl}(E_{kl}, B_p) = \partial X / \partial E_{kl} \qquad \tau_{kl} \equiv \tau_{kl}(E^{kl}, B^p) = \partial X / \partial E^{kl}$$
$$E^{kl} \equiv E^{kl}(\tau_{kl}, A_p) = \partial\Psi / \partial\tau_{kl} \qquad E_{kl} \equiv E_{kl}(\tau^{kl}, A^p) = \partial\Psi / \partial\tau^{kl}$$

Time Derivatives:

$$\tau T^{kl} \equiv \tau T^{kl}(E_{kl}, B_p) = (\partial X / \partial E_{kl})T \qquad \tau T_{kl} \equiv \tau T_{kl}(E^{kl}, B^p) = (\partial X / \partial E^{kl})T$$
$$ET^{kl} \equiv ET^{kl}(\tau_{kl}, A_p) = (\partial\Psi / \partial\tau_{kl})T \qquad ET_{kl} \equiv ET_{kl}(\tau^{kl}, A^p) = (\partial\Psi / \partial\tau^{kl})T$$

Or:

Stress/ Time = Stress * 1/ Time

$$\tau T^{kl} = \tau T^{kl}(E_{kl}, B_p) \qquad \tau T_{kl} = \tau T_{kl}(E^{kl}, B^p)\ \tau T_{kl}$$
$$= \rho\ eEE^{klmn}\ ET_{mn} + \rho\ eEB^{klr}\ BT_r \qquad = \rho\ eEE_{klmn}\ ET^{mn} + \rho\ eEB_{klr}\ BT^r$$
$$= \Xi EE^{klmn}\ ET_{mn} + \Xi EB^{klr}\ BT_r \qquad = \Xi EE_{klmn}\ ET^{mn} + \Xi EB_{klr}\ BT^r$$
$$= K^{klmn}\ ET_{mn} + \Pi^{klr}\ BT_r \qquad = K_{klmn}\ ET^{mn} + \Pi_{klr}\ BT^r$$

And:

1/ Time = 1 /Stress * Stress/ Time

$$ET^{kl} = ET^{kl}(\tau_{kl}, A_p) \qquad ET_{kl} = ET_{kl}(\tau^{kl}, A^p)$$
$$= \rho\ c\tau\tau^{klmn}\ \tau T_{mn} + \rho\ c\tau A^{klc}\ AT_c \qquad = \rho\ c\tau\tau_{klmn}\ \tau T^{mn} + \rho\ c\tau A_{klc}\ AT^c$$
$$= \Psi\tau\tau^{klmn}\ \tau T_{mn} + \Psi\tau A^{klc}\ AT_c \qquad = \Psi\tau\tau_{klmn}\ \tau T^{mn} + \Psi\tau A_{klc}\ AT^c$$
$$= \Gamma^{klmn}\ \tau T_{mn} + \Lambda^{klc}\ AT_c \qquad = \Gamma_{klmn}\ \tau T^{mn} + \Lambda_{klc}\ AT^c$$

Stress τ And Strain E:

$\tau^{kl} \equiv \tau^{kl}(E_{kl},B_p) \equiv P^k_i F^j_i P^k_j = \partial X/\partial E_{kl}$ $\quad\quad$ $\tau_{kl} \equiv \tau_{kl}(E^{kl},B^p) \equiv P^i_k F^i_j P^j_q = \partial X/\partial E^{kl}$

$E^{kl} \equiv E^{kl}(\tau_{kl},A_p) \equiv ½\,(Q^k_J\, Q^l_J) = \partial\Psi/\partial\tau_{kl}$ $\quad\quad$ $E_{kl} \equiv E_{kl}(\tau^{kl},A^p) \equiv ½\,(Q^J_k\, Q^J_l) = \partial\Psi/\partial\tau^{kl}$

Split Stress Rate τT And Strain Rate ET: Into: Elastic Part And Inelastic Part:

:

Stress Elastic Part

Stress/ Time = Stress * 1/ Time

τET^{kl} $\quad\quad$ τET_{kl}

And:

1/ Time = 1 /Stress * Stress/ Time

EET^{kl} $\quad\quad$ EET_{kl}

Stress/ Time = Stress * 1/ Time

$\tau T^{kl} = \tau ET^{kl} + \tau IT^{kl}$ $\quad\quad$ $\tau T_{kl} = \tau ET_{kl} + \tau IT_{kl}$

And:

1/ Time = 1 /Stress * Stress/ Time

$ET^{kl} = EET^{kl} + EIT^{kl}$ $\quad\quad$ $ET_{kl} = EET_{kl} + EIT_{kl} =$

(m . NL . n) Series : E Sum

Stress/ Time = Stress * 1/ Time

$\tau T^{kl} = \tau ET^{kl} + \tau IT^{kl}$ $\quad\quad$ $\tau T_{kl} = \tau ET_{kl} + \tau IT_{kl}$

And:

1/ Time = 1 /Stress * Stress/ Time

$ET^{kl} = EET^{kl} + EIT^{kl} =$ $\quad\quad$ $ET_{kl} = EET_{kl} + EIT_{kl} =$

Then:

Stress/ Time = Stress * 1/ Time

$$\begin{aligned}\tau T^{kl} &= \tau T^{kl}(E_{kl}, B_p) \\ &= \tau ET^{kl} + \tau IT^{kl} \\ &= \rho\, eEE^{klmn}\, ET_{mn} + \rho\, eEB^{klr}\, BT_r \\ &= \Xi EE^{klmn}\, ET_{mn} + \Xi EB^{klr}\, BT_r \\ &= K^{klmn}\, ET_{mn} + \Pi^{klr}\, BT_r\end{aligned}$$

$$\begin{aligned}\tau T_{kl} &= \tau T_{kl}(E^{kl}, B^p)\, \tau T_{kl} \\ &= \tau ET_{kl} + \tau IT_{kl} \\ &= \rho\, eEE_{klmn}\, ET^{mn} + \rho\, eEB_{klr}\, BT^r \\ &= \Xi EE_{klmn}\, ET^{mn} + \Xi EB_{klr}\, BT^r \\ &= K_{klmn}\, ET^{mn} + \Pi_{klr}\, BT^r\end{aligned}$$

And:

1/ Time = 1 /Stress * Stress/ Time

$$\begin{aligned}ET^{kl} &= ET^{kl}(\tau_{kl}, A_p) \\ &= EET^{kl} + EIT^{kl} = \\ &= \rho\, c\tau\tau^{klmn}\, \tau T_{mn} + \rho\, c\tau A^{klc}\, AT_c \\ &= \Psi\tau\tau^{klmn}\, \tau T_{mn} + \Psi\tau A^{klc}\, AT_c \\ &= \Gamma^{klmn}\, \tau T_{mn} + \Lambda^{klc}\, AT_c\end{aligned}$$

$$\begin{aligned}ET_{kl} &= ET_{kl}(\tau^{kl}, A^p) \\ &= EET_{kl} + EIT_{kl} = \\ &= \rho\, c\tau\tau_{klmn}\, \tau T^{mn} + \rho\, c\tau A_{klc}\, AT^c \\ &= \Psi\tau\tau_{klmn}\, \tau T^{mn} + \Psi\tau A_{klc}\, AT^c \\ &= \Gamma_{klmn}\, \tau T^{mn} + \Lambda_{klc}\, AT^c\end{aligned}$$

14.6.
Elastic and Inelastic Strain Rates:

Elasticity:

:

τT Is Not Split:

τT^{kl} $\quad\quad$ τT_{kl}

Naturally:

τ Is Not Split:

τ^{kl} $\quad\quad$ τ_{kl}

And:

ET Is Not Split:

$$ET^{kl} = EET^{kl} \qquad ET_{kl} = EET_{kl}$$

$$ET^{kl} = \Gamma^{klmn} \; \tau T_{mn} \qquad ET_{kl} = \Gamma_{klmn} \; \tau T^{mn}$$

Naturally:

$$E^{kl} = \Gamma^{klmn} \; \tau_{mn} \qquad E_{kl} = \Gamma_{klmn} \; \tau^{mn}$$

:

:

First Order D.E.:

… … Γ * τT … … = … … C * ET … …

When C=1 Get Elasticity:

… … Γ * τT … … = … … ET … …

Naturally:

… … Γ * τ … … = … … E … …

And:

… … Γ * ∫τ dt … … = … … ∫E dt … …

Inelasticity:

:

τT Is Not Split:

$$\tau T^{kl} \qquad \tau T_{kl}$$

Whereas:

Only ET Is Split Such That:

$$ET^{kl} = EET^{kl} + EIT^{kl} \qquad ET_{kl} = EET_{kl} + EIT_{kl}$$

$$= \qquad =$$

$$ET^{kl} = \Gamma^{klmn} \; \tau T_{mn} + \Lambda^{klc} \; AT_c \qquad ET_{kl} = \Gamma_{klmn} \; \tau T^{mn} + \Lambda_{klc} \; AT^c$$

:

:

First Order D.E.:

… … Γ * τT + Λ * τ … = … … C * ET … …

When C=1 Get Inelasticity:

… … Γ * τT + Λ * τ … = … … ET … …

Where ET Strain Rate.

:

:

Stress τ **Strain ε**

:

First Order D.E.:

… … Γ * τT + Λ * τ … = … … C * ET … …

When C=1 Get Viscosity

… … Γ * τT + Λ * τ … = … … ET … …

Where ET Strain Rate.

:

Inelastic Derivative:

$$\overset{\cdot\cdot}{(-2)} \; T\Gamma * \tau TT \qquad \overset{\cdot}{(-1)} \; T\Lambda * \tau T \qquad \overset{\cdot\cdot}{(-2)} \; TC * ETT$$

:

… Γ * τTT + Λ * τT … … = … C * ETT … … …

Inelastic:

. (-1)	(0)		. (-1)
……	1/Stiff		……
……	Flex		Damp
Γ	Λ		C
=	=		=
1/c	1/k		1/Γ
Γ	Λ		C
*	*		*
τT	τ		ET

:

…… Γ * τT + Λ * τ … = … … C * ET ……

: :

Inelastic Integral:

. (-1)	(0)	∫ (+1)	(0)
	Γ/T	Λ/T	C/T
	=	=	=
1/m/T	1/c/T	1/k/T	1/Γ/T
1/m/T	Γ/T	Λ/T	C/T
*	*	*	*
τT	τ	∫ τ dt	E

……… Γ/T * τ + Λ/T * ∫ τ dt = ……… C/T * E …

:

Stiffness (Resilience) View:

R

Differential equation:

Γ * τT + Λ * τ

=

C * ET + 0

τ_R

RττR = 1 **Stiffness (Resilience) RτeS**

R = 0

Γ/Λ τT + [1] τ

=

C/Λ * ET + 0

0ττ0 = [1] **Stiffness (Resilience) 0τeS**

Stiffness (Resilience) 0τe0 = 0

Stiffness (Resilience) 0τe1 = C/Λ

R = 1

[1] τT + Λ/Γ τ

=

C/Γ * ET + 0

0ττ0 = 1 **Stiffness (Resilience) 0τeS**

Stiffness (Resilience) 0τe0 = 0

Stiffness (Resilience) 0τe1 = C/Γ

Flexibility (Resilience) View:
S
Differential equation:

$$\Gamma * \tau T \;+\; \Lambda * \tau$$
$$=$$
$$C * ET \;+\; 0$$
$$e_S$$

Flexibility (Resilience) $[\, e\tau_S{}^R \,] \equiv Se\tau R$ $\qquad [\, ee_S{}^S \,] \equiv SeeS = 1$

There is No S=0

There is Only
S=1

$$\Gamma / C * \tau T \;+\; \Lambda / C * \tau$$
$$=$$
$$[1] * ET \;+\; 0$$

Flexibility (Resilience) $1e\tau R$ $\qquad [\, ee_1{}^1 \,] \equiv 1ee1 = 1$

Flexibility (Resilience) $1e\tau 0 = \Lambda / C$
Flexibility (Resilience) $1e\tau 1 = \Gamma / C$

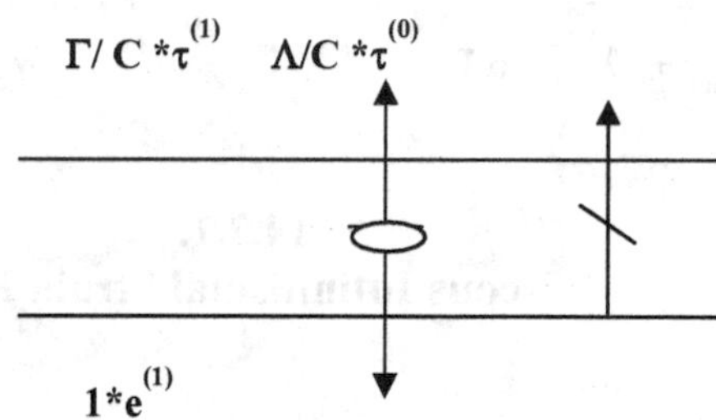

14.6.1.
Viscous Strain Rate:

τT Is Not Split:

$$\tau T^{kl} \qquad\qquad \tau T_{kl}$$

Whereas:
Only ET Is Split: Such That: Inelastic EIT = Viscous EVT :

$$ET^{kl} = EET^{kl} + EIT^{kl} \qquad ET_{kl} = EET_{kl} + EIT_{kl}$$
$$ET^{kl} = EET^{kl} + EVT^{kl} \qquad ET_{kl} = EET_{kl} + EVT_{kl}$$
$$ET^{kl} = \Gamma^{klmn} \; \tau T_{mn} + \Lambda^{klc} \; AT_c \qquad ET_{kl} = \Gamma_{klmn} \; \tau T^{mn} + \Lambda_{klc} \; AT^c$$
$$ET^{kl} = \Gamma^{klmn} \; \tau T_{mn} + \Lambda^{klmn} \; \tau_{mn} \qquad ET_{kl} = \Gamma_{klmn} \; \tau T^{mn} + \Lambda_{klmn} \; \tau^{mn}$$

14.6.2.
Plastic Strain Rate:

τT Is Not Split:

$$\tau T^{kl} \qquad \tau T_{kl}$$

Whereas:

Only ET Is Split: Such That: Inelastic EIT = Plastic EPT :

$$ET^{kl} = EET^{kl} + EIT^{kl} \qquad ET_{kl} = EET_{kl} + EIT_{kl}$$

$$ET^{kl} = EET^{kl} + EPT^{kl} \qquad ET_{kl} = EET_{kl} + EPT_{kl}$$

$$= \Gamma^{klmn}\ \tau T_{mn} + \Lambda^{klc}\ AT_c \qquad = \Gamma_{klmn}\ \tau T^{mn} + \Lambda_{klc}\ AT^c$$

$$= \Gamma^{klmn}\ \tau T_{mn} + \Lambda\ \sigma^{kl} \qquad = \Gamma_{klmn}\ \tau T^{mn} + \Lambda\ \sigma_{kl}$$

14.7.
Infinitesimal Strain Rates:

τT Is Not Split:

$$\tau T^{kl} \qquad \tau T_{kl}$$

Whereas:

Only εT Is Split Such That:

$$\varepsilon T^{kl} = \varepsilon ET^{kl} + \varepsilon IT^{kl} \qquad \varepsilon T_{kl} = \varepsilon ET_{kl} + \varepsilon IT_{kl}$$

$$= \qquad =$$

$$\gamma^{klmn}\ \tau T_{mn} + \lambda^{klc}\ \alpha T_c \qquad \gamma_{klmn}\ \tau T^{mn} + \lambda_{klc}\ \alpha T^c$$

14.7.1.
Viscous Infinitismal Strain Rate:

τT Is Not Split:

$$\tau T^{kl} \qquad \tau T_{kl}$$

Whereas:

Only εT Is Split: Such That: Inelastic εIT = Viscous εVT :

$$\varepsilon T^{kl} = \varepsilon ET^{kl} + \varepsilon IT^{kl} \qquad \varepsilon T_{kl} = \varepsilon ET_{kl} + \varepsilon IT_{kl}$$

$$\varepsilon T^{kl} = \varepsilon ET^{kl} + \varepsilon VT^{kl} \qquad \varepsilon T_{kl} = \varepsilon ET_{kl} + \varepsilon PT_{kl}$$

$$\varepsilon T^{kl} = \gamma^{klmn}\ \tau T_{mn} + \lambda^{klc}\ \alpha T_c \qquad \varepsilon T_{kl} = \gamma_{klmn}\ \tau T^{mn} + \lambda_{klc}\ \alpha T^c$$

$$\varepsilon T^{kl} = \gamma^{klmn}\ \tau T_{mn} + \lambda^{klmn}\ \tau_{mn} \qquad \varepsilon T_{kl} = \gamma_{klmn}\ \tau T^{mn} + \lambda_{klmn}\ \tau^{mn}$$

14.7.2.
Plastic Infinitismal Strain Rate:

τT Is Not Split:

$$\tau T^{kl} \qquad \tau T_{kl}$$

Whereas:

Only εT Is Split: Such That: Inelastic εIT = Plastic εPT :

$$\varepsilon T^{kl} = \varepsilon ET^{kl} + \varepsilon IT^{kl} \qquad \varepsilon T_{kl} = \varepsilon ET_{kl} + \varepsilon IT_{kl}$$

$$\varepsilon T^{kl} = \varepsilon ET^{kl} + \varepsilon PT^{kl} \qquad \varepsilon T_{kl} = \varepsilon ET_{kl} + \varepsilon PT_{kl}$$

$$\varepsilon T^{kl} = \gamma^{klmn} \ \tau T_{mn} + \lambda^{klc} \ \alpha T_c \qquad \varepsilon T_{kl} = \gamma_{klmn} \ \tau T^{mn} + \lambda_{klc} \ \alpha T^c$$

$$\varepsilon T^{kl} = \gamma^{klmn} \ \tau T_{mn} + \lambda \ \sigma^{kl} \qquad \varepsilon T_{kl} = \gamma_{klmn} \ \tau T^{mn} + \lambda \ \sigma_{kl}$$

14.8.
Specific Power Expansion Series:

Series e ≡ e (x):

$$e(x) = e0 + e1^I{}_m \ x_I{}^{,m} + 1/2! \ e2^{IJ}{}_{mn} \ x_I{}^{,m} \ x_J{}^{,n} + 1/3! \ e3^{IJK}{}_{mnp} \ x_I{}^{,m} \ x_J{}^{,n} \ x_K{}^{,p} + \ldots$$

$$e(x) = e0 + e1_I{}^m \ x^I{}_{,m} + 1/2! \ e2_{IJ}{}^{mn} \ x^I{}_{,m} \ x^J{}_{,n} + 1/3! \ e3_{IJK}{}^{mnp} \ x^I{}_{,m} \ x^J{}_{,n} \ x^K{}_{,p} + \ldots$$

Derivative of Series e ≡ e (x): ∂e(x)/∂x :

$$\partial e(x)/\partial x_I{}^{,m} = e1^I{}_m + 1/2! \ e2^{IJ}{}_{mn} \ x_J{}^{,n} + 1/3! \ e3^{IJK}{}_{mnp} \ x_J{}^{,n} \ x_K{}^{,p} + \ldots$$

$$\partial e(x)/\partial x^I{}_{,m} = e1_I{}^m + 1/2! \ e2_{IJ}{}^{mn} \ x^J{}_{,n} + 1/3! \ e3_{IJK}{}^{mnp} \ x^J{}_{,n} \ x^K{}_{,p} + \ldots$$

Multiplying By ρ To Get: ρ * ∂e/∂x :

$$\rho * \partial e(x)/\partial x_I{}^{,m} = \rho * e1^I{}_m + \rho * 1/2! \ e2^{IJ}{}_{mn} \ x_J{}^{,n} + \rho * 1/3! \ e3^{IJK}{}_{mnp} \ x_J{}^{,n} \ x_K{}^{,p} + \ldots$$

$$\rho * \partial e(x)/\partial x^I{}_{,m} = \rho * e1_I{}^m + \rho * 1/2! \ e2_{IJ}{}^{mn} \ x^J{}_{,n} + \rho * 1/3! \ e3_{IJK}{}^{mnp} \ x^J{}_{,n} \ x^K{}_{,p} + \ldots$$

Rewritten:

$$\rho * \partial e(x)/\partial x_I{}^{,m} = C1^I{}_m + C2^{IJ}{}_{mn} \ x_J{}^{,n} + C3^{IJK}{}_{mnp} \ x_J{}^{,n} \ x_K{}^{,p} + \ldots$$

$$\rho * \partial e(x)/\partial x^I{}_{,m} = C1_I{}^m + C2_{IJ}{}^{mn} \ x^J{}_{,n} + C3_{IJK}{}^{mnp} \ x^J{}_{,n} \ x^K{}_{,p} + \ldots$$

Time Derivative Of e(x) Is: eT (x):

$$eT(x) = \partial e/\partial (x_I{}^{,m}) \ * \ d(x_I{}^{,m})/dt = \partial e/\partial (x_I{}^{,m}) \ * \ xT_I{}^m$$

$$eT(x) = \partial e/\partial (x^I{}_{,m}) \ * \ d(x^I{}_{,m})/dt = \partial e/\partial (x^I{}_{,m}) \ * \ xT^I{}_{,m}$$

Multiplied By ρ To Get: ρ * eT ≡ ρ * eT (x):

$$\rho * eT(x) = \rho * \partial e/\partial (x_I{}^{,m}) \ * \ d(x_I{}^{,m})/dt = \rho * \partial e/\partial (x_I{}^{,m}) \ * \ xT_I{}^m$$

$$\rho * eT(x) = \rho * \partial e/\partial (x^I{}_{,m}) \ * \ d(x^I{}_{,m})/dt = \rho * \partial e/\partial (x^I{}_{,m}) \ * \ xT^I{}_{,m}$$

Substituting ρ * ∂e/∂x : From Above: To Get:
Series of Internal Power/Volume: (x) Specific Power Series:

$$\Xi T \equiv \rho * eT(x) = [C1^I{}_m + C2^{IJ}{}_{mn} \ x_J{}^{,n} + C3^{IJK}{}_{mnp} \ x_J{}^{,n} \ x_K{}^{,p} + \ldots] \ * \ xT_I{}^m$$

$$\Xi T \equiv \rho * eT(x) = [C1_I{}^m + C2_{IJ}{}^{mn} \ x^J{}_{,n} + C3_{IJK}{}^{mnp} \ x^J{}_{,n} \ x^K{}_{,p} + \ldots] \ * \ xT^I{}_{,m}$$

14.9.
Constitution Series:

Stress/Time:
Specific Power Minus Specific Heat Supply : ΞT-H:

$$F^I{}_J \ v_I{}^{,J} \equiv ½ \, (F^i{}_j \ v_i{}^{,j} + F^i{}_j \ v_i{}^{,j}) \equiv \ ET^{pq} \ \tau_{pq} \equiv (ET^{pq}) \, (\rho/\rho0 \ \tau0_{pq})$$

$$= \Xi T - H \equiv \rho \ eT - H =$$

$$F_I{}^J \ v^I{}_J \equiv ½ \, (F_i{}^j \ v^i{}_j + F_i{}^{,j} \ v^i{}_j) \equiv \ ET_{pq} \ \tau^{pq} \equiv (ET_{pq}) \, (\rho/\rho0 \ \tau0^{pq})$$

Or:

$$F^I{}_n \ v_I{}^{;m} \ P^n{}_m = F^I{}_n \ v_I{}^{,n} \equiv ET^{pq} \ \tau_{pq} = \Xi T - H = ET_{pq} \ \tau^{pq} \equiv F_I{}^n \ v^I{}_{,n} = F_I{}^n \ v^I{}_{;m} \ P^m{}_n$$

(x) : Stress:
Constitution Series Time Derivative: Stress/Time:
Stress/Time: [Unitless]*{ Stress/Time }:
Specific Power Series Minus Specific Heat Supply :

$$ET^{pq}\ \tau_{pq} = \Xi T - H = ET_{pq}\ \tau^{pq}$$

$$ET^{pq}\ \tau_{pq} = [C1^{I}{}_{m} + C2^{IJ}{}_{mn}\, x_{J}{}^{,n} + C3^{IJK}{}_{mnp}\, x_{J}{}^{,n}\, x_{K}{}^{,p} + \ldots]^{*}\ \{xT_{I}{}^{m}\} - H = ET_{pq}\ \tau^{pq}$$

$$ET^{pq}\ \tau_{pq} = [C1_{I}{}^{m} + C2_{IJ}{}^{mn}\, x^{J}{}_{,n} + C3_{IJK}{}^{mnp}\, x^{J}{}_{,n}\, x^{K}{}_{,p} + \ldots]^{*}\ \{xT^{I}{}_{m}\} - H = ET_{pq}\ \tau^{pq}$$

And:
(v) Unsymmetric Constitution Series Time Derivative: e[v(q,t)]:
Local Velocity

$$v_{N} \equiv qt_{N} \equiv \partial q_{N}/\partial t \qquad\qquad v^{N} \equiv qt^{N} \equiv \partial q^{N}/\partial t$$

Local Gradient of Local Velocity

$$v_{N}{}^{,n} = v_{N}{}^{;m}\ pq_{m}{}^{n} \equiv QT_{N}{}^{m}\ P_{m}{}^{n} \qquad\qquad v^{N}{}_{,n} = v^{N}{}_{;m}\ pq^{m}{}_{n} \equiv QT^{N}{}_{m}\ P^{m}{}_{n}$$

$$L_{N}{}^{n} \equiv vq_{N}{}^{n} \equiv vp_{N}{}^{K}\ pq_{K}{}^{n} = K_{N}{}^{K}\ P_{K}{}^{n} \qquad\qquad L^{N}{}_{n} \equiv vq^{N}{}_{n} \equiv vp^{N}{}_{K}\ pq^{K}{}_{n} = K^{N}{}_{K}\ P^{K}{}_{n}$$

$$v_{J}{}^{,I} = v_{J}{}^{;p}\ P_{p}{}^{I} = [v_{k}{}^{;p}\ \delta_{k}{}^{J}]\ P_{p}{}^{I}$$

$$v_{I}{}^{,J} = v_{I}{}^{;q}\ P_{q}{}^{J} = [v_{k}{}^{;q}\ \delta_{k}{}^{I}]\ P_{q}{}^{J}$$

Rewriting $\delta_{k}{}^{j}$ and $\delta_{k}{}^{i}$ In Two Reciprocal Forms:

$$\delta_{k}{}^{j} = (Q_{k}{}^{q}\ P_{q}{}^{j})$$

$$\delta_{k}{}^{i} = (P_{p}{}^{i}\ Q_{k}{}^{p})$$

Then:

$$v_{J}{}^{,I} = [v_{k}{}^{;p}\ (Q_{k}{}^{q}\ P_{q}{}^{J})]\ P_{p}{}^{I} \equiv [\,QT_{k}{}^{p}\ (Q_{k}{}^{q}\ P_{q}{}^{J})]\ P_{p}{}^{I}$$

$$v_{I}{}^{,J} = [v_{k}{}^{;q}\ (P_{p}{}^{I}\ Q_{k}{}^{p})]\ P_{q}{}^{J} \equiv [\,QT_{k}{}^{q}\ (P_{p}{}^{I}\ Q_{k}{}^{p})]\ P_{q}{}^{J}$$

The internal energy at any time, any location depends on the v at this time.
Stress/Time:
(v) Specific Power Series Minus Specific Heat Supply :

$$F^{I}{}_{n}\ v_{I}{}^{;m}\ P_{m}{}^{n} = F^{I}{}_{n}\ v_{I}{}^{,n} \equiv ET^{pq}\ \tau_{pq} = \Xi T - H = ET_{pq}\ \tau^{pq} \equiv F_{I}{}^{n}\, v^{I}{}_{,n} = F_{I}{}^{n}\, v^{I}{}_{;m}\ P^{m}{}_{n}$$

Stress/Time: [Stress*Time]*{1/Time/Time}:

$$F^{I}{}_{n}\ v_{I}{}^{,n} = [C1^{I}{}_{m} + C2^{IJ}{}_{mn}\, v_{J}{}^{,n} + C3^{IJK}{}_{mnp}\, v_{J}{}^{,n}\, v_{K}{}^{,p} + \ldots]^{*}\{vT_{I}{}^{m}\} - H = F_{I}{}^{n}\ v^{I}{}_{,n}$$

$$F^{I}{}_{n}\ v_{I}{}^{,n} = [C1_{I}{}^{m} + C2_{IJ}{}^{mn}\, v^{J}{}_{,n} + C3_{IJK}{}^{mnp}\, v^{J}{}_{,n}\, v^{K}{}_{,p} + \ldots]^{*}\{vT^{I}{}_{m}\} - H = F_{I}{}^{n}\ v^{I}{}_{,n}$$

Or:

$$F^{I}{}_{n}\, v_{I}{}^{;m}\ P_{m}{}^{n} = [\,[C1^{I}{}_{m} + C2^{IJ}{}_{mn}\, v_{J}{}^{,n} + C3^{IJK}{}_{mnp}\, v_{J}{}^{,n}\, v_{K}{}^{,p} + \ldots]^{*}\{vT_{I}{}^{m}\} - H = F_{I}{}^{n}\, v^{I}{}_{;m}\ P^{m}{}_{n}$$

$$F^{I}{}_{n}\, v_{I}{}^{;m}\ P_{m}{}^{n} = [C1_{I}{}^{m} + C2_{IJ}{}^{mn}\, v^{J}{}_{,n} + C3_{IJK}{}^{mnp}\, v^{J}{}_{,n}\, v^{K}{}_{,p} + \ldots]^{*}\{vT^{I}{}_{m}\} - H = F_{I}{}^{n}\, v^{I}{}_{;m}\ P^{m}{}_{n}$$

$$QT_{i}{}^{m}\ P_{m}{}^{n} \equiv v_{i}{}^{;m}\ pq_{m}{}^{n} = v_{i}{}^{,n} \qquad\qquad v^{i}{}_{,n} = v^{i}{}_{;m}\ pq^{m}{}_{n} \equiv QT^{i}{}_{m}\ P^{m}{}_{n}$$

Or:

$$F^{I}{}_{n}\ QT_{I}{}^{m}\ P_{m}{}^{n} = [\,[C1^{I}{}_{m} + C2^{IJ}{}_{mn}\, v_{J}{}^{,n} + C3^{IJK}{}_{mnp}\, v_{J}{}^{,n}\, v_{K}{}^{,p} + \ldots]^{*}\{vT_{I}{}^{m}\} - H = F_{I}{}^{n}\ QT^{I}{}_{m}\ P^{m}{}_{n}$$

$$F^{I}{}_{n}\ QT_{I}{}^{m}\ P_{m}{}^{n} = [C1_{I}{}^{m} + C2_{IJ}{}^{mn}\, v^{J}{}_{,n} + C3_{IJK}{}^{mnp}\, v^{J}{}_{,n}\, v^{K}{}_{,p} + \ldots]^{*}\{vT^{I}{}_{m}\} - H = F_{I}{}^{n}\ QT^{I}{}_{m}\ P^{m}{}_{n}$$

If H = 0:
Stress/Time: [Stress*Time]*{1/Time/Time}:

$$F^{I}{}_{m}\ v_{I}{}^{,m} = [C1^{I}{}_{m} + C2^{IJ}{}_{mn}\, v_{J}{}^{,n} + C3^{IJK}{}_{mnp}\, v_{J}{}^{,n}\, v_{K}{}^{,p} + \ldots]^{*}\ vT_{I}{}^{m} = \ldots$$

$$\ldots = [C1_{I}{}^{m} + C2_{IJ}{}^{mn}\, v^{J}{}_{,n} + C3_{IJK}{}^{mnp}\, v^{J}{}_{,n}\, v^{K}{}_{,p} + \ldots]^{*}\ vT^{I}{}_{m} = F_{I}{}^{n}\ v^{I}{}_{,n}$$

Stress/Time: [Stress*Time]*{1/Time/Time}:

$$F^I{}_m \; v_I{}^{,m} = [C1^I{}_m + C2^{IJ}{}_{mn} v_J{}^{,n} + C3^{IJK}{}_{mnp} v_J{}^{,n} v_K{}^{,p} + ...]^* \{\} v_I{}^{,m} = ...$$
$$... = [C1_I{}^m + C2_{IJ}{}^{mn} v^J{}_{,n} + C3_{IJK}{}^{mnp} v^J{}_{,n} v^K{}_{,p} + ...]^* \{\} v^I{}_{,m} = F_I{}^m \; v^I{}_{,m}$$

Or:

$$F^I{}_m = [C1^I{}_m + C2^{IJ}{}_{mn} v_J{}^{,n} + C3^{IJK}{}_{mnp} v_J{}^{,n} v_K{}^{,p} + ...] =$$
$$= [C1_I{}^m + C2_{IJ}{}^{mn} v^J{}_{,n} + C3_{IJK}{}^{mnp} v^J{}_{,n} v^K{}_{,p} + ...] = F_I{}^m$$

Stiffness (Resilience):

$$F^I{}_m = [C1^I{}_m + C2^{IJ}{}_{mn} v_J{}^{,n} + C3^{IJK}{}_{mnp} v_J{}^{,n} v_K{}^{,p} + ...] =$$
$$= [C1_I{}^m + C2_{IJ}{}^{mn} v^J{}_{,n} + C3_{IJK}{}^{mnp} v^J{}_{,n} v^K{}_{,p} + ...] = F_I{}^m$$

14.9.1.
Linear Constitution:

If Only $C2 \neq 0$ and $H = 0$ Then:

Stiffness (Resilience):

$$F^I{}_m = C2^{IJ}{}_{mn} \; v_J{}^{,n} = ...$$
$$... = C2_{IJ}{}^{mn} \; v^J{}_{,n} = F_I{}^m$$

14.9.2.
Infinitesimal Constitution Series:

Elastic Material is one for which the internal energy at any time, any location depends on the approximated Infinitesimal strain ε, at this time.

$$½ \, (q_I{}^{,m} + q_m{}^{,I}) \equiv \varepsilon_I{}^m$$
$$½ \, (qT_I{}^{,m} + qT_m{}^{,I}) \equiv ½ \, (v_I{}^{,m} + v_m{}^{,I}) \equiv \; \varepsilon T_I{}^m$$

Symmetric Deformation:

$$v_I{}^{,n} \equiv qT_I{}^{,n} = VT_I{}^n + WT_I{}^n = ½ \, (v_I{}^{,n} + v_n{}^{,I}) + ½ \, (v_I{}^{,n} - v_n{}^{,I})$$
$$v^I{}_{,n} \equiv qT^I{}_{,n} = VT^I{}_n + WT^I{}_n = ½ \, (v^I{}_{,n} + v^n{}_{,I}) + ½ \, (v^I{}_{,n} - v^n{}_{,I})$$

$$v_n{}^{,I} \equiv qT_n{}^{,I} = VT_n{}^I - WT_n{}^I = ½ \, (v_I{}^{,n} + v_n{}^{,I}) - ½ \, (v_I{}^{,n} - v_n{}^{,I})$$
$$v^n{}_{,I} \equiv qT^n{}_{,I} = VT^n{}_I - WT^n{}_I = ½ \, (v^I{}_{,n} + v^n{}_{,I}) - ½ \, (v^I{}_{,n} - v^n{}_{,I})$$

VT is Symmetric :

$$½ \, (v_I{}^{,n} + v_n{}^{,I}) \equiv ½ \, (qT_I{}^{,n} + qT_n{}^{,I}) \equiv VT_I{}^n$$
$$½ \, (v^I{}_{,n} + v^n{}_{,I}) \equiv ½ \, (qT^I{}_{,n} + qT^n{}_{,I}) \equiv VT^I{}_n$$

WT is Asymmetric :

$$½ \, (v_I{}^{,n} - v_n{}^{,I}) \equiv ½ \, (qT_I{}^{,n} - qT_n{}^{,I}) \equiv WT_I{}^n$$
$$½ \, (v^I{}_{,n} - v^n{}_{,I}) \equiv ½ \, (qT^I{}_{,n} - qT^n{}_{,I}) \equiv WT^I{}_n$$

Stress/Time:

Specific Power Minus Specific Heat Supply : ΞT-H:

$$F^I{}_J v_I{}^{,J} \equiv ½ \, (F^i{}_j v_i{}^{,j} + F^i{}_j v_i{}^{,j}) = \varepsilon T^{pq} \; \tau_{pq} = (ET^{pq}) \, (\rho/\rho0 \;\; \tau0_{pq})$$
$$= \Xi T - H = \rho \; eT - H \; =$$

$$F_I{}^J v^I{}_{,J} \equiv ½ (F_i{}^j v^i{}_j + F_i{}^{,j} v^i{}_j) = \varepsilon T_{pq}\ \tau^{pq} = (ET_{pq})(\rho/\rho 0\ \ \tau 0^{pq})$$

Or:

$$F^I{}_n v_I{}^{;m} P_m{}^n = F^I{}_n v_I{}^{,n} = \varepsilon T^{pq}\ \tau_{pq} = \Xi T - H = \varepsilon T_{pq}\ \tau^{pq} = F_I{}^n v^I{}_{,n} = F_I{}^n v^I{}_{;m} P^m{}_n$$

(ε) Constitution Series Time Derivative:
Stress/Time: [Stress]*{1/Time}:

$$\varepsilon T^{pq}\ \tau_{pq} = \Xi T - H = \varepsilon T_{pq}\ \tau^{pq}$$

$$\varepsilon T^{pq}\ \tau_{pq} = [C1^I{}_m + C2^{IJ}{}_{mn}\varepsilon_J{}^{,n} + C3^{IJK}{}_{mnp}\ \varepsilon_J{}^{,n}\varepsilon_K{}^{,p} + \ldots] * \varepsilon T_I{}^m - H = \varepsilon T_{pq}\ \tau^{pq}$$
$$\varepsilon T^{pq}\ \tau_{pq} = [C1_I{}^m + C2_{IJ}{}^{mn}\varepsilon^J{}_{,n} + C3_{IJK}{}^{mnp}\varepsilon^J{}_{,n}\ \varepsilon^K{}_{,p} + \ldots] * \varepsilon T^I{}_m - H = \varepsilon T_{pq}\ \tau^{pq}$$

(ε) Symmetric Constitution Series Time Derivative:

$$F^I{}_n\ \ v_I{}^{,n} = \Xi T - H \equiv \rho e T - H = F_I{}^n\ v^I{}_{,n}$$

$$F^I{}_m\ \ v_I{}^{,m} = [C1^I{}_m + C2^{IJ}{}_{mn}\varepsilon_J{}^{,n} + C3^{IJK}{}_{mnp}\ \varepsilon_J{}^{,n}\varepsilon_K{}^{,p} + \ldots] * \varepsilon T_I{}^m - H = F_I{}^m\ \ v^I{}_{,m}$$
$$F^I{}_m\ \ v_I{}^{,m} = [C1_I{}^m + C2_{IJ}{}^{mn}\varepsilon^J{}_{,n} + C3_{IJK}{}^{mnp}\varepsilon^J{}_{,n}\ \varepsilon^K{}_{,p} + \ldots] * \varepsilon T^I{}_m - H = F_I{}^m\ \ v^I{}_{,m}$$

Or:

$$F^I{}_m\ \ v_I{}^{,m} = [C1^I{}_m + C2^{IJ}{}_{mn}\varepsilon_J{}^{,n} + C3^{IJK}{}_{mnp}\ \varepsilon_J{}^{,n}\varepsilon_K{}^{,p} + \ldots] * v_I{}^{,m} - H = F_I{}^m\ \ v^I{}_{,m}$$
$$F^I{}_m\ \ v_I{}^{,m} = [C1_I{}^m + C2_{IJ}{}^{mn}\varepsilon^J{}_{,n} + C3_{IJK}{}^{mnp}\varepsilon^J{}_{,n}\ \varepsilon^K{}_{,p} + \ldots] * v^I{}_{,m} - H = F_I{}^m\ \ v^I{}_{,m}$$

If Only H = 0:

$$F^I{}_m\ \ v_I{}^{,m} = [C1^I{}_m + C2^{IJ}{}_{mn}\varepsilon_J{}^{,n} + C3^{IJK}{}_{mnp}\ \varepsilon_J{}^{,n}\varepsilon_K{}^{,p} + \ldots] * v_I{}^{,m} = \ldots\ \ldots\ \ldots$$
$$\ldots\ \ldots\ \ldots = [C1_I{}^m + C2_{IJ}{}^{mn}\varepsilon^J{}_{,n} + C3_{IJK}{}^{mnp}\varepsilon^J{}_{,n}\ \varepsilon^K{}_{,p} + \ldots] * v^I{}_{,m} = F_I{}^m\ \ v^I{}_{,m}$$

Or:

$$F^I{}_m = [C1^I{}_m + C2^{IJ}{}_{mn}\varepsilon_J{}^{,n} + C3^{IJK}{}_{mnp}\ \varepsilon_J{}^{,n}\varepsilon_K{}^{,p} + \ldots] = \ldots$$
$$\ldots = [C1_I{}^m + C2_{IJ}{}^{mn}\varepsilon^J{}_{,n} + C3_{IJK}{}^{mnp}\varepsilon^J{}_{,n}\ \varepsilon^K{}_{,p} + \ldots] = F_I{}^m$$

Stiffness (Resilience):

$$\tau^I{}_m = F^I{}_m = [C1^I{}_m + C2^{IJ}{}_{mn}\varepsilon_J{}^{,n} + C3^{IJK}{}_{mnp}\ \varepsilon_J{}^{,n}\varepsilon_K{}^{,p} + \ldots] = \ldots$$
$$\ldots = [C1_I{}^m + C2_{IJ}{}^{mn}\varepsilon^J{}_{,n} + C3_{IJK}{}^{mnp}\varepsilon^J{}_{,n}\ \varepsilon^K{}_{,p} + \ldots] = F_I{}^m = \tau_I{}^m$$

14.9.3.
Linear Infinitesimal Constitution:

If Only C2 ≠ 0 and H = 0 Then:
Stiffness (Resilience):

$$\tau^I{}_m = F^I{}_m = C2^{IJ}{}_{mn}\ \ \varepsilon_J{}^{,n} = \ldots\ \ldots\ \ldots$$
$$\ldots\ldots\ \ldots\ \ldots = C2_{IJ}{}^{mn}\ \varepsilon^J{}_{,n} = F_I{}^m = \tau_I{}^m$$

14.10.
Elasticity:

Elastic Flexibility (Resilience) :
1 / Time = 1 /Stress * Stress / Time

$$DET^{kl} = \Gamma^{klmn} \ \sigma T_{mn}$$

$$\sigma T_1 \equiv \sigma T_{11} + \sigma T_{22} + \sigma T_{33}$$

$$DET^{11} = \Gamma^{1111} \sigma T_{11} + \Gamma^{1122} \sigma T_{22} + \Gamma^{1133} \sigma T_{33}$$
$$DET^{22} = \Gamma^{2211} \sigma T_{11} + \Gamma^{2222} \sigma T_{22} + \Gamma^{2233} \sigma T_{33}$$
$$DET^{33} = \Gamma^{3311} \sigma T_{11} + \Gamma^{3322} \sigma T_{22} + \Gamma^{3333} \sigma T_{33}$$

$$DET^{1} \equiv DET^{11} + DET^{22} + DET^{33}$$

Normalized Bulk Constitution by a Value Θ:

$$[DET^{1} -3/\Theta \ \alpha h] = \Gamma^{1}/\Theta \ \sigma T_1$$

Where:

$$d[DET_1]/d[\sigma T_1] = \Gamma^{1}/\Theta$$

Since Elastic Strain is Funcion of Stress:

$$EET^{kl} = \Gamma^{klmn} \ \tau T_{mn}$$

$$DET^{kl} + EET^{ii}/3\ \delta^{kl} = \Gamma^{klmn} \ (\sigma T_{mn} + \tau T^{ii}/3\ \delta_{mn})$$

DET^{kl} Temperature-Independent Distortion:

Introducing Bulk Constitution:

$$EET^{ii}/3\ \delta^{kl} = \Gamma^{klmn} \ \tau T^{ii}/3\ \delta_{mn}$$

So:

Deviatoric Elastic Strain is funcion of Deviatoric Stress:

$$DET^{kl} = \Gamma^{klmn} \ \sigma T_{mn}$$

$$d(EET^{ii}) / d(\tau T^{ii})\ \delta^{kl} = \Gamma^{klmn} \ \delta_{mn}$$

Multiply DET^{kl} Flexibility (Resilience) by: σ_{kl}

$$DET^{kl} \sigma_{kl} = \Gamma^{klmn} \ \sigma T_{mn} \sigma_{kl} = (½\,\Gamma^{klmn} \ \sigma_{mn} \sigma_{kl})T$$

And:

<u>Infinitesimal</u> Elastic Flexibility (Resilience):

Since Elastic Strain is funcion of Stress:

$$\varepsilon ET^{kl} = (1+\nu)/E\ \tau T^{kl} - \nu/E\ \tau T^{ii}\ \delta^{kl}$$

$$\partial ET^{kl} + \varepsilon ET^{ii}/3\ \delta^{kl} = (1+\nu)/E\ (\sigma T^{kl} + \tau T^{ii}/3\ \delta^{kl}) - \nu/E\ \tau T^{ii}\ \delta^{kl}$$

Bulk Constitution:

$$\varepsilon ET^{ii}\ \delta^{kl} = (1+\nu)/E\ \tau T^{ii}/3\ \delta^{kl} - \nu/E\ \tau T^{ii}\ \delta^{kl} = (1-2\nu)/E\ \ \tau T^{ii}\ \delta^{kl}$$

So:

Deviatoric Elastic Strain is funcion of Deviatoric Stress:

$$\partial ET^{kl} = (1+\nu)/E\ \ \sigma T^{kl}$$

Multiply Flexibility (Resilience) by: σ_{kl}

$$\partial ET^{kl}\ \sigma_{kl} = (1+\nu)/E\ \ \sigma T^{kl}\ \sigma_{kl} = [½\,(1+\nu)/E\ \sigma^{kl}\ \sigma_{kl}]T$$
$$\partial ET^{kl}\ \sigma_{kl} = 1/(2G)\ \ \sigma T^{kl}\ \sigma_{kl} = [½\,1/(2G)\ \sigma^{kl}\ \sigma_{kl}]T$$

(v) Unsymmetric Constitution Series Time Derivative:

The internal energy at any time, any location depends on the v at this time.

e (v(q,t))

$$\rho eT=[C1^{i}_{m}+C2^{ij}_{nm}(v_j{}^{;n})](vT_i{}^{;m}) = v_i{}^{;m}\ pq_m{}^{n}\ F^{i}_{n} + H$$

$C1^{i}_{m} = 0 = -h_{n}{}^{,n}$ $\quad C2^{ij}_{nm}(v_{j}{}^{;n})\,(vT_{i}{}^{;m}) = v_{i}{}^{;m}\ \rho/\rho0\ \Phi^{i}_{m}$

Where: $\quad pq_{m}{}^{n}\ F^{i}_{n} \equiv \rho/\rho0\ \Phi^{i}_{m}$

(qp) Unsymmetric Constitution Series Time Derivative:

e (qp(q,t)) $\quad \rho eT = (C1^{i}_{m} + C2^{ij}_{nm}(qp_{j}{}^{;n}))(qpT_{i}{}^{;m}) = v_{i}{}^{;m}\ pq_{m}{}^{n}\ F^{i}_{n} + H$

Unsymmetric $\quad \rho eT = (C1^{i}_{m} + C2^{ij}_{nm}(qp_{j}{}^{;n}))(v_{i}{}^{;m}) = (v_{i}{}^{;m})\rho/\rho0\ \Phi^{i}_{m} + H$

$C1^{i}_{m} = 0 = H$ $\quad C2^{ij}_{nm}(qp_{j}{}^{;n})\,(v_{i}{}^{;m}) = (v_{i}{}^{;m})\ \rho/\rho0\ \Phi^{i}_{m}$

$C1^{i}_{m} = 0 = H$ $\quad C2^{ij}_{nm}(qp_{j}{}^{;n}) = \rho/\rho0\ \Phi^{i}_{m}$

Where: $\quad pq_{m}{}^{n}\ F^{i}_{n} \equiv \rho/\rho0\ \Phi^{i}_{m}$

$$qpT_{i}{}^{;m} \equiv v_{i}{}^{;m}$$

(ε) Symmetric Constitution Series Time Derivative:

Elastic Material is one for which the internal energy at any time, any location depends on the approximated Infinitesimal strain ε, at this time.

e (ε (q,t)) $\quad \rho eT = (C1^{i}_{m} + C2^{ij}_{nm}(\varepsilon_{j}{}^{;n}))(\varepsilon T_{i}{}^{m}) \approx v_{i}{}^{;m}\ pq_{m}{}^{n}\ F^{i}_{n} + H$

$C1^{i}_{m} = 0 = H$ $\quad C2^{ij}_{nm}(\varepsilon_{j}{}^{;n})\,(v_{i}{}^{;m}\ pq_{m}{}^{n}) \approx (v_{i}{}^{;m}\ pq_{m}{}^{n})\ F^{i}_{n}$

$C1^{i}_{m} = 0 = H$ $\quad C2^{ij}_{nm}(\varepsilon_{j}{}^{;n}) = \tau^{i}_{n} \approx F^{i}_{n}$

Where: $\quad \tfrac{1}{2}(v_{i}{}^{,m} + v_{m}{}^{,i}) \equiv \varepsilon T_{i}{}^{m}$

Symmetric Deformation:

$v_{i}{}^{,n} \equiv qT_{i}{}^{,n} = VT_{i}{}^{n} + WT_{i}{}^{n}$ $\quad v_{i}{}^{,n} = \tfrac{1}{2}(v_{i}{}^{,n} + v_{n}{}^{,i}) + \tfrac{1}{2}(v_{i}{}^{,n} - v_{n}{}^{,i})$

$v_{n}{}^{,i} \equiv qT_{n}{}^{,i} = VT_{n}{}^{i} - WT_{n}{}^{i}$ $\quad v_{n}{}^{,i} = \tfrac{1}{2}(v_{i}{}^{,n} + v_{n}{}^{,i}) - \tfrac{1}{2}(v_{i}{}^{,n} - v_{n}{}^{,i})$

$$\tfrac{1}{2}(v_{i}{}^{,m} + v_{m}{}^{,i}) \equiv \varepsilon T_{i}{}^{m}$$

Where vq is a sum of :

$$v_{i}{}^{,n} \equiv qT_{i}{}^{,n} = VT_{i}{}^{n} + WT_{i}{}^{n} = \tfrac{1}{2}(v_{i}{}^{,n} + v_{n}{}^{,i}) + \tfrac{1}{2}(v_{i}{}^{,n} - v_{n}{}^{,i})$$

$$v^{i}{}_{,n} \equiv qT^{i}{}_{,n} = VT^{i}{}_{n} + WT^{i}{}_{n} = \tfrac{1}{2}(v^{i}{}_{,n} + v^{n}{}_{,i}) + \tfrac{1}{2}(v^{i}{}_{,n} - v^{n}{}_{,i})$$

$$v_{n}{}^{,i} \equiv qT_{n}{}^{,i} = VT_{n}{}^{i} - WT_{n}{}^{i} = \tfrac{1}{2}(v_{i}{}^{,n} + v_{n}{}^{,i}) - \tfrac{1}{2}(v_{i}{}^{,n} - v_{n}{}^{,i})$$

$$v^{n}{}_{,i} \equiv qT^{n}{}_{,i} = VT^{n}{}_{i} - WT^{n}{}_{i} = \tfrac{1}{2}(v^{i}{}_{,n} + v^{n}{}_{,i}) - \tfrac{1}{2}(v^{i}{}_{,n} - v^{n}{}_{,i})$$

VT is symmetric :

$$\tfrac{1}{2}(v_{i}{}^{,n} + v_{n}{}^{,i}) \equiv \tfrac{1}{2}(qT_{i}{}^{,n} + qT_{n}{}^{,i}) \equiv VT_{i}{}^{n}$$

$$\tfrac{1}{2}(v^{i}{}_{,n} + v^{n}{}_{,i}) \equiv \tfrac{1}{2}(qT^{i}{}_{,n} + qT^{n}{}_{,i}) \equiv VT^{i}{}_{n}$$

WT is asymmetric :

$$\tfrac{1}{2}(v_{i}{}^{,n} - v_{n}{}^{,i}) \equiv \tfrac{1}{2}(qT_{i}{}^{,n} - qT_{n}{}^{,i}) \equiv WT_{i}{}^{n}$$

$$\tfrac{1}{2}(v^{i}{}_{,n} - v^{n}{}_{,i}) \equiv \tfrac{1}{2}(qT^{i}{}_{,n} - qT^{n}{}_{,i}) \equiv WT^{i}{}_{n}$$

Infinitesimal Elasticity:

$C1^{i}_{m} = 0 = H$ $\quad C2^{ij}_{nm}(\varepsilon_{j}{}^{;n}) = \tau^{i}_{n}$

$$\tau^{i}_{m} = C2^{ij}{}_{mn}\ \varepsilon_{j}{}^{n} \equiv C^{ij}{}_{mn}\ \varepsilon_{j}{}^{n}$$

There are: (3*3)(3*3) = 3^4 = 81 constants.

Since: 3 of τ^{i}_{m} are symmetric:

$$\tau^{i}_{m} = \tau^{m}_{i}$$

Then: (6 of τ^{i}_{m})(9 of ε_{j}^{n})=54 constants C^{ij}_{mn} :

81 constants C^{ij}_{mn} of which 3x(9) = 27 are redundants: (3*3)(3*3) – 27 = 54

$$C^{ij}_{mn} = C^{mj}_{in}$$

$\varepsilon_{xx} = dr_x / dx$ $\quad$ $1/2\ \gamma_{xy} = \varepsilon_{xy} = dr_x / dy$ $\quad$ $1/2\ \gamma_{xz} = \varepsilon_{xz} = dr_x / dz$

$\varepsilon_{yy} = dr_y / dy$ $\quad$ $1/2\ \gamma_{yz} = \varepsilon_{yz} = dr_y / dz$

$\varepsilon_{zz} = dr_z / dz$

Since:

$$\varepsilon_{j}^{n} = \varepsilon_{n}^{j} \qquad \tau^{i}_{m} = \tau^{m}_{i}$$

ε_1^{1} $\quad$ ε_2^{2} $\quad$ ε_3^{3} $\quad$ ε_1^{2} $\quad$ ε_2^{3} $\quad$ ε_3^{1}

τ_1^{1} $\quad$ τ_2^{2} $\quad$ τ_3^{3} $\quad$ τ_1^{2} $\quad$ τ_2^{3} $\quad$ τ_3^{1}

Then:

(6 of τ^{i}_{m})(6 of ε_{j}^{n})=36 constants C^{ij}_{mn} :

$$C^{ij}_{mn} = C^{in}_{mj}$$

$$C^{IJ}_{MN} = -\nu^{I}_{M} / E^{J}_{N} = -\nu^{I}_{M} / E^{M} = -\nu^{I}_{M} / E^{N}_{J} = C^{IN}_{MJ}$$

$1/E^1$	$-\nu^1_2/E^2$	$-\nu^1_3/E^3$	$-\nu^1_4/E^4$	$-\nu^1_5/E^5$	$-\nu^1_6/E^6$
$-\nu^2_1/E^1$	$1/E^2$	$-\nu^2_3/E^3$	$-\nu^2_4/E^4$	$-\nu^2_5/E^5$	$-\nu^2_6/E^6$
$-\nu^3_1/E^1$	$-\nu^3_2/E^2$	$1/E^3$	$-\nu^3_4/E^4$	$-\nu^3_5/E^5$	$-\nu^3_6/E^6$
$-\nu^4_1/E^1$	$-\nu^4_2/E^2$	$-\nu^4_3/E^3$	$1/E^4$	$-\nu^4_5/E^5$	$-\nu^4_6/E^6$
$-\nu^5_1/E^1$	$-\nu^5_2/E^2$	$-\nu^5_3/E^3$	$-\nu^5_4/E^4$	$1/E^5$	$-\nu^5_6/E^6$
$-\nu^6_1/E^1$	$-\nu^6_2/E^2$	$-\nu^6_3/E^3$	$-\nu^6_4/E^4$	$-\nu^6_5/E^5$	$1/E^6$

Furthermore Since:

$$C^{ij}_{mn} = \partial^2 e/\partial\varepsilon_{i}^{m} / \partial\varepsilon_{j}^{n} = \partial^2 e/\partial\varepsilon_{j}^{n} / \partial\varepsilon_{i}^{m} = C^{ji}_{nm}$$

Then: (6+1)6/2=21 independent constants C^{ij}_{mn} .

So:

36 constants C^{ij}_{mn} of which 5+4+3+2+1=15 are redundants:

(6)(6) – 15 = 21

C^{ij}_{mn} has 6*6=36 constants of which 21 independent constants

These are:

$$c^{i}_{m}$$

c^1_1	c^1_2	c^1_3	c^1_4	c^1_5	c^1_6
	c^2_2	c^2_3	c^2_4	c^2_5	c^2_6
		c^3_3	c^3_4	c^3_5	c^3_6
			c^4_4	c^4_5	c^4_6
	Symmetric			c^5_5	c^5_6
					c^6_6

14.11.
Orthotropic Constitution

Axials:

$$\varepsilon^{X}_{X} = [\ \tau^{X}_{X} - \nu^{Y}_{X}\ \tau^{Y}_{Y} - \nu^{Z}_{X}\ \tau^{Z}_{Z})] / E^{X}_{X} + \alpha_{X} *h$$

$$\varepsilon^{Y}_{Y} = [-\nu^{X}_{Y}\ \tau^{X}_{X} + \ \tau^{Y}_{Y} - \nu^{Z}_{Y}\ \tau^{Z}_{Z})] / E^{Y}_{Y} + \alpha_{Y} *h$$

$$\varepsilon^{Z}_{Z} = [-\nu^{X}_{Z}\ \tau^{X}_{X} - \nu^{Y}_{Z}\ \tau^{Y}_{Y} + \ \tau^{Z}_{Z})] / E^{Z}_{Z} + \alpha_{Z} *h$$

Shears:

$$\gamma^{Y}_{X} = \tau^{Y}_{X} / G^{Y}_{X} = \tau^{Y}_{X}\ 2(1+\nu^{Y}_{X}) / E^{X}_{X} = \gamma^{X}_{Y}$$

$$\gamma^{Z}_{Y} = \tau^{Z}_{Y} / G^{Z}_{Y} = \tau^{Z}_{Y}\ 2(1+\nu^{Z}_{Y}) / E^{Y}_{Y} = \gamma^{Y}_{Z}$$

$$\gamma^{X}_{Z} = \tau^{X}_{Z} / G^{X}_{Z} = \tau^{X}_{Z}\ 2(1+\nu^{X}_{Z}) / E^{Z}_{Z} = \gamma^{Z}_{X}$$

$1/E^1$	$-\nu^1{}_2/E^2$	$-\nu^1{}_3/E^3$	0	0	0
$-\nu^2{}_1/E^1$	$1/E^2$	$-\nu^2{}_3/E^3$	0	0	0
$-\nu^3{}_1/E^1$	$-\nu^3{}_2/E^2$	$1/E^3$	0	0	0
0	0	0	$1/E^4$	0	0
0	0	0	0	$1/E^5$	0
0	0	0	0	0	$1/E^6$

Axisymmetric Material:

$\varepsilon_r{}^r$	$\varepsilon_z{}^z$	$\varepsilon_3{}^3$	$\varepsilon_Z{}^R$	$\varepsilon_R{}^\Theta$	$\varepsilon_\Theta{}^Z$
$\tau_1{}^1$	$\tau_2{}^2$	$\tau_3{}^3$	$\tau_1{}^2$	$\tau_2{}^3$	$\tau_3{}^1$
$1/E^1$	$-\nu^1{}_2/E^2$	$-\nu^1{}_3/E^3$	$-\nu^1{}_4/E^4$	0	0
$-\nu^2{}_1/E^1$	$1/E^2$	$-\nu^2{}_3/E^3$	$-\nu^2{}_4/E^4$	0	0
$-\nu^3{}_1/E^1$	$-\nu^3{}_2/E^2$	$1/E^3$	$-\nu^3{}_4/E^4$	0	0
$-\nu^4{}_1/E^1$	$-\nu^4{}_2/E^2$	$-\nu^4{}_3/E^3$	$1/E^4$	0	0
0	0	0	0	$1/E^5$	$-\nu^5{}_6/E^6$
0	0	0	0	$-\nu^6{}_5/E^5$	$1/E^6$

	Esteel 210mPa		νsteel 0.30	Gsteel 80mPa	λsteel 120mPa	αsteel 0.000006	ρsteel 8000kg/m^3
	E/Esteel	ν/1	ν/νsteel	G/Gsteel	λ/λsteel	α/αsteel	ρ/ρsteel
Steel	1.	0.30	1.00	1.00	1.00	1.00	1.00
Aluminum	0.33	0.33	1.11	0.33	0.44	2.00	0.35
Concrete	0.13	0.20	0.66	0.15	0.06	0.92	0.30
Mercury	0	0.50	1.66	0	0.20	-	1.90
Ice	0.00003	0.49	1.60	0.00003	0.02	-	0.13
Water	0	0.50	1.66	0	0.02	-	0.13

Rock Mechanics Considerations

Elastic Energy of Finite Deformation assumes Constitutive Energy Function, Ξ such that:

$$\Xi = \Xi(E^{ii}, E^{ij} E^{ji}) = \Xi(E^{kr})$$

$$\Psi = \Psi(T_{ii}, T_{ij} T_{ji}) = \Psi(T_{kl})$$

$$\rho/\rho 0 \;\; d\Xi/dt \equiv \rho \;\; de/dt = F^{i}_{j} \;\; v_{i}^{,j} - h_{n}^{,n}$$

Symmetric Constitution:

$$d\Xi/dt = dE^{pq}/dt \;\; T_{pq} - h_{n}^{,n}$$

Where:

$$dE^{pq}/dt \equiv \tfrac{1}{2}(dF_{k}^{p}/dt \;\; F_{k}^{q} + F_{k}^{p} \;\; dF_{k}^{q}/dt)$$

$$\rho/\rho 0 \;\; T_{pq} \equiv P^{i}_{p} \;\; F^{i}_{j} \;\; P^{j}_{q}$$

Material Deformation Gradient F^{I}_{J} is given by:

$$\partial q^{I} / \partial p^{J} \equiv F^{I}_{J}$$

Spatial Deformation Gradient:

$$\partial p^{I} / \partial q^{J} \equiv P^{I}_{J}$$

$$v_{N} \equiv \partial q_{N} / \partial t$$

$$v_{i}^{,j} \equiv \partial v_{i} / \partial q_{j}$$

Assume Inelastic Variables: B^{p}, A_{p} :

So:

$$\Xi = \Xi(E^{ii}, E^{ij} E^{ji}) = \Xi(E^{kr}, B^{r})$$

$$\Psi = \Psi(T_{ii}, T_{ij} T_{ji}) = \Psi(T_{kl}, A_{r})$$

Stress Stiffness :

$$T_{kl} = T_{kl}(E^{kl}, B^{p})$$

Stress Stiffness Differential:

$$dT_{kl} = \partial T_{kl}/\partial E^{mn} \;\; dE^{mn} + \partial T_{kl}/\partial B^{r} \;\; dB^{r}$$

$$dT_{kl} = \partial \, \partial\Xi / \partial E^{kl} / \partial E^{mn} \;\; dE^{mn} + \partial \, \partial \Xi / \partial E^{kl} / \partial B^{r} \;\; dB^{r}$$

Then:

$$dT_{kl} = K_{klmn} \;\; dE^{mn} + \Pi_{klr} \;\; dB^{r}$$

And:

Strain Flexibility :

$$E^{kl} = E^{kl}(T_{kl}, A_{p})$$

Strain Flexibility Differential:

$$dE^{kl} = \partial E^{kl}/\partial T_{mn} \;\; dT_{mn} + \partial E^{kl}/\partial A_{c} \;\; dA_{c}$$

$$dE^{kl} = \partial \, \partial\Psi / \partial T_{kl} / \partial T_{mn} \;\; dT_{mn} + \partial \, \partial\Psi / \partial T_{kl} / \partial A_{c} \;\; dA_{c}$$

Internal Energy Differential:

$$d\Xi = dE^{kl} \;\; T_{kl} + \rho \text{ heat } \; dt - \rho 0 \;\; D \;\; dt$$

Complementary Energy Differential:

$$d\Psi = E^{kl} \;\; dT_{kl} - \rho \text{ heat } dt + \rho 0 \;\; D \, dt$$

This law:

$$dE^{kl} = dEE^{kl} + dEI^{kl}$$

Plastic Strain Hardening Material: Kinematic:

$$PPT \equiv \int EPT^{kl} \quad EPT_{kl}\ dt$$

$$\sigma = [(\sigma_1 + \sigma_3)/2 + (\sigma_1 - \sigma_3)/2 \quad \cos 2A]$$

$$\tau = \quad (\sigma_1 - \sigma_3)/2 \quad \sin 2A$$

$$\tau = C + \quad \sigma \quad \tan \Phi$$

$$\sigma_1 = 2\ C \tan\alpha + \sigma_3 \tan^2 \alpha$$

Stiffnes of ABAQUS

$$dT_{kl} = K_{klmn} \quad dE^{mn} + \Pi_{klr}\ dB^{r}$$

And:

$$ET^{kl} S_{kl} S^{pq} = \Gamma^{klmn} \quad TT_{mn} S_{kl} \quad S^{pq} + S^{pq} \quad \Lambda^{klMN} \quad S_{MN} S_{kl}$$

$$DT^{kl} S_{kl} S^{pq} = \Gamma^{klmn} \quad ST_{mn} S_{kl} S^{pq} + S^{pq} \quad \Lambda^{klMN} \quad S_{MN} S_{kl}$$

$$ET^{kl} S_{kl} S^{pq} = \quad \Gamma^{klmn} \quad TT_{mn} \ S_{kl} \ S^{pq} \quad + \quad S^{pq} \ \Lambda \ S^{2}$$

$$DT^{kl} S_{kl} S^{pq} = \quad \Gamma^{klmn} \quad ST_{mn} \ S_{kl} \ S^{pq} \quad + \quad S^{pq} \ \Lambda \ S^{2}$$

$(D^{kl}, E^{kl}, E^{q})(\partial_i^{\ j}, \varepsilon^{kl}, \varepsilon^{q})$ $\qquad$ $(S_{kl}, T_{kl}, T_q)(\sigma^{i}_{\ j}, \tau_{kl}, \tau_q)$

F Surface traction (force per area)

$(D^{kk} \equiv 0, E \equiv E^{kk})(\partial_k^{\ k} \equiv 0, \varepsilon \equiv \varepsilon^{kk})$ $\qquad$ $(S_{kk} \equiv 0, T \equiv T_{kk})(\sigma^{k}_{\ k} \equiv 0, \tau \equiv \tau_{kk})$

15. Infinitismal Elasticity of Two Constants:

(Two Constants Resilience)

15.1. 3-D Constitutions:

3-D Strain:

$\varepsilon^{IJ}(x) = ½ (u^{I}_{,J} + u^{J}_{,I})$ | $\varepsilon^{AA}(x) = u^{A}_{,A}$

ε^{IJ}	J = 1	J = 2	J = 3
I =1	$Strain_{XX}$	$Strain_{XY}$	$Strain_{XZ}$
I =2	$Strain_{YX}$	$Strain_{YY}$	$Strain_{YZ}$
I =3	$Strain_{ZX}$	$Strain_{ZY}$	$Strain_{ZZ}$

3-D STRESS:

:

τ^{IJ}	J = 1	J = 2	J = 3
I =1	$Stress_{XX}$	$Stress_{XY}$	$Stress_{XZ}$
I =2	$Stress_{YX}$	$Stress_{YY}$	$Stress_{YZ}$
I =3	$Stress_{ZX}$	$Stress_{ZY}$	$Stress_{ZZ}$

3-D Constitution:

E and ν are Independent Constants

:

$[E/(1+\nu)] \equiv 2 M$ $\qquad [E/(1+\nu)] * \nu/(1-2 \nu) \equiv \Lambda$ $\qquad E/(1-2 \nu) \equiv 2 M + 3 \Lambda \equiv \kappa$

***Temperature Dependent* Constitution Due to Heat and Strain:**

$\tau_{IJ} + E/(1-2\nu) \alpha h \delta_{IJ}$ $=$ $E/(1+\nu) \varepsilon_{IJ} + E/(1+\nu)*[\nu/(1-2 \nu)] \varepsilon_{1} \delta_{IJ}]$ $\qquad$ $\varepsilon_{IJ} - \alpha h \delta_{IJ}$ $=$ $(1+\nu)/E \tau_{IJ} - \nu/E \tau_{1}\delta_{IJ}$

Or:

$\tau_{IJ} + \kappa \alpha h \delta_{IJ} = 2 M \varepsilon_{IJ} + \Lambda \varepsilon_{1} \delta_{IJ}$ $\qquad$ $\varepsilon_{IJ} - \alpha h \delta_{IJ} = 1/(2 M) \tau_{IJ} - \nu/E \tau_{1} \delta_{IJ}$

Temperature Dependent Trace:

$\tau_{1} + 3 \kappa \alpha h = \kappa \varepsilon_{1}$ $\qquad$ $\varepsilon_{1} - 3 \alpha h = 1/\kappa \tau_{1}$

Temperature Independent Three Shears Constitution: xy, yz, zx:

$\tau_{I\neq J} = M \gamma_{I\neq J} = 2 M \varepsilon_{I\neq J}$ $\qquad$ $\varepsilon_{I\neq J} = 1/(2 M) \tau_{I\neq J}$

1/E	-ν /E	-ν /E	0	0	0
-ν /E	1/E	-ν /E	0	0	0

-ν /E	-ν /E	1/E	0	0	0
0	0	0	1/E	0	0
0	0	0	0	1/E	0
0	0	0	0	0	1/E

15.2.
3-D Distortion Constitutions:

Constitution is Two Components, Heat and Strain:

$\tau_{IJ} + E/(1-2\nu)\,\alpha h\,\delta_{IJ}$ = $E/(1+\nu)\,\varepsilon_{IJ} + E/(1+\nu) * \nu/(1-2\nu)\,\varepsilon_1\,\delta_{IJ}$	$\varepsilon_{IJ} - \alpha h\,\delta_{IJ}$ = $(1+\nu)/E\,\tau_{IJ} - \nu/E\,\tau_1\delta_{IJ}$

Its Trace is Two Components, Heat and Strain:

$\tau_1 + E/(1-2\nu)\,\alpha h * 3$ = $E/(1+\nu)\,\varepsilon_1 + E/(1+\nu) * \nu/(1-2\nu)\,\varepsilon_1\,3$	$\varepsilon_1 - \alpha h * 3$ = $(1+\nu)/E\,\tau_1 - \nu/E\,\tau_1 * 3$

$\delta_{IJ}/3$ * Trace:

$\delta_{IJ}/3 * \tau_1 + E/(1-2\nu)\,\alpha h\,\delta_{IJ}$ = $E/(1+\nu)\,\varepsilon_1\,\delta_{IJ}/3 + E/(1+\nu)\,\nu/(1-2\nu)\,\varepsilon_1\,\delta_{IJ}$	$\delta_{IJ}/3 * \varepsilon_1 - \alpha h\,\delta_{IJ}$ = $(1+\nu)/E\,\tau_1\,\delta_{IJ}/3 - \nu/E\,\tau_1\,\delta_{IJ}$

Distortion Defined by [Constitution - $\delta_{IJ}/3$ * Trace]:

Identical Constitution: Distortion and Shear are *Temperature Independent*:

$\tau_{IJ} - \tau_1/3\,\delta_{IJ} = E/(1+\nu) * (\varepsilon_{IJ} - \varepsilon_1/3\,\delta_{IJ})$	$\varepsilon_{IJ} - \varepsilon_1/3\,\delta_{IJ} = E/(1+\nu) * (\tau_{IJ} - \tau_1/3\,\delta_{IJ})$
$S_{IJ} = 2M * D_{IJ}$	$D_{IJ} = 1/[2M] * S_{IJ}$
$\tau_{I\neq J} = 2M * D_{IJ} = 2M\,\varepsilon_{I\neq J}$	$\varepsilon_{I\neq J} = 1/[2M] * \tau_{IJ} = 1/(2M)\,\tau_{I\neq J}$

Null Distortion Trace:

$\tau_I - \tau_1/3 * \delta_I = \tau_I - \tau_1/3 * 3 = 0$	$\varepsilon_I - \varepsilon_1/3 * \delta_I = \varepsilon_I - \varepsilon_1/3 * 3 = 0$

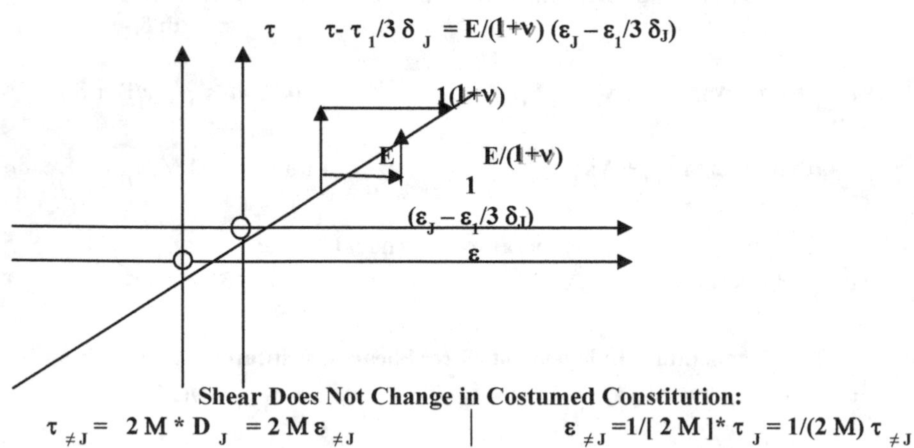

Shear Does Not Change in Costumed Constitution:

$\tau_{\neq J} = 2M * D_J = 2M\,\varepsilon_{\neq J}$	$\varepsilon_{\neq J} = 1/[2M] * \tau_J = 1/(2M)\,\tau_{\neq J}$

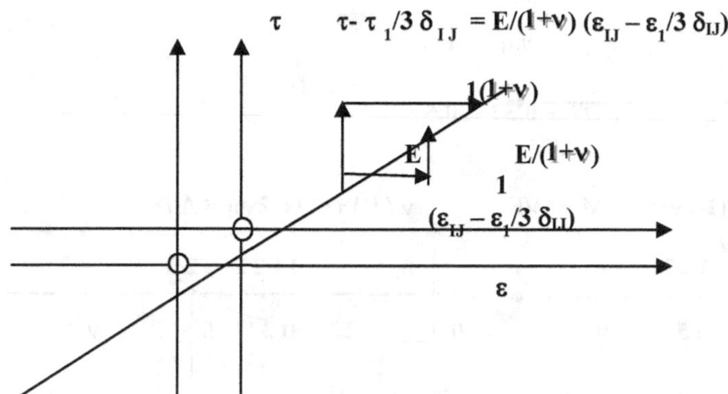

E and ν are Independent Constants

Numerical Values:

ν	0 <	0.25	0.366	0.40	< 0.500
$[1-\nu]$	1 >	0.75	0.633	0.60	> 0.500
$(1+\nu)$	1 <	1.25	1.366	1.40	< 1.500
$[1-\nu] / (1+\nu)$	1 >	0.60	0.462	0.424	> 0.333
$(1+\nu) / [1-\nu]$	1 <	1.666	2.164	2.334	< 3.000
$1/(1+\nu) \equiv 2\,M/E$	1 >	0.80	0.73	0.714	> 0.666
$1/2/(1+\nu) \equiv M/E$	0.5 >	0.40	0.36	0.357	> 0.333
$(1-2\,\nu)$	1 >	0.50	0.26	0.2	>0.000
$1/(1-2\,\nu)$	1<	2.00	3.73	5.00	< ∞
$\nu/(1-2\,\nu)$	0 <	0.50	1.37	2.00	< ∞
$(1+\nu)/(1-2\,\nu)$	1 <	2.50	5.12	7.00	< ∞
$1/(1+\nu)/(1-2\,\nu) \equiv \Lambda/\nu/E$	1 <	1.60	2.75	3.57	< ∞
$\nu/(1+\nu)/(1-2\,\nu) \equiv \Lambda/E$	0 <	0.40	1.00	1.42	< ∞
$[1-\nu]/(1+\nu)/(1-2\nu) = (2\,M+\Lambda)/E$	1 <	1.20	1.73	2.15	< ∞
$(1+\nu)*(1-2\nu)/[1-\nu] = E/(2\,M+\Lambda)$	1 >	0.83	0.57	0.46	> 0

Sum of the following:

$2\,\nu/(1+\nu)/(1-2\,\nu)] \equiv 2\,\Lambda/E$	0 <	0.80	2.00	2.84	< ∞
$[1-\nu]/(1+\nu)/(1-2\nu) = (2\,M+\Lambda)/E$	1 <	1.20	1.73	2.15	< ∞

Equal:

$1/(1-2\,\nu) \equiv (2\,M+3\,\Lambda)E \equiv \kappa/E$	1<	2.00	3.73	5.00	< ∞

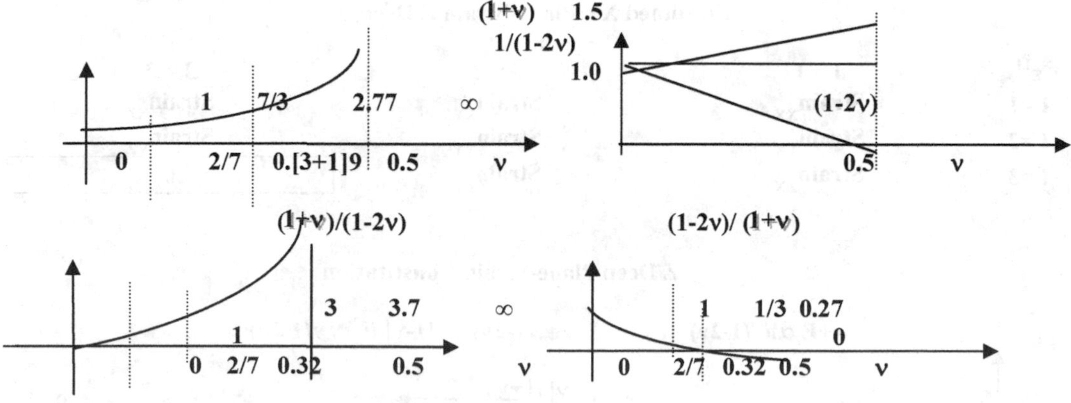

$1/(1+\nu) \equiv 2\,M/E$

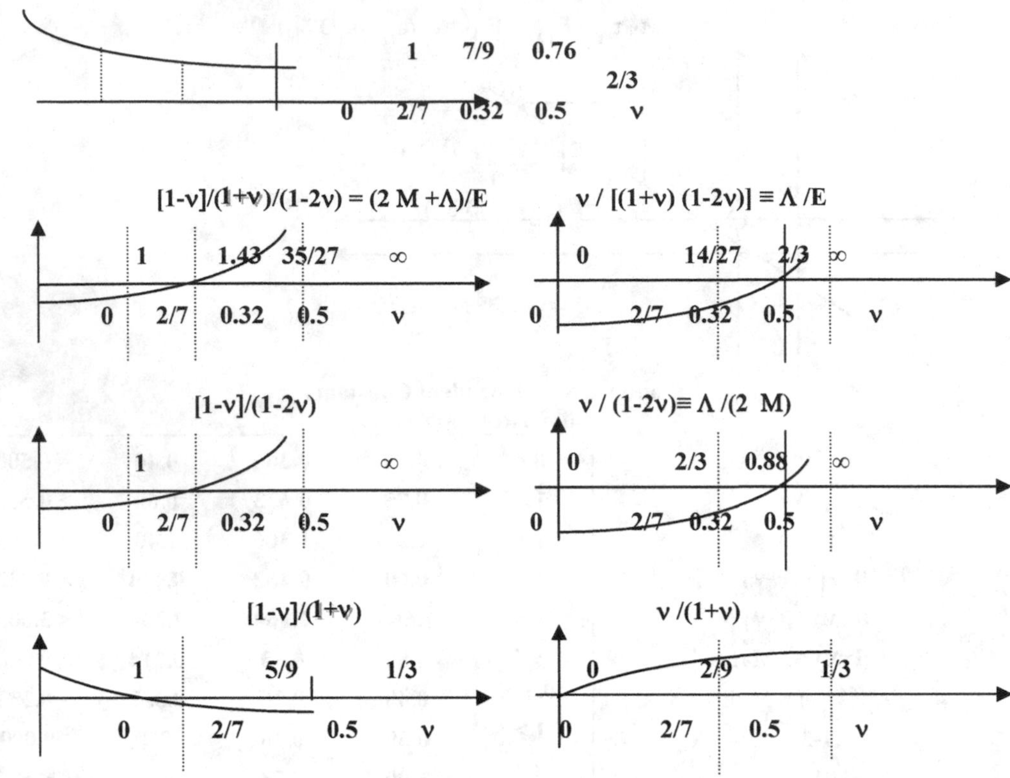

15.3.
3-D Distortion, Costumed Infinitesimal Elasticity:

15.3.1.
Strain Costumed:

15.3.1.1.
XY Plane-Strain Z Deep:

:

Costumed XY Plane-Strain Z Deep:

:

ε^{IJ}	J = 1	J = 2	J = 3
I =1	$Strain_{XX}$	$Strain_{XY}$	$Strain_{XZ}$
I =2	$Strain_{YX}$	$Strain_{YY}$	$Strain_{YZ}$
I =3	$Strain_{ZX}$	$Strain_{ZY}$	0

Z Deep Plane-Strain Constitution;

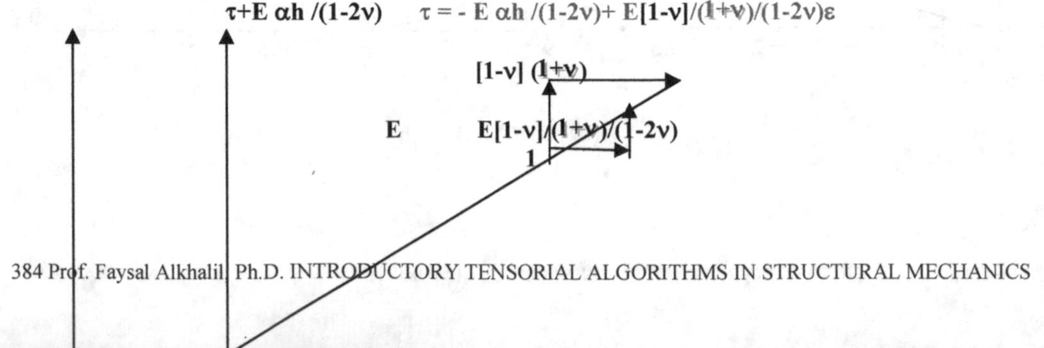

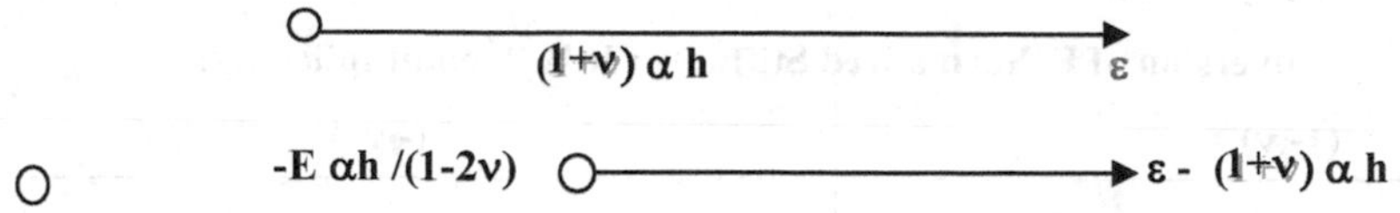

$(1+\nu)\,\alpha\, h$ ε

O $-E\,\alpha h/(1-2\nu)$ $\varepsilon - (1+\nu)\,\alpha\, h$

$\varepsilon_1 = \varepsilon_{XX} + \varepsilon_{YY}$ $\varepsilon_{ZZ} = 0$

Z Deep Plane-Strain Normal Constitution;

$\tau_{XX} + E\,\alpha h/(1-2\nu)$ = $E/(1+\nu)/(1-2\nu)\,[[1-\nu]\,\varepsilon_{XX} + \nu\,\varepsilon_{YY}]$	$\varepsilon_{XX} - (1+\nu)\ \alpha h$ = $[+[1-\nu]\,\tau_{XX} - \nu\,\tau_{YY}]\,(1+\nu)/E$
$\tau_{YY} + E\,\alpha h/(1-2\nu)$ = $E/(1+\nu)/(1-2\nu)\,[\nu\,\varepsilon_{XX} + [1-\nu]\,\varepsilon_{YY}]$	$\varepsilon_{YY} - (1+\nu)\ \alpha h$ = $[-\nu\,\tau_{XX} + [1-\nu]\,\tau_{YY}]\,(1+\nu)/E$

Reduction of Flexibility Equations from 3 to 2:
Cancel τ_{ZZ}, Find τ_{ZZ} in terms of τ_{XX} and τ_{YY}:
From Flexibility:

$$0 - E\,\alpha h = -\nu\,\tau_{XX} - \nu\,\tau_{YY} + \tau_{ZZ}$$

Get:

$$\tau_{ZZ} = \nu\,\tau_{XX} + \nu\,\tau_{YY} - E\,\alpha\,h$$

Substitute τ_{ZZ} in Flexibility Equations of ε_{XX} and ε_{YY} :

$$E\,\varepsilon_{XX} - E\,\alpha h = 1\,\tau_{XX} - \nu\,\tau_{YY} - \nu\,(\nu\,\tau_{XX} + \nu\,\tau_{YY} - \alpha h)$$
$$E\,\varepsilon_{XX} - E\,\alpha h = 1\,\tau_{XX} - \nu\nu\,\tau_{XX} - \nu\,\tau_{YY} - \nu\nu\,\tau_{YY} + \nu\,\alpha h$$

$$E\,\varepsilon_{YY} - E\,\alpha h = 1\,\tau_{YY} - \nu\,\tau_{XX} - \nu\,(\nu\,\tau_{YY} + \nu\,\tau_{XX} - \alpha h)$$
$$E\,\varepsilon_{YY} - E\,\alpha h = 1\,\tau_{YY} - \nu\nu\,\tau_{YY} - \nu\,\tau_{XX} - \nu\nu\,\tau_{XX} + \nu\,\alpha h$$

Symmetric Flexibility :
Plane-Strain Axial Strains:

$$E\,\varepsilon_{XX} - (1+\nu)\,E\,\alpha h = [+[1-\nu]\,\tau_{XX} - \nu\,\tau_{YY}]\,(1+\nu)$$

$$E\,\varepsilon_{YY} - (1+\nu)\,E\,\alpha h = [-\nu\,\tau_{XX} + [1-\nu]\,\tau_{YY}]\,(1+\nu)$$

Multiply First Flexibility Equation by [1-ν] and Second One by ν and Added:
Multiply First Flexibility Equation by ν and Second One by [1-ν] and Added:

$$[1-\nu]\,E\,\varepsilon_{XX} + \nu\,E\,\varepsilon_{YY} - (1+\nu)\,E\,\alpha h = (1+\nu)(1-2\nu)\,\tau_{XX}$$

$$[1-\nu]\,E\,\varepsilon_{YY} + \nu\,E\,\varepsilon_{XX} - (1+\nu)\,E\,\alpha h = (1+\nu)(1-2\nu)\,\tau_{YY}$$

Plane-Strain Constitution;

$\tau_{XX} + E\,\alpha h/(1-2\nu) =$ $E/(1+\nu)/(1-2\nu)\,[\,[1-\nu]\,\varepsilon_{XX} + \nu\,\varepsilon_{YY}]$	$\varepsilon_{XX} - (1+\nu)\ \alpha h =$ $[+[1-\nu]\,\tau_{XX} - \nu\,\tau_{YY}]\,(1+\nu)/E$
$\tau_{YY} + E\,\alpha h/(1-2\nu) =$ $E/(1+\nu)/(1-2\nu)\,[\nu\,\varepsilon_{XX} + [1-\nu]\,\varepsilon_{YY}]$	$\varepsilon_{YY} - (1+\nu)\ \alpha h =$ $[-\nu\,\tau_{XX} + [1-\nu]\,\tau_{YY}]\,(1+\nu)/E$
$\tau_{ZZ} + E\,\alpha h/(1-2\nu) =$ $E/(1+\nu)/(1-2\nu)\ \nu\,[\varepsilon_{XX} + \varepsilon_{YY}]$	$0 - \alpha h =$ $[-\nu\,\tau_{XX} - \nu\,\tau_{YY} + \tau_{ZZ}]/E$

Inversion of E-Normalized Flexibility : It Is Premultiplied by:

(1- ν)	(ν)

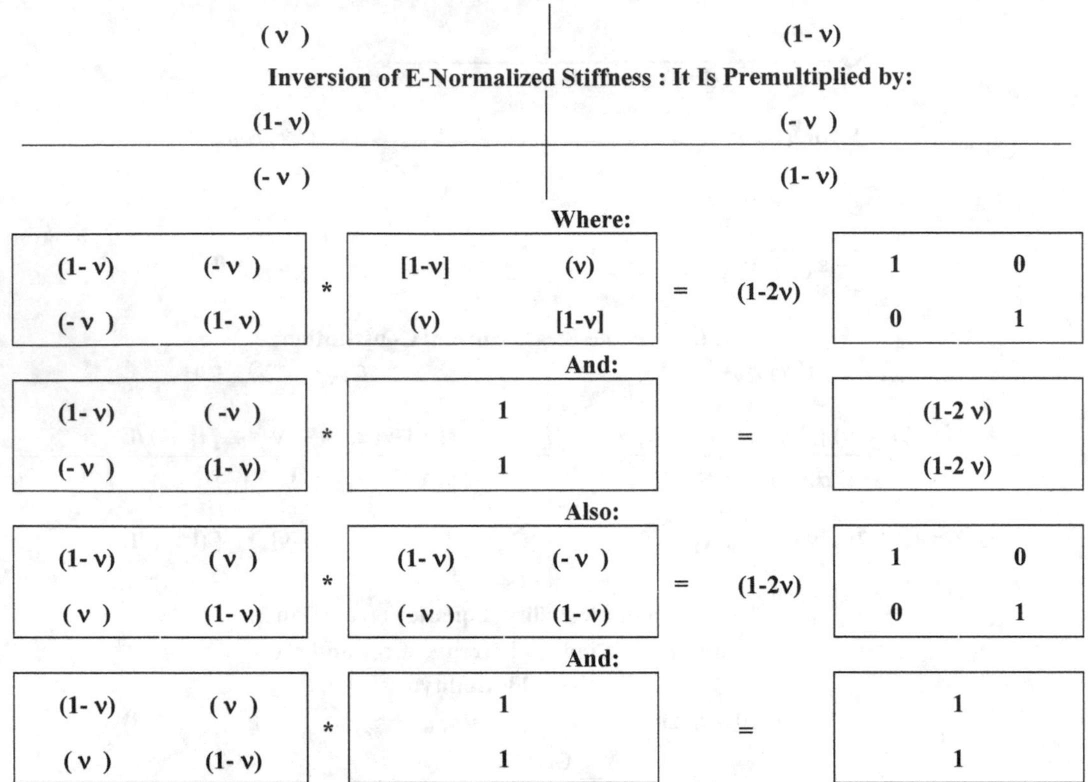

(ν) | (1- ν)
Inversion of E-Normalized Stiffness : It Is Premultiplied by:
(1- ν) | (- ν)
(- ν) | (1- ν)
Where:
(1- ν) (- ν) / (- ν) (1- ν) * [1-ν] (ν) / (ν) [1-ν] = (1-2ν) 1 0 / 0 1
And:
(1- ν) (-ν) / (- ν) (1- ν) * 1 / 1 = (1-2 ν) / (1-2 ν)
Also:
(1- ν) (ν) / (ν) (1- ν) * (1- ν) (- ν) / (- ν) (1- ν) = (1-2ν) 1 0 / 0 1
And:
(1- ν) (ν) / (ν) (1- ν) * 1 / 1 = 1 / 1

15.3.1.2.
1-Dmensional Strain:

Costumed X Lineal-Strain:
:

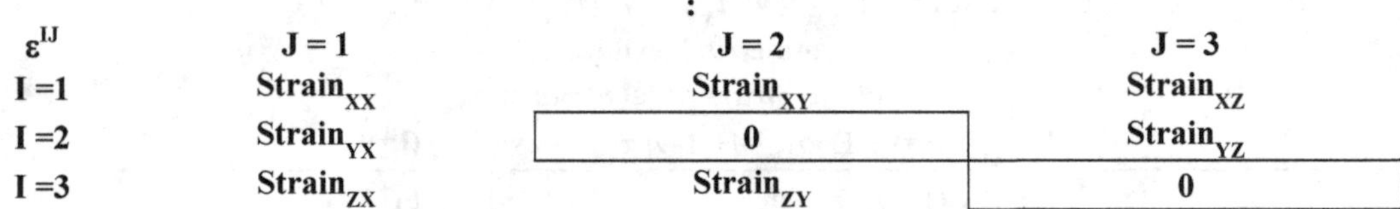

ε^{IJ}	J = 1	J = 2	J = 3
I =1	$Strain_{XX}$	$Strain_{XY}$	$Strain_{XZ}$
I =2	$Strain_{YX}$	0	$Strain_{YZ}$
I =3	$Strain_{ZX}$	$Strain_{ZY}$	0

From X Axial Stiffness :

$$\tau_{XX} + E\,\alpha h/(1-2\nu) = E[1-\nu]/(1+\nu)/(1-2\nu)\,\varepsilon_{XX}$$

Inverted to Get:

$$\varepsilon_{XX} - (1+\nu)/[1-\nu]\,\alpha h = (1+\nu)(1-2\nu)/[1-\nu]/E\ \tau_{XX}$$

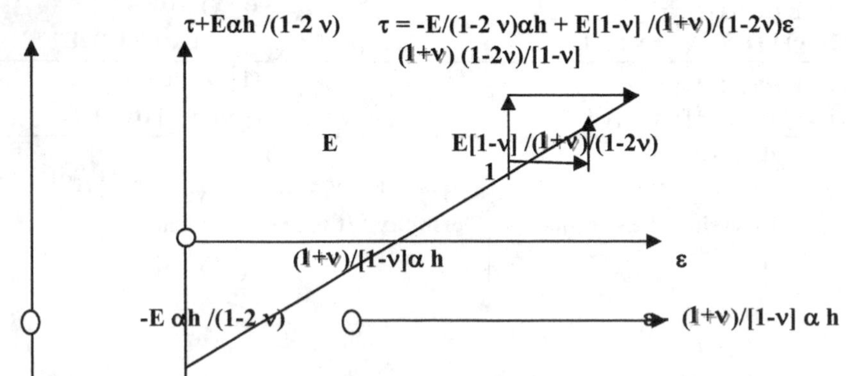

:

$$0 - E\alpha h = [-\nu\tau_{XX} + \tau_{YY} - \nu\tau_{ZZ}]$$

$$0 - E\alpha h = [-\nu\tau_{XX} - \nu\tau_{YY} + \tau_{ZZ}]$$

Subtracted to get:

$$\tau_{YY} = \tau_{ZZ}$$

Back Substituted:

$$0 - E\alpha h = [-\nu\tau_{XX} + (1-\nu)\tau_{YY}] = [-\nu\tau_{XX} + (1-\nu)\tau_{ZZ}]$$

Get:

$$\tau_{YY} + E\alpha h / [1-\nu] = \tau_{ZZ} + E\alpha h / [1-\nu] = \nu/[1-\nu]\tau_{XX}$$

But:

$$\nu/[1-\nu]\tau_{XX} = E\ \nu/(1+\nu)/(1-2\nu)\varepsilon_{XX} - E\nu/[1-\nu]/(1-2\nu)\alpha h$$

So:

$$\tau_{YY} + E\alpha h/[1-\nu] = \tau_{ZZ} + E\alpha h/[1-\nu] = E\nu/(1+\nu)/(1-2\nu)\varepsilon_{XX} - E\nu/[1-\nu]/(1-2\nu)\alpha h$$

Or:

$$\tau_{YY} + E\alpha h / (1-2\nu) = \tau_{ZZ} + E\alpha h / (1-2\nu) = E\ \nu/(1+\nu)/(1-2\nu)\varepsilon_{XX}$$

Constitution: Axial-Strain $\varepsilon_{YY} = \varepsilon_{ZZ} = 0$:

$\tau_{XX} + E\alpha h/(1-2\nu)$ = $E[1-\nu]/(1+\nu)/(1-2\nu)\varepsilon_{XX}$	$\varepsilon_{XX} - (1+\nu)/[1-\nu]\alpha h$ = $(1+\nu)(1-2\nu)/[1-\nu]\tau_{XX}/E$
$\tau_{YY} + E\alpha h/(1-2\nu)$ = $E\nu/(1+\nu)/(1-2\nu)\varepsilon_{XX}$	$\varepsilon_{YY} = 0$
$\tau_{ZZ} + E\alpha h/(1-2\nu)$ = $E\nu/(1+\nu)/(1-2\nu)\varepsilon_{XX}$	$\varepsilon_{XX} = 0$

$\tau\ \varepsilon$ Axial Strain Constitution:

$\tau + E/(1-2\nu)\alpha h$ = $(2M+\Lambda)\varepsilon$ = $E/(1+\nu)*[1-\nu]/(1-2\nu)\varepsilon$	$\varepsilon - (1+\nu)/[1-\nu]\alpha h$ = $1/(2M+\Lambda)\tau$ = $(1+\nu)/[1-\nu]*(1-2\nu)/E\,\tau$

:

1-Dimensional Strain Constitution:

:

$\tau + E/(1-2\nu)\alpha h = E*[1-\nu]/(1+\nu)/(1-2\nu)\varepsilon$	$\varepsilon - (1+\nu)/[1-\nu]\alpha h = (1+\nu)(1-2\nu)/[1-\nu]/E\,\tau$

15.3.2.
Stress Costumed:

15.3.2.1.
2-Dmensional Stress:

:

Costumed XY Plane-Stress Z Deep:

:

τ^{IJ}	J = 1	J = 2	J = 3
I =1	$Stress_{XX}$	$Stress_{XY}$	$Stress_{XZ}$
I =2	$Stress_{YX}$	$Stress_{YY}$	$Stress_{YZ}$

I =3	$Stress_{ZX}$	$Stress_{ZY}$	0

$Stress_{XX}$:

$\tau^{XX}_{(Heat)} = E\,\alpha h/[1-\nu]$	$\tau^{XX} = E/(1-\nu^2)(+\varepsilon_{XX}+\nu\,\varepsilon_{YY})$

Z Deep XY Plane-Stress:

$\tau + E\,\alpha h/[1-\nu]$ $\quad \tau = -E\,\alpha h/[1-\nu] + E/(1-\nu^2)\,\varepsilon$

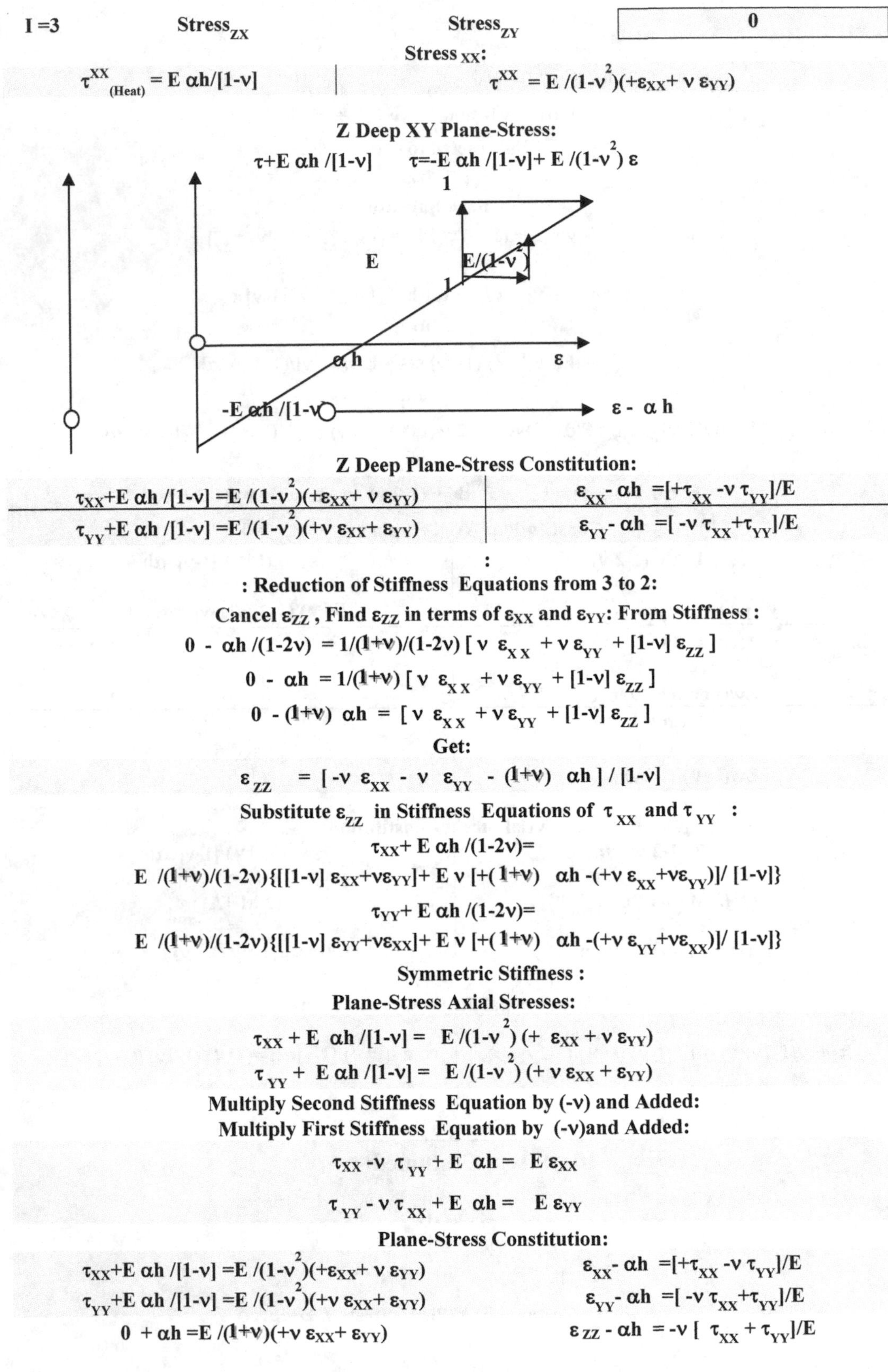

Z Deep Plane-Stress Constitution:

$\tau_{XX} + E\,\alpha h/[1-\nu] = E/(1-\nu^2)(+\varepsilon_{XX}+\nu\,\varepsilon_{YY})$	$\varepsilon_{XX} - \alpha h = [+\tau_{XX} - \nu\,\tau_{YY}]/E$
$\tau_{YY} + E\,\alpha h/[1-\nu] = E/(1-\nu^2)(+\nu\,\varepsilon_{XX}+\varepsilon_{YY})$	$\varepsilon_{YY} - \alpha h = [-\nu\,\tau_{XX}+\tau_{YY}]/E$

:

: Reduction of Stiffness Equations from 3 to 2:

Cancel ε_{ZZ}, Find ε_{ZZ} in terms of ε_{XX} and ε_{YY}: From Stiffness :

$$0 - \alpha h/(1-2\nu) = 1/(1+\nu)/(1-2\nu)\,[\,\nu\,\varepsilon_{XX} + \nu\,\varepsilon_{YY} + [1-\nu]\,\varepsilon_{ZZ}\,]$$

$$0 - \alpha h = 1/(1+\nu)\,[\,\nu\,\varepsilon_{XX} + \nu\,\varepsilon_{YY} + [1-\nu]\,\varepsilon_{ZZ}\,]$$

$$0 - (1+\nu)\,\alpha h = [\,\nu\,\varepsilon_{XX} + \nu\,\varepsilon_{YY} + [1-\nu]\,\varepsilon_{ZZ}\,]$$

Get:

$$\varepsilon_{ZZ} = [\,-\nu\,\varepsilon_{XX} - \nu\,\varepsilon_{YY} - (1+\nu)\,\alpha h\,]/[1-\nu]$$

Substitute ε_{ZZ} in Stiffness Equations of τ_{XX} and τ_{YY} :

$$\tau_{XX} + E\,\alpha h/(1-2\nu) =$$

$$E/(1+\nu)/(1-2\nu)\{[[1-\nu]\,\varepsilon_{XX}+\nu\varepsilon_{YY}] + E\,\nu\,[+(1+\nu)\;\alpha h - (+\nu\,\varepsilon_{XX}+\nu\varepsilon_{YY})]/[1-\nu]\}$$

$$\tau_{YY} + E\,\alpha h/(1-2\nu) =$$

$$E/(1+\nu)/(1-2\nu)\{[[1-\nu]\,\varepsilon_{YY}+\nu\varepsilon_{XX}] + E\,\nu\,[+(1+\nu)\;\alpha h - (+\nu\,\varepsilon_{YY}+\nu\varepsilon_{XX})]/[1-\nu]\}$$

Symmetric Stiffness :

Plane-Stress Axial Stresses:

$$\tau_{XX} + E\;\alpha h/[1-\nu] = E/(1-\nu^2)\,(+\,\varepsilon_{XX} + \nu\,\varepsilon_{YY})$$

$$\tau_{YY} + E\,\alpha h/[1-\nu] = E/(1-\nu^2)\,(+\,\nu\,\varepsilon_{XX} + \varepsilon_{YY})$$

Multiply Second Stiffness Equation by $(-\nu)$ and Added:

Multiply First Stiffness Equation by $(-\nu)$and Added:

$$\tau_{XX} - \nu\;\tau_{YY} + E\;\alpha h = E\,\varepsilon_{XX}$$

$$\tau_{YY} - \nu\,\tau_{XX} + E\;\alpha h = E\,\varepsilon_{YY}$$

Plane-Stress Constitution:

$\tau_{XX} + E\,\alpha h/[1-\nu] = E/(1-\nu^2)(+\varepsilon_{XX}+\nu\,\varepsilon_{YY})$	$\varepsilon_{XX} - \alpha h = [+\tau_{XX} - \nu\,\tau_{YY}]/E$
$\tau_{YY} + E\,\alpha h/[1-\nu] = E/(1-\nu^2)(+\nu\,\varepsilon_{XX}+\varepsilon_{YY})$	$\varepsilon_{YY} - \alpha h = [-\nu\,\tau_{XX}+\tau_{YY}]/E$
$0 + \alpha h = E/(1+\nu)(+\nu\,\varepsilon_{XX}+\varepsilon_{YY})$	$\varepsilon_{ZZ} - \alpha h = -\nu\,[\;\tau_{XX} + \tau_{YY}]/E$

15.3.2.2.
1-Dmensional Stress X Lineal-Stress:

Costumed X Lineal-Stress:

:

τ^{IJ}	J = 1	J = 2	J = 3
I =1	$Stress_{XX}$	$Stress_{XY}$	$Stress_{XZ}$
I =2	$Stress_{YX}$	0	$Stress_{YZ}$
I =3	$Stress_{ZX}$	$Stress_{ZY}$	0

Axial Stress Constitution:

$\tau + E\,\alpha h$ $\tau = -E\,\alpha h + E\varepsilon$

1

E E

1

αh ε

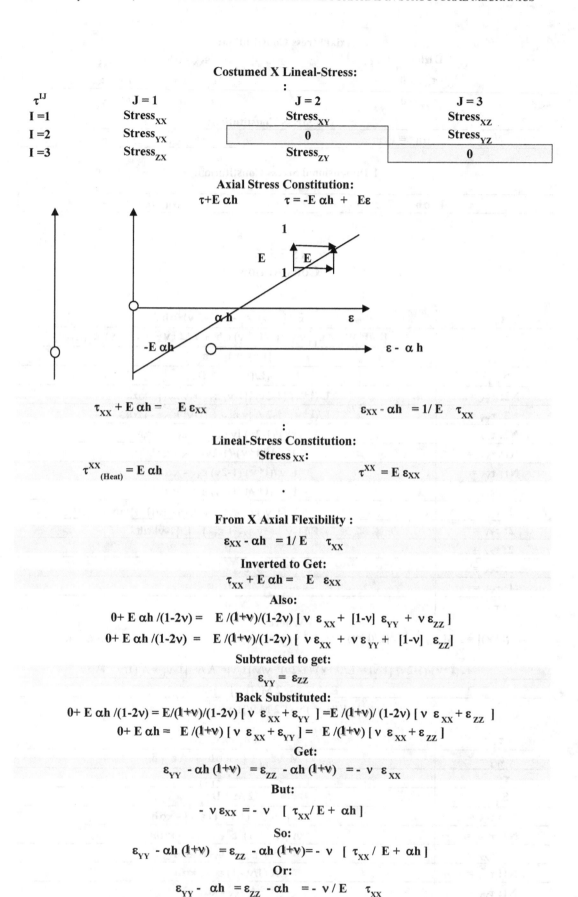

$-E\,\alpha h$ $\varepsilon - \alpha h$

$\tau_{XX} + E\,\alpha h = E\,\varepsilon_{XX}$ $\varepsilon_{XX} - \alpha h = 1/E \quad \tau_{XX}$

:

Lineal-Stress Constitution:

Stress $_{XX}$:

$\tau^{XX}_{(Heat)} = E\,\alpha h$ $\tau^{XX} = E\,\varepsilon_{XX}$

.

From X Axial Flexibility :

$$\varepsilon_{XX} - \alpha h = 1/E \quad \tau_{XX}$$

Inverted to Get:

$$\tau_{XX} + E\,\alpha h = E\,\varepsilon_{XX}$$

Also:

$$0 + E\,\alpha h/(1-2\nu) = E/(1+\nu)/(1-2\nu)\,[\,\nu\,\varepsilon_{XX} + [1-\nu]\,\varepsilon_{YY} + \nu\,\varepsilon_{ZZ}\,]$$

$$0 + E\,\alpha h/(1-2\nu) = E/(1+\nu)/(1-2\nu)\,[\,\nu\,\varepsilon_{XX} + \nu\,\varepsilon_{YY} + [1-\nu]\,\varepsilon_{ZZ}]$$

Subtracted to get:

$$\varepsilon_{YY} = \varepsilon_{ZZ}$$

Back Substituted:

$$0 + E\,\alpha h/(1-2\nu) = E/(1+\nu)/(1-2\nu)\,[\,\nu\,\varepsilon_{XX} + \varepsilon_{YY}\,] = E/(1+\nu)/(1-2\nu)\,[\,\nu\,\varepsilon_{XX} + \varepsilon_{ZZ}\,]$$

$$0 + E\,\alpha h = E/(1+\nu)\,[\,\nu\,\varepsilon_{XX} + \varepsilon_{YY}\,] = E/(1+\nu)\,[\,\nu\,\varepsilon_{XX} + \varepsilon_{ZZ}\,]$$

Get:

$$\varepsilon_{YY} - \alpha h\,(1+\nu) = \varepsilon_{ZZ} - \alpha h\,(1+\nu) = -\,\nu\,\varepsilon_{XX}$$

But:

$$-\,\nu\,\varepsilon_{XX} = -\,\nu\,[\,\tau_{XX}/E + \alpha h\,]$$

So:

$$\varepsilon_{YY} - \alpha h\,(1+\nu) = \varepsilon_{ZZ} - \alpha h\,(1+\nu) = -\,\nu\,[\,\tau_{XX}/E + \alpha h\,]$$

Or:

$$\varepsilon_{YY} - \alpha h = \varepsilon_{ZZ} - \alpha h = -\,\nu/E \quad \tau_{XX}$$

Axial Stress Constitution:

$\tau_{XX} + E\,\alpha h = \quad E\,\varepsilon_{XX}$	$\varepsilon_{XX} - \alpha h \;= 1/E \quad \tau_{XX}$
$\tau_{YY} = 0$	$\varepsilon_{YY} - \alpha h \;= -\;\nu/E \;\; \tau_{XX}$
$\tau_{ZZ} = 0$	$\varepsilon_{ZZ} - \alpha h \;= -\;\nu/E \;\; \tau_{XX}$

$\tau\,\varepsilon$ Axial Stress Constitution:

$\tau \;\; + E \;\alpha h \;\; = E \;\varepsilon$	$\varepsilon \;\; - \alpha h \;\; = 1/E \;\; \tau$

1-Dimensional Stress Constitution:

:

$\tau \;\; + E \;\alpha h \;\; = E \;\varepsilon$	$\varepsilon \;\; - \alpha h \;\; = 1/E \;\; \tau$

15.3.3.
All Constitutions:

$\tau_1 =$	$E/(1\text{-}2\,\nu)\;\varepsilon_1 - E/(1\text{-}2\,\nu)\;\alpha h\;3$
$\tau_{IJ} =$	$[E/(1+\nu)]\;*\,\varepsilon_{IJ} + [\,-\,E/(1\text{-}2\,\nu)\,\alpha h\; + E/(1+\nu)*\,\nu/(1\text{-}2\,\nu)\;\varepsilon_1]\,\delta_{IJ}$
$\tau_{I\neq J} =$	$[E/(1+\nu)]\;*\,\varepsilon_{I\neq J}$
$S_{IJ} =$	$[E/(1+\nu)]\;*\,D_{IJ}$
N2: $\tau_{XX} =$	$E/(1+\nu)/(1\text{-}2\nu)\;[[1\text{-}\nu]\,\varepsilon_{XX} + \nu\,\varepsilon_{YY}] - \kappa\alpha h$
N2: $\tau_{YY} =$	$E/(1+\nu)/(1\text{-}2\nu)\;[\nu\,\varepsilon_{XX} + [1\text{-}\nu]\,\varepsilon_{YY}] - \kappa\alpha h$
N2: $\tau_{ZZ} =$	$E/(1+\nu)/(1\text{-}2\nu)\;[\nu\,\varepsilon_{XX} + \nu\,\varepsilon_{YY}] - \kappa\alpha h$
N1: $\tau_{XX} =$	$E\;[1\text{-}\nu]\;/(1+\nu)/(1\text{-}2\nu)\;\varepsilon_{XX} - \kappa\alpha h$
N1: $\tau_{YY} =$	$E\;\nu\;/(1+\nu)/(1\text{-}2\nu)\;\varepsilon_{XX} - \kappa\alpha h$
N1: $\tau_{ZZ} =$	$E\;\nu\;/(1+\nu)/(1\text{-}2\nu)\;\varepsilon_{XX} - \kappa\alpha h$
2: $\tau_{XX} =$	$E/(1\text{-}\nu^2)\;(+\,\varepsilon_{XX} + \nu\,\varepsilon_{YY}) - E/[1\text{-}\nu]\;\alpha h$
2: $\tau_{YY} =$	$E/(1\text{-}\nu^2)\;(+\nu\,\varepsilon_{XX} + \varepsilon_{YY}) - E/[1\text{-}\nu]\;\alpha h$
2: $\tau_{ZZ} =$	0
1: $\tau_{XX} =$	$E\,\varepsilon_{XX} - E\;\alpha h$
1: $\tau_{YY} =$	0
1: $\tau_{ZZ} =$	0

Where:

$[E/(1+\nu)] \equiv 2\,M$	$[E/(1+\nu)]*\;\nu/(1\text{-}2\,\nu) \equiv \Lambda$	$E/(1\text{-}2\,\nu) \equiv 2\,M + 3\,\Lambda \equiv \kappa$

:

$$E/(1+\nu)/(1\text{-}2\nu)\;[1\text{-}\nu] = E/(1+\nu)/(1\text{-}2\nu)*\;\nu/\nu\;[1\text{-}\nu] = \Lambda/\nu\;\;[1\text{-}\nu] = \Lambda\,(1/\nu - 1)$$

$$E/(1\text{-}\nu^2) \;= 2\,M\,/[1\text{-}\nu]$$

$$E/(1\text{-}\nu^2)\;\nu = \;2\,M\;\nu/[1\text{-}\nu]$$

:

$\tau_1 =$	$\kappa\,\varepsilon_1 - \kappa\,\alpha h\;3$
$\tau_{IJ} =$	$2\,M * \varepsilon_{IJ} + [\,-\,\kappa\,\alpha h\; + \;\Lambda\,\varepsilon_1]\,\delta_{IJ}$
$\tau_{I\neq J} =$	$2\,M * \varepsilon_{I\neq J}$
$S_{IJ} =$	$2\,M * D_{IJ}$
N2: $\tau_{XX} =$	$\Lambda\,[\;\varepsilon_{XX} + (1/\nu - 1)\,\varepsilon_{YY}] - \kappa\alpha h$
N2: $\tau_{YY} =$	$\Lambda\,[(1/\nu - 1)\;\;\varepsilon_{XX} + \varepsilon_{YY}] - \kappa\alpha h$
N2: $\tau_{ZZ} =$	$\Lambda\,[\;\;\varepsilon_{XX} + \;\varepsilon_{YY}] - \kappa\alpha h$
N1: $\tau_{XX} =$	$\Lambda\;(1/\nu - 1)\;\varepsilon_{XX} - \kappa\alpha h$
N1: $\tau_{YY} =$	$\Lambda\;\varepsilon_{XX} - \kappa\alpha h$

N1: τ_{ZZ} =	$\Lambda\, \varepsilon_{XX}$ - $\kappa\alpha h$
2: τ_{XX} =	2 M /[1-ν] (+ ε_{XX} + ν ε_{YY}) - E/[1-ν] αh
2: τ_{YY} =	2 M /[1-ν] (+ν ε_{XX} + ε_{YY}) - E/[1-ν] αh
2: τ_{ZZ} =	0
1: τ_{XX} =	E ε_{XX} - E αh
1: τ_{YY} =	0
1: τ_{ZZ} =	0

Substituting:

$\varepsilon^{IJ}(x) = ½ (u^I_{,J} + u^J_{,I})$ | $\varepsilon^{AA}(x) = u^A_{,A}$

:

τ_1 =	$\kappa\, u^A_{,A} - \kappa\alpha h\, 3$
τ_{IJ} =	$2 M * ½ (u^I_{,J} + u^J_{,I}) + [- \kappa\alpha h + \Lambda u^A_{,A}]\, \delta_{IJ}$
$\tau_{I \neq J}$ =	$2 M * ½ (u^I_{,J} + u^J_{,I})$
$\tau_{IJ} - \tau_1 /3$ =	$2 M * [½ (u^I_{,J} + u^J_{,I}) - u^A_{,A} /3]D_{IJ}$
N2: τ_{XX} =	$\Lambda [u^X_{,X} + (1/\nu - 1) u^Y_{,Y}] - \kappa\alpha h$
N2: τ_{YY} =	$\Lambda [(1/\nu - 1) u^X_{,X} + u^Y_{,Y}] - \kappa\alpha h$
N2: τ_{ZZ} =	$\Lambda [u^X_{,X} + u^Y_{,Y}] - \kappa\alpha h$
N1: τ_{XX} =	$\Lambda (1/\nu - 1) u^X_{,X} - \kappa\alpha h$
N1: τ_{YY} =	$\Lambda u^X_{,X} - \kappa\alpha h$
N1: τ_{ZZ} =	$\Lambda u^X_{,X} - \kappa\alpha h$
2: τ_{XX} =	$2 M /[1-\nu] (+ u^X_{,X} + \nu\, u^Y_{,Y}) - E/[1-\nu]\, \alpha h$
2: τ_{YY} =	$2 M /[1-\nu] (+\nu\, u^X_{,X} + u^Y_{,Y}) - E/[1-\nu]\, \alpha h$
2: τ_{ZZ} =	0
1: τ_{XX} =	$E\, u^X_{,X} - E\, \alpha h$
1: τ_{YY} =	0
1: τ_{ZZ} =	0

Plane-Strain Constitution;

τ_{XX}+ E αh /(1-2ν)= E/(1+ν)/(1-2ν) [[1-ν] ε_{XX} + ν ε_{YY}]	ε_{XX} - (1+ν) αh = [+[1-ν] τ_{XX} - ν τ_{YY}] (1+ν) /E
τ_{YY}+ E αh /(1-2ν)= E/(1+ν)/(1-2ν) [ν ε_{XX} +[1-ν] ε_{YY}]	ε_{YY} - (1+ ν) αh = [- ν τ_{XX} + [1-ν] τ_{YY}] (1+ν) /E
τ_{ZZ}+ E αh /(1-2ν) = E /(1+ν)/(1-2ν) ν [ε_{XX} + ε_{YY}]	0 - αh = [- ν τ_{XX} - ν τ_{YY} + τ_{ZZ}] /E

1-Dimensional Strain State Constitution

τ_{XX} + E αh /(1-2 ν) = E [1-ν] /(1+ν)/(1-2ν) ε_{XX}	ε_{XX} - (1+ν)/[1-ν] αh = (1+ν) (1-2ν)/[1-ν] τ_{XX} / E
τ_{YY} + E αh/(1-2ν) = Eν/(1+ν)/(1-2ν) ε_{XX}	ε_{YY} = 0
τ_{ZZ} + E αh / (1-2ν)= Eν/(1+ν)/(1-2ν) ε_{XX}	ε_{XX} = 0

Plane-Stress Constitution:

τ_{XX}+E αh /[1-ν] =E /(1-ν^2)(+ε_{XX}+ ν ε_{YY})	ε_{XX}- αh =[+τ_{XX} -ν τ_{YY}]/E
τ_{YY}+E αh /[1-ν] =E /(1-ν^2)(+ν ε_{XX}+ ε_{YY})	ε_{YY}- αh =[-ν τ_{XX}+τ_{YY}]/E

$0 + \alpha h = E/(1+\nu)(+\nu\, \varepsilon_{XX} + \varepsilon_{YY})$	$\varepsilon_{ZZ} - \alpha h = -\nu\, [\tau_{XX} + \tau_{YY}]/E$

Axial Stress Constitution:

$\tau_{XX} + E\,\alpha h = E\,\varepsilon_{XX}$	$\varepsilon_{XX} - \alpha h = 1/E\ \tau_{XX}$
$\tau_{YY} = 0$	$\varepsilon_{YY} - \alpha h = -\nu/E\ \tau_{XX}$
$\tau_{ZZ} = 0$	$\varepsilon_{ZZ} - \alpha h = -\nu/E\ \tau_{XX}$

15.3.4.
Comparison of Plane-Strain and Plane-Stress:

Plane-Strain:

$$\tau_{XX} + E\,\alpha h/(1-2\nu) = E/(1+\nu)/(1-2\nu)\ [[1-\nu]\ \varepsilon_{XX} + \nu\, \varepsilon_{YY} + 0]$$

Plane-Stress:

$$\tau_{XX} + E_S\, \alpha h/(1-\nu_S) = E_S/(1-\nu_S^2)\ (\varepsilon_{XX} + \nu_S\, \varepsilon_{YY})$$

Let:

1	2	3
$E\alpha/(1-2\nu)$ = $E_S\, \alpha_S/(1-\nu_S)$	$E\,[1-\nu]/(1+\nu)/(1-2\nu)$ = $E_S/(1-\nu_S^2)$	$E\,\nu/(1+\nu)/(1-2\nu)$ = $E_S\, \nu_S/(1-\nu_S^2)$

Divide 2 by 3 to get:

$$[1-\nu]\ /\nu = 1/\nu_S$$

Plane-Stress E_S and ν_S in terms of Plane-Strain E and ν:

$$\nu_S = \nu/[1-\nu]$$

$$(1-\nu_S) = 1 - \nu/[1-\nu] = (1-2\nu)/[1-\nu]$$

$$1-\nu_S^2 = 1 - \nu^2/[1-\nu]^2 = (1-2\nu)/[1-\nu]^2$$

$$\nu_S/(1-\nu_S^2) = \nu/[1-\nu]\ [1-\nu]^2/(1-2\nu) = \nu\,[1-\nu]\ /(1-2\nu)$$

Substitute $(1-\nu_S^2)$ in 2 to get:

$$E_S/(1-\nu_S^2) = E_S\,[1-\nu]^2/(1-2\nu) = E\,[1-\nu]/(1+\nu)/(1-2\nu)$$

$$E_S = E/(1-\nu^2)$$

If we Substitute $\nu_S/(1-\nu_S^2)$ in 3 to get same result:

$$E_S\, \nu_S/(1-\nu_S^2) = E_S\ \nu\,[1-\nu]\ /(1-2\nu) = E\,\nu/(1+\nu)/(1-2\nu)$$

$$E_S = E/(1-\nu^2)$$

Substitute E_S and $(1-\nu_S^2)$ in right side of 2 to get 2:

$$E_S/(1-\nu_S^2) = E/(1-\nu^2)\ [1-\nu]^2/(1-2\nu) = E\,[1-\nu]/(1+\nu)/(1-2\nu)\ \text{True}$$

Substitute E_S and $\nu_S/(1-\nu_S^2)$ in right side of 3 to get 3:

$$E_S\, \nu_S/(1-\nu_S^2) = E/(1-\nu^2)\ \nu\,[1-\nu]\ /(1-2\nu) = E\,\nu/(1+\nu)/(1-2\nu)\ \text{True}$$

Substitute E_S and $(1-\nu_S)$ in 1 to get:

$$E_S\, \alpha_S/(1-\nu_S) = E/(1-\nu^2)\ \alpha_S\,[1-\nu]/(1-2\nu) = E\alpha/(1-2\nu)$$

$$\alpha_S = \alpha\ (1+\nu)$$

Substitute E_S, ν_S and α_S in right side of 1 to get 1:

$$E_S\, \alpha_S/(1-\nu_S) = E/(1-\nu^2)\ \alpha\ (1+\nu)\,[1-\nu]/(1-2\nu) = E\alpha/(1-2\nu)\ \text{True}$$

So:

$\alpha_S = \alpha\ (1+\nu)$	$E_S = E/(1-\nu^2)$	$\nu_S = \nu/[1-\nu]$

Plane-Strain E and ν in terms of Plane-Stress E_S and ν_S:

$$1/\nu = 1 + 1/\nu_S = (1+\nu_S)/\nu_S$$

$$\nu = \nu_S/(1+\nu_S)$$

$$(1-\nu) = 1-\nu_S/(1+\nu_S) = 1/(1+\nu_S)$$

$$(1+\nu) = 1+\nu_S/(1+\nu_S) = (1+2\nu_S)/(1+\nu_S)$$

$$(1-2\nu) = 1-2\nu_S/(1+\nu_S) = (1-\nu_S)/(1+\nu_S)$$

Substitute [1-ν] / (1+ν)/ (1-2ν) in 2 to get:

$$E[1-\nu]/(1+\nu)/(1-2\nu)=E/(1+\nu_S)(1+\nu_S)/(1+2\nu_S)(1+\nu_S)/(1-\nu_S)=E_S/(1-\nu_S^2)$$

$$E=E_S(1+2\nu_S)/(1+\nu_S)^2$$

If we Substitute ν/ (1+ν)/ (1-2ν) in 3 to get same result:

$$E\nu/(1+\nu)/(1-2\nu)=E\nu_S/(1+\nu_S)(1+\nu_S)/(1+2\nu_S)(1+\nu_S)/(1-\nu_S)=E_S\nu_S/(1+\nu_S)/(1-\nu_S)$$

$$E=E_S(1+2\nu_S)/(1+\nu_S)^2$$

Substitute E and [1-ν] / (1+ν)/ (1-2ν) in left side of 2 to get 2:

$$E[1-\nu]/(1+\nu)/(1-2\nu)=E_S(1+2\nu_S)/(1+\nu_S)^2/(1+\nu_S)(1+\nu_S)/(1+2\nu_S)(1+\nu_S)/(1-\nu_S)$$

$$=$$

$$E_S/(1+\nu_S)/(1-\nu_S) = -E_S/(1+\nu_S)/(-1+\nu_S)=E_S/(1-\nu_S^2)\text{True}$$

Substitute E and ν/ (1+ν)/ (1-2ν) in left side of 3 to get 3:

$$E\nu/(1+\nu)/(1-2\nu)= E_S(1+2\nu_S)/(1+\nu_S)^2\,\nu_S/(1+\nu_S)(1+\nu_S)/(1+2\nu_S)(1+\nu_S)/(1-\nu_S)$$

$$=$$

$$E_S/(1+\nu_S)\nu_S/(1-\nu_S) = E_S\nu_S/(1-\nu_S^2)\text{True}$$

Substitute E and (1-2 ν) in 1 to get:

$$E\alpha/(1-2\nu)= E_S(1+2\nu_S)/(1+\nu_S)^2\,\alpha(1+\nu_S)/(1-\nu_S) = E_S\alpha_S/(1-\nu_S)$$

$$\alpha=\alpha_S(1+\nu_S)/(1+2\nu_S)$$

Substitute E, ν and α in left side of 1 to get 1:

$$E\alpha/(1-2\nu) = E_S(1+2\nu_S)/(1+\nu_S)^2\,\alpha_S(1+\nu_S)/(1+2\nu_S)(1+\nu_S)/(1-\nu_S)$$

$$=$$

$$E_S\alpha_S/(1-\nu_S)\text{ True}$$

So:

$\alpha=\alpha_S(1+\nu_S)/(1+2\nu_S)$	$E=E_S(1+2\nu_S)/(1+\nu_S)^2$	$\nu=\nu_S/(1+\nu_S)$

A Check Comparison:

$\alpha_S=\alpha(1+\nu)$	$E_S=E/(1-\nu^2)$	$\nu_S=\nu/[1-\nu]$
$\alpha=\alpha_S(1+\nu_S)/(1+2\nu_S)$	$E=E_S(1+2\nu_S)/(1+\nu_S)^2$	$\nu=\nu_S/(1+\nu_S)$

If multiplied:

$1=(1+\nu)(1+\nu_S)/(1+2\nu_S)$	$1=(1+2\nu_S)/(1-\nu^2)/(1+\nu_S)^2$	$1=1/[1-\nu]/(1+\nu_S)$
	$1=(1+2\nu_S)/(1+\nu)/(1+\nu_S)$	

Notice that if Poisson ratio=0 then all values are equal.

Plane-Strain:

$$\tau_{XX} + E\alpha h/(1-2\nu) = E/(1+\nu)/(1-2\nu)\ [(1-\nu)\varepsilon_{XX} + \nu\varepsilon_{YY} + 0]$$

Plane-Stress:

$$\tau_{XX} + E_S\alpha_S h/(1-\nu_S) = E_S/(1-\nu_S^2)(\varepsilon_{XX}+\nu_S\varepsilon_{YY})$$

:

$\alpha_S=\alpha(1+\nu)$	$E_S=E/(1-\nu^2)$	$\nu_S=\nu/(1-\nu)$
$\alpha=\alpha_S(1+\nu_S)/(1+2\nu_S)$	$E=E_S(1+2\nu_S)/(1+\nu_S)^2$	$\nu=\nu_S/(1+\nu_S)$

If Poisson ratio=0 then all values are equal.

Notice that if Poisson ratio $\nu_S = 0.5$ then:

$\alpha_S=\alpha\ 1.33$	$E_S=E/(8/9)$	$\nu_S=0.5$

$\alpha = \alpha_S\ 1.5/2$	$E = E_S\ 2/1.5^2$	$\nu = 0.5/1.5 = 0.33$

It is not applicable to Poisson ratio ν larger than 1/3, for example ν = 0.5 then:

$\alpha_S = \alpha\ 1.5$	$E_S = E/0.75$	$\nu_S = 0.5/0.5 = 1.0$
$\alpha = \alpha_S\ 2/3$	$E = E_S\ 2/1.5^2$	$\nu = 0.5$

An Application on Steel Two Dimensional Values :

If α=1.2E-05, Modulus E=2E+05 mN/m², Poisson 0.3 Then:

α_S = 1.2E-05*1.3	E_S = 2E+05/0.91	$\nu_S = 0.3/0.7 = 0.429$

If α_S =1.2E-05, Modulus E_S =2E+05 mN/m², Poisson 0.3 Then:

α=1.2E-05*1.3/1.6	$E = 2E{+}05 \cdot 1.6/1.3^2$	$\nu = 0.3/1.3 = 0.231$

No Heat Comparison of
Plane-Strain and Plane-Stress:

No Heat Plane-Strain XX Normal Constitution;

$\tau_{XX}=E^N[(1-\nu^N)\varepsilon_{XX}+\nu^N\varepsilon_{YY}]/[(1+\nu^N)(1-2\nu^N)]$ | $\varepsilon_{XX}=[+(1-\nu^N)\tau_{XX}-\nu^N\tau_{YY}](1+\nu^N)/E^N$

Plane-Strain Shear Constitution:

$\tau_{XY}=E^N/(1+\nu^N)\varepsilon_{XY}=E^N/(1+\nu^N)/2\gamma_{XY}$ | $\gamma_{XY}=2\varepsilon_{XY}=2(1+\nu^N)/E^N\ \tau_{XY}$

No Heat Plane-Stress XX Normal Constitution:

$\tau_{XX}=E/(1-\nu^2)(+\varepsilon_{XX}+\nu\varepsilon_{YY})$ | $\varepsilon_{XX}=[+\tau_{XX}-\nu\tau_{YY}]/E$

Plane-Stress Shear Constitution:

$\tau_{XY}=E/(1+\nu)\varepsilon_{XY}=E/(1+\nu)/2\ \gamma_{XY}$ | $\gamma_{XY}=2\varepsilon_{XY}=2(1+\nu)/E\ \tau_{XY}$

Directly Shear Comparison:

$$E^N/(1+\nu^N)=E/(1+\nu)$$

Normal Comparison:

$$E^N[(1-\nu^N)\varepsilon_{XX}+\nu^N\varepsilon_{YY}]/[(1+\nu^N)(1-2\nu^N)] = E/(1-\nu^2)(+\varepsilon_{XX}+\nu\varepsilon_{YY})$$

$$[+(1-\nu^N)\tau_{XX}-\nu^N\tau_{YY}](1+\nu^N)/E^N = [+\tau_{XX}-\nu\tau_{YY}]/E$$

Using Shear Comparison Then: E^N/E =

$$(1+\nu^N)/(1+\nu) = (+\varepsilon_{XX}+\nu\varepsilon_{YY})/[(1-\nu^N)\varepsilon_{XX}+\nu^N\varepsilon_{YY}]\ [(1+\nu^N)(1-2\nu^N)]/(1-\nu^2)$$

$$(1+\nu^N)[(1-\nu^N)\tau_{XX}-\nu^N\tau_{YY}]/[\tau_{XX}-\nu\tau_{YY}] = (1+\nu^N)/(1+\nu)$$

Considering $\varepsilon_{YY}=0$, Then:

$$(1+\nu^N)/(1+\nu) = 1/(1-\nu^N)[(1+\nu^N)(1-2\nu^N)]/(1-\nu^2)$$

$$(1+\nu^N)[(1-\nu^N)] = (1+\nu^N)/(1+\nu)$$

Considering $\varepsilon_{XX}=0$, Then:

$(1+\nu^N)/(1+\nu)=\nu/[+\nu^N][(1+\nu^N)(1-2\nu^N)]/(1-\nu^2)$ | $(1+\nu^N)[-\nu^N]/[-\nu]=(1+\nu^N)/(1+\nu)$

Compared:

$$\nu=\nu^N/(1-\nu^N)$$

Substituted in Shear Comparison:

$$E^N/E=(1+\nu^N)/(1+\nu)=(1+\nu^N)/[1+\nu^N/(1-\nu^N)]=(1+\nu^N)(1-\nu^N)=(1-\nu^N\nu^N)$$

Or:

$\nu=\nu^N/(1-\nu^N)$ | $E=E^N/\{1-[\nu^N]^2\}$

Also:

$\nu-\nu\nu^N=\nu^N$	$\nu^N+\nu\nu^N=\nu$	$E(1-\nu^N\nu^N)=E^N$
$\nu=\nu^N/(1-\nu^N)$	$\nu^N=\nu/(1+\nu)$	$E^N=E(1+2\nu)/[(1+\nu)]^2$
$(1-\nu^N)=\nu^N/\nu=$	$=\nu^N/\nu=1/(1+\nu)$	

Where:

$$E^N/E=1-[\nu^N]^2=1-[\nu]^2/[(1+\nu)]^2=[(1+\nu)]^2-[\nu]^2/[(1+\nu)]^2=(1+2\nu)/[(1+\nu)]^2$$

So:

$\nu^N < \nu = \nu^N/(1-\nu^N)$	$E^N < E = E^N/\{1-[\nu^N]^2\}$
$\nu^N = \nu/(1+\nu) < \nu$	$E^N = E(1+2\nu)/(1+\nu)^2 < E$

And:

$$E^N/(1+\nu^N) = E/(1+\nu)$$

Check:

$$E^N/(1+\nu^N)=E^N/[(1+2\nu)/(1+\nu)]=E(1+2\nu)/(1+\nu)^2/[(1+2\nu)/(1+\nu)]=E/(1+\nu)$$

O.K.

15.3.5.
Comparison of Axial-Strain and Axial-Stress:

1-Dimensional Constitution:

Strain:

$$\tau + E/(1-2\nu)\alpha h = E*[1-\nu]/(1+\nu)/(1-2\nu)\ \varepsilon \equiv (2M+\Lambda)\ \varepsilon$$

Stress:

$$\tau + E_S\ \alpha_S\ h = E_S\ \varepsilon$$

Solve These:

$$E\alpha/(1-2\nu)= E_S\ \alpha_S \qquad E[1-\nu]/(1+\nu)/(1-2\nu)= E_S$$

To Get:

$$\alpha_S = (1+\nu)/[1-\nu]\ \alpha \qquad E_S = E[1-\nu]/(1+\nu)/(1-2\nu)$$

$$\alpha = [1-\nu]/(1+\nu)\ \alpha_S \qquad E = E_S*(1+\nu)*(1-2\nu)/[1-\nu]$$

No Heat Comparison

$$E_S = E[1-\nu]/(1+\nu)/(1-2\nu)$$

$$E = E_S*(1+\nu)*(1-2\nu)/[1-\nu]$$

15.3.6.
Matrix Representation:

Constitutive Equations of only two independent constants, E and ν:

Axial Flexibility :

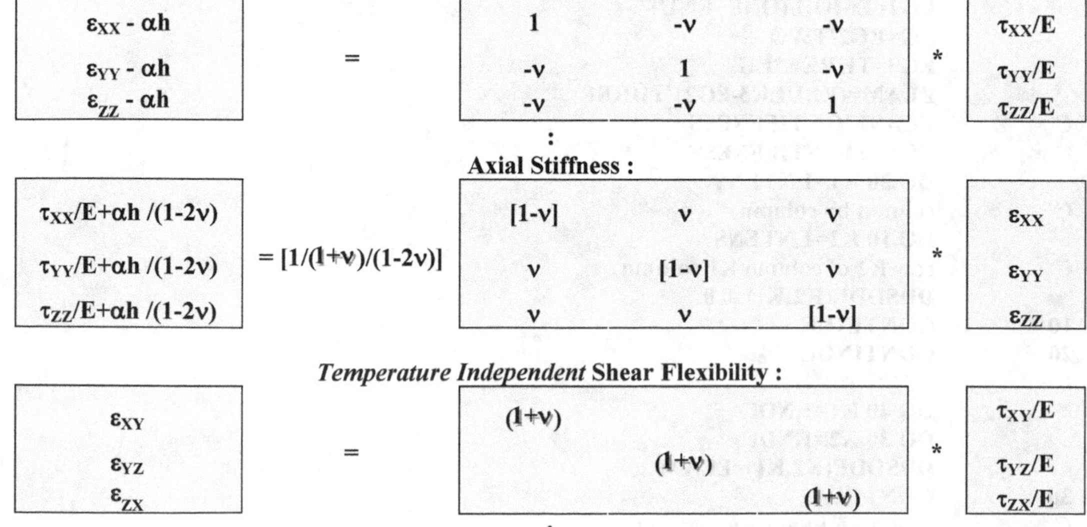

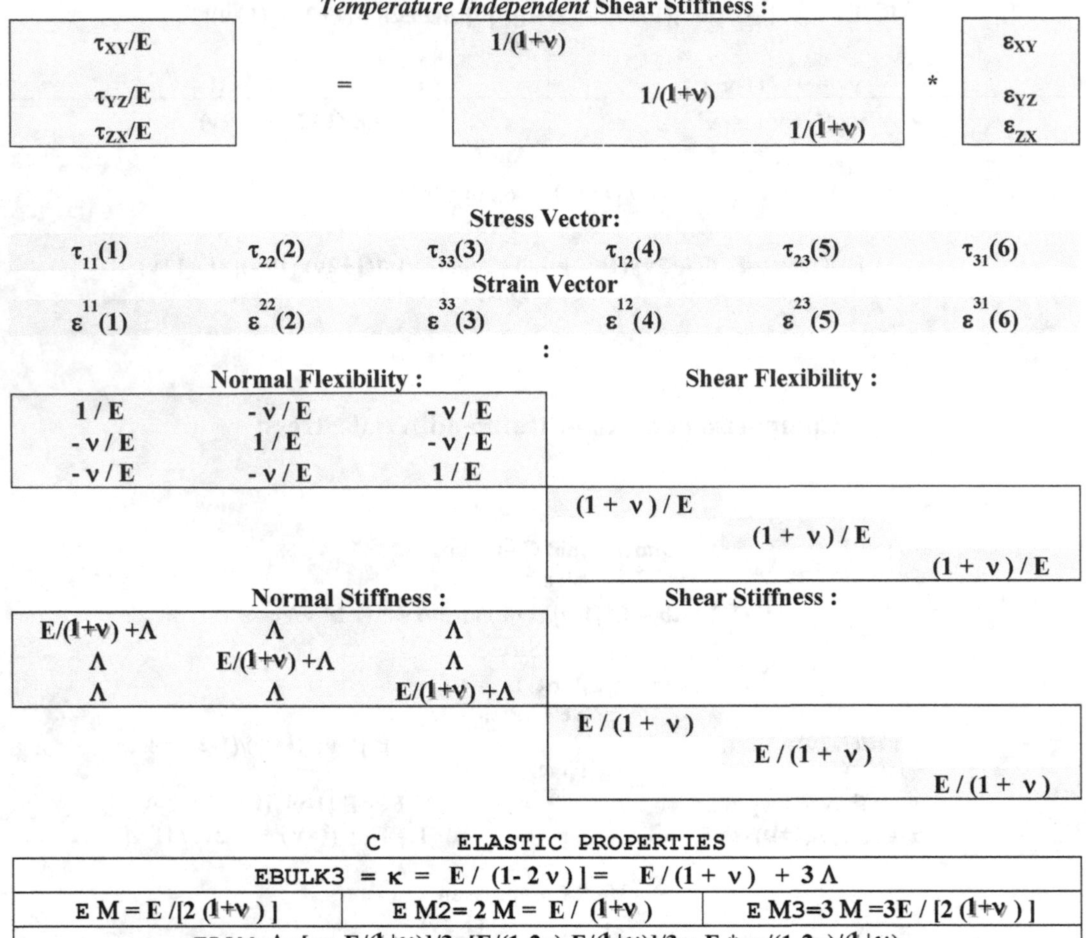

C ELASTIC PROPERTIES

EBULK3 = κ = E/ (1-2ν)] = E/(1+ν) + 3Λ		
E M = E /[2 (1+ν)]	E M2= 2 M = E/ (1+ν)	E M3=3 M =3E / [2 (1+ν)]
ELAM=Λ=[κ -E/(1+ν)]/3=[E/(1-2ν)-E/(1+ν)]/3= E * ν /(1-2ν)/(1+ν)		

FORTRAN Program:

```
      EBULK3=EMOD/(ONE-TWO*ENU)
      EG2=EMOD/(ONE+ENU)
      EG=EG2/TWO
      EG3=THREE*EG
      ELAM=(EBULK3-EG2)/THREE
C     ELASTIC STIFFNESS
C     ELASTIC STIFFNESS
      DO 20 K1=1,NTENS
C     column by column
      DO 10 K2=1,NTENS
C     row K2 of column K1 column
      DDSDDE(K2,K1)=0.0
10    CONTINUE
20    CONTINUE
C
      DO 40 K1=1,NDI
      DO 30 K2=1,NDI
      DDSDDE(K2,K1)=ELAM
30    CONTINUE
C     column K1 values are all set
```

```
          DDSDDE(K1,K1)=EG2+ELAM
40        CONTINUE
C
C         ELASTIC SHEAR STIFFNESS
          DO 50 K1=NDI+1,NTENS
C         only diagonal K1,K1
          DDSDDE(K1,K1)=EG
50        CONTINUE
C
C         CALCULATE STRESS FROM ELASTIC STRAINS
          DO 70 K1=1,NTENS
          DO 60 K2=1,NTENS
          STRESS(K2)=STRESS(K2)+DDSDDE(K2,K1)*DSTRAN(K1)
60        CONTINUE
70        CONTINUE
```

15.4.
3-D Virtual Concept:

15.4.1.
3-D Virtual Strain:

Strain:

$\varepsilon^{IJ}(x) = \frac{1}{2}(u^I{}_{,J} + u^J{}_{,I})$ $\qquad$ $\varepsilon^{AA}(x) = u^A{}_{,A}$

Virtual Strain:

$\varepsilon_{IJ} = \frac{1}{2}[u_I{}^{,J} + u_J{}^{,I}]$	$(\varepsilon_{BB}) = [u_B{}^{,B}]$

:

3-D VIRTUAL WORK:

:

$$\tau_{IJ} = 2G * \frac{1}{2}(u^I{}_{,J} + u^J{}_{,I}) + [-\kappa\alpha h + \Lambda u^A{}_{,A}]\,\delta_{IJ}$$

:

Pre-multiplied By: ε^{IJ}

:

$$\frac{1}{2}[u_I{}^{,J} + u_J{}^{,I}]\,\tau_{IJ}$$

$$=$$

$$\frac{1}{2}[u_I{}^{,J} + u_J{}^{,I}]\,2G * \frac{1}{2}(u^I{}_{,J} + u^J{}_{,I}) + \frac{1}{2}(u^I{}_{,J} + u^J{}_{,I})[-\kappa\alpha h + \Lambda u^A{}_{,A}]\,\delta_{IJ}$$

15.4.2.
3-D Virtual Distortion Work:

:

$$\tau_{IJ} - \tau_I/3$$

$$=$$

$$2 G * [½ (u^{I}_{,J} + u^{J}_{,I}) - u^{A}_{,A} /3]$$

:

Pre-multiplied By: ε^{IJ}

:

$$½[u_{I}{}^{,J} + u_{J}{}^{,I}] [\tau_{IJ} - \tau_{I} /3]$$
$$=$$
$$½[u_{I}{}^{,J} + u_{J}{}^{,I}] 2 G * [½ (u^{I}_{,J} + u^{J}_{,I}) - u^{A}_{,A} /3]$$

15.5. Minimum Potential Capacity and Energy:

15.5.1. Minimum Potential Capacity:

U Strain Energy **W External Work Energy**

Define Potential Energy:

$$P = U - 2 W = (U - W) - W$$

In Terms of Isoperimetrics: r^{J} **, I = 1,…M**

$$\partial P/\partial r^{J} = 0$$

Lead to:

$$K^{I}_{J} \; r^{J} = R^{I}$$

In Finite Elements r^{J} **are Displacements.**

$$V = \int_{volume} \int_{0}^{\varepsilon} \sigma_{J} \; d\varepsilon^{J} \; (dVolume) - \int f_{J} \; u^{J} \; (dVolume) - P_{J} \; u^{J}$$
$$V = 1/2 \; u_{I} \; K^{I}_{J} \; u^{J} - u_{I} B^{I} - u_{I} P^{I}$$
$$V = 1/2 \; u_{I} \; K^{I}_{J} \; u^{J} - u_{I} R^{I}$$
$$\partial V/\partial u_{I} = 0$$

$$K^{I}_{J} \; u^{J} - R^{I} = 0$$

$$d (\partial L / \partial uT_{I}) / dt - \partial L / \partial u_{I} = 0$$

T is function of (uT_{I}) **Only**

V is function of (u_{I}) **Only**

$$L \equiv T - V$$
$$L = T(uT_{I}) - V(u_{I})$$

$$T = \int_{volume} dT = \int_{volume} 1/2 \; uT_{I}(x,y,z) \;\; \rho \;\; uT^{I}(x,y,z) \;\; (dVolume)$$

$$uT^{K} (x,y,z) = B^{K}_{J}(x,y,z) \;\; UT^{J} \quad \text{where } K=x,y,z$$

$$T = UT_{I} \int_{volume} 1/2 \; B^{I}_{K}(x,y,z) \;\; \rho \;\; B^{K}_{J}(x,y,z) \;\; (dVolume)] \;\; UT^{J}$$

$$M^{I}_{J} = \int_{volume} B^{I}_{K}(x,y,z) \;\; \rho \;\; B^{K}_{J}(x,y,z) \;\; (dVolume)$$

$$T = 1/2 \; UT_{I} \;\; M^{I}_{J} \;\; UT^{J}$$
$$(\partial T / \partial UT_{I}) = M^{I}_{J} \;\; UT^{J}$$
$$d (\partial T / \partial UT_{I}) / dt = M^{I}_{J} \;\; UTT^{J}$$

$$\partial L / \partial u_I = 0 - \partial V / \partial u_I = - K^I{}_J\ u^J + R^I$$
$$d(\partial L / \partial uT_I) / dt - \partial L / \partial u_I = M^I{}_J\ UTT^J + K^I{}_J\ u^J - R^I = 0$$

15.5.2.
Potential Energy:

Define Internal Strain Energy:

$$W_q = \int_{volume} \{ \varepsilon_I\ [1/2 * C^I{}_J\ \varepsilon^J{}_q + \tau^I{}_q(0)] \} (dVolume) + Order([u_M]^2)$$
$$W_q = \int_{volume} [\varepsilon_I * 1/2\ C^I{}_J\ \varepsilon^J{}_q] (dVolume)_q + \int_{volume} [\varepsilon_I\ \tau^I{}_q(0) \} (dVolume)_q$$

Writing:

$$\varepsilon_I = u_M\ a^M{}_I$$

Then:

$$W_q = 1/2 u_M [\int_{volumeq} a^M{}_I C^J a^J{}_{Nq} (dVolume)_q] u^N + u_M [\int_{volumeq} a^M{}_I \tau^I{}_q(0) (dVolume)_q]$$

Or:

$$W_q = 1/2\ u_M [K_{Elastic}]^M{}_{Nq}\ u^N + u_M [B]^M{}_q$$

If Higher Order Values are Inccluded:

$$W_q = 1/2\ u_M [K_{Elastic}]^M{}_{Nq}\ u^N + u_M [B]^M{}_q + Order([u_M]^3) + Order([u_M]^4)$$

Where:

$$[K_{Elastic}]^M{}_{Nq} = [\int_{volumeq} a^M{}_I C^J a^J{}_{Nq} (dVolume)_q]$$

And:

$$[B]^M{}_q = [\int_{volumeq} a^M{}_I \tau^I{}_q(0) (dVolume)_q]$$

Adding Large Deflection Geometric Energy:

$$W_q = 1/2\ u_M [K_{Elastic}]^M{}_{Nq}\ u^N + 1/2\ u_M [K_{Geometric}]^M{}_{Nq}\ u^N + u_M [B]^M{}_{Nq}$$

$$[K_{Elastic}]^M{}_N = \Sigma_q\ [K_{Elastic}]^M{}_{Nq} = \Sigma_q\ [\int_{volumeq} a^M{}_I C^J a^J{}_{Nq} (dVolume)_q]$$

$$[K_{Geometric}]^M{}_N = \Sigma_q\ [K_{Geometric}]^M{}_{Nq} = \Sigma_q\ [\int_{volumeq} a^M{}_I C^J a^J{}_{Nq} (dVolume)_q]$$

$$[B]^M{}_N = \Sigma_q\ [B]^M{}_{Nq} = \Sigma_q\ [\int_{volumeq} a^M{}_I \tau^I{}_q(0) (dVolume)_q]$$

$$V = \Sigma_q\ V_q = \Sigma_q\ W_q - u_M [R]^M$$
$$V = 1/2\ u_M [K_{Elastic}]^M{}_N\ u^N + u_M [B]^M + 1/2\ u_M [K_{Geometric}]^M{}_N\ u^N - u_M [R]^M$$

Let:

$$\partial V / \partial u^I = 0$$

Then:

$$\partial V / \partial u^I = [K_{Elastic}]^M{}_N\ u^N + [B]^M + [K_{Geometric}]^M{}_N\ u^N - [R]^M = 0$$

Stability:

$$[K_{Elastic} + K_{Geometric}]^M{}_N\ u^N = 0$$

15.6.
Conservation of Infinitesimal Elasticity of Two Constants:

15.6.1.
First Derivative of Constitutive Equations:

$$\tau_{ij} = 2G\ \varepsilon_{ij} + \Lambda\ \varepsilon_{kk}\ \delta_{ij}$$
$$\tau_{ij,j} = 2G\ \varepsilon_{ij,j} + \Lambda\ \varepsilon_{kk,j}\ \delta_{ij}$$
$$\tau_{ij,j} = 2G\ ½\,(r_{i,jj} + r_{j,ij}) + \Lambda\ r_{k,kj}\ \delta_{ij}$$
$$\tau_{ij,j} = G\ r_{i,jj} + (G+\Lambda)\ r_{k,kj}\ \delta_{ij} = G\ r_{i,jj} + (G+\Lambda)\ r_{k,ki}$$

Conservation of Momentum:

$$\rho\ vT_m \equiv \rho\ dv_m / dt = \rho\ b_m + \partial(\tau_m^{\ n}) / \partial q_n \equiv \rho\ b_m + \tau_m^{\ n}{}_{,n}$$

So

$$\rho\ vT_m = \rho\ b_m + \tau_m^{\ n}{}_{,n} = \rho\ b_m + G\ r_m^{\ n}{}_{,n} + (G+\Lambda)\ r_k^{\ k}{}_{,m}$$

First Derivative of Conservation of Momentum:

$$\rho\ vT_m^{\ ,m} = \rho\ b_m^{\ ,m} + \tau_m^{\ n}{}_{,n}^{\ ,m} = \rho\ b_m^{\ ,m} + G\ r_m^{\ ,n}{}_{,n}^{\ ,m} + (G+\Lambda)\ r_k^{\ ,k}{}_{,m}^{\ ,m}$$
$$\rho\ vT_m^{\ ,m} = \rho\ b_m^{\ ,m} + \tau_m^{\ n}{}_{,n}^{\ ,m} = \rho\ b_m^{\ ,m} + (2G+\Lambda)\ r_k^{\ ,k}{}_{,m}^{\ ,m}$$

Second Derivative of Conservation of Momentum:

$$\rho\ vT_m^{\ ,mq} = \rho\ b_m^{\ ,mq} + \tau_m^{\ n}{}_{,n}^{\ ,mq} = \rho\ b_m^{\ ,mq} + G\ r_m^{\ ,n}{}_{,n}^{\ ,mq} + (G+\Lambda)\ r_k^{\ ,k}{}_{,m}^{\ ,mq}$$
$$\rho\ vT_m^{\ ,mq} = \rho\ b_m^{\ ,mq} + (2G+\Lambda)\ r_k^{\ ,k}{}_{,m}^{\ ,mq}$$

Orthonormal

If: $\underline{v}T = 0$ **and** $\underline{b} = 0$ **then** $\tau_{mn,n} = 0$

$$\tau_{ij,j} = G\ r_{i,jj} + (G+\Lambda)\ r_{k,ki} = 0$$
$$\tau_{ij,ji} = G\ r_{i,jji} + (G+\Lambda)\ r_{k,kii} = (2G+\Lambda)\ r_{k,kmm} = 0$$
$$\nabla^2\ r_{k,k} \equiv r_{k,kmm} = 0$$
$$\tau_{ij,jmm} = G\ r_{i,jjmm} + (G+\Lambda)\ r_{k,kimm} = 0$$
$$r_{k,kmm} = 0$$
$$\tau_{ij,jmm} = G\ r_{i,jjmm} = 0$$
$$\nabla^4\ r_i \equiv r_{i,jjmm} = 0$$

Summary of Results:

$$\varepsilon_{ij} = (1+\nu)/E\ \tau_{ij} - \nu/E\ \tau_{kk}\ \delta_{ij}$$
$$\tau_{ij} = E/(1+\nu)\varepsilon_{ij} + \nu/E/((1-2\nu)(1+\nu))\ \varepsilon_{kk}\ \delta_{ij}$$

$$\varepsilon_{ij} = (2G)^{-1}\tau_{ij} - 1/3[(2G)^{-1} - (2G+3\Lambda)^{-1}]\tau_{kk}\delta_{ij}$$
$$\tau_{ij} = 2G\ \varepsilon_{ij} + \Lambda\ \varepsilon_{kk}\ \delta_{ij}$$

$$E_1 = (1-2\nu)/E\ \ T_1$$
$$E_1 = T_1\ /K$$
$$T_1 = E\ [(1-2\nu)+3\nu)/((1+\nu)(1-2\nu)]\ E_1$$

$$E_1 = [1/(2G) - (1/(2G) - 1/(2G+3\Lambda)]\ T_1$$
$$E_1 = 1/(2G + 3\Lambda)\ T_1$$
$$T_1 = (2G + 3\Lambda)\,E_1$$

$$E/(1-2\nu) \equiv K = 2G + 3\Lambda$$
$$G \equiv ½\,E/(1+\nu)$$
$$\Lambda = E\ \nu\ /\ [(1-2\nu)(1+\nu)] = K\ \nu\ /\ (1+\nu)$$

$$\tau_{kk} \equiv T_1 \qquad T_1 = K\ E_1 \qquad \varepsilon_{ii} \equiv E_1$$

$$\tau_{ij,j} = 0 \qquad \tau_{ij,ji} = 0 \qquad \tau_{ij,jmm} = 0$$
$$\nabla^2\ r_{k,k} \equiv r_{k,kmm} = 0 \qquad \nabla^4\ r_i \equiv r_{i,jjmm} = 0$$

15.6.2.
General and Principal Internal Energy per Unit Mass:

$$e_{internal} = \int_{\varepsilon} \tau_{kl}\, d\varepsilon_{kl}$$

Internal Energy per Unit Mass for Linear Stress Strain relation:

$$e_{internal} = ½\ \tau_{kl}\ \varepsilon_{kl}$$

Divide right hand side into two parts: k = l and k ≠ l :

$$e_{internal} = ½\ \tau_{\underline{m}m}\ \varepsilon_{\underline{m}m} + ½\ \tau_{kl}\ \varepsilon_{kl\neq k}$$

Considering First Part: k = l = m :

$$½\ \tau_{\underline{m}m}\ \varepsilon_{\underline{m}m} = ½(\tau_{11}\ \varepsilon_{11} + \tau_{22}\,\varepsilon_{22} + \tau_{33}\,\varepsilon_{33})$$

Considering Second Part: k ≠ l :

Define:

$$\gamma_{kl} \equiv 2\ \varepsilon_{kl}$$

Then:

$$\tau_{kl}\ \varepsilon_{kl\neq k} = \tau_{kl}\ \gamma_{kl<k} = \tau_{kl}\ \gamma_{kl>k}$$

$$\tau_{12}\,\varepsilon_{12} + \tau_{23}\,\varepsilon_{23} + \tau_{31}\,\varepsilon_{31} + \tau_{21}\,\varepsilon_{21} + \tau_{32}\,\varepsilon_{32} + \tau_{13}\,\varepsilon_{13} = \tau_{12}\,\gamma_{12} + \tau_{23}\,\gamma_{23} + \tau_{31}\,\gamma_{31}$$

$$\tau_{kl}\ \varepsilon_{kl\neq k} = 2(\tau_{12}\,\varepsilon_{12} + \tau_{23}\,\varepsilon_{23} + \tau_{31}\,\varepsilon_{31}) = (\tau_{12}\,\gamma_{12} + \tau_{23}\,\gamma_{23} + \tau_{31}\,\gamma_{31})$$

Or:

$$½\ \tau_{kl}\ \varepsilon_{kl\neq k} = (\tau_{12}\,\varepsilon_{12} + \tau_{23}\,\varepsilon_{23} + \tau_{31}\,\varepsilon_{31}) = ½(\tau_{12}\,\gamma_{12} + \tau_{23}\,\gamma_{23} + \tau_{31}\,\gamma_{31})$$

So:

$$e_{internal} = ½\ \tau_{\underline{m}m}\ \varepsilon_{\underline{m}m} + ½\ \tau_{kl}\ \varepsilon_{kl\neq k}$$

$$e_{internal} = ½\ \tau_{\underline{m}m}\ \varepsilon_{\underline{m}m} + ½\ \tau_{kl}\ \gamma_{kl<k}$$

Or:

$$e_{internal} = ½(\tau_{11}\ \varepsilon_{11} + \tau_{22}\,\varepsilon_{22} + \tau_{33}\,\varepsilon_{33}) + (\tau_{12}\,\varepsilon_{12} + \tau_{23}\,\varepsilon_{23} + \tau_{31}\,\varepsilon_{31})$$

$$e_{internal} = ½(\tau_{11}\ \varepsilon_{11} + \tau_{22}\,\varepsilon_{22} + \tau_{33}\,\varepsilon_{33}) + ½(\tau_{12}\,\gamma_{12} + \tau_{23}\,\gamma_{23} + \tau_{31}\,\gamma_{31})$$

Substituting Values of ε_{ij} :

$$\varepsilon_{ij} = (1+\nu)/E\ \tau_{ij} - \nu/E\ \tau_{kk}\ \delta_{ij}$$

In:

$$e_{internal} = ½\ \tau_{\underline{m}m}\ \varepsilon_{\underline{m}m} + ½\ \tau_{kl}\ \varepsilon_{kl\neq k}$$

$$e_{internal} = ½\ \tau_{\underline{m}m}\ \varepsilon_{\underline{m}m} + ½\ \tau_{kl}\ \gamma_{kl<k}$$

First Part: k = l = m :

$$\varepsilon_{\underline{m}m} = (1+\nu)/E\ \tau_{\underline{m}m} - \nu/E\ \tau_{kk} = 1/E\ \tau_{\underline{m}m} - \nu/E\ \tau_{nn\neq m}$$

So:

$$½\ \tau_{\underline{m}m}\ \varepsilon_{\underline{m}m} = ½\ 1/E\ (\tau_{11}^2 + \tau_{22}^2 + \tau_{33}^2) - \nu/E\ (\tau_{11}\,\tau_{22} + \tau_{22}\,\tau_{33} + \tau_{33}\,\tau_{11})$$

Second Part: k ≠ l : or l < k :

$$\varepsilon_{kl} \equiv ½\ \gamma_{kl} = (1+\nu)/E\ \tau_{kl}$$

So:

$$½\ \tau_{kl}\ \varepsilon_{kl\neq k} = ½\ (1+\nu)/E\ (\tau_{12}^2 + \tau_{23}^2 + \tau_{31}^2 + \tau_{21}^2 + \tau_{32}^2 + \tau_{13}^2)$$

$$\tau_{kl}\ \gamma_{kl<k} = (1+\nu)/E\ (\tau_{12}^2 + \tau_{23}^2 + \tau_{31}^2)$$

$$½\ (1+\nu)/E\ (\tau_{12}^2 + \tau_{23}^2 + \tau_{31}^2 + \tau_{21}^2 + \tau_{32}^2 + \tau_{13}^2) = (1+\nu)/E\ (\tau_{12}^2 + \tau_{23}^2 + \tau_{31}^2)$$

Sum:

$$e_{internal} = ½\,\tau_{\underline{m}m}\ (1/E\ \tau_{\underline{m}m} - \nu/E\ \tau_{nn\neq m}) + ½\ \tau_{kl}\ (1+\nu)/E\ \tau_{kl\neq k}$$

$$e_{internal} = ½\,\tau_{\underline{m}m}\ (1/E\ \tau_{\underline{m}m} - \nu/E\ \tau_{nn\neq m}) + \tau_{kl}\ (1+\nu)/E\ \tau_{kl<k}$$

Or:

$$e_{internal} = ½\ 1/E(\tau_{11}^2 + \tau_{22}^2 + \tau_{33}^2) - \nu/E(\tau_{11}\,\tau_{22} + \tau_{22}\,\tau_{33} + \tau_{33}\,\tau_{11}) + (1+\nu)/E(\tau_{12}^2 + \tau_{23}^2 + \tau_{31}^2)$$

15.6.3.
Internal Energy per Unit Mass:

If Principal Stress: If shear is zero: $\tau_{k l \neq k} = 0$:
Principal Internal Energy per Unit Mass:

$\tau_{k \neq l} = 0$ $$e_{principal} = ½\ 1/E\ \tau_{\underline{m}\,m}\ \tau_{\underline{m}\,m} - ½\ \nu/E\ \tau_{\underline{m}\,m}\ \tau_{k\,k \neq m}$$

$\tau_{k \neq l} = 0$ $$e_{principal} = ½\ 1/E(\tau_{11}^2 + \tau_{22}^2 + \tau_{33}^2) - \nu/E(\tau_{11}\ \tau_{22} + \tau_{22}\ \tau_{33} + \tau_{33}\ \tau_{11})$$

$\tau_{k \neq l} = 0$ $$e_{principal} = ½\ 1/E(\tau_{11} + \tau_{22} + \tau_{33})^2 - (1+\nu)/E(\tau_{11}\ \tau_{22} + \tau_{22}\ \tau_{33} + \tau_{33}\ \tau_{11})$$

$\tau_{k \neq l} = 0$ $$e_{principal} = ½\ 1/E\ (\ T_1^2 - 2\ T_2) - \nu/E\ (T_2)$$

15.6.3.1.
Spherical Stress:
Spherical Internal Energy per Unit Mass:

$$\varepsilon_{\underline{L}\,L} = \varepsilon_{11} = \varepsilon_{22} = \varepsilon_{33}$$
$$\tau_{\underline{L}\,L} = \tau_{11} = \tau_{22} = \tau_{33}$$

$$E_1 \equiv 3\ \varepsilon_{\underline{L}\,L} \equiv \varepsilon_{i\,i}$$
$$T_1 \equiv 3\ \tau_{\underline{L}\,L} \equiv \tau_{i\,i}$$

$$\varepsilon_{i\,i} = 1/K\ \ \tau_{k\,k}$$
$$E_1 = 1/K\ \ T_1$$
$$3\ \varepsilon_{\underline{L}\,L} = 1/K\ \ 3\ \tau_{\underline{L}\,L}$$

$$\varepsilon_{\underline{L}\,L} = 1/K\ \ \tau_{\underline{L}\,L}$$
$$E_1/3 = 1/K\ \ T_1/3$$

$$3\ \varepsilon_{\underline{L}\,L} = 1/K\ \ T_1$$
$$E_1 = 1/K\ \ 3\ \tau_{\underline{L}\,L}$$

$$E\,/\,(1 - 2\ \nu) = K = 2\,G + 3\,\Lambda$$

$$e_{spherical} = ½\ \tau_{\underline{m}\,m}\ \ \varepsilon_{\underline{m}\,m} = ½\ (\tau_{11}\ \tau_{11}/K + \tau_{22}\ \tau_{22}/K + \tau_{33}\ \tau_{33}\ /K) = ½\ 3/K\ \ \tau_{\underline{L}\,L}^2$$

$$e_{spherical} = ½\ 3(1 - 2\ \nu)/E\ \ \tau_{\underline{L}\,L}^2 = ½\ 3/K\ \ \tau_{\underline{L}\,L}^2 = ½\ 3/(2\ G + 3\ \Lambda)\ \ \tau_{\underline{L}\,L}^2$$

$$e_{spherical} = ½\ 3(1 - 2\ \nu)/E\ (T_1/3)^2 = ½\ 3/K\ (T_1/3)^2 = ½\ 3/(2\ G + 3\ \Lambda)\ (T_1/3)^2$$

$$e_{spherical} = 1/6\ (1 - 2\ \nu)/E\ T_1^2 = 1/6\ \ 1/K\ \ T_1^2 = 1/6\ \ 1/(2\ G + 3\ \Lambda)\ T_1^2$$

15.6.3.2.
Distortional Internal Energy per Unit Mass: Principal Minus Spherical:

$$e_{Distortional} \equiv e_{principal} - e_{spherical}$$

Subtracting:

$$e_{principal} = ½\ 1/E\ (T_1^2 - 2T_2) - \nu/E\ (T_2) = 1/6\ \ 3/E\ T_1^2 - (1+\nu)/E\ T_2$$

$$e_{spherical} = 1/6\ (1 - 2\nu)/E\ T_1^2 = 1/6\ \ 1/K\ \ T_1^2 = 1/6\ \ 1/(2G + 3\Lambda)\ T_1^2$$

$$e_{Distortional} = 1/6\ \ 3/E\ T_1^2 - (1+\nu)/E\ T_2 - 1/6\ (1 - 2\nu)/E\ T_1^2$$

$$e_{Distortional} = 1/6\ \ (2 + 2\nu)/E\ T_1^2 - (1+\nu)/E\ T_2$$

$$e_{Distortional} = 1/(6G)\ \ T_1^2 - 1/(2G)\ T_2$$

$$e_{Distortional} = 1/(6G)\ \ T_1^2 - 3/(6G)\ T_2$$

$$e_{Distortional} = 1/(6G)\ (T_1^2 - 3T_2)$$

$$e_{Distortional} = 1/(6G)\ \ ½\ \ ((\tau_{11}-\tau_{22})^2 + (\tau_{22}-\tau_{33})^2 + (\tau_{33}-\tau_{11})^2)$$

$$\text{If}: T_{11} = T_{22} = T_{33} \quad \text{and} \quad T_{12} = T_{23} = T_{31} = 0$$

$$e_{Distortional} = 0$$

Distortional Internal Energy per Unit Mass:

$$e_{Distortional} = ½\, s_{kl}\, e_{kl} = ½\,(\tau_{ij} - \tau_{kk}/3\ \ \delta_{ij})(\varepsilon_{ij} - \varepsilon_{kk}/3\ \ \delta_{ij})$$

$$e_{Distortional} = ½\, s_{kl}\, e_{kl} = ½\,(\tau_{ij} - \Pi\ \ \delta_{ij})(\varepsilon_{ij} - \varepsilon\ \ \delta_{ij})$$

$$e_{Distortional} = ½\,(\tau_{ij}\,\varepsilon_{ij} - \tau_{ij}\,\varepsilon_{kk}/3\,\delta_{ij} - \tau_{kk}/3\,\delta_{ij}\,\varepsilon_{ij} + \tau_{kk}/3\ \ \delta_{ij}\,\varepsilon_{ll}/3\ \ \delta_{ij})$$

$$e_{Distortional} = ½\,(\tau_{ij}\,\varepsilon_{ij} - \tau_{mm}\,\varepsilon_{kk}/3 - \tau_{kk}/3\ \varepsilon_{mm} + \tau_{kk}/3\ \ \varepsilon_{ll}/3\ \ 3)$$

$$e_{Distortional} = ½\,(\tau_{ij}\,\varepsilon_{ij} - \tau\,\varepsilon/3 - \tau/3\ \varepsilon + \ \tau/3\ \ \varepsilon/3\ \ 3)$$

$$e_{Distortional} = ½\ (\tau_{ij}\,\varepsilon_{ij} - \tau_{mm}\ \varepsilon_{kk}/3) = ½\ (\tau_{ij}\,\varepsilon_{ij} - \tau\ \ \varepsilon/3)$$

Where:

$$e_{internal} = ½\ \ \tau_{ij}\,\varepsilon_{ij}$$

$$e_{spherical} = ½\ \ \tau_{mm}\ \varepsilon_{kk}/3 = ½\ T_1\ \ E_1/3 = ½\ 3\ \ T_1\ \ E_1 = ½\ 3/K\ \ T_1^2$$

$$e_{spherical} = 1/6\ (1 - 2\nu)/E\ T_1^2 = 1/6\ \ 1/K\ \ T_1^2 = 1/6\ \ 1/(2G + 3\Lambda)\ T_1^2$$

Summary of
Internal Energy per Unit Mass for Linear Stress Strain relation:

$$e_{internal} = ½\ 1/E(\tau_{11}^2 + \tau_{22}^2 + \tau_{33}^2) - \nu/E(\tau_{11}\,\tau_{22} + \tau_{22}\,\tau_{33} + \tau_{33}\,\tau_{11}) + (1+\nu)/E\ (\tau_{12}^2 + \tau_{23}^2 + \tau_{31}^2)$$

Subtracting:

$$e_{principal} = ½\ 1/E\ (T_1^2 - 2T_2) - \nu/E\ (T_2) = 1/6\ \ 3/E\ T_1^2 - (1+\nu)/E\ T_2$$

$$e_{spherical} = 1/6\ (1 - 2\nu)/E\ T_1^2 = 1/6\ \ 1/K\ \ T_1^2 = 1/6\ \ 1/(2G + 3\Lambda)\ T_1^2$$

$$e_{Distortional} \equiv e_{principal} - e_{spherical} = 1/(6G)\ \ (T_1^2 - 3\,T_2)$$

15.7.
Stress View:

15.7.1.
Stress Invariants:

	Not Principal (Off-Diagonals)	**Principal (Only Diagonals)**
$T_{ii} \setminus T_{ij}$ **General** $T_{ii} \neq 0$	**Not Principal** ($T_{ij} \neq 0$ if $i \neq j$) $T_{ii} \neq 0$	**Principal** ($T_{ij} = 0$ if $i \neq j$) $T_{ii} \neq 0$

Spherical T_{spher}	**NA**	**$T_{ii} = 3\, T_{spher}$**
Deviatoric $T_{ii}=0$	**$T_{ii} = 0$**	**$T_{ii} = 0$**

T1 ≡ Trace of Values ≡ Diagonals Sum

$$T1 \equiv T_{ii} = T_{11} + T_{22} + T_{33}$$

Then:

T_1	**Not Principal ($T_{ij} \neq 0$ if $i \neq j$)**	**Principal ($T_{ij} = 0$ if $i \neq j$)**
General $T_{ii}\neq 0$	$T_{11} + T_{22} + T_{33}$	$T_{11} + T_{22} + T_{33}$
Spherical T_{spher}	NA	$3\, T_{spher}$
Deviatoric $T_{ii}=0$	0	0

T2 ≡ ½ [(Diagonals Sum)² - Squared Values Sum]

$$T2 = \tfrac{1}{2}\,(T_{ii}\,T_{jj} - T_{ik}\;T_{ki})$$

$$T2 = \tfrac{1}{2}\,\{\,(T_{11} + T_{22} + T_{33})^2 - [(T_{11}^2 + T_{22}^2 + T_{33}^2) + 2(\,T_{12}^2 + T_{23}^2 + T_{31}^2)]\}$$

Then:

T2	**Not Principal ($T_{ij} \neq 0$ if $i \neq j$)**	**Principal ($T_{ij} = 0$ if $i \neq j$)**
General $T_{ii}\neq 0$	$\tfrac{1}{2}\,(T_{ii}\,T_{jj} - T_{ik}\;T_{ki})$	$T_{11}\,T_{22} + T_{22}\,T_{33} + T_{33}\,T_{11}$
Spherical T_{spher}	NA	$3\, T_{spher}^2$
Deviatoric $T_{ii}=0$	$\tfrac{1}{2}\,(\,-T_{ik}\;T_{ki})$	$\tfrac{1}{2}\,\{\,-[(T_{11}^2 + T_{22}^2 + T_{33}^2)]\}$

$$T3 = \det(T_{ij})$$

T3=1/3 $T_{ij}\,T_{jk}\,T_{ki}$ -½ Diagonals Sum*Squared Values Sum +1/6(Diagonals Sum)³

$$T3 = 1/3\; T_{ij}\;T_{jk}\;T_{ki} - 1/2\;\;T_{ii}\;\;T_{jk}\;\;T_{kj} + 1/6\;T_{ii}\;\;T_{jj}\;\;T_{kk}$$

$$T3 = 1/3\;T_{ij}\,T_{jk}\,T_{ki} - 1/2\;(T_{ii})^*[(T_{11}^2 + T_{22}^2 + T_{33}^2) + 2(\,T_{12}^2 + T_{23}^2 + T_{31}^2)] + 1/6\;(T_{ii})^3$$

$$T3 = 1/3\;T_{ij}\;\;T_{jk}\;\;T_{ki} - 1/2\;\;T_1\;\;T_{jk}\;\;T_{kj} + 1/6\;T_1\;T_1\;T_1$$

Then:

T_3	**Not Principal ($T_{ij} \neq 0$ if $i \neq j$)**	**Principal ($T_{ij} = 0$ if $i \neq j$)**
General $T_{ii}\neq 0$	$\det(T_{ij})$	$T_{11}\;T_{22}\;T_{33}$
Spherical T_{spher}	NA	T_{spher}^3
Deviatoric $T_{ii}=0$	$\det(T_{ij})$	$T_{11}\;T_{22}\;T_{33}$

So:

$$F \equiv \det(T_{mn} - T\,\delta_{mn}) \equiv T3 - \;T2\;\;T\; + \;T1\;\;T^2 - \;T0\;\;T^3$$

Then:

$T_{ii} \setminus T_{ij}$	**Not Principal ($T_{ij} \neq 0$ if $i \neq j$)**	**Principal ($T_{ij} = 0$ if $i \neq j$)**
General $T_{ii}\neq 0$	$\det(T_{ij}) - \tfrac{1}{2}(T_{ii}T_{jj} - T_{ik}T_{ki})T + (T_{ii})T^2 - T^3$	$T_{11}T_{22}T_{33} - (T_{11}T_{22} + T_{22}T_{33} + T_{33}T_{11})T + (T_{ii})T^2 - T^3$
Spherical T_{spher}	NA	$T_{spher}^3 - (3T_{spher}^2)T + (3T_{spher})T^2 - T^3$
Deviatoric $T_{ii}=0$	$\det(T_{ij}) - \tfrac{1}{2}\,(-T_{ik}\,T_{ki})T - T^3$	$T_{11}T_{22}T_{33} - (T_{11}T_{22} + T_{22}T_{33} + T_{33}T_{11})T - T^3$

15.7.2.
Eigen-Values:

Eigen-Values, T of F are Roots of: F:

$$F \equiv \det(T_{mn} - T\,\delta_{mn}) = 0$$

$$T3 \;-\; T2\;\;T \;+\; T1\;\;T^2 - \;T0\;\;T^3 = 0$$

$$\det(T_{ij}) - \;\tfrac{1}{2}\,[(T_{ii})^2 - T_{ik}\;\;T_{ki}\,]\;T \;+\; T_{ii}\,T^2 - \;T^3 = 0$$

F = 0	**Not Principal ($T_{ij} \neq 0$ if $i \neq j$)**	**Principal ($T_{ij} = 0$ if $i \neq j$)**
General $T_{ii}\neq 0$	$\det(T_{ij}) - \tfrac{1}{2}(T_{ii}T_{jj} - T_{ik}T_{ki})\;T + (T_{ii})\;T^2 - T^3 = 0$	$T_{11}T_{22}T_{33} - (T_{11}T_{22} + T_{22}T_{33} + T_{33}T_{11})T + (T_{ii})T^2 - T^3 = 0$

Spherical T_{spher}	**NA**	$T_{spher}^3 - 3T_{spher}^2 T + 3T_{spher} T^2 - T^3 = 0$
Deviatoric $T_{ii}=0$	$\det(T_{ij})$ $-\tfrac{1}{2}(-T_{ik}T_{ki})T - T^3 = 0$	$T_{11}T_{22}T_{33}$ $-(T_{11}T_{22}+T_{22}T_{33}+T_{33}T_{11})T - T^3 = 0$

15.7.3.
Squared Values Sum:

$T_{ii} \setminus T_{ij}$	Not Principal ($T_{ij} \neq 0$ if $i \neq j$)	Principal ($T_{ij} = 0$ if $i \neq j$)
General $T_{ii} \neq 0$	$T1^2 - 2\,T2$	$T_{11}^2 + T_{22}^2 + T_{33}^2$
Spherical T_{spher}	NA	$3\,T_{spher}^2$
Deviatoric $T_{ii}=0$	$-2\,T2$	$T_{11}^2 + T_{22}^2 + T_{33}^2$

15.7.4.
Distortion:

$T_{ii} \setminus T_{ij}$	Not Principal ($T_{ij} \neq 0$ if $i \neq j$)	Principal ($T_{ij} = 0$ if $i \neq j$)
General $T_{ii} \neq 0$	$T_{ii} \neq 0$	$T_{ii} \neq 0$
Spherical T_{spher}	NA	$T_{ii} = 3\,T_{spher}$
Deviatoric $T_{ii}=0$	$T_{ii} = 0$	$T_{ii} = 0$

Distortional Property:

$$6G\,e_{Distortional} = T_1^2 - 3T_2$$

$$6G\,e_{Distortional} = (T_{11}^2 + T_{22}^2 + T_{33}^2) + 3(T_{12}^2 + T_{23}^2 + T_{31}^2) - (T_{11}T_{22} + T_{22}T_{33} + T_{33}T_{11})$$

$$6G\,e_{Distortional} = \tfrac{1}{2}\,[(T_{11} - T_{22})^2 + (T_{22} - T_{33})^2 + (T_{33} - T_{11})^2] + 3(T_{12}^2 + T_{23}^2 + T_{31}^2)$$

$T_1^2 - 3T_2$	Not Principal ($T_{ij} \neq 0$ if $i \neq j$)	Principal ($T_{ij} = 0$ if $i \neq j$)
$T_{ii} \neq 0$	$\tfrac{1}{2}[(T_{11}-T_{22})^2+(T_{22}-T_{33})^2+(T_{33}-T_{11})^2] + 3(T_{12}^2 + T_{23}^2 + T_{31}^2)$	$\tfrac{1}{2}[(T_{11}-T_{22})^2+(T_{22}-T_{33})^2+(T_{33}-T_{11})^2]$
T_{spher}	NA	0
$T_{ii}=0$	$\tfrac{1}{2}[(T_{11}-T_{22})^2+(T_{22}-T_{33})^2+(T_{33}-T_{11})^2] + 3(T_{12}^2 + T_{23}^2 + T_{31}^2)$	$\tfrac{1}{2}[(T_{11}-T_{22})^2+(T_{22}-T_{33})^2+(T_{33}-T_{11})^2]$

Principal Distortional Property:

$T_{ii} \setminus T_{ij}$	Principal ($T_{ij} = 0$ if $i \neq j$)
General $T_{ii} \neq 0$	$T_{ii} \neq 0$

$$T_{11}T_{22}T_{33} - (T_{11}T_{22} + T_{22}T_{33} + T_{33}T_{11})T + (T_{11} + T_{22} + T_{33})T^2 - T^3 = 0$$

$$(T_{11} - T)(T_{22} - T)(T_{33} - T) = 0$$

$$T_2 = (T_{11}T_{22} + T_{22}T_{33} + T_{33}T_{11})$$

$$6G\,e_{Distortional} = (T_{11}^2 + T_{22}^2 + T_{33}^2) - (T_{11}T_{22} + T_{22}T_{33} + T_{33}T_{11})$$

$$6G\,e_{Distortional} = \tfrac{1}{2}\,[(T_{11} - T_{22})^2 + (T_{22} - T_{33})^2 + (T_{33} - T_{11})^2]$$

$$F = \tfrac{1}{2}[\,(T_{11} - T_{22})^2 + (T_{22} - T_{33})^2 + (T_{33} - T_{11})^2\,]$$

T_{33}

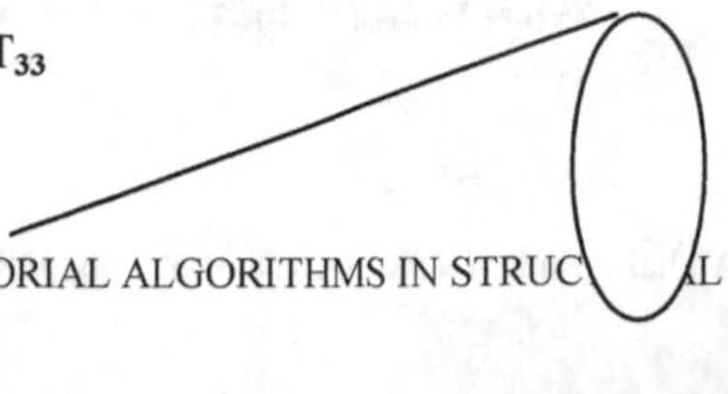

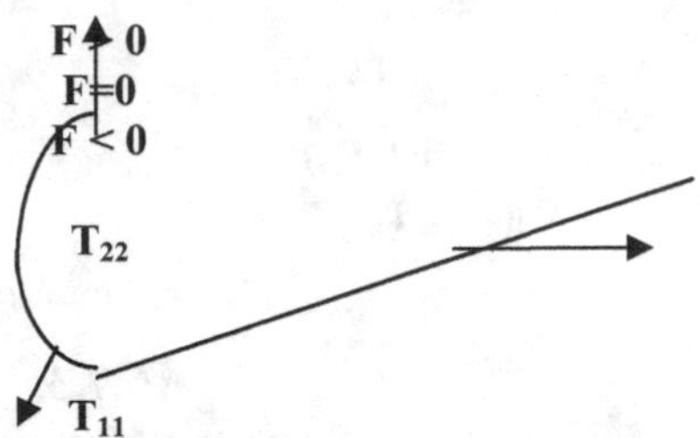

15.8.
Infinitesimal:

General:

$$\det(\tau_{mn} - \tau\,\delta_{mn}) \equiv \det(\tau_{ij}) - \tfrac{1}{2}(\tau_{ii}\tau_{kk} - \tau_{ik}\tau_{ki})\,\tau + \tau_{ii}\,\tau^2 - \tau^3 = 0$$

$$\tau_2 = (\tau_{11}\tau_{22} + \tau_{22}\tau_{33} + \tau_{33}\tau_{11}) - (\tau_{12}^2 + \tau_{23}^2 + \tau_{31}^2)$$

Principal ($\tau_{ij} = 0$ if $i \neq j$):

$$(\tau_{11}-\tau)(\tau_{22}-\tau)(\tau_{33}-\tau) = \tau_{11}\tau_{22}\tau_{33} - (\tau_{11}\tau_{22} + \tau_{22}\tau_{33} + \tau_{33}\tau_{11})\,\tau + (\tau_{11}+\tau_{22}+\tau_{33})\,\tau^2 - \tau^3 = 0$$

$$\tau_2 = (\tau_{11}\tau_{22} + \tau_{22}\tau_{33} + \tau_{33}\tau_{11})$$

Deviatoric $T_{ii} = 0$:

$$\det(T_{mn} - T\,\delta_{mn}) \equiv \det(T_{ij}) - \tfrac{1}{2}(0 - T_{ik}T_{ki})\,T \qquad - T^3 = 0$$

$$T_2 = (T_{11}T_{22} + T_{22}T_{33} + T_{33}T_{11}) - (T_{12}^2 + T_{23}^2 + T_{31}^2)$$

$$T_2 = -\tfrac{1}{2}(T_{11}^2 + T_{22}^2 + T_{33}^2) - (T_{12}^2 + T_{23}^2 + T_{31}^2)$$

Deviatoric $T_{ii} = 0$:and Principal ($T_{ij} = 0$ if $i \neq j$):

$$(T_{11}-T)(T_{22}-T)(T_{33}-T) = T_{11}T_{22}T_{33} - (T_{11}T_{22} + T_{22}T_{33} + T_{33}T_{11})\,T \qquad - T^3 = 0$$

$$T_2 = (T_{11}T_{22} + T_{22}T_{33} + T_{33}T_{11})$$

$$T_2 = -\tfrac{1}{2}(T_{11}^2 + T_{22}^2 + T_{33}^2)$$

$$T_2 = 1/6\ [(\tau_{11}-\tau_{22})^2 + (\tau_{22}-\tau_{33})^2 + (\tau_{33}-\tau_{11})^2]$$

$$T_2 = 1/6\ [2(\tau_{11}^2 + \tau_{22}^2 + \tau_{33}^2) - 2(\tau_{11}\tau_{22} + \tau_{22}\tau_{33} + \tau_{33}\tau_{11})]$$

$$T_2 = 1/6\ [(\tau_{11}-\tau_{22})^2 + (\tau_{22}-\tau_{33})^2 + (\tau_{33}-\tau_{11})^2]$$

15.8.1.
Physics of Invariants:

15.8.1.1.
Equa-angle Principal (Octahedral) Plane-Stresses:

Octahedral Plane makes equal angles with each of the principal axes: $n_1 = n_2 = n_3$:

$n_1 = 1/3^{1/2}$ $\qquad$ $n_2 = 1/3^{1/2}$ $\qquad$ $n_3 = 1/3^{1/2}$

The Octahedral Axial Stress:

$$\tau_{NO} \equiv \tau_{Normal} = n_i\,\tau_{ij}\,n_j = \tau_{11}\,n_1^2 + \tau_{22}\,n_2^2 + \tau_{33}\,n_3^2$$

Or:

$$\boxed{\tau_{NO}} = \boxed{1/3^{1/2} \quad 1/3^{1/2} \quad 1/3^{1/2}} \begin{bmatrix} \tau_{11} & & \\ & \tau_{22} & \\ & & \tau_{33} \end{bmatrix} \begin{bmatrix} 1/3^{1/2} \\ 1/3^{1/2} \\ 1/3^{1/2} \end{bmatrix}$$

Or:

$$\tau_{NO} = 1/3\,(\tau_{11} + \tau_{22} + \tau_{33}) = \tau_{spherical}$$

The Square of Octahedral Shear Stress:

$$\tau_{SO}^2 \equiv \tau_{Shear}^2 = (\tau_{11}\, n_1)^2 + (\tau_{22}\, n_2)^2 + (\tau_{33}\, n_3)^2 - (\tau_{Normal})^2$$

$$\tau_{SO}^2 = 1/3\,(\tau_{11}^2 + \tau_{22}^2 + \tau_{33}^2) - [1/3\,(\tau_{11} + \tau_{22} + \tau_{33})]^2$$

$$\tau_{SO}^2 = 1/9\,[(\tau_{11} - \tau_{22})^2 + (\tau_{22} - \tau_{33})^2 + (\tau_{33} - \tau_{11})^2]$$

If Compared With:

Deviatoric $T_{ii} = 0$:Principal: $T_{ij} = 0$ if $i \neq j$: Half Squares Sum:

$$0 \leq -T_2 = 1/6\,[(\tau_{11} - \tau_{22})^2 + (\tau_{22} - \tau_{33})^2 + (\tau_{33} - \tau_{11})^2]$$

To get:

$$\tau_{SO}^2/(-T_2) = 2/3$$

15.8.1.2.
Principal Shear Stresses:

$$\tau 1_{Shear} = 1/2\,|\tau_{22} - \tau_{33}| \qquad \tau 2_{Shear} = 1/2\,|\tau_{33} - \tau_{11}| \qquad \tau 3_{Shear} = 1/2\,|\tau_{11} - \tau_{22}|$$

Where:

$$\tau_{MaximumShear} \equiv Max(\tau 1_{Shear}, \tau 2_{Shear}, \tau 3_{Shear})$$

From:

Previous Paragraph:

$$\tau_{SO}^2 = 1/9\,[(\tau_{11} - \tau_{22})^2 + (\tau_{22} - \tau_{33})^2 + (\tau_{33} - \tau_{11})^2]$$

So:

$$\tau_{SO}^2 = 4/9\,[(\tau 1_{Shear})^2 + (\tau 2_{Shear})^2 + (\tau 3_{Shear})^2]$$

But From:

Deviatoric $T_{ii} = 0$:Principal: $T_{ij} = 0$ if $i \neq j$: Half Squares Sum:

$$(-T_2) = 1/6\,[(\tau_{11} - \tau_{22})^2 + (\tau_{22} - \tau_{33})^2 + (\tau_{33} - \tau_{11})^2]$$

$$(-T_2) = 2/3\,[(1/2\,|\tau_{11} - \tau_{22}|)^2 + (1/2\,|\tau_{22} - \tau_{33}|)^2 + (1/2\,|\tau_{33} - \tau_{11}|)^2]$$

$$(-T_2) = 2/3\,[(\tau 1_{Shear})^2 + (\tau 2_{Shear})^2 + (\tau 3_{Shear})^2]$$

So:

$$\tau_{SO}^2 = 4/9\,[(\tau 1_{Shear})^2 + (\tau 2_{Shear})^2 + (\tau 3_{Shear})^2]$$

And:

$$(-T_2) = 2/3\,[(\tau 1_{Shear})^2 + (\tau 2_{Shear})^2 + (\tau 3_{Shear})^2]$$

Such That:

$$\tau_{SO}^2/(-T_2) = 2/3$$

As It Was Shown In Previous Paragraph:

In Order To Determine Bounds of:

$$2/3 \leq (\tau_{SO}^2)/(\tau_{MaximumShear}^2) \leq 8/9$$

Introduce:

Order:

$$\tau_{33} \leq \tau_{22} \leq \tau_{11}$$

$$\tau_{MaximumShear} \equiv Max(\tau 1_{Shear}, \tau 2_{Shear}, \tau 3_{Shear}) = 1/2\,(\tau_{11} - \tau_{33}) = \tau 2_{Shear}$$

So:

$0 \leq \tau 1_{Shear} = (\tau_{22} - \tau_{33})/2$ $\quad 0 \leq \tau 2_{Shear} = (\tau_{11} - \tau_{33})/2$ $\quad 0 \leq \tau 3_{Shear} = (\tau_{11} - \tau_{22})/2$

Then:

$$\tau 1_{Shear} + \tau 3_{Shear} = \tau 2_{Shear}$$

$0 \leq \tau 1_{Shear} \leq \tau 2_{Shear}$ $\quad 0 \leq \tau 3_{Shear} \leq \tau 2_{Shear}$

So:

Summed Principal Shear Stresses:

$$\tau 1_{Shear} + \tau 2_{Shear} + \tau 3_{Shear} = 2\, \tau 2_{Shear}$$

Squared Summed Principal Shear Stresses:

$$(\tau 1_{Shear} + \tau 2_{Shear} + \tau 3_{Shear})^2 = 4\, \tau 2_{Shear}^2$$

$$\tau 1_{Shear}^2 + \tau 2_{Shear}^2 + \tau 3_{Shear}^2 + 2\tau 1_{Shear}\tau 2_{Shear} + 2\tau 2_{Shear}\tau 3_{Shear} + 2\tau 3_{Shear}\tau 1_{Shear} = 4\tau 2_{Shear}^2$$

$$\tau 1_{Shear}^2 + \tau 2_{Shear}^2 + \tau 3_{Shear}^2 + 2\tau 2_{Shear}(\tau 1_{Shear} + \tau 3_{Shear}) + 2\tau 3_{Shear}\tau 1_{Shear} = 4\tau 2_{Shear}^2$$

$$\tau 1_{Shear}^2 + \tau 2_{Shear}^2 + \tau 3_{Shear}^2 + 2\tau 2_{Shear}^2 + 2\tau 3_{Shear}\tau 1_{Shear} = 4\tau 2_{Shear}^2$$

So: Summed Principal Shear Stresses Squares:

$$\tau 1_{Shear}^2 + \tau 2_{Shear}^2 + \tau 3_{Shear}^2 = 4\tau 2_{Shear}^2 - 2(\tau 1_{Shear}\tau 2_{Shear} + \tau 2_{Shear}\tau 3_{Shear} + \tau 3_{Shear}\tau 1_{Shear})$$

Or:

$$(\tau 1_{Shear} + \tau 3_{Shear}) = \tau 2_{Shear}$$

$$(\tau 1_{Shear} + \tau 3_{Shear})^2 = \tau 2_{Shear}^2$$

$$\tau 1_{Shear}^2 + \tau 3_{Shear}^2 + 2\, \tau 1_{Shear}\, \tau 3_{Shear} = \tau 2_{Shear}^2$$

$$\tau 1_{Shear}^2 + \tau 2_{Shear}^2 + \tau 3_{Shear}^2 + 2\, \tau 1_{Shear}\, \tau 3_{Shear} = 2\, \tau 2_{Shear}^2$$

So: Summed Principal Shear Stresses Squares:

$$\tau 1_{Shear}^2 + \tau 2_{Shear}^2 + \tau 3_{Shear}^2 = 2\, [\tau 2_{Shear}^2 - \tau 1_{Shear}\, \tau 3_{Shear}]$$

Now:

To Determine An Upper Bound:

$$(\tau_{SO}^2)/(\tau 2_{Shear})^2 \leq 8/9$$

Substitute:

$$\tau 1_{Shear}^2 + \tau 2_{Shear}^2 + \tau 3_{Shear}^2 = 2\, [\tau 2_{Shear}^2 - \tau 1_{Shear}\, \tau 3_{Shear}]$$

In:

$$\tau_{SO}^2 = 4/9\, [(\tau 1_{Shear})^2 + (\tau 2_{Shear})^2 + (\tau 3_{Shear})^2]$$

To Get:

$$\tau_{SO}^2 = 2 * 4/9\, [\tau 2_{Shear}^2 - \tau 1_{Shear}\, \tau 3_{Shear}]$$

But:

$$0 \leq \tau 1_{Shear}\, \tau 3_{Shear}$$

So:

An Upper Bound:

$$(\tau_{SO}^2)/(\tau 2_{Shear})^2 \leq 8/9$$

To Determine A Lower Bound:

$$2/3 \leq (\tau_{SO}^2)/(\tau 2_{Shear})^2$$

Substitute:

$$\tau 1_{Shear}^2 + \tau 2_{Shear}^2 + \tau 3_{Shear}^2 = 2\, [\tau 2_{Shear}^2 - \tau 1_{Shear}\, \tau 3_{Shear}]$$

In:

$$0 \leq 2/3\, [(\tau 1_{Shear})^2 + (\tau 2_{Shear})^2 + (\tau 3_{Shear})^2] = (-T_2)$$

To Get:

$$0 \leq 4/3\, [\tau 2_{Shear}^2 - \tau 1_{Shear}\, \tau 3_{Shear}] = (-T_2)$$

Now:

Since:

$$0 \leq [\tau 2_{Shear}^2 - \tau 1_{Shear}\, \tau 3_{Shear}]$$

And:

$0 \leq \tau 1_{Shear}$ $\quad 0 \leq \tau 2_{Shear}$ $\quad 0 \leq \tau 3_{Shear}$

The Following Case Is Impossible:

$$\tau 1_{Shear}\, \tau 3_{Shear} \leq 0$$

So Only The Following Case Is Possible:

$$0 \leq \tau1_{Shear}\,\tau3_{Shear} \leq \tau2_{Shear}^{2}$$

But:

$$(\tau1_{Shear} + \tau3_{Shear}) = \tau2_{Shear}$$

$$(\tau1_{Shear}^{2} + 2\tau1_{Shear}\,\tau3_{Shear} + \tau3_{Shear}^{2}) = (\tau1_{Shear} + \tau3_{Shear})^{2} = \tau2_{Shear}^{2}$$

And:

$$0 \leq (\tau1_{Shear}^{2} - 2\tau1_{Shear}\,\tau3_{Shear} + \tau3_{Shear}^{2}) = (\tau1_{Shear} - \tau3_{Shear})^{2}$$

$$2\,\tau1_{Shear}\,\tau3_{Shear} \leq (\tau1_{Shear}^{2} + \tau3_{Shear}^{2})$$

Sum of:

$$(\tau1_{Shear}^{2} + 2\tau1_{Shear}\,\tau3_{Shear} + \tau3_{Shear}^{2}) = \tau2_{Shear}^{2}$$

$$2\,\tau1_{Shear}\,\tau3_{Shear} \leq (\tau1_{Shear}^{2} + \tau3_{Shear}^{2})$$

Is:

$$4\,\tau1_{Shear}\,\tau3_{Shear} \leq \tau2_{Shear}^{2}$$

Or:

$$0 \leq \tau2_{Shear}^{2} - 4\,\tau1_{Shear}\,\tau3_{Shear}$$

Then:

$$3\,(\tau2_{Shear})^{2} \leq 4\,\tau2_{Shear}^{2} - 4\,\tau1_{Shear}\,\tau3_{Shear}$$

So:

$$0 \leq (\tau2_{Shear})^{2} \leq 4/3\,[\tau2_{Shear}^{2} - \tau1_{Shear}\,\tau3_{Shear}] = (-T_2)$$

$$0 \leq 1/(-T_2) \leq 1/(\tau2_{Shear})^{2}$$

And Recall:

$$2/3\,(-T_2) = \tau_{SO}^{2}$$

If Last Equations are Multiplied:

$$2/3 \leq (\tau_{SO}^{2})/(\tau2_{Shear})^{2}$$

So Both Bounds:

$$2/3 \leq (\tau_{SO}^{2})/(\tau_{MaximumShear})^{2} \leq 8/9$$

15.8.1.3.
Evaluation of Principal Stresses:

Introduce Trigonometric Paragraph

$$4\cos^3\alpha - 3\cos\alpha = \cos 3\alpha$$

equals:

$$4\cos\alpha\,(1 - \sin^2\alpha) - 3\cos\alpha = \cos(\alpha + 2\alpha)$$

equals:

$$\cos\alpha - 4\cos\alpha\,\sin^2\alpha = \cos\alpha\cos 2\alpha - \sin\alpha\,\sin 2\alpha$$

equals:

$$\cos\alpha - 4\cos\alpha\,\sin^2\alpha = \cos\alpha\,(1 - 2\sin^2\alpha) - \sin\alpha\,(2\sin\alpha\cos\alpha)$$

Write:

$$4\cos^3\alpha - 3\cos\alpha = \cos 3\alpha$$

in this form:

$$+\,1/4\cos 3\alpha + 3/4\cos\alpha - \cos^3\alpha = 0$$

Now Let:

$$T = \rho\cos\alpha$$

In:

$$F \equiv T_3 - T_2\,T + \qquad - T^3 = 0$$

To Get:

$$T_3/\rho^{3} - T_2/\rho^{2}\cos\alpha \qquad - \cos^{3}\alpha = 0$$

Compared terms with:

$$+ 1/4 \cos 3\alpha + 3/4 \cos\alpha - \cos^3\alpha = 0$$

To Get:

$$T_3/\rho^3 = 1/4 \cos 3\alpha$$

$$- T_2/\rho^2 = 3/4$$

Second Leads To:

$$\rho^2 = 4 (-\Sigma_2/3)$$

$$\rho = 2 (-T_2/3)^{1/2}$$

$$\rho^3 = 8 (-T_2/3)^{3/2}$$

First Leads To:

$$\cos 3\alpha = 4 T_3/\rho^3 = ½ T_3/(-T_2/3)^{3/2}$$

Where:First Root: $0 \le 3\alpha_1 \le \pi$

$$0 \le \alpha_1 \le \pi/3$$

All Roots:

$$\alpha_1$$

$$\alpha_2 = \alpha_1 + 2\pi/3$$

$$\alpha_3 = \alpha_2 + 2\pi/3 = \alpha_1 + 4\pi/3$$

$$\alpha_1 = \alpha_3 + 2\pi/3 = \alpha_2 + 4\pi/3 = \alpha_1 + 6\pi/3$$

Principal Stresses:

$$T_{11} = \tau_{11} - \tau_{spherical} = 2 (-T_2/3)^{1/2} \cos\alpha_1$$

$$T_{22} = \tau_{22} - \tau_{spherical} = 2 (-T_2/3)^{1/2} \cos\alpha_2$$

$$T_{33} = \tau_{33} - \tau_{spherical} = 2 (-T_2/3)^{1/2} \cos\alpha_3$$

$$T_{33} \le T_{22} \le T_{11}$$

Where Invariants:

$$0 \le \alpha_1 \le \pi/3$$

$\tau_{spherical} = \tau_{m\,m}/3 = \tau_1/3$ **Or Invariant:** τ_1

$(-T_2/3)^{1/2}$ **Or Invariant:** $(-T_2)^{1/2}$ **Or Invariant:** T_2

Various Failure Criteria:

$$F(\alpha_1, \tau_1, T_2) = 0$$

15.8.1.4.
Physical Interpretation of Invariants:

Invariants:

$$0 \le \alpha_1 \le \pi/3$$

$\tau_{spherical} = \tau_{m\,m}/3 = \tau_1/3$ **Or Invariant:** τ_1

$(-T_2/3)^{1/2}$ **Or Invariant:** $(-T_2)^{1/2}$ **Or Invariant:** T_2

Various Failure Criteria:

$$F(\alpha_1, \tau_1, T_2) = 0$$

Octahedral Stresses:

Normal:

$$\tau_{NO} = 1/3 (\tau_{11} + \tau_{22} + \tau_{33}) = \tau_{spherical}$$

Shear:

$$\tau_{SO}^2 = 1/9 [(\tau_{11} - \tau_{22})^2 + (\tau_{22} - \tau_{33})^2 + (\tau_{33} - \tau_{11})^2]$$

$$\tau_{SO}^2 = 2/3 (-T_2)$$

Or:

$$(-T_2) = 3/2 \tau_{SO}^2$$

Substituted In:

$$\cos 3\alpha = 4 T_3/\rho^3 = ½ T_3/(-T_2/3)^{3/2}$$

To Get:

$$\cos 3\alpha = \tfrac{1}{2}\, T_3 / (1/2\, \tau_{SO}^{2})^{3/2} = 2^{1/2}\, T_3 / \tau_{SO}^{3}$$

Various Failure Criteria:

$$F(\alpha_1, \tau_{NO}, \tau_{SO}) = 0$$

Elastic Strain Energy: Dilatation Energy Plus Distortion Energy.

15.9.
Elastic Internal Energy per Unit Mass:

General Assumption:

$$\rho e = \rho e\,(I1, I2, I3) = \rho e\,(E^{ii}, E^{ij}E^{ji}, E^{ij}E^{jk}E^{ki})$$
$$\rho c = \rho c\,(J1, J2, J3) = \rho c\,(T_{ii}, T_{ij}T_{ji}, T_{ij}T_{jk}T_{ki})$$

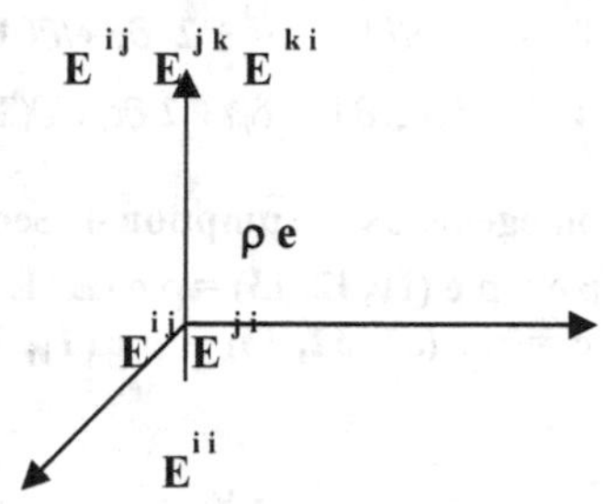

$T_{ij}\,T_{jk}\,T_{ki}$

ρc

$T_{ij}\ T_{ji}$

T_{ii}

$$T_{ij} = \rho\ \partial e / \partial E^{ij}$$
$$=$$
$$\partial\rho e/\partial E^{ii}\ \partial E^{ii}/\partial E^{ij} + \partial\rho e/\partial(E^{ij}E^{ji})\ \partial(E^{ij}E^{ji})/\partial E^{ij} + \partial\rho e/\partial(E^{ij}E^{jk}E^{ki})\ \partial(E^{ij}E^{jk}E^{ki})/\partial E^{ij}$$

$$E^{ij}\ \rho\ \partial c / \partial T_{ij}$$
$$=$$
$$\partial\rho c/\partial T_{ii}\ \partial T_{ii}/\partial T_{ij} + \partial\rho c/\partial(T_{ij}T_{ji})\ \partial(T_{ij}T_{ji})/\partial T_{ij} + \partial\rho c/\partial(T_{ij}T_{jk}T_{ki})\ \partial(T_{ij}T_{jk}T_{ki})/\partial T_{ij}$$

Or:

$$T_{ij} = \partial\rho e/\partial E^{ii}\ \delta^{ij} + 2\ \partial\rho e/\partial(E^{ij}E^{ji})\ E^{ij} + 3\ \partial\rho e/\partial(E^{ij}E^{jk}E^{ki})\ E^{jk}E^{ki}$$

$$E^{ij} = \partial\rho c/\partial T_{ii}\ \delta_{ij} + 2\ \partial\rho c/\partial(T_{ij}T_{ji})\ T_{ij} + 3\ \partial\rho c/\partial(T_{ij}T_{jk}T_{ki})\ T_{jk}T_{ki}$$

Linear Assumption:

$$\rho\, e = \rho\, e\,(I1, I2, I3) = \rho\, e\,(E^{ii}, E^{ij}E^{ji})$$
$$\rho\, c = \rho\, c\,(J1, J2, J3) = \rho\, c\,(T_{ii}, T_{ij}T_{ji})$$

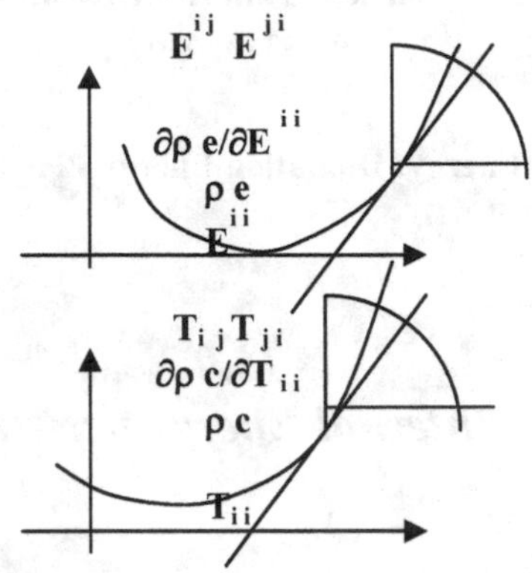

$$T_{ij} = \rho\ \partial e / \partial E^{ij} = \partial \rho\, e / \partial E^{ii}\ \partial E^{ii} / \partial E^{ij} + \partial \rho\, e / \partial (E^{ij}E^{ji})\ \partial (E^{ij}E^{ji}) / \partial E^{ij}$$
$$E^{ij} = \rho\ \partial c / \partial T_{ij} = \partial \rho\, c / \partial T_{ii}\ \partial T_{ii} / \partial T_{ij} + \partial \rho\, c / \partial (T_{ij}T_{ji})\ \partial (T_{ij}T_{ji}) / \partial T_{ij}$$

Or:

$$T_{ij} = \partial \rho\, e / \partial E^{ii}\ \delta^{ij} + 2\ \partial \rho\, e / \partial (E^{ij}E^{ji})\ E^{ij}$$
$$E^{ij} = \partial \rho\, c / \partial T_{ii}\ \delta_{ij} + 2\ \partial \rho\, c / \partial (T_{ij}T_{ji})\ T_{ij}$$

Homogeneous Assumption of Second Degree:

$$\rho\, e = \rho\, e\,(I1, I2, I3) = \rho\, e\,(E^{ii}E^{jj}, E^{ij}E^{ji})$$
$$\rho\, c = \rho\, c\,(J1, J2, J3) = \rho\, c\,(T_{ii}T_{jj}, T_{ij}T_{ji})$$

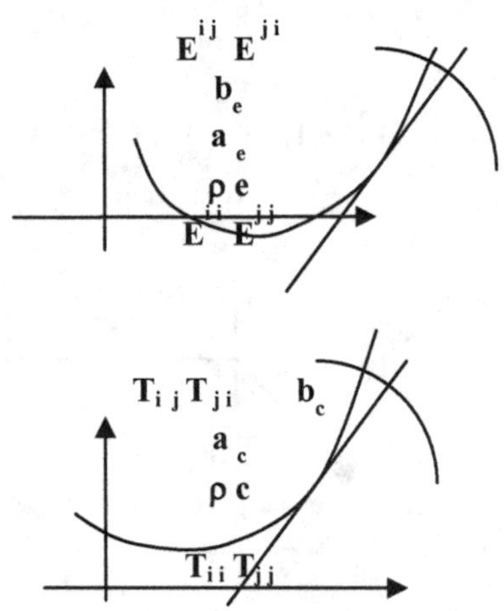

$$T_{ij} = \rho\ \partial e / \partial E_{ij} = \partial \rho\, e / \partial (E^{ii}E^{jj})\ \partial (E^{ii}E^{jj}) / \partial E^{ij} + \partial \rho\, e / \partial (E^{ij}E^{ji})\ \partial (E^{ij}E^{ji}) / \partial E^{ij}$$
$$E^{ij} = \rho\ \partial c / \partial T_{ij} = \partial \rho\, c / \partial (T_{ii}T_{jj})\ \partial (T_{ii}T_{jj}) / \partial T^{ij} + \partial \rho\, c / \partial (T_{ij}T_{ji})\ \partial (T_{ij}T_{ji}) / \partial T^{ij}$$

Or:

$$\rho\, e = a_e * (E^{ii}E^{jj}) + b_e * (E^{ij}E^{ji})$$
$$\rho\, c = a_c * (T_{ii}T_{jj}) + b_c * (T_{ij}T_{ji})$$

Or:

$$T_{ij} = 2\ a_e * E^{ii}\delta^{ij} + 2\ b_e * E^{ij}$$

$$E^{ij} = 2\,a_c * T_{ii}\,\delta_{ij} + 2\,b_c * T_{ij}$$

Infinitesimal:

$$\tau_{ij} = 2\,a_e * \varepsilon^{ii}\,\delta^{ij} + 2\,b_e * \varepsilon^{ij}$$

$$\varepsilon^{ij} = 2\,a_c * \tau_{ii}\,\delta_{ij} + 2\,b_c * \tau_{ij}$$

A theory due to Euler for Homogeneous Assumption :

If $F(T_{ij})$ is homogeneous in T_{ij} of degree n then:

$$\partial F/\partial T_{ij}\;T_{ij} \equiv FT^{ij}\;T_{ij} = n\;F$$

15.10.
Elastically Bounded Stress Yield Surface:
Eastic Internal Energy per Unit Mass:

Series Derivative $\rho\partial e/\partial(x_i{}^{,m}) = C1^i{}_m + C2^{ij}{}_{mn}(x_j{}^{,n}) + C3^{ij}{}_{mn}(x_i{}^{,m})(x_j{}^{,n}) + \ldots$

Time derivative $\rho\;eT = \rho\;\;\partial e/\partial(x_i{}^{,m})\;\;(xT_i{}^{,m})$

x Series Time Derivative $\rho\,eT = (C1^i{}_m + C2^{ij}{}_{mn}\,x_j{}^{,n} + C3^{ij}{}_{mn}\,x_i{}^{,m}\,x_j{}^{,n} + \ldots)(xT_i{}^{,m})$

Elasticity:

e Conservation $\rho\,eT = v_i{}^{,n}\,F^i{}_n - h_n{}^{,n} = ET^{pq}\,\rho/\rho 0\;T_{pq} - h_n{}^{,n}$

x Series Time Derivative $\rho\;eT = (C1^i{}_m + C2^{ij}{}_{mn}\;\;(x_j{}^{,n}))\;\;(xT_i{}^{,m})$

x Constit. Series T Deriv $\rho eT = (C1^i{}_m + C2^{ij}{}_{mn}(x_j{}^{,n}))(xT_i{}^{,m}) = ET^{pq}\;\rho/\rho 0\;T_{pq} - h_n{}^{,n}$

$$\rho\;\;eT\;=\;ET^{pq}\;\;T_{pq}$$

Introduce Elastically Bounded Stress Yield Surface:

Elastic: $F < 0$ **Yield Surface:$F(J_1,J_2) = 0$** **Impossible: $0 < F$**

Introduce Simplest Stress Yield Surface: Von Mises:

$$F(J_1, J_2) = F(T_{kl}, T_p) = F(S_{kl}, S_q) = F(J_2) = 0$$

$$F = F(S_{kl}, S_q) = F(J_2) = F(S_{ij}S_{ji}) = -J_2 - ½S^2 = ½S^{ik}S_{ki} - ½S^2 = 0$$

$$½S^{ik}S_{ki} = ½S^2 \equiv K^2$$

In simple shear $T1_2 = T_{21} = Y_s$

$$T_{ii}/3 = 0$$

$$S_{ij} = T_{ij} - T_{ii}/3\;\;\delta_{ij} = T_{ij} - 0 = T_{ij}$$

0	**$S_{12} = Y_s$**	**0**
$S_{21} = Y_s$	**0**	**0**
0	**0**	**0**

	$S_{ik}S_{ki}$	
$S_{1k}S_{k1} = Y_s^2$	**0**	**0**
0	**$S_{2k}S_{k2} = Y_s^2$**	**0**
0	**0**	**0**

$$½ S_{ik} S_{ki} = ½ (2 Y_s^2) = Y_s^2 = ½ S^2 = K^2$$
$$½ S^2 = Y_s^2 \equiv K^2$$
$$S = \sqrt{2}\ Y_s = \sqrt{2}\ K$$

So:

$$F (S_{kl}, S_q) = ½ S_{ik} S_{ki} - Y_s^2 = ½ S_{ik} S_{ki} - ½ S^2 = ½ S_{ik} S_{ki} - K^2 = 0$$

In simple tension $T_{11} = Y_t$

$$T_{ii} / 3 = Y_t / 3$$

$$S_{ij} = T_{ij} - T_{ii} / 3 \quad \delta_{ij} = T_{ij} - Y_t / 3 \quad \delta_{ij}$$

$S_{11} = 2/3\ Y_t$	0	0
0	$-1/3\ Y_t$	0
0	0	$-1/3\ Y_t$

$S_{ik} S_{ki}$

$S_{1k} S_{k1} = 4/9\ Y_t^2$	0	0
0	$S_{2k} S_{k2} = 1/9\ Y_t^2$	0
0	0	$S_{3k} S_{k3} = 1/9\ Y_t^2$

$$½ S_{ik} S_{ki} = ½ (4/9 + 1/9 + 1/9)\ Y_t^2 = ½ S^2 \equiv K^2$$
$$½ S^2 = 1/3\ Y_t^2 = K^2$$
$$½ S^2 = Y_s^2 \equiv K^2$$

$$1/3\ Y_t^2 = 1/2\ S^2 = Y_s^2 \equiv K^2$$
$$3\ K^2 \equiv 3\ Y_s^2 = 3/2\ S^2 = Y_t^2$$

$$2/3\ Y_t^2 = S^2 = 2\ Y_s^2 \equiv 2\ K^2$$
$$\sqrt{2} / \sqrt{3}\ Y_t = S = \sqrt{2}\ Y_s \equiv \sqrt{2}\ K$$

So:

$$F (S_{kl}, S_q) = ½ S_{ik} S_{ki} - 1/3\ Y_t^2 = ½ S_{ik} S_{ki} - ½ S^2 = ½ S_{ik} S_{ki} - K^2 = 0$$

Principal T_i

$$T_{ii} / 3 = (T_1 + T_2 + T_3) / 3$$

$$S_i = T_i - (T_1 + T_2 + T_3) / 3$$

$S_1 = T_1 - (T_1 + T_2 + T_3)/3$	0	0
0	$S_2 = T_2 - (T_1 + T_2 + T_3)/3$	0
0	0	$S_3 = T_3 - (T_1 + T_2 + T_3)/3$

$S_{ik} S_{ki}$

$(S_1)^2$	0	0
0	$(S_2)^2$	0
0	0	$(S_3)^2$

$$½ S_{ik} S_{ki} = ½ [(T_1-(T_1+T_2+T_3)/3)^2 + (T_2-(T_1+T_2+T_3)/3)^2 + (T_3-(T_1+T_2+T_3)/3)^2] = ½ S^2$$
$$½ S_{ik} S_{ki} = ½ [(T_1^2+T_2^2+T_3^2+(T_1+T_2+T_3)^2/3 - 2(T_1+T_2+T_3)(T_1+T_2+T_3)/3] = ½ S^2$$

$$\tfrac{1}{2} S_{ik} S_{ki} = \tfrac{1}{2}\,[(T_1^2+T_2^2+T_3^2) - (T_1+T_2+T_3)(T_1+T_2+T_3)/3] = \tfrac{1}{2}S^2$$
$$\tfrac{1}{2} S_{ik} S_{ki} = \tfrac{1}{2}\,[(T_1^2+T_2^2+T_3^2) - (T_1^2+T_2^2+T_3^2)/3 - 2(T_1 T_2+T_2 T_3+T_3 T_1)/3] = \tfrac{1}{2}S^2$$
$$\tfrac{1}{2} S_{ik} S_{ki} = 1/2\ \ 1/3\,[2(T_1^2+T_2^2+T_3^2) - 2(T_1 T_2+T_2 T_3+T_3 T_1)] = \tfrac{1}{2}S^2$$
$$(T_1 - T_2)^2 + (T_2 - T_3)^2 + (T_3 - T_1)^2 = 2(T_1^2+T_2^2+T_3^2) - 2(T_1 T_2+T_2 T_3+T_3 T_1)$$
$$\tfrac{1}{2} S_{ik} S_{ki} = 1/2\ \ 1/3\ [(T_1 - T_2)^2 + (T_2 - T_3)^2 + (T_3 - T_1)^2] = \tfrac{1}{2}S^2$$

But:

$$\tfrac{1}{2} S^2 = 1/3\ Y_t^2 = Y_s^2 \equiv K^2$$

So:

$$F(S_{kl},S_q) = \tfrac{1}{2} S_{ik} S_{ki} = 1/2\ \ 1/3[(T_1-T_2)^2+(T_2-T_3)^2+(T_3-T_1)^2] = K^2 \equiv Y_s^2 = \tfrac{1}{2} S^2 = 1/3\ Y_t^2$$

Or:

$$(T_1 - T_2)^2 + (T_2 - T_3)^2 + (T_3 - T_1)^2 = 6K^2 \equiv 6Y_s^2 = 3S^2 = 2\ Y_t^2$$

Distortioal Energy $e_{Distortional}$ Stress Yield Surface:

Elastic: $F < 0$ $\qquad F(J_1, J_2) = 0 \qquad$ **Impossible: $0 < F$**

Where Distortioal Energy $e_{Distortional}$:

$$6\,G\,e_{Distortional} = (J_1^2 - 3J_2)$$

General:

$$6G\,e_{Distortional} \equiv J_1^2 - 3\,J_2$$
$$J_1^2 - 3\,J_2 \equiv (T_{11}^2+T_{22}^2+T_{33}^2) - (T_{11} T_{22} + T_{22} T_{33} + T_{33} T_{11}) + 3(T_{12}^2+T_{23}^2+T_{31}^2)$$
$$J_1^2 - 3\,J_2 = \tfrac{1}{2}\,[(T_1 - T_2)^2 + (T_2 - T_3)^2 + (T_3 - T_1)^2] \ \ + 3(T_{12}^2+T_{23}^2+T_{31}^2)$$

Principal:

$$6G\,e_{Distortional} \equiv J_1^2 - 3\,J_2$$
$$J_1^2 - 3\,J_2 \equiv (T_1^2+T_2^2+T_3^2) - (T_1 T_2 + T_2 T_3 + T_3 T_1)$$
$$J_1^2 - 3\,J_2 = \tfrac{1}{2}\,[(T_1 - T_2)^2 + (T_2 - T_3)^2 + (T_3 - T_1)^2]$$

General Deviator:

$$6G\,e_{Distortional} \equiv J_1^2 - 3\,J_2 = -3\,J_2$$
$$-3\,J_2 = -3\,(S_{11} S_{22}+S_{22} S_{33}+S_{33} S_{11}) + 3\,(S_{12}^2+S_{23}^2+S_{31}^2)$$
$$-3\,J_2 = 3/2\,(S_{11}^2+S_{22}^2+S_{33}^2) + 3\,(S_{12}^2+S_{23}^2+S_{31}^2)$$
$$J_1^2 - 3J_2 = (S_{11}^2+S_{22}^2+S_{33}^2) - (S_{11} S_{22} + S_{22} S_{33} + S_{33} S_{11}) + 3(S_{12}^2+S_{23}^2+S_{31}^2)$$
$$J_1^2 - 3J_2 = (S_{11}^2+S_{22}^2+S_{33}^2) + \tfrac{1}{2}(S_{11}^2+S_{22}^2+S_{33}^2) + 3(S_{12}^2+S_{23}^2+S_{31}^2)$$
$$J_1^2 - 3\,J_2 = \tfrac{1}{2}\,[(S_1 - S_2)^2 + (S_2 - S_3)^2 + (S_3 - S_1)^2] \ \ + 3(S_{12}^2+S_{23}^2+S_{31}^2)$$

Principal Deviator:

$$6G\,e_{Distortional} \equiv J_1^2 - 3\,J_2 = -3\,J_2$$
$$-3\,J_2 = -3\,(S_{11} S_{22}+S_{22} S_{33}+S_{33} S_{11})$$
$$-3\,J_2 = 3/2\,(S_{11}^2+S_{22}^2+S_{33}^2)$$
$$J_1^2 - 3J_2 = (S_{11}^2+S_{22}^2+S_{33}^2) - (S_{11} S_{22} + S_{22} S_{33} + S_{33} S_{11})$$
$$J_1^2 - 3J_2 = (S_{11}^2+S_{22}^2+S_{33}^2) + \tfrac{1}{2}(S_{11}^2+S_{22}^2+S_{33}^2)$$
$$J_1^2 - 3\,J_2 = \tfrac{1}{2}\,[(S_1 - S_2)^2 + (S_2 - S_3)^2 + (S_3 - S_1)^2]$$

Apply these to Von Mises:

Von Mises General Distortional energy:

$$1/3\ Y_t^2 = 1/2\ S^2 = Y_s^2 \equiv K^2$$

$$6G\, e_{Distortional} \equiv J_1^2 - 3J_2$$
$$J_1^2 - 3J_2 \equiv (T_{11}^2 + T_{22}^2 + T_{33}^2) - (T_{11}T_{22} + T_{22}T_{33} + T_{33}T_{11}) + 3(T_{12}^2 + T_{23}^2 + T_{31}^2)$$
$$J_1^2 - 3J_2 = ½\,[(T_1 - T_2)^2 + (T_2 - T_3)^2 + (T_3 - T_1)^2] \;\; + 3(T_{12}^2 + T_{23}^2 + T_{31}^2)$$

Von Mises General Distortional energy in simple tension, Y_t:

$$J_1^2 - 3J_2 = (Y_t)^2 = 3K^2$$

So:

$$(T_{11}^2 + T_{22}^2 + T_{33}^2) - (T_{11}T_{22} + T_{22}T_{33} + T_{33}T_{11}) + 3(T_{12}^2 + T_{23}^2 + T_{31}^2) = Y_t^2$$
$$½\,[(T_1 - T_2)^2 + (T_2 - T_3)^2 + (T_3 - T_1)^2] \;\; + 3(T_{12}^2 + T_{23}^2 + T_{31}^2) = Y_t^2$$

If : $T_{33} = T$:

$$(T_{11}^2 + T_{22}^2 + T^2) - (T_{11}T_{22} + T_{22}T + TT_{11}) + 3(T_{12}^2 + T_{23}^2 + T_{31}^2) = Y_t^2$$
$$½\,[(T_1 - T_2)^2 + (T_2 - T)^2 + (T - T_1)^2] \;\; + 3(T_{12}^2 + T_{23}^2 + T_{31}^2) = Y_t^2$$

$$T_{11}^2 + T_{22}^2 + T^2 - T_{11}T_{22} - T_{22}T - TT_{11} + 3(T_{12}^2 + T_{23}^2 + T_{31}^2) = Y_t^2$$
$$T^2 - (T_{11} + T_{22})T + \;\; T_{11}^2 + T_{22}^2 - T_{11}T_{22} + 3(T_{12}^2 + T_{23}^2 + T_{31}^2) = Y_t^2$$

If : $T_{33} = T = 0$:

$$T_{11}^2 + T_{22}^2 - T_{11}T_{22} + 3(T_{12}^2 + T_{23}^2 + T_{31}^2) = Y_t^2$$

If : $T_{33} = T_{32} = T_{31} = 0$:

$$T_{11}^2 + T_{22}^2 + T_{12}^2 - T_{11}T_{22} = Y_t^2$$

Von Mises Principal Distortional energy:

$$6G\, e_{Distortional} \equiv J_1^2 - 3J_2$$
$$J_1^2 - 3J_2 \equiv (T_1^2 + T_2^2 + T_3^2) - (T_1T_2 + T_2T_3 + T_3T_1)$$
$$J_1^2 - 3J_2 = ½\,[(T_1 - T_2)^2 + (T_2 - T_3)^2 + (T_3 - T_1)^2]$$

Von Mises Principal Distortional energy in simple tension:

$$J_1^2 - 3J_2 = ½\,((Y_t - 0)^2 + (0-0)^2 + (0 - Y_t)^2) = ½\;\; 2Y_t^2$$

So:

$$(T_1^2 + T_2^2 + T_3^2) - (T_1T_2 + T_2T_3 + T_3T_1) \;=\; Y_t^2$$
$$½\,[(T_1 - T_2)^2 + (T_2 - T_3)^2 + (T_3 - T_1)^2] \;=\; Y_t^2$$

If : $T_3 = T$:

$$½\,((T_1 - T_2)^2 + (T_2 - T)^2 + (T . T_1)^2) = Y_t^2$$
$$T_1^2 + T_2^2 \;\; - T_1T_2 \; - \; (T_1 + T_2)\; T \;\; + T^2 \;=\; Y_t^2$$
$$(T_1/Y_t)^2 \; + (T_2/Y_t)^2 - (T_1/Y_t)(T_2/Y_t) - (T_1 + T_2)/Y_t \;\; T/Y_t + (T/Y_t)^2 = 1$$

If : $T_{33} = 0$:

$$(T_1)^2 + (T_2)^2 - (T_1T_2) = Y_t^2$$
$$(T_1/Y_t)^2 \; + (T_2/Y_t)^2 - (T_1/Y_t)(T_2/Y_t) = 1$$

Von Mises General Deviator Distortional energy:

$$6G\, e_{Distortional} \equiv J_1^2 - 3J_2 = -3J_2$$
$$-3J_2 \;\; = -3(S_{11}S_{22} + S_{22}S_{33} + S_{33}S_{11}) + 3(S_{12}^2 + S_{23}^2 + S_{31}^2)$$
$$-3J_2 \; = 3/2\,(S_{11}^2 + S_{22}^2 + S_{33}^2) + 3(S_{12}^2 + S_{23}^2 + S_{31}^2)$$
$$J_1^2 - 3J_2 = (S_{11}^2 + S_{22}^2 + S_{33}^2) - (S_{11}S_{22} + S_{22}S_{33} + S_{33}S_{11}) + 3(S_{12}^2 + S_{23}^2 + S_{31}^2)$$
$$J_1^2 - 3J_2 = (S_{11}^2 + S_{22}^2 + S_{33}^2) + ½\,(S_{11}^2 + S_{22}^2 + S_{33}^2) + 3(S_{12}^2 + S_{23}^2 + S_{31}^2)$$
$$J_1^2 - 3J_2 \; = ½\,[(S_1 - S_2)^2 + (S_2 - S_3)^2 + (S_3 - S_1)^2] \;\; + 3(S_{12}^2 + S_{23}^2 + S_{31}^2)$$

Von Mises General Deviator Distortional energy in simple tension:

$$3/2(\ (2/3\ Y_t)^2 + (1/3\ Y_{te})^2 + (1/3\ Y_{te})^2\ ) = (Y_t)^2$$

So:

$$-3\,(S_{11}\,S_{22}+S_{22}\,S_{33}+S_{33}\,S_{11}) + 3\,(S_{12}^2+S_{23}^2+S_{31}^2) = (Y_t)^2$$

$$3/2\,(S_{11}^2+S_{22}^2+S_{33}^2) + 3\,(S_{12}^2+S_{23}^2+S_{31}^2) = (Y_t)^2$$

$$(S_{11}^2+S_{22}^2+S_{33}^2) - (S_{11}\,S_{22} + S_{22}\,S_{33} + S_{33}\,S_{11}) + 3(S_{12}^2+S_{23}^2+S_{31}^2) = (Y_t)^2$$

$$(S_{11}^2+S_{22}^2+S_{33}^2) + ½\,(S_{11}^2+S_{22}^2+S_{33}^2) + 3(S_{12}^2+S_{23}^2+S_{31}^2) = (Y_t)^2$$

If : $S_{33} = S$:

$$3/2(S_{11}^2 + S_{22}^2 + S^2) + 3\,(S_{12}^2+S_{23}^2+S_{31}^2) = (Y_t)^2$$

If : $S_{33} = 0$:

$$3/2(S_{11}^2 + S_{22}^2\ ) + 3(S_{12}^2+S_{23}^2+S_{31}^2) = (Y_t)^2$$

If : $S_{33} = S_{32} = S_{31} = 0$:

$$3/2(S_{11}^2 + S_{22}^2\ ) + 3\,S_{12}^2 = (Y_t)^2$$

Von Mises Principal Deviatoric Distortional energy :

$$6G\ e_{Distortional} \equiv J_1^2 - 3\,J_2 = -3\,J_2$$

$$-3\,J_2 = -3\,(S_{11}\,S_{22}+S_{22}\,S_{33}+S_{33}\,S_{11})$$

$$-3\,J_2 = 3/2\,(S_{11}^2+S_{22}^2+S_{33}^2)$$

$$J_1^2 - 3J_2 = (S_{11}^2+S_{22}^2+S_{33}^2) - (S_{11}\,S_{22} + S_{22}\,S_{33} + S_{33}\,S_{11})$$

$$J_1^2 - 3J_2 = (S_{11}^2+S_{22}^2+S_{33}^2) + ½\,(S_{11}^2+S_{22}^2+S_{33}^2)$$

$$J_1^2 - 3\,J_2 = ½\,[\,(S_1 - S_2)^2 + (S_2 - S_3)^2 + (S_3 - S_1)^2]$$

Von Mises Principal Deviator Distortional energy in simple tension:

$$3/2\,(\ (2/3\ Y_t)^2 + (1/3\ Y_{te})^2 + (1/3\ Y_{te})^2\ ) = 3/2\ \ 2/3\,(Y_t)^2$$

So:

$$-3\,J_2 = -3\,(S_{11}\,S_{22}+S_{22}\,S_{33}+S_{33}\,S_{11}) = (Y_t)^2$$

$$-3\,J_2 = 3/2\,(S_{11}^2+S_{22}^2+S_{33}^2) = (Y_t)^2$$

$$J_1^2 - 3J_2 = (S_{11}^2+S_{22}^2+S_{33}^2) - (S_{11}\,S_{22} + S_{22}\,S_{33} + S_{33}\,S_{11}) = (Y_t)^2$$

$$J_1^2 - 3J_2 = (S_{11}^2+S_{22}^2+S_{33}^2) + ½\,(S_{11}^2+S_{22}^2+S_{33}^2) = (Y_t)^2$$

$$J_1^2 - 3\,J_2 = ½\,[\,(S_1 - S_2)^2 + (S_2 - S_3)^2 + (S_3 - S_1)^2] = (Y_t)^2$$

If : $S_{33} = S$:

$$S_{11}^2 + S_{22}^2 + S^2 = 2/3\,(Y_t)^2$$

If : $S_3 = 0$:

$$S_{11}^2 + S_{22}^2 = 2/3\,(Y_t)^2$$

Von Mises: $6\,G\,e_{Distortional} = (J_1^2 - 3J_2) = [½\,(\,3\,T_{ik}\,T_{ki} - J_1^2\,)\,]$

Equal:

General:

$$6Ge_{Distortional} = (T_{11}^2+T_{22}^2+T_{33}^2) - (T_{11}\,T_{22}+T_{22}\,T_{33}+T_{33}\,T_{11}) + 3(T_{12}^2 + T_{23}^2 + T_{31}^2)$$

$$= ½\,[\,(T_1 - T_2)^2 + (T_2 - T_3)^2 + (T_3 - T_1)^2] + 3\,(T_{12}^2 + T_{23}^2 + T_{31}^2)$$

Principal:

$$6G\ e_{Distortional} = ½\,((T_1 - T_2)^2 + (T_2 - T_3)^2 + (T_3 - T_1)^2)$$

General Deviator:

$$6G\ e_{Distortional} = [3/2(S_{11}^2 + S_{22}^2 + S_{33}^2) + 3(S_{12}^2+S_{23}^2+S_{31}^2\ )]$$

Principal Deviator:

$$6G\ e_{Distortional} = 3/2(S_1^2 + S_2^2 + S_3^2) = -3(S_1\ S_2 + S_2\,S_3 + S_3\ S_1)$$

$(D^{kl}, E^{kl}, E^{q})\ (\partial_i^{\ j}, \varepsilon^{kl}, \varepsilon^{q})$ $(S_{kl}, T_{kl}, T_q)\ (\sigma^i_j, \tau_{kl}, \tau_q)$

F Surface traction (force per area)

$(D^{kk} \equiv 0, E \equiv E^{kk})\ (\partial_k^{\ k} \equiv 0, \varepsilon \equiv \varepsilon^{kk})$ $(S_{kk} \equiv 0, T \equiv T_{kk})\ (\sigma^k_{\ k} \equiv 0, \tau \equiv \tau_{kk})$

Recall:

Elastic Flexibility :

1 / Time = 1 /Stress * Stress / Time

Since Elastic Strain is function of Stress:

$$EET^{kl} = \Gamma^{klmn}\ TT_{mn}$$

$$DET^{kl} + EET^{ii}/3\ \delta^{kl} = \Gamma^{klmn}\ (ST_{mn} + TT^{ii}/3\ \delta_{mn})$$

But: Introducing Bulk Constitution:

$$EET^{ii}/3\ \delta^{kl} = \Gamma^{klmn}\ TT^{ii}/3\ \delta_{mn}$$

So:

Deviatoric Elastic Strain is function of Deviatoric Stress:

$$DET^{kl} = \Gamma^{klmn}\ ST_{mn}$$

Multiply Flexibility by: S_{kl}

$$DET^{kl} S_{kl} = \Gamma^{klmn}\ ST_{mn}\ S_{kl} = (½\,\Gamma^{klmn}\ S_{mn} S_{kl})T$$

And:

Infinitesimal Elastic Flexibility :

Since Elastic Strain is function of Stress:

$$\varepsilon ET^{kl} = (1+\nu)/E\ \tau T^{kl} - \nu/E\ \tau T^{ii}\ \delta^{kl}$$

$$\partial ET^{kl} + \varepsilon ET^{ii}/3\ \delta^{kl} = (1+\nu)/E\ (\sigma T^{kl} + \tau T^{ii}/3\ \delta^{kl}) - \nu/E\ \tau T^{ii}\ \delta^{kl}$$

Bulk Constitution:

$$\varepsilon ET^{ii}\ \delta^{kl} = (1+\nu)/E\ \tau T^{ii}/3\ \delta^{kl} - \nu/E\ \tau T^{ii}\ \delta^{kl} = (1-2\nu)/E\ \ \tau T^{ii}\ \delta^{kl}$$

So:

Deviatoric Elastic Strain is function of Deviatoric Stress:

$$\partial ET^{kl} = (1+\nu)/E\ \ \sigma T^{kl}$$

Multiply Flexibility by: σ_{kl}

$$\partial ET^{kl}\ \sigma_{kl} = (1+\nu)/E\ \ \sigma T^{kl}\ \sigma_{kl} = (½\,(1+\nu)/E\ \sigma^{kl}\ \sigma_{kl})T$$

$$\partial ET^{kl}\ \sigma_{kl} = 1/(2G)\ \ \sigma T^{kl}\ \sigma_{kl} = (½1/(2G)\ \sigma^{kl}\ \sigma_{kl})T$$

:

15.11.
Infinitesimal Elasticity:

$C1^i_{\ m} = 0 = H$ $C2^{ij}_{\ mn}(\varepsilon_j^{\ n}) = \tau^i_{\ n}$

$$\tau^i_{\ m} = C2^{ij}_{\ mn}\ \varepsilon_j^{\ n} \equiv C^{ij}_{\ mn}\ \varepsilon_j^{\ n}$$

There are: (3*3)(3*3) = 3^4 = 81 constants.

Since: 3 of $\tau^i_{\ m}$ **are symmetric:**

$$\tau^i_{\ m} = \tau^m_{\ i}$$

Then: (6 of τ^{i}_{m})(9 of ε_{j}^{n})=54 constants C^{ij}_{mn} :

81 constants C^{ij}_{mn} of which 3x(9) = 27 are redundants: (3*3)(3*3) – 27 = 54

$$C^{ij}_{mn} = C^{mj}_{in}$$

$\varepsilon_{xx} = dr_x / dx$	$1/2\ \gamma_{xy} = \varepsilon_{xy} = dr_x / dy$	$1/2\ \gamma_{xz} = \varepsilon_{xz} = dr_x / dz$
	$\varepsilon_{yy} = dr_y / dy$	$1/2\ \gamma_{yz} = \varepsilon_{yz} = dr_y / dz$
		$\varepsilon_{zz} = dr_z / dz$

Since:

$\varepsilon_{j}^{n} = \varepsilon_{n}^{j}$ $\qquad$ $\tau^{i}_{m} = \tau^{m}_{i}$

ε_{1}^{1}	ε_{2}^{2}	ε_{3}^{3}	ε_{1}^{2}	ε_{2}^{3}	ε_{3}^{1}
τ_{1}^{1}	τ_{2}^{2}	τ_{3}^{3}	τ_{1}^{2}	τ_{2}^{3}	τ_{3}^{1}

Then:

(6 of τ^{i}_{m})(6 of ε_{j}^{n})=36 constants C^{ij}_{mn} :

$$C^{ij}_{mn} = C^{in}_{mj}$$

$$C^{IJ}_{MN} = -\nu^{I}_{M} / E^{J}_{N} = -\nu^{I}_{M} / E^{M} = -\nu^{I}_{M} / E^{N}_{J} = C^{IN}_{MJ}$$

$1/E^1$	$-\nu^{1}_{2}/E^2$	$-\nu^{1}_{3}/E^3$	$-\nu^{1}_{4}/E^4$	$-\nu^{1}_{5}/E^5$	$-\nu^{1}_{6}/E^6$
$-\nu^{2}_{1}/E^1$	$1/E^2$	$-\nu^{2}_{3}/E^3$	$-\nu^{2}_{4}/E^4$	$-\nu^{2}_{5}/E^5$	$-\nu^{2}_{6}/E^6$
$-\nu^{3}_{1}/E^1$	$-\nu^{3}_{2}/E^2$	$1/E^3$	$-\nu^{3}_{4}/E^4$	$-\nu^{3}_{5}/E^5$	$-\nu^{3}_{6}/E^6$
$-\nu^{4}_{1}/E^1$	$-\nu^{4}_{2}/E^2$	$-\nu^{4}_{3}/E^3$	$1/E^4$	$-\nu^{4}_{5}/E^5$	$-\nu^{4}_{6}/E^6$
$-\nu^{5}_{1}/E^1$	$-\nu^{5}_{2}/E^2$	$-\nu^{5}_{3}/E^3$	$-\nu^{5}_{4}/E^4$	$1/E^5$	$-\nu^{5}_{6}/E^6$
$-\nu^{6}_{1}/E^1$	$-\nu^{6}_{2}/E^2$	$-\nu^{6}_{3}/E^3$	$-\nu^{6}_{4}/E^4$	$-\nu^{6}_{5}/E^5$	$1/E^6$

Furthermore Since:

$$C^{ij}_{mn} = \partial^2 e / \partial\varepsilon_{i}^{m} / \partial\varepsilon_{j}^{n} = \partial^2 e / \partial\varepsilon_{j}^{n} / \partial\varepsilon_{i}^{m} = C^{ji}_{nm}$$

Then: (6+1)6/2=21 independent constants C^{ij}_{mn} .

So:

36 constants C^{ij}_{mn} of which 5+4+3+2+1=15 are redundants:

(6)(6) – 15 = 21

C^{ij}_{mn} has 6*6=36 constants of which 21 independent constants

These are:

c^{i}_{m}

c^{1}_{1}	c^{1}_{2}	c^{1}_{3}	c^{1}_{4}	c^{1}_{5}	c^{1}_{6}
	c^{2}_{2}	c^{2}_{3}	c^{2}_{4}	c^{2}_{5}	c^{2}_{6}
		c^{3}_{3}	c^{3}_{4}	c^{3}_{5}	c^{3}_{6}
			c^{4}_{4}	c^{4}_{5}	c^{4}_{6}
	Symmetric			c^{5}_{5}	c^{5}_{6}
					c^{6}_{6}

15.12.
Orthotropic Constitution

Axials:

$$\varepsilon_{X}^{X} = [\ \tau^{X}_{X} - \nu_{X}^{Y}\ \tau^{Y}_{Y} - \nu_{X}^{Z}\ \tau^{Z}_{Z})] / E_{X}^{X} + \alpha_{X} * h$$

$$\varepsilon_{Y}^{Y} = [- \nu_{Y}^{X}\ \tau^{X}_{X} + \tau^{Y}_{Y} - \nu_{Y}^{Z}\ \tau^{Z}_{Z})] / E_{Y}^{Y} + \alpha_{Y} * h$$

$$\varepsilon_{Z}^{Z} = [- \nu_{Z}^{X}\ \tau^{X}_{X} - \nu_{Z}^{Y}\ \tau^{Y}_{Y} + \tau^{Z}_{Z})] / E_{Z}^{Z} + \alpha_{Z} * h$$

Shears:

$$\gamma_{X}^{Y} = \tau^{Y}_{X} / G_{X}^{Y} = \tau^{Y}_{X}\ 2(1+\nu_{X}^{Y}) / E_{X}^{X} = \gamma_{Y}^{X}$$

$$\gamma_{Y}^{Z} = \tau^{Z}_{Y} / G_{Y}^{Z} = \tau^{Z}_{Y}\ 2(1+\nu_{Y}^{Z}) / E_{Y}^{Y} = \gamma_{Z}^{Y}$$

$$\gamma_{Z}^{X} = \tau^{X}_{Z} / G_{Z}^{X} = \tau^{X}_{Z}\ 2(1+\nu_{Z}^{X}) / E_{Z}^{Z} = \gamma_{X}^{Z}$$

$1/E^1$	$-\nu^1{}_2/E^2$	$-\nu^1{}_3/E^3$	0	0	0
$-\nu^2{}_1/E^1$	$1/E^2$	$-\nu^2{}_3/E^3$	0	0	0
$-\nu^3{}_1/E^1$	$-\nu^3{}_2/E^2$	$1/E^3$	0	0	0
0	0	0	$1/E^4$	0	0
0	0	0	0	$1/E^5$	0
0	0	0	0	0	$1/E^6$

15.13.
Axisymmetric Material:

$\varepsilon_r^{\ r}$	$\varepsilon_z^{\ z}$	$\varepsilon_3^{\ 3}$	$\varepsilon_Z^{\ R}$	$\varepsilon_R^{\ \Theta}$	$\varepsilon_\Theta^{\ Z}$
$\tau_1^{\ 1}$	$\tau_2^{\ 2}$	$\tau_3^{\ 3}$	$\tau_1^{\ 2}$	$\tau_2^{\ 3}$	$\tau_3^{\ 1}$
$1/E^1$	$-\nu^1{}_2/E^2$	$-\nu^1{}_3/E^3$	$-\nu^1{}_4/E^4$	0	0
$-\nu^2{}_1/E^1$	$1/E^2$	$-\nu^2{}_3/E^3$	$-\nu^2{}_4/E^4$	0	0
$-\nu^3{}_1/E^1$	$-\nu^3{}_2/E^2$	$1/E^3$	$-\nu^3{}_4/E^4$	0	0
$-\nu^4{}_1/E^1$	$-\nu^4{}_2/E^2$	$-\nu^4{}_3/E^3$	$1/E^4$	0	0
0	0	0	0	$1/E^5$	$-\nu^5{}_6/E^6$
0	0	0	0	$-\nu^6{}_5/E^5$	$1/E^6$

	Esteel 210mPa		νsteel 0.30	Gsteel 80mPa	λsteel 120mPa	αsteel 0.000006	ρsteel 8000kg/m³
	E/Esteel	ν/1	ν/νsteel	G/Gsteel	λ/λsteel	α/αsteel	ρ/ρsteel
Steel	1.	0.30	1.00	1.00	1.00	1.00	1.00
Aluminum	0.33	0.33	1.11	0.33	0.44	2.00	0.35
Concrete	0.13	0.20	0.66	0.15	0.06	0.92	0.30
Mercury	0	0.50	1.66	0	0.20	-	1.90
Ice	0.00003	0.49	1.60	0.00003	0.02	-	0.13
Water	0	0.50	1.66	0	0.02	-	0.13

16.
Viscosity:

16.1.
Viscoelasticity, EVT:
ET = EET + *EVT*

Assume:
(First Order Tensor, A Is Associated With Second Order Tensor, T)
And
(First Order Tensor, B Is Associated With Second Order Tensor, E)

(First Order Tensor, A is Associated With Second Order Tensor, τ)
And
(First Order Tensor, B is Associated With Second Order Tensor, E)

16.1.1.
Stress Strain Resilience:

Stress Stiffness:
Stress = Stress * 1

$$T^{kl}=T^{kl}(E_{kl},B_p) \qquad T_{kl}=T_{kl}(E^{kl},B^p)$$
$$\tau^{kl}=\tau^{kl}(E_{kl},B_p) \qquad \tau_{kl}=\tau_{kl}(E^{kl},B^p)$$

Strain Flexibility:
1 = 1 /Stress * Stress

$$E^{kl}=E^{kl}(T_{kl},A_p) \qquad E_{kl}=E_{kl}(T^{kl},A^p)$$
$$E^{kl}=E^{kl}(\tau_{kl},A_p) \qquad E_{kl}=E_{kl}(\tau^{kl},A^p)$$

Stress and Strain are Reciprocal functions of each other
Time Rate of Stress Elastic Stiffness:
Stress/ Time = Stress * 1/ Time

$$TT^{kl}=TT^{kl}(E_{kl},B_p) \qquad TT_{kl}=TT_{kl}(E^{kl},B^p)$$
$$\tau T^{kl}=\tau T^{kl}(E_{kl},B_p) \qquad \tau T_{kl}=\tau T_{kl}(E^{kl},B^p)$$

Time Rate of Strain Elastic Flexibility:
1/ Time = 1 /Stress * Stress/ Time

$$ET^{kl}=ET^{kl}(T_{kl},A_p) \qquad ET_{kl}=ET_{kl}(T^{kl},A^p)$$
$$ET^{kl}=ET^{kl}(\tau_{kl},A_p) \qquad ET_{kl}=ET_{kl}(\tau^{kl},A^p)$$

Differential:

$$dT^{kl}=\partial T^{kl}/\partial E_{mn}\, dE_{mn}+\partial T^{kl}/\partial B_r\, dB_r \qquad dT_{kl}=\partial T_{kl}/\partial E^{mn}\, dE^{mn}+\partial T_{kl}/\partial B^r\, dB^r$$
$$dE^{kl}=\partial E^{kl}/\partial T_{mn}\, dT_{mn}+\partial E^{kl}/\partial A_c\, dA_c \qquad dE_{kl}=\partial E_{kl}/\partial T^{mn}\, dT^{mn}+\partial E_{kl}/\partial A^c\, dA^c$$

$$d\tau^{kl}=\partial \tau^{kl}/\partial E_{mn}\, dE_{mn}+\partial \tau^{kl}/\partial B_r\, dB_r \qquad d\tau_{kl}=\partial \tau_{kl}/\partial E^{mn}\, dE^{mn}+\partial \tau_{kl}/\partial B^r\, dB^r$$
$$dE^{kl}=\partial E^{kl}/\partial \tau_{mn}\, d\tau_{mn}+\partial E^{kl}/\partial A_c\, dA_c \qquad dE_{kl}=\partial E_{kl}/\partial \tau^{mn}\, d\tau^{mn}+\partial E_{kl}/\partial A^c\, dA^c$$

••• ••• •••

17. Plasticity (Plastic Inelasticity):

17.1. Plastic Strain Rates

Plastic Strain Rate:

$$ET^{kl} = (\Gamma^{klmn} \; \tau T_{mn} + + \Lambda \; S^{kl})$$

Strain:

$$E^{kl} = \int ET^{kl} \; dt = \int (\Gamma^{klmn} \; \tau T_{mn} + \Lambda \; S^{kl}) \; dt$$

$$E^{kl} = \quad (\Gamma^{klmn} \; \tau_{mn}) + \Lambda \; (\int S_{kl}) \; dt$$

$$\tau^{kl} = (\Gamma^{klmn} \tau T_{mn} + \Lambda \; \sigma^{kl})T * 0K2 + (\Gamma^{klmn} \tau T_{mn} + \Lambda \; \sigma^{kl}) * 0K1 + \int (\Gamma^{klmn} \; \tau T_{mn} + \Lambda \; \sigma^{kl}) \; dt * 0K0$$

$$\tau T^{kl} = (\Gamma^{klmn} \tau T_{mn} + \Lambda \; \sigma^{kl})T * 1K2 + (\Gamma^{klmn} \tau T_{mn} + \Lambda \; \sigma^{kl}) * 1K1 + \int (\Gamma^{klmn} \; \tau T_{mn} + \Lambda \; \sigma^{kl}) \; dt * 1K0$$

$$\tau TT^{kl} = (\Gamma^{klmn} \tau T_{mn} + \Lambda \; \sigma^{kl})T * 2K2 + (\Gamma^{klmn} \tau T_{mn} + \Lambda \; \sigma^{kl}) * 2K1 + \int (\Gamma^{klmn} \; \tau T_{mn} + \Lambda \; \sigma^{kl}) \; dt * 2K0$$

Combinations of Two values of m with Three Values of n

$$\tau^{kl} + \tau T^{kl} = (\Gamma^{klmn} \; \tau T_{mn} + \Lambda \; \sigma^{kl})T * (0K2 + 1K2) + (\Gamma^{klmn} \; \tau T_{mn} + \Lambda \; \sigma^{kl}) * (0K1 + 1K1) + \int (\Gamma^{klmn} \; \tau T_{mn} + \Lambda \; \sigma^{kl}) \; dt * (0K0 + 1K0)$$

••• ••• ••• (not published due to page number requirement)

18.
Structural Dynamics:

18.1.
Two Functions of a Variable:

Normalized Two Functions:

Let :
Time to be an Independent Variable and T ≡ Unit of Time.
Unit of x a Time Dependent Variable : [x] ≡ xU ≡ Unit of x

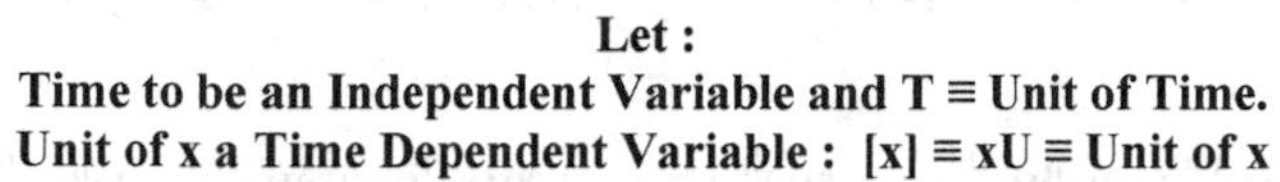

:

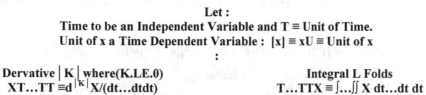

Dervative | K | where(K.LE.0)
XT…TT ≡ $d^{|K|}$X/(dt…dtdt)

Integral L Folds
T…TTX ≡ ∫…∫∫ X dt…dt dt

Unitless General Two Variables Differential equation:

$$\tau^{\underline{M}}\,\tau^{(M)} = \tau^{(M)} / \tau_M = \varepsilon^{(N)} / \varepsilon_N = \varepsilon^N\ \ \varepsilon^{(N)}$$

Factors of Derivatives:

$$\tau^M \equiv 1 / \tau_M \qquad\qquad \varepsilon^N \equiv 1 / \varepsilon_N$$

:

Expanded Unitless General Two Variables Differential equation:

$$\{\tau^m * \tau^{(m)} + \tau^{m+1} * \tau^{(m+1)} + \ldots\ \tau^R * \tau^{(R)} + \ldots + \tau^p * \tau^{(p)}\}$$
$$=$$
$$\{\varepsilon^n * \varepsilon^{(n)} + \varepsilon^{n+1} * \varepsilon^{(n+1)} + \ldots + \varepsilon^S * \varepsilon^{(S)} + \ldots + \varepsilon^q * \varepsilon^{(q)}\}$$

$$\{\tau^{(m)} / \tau_m + \tau^{(m+1)} / \tau_{m+1} + \ldots\ \tau^{(R)} / \tau_R + \ldots + \tau^{(p)} / \tau_p\}$$
$$=$$
$$\{\varepsilon^{(n)} / \varepsilon_n + \varepsilon^{(n+1)} / \varepsilon_{n+1} + \ldots + \varepsilon^{(S)} / \varepsilon_S + \ldots + \varepsilon^{(q)} / \varepsilon_q\}$$

$$(m) \le M \le (p) \qquad\qquad (n) \le N \le (q)$$

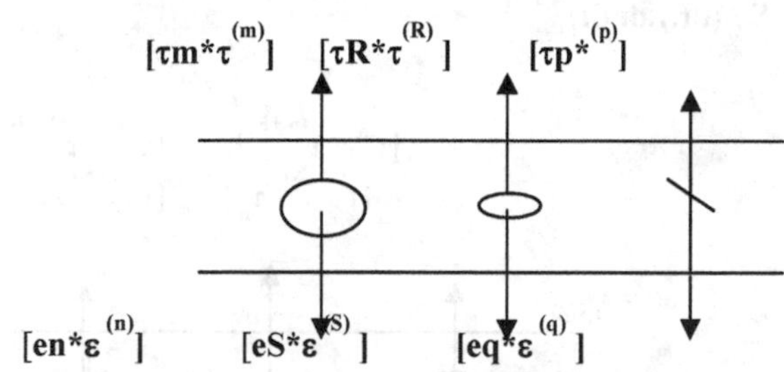

Where :

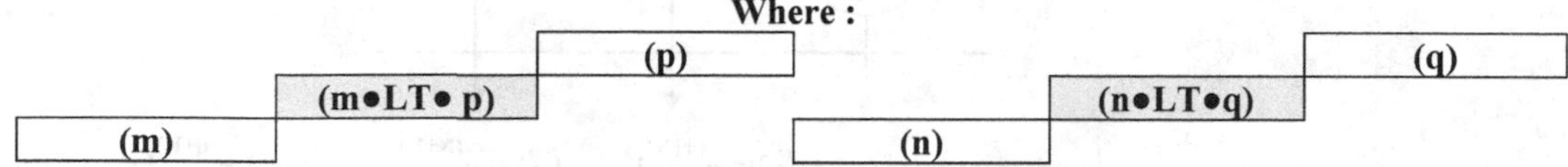

Assumed Uniformly Normalized Differential equation:

Units:

$[\tau^{(m)}] = [\tau_m \equiv 1/\tau^m]$ $[\tau^{(R)}] = [\tau_R \equiv 1/\tau^R]$ $[\tau^{(p)}] = [\tau_p \equiv 1/\tau^P]$ $[\varepsilon^{(n)}] = [\varepsilon_n \equiv 1/\varepsilon^n]$ $[\varepsilon^{(S)}] = [\varepsilon_S \equiv 1/\varepsilon^S]$ $[\varepsilon^{(q)}] = [\varepsilon_q \equiv 1/\varepsilon^q]$

Units of Derivatives:

$[\tau^{(M)}] = \tau U / [T]^M$ $[\varepsilon^{(N)}] = \varepsilon U / [T]^N$

Units of Factors:

$[\tau^M] = [T]^M / \tau U$ $[\varepsilon^N] = [T]^N / \varepsilon U$

Such that:

$[\tau^M] = 1/[\tau^{(M)}]$ $[\varepsilon^N] = 1/[\varepsilon^{(N)}]$

$[T]^M / [\tau^M] = [\tau^{(M)}] * [T]^M = \tau U$ $[T]^N / [\varepsilon^N] = [\varepsilon^{(N)}] * [T]^N = \varepsilon U$

So that Products of Factors and Derivatives are Unitless:

$[\tau^{\underline{m}}][\tau^{(m)}]=1$ $[\tau^{\underline{R}}][\tau^{(R)}]=1$ $[\tau^{\underline{p}}][\tau^{(p)}]=1$ $[\varepsilon^{\underline{n}}][\varepsilon^{(n)}]=1$ $[\varepsilon^{\underline{S}}][\varepsilon^{(S)}]=1$ $[\varepsilon^{\underline{q}}][\varepsilon^{(q)}]=1$

:

$$[\tau^{\underline{m}}][\tau^{(m)}] = [\tau^{\underline{R}}][\tau^{(R)}] = [\tau^{\underline{p}}][\tau^{(p)}] = 1 = [\varepsilon^{\underline{n}}][\varepsilon^{(n)}] = [\varepsilon^{\underline{S}}][\varepsilon^{(S)}] = [\varepsilon^{\underline{q}}][\varepsilon^{(q)}]$$

Ratios:

$[\tau^{(R)}] / [\tau^{(P)}] = [\tau_R/\tau_P] = 1 / [\tau^R/\tau^P] = 1/[T]^{R-P}$ $[\varepsilon^{(S)}] / [\varepsilon^{(Q)}] = [\varepsilon_S/\varepsilon_Q] = 1/[\varepsilon^S/\varepsilon^Q] = 1/[T]^{S-Q}$

$[\tau^R/\tau^P] = [T]^{R-P}$ $[\varepsilon^S/\varepsilon^Q] = [T]^{S-Q}$

18.1.1.
Dervative to No Positive Operations:
(No Variable Integral)

:

Defined

$T \equiv$ Unit of Time

$[x] \equiv$ Unit of x

Dervative $|K|$ where (K.LE.0)

$X^{(K)} \equiv d^{|K|} X / (dt\ldots dt\ dt)$

$[\tau^m * \tau^{(m+K)}]$ $[\tau^{(m+K)}/\tau_m]$ $[\tau^M * \tau^{(M+K)}]$ $[\tau^{(M+K)}/\tau_M]$ $[\tau^p * \tau^{(p+K)}]$ $[\tau^{(p+K)}/\tau_p]$

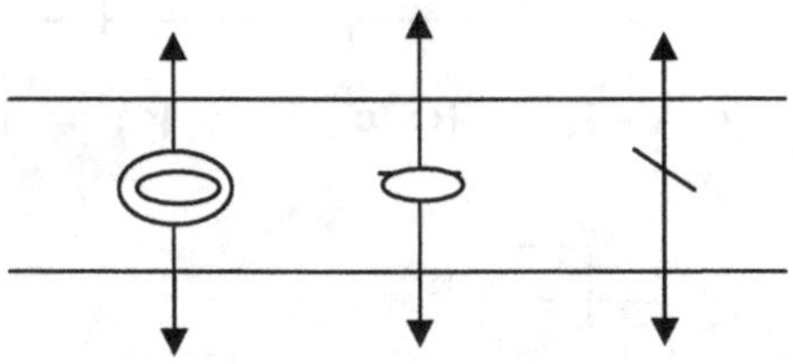

$[\varepsilon^n * \varepsilon^{(n+K)}]$ $[\varepsilon^{(n+K)}/\varepsilon_n]$ $[\varepsilon^N * \varepsilon^{(N+K)}]$ $[\varepsilon^{(N+K)}/\varepsilon_N]$ $[\varepsilon^q * \varepsilon^{(q+K)}]$ $[\varepsilon^{(q+K)}/\varepsilon_q]$

$$[\tau^M * \tau^{(M+K)}] = \tau^{(M+K)}/\tau_M = [\varepsilon^N * \varepsilon^{(N+K)}] = \varepsilon^{(N+K)}/\varepsilon_N$$

Expanded :

$$[\tau^m * \tau^{(m+K)}] + [\tau^M * \tau^{(M+K)}] + [\tau^p * \tau^{(p+K)}]$$

$$=$$

$$[\tau^{(m+K)}/\tau_m] + [\tau^{(M+K)}/\tau_M] + [\tau^{(p+K)}/\tau_p]$$

$$=$$

$$[\varepsilon^m * \varepsilon^{(m+K)}] + [\varepsilon^M * \varepsilon^{(M+K)}] + [\varepsilon^p * \varepsilon^{(p+K)}]$$

$$=$$

$$[\varepsilon^{(m+K)}/\varepsilon_m] + [\varepsilon^{(M+K)}/\varepsilon_M + [\varepsilon^{(p+K)}/\varepsilon_p]$$

Two Cases of Dervative Operations:

$(m) \leq M \leq (p)$ **$(n) \leq N \leq (q)$**

:

First Case:

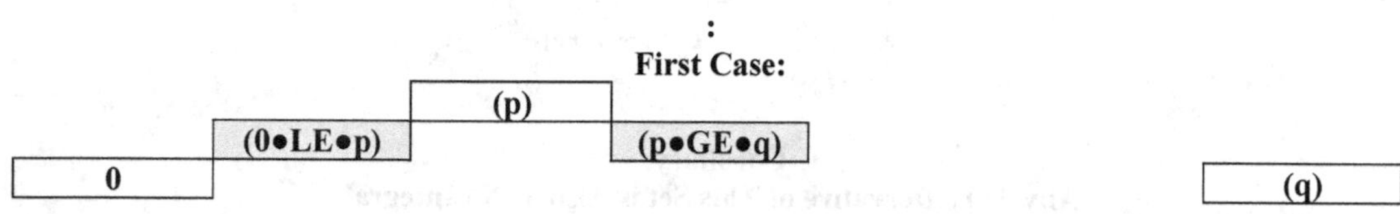

(0•LE•p) **(q•LE•p)**

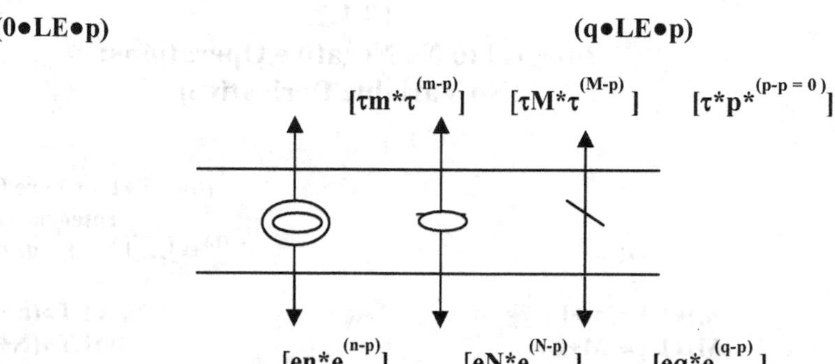

$$\tau^{(M-p)}/\tau_M = e^{(N-p)}/e_N$$

Expanded :

$$\tau^{(m-p)}/\tau_m + \tau^{(m+1-p)}/\tau_{m+1} + \ldots + \tau^{(p-p)}/\tau_p$$

$$=$$

$$[e^{(n-p)}]/e_n + [e^{(n+1-p)}]/e_{n+1} + \ldots + [e^{(q-p)}]/e_q$$

Second Case:

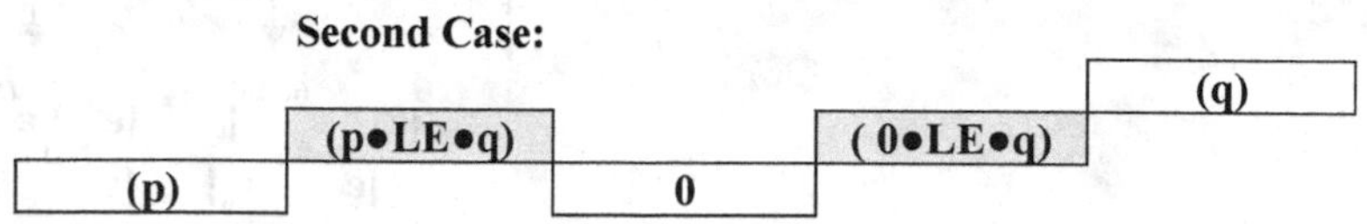

(p•LE•q) **(0•LE•q)**

$$[\tau m*\tau^{(m-q)}] \quad [\tau M*\tau^{(M-q)}] \quad [\tau*p*^{(p-q)}]$$

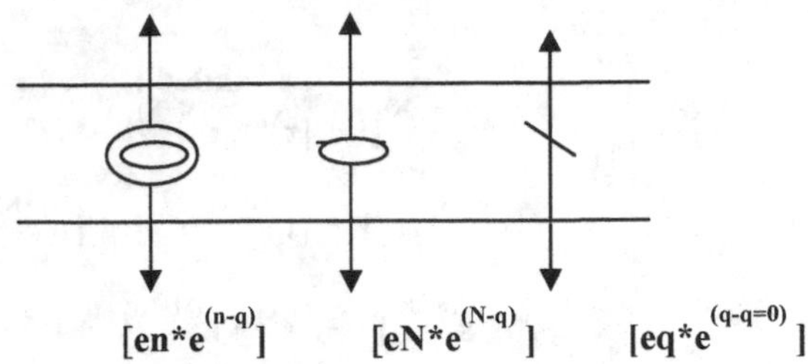

$$\tau^{(M-q)}/\tau_M = e^{(N-q)}/e_N$$

Expanded :

$$\tau^{(m-q)}/\tau_m + \tau^{(m+1-q)}/\tau_{m+1} + \ldots + \tau^{(q-q)}/\tau_q$$

$$=$$

$$[e^{(n-q)}]/e_n + [e^{(n+1-q)}]/e_{n+1} + \ldots + [e^{(q-q)}]/e_q$$

Corollary:
Any |Q| Dervative of This Set is Also of No Integral

18.1.2.
Integral to No Negative Operations:
(No Variable Derivative)

(0•LT•L) where (L.GE.0)
Integral L
$X^{(L)} \equiv \int\ldots\iint X\, dt\ldots dt\, dt$ (L Folds)

(m)•LT•(m+L)
(M)•LT•(M+L)
(q+L)•LT•(q)

(n)•LT•(n+L)
(N)•LT•(N+L)
(q+K)•LT•(q)

$[\tau^m * \tau^{(m+L)}]$ $[\tau^M * \tau^{(M+L)}]$ $[\tau^p * \tau^{(p+L)}]$

$[\tau^{(m+L)}/\tau_m]$ $[\tau^{(M+L)}/\tau_M]$ $[\tau^{(p+L)}/\tau_p]$

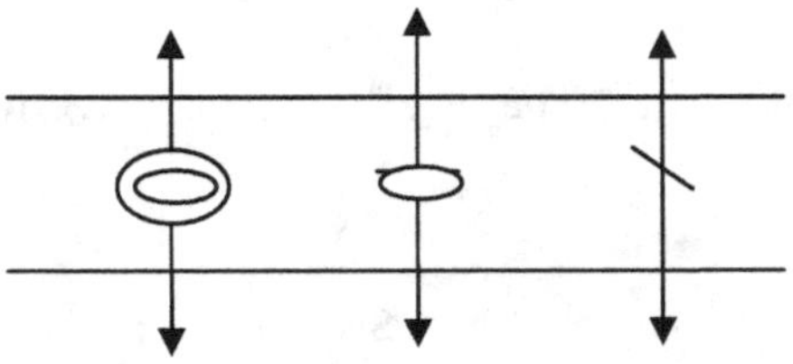

$[\varepsilon^n * \varepsilon^{(n+L)}]$ $[\varepsilon^N * \varepsilon^{(N+L)}]$ $[\varepsilon^q * \varepsilon^{(q+L)}]$

$[\varepsilon^{(n+L)}/\varepsilon_n]$ $[\varepsilon^{(N+L)}/\varepsilon_N]$ $[\varepsilon^{(q+L)}/\varepsilon_q]$

$$[\tau^M * \tau^{(M+L)}] = \tau^{(M+L)}/\tau_M = [\varepsilon^N * \varepsilon^{(N+L)}] = \varepsilon^{(N+L)}/\varepsilon_N$$

Expanded :

$$[\tau^{m} * \tau^{(m+L)}] + [\tau^{M} * \tau^{(M+L)}] + [\tau^{p} * \tau^{(p+L)}]$$

$$=$$

$$[\tau^{(m+L)} / \tau_{m}] + [\tau^{(M+L)} / \tau_{M}] + [\tau^{(p+L)} / \tau_{p}]$$

$$=$$

$$[\varepsilon^{m} * \varepsilon^{(m+L)}] + [\varepsilon^{M} * \varepsilon^{(M+L)}] + [\varepsilon^{p} * \varepsilon^{(p+L)}]$$

$$=$$

$$[\varepsilon^{(m+L)} / \varepsilon_{m}] + [\varepsilon^{(M+L)} / \varepsilon_{M}] + [\varepsilon^{(p+L)} / \varepsilon_{p}]$$

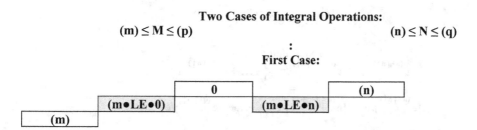

Two Cases of Integral Operations:

(m) ≤ M ≤ (p) **(n) ≤ N ≤ (q)**

:

First Case:

(m) (m•LE•0) 0 (m•LE•n) (n)

(m•LE•0) **(m•LE•n)**

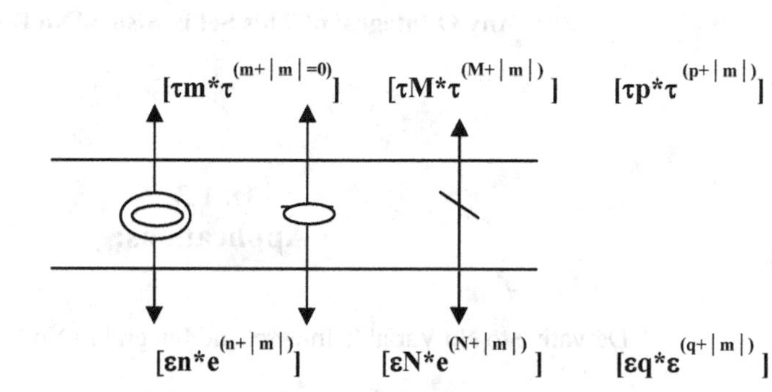

$$\tau^{(M+|m|)} / \tau_{M} = e^{(N+|m|)} / \varepsilon_{N}$$

Expanded :

$$\tau^{(m+|m|)} / \tau_{m} + \tau^{(m+1+|m|)} / \tau_{m+1} + \ldots + \tau^{(p+|m|)} / \tau_{p}$$

$$=$$

$$[e^{(n+|m|)}] / e_{n} + [e^{(n+1+|m|)}] / e_{n+1} + \ldots + [e^{(q+|m|)}] / e_{q}$$

Second Case:

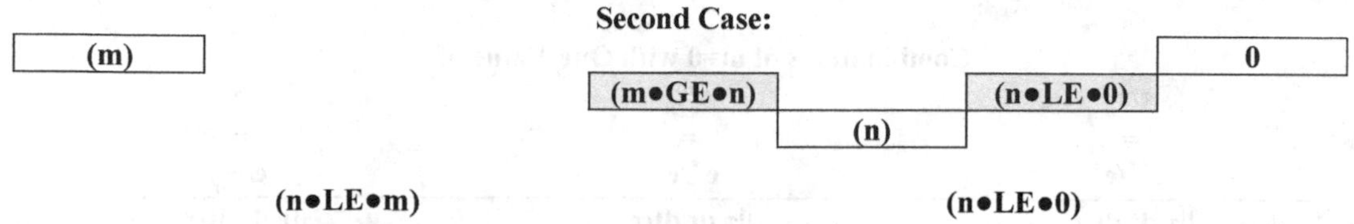

(n•LE•m) **(n•LE•0)**

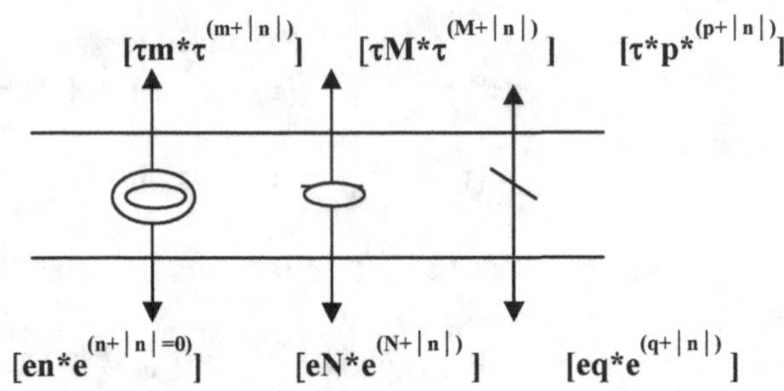

$$\tau^{(M+|n|)}/\tau_M = e^{(N+|n|)}/e_N$$

Expanded :

$$\tau^{(n+|n|)}/\tau_n + \tau^{(n+1+|n|)}/\tau_{n+1} + \ldots + \tau^{(p+|n|)/}\tau_p$$

$$=$$

$$[e^{(n+|n|)}]/e_n + [e^{(n+1+|n|)}]/e_{n+1} + \ldots + [e^{(q+|n|)}]/e_q$$

Corollary:
Any Q Integral of This Set is Also of No Derivative

18.1.3.
Applications:

Dervative to No Variable Integral and Integral to No Variable Derivative:

Two Variables of (m=2) and (n=2)

$$\tau^{\bullet\bullet}/\tau_2 + \tau^{\bullet}/\tau_1 + \tau/\tau_0 = e^{\bullet\bullet}/e_2 + e^{\bullet}/e_1 + e/e_0$$

No Case of Derivative Transfer to
(No Positive Operations ≡ No Integral Variables):
All Variables and Not Integrals
Integral Transfer to (No Derivative Variables):
:
Two Cases :
No Integrals & No Derivatives:

$$\tau^{\bullet\bullet}/\tau_2 + \tau^{\bullet}/\tau_1 + \tau/\tau_0$$

$$=$$

$$e^{\bullet\bullet}/e_2 + e^{\bullet}/e_1 + e/e_0$$

$$\tau/\tau_2 + \int \tau\, dt/\tau_1 + \int\int \tau\, dt\, dt/\tau_0$$

$$=$$

$$e/e_2 + \int e\, dt/e_1 + \int\int e\, dt\, dt/e_0$$

Combinations of m=0 with One Value of n:

τ/τ_0	τ/τ_0	τ/τ_0
=	=	=
$e^{\bullet\bullet}/e_2$	$e^{\bullet}/e_1$	e/e_0
$\int\int\tau\, dt\, dt/\tau_0$	$\int\int\tau\, dt\, dt/\tau_0$	$\int\int\tau\, dt\, dt/\tau_0$
=	=	=

e/e_2	$\int e\,dt/e_1$	$\iint e\,dt\,dt/e_0$

Combinations of m=0 with Two Values of n:

$\tau/\tau_0 = e^{\bullet\bullet}/e_2 + e^{\bullet}/e_1$	$\tau/\tau_0 = e^{\bullet\bullet}/e_2 + e/e_0$	$\tau/\tau_0 = e^{\bullet}/e_1 + e/e_0$
$\int\int \tau\,dt\,dt/\tau_0 = e/e_2 + \int e\,dt/e_1$	$\int\int \tau\,dt\,dt/\tau_0 = e/e_2 + \int\int e\,dt\,dt/e_0$	$\iint \tau\,dt\,dt/\tau_0 = \int e\,dt/e_1 + \int\int e\,dt\,dt/e_0$

One Combination of m=0 with Three Values of n:

$$\tau/\tau_0 = e^{\bullet\bullet}/e_2 + e^{\bullet}/e_1 + e/e_0$$

$$\int\int \tau\,dt\,dt/\tau_0 = e/e_2 + \int e\,dt/e_1 + \int\int e\,dt\,dt/e_0$$

Combinations of m=1 with One Value of n:

$\tau^{\bullet}/\tau_1 = e^{\bullet\bullet}/e_2$	$\tau^{\bullet}/\tau_1 = e^{\bullet}/e_1$	$\tau^{\bullet}/\tau_1 = e/e_0$
$\int \tau\,dt/\tau_1 = e/e_2$	$\int \tau\,dt/\tau_1 = \int e\,dt/e_1$	$\int \tau\,dt/\tau_1 = \iint e\,dt\,dt/e_0$

Combinations of m=1 with Two Values of n:

$\tau^{\bullet}/\tau_1 = e^{\bullet\bullet}/e_2 + e^{\bullet}/e_1$	$\tau^{\bullet}/\tau_1 = e^{\bullet\bullet}/e_2 + e/e_0$	$\tau^{\bullet}/\tau_1 = e^{\bullet}/e_1 + e/e_0$
$\int \tau\,dt/\tau_1 = e/e_2 + \int e\,dt/e_1$	$\int\int \tau\,dt\,dt/\tau_0 = e/e_2 + \int\int e\,dt\,dt/e_0$	$\iint \tau\,dt\,dt/\tau_0 = \int e\,dt/e_1 + \int\int e\,dt\,dt/e_0$

One Combination of m=1 with Three Values of n:

$$\tau^{\bullet}/\tau_1 = e^{\bullet\bullet}/e_2 + e^{\bullet}/e_1 + e/e_0$$

$$\int \tau\,dt/\tau_1 = e/e_2 + \int e\,dt/e_1 + \int\int e\,dt\,dt/e_0$$

Combinations of m=2 with One Value of n:

$\tau^{\bullet\bullet}/\tau_2 = e^{\bullet\bullet}/e_2$	$\tau^{\bullet\bullet}/\tau_2 = e^{\bullet}/e_1$	$\tau^{\bullet\bullet}/\tau_2 = e/e_0$
$\tau/\tau_2 = e/e_2$	$\tau/\tau_2 = \int e\,dt/e_1$	$\tau/\tau_2 = \iint e\,dt\,dt/e_0$

Combinations of m=2 with Two Values of n:

$\tau^{\bullet\bullet}/\tau_2 = e^{\bullet\bullet}/e_2 + e^{\bullet}/e_1$	$\tau^{\bullet\bullet}/\tau_2 = e^{\bullet\bullet}/e_2 + e/e_0$	$\tau^{\bullet\bullet}/\tau_2 = e^{\bullet}/e_1 + e/e_0$
$\tau/\tau_2 = e/e_2 + \int e\,dt/e_1$	$\tau/\tau_2 = e/e_2 + \iint e\,dt\,dt/e_0$	$\tau/\tau_2 = \int e\,dt/e_1 + \int\int e\,dt\,dt/e_0$

One Combination of m=2 with Three Values of n

$$\frac{\tau^{\bullet\bullet}/\tau_2 = e^{\bullet\bullet}/e_2 + e^{\bullet}/e_1 + e/e_0}{\tau/\tau_2 = e/e_2 + \int e\,dt/e_1 + \int\int e\,dt\,dt/e_0}$$

Combinations of m=0 and m=1 with One Value of n:

$$\frac{\tau^{\bullet}/\tau_1 + \tau/\tau_0 = e^{\bullet\bullet}/e_2}{\int \tau\,dt/\tau_1 + \int\int \tau\,dt\,dt/\tau_0 = e/e_2} \qquad \frac{\tau^{\bullet}/\tau_1 + \tau/\tau_0 = e^{\bullet}/e_1}{\int \tau\,dt/\tau_1 + \int\int \tau\,dt\,dt/\tau_0 = \int e\,dt/e_1} \qquad \frac{\tau^{\bullet}/\tau_1 + \tau/\tau_0 = e/e_0}{\int \tau\,dt/\tau_1 + \int\int \tau\,dt\,dt/\tau_0 = \int\int e\,dt\,dt/e_0}$$

m=0 and n=1 with Two Values of n:

$$\frac{\tau^{\bullet}/\tau_1 + \tau/\tau_0 = e^{\bullet\bullet}/e_2 + e^{\bullet}/e_1}{\int \tau\,dt/\tau_1 + \int\int \tau\,dt\,dt/\tau_0 = e/e_2 + \int e\,dt/e_1} \qquad \frac{\tau^{\bullet}/\tau_1 + \tau/\tau_0 = e^{\bullet\bullet}/e_2 + e/e_0}{\int \tau\,dt/\tau_1 + \int\int \tau\,dt\,dt/\tau_0 = e/e_2 + \int\int e\,dt\,dt/e_0} \qquad \frac{\tau^{\bullet}/\tau_1 + \tau/\tau_0 = e^{\bullet}/e_1 + e/e_0}{\int \tau\,dt/\tau_1 + \int\int \tau\,dt\,dt/\tau_0 = \int e\,dt/e_1 + \int\int e\,dt\,dt/e_0}$$

One Combination of m=0 and m=1 with Three Values of n:

$$\frac{\tau^{\bullet}/\tau_1 + \tau/\tau_0 = e^{\bullet\bullet}/e_2 + e^{\bullet}/e_1 + e/e_0}{\int \tau\,dt/\tau_1 + \int\int \tau\,dt\,dt/\tau_0 = e/e_2 + \int e\,dt/e_1 + \int\int e\,dt\,dt/e_0}$$

Combinations of m=0 and m=2 with One Value of n:

$$\frac{\tau^{\bullet\bullet}/\tau_2 + \tau/\tau_0 = e^{\bullet\bullet}/e_2}{\tau/\tau_2 + \int\int \tau\,dt\,dt/\tau_0 = e/e_2} \qquad \frac{\tau^{\bullet\bullet}/\tau_2 + \tau/\tau_0 = e^{\bullet}/e_1}{\tau/\tau_2 + \int\int \tau\,dt\,dt/\tau_0 = \int e\,dt/e_1} \qquad \frac{\tau^{\bullet\bullet}/\tau_2 + \tau/\tau_0 = e/e_0}{\tau/\tau_2 + \int\int \tau\,dt\,dt/\tau_0 = \int\int e\,dt\,dt/e_0}$$

Combinations of m=0 and m=2 with Two Values of n:

$$\frac{\tau^{\bullet\bullet}/\tau_2 + \tau/\tau_0 = e^{\bullet\bullet}/e_2 + e^{\bullet}/e_1}{\tau/\tau_2 + \int\int \tau\,dt\,dt/\tau_0 = e/e_2 + \int e\,dt/e_1} \qquad \frac{\tau^{\bullet\bullet}/\tau_2 + \tau/\tau_0 = e^{\bullet\bullet}/e_2 + e/e_0}{\tau/\tau_2 + \int\int \tau\,dt\,dt/\tau_0 = e/e_2 + \int\int e\,dt\,dt/e_0} \qquad \frac{\tau^{\bullet\bullet}/\tau_2 + \tau/\tau_0 = e^{\bullet}/e_1 + e/e_0}{\tau/\tau_2 + \int\int \tau\,dt\,dt/\tau_0 = \int e\,dt/e_1 + \int\int e\,dt\,dt/e_0}$$

One Combination of m=0 and m=2 with Three Values of n:

$$\frac{\tau^{\bullet\bullet}/\tau_2 + \tau/\tau_0 = e^{\bullet\bullet}/e_2 + e^{\bullet}/e_1 + e/e_0}{\tau/\tau_2 + \int\int \tau\,dt\,dt/\tau_0 = e/e_2 + \int e\,dt/e_1 + \int\int e\,dt\,dt/e_0}$$

Combinations of m=1 and m=2 with One Value of n:

$$\tau^{\bullet\bullet}/\tau_2 + \tau^{\bullet}/\tau_1 \qquad \tau^{\bullet\bullet}/\tau_2 + \tau^{\bullet}/\tau_1 \qquad \tau^{\bullet\bullet}/\tau_2 + \tau^{\bullet}/\tau_1$$

$=$ $\quad$ $=$ $\quad$ $=$

$e^{\bullet\bullet}/e_2$ $\quad$ $e^{\bullet}/e_1$ $\quad$ e/e_0

$\tau/\tau_2 + \int\int \tau \, dt \, dt/\tau_0$ $\quad$ $\tau/\tau_2 + \int\int \tau \, dt \, dt/\tau_0$ $\quad$ $\tau/\tau_2 + \int\int \tau \, dt \, dt/\tau_0$

$=$ $\quad$ $=$ $\quad$ $=$

e/e_2 $\quad$ $\int e \, dt/e_1$ $\quad$ $\int\int e \, dt \, dt/e_0$

Combinations of m=1 and m=2 with Two Values of n:

$\tau^{\bullet\bullet}/\tau_2 + \tau^{\bullet}/\tau_1$ $\quad$ $\tau^{\bullet\bullet}/\tau_2 + \tau^{\bullet}/\tau_1$ $\quad$ $\tau^{\bullet\bullet}/\tau_2 + \tau^{\bullet}/\tau_1$

$=$ $\quad$ $=$ $\quad$ $=$

$e^{\bullet\bullet}/e_2 + e^{\bullet}/e_1$ $\quad$ $e^{\bullet\bullet}/e_2 + e/e_0$ $\quad$ $e^{\bullet}/e_1 + e/e_0$

$\tau/\tau_2 + \int\int \tau \, dt \, dt/\tau_0$ $\quad$ $\tau/\tau_2 + \int\int \tau \, dt \, dt/\tau_0$ $\quad$ $\tau/\tau_2 + \int\int \tau \, dt \, dt/\tau_0$

$=$ $\quad$ $=$ $\quad$ $=$

$e/e_2 + \int e \, dt/e_1$ $\quad$ $e/e_2 + \int\int e \, dt \, dt/e_0$ $\quad$ $\int e \, dt/e_1 + \int\int e \, dt \, dt/e_0$

One Combination of m=1 and m=2 with Three Values of n:

$$\tau^{\bullet\bullet}/\tau_2 + \tau^{\bullet}/\tau_1$$

$$=$$

$$e^{\bullet\bullet}/e_2 + e^{\bullet}/e_1 + e/e_0$$

$$\tau/\tau_2 + \int\int \tau \, dt \, dt/\tau_0$$

$$=$$

$$e/e_2 + \int e \, dt/e_1 + \int\int e \, dt \, dt/e_0$$

Combinations of m=0, m=1 and m=2 with One Value of n:

$\tau^{\bullet\bullet}/\tau_2 + \tau^{\bullet}/\tau_1 + \tau/\tau_0$ $\quad$ $\tau^{\bullet\bullet}/\tau_2 + \tau^{\bullet}/\tau_1 + \tau/\tau_0$ $\quad$ $\tau^{\bullet\bullet}/\tau_2 + \tau^{\bullet}/\tau_1 + \tau/\tau_0$

$=$ $\quad$ $=$ $\quad$ $=$

$e^{\bullet\bullet}/e_2$ $\quad$ $e^{\bullet}/e_1$ $\quad$ e/e_0

$\tau/\tau_2 + \int\tau \, dt/\tau_1 + \int\int\tau \, dtdt/\tau_0$ $\quad$ $\tau/\tau_2 + \int\tau \, dt/\tau_1 + \int\int\tau \, dtdt/\tau_0$ $\quad$ $\tau/\tau_2 + \int\tau \, dt/\tau_1 + \int\int\tau \, dtdt/\tau_0$

$=$ $\quad$ $=$ $\quad$ $=$

e/e_2 $\quad$ $\int e \, dt/e_1$ $\quad$ $\int\int e \, dt \, dt/e_0$

Combinations of m=0, m=1 and m=2 with Two Values of n:

$\tau^{\bullet\bullet}/\tau_2 + \tau^{\bullet}/\tau_1 + \tau/\tau_0$ $\quad$ $\tau^{\bullet\bullet}/\tau_2 + \tau^{\bullet}/\tau_1 + \tau/\tau_0$ $\quad$ $\tau^{\bullet\bullet}/\tau_2 + \tau^{\bullet}/\tau_1 + \tau/\tau_0$

$=$ $\quad$ $=$ $\quad$ $=$

$e^{\bullet\bullet}/e_2 + e^{\bullet}/e_1$ $\quad$ $e^{\bullet\bullet}/e_2 + e/e_0$ $\quad$ $e^{\bullet}/e_1 + e/e_0$

$\tau/\tau_2 + \int\tau \, dt/\tau_1 + \int\int\tau \, dtdt/\tau_0$ $\quad$ $\tau/\tau_2 + \int\tau \, dt/\tau_1 + \int\int\tau \, dtdt/\tau_0$ $\quad$ $\tau/\tau_2 + \int\tau \, dt/\tau_1 + \int\int\tau \, dtdt/\tau_0$

$=$ $\quad$ $=$ $\quad$ $=$

$e/e_2 + \int e \, dt/e_1$ $\quad$ $e/e_2 + \int\int e \, dt \, dt/e_0$ $\quad$ $\int e \, dt/e_1 + \int\int e \, dt \, dt/e_0$

One Combination of m=0, m=1 and m=2 with Three Values of n

$$\tau^{\bullet\bullet}/\tau_2 + \tau^{\bullet}/\tau_1 + \tau/\tau_0$$

$$=$$

$$e^{\bullet\bullet}/e_2 + e^{\bullet}/e_1 + e/e_0$$

$$\tau/\tau_2 + \int \tau \, dt/\tau_1 + \int\int \tau \, dt \, dt/\tau_0$$

$$=$$

$$e/e_2 + \int e \, dt/e_1 + \int\int e \, dt \, dt/e_0$$

:

Rewrite

Combinations of One value of m with Three Values of n

$$\tau/\tau_0 = e^{\bullet\bullet}/e_2 + e^{\bullet}/e_1 + e/e_0$$

$$\tau^{\bullet}/\tau_1 = e^{\bullet\bullet}/e_2 + e^{\bullet}/e_1 + e/e_0$$

$$\tau^{\bullet\bullet}/\tau_2 = e^{\bullet\bullet}/e_2 + e^{\bullet}/e_1 + e/e_0$$

Or:

$$\tau = e^{\bullet\bullet} \; \tau_0/e_2 + e^{\bullet} \; \tau_0/e_1 + e \; \tau_0/e_0$$

$$\tau^{\bullet} = e^{\bullet\bullet} \; \tau_1/e_2 + e^{\bullet} \; \tau_1/e_1 + e \; \tau_1/e_0$$

$$\tau^{\bullet\bullet} = e^{\bullet\bullet} \; \tau_2/e_2 + e^{\bullet} \; \tau_2/e_1 + e \; \tau_2/e_0$$

Or:

$$\tau = e^{\bullet\bullet}\ 0\tau e2 + e^{\bullet}\ 0\tau e1\ +\ e\ 0\tau e0$$
$$\tau^{\bullet} = e^{\bullet\bullet}\ 1\tau e2 + e^{\bullet}\ 1\tau e1\ +\ e\ 1\tau e2$$
$$\tau^{\bullet\bullet} = e^{\bullet\bullet}\ 2\tau e2 + e^{\bullet}\ 2\tau e1\ +\ e\ 2\tau e0$$

$$\tau^{\bullet} \equiv \tau T$$
$$\tau^{\bullet\bullet} \equiv \tau T^{\bullet} \equiv \tau TT$$

$$(eT) \equiv e^{\bullet} \equiv eT$$
$$e^{\bullet\bullet} \equiv eT^{\bullet} \equiv eTT$$

τ_0 / e_2	τ_0/e_1	τ_0/e_0
τ_1 / e_2	τ_1/e_1	τ_1/e_0
τ_2 / e_2	τ_2/e_1	τ_2/e_0

Example:

V Volt	**Electric Circuits**	**I Ampere**
$I = 1/H \int V\,dt$	**H Inductance (Henry)**	$V = H\ d(I)/dt$
$I = 1/R\ V$	**R Resistance (Ohm)**	$V = R\ I$
$I = D\ dV/dt$	**D Capacitance (Farad)**	$V = 1/D \int I\ dt$

(-2)	(-1)	(0)	(-2)	(-1)	(0)
C	1/R	1/H	H	R	1/D
*	*	*	*	*	*
VTT	VT	V	ITT	IT	I
1/m	Γ	Λ	M	C	K
*	*	*	*	*	*
τTT	τT	τ	ETT	ET	E

:

... C * VTT + 1/R * VT + 1/H * V ... = ... H * ITT + R * IT + 1/C * I ...

18.1.4.
Stiffness View:

Normalized Differential Equation by $[\tau_R]$

$$\tau^{(M)}/\ \tau_M = \tau^{(M)}/\ \tau_M = e^{(N)}\ /e_N$$
$$*$$
$$(\tau_R)$$
$$=$$
$$\tau_R\ /\tau_M\ *\ [\tau^{(M)}] = \tau_R/\ e_N * [\ e^{(N)}\ ]$$
$$[\tau\tau_R^{\ M}\ ] * [\tau^{(M)}] = [\tau e_R^{\ N}] * [\ e^{(N)}\ ]$$

$$[\tau\tau_R^{\ M}\ ] = \tau_R\ /\tau_M \qquad [\tau e_R^{\ N}] = \tau_R/\ e_N$$

Expanded :

$$\tau\tau_R^{\ m} * \tau^{(m)} + \tau\tau_R^{\ m+1} * \tau^{(m+1)} + \ldots + [\tau\tau_R^{\ R} = 1\ ] * [\tau^{(R)}] + \ldots + \tau\tau_R^{\ p} * \tau^{(p)}$$
$$=$$
$$\tau e_R^{\ n} * e^{(n)} + \tau e_R^{\ n+1} * e^{(n+1)} + \ldots + [\ \tau e_R^{\ S}\ ] * [\ e^{(S)}\ ] + \ldots + \tau e_R^{\ q} * e^{(q)}$$

$$\tau\tau_R^{\ R} \equiv R\tau\tau R = 1 \qquad \text{Stiffness } [\tau e_R^{\ S}] \equiv R\tau eS$$

$$\tau\tau_R^{\ R} \equiv R\tau\tau R = 1$$

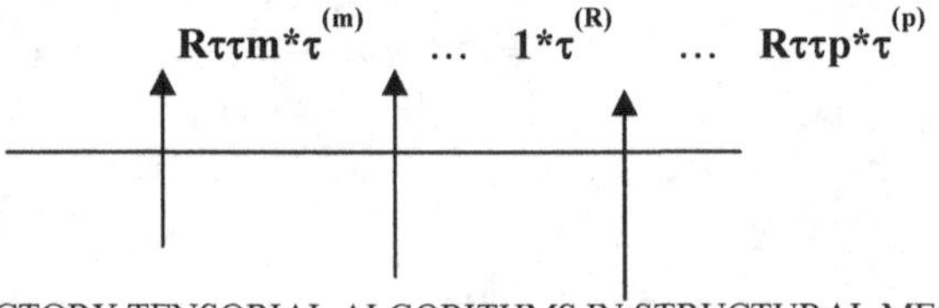

$$R\tau\tau m*\tau^{(m)}\ \ \ldots\ \ 1*\tau^{(R)}\ \ \ldots\ \ R\tau\tau p*\tau^{(p)}$$

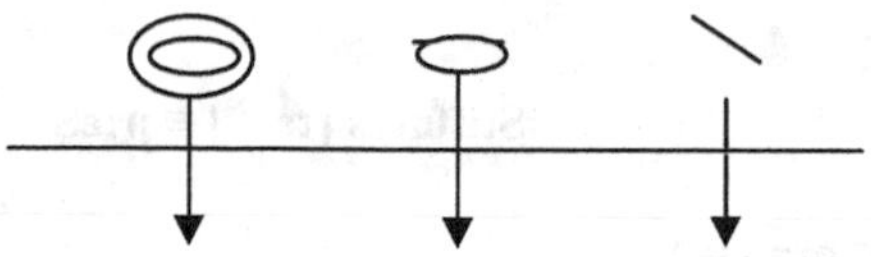

$R\tau en*e^{(n)}$ ... $R\tau eS*e^{(S)}$ ... $R\tau eq*e^{(q)}$

$\tau\tau_R{}^R \equiv R\tau\tau R = 1$ Stiffness $[\tau e_R{}^S] \equiv R\tau eS$

$R = m$

$\tau^{(M)}/\tau_M = e^{(N)}/e_N$

$*$

(τ_m)

$=$

$[\tau\tau_m{}^M] * [\tau^{(M)}] = [\tau e_m{}^N] * [e^{(N)}]$

$=$

$\tau\tau_m{}^m \equiv m\tau\tau m = 1$ Stiffness $[\tau e_m{}^S] \equiv m\tau eS$

$\tau\tau_m{}^m \equiv m\tau\tau m = 1$

$[1*\tau^{(m)}]$ ... $[m\tau\tau R*\tau^{(R)}]$ ... $[m\tau\tau p*\tau^{(p)}]$

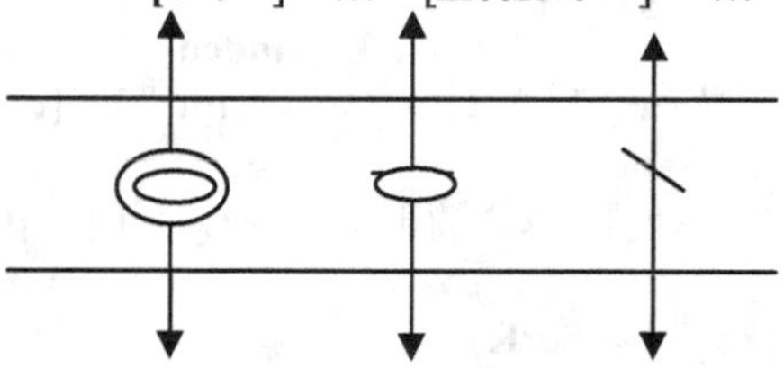

$[m\tau en*e^{(n)}]$ ... $[m\tau eS*e^{(S)}]$ ... $[m\tau eq*e^{(q)}]$

$\tau\tau_m{}^m \equiv m\tau\tau m = 1$ Stiffness $[\tau e_m{}^S] \equiv m\tau eS$

:

$R = p$

$\tau^{(M)}/\tau_M = e^{(N)}/e_N$

$*$

(τ_p)

$=$

$[\tau\tau_p{}^M] * [\tau^{(M)}] = [\tau e_p{}^N] * [e^{(N)}]$

$\tau\tau_p{}^p \equiv p\tau\tau p = 1$ Stiffness $[\tau e_p{}^S] \equiv p\tau eS$

$\tau\tau_p{}^p \equiv p\tau\tau p = 1$

$[p\tau\tau m*\tau^{(m)}]$ ... $[p\tau\tau R*\tau^{(R)}]$ ... $[1*\tau^{(p)}]$

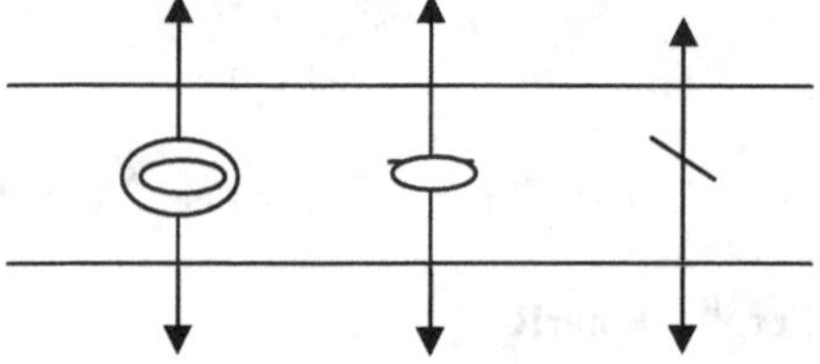

$[p\tau en*e^{(n)}]$ ... $[p\tau eS*e^{(S)}]$ ... $[p\tau e0*e^{(q)}]$

Stiffness $[\tau e_p^{\ S}] \equiv p\tau eS$

$\tau\tau_p^{\ p} \equiv p\tau\tau p = 1$

Stiffness $[\tau e_p^{\ S}] \equiv p\tau eS$

18.1.5.
Flexibility View:

$$\tau^{(M)} / \tau_M = e^{(N)} / e_N$$
$$*$$
$$(e_S)$$
$$=$$
$$\frac{(e_S)/\tau_M \ * \ \tau^{(M)} = (e_S)/e_N * [e^{(N)}]}{[e\tau_S^{\ M}] \ * [\tau^{(M)}] = [ee_S^{\ N}] \ * \ [e^{(N)}]}$$

$[e\tau_S^{\ M}] = (e_S)/\tau_M$ $[ee_S^{\ N}] = (e_S)/e_N$

Expanded :

$$e\tau_S^{\ m} * \tau^{(m)} + e\tau_S^{\ m+1} * \tau^{(m+1)} + \ldots + [e\tau_S^{\ R}] * [\tau^{(R)}] + \ldots + e\tau_S^{\ p} * \tau^{(p)}$$
$$=$$
$$ee_S^{\ n} * e^{(n)} + ee_S^{\ n+1} * e^{(n+1)} + \ldots + [ee_S^{\ S} = 1] * [e^{(S)}] + \ldots + ee_S^{\ q} * e^{(q)}$$

Flexibility $[e\tau_S^{\ R}] \equiv Se\tau R$ $[ee_S^{\ S}] \equiv SeeS = 1$

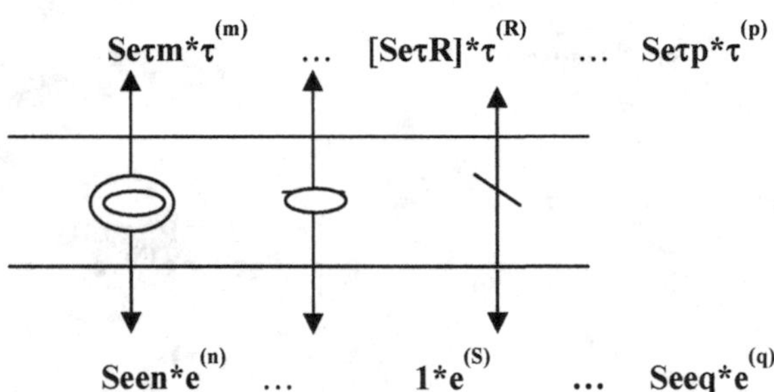

$[ee_S^{\ S}] \equiv SeeS = 1$

Flexibility $[e\tau_S^{\ R}] \equiv Se\tau R$ $[ee_S^{\ S}] \equiv SeeS = 1$

$$S = n$$
$$\tau^{(M)} / \tau_M = e^{(N)} / e_N$$
$$*$$
$$(e_n)$$
$$=$$
$$[e\tau_n^{\ M}] \ * [\tau^{(M)}] = [ee_n^{\ N}] \ * \ [e^{(N)}]$$

Flexibility $[e\tau_n^{\ R}] \equiv ne\tau R$ $[ee_n^{\ n}] \equiv [neen] = 1$

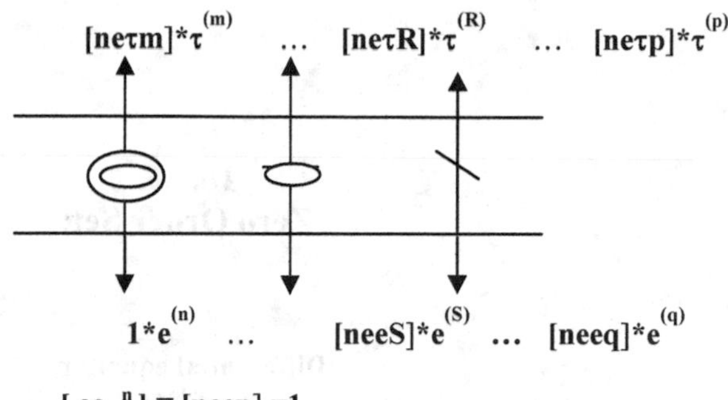

$[\, ee_{n}{}^{n}\,] \equiv [neen] = 1$

Flexibility $[\, e\tau_{n}{}^{R}\,] \equiv ne\tau R$ $\quad\quad$ $[\, ee_{n}{}^{n}\,] \equiv [neen] = 1$

$$\vdots$$

$$S = q$$

$$\tau^{(M)} / \tau_{M} = e^{(N)} / e_{N}$$

$$*$$

$$(\, e_{q}\,)$$

$$=$$

$$[\, e\tau_{q}{}^{M}\,] \;*\; [\tau^{(M)}] = [\, ee_{q}{}^{N}\,] \;*\; [e^{(N}\,]$$

Flexibility $[\, e\tau_{q}{}^{R}\,] \equiv qe\tau R$ $\quad\quad$ $[\, ee_{q}{}^{q}\,] \equiv [qeeq] = 1$

Flexibility $[e\tau_{q}{}^{R}] \equiv [qe\tau R\,]$

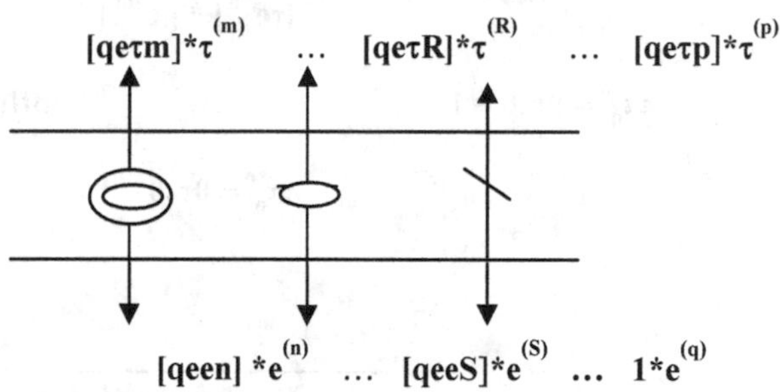

$[\, ee_{q}{}^{q}\,] \equiv [qeeq] = 1$

Flexibility $[\, e\tau_{q}{}^{R}\,] \equiv qe\tau R$ $\quad\quad$ $[\, ee_{q}{}^{q}\,] \equiv [qeeq] = 1$

So:

$$[\tau\tau_{R}{}^{M}\,] * [\tau^{(M)}] = [\tau e_{R}{}^{N}\,] \;*\; [\, e^{(N)}\,]$$

$$[\, e\tau_{S}{}^{M}\,] \;*\; [\tau^{(M)}] = [\, ee_{S}{}^{N}\,] \;*\; [e^{(N}\,]$$

Where :

$$e\tau_{S}{}^{M} \;\; \delta_{M}{}^{R} \;\; \tau e_{R}{}^{N} = (\, e_{S}\,) / \tau_{M} \;\; \delta_{M}{}^{R} \; \tau_{R} / \, e_{N} = \delta_{S}{}^{N}$$

$$\tau e_{R}{}^{N} \, \delta_{N}{}^{S} \;\; e\tau_{S}{}^{M} = \tau_{R} / \, e_{N} \;\; \delta_{N}{}^{S} \;\; (\, e_{S}\,) / \tau_{M} \; = \; \delta_{R}{}^{M}$$

Diagonal Terms:

$$e\tau_{\underline{S}}{}^{P} \quad \tau e_{P}{}^{\underline{S}} = 1$$

$$\tau e_{\underline{R}}{}^{N} \quad e\tau_{S}{}^{\underline{R}} = 1$$

18.1.6.
Zero Order Set:

Differential equation:

$\tau^{(0)} / \tau_0$

=

$[e^{(0)} / e_0]$

Variable Operations:

$\tau^{(0)}$

$e^{(0)}$

:

Stiffness View:

Only R = 0

Differential equation:

$\tau^{(0)} / \tau_0$

=

$[e^{(0)} / e_0]$

*

τ_0

=

$[\tau\tau_0^{\ 0} = 1] * [\tau^{(0)}] = [1] * [\tau^{(0)}]$

=

$[\tau e_0^{\ 0}] * [e^{(0)}]$

$\tau\tau_0^{\ 0} \equiv 0\tau\tau 0 = 1$ **Stiffness** $[\tau e_0^{\ S}] \equiv 0\tau eS$ **Only S=0**

$\tau\tau_0^{\ 0} \equiv 0\tau\tau 0 = 1$

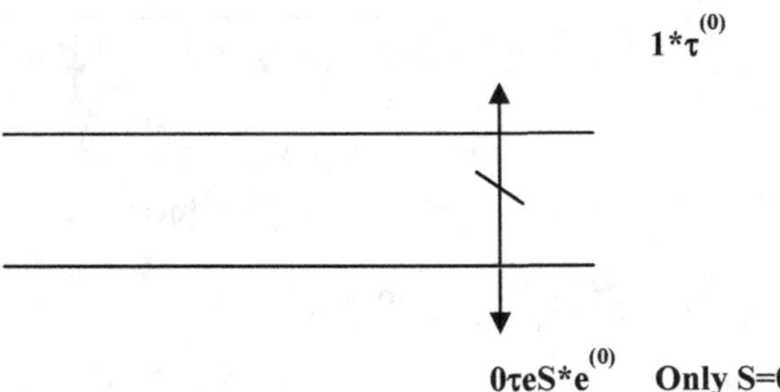

:

Flexibility View:

Only S = 0

Differential equation:

$[\tau^{(0)}] / \tau_0$

=

$[e^{(0)} / e_0]$

*

e_0

=

$$[e\tau_0^{\ 0}] * [\tau^{(0)}]$$
$$=$$
$$[ee_0^{\ 0} = 1] * [e^{(0)}] = [1] * [e^{(0)}]$$

Flexibility $[e\tau_0^{\ R}] \equiv 0e\tau R$ **Only R=0** $[ee_0^{\ 0}] \equiv 0ee0 = 1$

$[0e\tau R] * \tau^{(R)}$ **Only R=0**

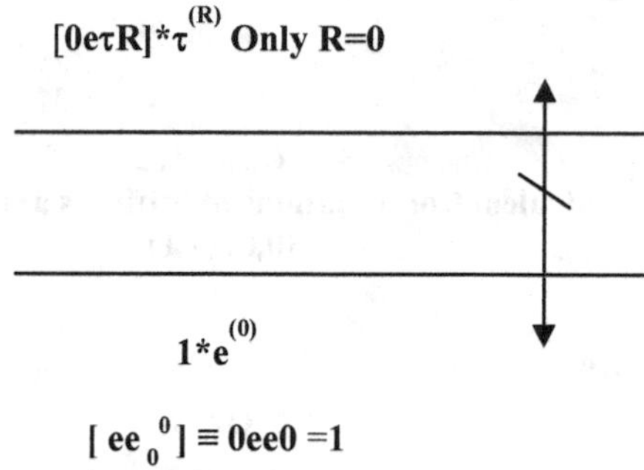

$1 * e^{(0)}$

$[ee_0^{\ 0}] \equiv 0ee0 = 1$

18.1.7.
Stiffness and Flexibility of First Order Set:

Differential equation:

$$\tau^{(M)} / \tau_M = e^{(N)} / e_N$$

:

Combinations of Variable Operations:

:

$C^2_{\ 0} * 2 = 1 * 2 = 2$ **Cases of 0**

:

(0,0) (0,0)

0 **0**

(0,0) (0,0)

0 **0**

4 Cases of 1
equivalent Combinations of Stiffness and Flexibility:
(0,0) (0,1)

1/e

Or :

(0,1) (0,0)

$1/\tau$

(0,0) (1,0)

1/e

Or :

(1,0) (0,0)

1/τ

2 Cases of 2
equivalent Combinations of Stiffness and Flexibility:
(0,0) (1,1)

1/e **1/e**

Or :

(1,1) (0,0)

1/τ **1/τ**

:

4 Cases of 3
equivalent Combinations of Stiffness and Flexibility:
(0,1) (1,1)

1/τ

1/e **1/e**

Or :

(1,1) (0,1)

1/τ **1/τ**

1/e

(1,0) (1,1)

1/τ

1/e **1/e**

Or :

(1,1) (1,0)

1/τ **1/τ**

1/e

$$\left\{ \left[\left(C^2_0 + C^2_1 + C^2_2 \right) \right] + \left[C^2_2(1/\tau) * C^2_1(1/e) \right] \right\} * 2 =$$

$$\{ [1+2+1] + [1*2] \} *2 = \{ [4] + [2] \} *2 = 12 \text{ Cases}$$

Due to Symmetry of (1/τ) and (1/e) 6 Cases:
{ [1+ 2 +1] + [1*2]} = { [4] + [2]} = 6
:
Null Case C^2_0 =1 Case of 0
(0,0) (0,0)
C^2_1 =2 Cases of 1
(0,0) (0,1) or (0,1) (0,0)
(0,0) (1,0) or (1,0) (0,0)
C^2_2 =1 Cases of 2
(0,0) (1,1) or (0,0) (0,0)
:
C^2_2 (1/τ) * C^2_1(1/e)= 1*2 =2 Cases of 3
(0,1) (1,1) or (1,1) (0,1)
(1,0) (1,1) or (1,1) (1,0)
:
Stiffness View:
R
Differential equation:

$$\tau^{(1)}/\tau_1 + \tau^{(0)}/\tau_0$$
$$=$$
$$e^{(1)}/e_1 + e^{(0)}/e_0$$
$$*$$
$$(\tau_R)$$
$$=$$

$\tau\tau_R{}^R \equiv R\tau\tau R = 1$ **Stiffness $[\tau e_R{}^S] \equiv R\tau eS$**

R = 0

$$\tau\tau_0{}^1 * \tau^{(1)} + [1] * [\tau^{(0)}]$$
$$=$$
$$[\tau e_0{}^1] * [e^{(1)}] + [\tau e_0{}^0] * [e^{(0)}]$$

$\tau\tau_0{}^0 \equiv 0\tau\tau 0 = 1$ **Stiffness $[\tau e_0{}^S] \equiv 0\tau eS$**

Stiffness $[\tau e_0{}^0] \equiv 0\tau e0$
Stiffness $[\tau e_0{}^1] \equiv 0\tau e1$

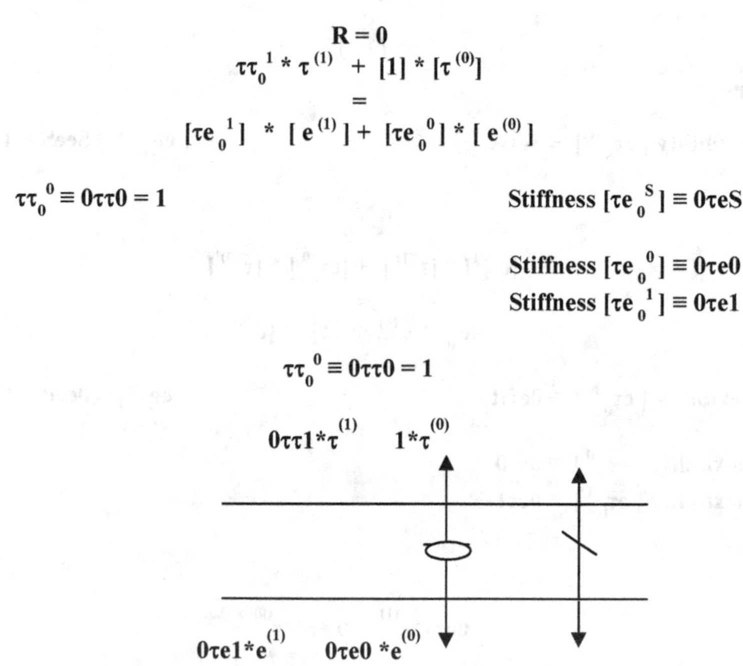

R = 1

$$[1] * [\tau^{(1)}] + \tau\tau_1{}^0 * \tau^{(0)}$$
$$=$$
$$[\tau e_1{}^1] * [e^{(1)}] + [\tau e_1{}^0] * [e^{(0)}]$$

$\tau\tau_0{}^0 \equiv 0\tau\tau 0 = 1$

Stiffness $[\tau e_0{}^S] \equiv 0\tau eS$

Stiffness $[\tau e_0{}^0] \equiv 0\tau e0$
Stiffness $[\tau e_0{}^1] \equiv 0\tau e1$

$\tau\tau_R{}^R \equiv R\tau\tau R = 1$

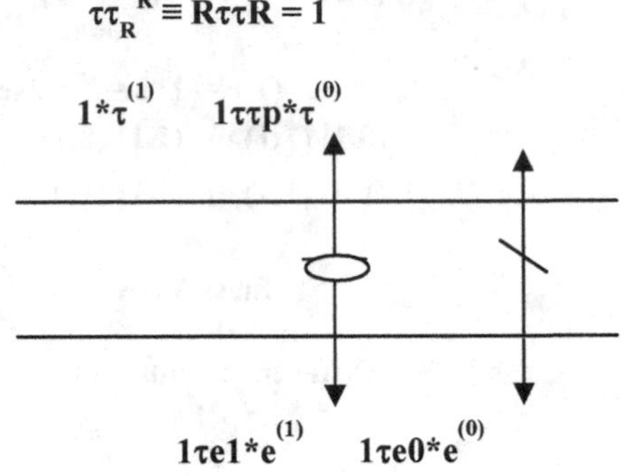

Flexibility View:
S
Differential equation:

$$\tau^{(1)}/\tau_1 + \tau^{(0)}/\tau_0$$
$$=$$
$$e^{(1)}/e_1 + e^{(0)}/e_0$$
$$*$$
$$(e_S)$$
$$=$$

Flexibility $[e\tau_S{}^R] \equiv Se\tau R$

$[ee_S{}^S] \equiv SeeS = 1$

S=0

$$[e\tau_0{}^1] * [\tau^{(1)}] + [e\tau_0{}^0] * [\tau^{(0)}]$$
$$=$$
$$(ee_0{}^1) * e^{(1)} + [1] * [e^{(0)}$$

Flexibility $[e\tau_0{}^R] \equiv 0e\tau R$

$[ee_0{}^0] \equiv 0ee0 = 1$

Flexibility $[e\tau_0{}^0] \equiv 0e\tau 0$
Flexibility $[e\tau_0{}^1] \equiv 0e\tau 1$

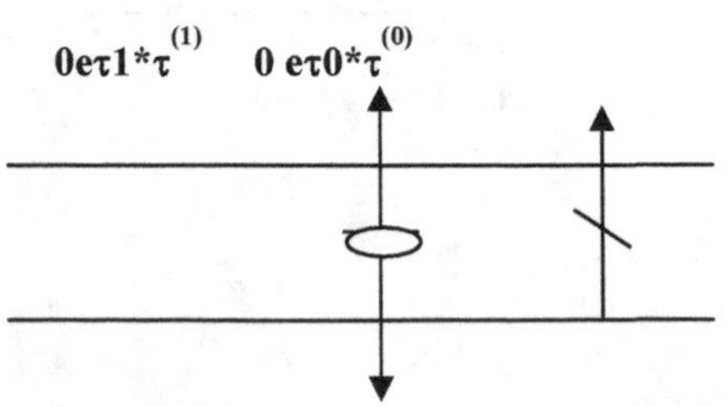

$0ee1*e^{(1)}$ $[1]*e^{(0)}$
$[ee_0^{\ 0}] \equiv 0ee0 = 1$

S=1
$[e\tau_1^{\ 1}] * [\tau^{(1)}] + [e\tau_1^{\ 0}] * [\tau^{(0)}]$
=
$[ee_1^{\ 1}] * e^{(1)} + [1] * [e^{(0)}]$

Flexibility $[e\tau_1^{\ R}] \equiv 1e\tau R$ $[ee_1^{\ 1}] \equiv 1ee1 = 1$

Flexibility $[e\tau_0^{\ 0}] \equiv 1e\tau 0$
Flexibility $[e\tau_0^{\ 1}] \equiv 1e\tau 1$

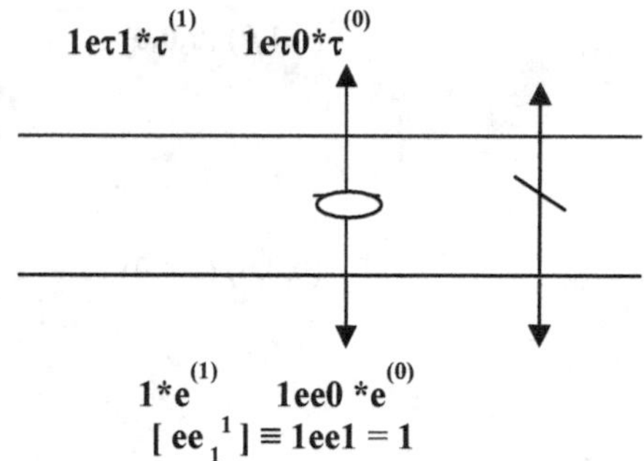

18.1.8.
Second Order Set:

Differential equation:
$\tau^{(M)}/\tau_M = e^{(N)}/e_N$
$\tau^{(2)}/\tau_2 + \tau^{(1)}/\tau_1 + \tau^{(0)}/\tau_0$
=
$[e^{(2)}/e_2] + [e^{(1)}/e_1] + [e^{(0)}/e_0]$
:
Combinations of Variable Operations:

:
C^3_0 ***2 =1*2=2 Cases of 0**
Identical Tow equivalent Combinations of Stiffness and Flexibility:
(0,0,0) (0,0,0)

0 **0** **0**

(0,0,0) (0,0,0)
0 **0** **0**

C^3_1 *2 =3*2= 6 Cases of 1
equivalent Combinations of Stiffness and Flexibility:
(0,0,0) (0,0,1)

1/e

(0,0,1) (0,0,0)

1/τ

(0,0,0) (0,1,0)

1/e

(0,1,0) (0,0,0)

1/τ

(0,0,0) (1,0,0)

1/e

(1,0,0) (0,0,0)

1/τ

C^3_2 *2 =3*2=6 Cases of 2
equivalent Combinations of Stiffness and Flexibility:
(0,0,0) (0,1,1)

1/e **1/e**

(0,1,1) (0,0,0)

1/τ **1/τ**

(0,0,0) (1,1,0)

1/e **1/e**

(1,1,0) (0,0,0)

1/τ **1/τ**

(0,0,0) (1,0,1)

1/e **1/e**

(1,0,1) (0,0,0)

1/τ **1/τ**

C^3_3 *2 =1*2=2 Cases of 3
equivalent Combinations of Stiffness and Flexibility:
(0,0,0) (1,1,1)

1/e **1/e** **1/e**

(1,1,1) (0,0,0)

1/τ **1/τ** **1/τ**

:

:

:

C^3_3 (1/e) * 2 * C^3_1(1/τ)= 1*2*3 =6 Cases of 4
equivalent Combinations of Stiffness and Flexibility:
(0,0,1) (1,1,1)

1/τ
1/e **1/e** **1/e**

(1,1,1) (0,0,1)

1/τ **1/τ** **1/τ**
1/e

(0,1,0) (1,1,1)

1/τ
1/e **1/e** **1/e**

(1,1,1) (0,1,0)

1/τ **1/τ** **1/τ**
1/e

(1,0,0) (1,1,1)

1/τ
1/e **1/e** **1/e**

(1,1,1) (1,0,0)

1/τ **1/τ** **1/τ**

1/e

:
:
:

C^3_3 (1/e) * 2 * C^3_2(1/τ)= 1*2*3 =6 Cases of 5 equivalent Combinations of Stiffness and Flexibility:

(0,1,1) (1,1,1)

	1/τ	**1/τ**
1/e	**1/e**	**1/e**

(1,1,1) (0,1,1)

1/τ	**1/τ**	**1/τ**
	1/e	**1/e**

(1,1,0) (1,1,1)

1/τ	**1/τ**	
1/e	**1/e**	**1/e**

(1,1,1) (1,1,0)

1/τ	**1/τ**	**1/τ**
1/e	**1/e**	

(1,0,1) (1,1,1)

1/τ		**1/τ**
1/e	**1/e**	**1/e**

(1,1,1) (1,0,1)

1/τ	**1/τ**	**1/τ**
1/e		**1/e**

:

{

[(C^3_0 + C^3_1 + C^3_2 + C^3_3)]

+

[C^3_3(1/τ) * C^3_1(1/e) + C^3_3(1/ τ) * C^3_2(1/e)]

}

*****

2

=

{ [1+ 3 +3+1] + [1*3+1*3]} *2 = { [8] + [6]} *2 = 14*2

=

28 Cases

Due to Symmetry of (1/τ) and (1/e) 14 Cases:
{ [1+ 3 +3+1] + [1*3+1*3]} = { [8] + [6]} = 14

:

Null Case C^3_0 =1 Case of 0

(0,0,0) (0,0,0)

C^3_1 =3 Cases of 1

(0,0,0) (0,0,1) or (0,0,1) (0,0,0)

(0,0,0) (0,1,0) or (0,1,0) (0,0,0)

(0,0,0) (1,0,0) or (1,0,0) (0,0,0)

C^3_2 =3 Cases of 2

(0,0,0) (0,1,1) or (0,1,1) (0,0,0)

(0,0,0) (1,1,0) or (1,1,0) (0,0,0)

(0,0,0) (1,0,1) or (1,0,1) (0,0,0)

C^3_3 =1 Case of 3

(0,0,0) (1,1,1) or (1,1,1) (0,0,0)

:

$C^3_3(1/\tau) * C^3_1(1/e) = 1*3 = 3$ Cases of 4

(0,0,1) (1,1,1) or (1,1,1) (0,0,1)

(0,1,0) (1,1,1) or (1,1,1) (0,1,0)

(1,0,0) (1,1,1) or (1,0,0) (1,0,0)

$C^3_3(1/\tau) * C^3_2(1/e) = 1*3 = 3$ Cases of 5

(0,1,1) (1,1,1) or (1,1,1) (0,1,1)

(1,1,0) (1,1,1) or (1,1,1) (1,1,0)

(1,0,1) (1,1,1) or (1,1,1) (1,0,1)

18.2. Second Order Dynamic Set:

18.2.1. Normalized Second Order Dynamic Set:

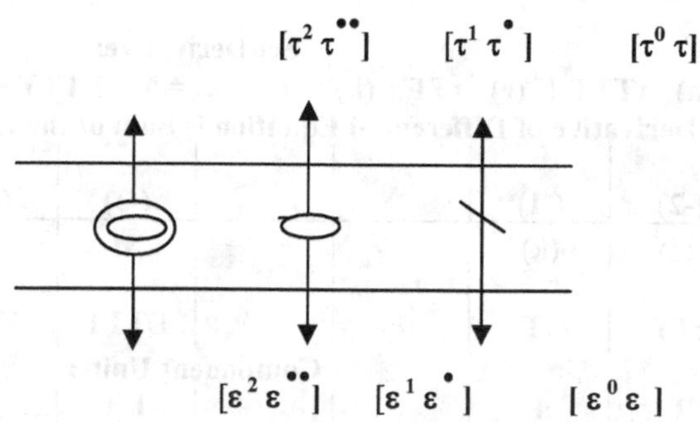

Unitless Second Order Dynamic Set:

$$(\tau^2\ \ddot{\tau}) + (\tau^1\ \dot{\tau}) + (\tau^0\ \tau) = (\varepsilon^2\ \ddot{\varepsilon}) + (\varepsilon^1\ \dot{\varepsilon}) + (\varepsilon^0\ \varepsilon)$$

$$(m)\ \tau TT + (c)\ \tau T + (k)\ \tau = M\ \varepsilon TT + C\ \varepsilon T + K\ \varepsilon$$

$$(\ddot{\tau} / \tau_2) + (\dot{\tau} / \tau_1) + (\tau / \tau_0) = (\ddot{\varepsilon} / \varepsilon_2) + (\dot{\varepsilon} / \varepsilon_1) + (\varepsilon / \varepsilon_0)$$

$\tau^2 = 1/\tau_2$	$\tau^1 = 1/\tau_1$	$\tau^0 = 1/\tau_0$	$\varepsilon^2 = 1/\varepsilon_2$	$\varepsilon^1 = 1/\varepsilon_1$	$\varepsilon^0 = 1/\varepsilon_0$
=	=	=	=	=	=
(m)	(c)	(k)	(M)	(C)	(K)

Let Units of Variables:

[m] = TT	[c] = T	[k] = 1	[M] = TT	[C] = T	[K] = 1

Units of Derivatives:

$[\tau^{(M)}] = \tau U / [T]^M$	$[\varepsilon^{(N)}] = \varepsilon U / [T]^N$

:

$[\ddot{\tau}]$	$[\dot{\tau}]$	$[\tau]$	$[\ddot{\varepsilon}]$	$[\dot{\varepsilon}]$	$[\varepsilon]$
=	=	=	=	=	=
$[\tau_2=1/m=1/\tau^2]$	$[\tau_1=1/c=1/\tau^1]$	$[\tau_0=1/k=1/\tau^0]$	$[\varepsilon_2=1/M=1/\varepsilon^2]$	$[\varepsilon_1=1/C=1/\varepsilon^1]$	$[\varepsilon_0=1/K=1/\varepsilon^0]$
=	=	=	=	=	=
$\tau U / [T]^2$	$\tau U / [T]^1$	$\tau U / [T]^0$	$\varepsilon U / [T]^2$	$\varepsilon U / [T]^1$	$1/ [T]^0$

Units of Factors:

$[\tau^M] = [T]^M / \tau U$	$[\varepsilon^N] = [T]^N / \varepsilon U$

:

$[\tau^2]$	$[\tau^1]$	$[\tau^0]$	$[\varepsilon^2]$	$[\varepsilon^1]$	$[\varepsilon^0]$
=	=	=	=	=	=
$[T]^2 / \tau U$	$[T]^1 / \tau U$	$[T]^M / \tau U$	$[T]^2 / \varepsilon U$	$[T]^1 / \varepsilon U$	$[T]^0 / \varepsilon U$

Such that:

$[\tau^M] = 1/ [\tau^{(M)}]$	$[\varepsilon^N]= 1/ [\varepsilon^{(N)}]$
$[T]^M / [\tau^M] = [\tau^{(M)}]\ [T]^M = \tau U$	$[T]^N / [\varepsilon^N] = [\varepsilon^{(N)}]\ [T]^N = \varepsilon U$

So that Products of Factors and Derivatives are Unitless:

$[\tau^2][\ddot{\tau}]=1$	$[\tau^1][\dot{\tau}]=1$	$[\tau^0][\tau]=1$	$[\varepsilon^2][\ddot{\varepsilon}]=1$	$[\varepsilon^1][\dot{\varepsilon}]=1$	$[\varepsilon^0][\varepsilon]=1$

:

$$[\tau^2][\ddot{\tau}] = [\tau^1][\dot{\tau}] = [\tau^0][\tau] = 1 = [\varepsilon^2][\ddot{\varepsilon}] = [\varepsilon^1][\dot{\varepsilon}] = [\varepsilon^0][\varepsilon]$$

Ratios:

$[\tau^{(R)}] / [\tau^{(P)}] = [\tau_R/\tau_P] = 1 / [\tau^R/\tau^P]$ = $1/ [T]^{R-P}$	$[\varepsilon^{(S)}] / [\varepsilon^{(Q)}] = [\varepsilon_S/\varepsilon_Q] = 1/ [\varepsilon^S/\varepsilon^Q]$ = $1/ [T]^{S-Q}$
$[\tau^R/\tau^P] = [T]^{R-P}$	$[\varepsilon^S/\varepsilon^Q] = [T]^{S-Q}$

Set Derivative:

(m) τTTT + (c) τTT + (k) τT …… = M ETTT + C ETT + K ET ……

Derivative of Differential Equation is Sum of the following Components:

(-3)	(-2)	(-1)			(-3)	(-2)	(-1)		
(m)	(c)	(k)			M	C	K		
τTTT	τTT	τT			ETTT	ETT	ET		

Component Units:

T T	T	1			T T	T	1		
1/TTT	1/TT	1/T			1/TTT	1/TT	1/T		

So Unit is (1/T):

[(TT)] [1/TTT]=[(T)] [1/TT]=[1] [1/T]=1/T=[TT] [1/TTT]=[T] [1/TT]=[1 1/T]

:

Set:

(m) τTT + (c) τ T + (k) τ = M εTT + C εT + K ε

Differential Equation is Sum of the following Components:

(-2)	(-1)			(-2)	(-1)	
(m)	(c)	(k)		M	C	K
τTT	τT	τ		εTT	εT	ε

Unitless Components:

TT	T	1	TT	T	1
1/TT	1/T	1	1/TT	1/T	1

Unitless:

[TT] [1/ TT] = [T] [1/T] = [1] [1/1] = 1 = [TT] [1/ TT] = [T] [1/T] = [1 1]

:

Set Integral:

… … (m) τT + (c) τ + (k) ∫τ dt = … … M ET + C E + K ∫E dt

Integral of Differential Equation is Sum of the following Components:

(-1)		∫ (+1)	(-1)		∫ (+1)
(m)	(c)	(k)	M	C	K
τT	τ	∫τ dt	ET	E	∫E dt

Component Units:

TT	T	1	TT	T	1
1/T	1/1	T	1/T	1/1	T

So Unit is (T):

[TT] [1/ T] = [T] [1/1] = [1] [T/1] = 1 = [TT] [1/ T] = [T] [1/1] = [1 T]

18.2.1.1.
Stress and Strain Units:

Unit of τ = Stress | **Unit of ε = Strain**

Normalize by their Units:

{ τU}/[Stress] | **{ εU}/[Strain]**

Then:

{ m τTT + c τ T + k τ} / [τU]

=

{ M ε TT + C εT + K ε } / [εU]

{ TT τU/TT = T τU/T = 1 τU/1 } / [Stress]

=

{ TT εU/TT = T εU/T = 1 εU/1 } / [Strain]

{ τU } / [Stress] = 1 = { εU } / [1]

18.2.1.2.
Force and Deformation Units:

Unit of τ = Force | **Unit of ε = Deformation**

Normalize by their Units:

{ τU}/[Force] | **{ εU}/[Deformation]**

Then:

{ m τTT + c τ T + k τ} / [τU]

=

{ M ε TT + C εT + K ε } / [εU]

{ TT τU/TT = T τU/T = 1 τU/1 } / [Force]

=

{ TT εU/TT = T εU/T = 1 εU/1 } / [Deformation]

{ τU } / [Force] = 1 = { εU}/[Deformation]

18.2.2.
Stiffness (Power Resilience) and Flexibility (Form Resilience):

18.2.2.1.
Stiffness (Power Resilience) of Dynamic Structural Set:

Dynamic Structural Set Equation:

$$(m)\ \tau^{\bullet\bullet} + (c)\ \tau^{\bullet} + (k)\ \tau = (M)\ \varepsilon^{\bullet\bullet} + (C)\ \varepsilon^{\bullet} + (K)\ \varepsilon$$

$$\tau_R$$
$$=$$

$\tau\tau_R^{\ R} \equiv R\tau\tau R = 1$	Stiffness (Power Resilience) $(\tau e_R^{\ S}) \equiv R\tau eS$

$$(\tau e_R^{\ S})$$

$$\tau\tau_R^{\ 2}\ \tau^{\bullet\bullet} + \tau\tau_R^{\ 1}\ \tau^{\bullet} + \tau\tau_R^{\ 0}\ \tau = (\tau e_R^{\ 2})\ (e^{\bullet\bullet}) + (\tau e_R^{\ 1})\ (e^{\bullet}) + (\tau e_R^{\ 0})\ (e)$$

Applied:
For R = 0

$$\tau_{R=0} = (1/k)$$

:

$$(m/k)\ \tau^{\bullet\bullet} + (c/k)\ \tau^{\bullet} + (1)\ \tau = (M/k)\ \varepsilon^{\bullet\bullet} + (C/k)\ \varepsilon^{\bullet} + (K/k)\ \varepsilon$$

$$(\tau e_0^{\ S})$$

$\tau\tau_0^{\ 0} \equiv 0\tau\tau 0 = 1$ Stiffness (Power Resilience) $(\tau e_0^{\ S}) \equiv 0\tau eS$

$$(\tau e_0^{\ 0}) \equiv 0\tau e0 = (K/k)$$
$$(\tau e_0^{\ 1}) \equiv 0\tau e1 = (C/k)$$
$$(\tau e_0^{\ 2}) \equiv 0\tau e2 = (M/k)$$

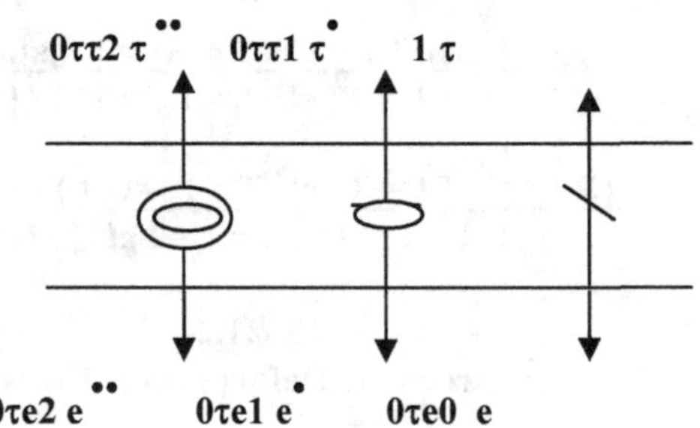

Applied:
For R = 1

$$\tau_{R=1} = (c)$$

:

$$(m/c)\ \tau^{\bullet\bullet} + (1)\ \tau^{\bullet} + (k/c)\ \tau = (M/c)\ \varepsilon^{\bullet\bullet} + (C/c)\ \varepsilon^{\bullet} + (K/c)\ \varepsilon$$

$$(\tau e_1^{\ S})$$

$\tau\tau_1^{\ 1} \equiv 1\tau\tau 1 = 1$ Stiffness (Power Resilience) $(\tau e_1^{\ S}) \equiv 1\tau eS$

$$(\tau e_1^{\ 0}) \equiv 1\tau e0 = (K/c)$$
$$(\tau e_1^{\ 1}) \equiv 1\tau e1 = (C/c)$$
$$(\tau e_1^{\ 2}) \equiv 1\tau e2 = (M/c)$$

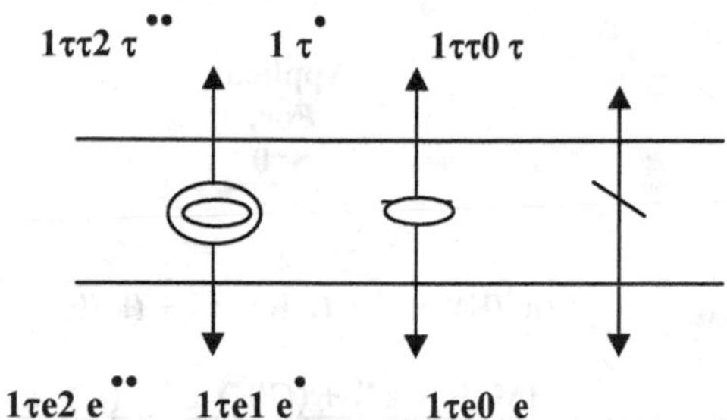

Applied:
For R = 2

$\tau_{R=2} = (1/m)$

:

$$(1)\ \tau^{\bullet\bullet} + (c/m)\ \tau^{\bullet} + (k/m)\ \tau = (M/m)\ \varepsilon^{\bullet\bullet} + (C/m)\ \varepsilon^{\bullet} + (K/m)\ \varepsilon$$

$(\tau e_2^{\ S})$

$\tau\tau_2^{\ 2} \equiv 2\tau\tau 2 = 1$

Stiffness (Power Resilience) $(\tau e_2^{\ S}) \equiv 2\tau eS$

$(\tau e_2^{\ 0}) \equiv 2\tau e0 = (K/m)$

$(\tau e_2^{\ 1}) \equiv 2\tau e1 = (C/m)$

$(\tau e_2^{\ 2}) \equiv 2\tau e2 = (M/m)$

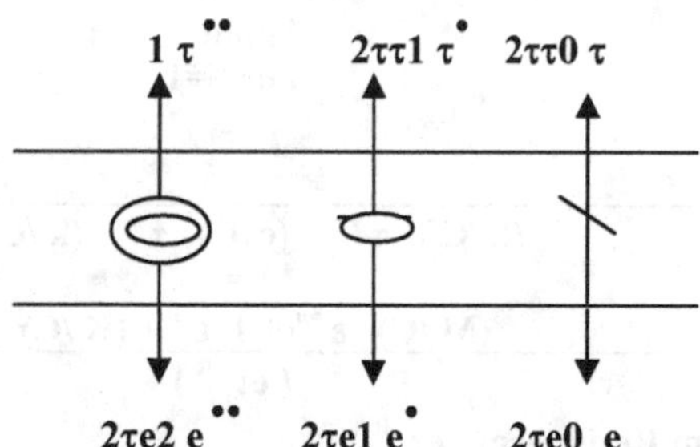

18.2.2.2.
Flexibility (Form Resilience) of Dynamic Structural Set:

:

Dynamic Structural Set Equation:

$$(m)\ \tau^{\bullet\bullet} + (c)\ \tau^{\bullet} + (k)\ \tau = (M)\ \varepsilon^{\bullet\bullet} + (C)\ \varepsilon^{\bullet} + (K)\ \varepsilon$$

(e_S)

=

Flexibility (Form Resilience) $(e\tau_S^{\ R}) \equiv Se\tau R$	$(ee_S^{\ S}) \equiv SeeS = 1$

$(e\tau_S^{\ R})$

$$(e\tau_S^{\ 2})\ (\tau^{\bullet\bullet}) + (e\tau_S^{\ 1})\ (\tau^{\bullet}) + (e\tau_S^{\ 0})\ (\tau) = ee_S^{\ 2}\ e^{\bullet\bullet} + ee_S^{\ 1}\ e^{\bullet} + ee_S^{\ 0}\ (e)$$

Applied:
For:
S=0

$\varepsilon_{S=0} = (1/K)$

:

$$(m/K)\ \tau^{\bullet\bullet} + (c/K)\ \tau^{\bullet} + (k/K)\ \tau = (M/K)\ \varepsilon^{\bullet\bullet} + (C/K)\ \varepsilon^{\bullet} + (1)\ \varepsilon$$

$(e\tau_0{}^R)$

Flexibility (Form Resilience) $(e\tau_0{}^R)$

$(ee_0{}^0) \equiv 0ee0 = 1$

$(e\tau_0{}^0) \equiv 0e\tau0 = (k/K)$
$(e\tau_0{}^1) \equiv 0e\tau1 = (c/K)$
$(e\tau_0{}^2) \equiv 0e\tau2 = (m/K)$

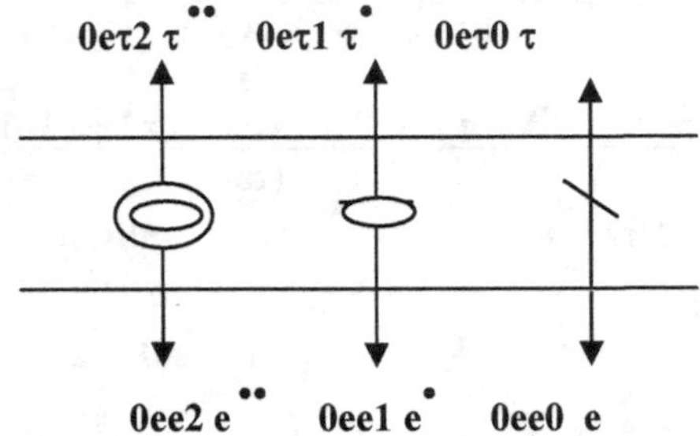

Applied:
For S=1

$\varepsilon_{S=1} = (1/C)$

$$(k/C)\ \tau^{\bullet\bullet} + (c/C)\ \tau^{\bullet} + (k/C)\ \tau = (M/C)\ \varepsilon^{\bullet\bullet} + 1\ \varepsilon^{\bullet} + (K/C)\ \varepsilon$$

$(e\tau_1{}^R)$

Flexibility (Form Resilience) $(e\tau_1{}^R)$

$(ee_1{}^1) \equiv 1ee1 = 1$

$(e\tau_1{}^0) \equiv 1e\tau0 = (k/C)$
$(e\tau_1{}^1) \equiv 1e\tau1 = (c/C)$
$(e\tau_1{}^2) \equiv 1e\tau2 = (k/C)$

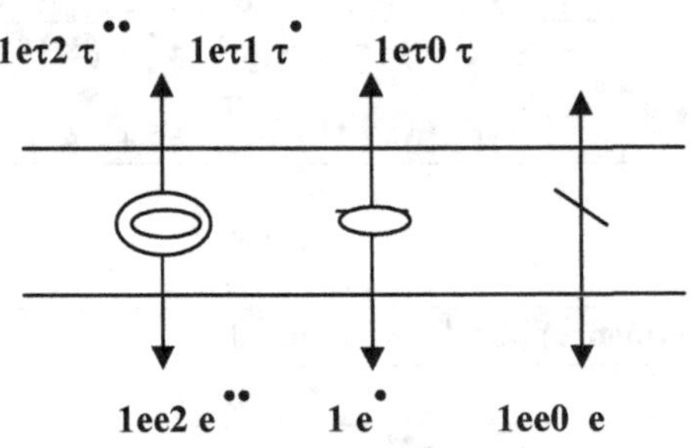

Applied:
For S=2

$$\varepsilon_{S=2} = (1/M)$$

$$(m/M)\ \tau^{\bullet\bullet} + (c/M)\ \tau^{\bullet} + (k/M)\ \tau = (1)\ \varepsilon^{\bullet\bullet} + (C/M)\ \varepsilon^{\bullet} + (K/M)\ \varepsilon$$

$(e\tau_2{}^R)$

Flexibility (Form Resilience) $(e\tau_2{}^R)$ $\quad (ee_2{}^2) \equiv 2ee2 = 1$

$(e\tau_2{}^0) \equiv 2e\tau 0 = (k/M)$
$(e\tau_2{}^1) \equiv 2e\tau 1 = (c/M)$
$(e\tau_2{}^2) \equiv 2e\tau 2 = (m/M)$

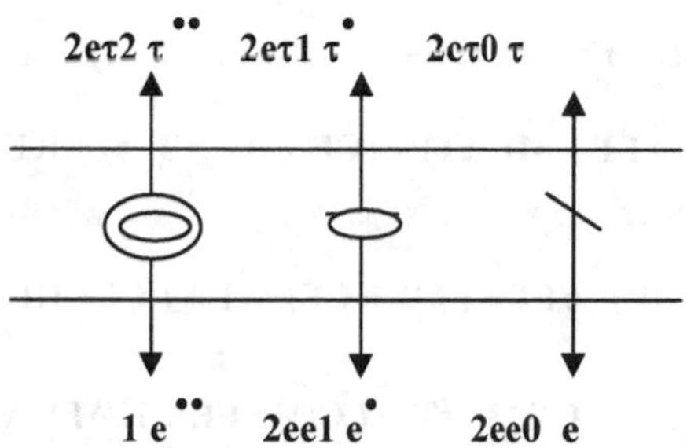

18.2.2.3.
Logging Results of Stiffness (Power Resilience) and Flexibility (Form Resilience):

Logging Results:

Flexibility (Form Resilience) $(e\tau_{\underline{S}}{}^{\underline{R}})$	Stiffness (Power Resilience) $(\tau e_{\underline{R}}{}^{\underline{S}})$
1:1 $(e\tau_0{}^0) \equiv 0e\tau 0 = (k/K)$	1:1 $(\tau e_0{}^0) \equiv 0\tau e0 = (K/k)$
T:1 $(e\tau_0{}^1) \equiv 0e\tau 1 = (c/K)$	T:1 $(\tau e_0{}^1) \equiv 0\tau e1 = (C/k)$
TT:1 $(e\tau_0{}^2) \equiv 0e\tau 2 = (m/K)$	TT:1 $(\tau e_0{}^2) \equiv 0\tau e2 = (M/k)$
1:T $(e\tau_1{}^0) \equiv 1e\tau 0 = (k/C)$	1:T $(\tau e_1{}^0) \equiv 1\tau e0 = (K/c)$
T:T $(e\tau_1{}^1) \equiv 1e\tau 1 = (c/C)$	T:T $(\tau e_1{}^1) \equiv 1\tau e1 = (C/c)$
1:T $(e\tau_1{}^2) \equiv 1e\tau 2 = (k/C)$	TT:T $(\tau e_1{}^2) \equiv 1\tau e2 = (M/c)$
1:TT $(e\tau_2{}^0) \equiv 2e\tau 0 = (k/M)$	1:TT $(\tau e_2{}^0) \equiv 2\tau e0 = (K/m)$
T:TT $(e\tau_2{}^1) \equiv 2e\tau 1 = (c/M)$	T:TT $(\tau e_2{}^1) \equiv 2\tau e1 = (C/m)$
TT:TT $(e\tau_2{}^2) \equiv 2e\tau 2 = (m/M)$	TT:TT $(\tau e_2{}^2) \equiv 2\tau e2 = (M/m)$

Where:

$$(e\tau_{\underline{S}}{}^{\underline{R}})\ (\tau e_{\underline{R}}{}^{\underline{S}}) = 1$$

18.3.
Imposed Movement:

$$(m)\ \tau TT + (c)\ \tau T + (k)\ \tau = M\ UTT + C\ UT + K\ U$$

Discussion of Right Hand Side:

M UTT + C UT + K U

(UT) Is Split into Essential and Imposed Movements Such That:

(UT) = (UET) + (UIT)

And:

(UET) = (E τT)

(UIT) = (+ j τ)

Then:

(UT) = (E τT + I τ)

Substituted in Right Hand Side :

M UTT + C UT + K U

=

M (UT)T + C (UT) + K ∫(UT) dt

=

M (E τT + I τ)T + C (E τT + I τ) + K ∫(E τT + I τ) dt

=

(ME τTT + MI τT) + (CE τT + CI τ) + KE ∫τT dt + KI ∫τ dt

:

(ME) τTT + (MI + CE) τT + (CI + KE) τ + KI ∫τ dt

:

τTT + (I/E + C/M) τT + (C/M I/E + K/M) τ + (K/M I/E) ∫τ dt

:

In Case of Essential movement:

(UT) = (E τT + 0)

:

(ME) τTT + (M0 + CE) τT + (C0 + KE) τ + K0 ∫τ dt

So:

(ME) τTT + (CE) τT + (KE) τ = M UTT + C UT + K U

τTT + (C/M) τT + (K/M) τ = 1/ E UTT + C/ M/ E UT + K/ M/ E U

:

In Case of Imposed movement:

(UT) = (0 + I τ)

:

(M0) τTT + (MI + C0) τT + (CI + K0) τ + KI ∫τ dt

(MI) τT + (CI) τ + KI ∫τ dt = M UTT + C UT + K U

τT + (C/M) τ + (K/M) ∫τ dt = 1/I UTT + C/M/I UT + K/M/I U

:

Logging Essential and Imposed:

(ME) τTT + (MI + CE) τT + (CI + KE) τ + KI ∫τ dt

=

M UTT + C UT + K U

Essential I=0:

(ME) τTT + (… + CE) τT + (… + KE) τ + …..……..

Imposed E=0:

…..…………….. (MI …) τT + (CI … …) τ + KI ∫τ dt

:

Derivative:

$$\mathbf{(ME)}\ \tau\mathbf{TTT} + (\mathbf{MI} + \mathbf{CE})\ \tau\mathbf{TT} + (\mathbf{CI} + \mathbf{KE})\ \tau\mathbf{T} + \mathbf{(KI)}\ \tau$$
$$=$$
$$\mathbf{M}\ \ \mathbf{UTTT} + \mathbf{C}\ \ \mathbf{UTT} + \mathbf{K}\ \ \mathbf{UT}$$

Essential and Imposed Set:

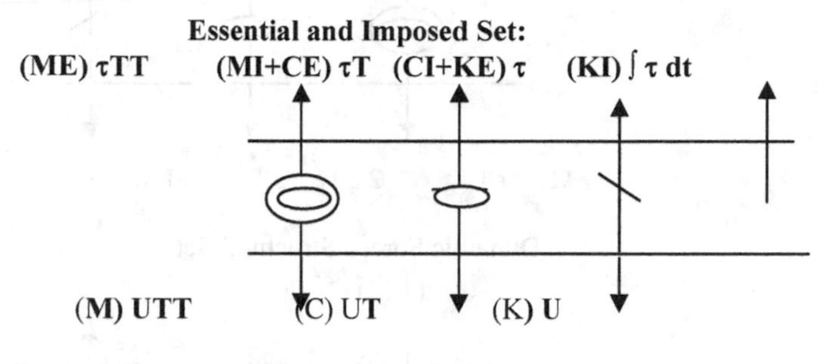

Derivative of Essential and Imposed Set:

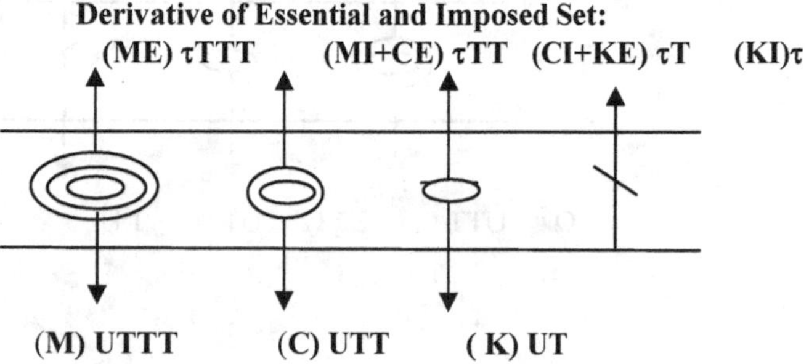

18.4.
Normalized Forced Set:

Define:

$\beta^2 \equiv \omega^2 / \Omega^2$	$-\xi\,\Omega \pm i\,\sqrt{(1-\xi^2)}\,\Omega \equiv -\Omega_\xi \pm i\,\Omega_\eta$	$C/M \equiv 2\,\xi\,\Omega\ \equiv 2\,\Omega_\xi$
$\gamma^2 \equiv 1 - \beta^2$	$\xi^2\ \leq 1$	$K/\,M \equiv \Omega^2$
$(\gamma^2 + j\ \ 2\xi\beta) \equiv \rho^2\ \exp(+\,j\ \theta)$	$\eta^2 \equiv 1 - \xi^2 \equiv\ \Omega_\eta{}^2 / \Omega^2$	$C^2 / (M\,K) =\ 4\,\xi^2\ \leq 4 * 1$
	Unit-wise:	
$[\{\alpha\}] = [U]/TT$	$[\exp(+\,j\ \text{angle})] = 1$	$[\Omega_\xi]=[\ \Omega_\eta]=[\Omega]=1/T$

$$(m)\ \ \tau TT + (c)\ \ \tau T + (k)\ \ \tau$$
$$=$$
$$(M)\ \ \varepsilon TT + (C)\ \ \varepsilon T + (K)\ \ \varepsilon$$

18.4.1.
Normalize by (K):

$$(m\,/K)\ \ \tau TT + (c/K)\ \ \tau T + (k\,/K)\ \ \tau$$
$$=$$
$$(M\,/K)\ \ \varepsilon TT\ + (C/K)\ \ \varepsilon T + \ 1\ \ \varepsilon$$

$$(m/M)\,\Omega^{-2}\ \ \tau TT +\ (c/M)\,\Omega^{-2}\ \ \tau T +\ (k/M)\,\Omega^{-2}\ \ \tau$$
$$=$$
$$\Omega^{-2}\ \ \varepsilon TT + 2\,\xi\,\Omega^{-1}\ \ \varepsilon T +\ \ 1\ \ \varepsilon$$

$$(k/K) = (k/M)\ M/K =\ (k/M)\,\Omega^{-2}$$
$$(c/K) = (c/M)\ M/K =\ (c/M)\,\Omega^{-2}$$
$$(m/K) = (m/M)\ M/K =\ (m/M)\,\Omega^{-2}$$

$$(K/K) = 1$$
$$(C/K) = (C/M)\ M/K = 2\,\xi\,\Omega^{-1}$$
$$(M/K) = \Omega^{-2}$$

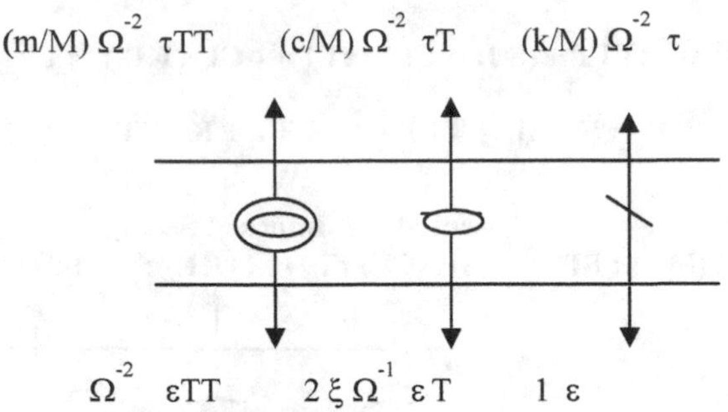

Dynamic Forced Structural Set:

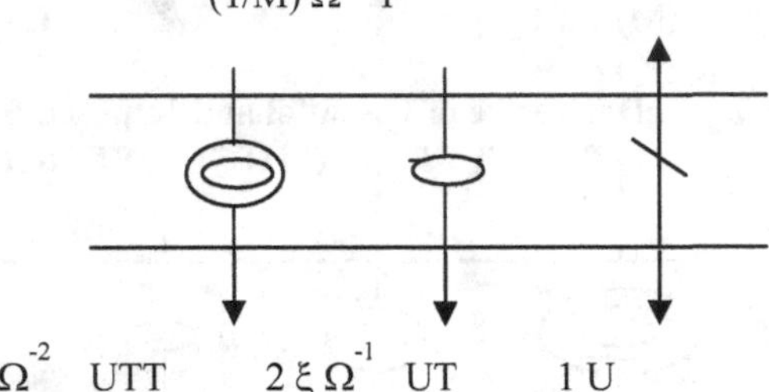

18.4.2.
Normalize by (C):

$$(k/C)\ \tau TT + (c/C)\ \tau T + (k/C)\ \tau = (M/C)\ \varepsilon TT + 1\ \varepsilon T + (K/C)\ \varepsilon$$

$$(m/M)\,\Omega^{-1}\ 1/(2\,\xi)\ \tau TT + (c/C)\ \tau T + (k/K)\,\Omega^{+1}\ 1/(2\,\xi)\,\tau = \Omega^{-1}\ 1/(2\,\xi)\ \varepsilon TT + 1\ \varepsilon T + \ \Omega^{+1}\ 1/(2\,\xi)\ \varepsilon$$

$(k/C) = (k/K)\,K/C = (k/K)\,\Omega^{+1}\ 1/(2\,\xi)$	$(K/C) = (K/M)\,M/C = \Omega^{+1}\ 1/(2\,\xi)$
(c/C)	$(C/C) = 1$
$(m/C) = (m/M)\,M/C = (m/M)\,\Omega^{-1}\ 1/(2\,\xi)$	$(M/C) = (M/K)\,K/C = \Omega^{-1}\ 1/(2\,\xi)$

$(m/M)\,\Omega^{-1}\ 1/(2\,\xi)\ \tau TT$ $\quad (c/C)\ \tau T$ $\quad (k/K)\,\Omega^{+1}\ 1/(2\,\xi)\,\tau$

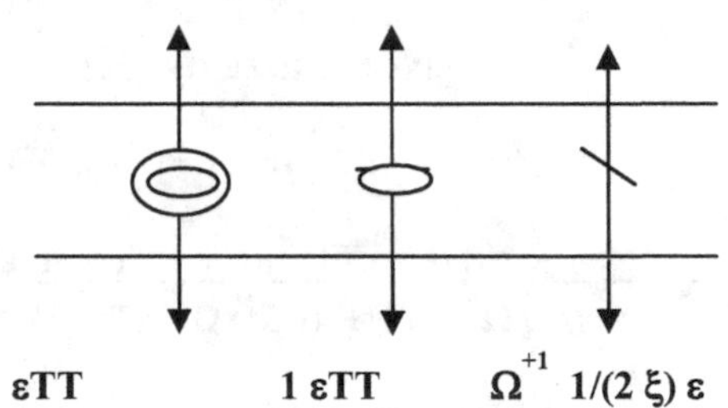

Dynamic Forced Structural Set:

$(1/K)\Omega^{+1}1/(2\xi)F$

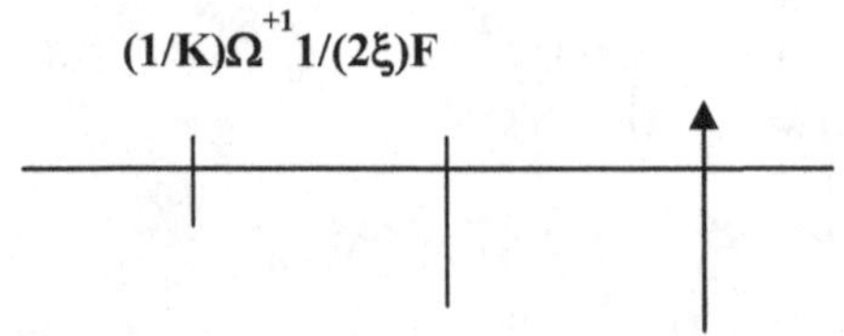

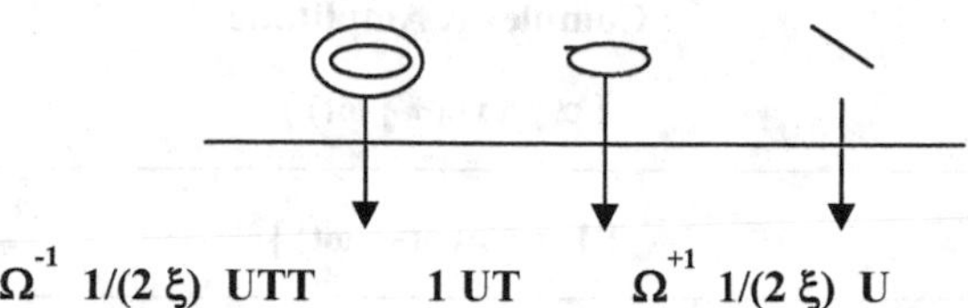

18.4.3.
Normalize by (M):

$$(m/M)\ \tau TT + (c/M)\ \tau T + (k/M)\ \tau$$
$$=$$
$$1\ \varepsilon TT + (C/M)\ \varepsilon T + (K/M)\ \varepsilon$$

$$(m/M)\ \tau TT + (c/K)2\xi\Omega^{+1}\ \tau TT + (k/K)\Omega^{+2}\ \tau$$
$$=$$
$$1\ \varepsilon TT + 2\,\xi\,\Omega^{+1}\ \varepsilon T + \Omega^{+2}\ \varepsilon$$

$(k/M) = (k/K)\ K/M = (k/K)\ \Omega^2$	$(K/M) = \Omega^2$
$(c/M) = (c/C)\ C/M = (c/K)\ 2\xi\Omega$	$(C/M) = 2\,\xi\,\Omega$
$(m/M) = (m/M)\ M/M = (m/M)\ 1$	$(M/M) = 1$

$(m/M)\ \tau TT$ $\quad (c/K)2\xi\Omega\ \tau T$ $\quad (k/K)\Omega^2\ \tau$

$1\ \varepsilon TT$ $\quad 2\,\xi\,\Omega\ \varepsilon T$ $\quad \Omega^2\ \varepsilon$

Dynamic Forced Structural Set:

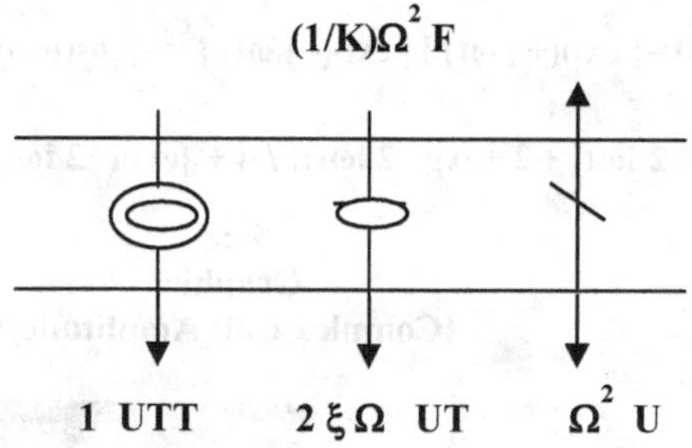

18.5.
Exponential Forms:

Trigeometric Relations:

{ Complex α Amplitude)

{ α * exp(+ j ωt) }

{ 1/α * exp(- j ωt) }

{ α * exp(+ j ωt) } * { 1/α * exp(- j ωt) } = 1

:

Graphic:

{ Complex α Amplitude)

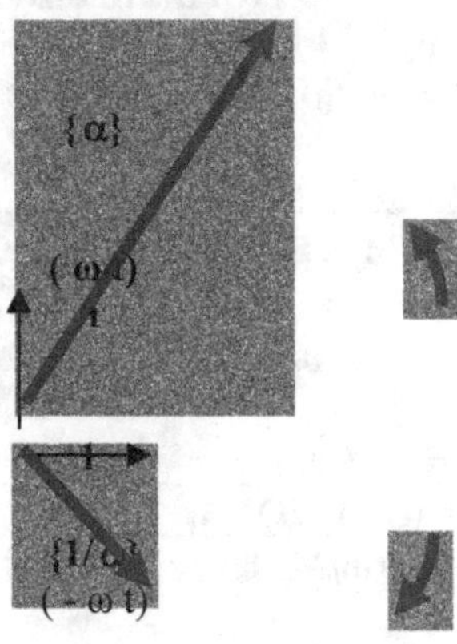

Even Symmetry: **Odd Symmetry:**

{Complex Unit Amplitude)

{ exp(+ j ωt) } = {cos(ωt)} + i {sin(ωt)} **{ exp(- j ωt) } = {cos(ωt)} - i {sin(ωt)}**

{cos(ωt)} = {exp(+ jωt)+exp(-iωt)} / 2 **{sin(ωt)} = {exp(+ jωt)-exp(-iωt)}/(2i)**

:

1={ exp(+ j ωt) }{ exp(- j ωt) } = {cos(ωt)}^{2} + {sin(ωt)}^{2}

1= {exp(+2 iωt) + 2 +exp(-2 iωt)} / 4 + {exp(+2 iωt) – 2 +exp(-2 iωt)} / (-4)

:

Graphic:

{Complex Unit Amplitude):

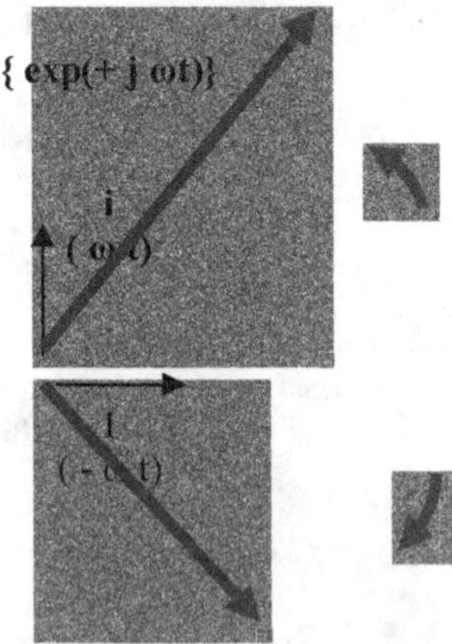

{ exp(- j ωt)}

{Real Unit Amplitude)

{ exp(+ ωt) } = {cosh(ωt)} + {sinh(ωt)} **{ exp(- ωt) } = {cosh(ωt)} - {sinh(ωt)}**

{cosh(ωt)} = {exp(+ ωt) + exp(-ωt)} / 2 **{sinh(ωt)} = {exp(+ωt)-exp(-ωt)}/2**

:

$$1=\{ \exp(+ \omega t) \}\{ \exp(- \omega t) \} = \{\cosh(\omega t)\}^2 - \{\sin(\omega t)\}^2$$

$$1= \{\exp(+2 \omega t) + 2 +\exp(-2 \omega t)\} / 4 - \{\exp(+2 \omega t) - 2 +\exp(-2 \omega t)\} / (4)$$

:

Graphic:
{Real Unit Amplitude)

{ exp(+ ωt)}

{sinh(ωt)}
(ωt)
{cosh(ωt)}
(- ωt)

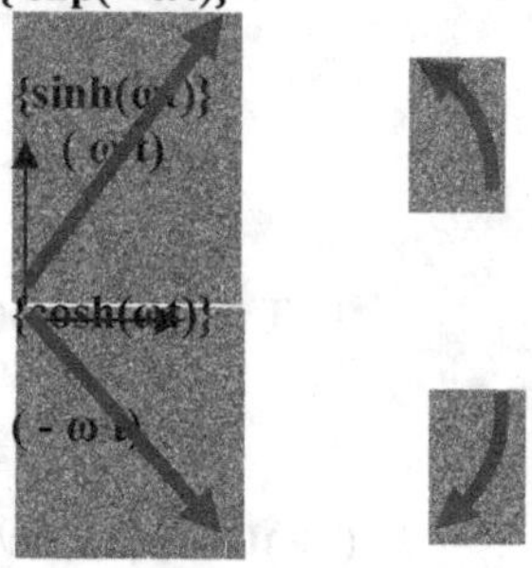

{ exp(- ωt)}

:

And:

$\cos(\alpha+\beta) = \cos(\alpha) \cos(\beta) - \sin(\alpha) \sin(\beta)$ $\sin(\alpha+\beta) = \sin(\alpha) \cos(\beta) + \cos(\alpha) \sin(\beta)$

$\cos(\alpha-\beta) = \cos(\alpha) \cos(\beta) + \sin(\alpha) \sin(\beta)$ $\sin(\alpha-\beta) = \sin(\alpha) \cos(\beta) - \cos(\alpha) \sin(\beta)$

Summed:

$\cos(\alpha + \beta) + \cos(\alpha-\beta) = 2 \cos(\alpha) \cos(\beta)$ $\sin(\alpha + \beta) + \sin(\alpha-\beta) = 2 \sin(\alpha) \cos(\beta)$

Subtracted:

$\cos(\alpha + \beta) - \cos(\alpha-\beta) = - 2 \sin(\alpha) \sin(\beta)$ $\sin(\alpha + \beta) - \sin(\alpha-\beta) = + 2 \cos(\alpha) \sin(\beta)$

Let:

$\gamma = (\alpha + \beta)/2$ $\delta = (\alpha-\beta)/2$

$[\gamma + \delta] = \alpha$ $[\gamma-\delta] = \beta$

So that:

$\cos(2\gamma)=\cos(\gamma + \delta)\cos(\gamma-\delta)-\sin(\gamma + \delta)\sin(\gamma-\delta)$ $\sin(2\gamma)=\sin(\gamma + \delta)\cos(\gamma-\delta)+\cos(\gamma + \delta)\sin(\gamma-\delta)$

$\cos(2\delta)=\cos(\gamma + \delta)\cos(\gamma-\delta)+\sin(\gamma + \delta)\sin(\gamma-\delta)$ $\sin(2\delta)=\sin(\gamma + \delta)\cos(\gamma-\delta)-\cos(\gamma + \delta)\sin[(\gamma-\delta)$

Summed:

$\cos(2\gamma) + \cos(2\delta) = 2 \cos(\gamma + \delta) \cos(\gamma-\delta)$ $\sin(2\gamma) + \sin(2\delta) = 2 \cos(\gamma + \delta) \sin(\gamma + \delta)$

Subtracted:

$\cos(2\gamma) - \cos(2\delta) = - 2 \sin(\gamma + \delta) \sin(\gamma-\delta)$ $\sin(2\gamma) - \sin(2\delta) = 2 \sin(\gamma + \delta) \cos(\gamma-\delta)$

And:

$\cos(2\theta) = 2\cos^2(\theta) - 1$ $\cos(2\theta)=\cos^2(\theta)-\sin^2(\theta)$ $1-2\sin^2(\theta) = \cos(2\theta)$

Divided by $\cos^2(\theta)$ and $\sin^2(\theta)$ respectively:

$\cos(2\theta)\cos^{-2}(\theta) = 2 - \cos^{-2}(\theta) = 1-\tan^2(\theta)$ $\tan^{-2}(\theta) - 1 = \sin^{-2}(\theta) - 2 = \cos(2\theta)\sin^{-2}(\theta)$

$[1 + \cos(2\theta)]/2=\cos^2(\theta)$	$[1-\cos(2\theta)]/2=\sin^2(\theta)$
$\cos(2\theta)\cos^{-2}(\theta) = 2 - \cos^{-2}(\theta)$	$\sin^{-2}(\theta) - 2 = \cos(2\theta)\sin^{-2}(\theta)$
$\cos(2\theta)\cos^{-2}(\theta) = 1 - \tan^2(\theta)$	$\tan^{-2}(\theta) - 1=\cos(2\theta)\sin^{-2}(\theta)$
$2- \cos^{-2}(\theta) = 1 - \tan^2(\theta)$	$\tan^{-2}(\theta) - 1= \sin^{-2}(\theta) - 2$
$1- \cos^{-2}(\theta) = - \tan^2(\theta)$	$\tan^{-2}(\theta) = \sin^{-2}(\theta) - 1$
$[1 + \tan^2(\theta)]^{-1} = \cos^2(\theta)$	$[1 + \tan^{-2}(\theta)]^{-1} = \sin^2(\theta)$

18.5.1.
Positive Exponential Forms:

$$U_\omega TT \equiv (-\omega^2) U \exp(+ j\omega t) = (-\omega^2) U_\omega = (+\omega^2) U \exp\{+i (+\pi+\omega t)\}$$

$$U_\omega T \equiv (i\omega) U \exp(+ j\omega t) = (+ j\omega) U_\omega = (+\omega) U \exp\{+i (+\pi/2+\omega t)\}$$

$$U_\omega \equiv U \exp(+ j\omega t)$$

$U_\omega T =(+\omega) U \exp\{+i (+\pi/2+\omega t)\}$

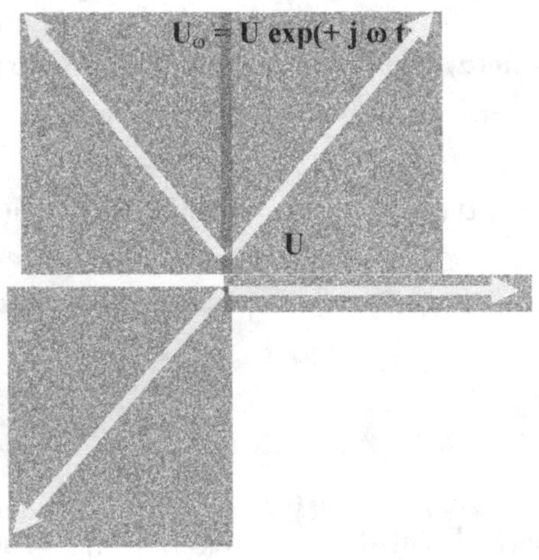

$U_\omega TT = (+\omega^2) U \exp\{+i (\pi+\omega t)\}$

18.5.2.

Negative Exponential Forms:

$$U_\omega TT \equiv (-\omega^2) U \exp(-i\omega t) = (-\omega^2)\, U_\omega = (+\omega^2)\, U \exp\{-i\,(+\pi+\omega t)\}$$

$$U_\omega T \equiv (i\omega) U \exp(-i\omega t) = (-j\,\omega)\, U_\omega = (+\omega) U \exp\{-i\,(+\pi/2+\omega t)\}$$

$$U_\omega \equiv U \exp(-i\omega t)$$

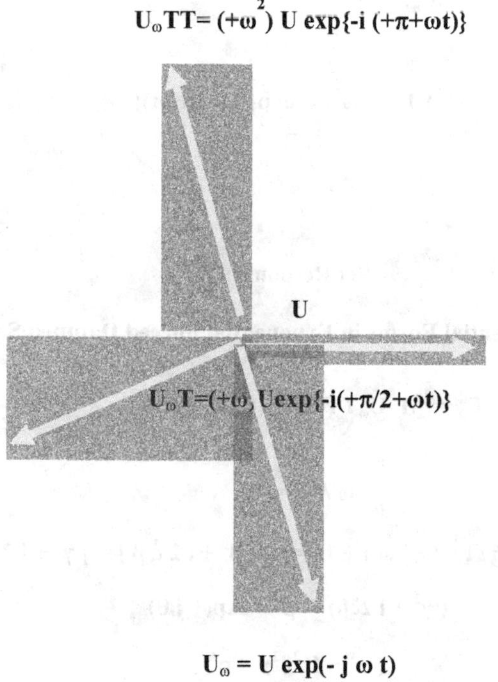

18.6.
Response of Forced Set:

Given Force Initial Value:

$F = F_0\ \exp(iwt_0)$ $\qquad\qquad$ $U = U_0\ \exp(iwt_0)$

18.6.1.
Positive Exponential Force Forms:

Let Set Response Given Exponential Force be in the Same Form:

$F_\omega = F\ \exp(+\ j\ \omega t)$ $\qquad$ $F_\omega = M\ U_\omega TT + C\ U_\omega T + K\ U_\omega$ $\qquad$ $U_\omega = U\ \exp(+\ j\ \omega t)]$

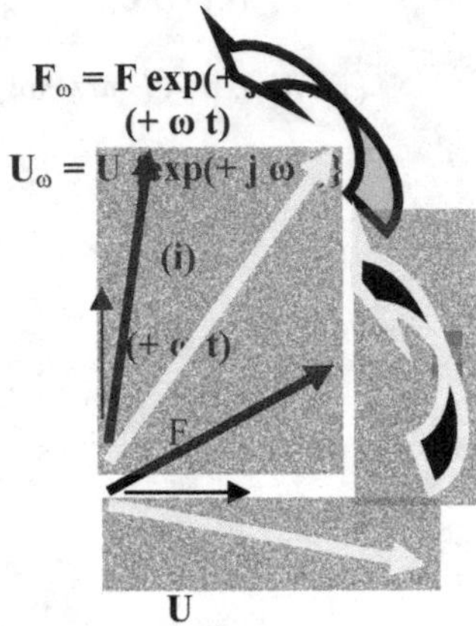

$$U_\omega TT = (+\omega^2)\ U\ \exp\{-i\ (+\pi+\omega t)\}$$

Set Response:

Substitute Exponential Forms in Exponential Forced Damped Set to Get:

$$F_\omega = M\ U_\omega TT + C\ U_\omega T + K\ U_\omega$$

$$\omega / \Omega = \beta$$

$$[-\Omega^{-2}\ \omega^2 + 2\,\xi\,\Omega^{-1}\ (+\ j\ \omega) + 1] = [\ 1 - \beta^2 + i\ 2\ \xi\ \beta\ ] = [\ \gamma^2 + i\ 2\ \xi\ \beta\ ]$$

$$(\ \gamma^2 + i\ 2\xi\beta) \equiv \rho^2\ \ \exp(+\ j\ \theta)$$

$$(\ \gamma^4 + 4\xi^2\beta^2) \equiv \rho^4$$

Then:

$$F = M\ U\ \{-\ \omega^2 + 2\ \xi\ \Omega^1\ (+\ j\ \omega) + \Omega^2\} = K\ U\ \{-\Omega^{-2}\ \omega^2 + 2\ \xi\ \Omega^{-1}\ (+\ j\ \omega) + 1\}$$

$$F = M\ U\ \Omega^2\ \rho^2 \exp(+\ j\ \theta) = K\ U\ \ \rho^2 \exp(+\ j\ \theta)$$

$$F / U = M\ \ \Omega^2\ \rho^2 \exp(+\ j\ \theta) = K\ \ \ \rho^2 \exp(+\ j\ \theta)$$

$$F / \{(U\ \rho^2\ \ \exp(+\ j\ \theta)\} = M\ \Omega^2\ = \ K$$

$$\alpha / \{ (U \rho^2 \exp(+ j \theta) \} = \Omega^2 = K / M$$

Also written:

$$\alpha D / \{ (U \exp(+ j \theta) \} = \Omega^2 = K / M$$

Where:

$$D \equiv 1 / \rho^2$$

:

Inverted to Get Response:

$$U_\omega = (1/M) F_\omega / [\rho^2 \Omega^2 \exp(+ j \theta)] = (1 /K) F_\omega / [\rho^2 \exp(+ j \theta)]$$

$$U = (1/M) F / [\rho^2 \Omega^2 \exp(+ j \theta)] = (1 /K) F / [\rho^2 \exp(+ j \theta)]$$

$$U / F = (1/M) / [\rho^2 \Omega^2 \exp(+ j \theta)] = (1 /K) / [\rho^2 \exp(+ j \theta)]$$

$$\{(U \rho^2 \exp(+ j \theta) \} / F = 1 / \{ M \Omega^2 \} = 1 / K$$

$$\{ (U \rho^2 \exp(+ j \theta) \} / \alpha = 1 / \Omega^2 = M / K$$

Also written:

$$U 1 / D \exp(+ j \theta) \} / \alpha = 1 / \Omega^2 = M / K$$

Where:

$$1 / D \equiv \rho^2$$

:

So:

$$\alpha / \{ (U \rho^2 \exp(+ j \theta) \} \equiv \alpha D / \{ (U \exp(+ j \theta) \} = \Omega^2 = K / M$$

$$\{ (U \rho^2 \exp(+ j \theta) \} / \alpha \equiv U 1 / D \exp(+ j \theta) \} / \alpha = 1 / \Omega^2 = M / K$$

:

Amplitude of Dynamic Stiffness (Power Resilience):

$D / (U) = 1 / \rho^2 / (U)$ **(-1)Angle θ**

Amplitude of Dynamic Flexibility (Form Resilience):

$(U)\ 1 / D = \rho^2\ (U)$ **Angle θ**

:

$$F = K U \{ -\Omega^{-2} \omega^2 + 2 \xi \Omega^{-1} (+ j \omega) + 1 \}$$

$$F = -\beta^2 K U + 2 \xi (+ j \beta) K U + K U$$

F = Inertia + Damping + Stiffness (Power Resilience)

F /M (Power Resilience) = Unit Mass Inertia + Unit Mass Damping + Unit Mass Stiffness

:

$F/M = \{\alpha\} = \Omega^2 \rho^2 \exp(+ j \theta) U$ $\quad$ $U= \{\alpha\}/[\Omega^2 \rho^2 \exp(+ j \theta)]= \exp(-i \theta) \{\alpha\}/[\Omega^2 \rho^2]$

:

Response:		Angles with Respect to	
		U	F
Displacement Response:	U	0	$-\theta$
Stiffness (Power Resilience):	K U	0	$-\theta$

Damping:	$2\,\xi\,\beta\;K\,U$	$+\pi/2$	$-\theta+\pi/2$
Inertia:	$-\beta^2\,K\,U$	0	$-\theta$
Inertia & Stiffness (Power Resilience):	$(-\beta^2+1)\,K\;U$	0	$-\theta$

Low Frequencies: $\beta^2 \leq 1$

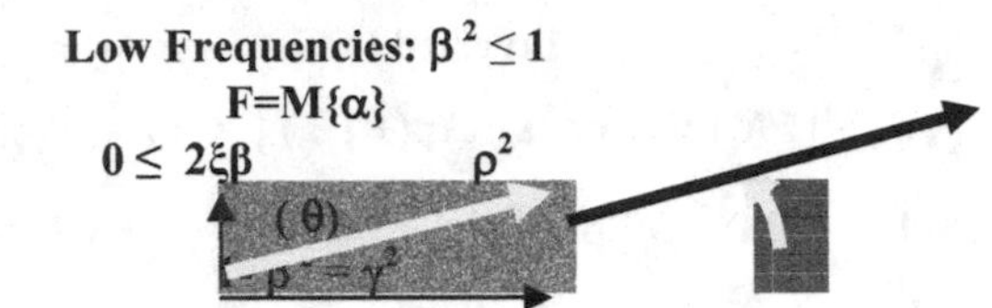

High Frequencies: $1 \leq \beta^2$

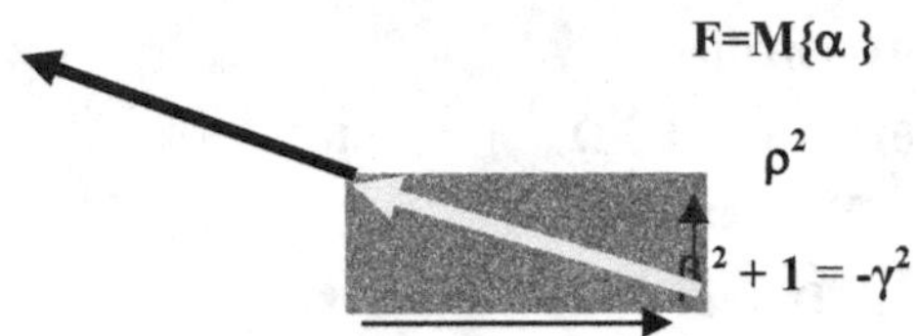

Seperated Forces :

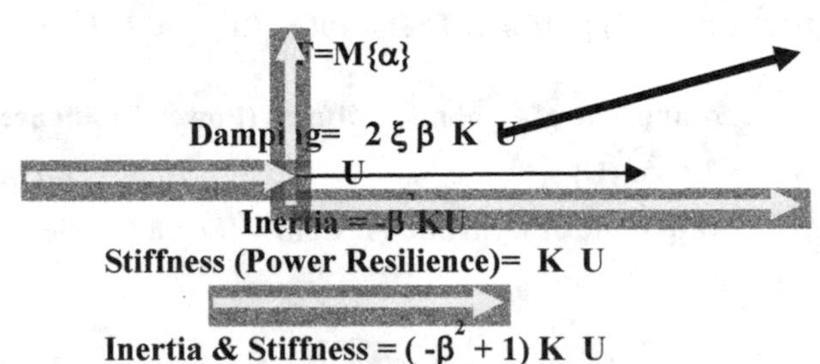

Stiffness (Power Resilience)= K U

Inertia & Stiffness = ($-\beta^2$ + 1) K U

18.6.2.
Negative Exponential Force Forms:

Let Set Response Given Exponential Force be in the Same Form:

$F_\omega = F\exp(-j\,\omega t)$ **$F_\omega = M\,U_\omega TT + C\,U_\omega T + K\,U_\omega$** **$U_\omega = U\exp(-j\,\omega t)]$**

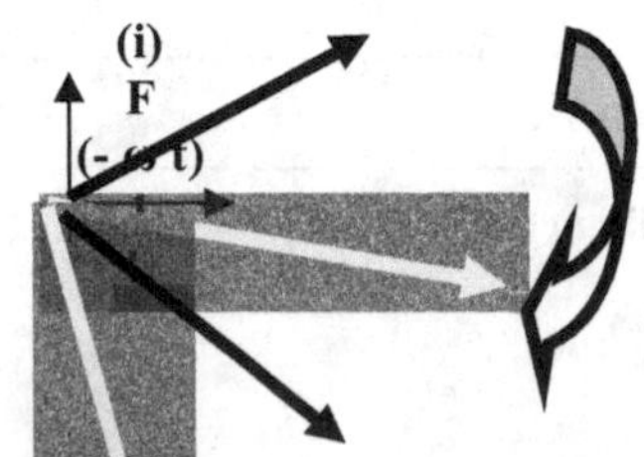

$F_{\omega} = F \exp(- j \omega t)$

(- ω t)

$U_{\omega} = U \ [\exp(- j \omega$

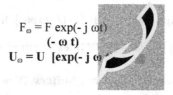

Set Response:

Substitute Exponential Forms in Exponential Forced Damped Set to Get:

$$F_{\omega} = M\ U_{\omega}TT + C\ U_{\omega}T + K\ U_{\omega}$$

$$\omega / \Omega = \beta$$

$$[-\Omega^{-2}\ \omega^2 + 2 \xi \Omega^{-1}\ (+ j \omega) + 1] = [1 - \beta^2 - i\, 2 \xi \beta] = [\gamma^2 - i\, 2 \xi \beta]$$

$$(\gamma^2 - i\, 2\xi\beta) \equiv \rho^2\ \exp(- j \theta)$$

$$(\gamma^4 + 4\xi^2\beta^2) \equiv \rho^4$$

Then:

$$F = M\ U\ \{ - \omega^2 + 2 \xi \Omega^{1} (- j \omega) + \Omega^{2} \} = K\ U\ \{ -\Omega^{-2}\ \omega^2 + 2 \xi \Omega^{-1} (- j \omega) + 1 \}$$

$$F = M\ U\ \Omega^{2}\ \rho^2 \exp(- j \theta) = K\, U\ \rho^2 \exp(- j \theta)$$

$$F / U = M\ \ \Omega^{2}\ \rho^2 \exp(- j \theta) = K\ \ \rho^2 \exp(- j \theta)$$

$$F / \{(U \rho^2\ \exp(- j \theta) \} = M\ \Omega^{2} = K$$

$$\alpha / \{ (U \rho^2\ \exp(- j \theta) \} = \Omega^{2} = K / M$$

Also written:

$$\alpha D / \{ (U\ \ \exp(- j \theta) \} = \Omega^{2} = K / M$$

Where:

$$D \equiv 1 / \rho^2$$

:

Inverted to Get Response:

$$U_{\omega} = (1/M) F_{\omega} / [\rho^2 \Omega^{2} \exp(- j \omega) \theta)] = (1 /K) F_{\omega} / [\rho^2 \exp(- j \theta)]$$

$$U = (1/M)\ F / [\rho^2 \Omega^{2} \exp(- j \theta)] = (1 /K)\ F / [\rho^2 \exp(- j \theta)]$$

$$U / F = (1/M)\ / [\rho^2 \Omega^{2} \exp(- j \theta)] = (1 /K)\ / [\rho^2 \exp(- j \theta)]$$

$$\{(U \rho^2\ \exp(- j \theta) \} /\ F = 1 / \{ M\ \Omega^{2} \} = 1 / K$$

$$\{ (U \rho^2\ \exp(- j \theta) \} / \alpha = 1 / \Omega^{2} = M / K$$

Also written:

$$U\, 1 / D\ \exp(- j \theta) \} / \alpha = 1 / \Omega^{2} = M / K$$

Where:

$$1 / D \equiv \rho^2$$

:

So:

$$\alpha / \{ (U \rho^2 \exp(-j\theta) \} \equiv \alpha D / \{ (U \exp(-j\theta) \} = \Omega^2 = K / M$$

$$\{ (U \rho^2 \exp(-j\theta) \} / \alpha \equiv U 1 / D \exp(-j\theta) \} / \alpha = 1 / \Omega^2 = M / K$$

:

Amplitude of Dynamic Stiffness (Power Resilience):

$D / (U) = 1 / \rho^2 / (U)$ **Angle θ**

Amplitude of Dynamic Flexibility (Form Resilience):

$(U)\ 1 / D = \rho^2\ (U)$ **(-1)Angle θ**

:

$$F = K\ U\ \{ -\Omega^{-2} \omega^2 + 2\xi \Omega^{-1} (-j\omega) + 1 \}$$

$$F = -\beta^2 K\ U + 2\xi\ (-j\beta)\ K\ U + K\ U$$

F = Inertia + Damping + Stiffness

F/M (Power Resilience) = Unit Mass Inertia + Unit Mass Damping + Unit Mass Stiffness

$F/M = \{\alpha\} = \Omega^2 \rho^2 \exp(-j\theta)\ U$ $U = \{\alpha\}/[\Omega^2 \rho^2 \exp(-j\theta)] = \exp(+i\theta)\ \{\alpha\}/[\Omega^2 \rho^2]$

:

Response:		Angles with Respect to	
		U	F
Displacement Response:	U	0	+θ
Stiffness (Power Resilience):	K U	0	+θ
Damping:	2 ξ β K U	-π/2	+θ - π/2
Inertia:	$-\beta^2$ K U	0	+θ
Inertia & Stiffness (Power Resilience):	$(-\beta^2 + 1)$ K U	0	+θ

Low Frequencies: $\beta^2 \leq 1$

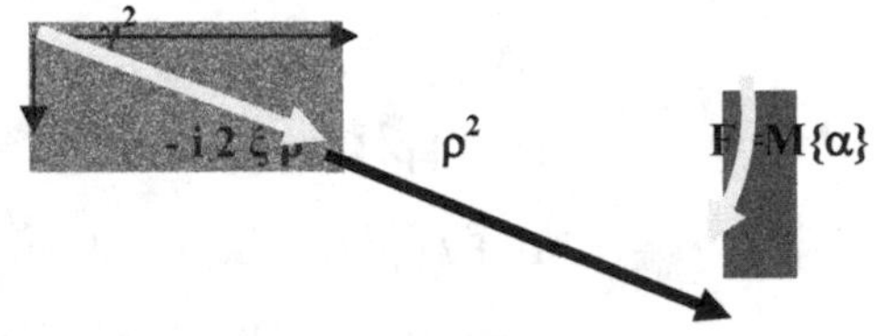

High Frequencies: $1 \leq \beta^2$

$-\beta^2 + 1 = -\gamma^2$

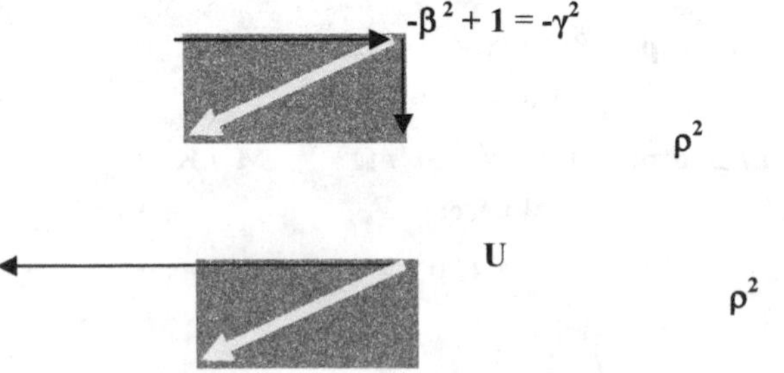

Seperated Forces :

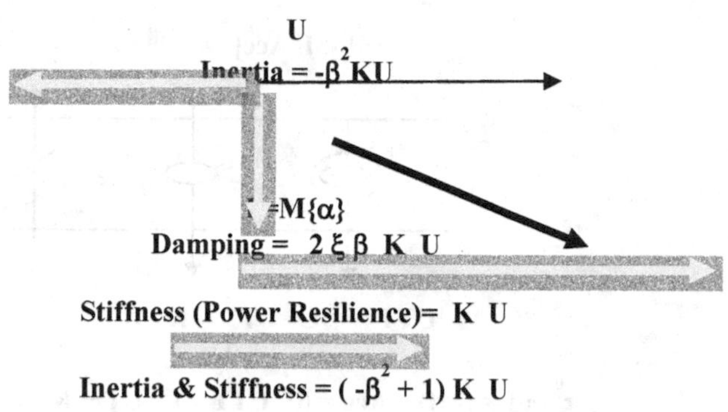

18.7. Structural Base Motion:

Given:

Displacement = Ground Displacement + Node Relative Displacement U

Velocity = Ground Velocity + Node Relative Velocity UTT

Acceleration = Ground Acceleration + Node Relative Acceleration UTT

Shaked Base by Acceleration is Normalized by (M):

$$(m/M) \quad movTT = (c/K)\, 2\xi\Omega^{+1} \quad velT = (k/K)\, \Omega^{+2} \quad Acc$$

$$=$$

$$U\,TT + 2\,\Omega_{\xi}\,UT + \Omega^{2}\,U$$

Ground Acceleration = movTT = velT = Acc

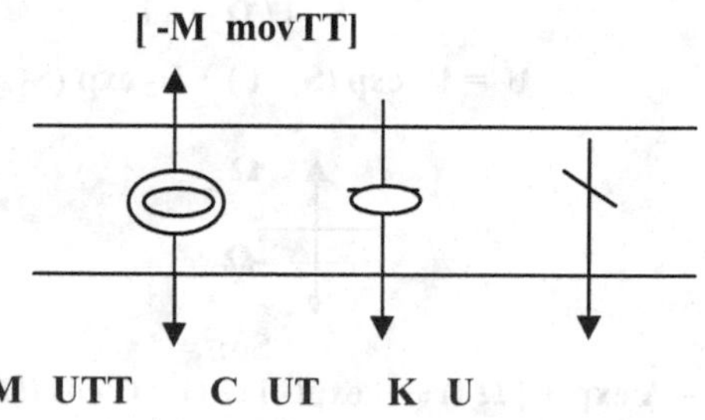

$$\tau^{2} \quad \tau TT = (-M) \quad movTT = M \;\; UTT + C \;\; UT + K \;\; U$$

[-M velT]

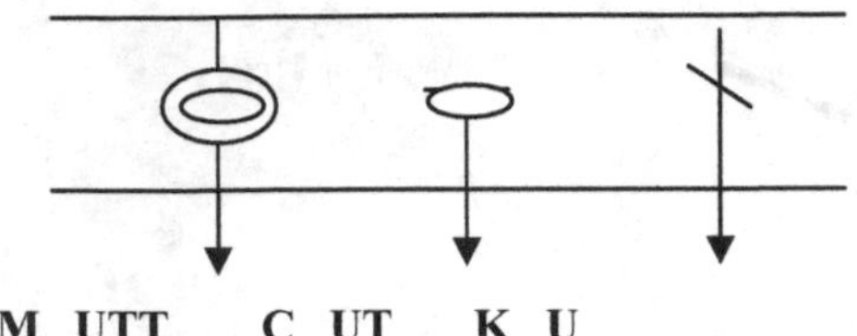

$$\tau^{1}\ \tau T = (\text{-}M)\ \ velT = M\ \ UTT + C\ \ UT + K\ \ U$$

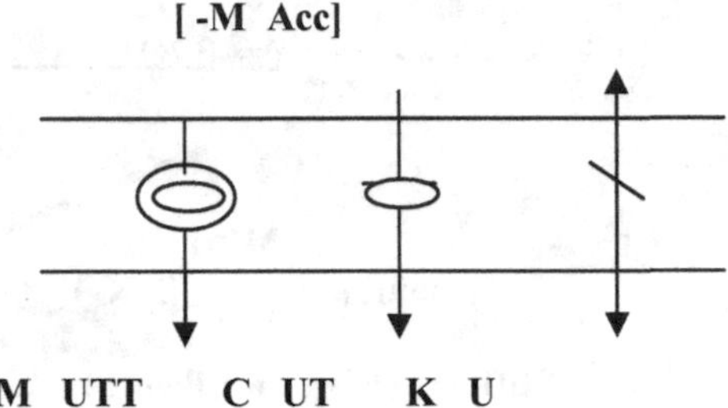

$$\tau^{0}\ \tau T = (\text{-}M)\ \ Acc = M\ \ UTT + C\ \ UT + K\ \ U$$

:

$$\{movTT\} = \{velT\} = \{Acc\} = UTT + 2\ \Omega_{\xi}\ UT + \Omega^{2}\ U$$

18.7.1.
Free Set:

18.7.1.1.
Free Undamped Set:

Free Base Movement: :

$$0 = M\ \ UTT + K\ \ U$$
$$0 = UTT + K/\ M\ \ U$$

Let :

$$(U) = A\ \ \exp(S\ \ t)$$

Then:

$$0 = (S\ S + \Omega^{2})\ \ \exp(S\ t)$$
$$0 = S\ S + \Omega^{2}$$

$$S_{1,2} = \pm\ i\ \Omega$$
$$S_{1,2}\ /\ \Omega = \pm\ i$$

$$U\ = U_{1}\exp(S_{1}\ \ t) + U_{2}\exp(S_{2}\ \ t)$$

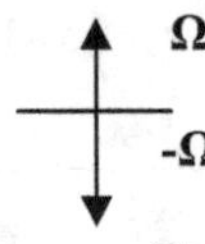

So:

$$U = A\ \exp(+\ j\ \Omega\ t) + B\ \exp(-\ j\ \Omega\ t) = C\ \cos(\Omega\ t) + D\ \sin(\Omega\ t)$$

AssUme Initial Zero Displacement:

$U = 0$ **at** $t=0$ **Then:** $C = 0$

So:

$$U = D\ \sin(\ \Omega\ t)$$

And:

UT = D Ω cos (Ω t)
AssUme Initial Zero Velocity:
U_0T at t=0 Then: U_0T = D Ω
So:
U = U_0T/Ω sin (Ω_η t)

Notice:

$$UTT = -\Omega^2 [C \cos(\Omega t) + D \sin(\Omega t)] = -\Omega^2 [A \exp(+ j \Omega t) + B \exp(- j \Omega t)]$$

Verifying:

$$0 = UTT + \Omega^2 U$$

18.7.1.2.
Free Damped Set:

Free Base Movement:

$$0 = M \ \ UTT + C \ \ UT + K \ \ U$$
$$0 = UTT + C/M \ \ UT + K/M \ \ U$$

Let :

$$(U) = A \ \ \exp(S \ \ t)$$

Then:

$$0 = (S S + 2 \xi \Omega \ \ S + \Omega^2) \ \ \exp(S t)$$
$$0 = S S + 2 \xi \Omega \ \ S + \Omega^2$$
$$S_{1,2} = - \xi \Omega \pm i \sqrt{(1- \xi^2)} \ \ \Omega$$
$$S_{1,2} = - \Omega_\xi \pm i \ \ \Omega_\eta$$

$$S_{1,2} / \Omega = -\xi \pm i\sqrt{(1-\xi^2)} = -\xi \pm i \eta$$

$$U = U_1 \exp(S_1 \ \ t) + U_2 \exp(S_2 \ \ t)$$

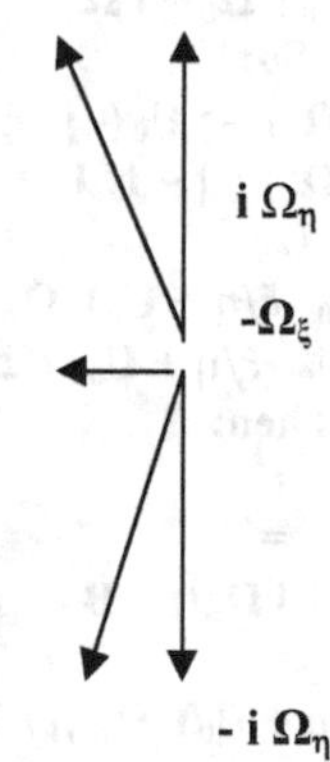

And:

$$U = A \ \ \exp[(- \xi \Omega + i \eta \Omega) t] + B \ \ \exp[(- \xi \Omega - i \eta \Omega) t]$$
$$U = \exp(-\Omega_\xi t) [C \cos(\Omega_\eta t) + D \sin(\Omega_\eta t)]$$

Assume Initial Zero Displacement:
U= 0 at t=0 Then: C =0
So:

$$U = D \ \ \exp(-\Omega_\xi t) \ \ \sin(\Omega_\eta t)$$

And:

$$U^\bullet = D \ \ \exp(-\Omega_\xi t) \ \ [(-\Omega_\eta) \ \ \sin(\Omega_\eta t) + \Omega_\eta \ \ \cos(\Omega_\eta t)]$$

Assume Initial Zero Velocity:

U_0T at t=0 Then: $U_0T = D \quad \Omega_\eta$

So:

$$U = U_0T/\Omega_\eta \quad \exp(-\Omega_\xi t) \quad \sin(\Omega_\eta t)$$

When $\xi = 0$ or $\eta = 1$ It becomes Free Undamped Set Where:

0 1 $\quad U = [A \quad \exp(+i\,\Omega\,t) + B \quad \exp(-i\Omega\,t)]$

ξ η $\quad U = \exp(-\Omega_\xi t) \quad [A \quad \exp(+i\,\Omega_\eta t) + B \quad \exp(-i\Omega_\eta t)]$

Or:

0 1 $\quad U = A \quad \exp[(+i\,\Omega)\,t] + B \quad \exp[(-i\,\Omega)\,t]$

ξ η $\quad U = A \quad \exp[(-\xi\,\Omega + i\,\eta\,\Omega)\,t] + B \quad \exp[(-\xi\,\Omega - i\,\eta\,\Omega)\,t]$

Velocity:

$$UT$$
$$=$$
$$(-\Omega_\xi) \quad \exp(-\Omega_\xi t) \quad [A \ \exp(+i\,\Omega_\eta t) + B \ \exp(-i\Omega_\eta t)]$$
$$+$$
$$\exp(-\Omega_\xi t) \quad [A \ (+i\,\Omega_\eta)\exp(+i\,\Omega_\eta t) + B \ (-i\Omega_\eta)\exp(-i\Omega_\eta t)]$$

$$UT/\exp(-\Omega_\xi\,t)$$
$$=$$
$$A\,[(-\Omega_\xi) \quad \exp(+i \quad \Omega_\eta\,t) + (+i \quad \Omega_\eta) \quad \exp(+i \quad \Omega_\eta\,t)]$$
$$+$$
$$B \quad [(-\Omega_\xi) \quad \exp(-i \quad \Omega_\eta\,t) + (-i \quad \Omega_\eta) \quad \exp(-i \quad \Omega_\eta\,t)]$$

Assume Initial Displacement:

$$U_0 = A + B$$

Assume Initial Zero Velocity:

$$U_0T = A\,(-\Omega_\xi + i\,\Omega_\eta) + B\,(-\Omega_\xi - i\,\Omega_\eta)$$

Determinants:

$$\begin{vmatrix} 1 & 1 \\ (-\Omega_\xi + i\,\Omega_\eta) & (-\Omega_\xi - i\,\Omega_\eta) \end{vmatrix} = (-2\,i\,\Omega_\eta)$$

$$\begin{vmatrix} U_0 & 1 \\ U_0T & (-\Omega_\xi - i\,\Omega_\eta) \end{vmatrix} = U_0\,(-\Omega_\xi - i\,\Omega_\eta) - U_0T$$

$$\begin{vmatrix} 1 & U_0 \\ (-\Omega_\xi + i\,\Omega_\eta) & U_0T \end{vmatrix} = U_0T - U_0\,(-\Omega_\xi + i\,\Omega_\eta)$$

So:

$$A = [U_0\,(-\Omega_\xi - i\,\Omega_\eta) - U_0T\,]/(-2i\,\Omega_\eta) = [\,U_0\,(-j \quad \xi/\eta + 1) - i \quad U_0T/\Omega_\eta\,]/2$$
$$B = [U_0T - U_0\,(-\Omega_\xi + i\,\Omega_\eta)]/(-2i\,\Omega_\eta) = [\,+i \quad U_0T/\Omega_\eta - U_0\,(-j \quad \xi/\eta - 1)]/2$$

Or:

$$A = [U_0 - i \quad (U_0 \quad \xi/\eta + U_0T/\Omega_\eta\,)]/2$$
$$B = [U_0 + i \quad (\,U_0 \quad \xi/\eta + U_0T/\Omega_\eta)]/2$$

Then:

$$U$$
$$=$$
$$\exp(-\Omega_\xi t)\,[A\,\exp(+i\,\Omega_\eta t) + B\,\exp(-i\Omega_\eta t)]$$
$$=$$
$$\exp(-\Omega_\xi t) \quad \{[U_0 - i \quad (U_0\,\xi/\eta + U_0T/\Omega_\eta\,)]\,/\,2 \quad \exp(+i\,\Omega_\eta t)$$
$$+$$
$$[U_0 + i \quad (\,U_0\,\xi/\eta + U_0T/\Omega_\eta)]/2 \quad \exp(-i\Omega_\eta t)\}$$

From Rest: If $U_0 = 0$ then:

$$U = \exp(-\Omega_\xi t)[-i \quad (U_0T/\Omega_\eta\,)/2\,\exp(+i\,\Omega_\eta t) + U_0T/\Omega_\eta)/2\,\exp(-i\Omega_\eta t)]$$
$$U = (U_0T/\Omega_\eta\,) \quad \exp(-\Omega_\xi t)[\,+\exp(-i\Omega_\eta t) - i\,\exp(+i\,\Omega_\eta t)]\,/2$$
$$U = (U_0T/\Omega_\eta\,) \quad \exp(-\Omega_\xi t)[\,+i\,\exp(-i\Omega_\eta t) + \exp(+i\,\Omega_\eta t)]\,/(2i)$$

That is:

$$U = (U_0T/\Omega_\eta\,)\,\exp(-\Omega_\xi t)\,\sin(\Omega_\eta t)$$

From Rest:

0 1 $\quad U = (U_0T/\Omega\,) \quad [\exp(+i\,\Omega\,t) + i\,\exp(-i\Omega\,t)]\,/(2i)$

ξ η $\quad U = (U_0T/\Omega_\eta\,) \quad \exp(-\Omega_\xi t)\,[\exp(+i\,\Omega_\eta t) + i\,\exp(-i\Omega_\eta t)]\,/(2i)$

Similarly:

$$U= \exp(-\Omega_\xi t) [C \cos(\Omega_\eta t) + D \sin(\Omega_\eta t)]$$

$$U_0T = (-\Omega_\xi) \exp(-\Omega_\xi t) [C \cos(\Omega_\eta t) + D \sin(\Omega_\eta t)] + (\Omega_\eta) \exp(-\Omega_\xi t) [-C \sin(\Omega_\eta t) + D \cos(\Omega_\eta t)]$$

$$U_0T = \exp(-\Omega_\xi t) \{(-\Omega_\xi) [C \cos(\Omega_\eta t) + D \sin(\Omega_\eta t)] + (\Omega_\eta) [-C \sin(\Omega_\eta t) + D \cos(\Omega_\eta t)]\}$$

$$U_0T = \exp(-\Omega_\xi t) \{C[(-\Omega_\xi) \cos(\Omega_\eta t) -(\Omega_\eta) \sin(\Omega_\eta t)] + D [(-\Omega_\xi) D \sin(\Omega_\eta t) + (\Omega_\eta) \cos(\Omega_\eta t)]\}$$

From Rest: If $U_0=0$ then:

$$C=0$$

So:

$$U= D \exp(-\Omega_\xi t) \sin(\Omega_\eta t)$$

And:

$$UT = D \exp(-\Omega_\xi t) [(-\Omega_\eta) \sin(\Omega_\eta t) + \Omega_\eta \cos(\Omega_\eta t)]$$

Initial Velocity:

U_0T at t=0 Then: $U_0T = D\,\Omega_\eta$

So From Rest:

0 1 $$U= (U_0T/\Omega) \sin(\Omega_\eta t)$$

ξ η $$U= (U_0T/\Omega_\eta) \exp(-\Omega_\xi t) \sin(\Omega_\eta t)$$

18.7.2.
Pulse Forced Set:

Assume at t=0:

$\{\dot{h}(0^-)\}=0$ $\{h(0^-)\} = \{h(0^+)\}$ $\int h(0)\, dt = 0$

18.7.2.1.
Pulse Forced Undamped Set:

Let a Forced Undamped Dynamic Pulse: $\delta(t=0)$:

$$\delta = \ddot{h} + \Omega^2 h$$

Then:

$$\int\delta\, dt = \int\ddot{h}\, dt + \Omega^2 \int h\, dt$$

Integrated from $t = 0^-$ To $t = 0^+$

$$1 = [\dot{h}(0^+) - \dot{h}(0^-)] + \Omega^2 \int h\, dt$$

$$1 = \dot{h}(0^+) + \Omega^2 (0)$$

Concluding:

Main Initial Condition:

$\{\dot{h}(0^+)\}=1$

0 < t Free Undamped Dynamic:

$$0 = + \ddot{h} + \Omega^2 h$$

So:

$$\{h(t)\} = \{\sin(\Omega\ \ t) / \Omega \}$$

With its Properties at t=0:

$\{\dot{h}(0^-)\}=0$ $\{\dot{h}(0^+)\}=1$ $\{h(0^-)\} = \{h(0^+)\}$ $\int h(0)\, dt = 0$

No Decay Term:

Single Term $\{\sin(\Omega t)\} = \{h(t)\}\ \Omega$ Unitless

Time Scaled $1/\Omega$ Single Unitless Term $\sin(\Omega t) / \Omega = \{h(t)\}$

18.7.2.2.

Pulse Forced Damped Set:

Let a Forced Damped Dynamic Pulse: $\delta(t=0)$:

$$\delta = \ddot{h} + 2\,\Omega_\xi\,\dot{h} + \Omega^2 h$$

Then:

$$\int\delta\, dt = \int\ddot{h}\, dt + 2\,\Omega_\xi \int\dot{h}\, dt + \Omega^2 \int h\, dt$$

Integrated from $t = 0^-$ To $t = 0^+$

$$1 = [\,\dot{h}(0^+) - \dot{h}(0^-)] + 2\,\Omega_\xi\,[\,h(0^+) - h(0^-)] + \Omega^2 \int h\, dt$$

$$1 = \dot{h}(0^+) + \Omega^2\,(0)$$

Concluding:

Main Initial Condition:

$\{\dot{h}(0^+)\}=1$

0 < t Free Damped Dynamic:

$$0 = +\ddot{h} + 2\,\Omega_\xi\,\dot{h} + \Omega^2 h$$

So:

$$\{h(t)\} = \exp(-\Omega_\xi t)\,\sin(\Omega_\eta t)\,/\,\Omega_\eta$$

With its Properties at t=0:

$\{\dot{h}(0^-)\}=0$ $\{\dot{h}(0^+)\}=1$ $\{h(0^-)\} = \{h(0^+)\}$ $\int h(0)\, dt = 0$

Damped Set Unitless Decay Term $[\exp(-\Omega_\xi\, t)]$:

Two Unitless Terms $\{\exp(-\Omega_\xi\, t)\}\,\{\sin(\Omega_\eta\, t)\} = \{h(t)\}\,\Omega$

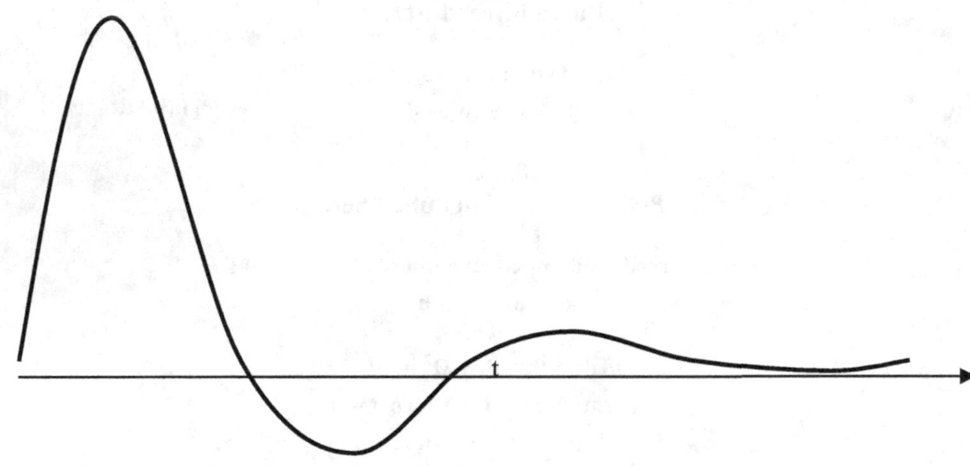

18.6.2.3.

Approximations:

Damped Set Unitless Decay Term $[\exp(-\Omega_\xi\, t)]$ is Not Approximated:

$\boldsymbol{\Omega_\xi}$

Approximated Terms:

Ω_η $\approx$ Approximate $\Omega_\eta \equiv \eta\,\Omega \approx \Omega$

$\{\sin[\Omega_\eta t]\}$	$\approx$	$\{\sin(\Omega t)\}$
$\{\exp[-\Omega_\xi t]\sin[\Omega_\eta t]/\Omega_\eta\}$	$=\{h(t)\}\approx$	$\{\exp(-\Omega_\xi t)\}\{\sin(\Omega t)\}/\Omega$
$\int\text{-}Acc(\tau)\exp[-\Omega_\xi(t-\tau)]\sin[\Omega_\eta(t-\tau)]/\Omega_\eta\,d\tau$	$=U\approx$	$\int\text{-}Acc(\tau)\exp(-\Omega_\xi(t-\tau))\}\sin(\Omega(t-\tau))/\Omega\,d\tau$
$\{-\Omega_\xi U+\int\text{-}Acc(\tau)\exp[-\Omega_\xi(t-\tau)]\cos[\Omega_\eta(t-\tau)]\,d\tau\}$	$=UT\approx$	$-\Omega_\xi U+\int\text{-}Acc(\tau)\exp(-\Omega_\xi(t-\tau))\cos(\Omega(t-\tau))\,d\tau$
Spectral Displacement		$U_\Omega=V_\Omega/\Omega=A_\Omega/\Omega^2$
Spectral Velocity		$\Omega U_\Omega=V_\Omega=A_\Omega/\Omega$
Spectral Acceleration		$\Omega^2U_\Omega=\Omega V_\Omega=A_\Omega$

18.7.3.
Convolution of Base Accelerated Set:

External Acceleration, {Acc(t)} is written:

$$\{Acc(t)\}=\int\{Acc(\tau)\}\ \ \delta(t-\tau)\,d\tau$$

And Set Response:

$$U(t)=\int\{Acc(\tau)\}\ \ h(t-\tau)\,d\tau$$

Accelerated Set:

$$\{Acc(t)\}=\int\{Acc(\tau)\}\ \ \delta(t-\tau)\,d\tau=U(t)TT+2\,\Omega_\xi\,U(t)\,T+\Omega^2\,U(t)$$

18.7.3.1.
Convolution of Base Accelerated of Undamped Set:

$$U_a(t)=\int\{Acc(\tau)\}/\Omega\ \ \sin[\Omega\,(t-\tau)]\,d\tau$$

Undamped Set

$$-Acc\ =UTT+\Omega^2\,U$$

Is written:

$$-Acc(t)=\int-Acc(\tau)\,\delta(t-\tau)\,d\tau=+U(t)TT+\Omega^2\,U(t)$$

Pulse at t- τ =0 that is at t = τ:

$$\delta(t-\tau)=U(t-\tau)TT+\Omega^2\,U(t-\tau)$$

:

Defined:

$$\{h(t-\tau)\}\equiv\{\{\exp[+i\,\Omega(t-\tau)]+i\ \exp[-i\Omega(t-\tau)]\}/(2i\,\Omega)$$

$$\{h(t-\tau)\}=\ \sin[\Omega_\eta\,(t-\tau)]\,/\,\Omega_\eta$$

Unit-wise:

$$[h(t-\tau)]=\ [1/\Omega_\eta]\ =T$$

Set Undamped Response:

$$U=\int-Acc(\tau)\,h(t-\tau)\,d\tau=\int-Acc(\tau)\sin[\Omega(t-\tau)]\,/\Omega\,d\tau$$

First Derivative, UT:

$$UT=\int-Acc(\tau)\ \ \cos\,[\Omega\,(t-\tau)\,d\tau$$

UTT + Acc:

$$UTT+Acc\ =-\ \ \Omega^2\,U$$

18.7.3.2.
Convolution of Base Acceleration of Damped Set:

$$U_a(t)=\int\{Acc(\tau)/\}\Omega_\eta\ \ \exp\,[-\Omega_\xi\,(t-\tau)]\,\sin\,[\Omega_\eta\,(t-\tau)]\,d\tau$$

Damped Set

Base Shaked by Acceleration (-Acc(t)) then Damped Set:

$$-Acc = UTT + 2\,\Omega_\xi\,UT + \Omega^2\,U$$

Is written:

$$-Acc(t) = \int -Acc(\tau)\,\delta(t-\tau)\,d\tau = U(t)TT + 2\,\Omega_\xi\,U(t)T + \Omega^2\,U(t)$$

:

Defined:

$$\{h(t-\tau)\} \equiv \{\exp[-\Omega_\xi(t-\tau)]\}\ \ \{\exp[+i\,\Omega_\eta(t-\tau)] + i\ \exp[-i\Omega_\eta(t-\tau)]\}/(2i\,\Omega_\eta)$$

$$\{h(t-\tau)\} \equiv \{\exp[(-\Omega_\xi + i\,\Omega_\eta)(t-\tau)] - \exp[(-\Omega_\xi - i\,\Omega_\eta)(t-\tau)]\}\ \ 1/(2i)\ \ \ 1/\Omega_\eta$$

$$\{h(t-\tau)\} = \{\exp[-\Omega_\xi(t-\tau)]\}\ \ \ \sin[\Omega_\eta(t-\tau)]\,/\,\Omega_\eta$$

Unit-wise:

$$[h(t-\tau)] = [\,1/\Omega_\eta] = T$$

:

Set Damped Response:

$$U = \int -Acc(\tau)\,h(t-\tau)\,d\tau = \int -Acc(\tau)\,\exp[-\,\Omega_\xi\,(t-\tau)]\,\sin[\Omega_\eta(t-\tau)]\,/\Omega_\eta\,d\tau$$

First Derivative, UT:

$$UT = -\Omega_\xi\,U + \int -Acc(\tau)\,\exp[-\Omega_\xi\,(t-\tau)]\,\cos[\Omega_\eta\,(t-\tau)]\,d\tau$$

$$UT = \int -Acc(\tau)\ \ \ \exp[-\,\Omega_\xi\,(t-\tau)]\ \ \{\Omega_\xi/\Omega_\eta\,\sin[\Omega_\eta\,(t-\tau)] - \cos[\Omega_\eta\,(t-\tau)]\,\}\,d\tau$$

$$UT = \int -Acc(\tau)\,\exp[-\,\Omega_\xi\,(t-\tau)]\,\{\xi/\eta\,\sin[\Omega_\eta\,(t-\tau)] - \cos[\Omega_\eta\,(t-\tau)]\,\}\,d\tau$$

$$UT = \int -Acc(\tau)\,\exp[-\,\Omega_\xi\,(t-\tau)]\,\sin[\Omega_\eta(t-\tau) - \arctan(\eta/\xi)]\,d\tau$$

UTT + Acc:

$$UTT + Acc = -\,2\,\Omega_\xi\,UTT\ -\ \Omega^2\,U$$

18.7.4.
Base Accelerated (Earthquake) Convolution Spectrals:

Convolution Spectral

$$A_\Omega = \Omega\,V_\Omega = \Omega^2\,U_\Omega$$

Convolution Spectrals of $-\,Acc(\tau) = UTT + 2\,\Omega_\xi\ \ UT + \Omega^2\ \ U\,)$

Spectral Displacement $A_\Omega/\Omega^2 = V_\Omega/\Omega = U_\Omega$	**TT[U]**	$U_\Omega\ (\xi,\Omega)$ = $1\,\{\int -Acc(\tau)\,H(t-\tau)\,d\tau\}_{max}/\,\Omega = 1\,V_\Omega\,/\,\Omega$
Spectral Velocity $A_\Omega/\Omega = V_\Omega = \Omega\,U_\Omega$	**T[U]**	$V_\Omega\,(\xi,\Omega) \approx \{UT + \Omega_\xi\,U\}_{max}$ = $1\,\{\int -Acc(\tau)\,H(t-\tau)\,d\tau\}_{max}/\,1 = 1\,V_\Omega\,/\,1$
Spectral Acceleration $A_\Omega = \Omega\,V_\Omega = \Omega^2\,U_\Omega$	**[U]**	$A_\Omega\ (\xi,\Omega)$ = $\Omega\,\{\int -Acc(\tau)\,H(t-\tau)\,d\tau\}_{max}/\,1 = \Omega\,V_\Omega\,/\,1$

:

Unit of U_Ω **is Equal Unit of U multiplied by Square of Time Unit:**

$$[U_\Omega] = [\,U\,]\ \ [\int H(t)\,dt\,/\,\Omega] = [\,U\,]\ \ \ TT$$

:

U_Ω or $T=2\pi/\Omega$ $\qquad$ V_Ω $\qquad$ A_Ω

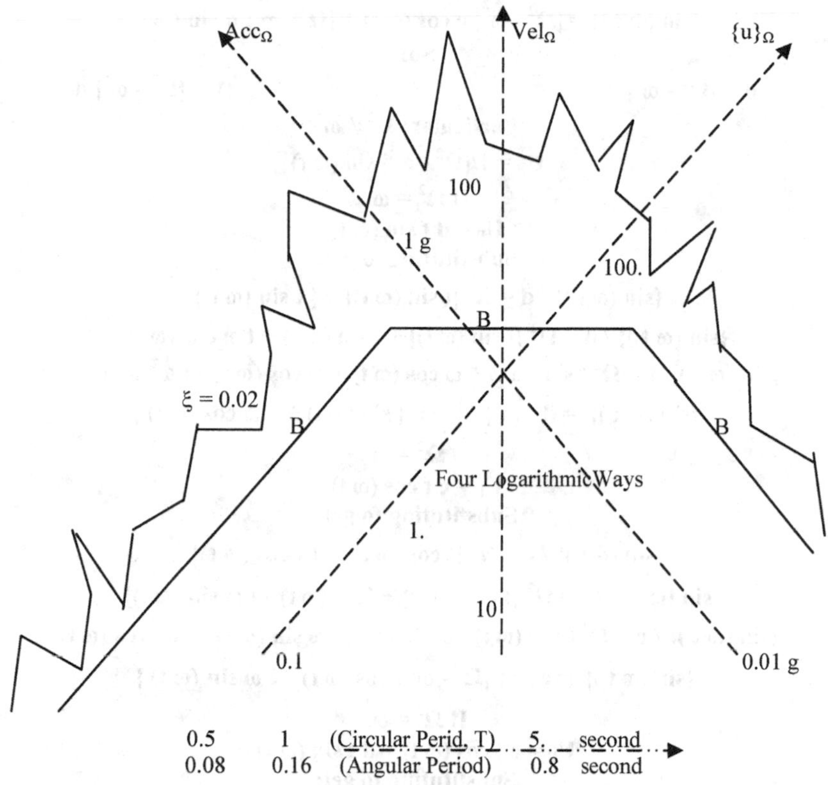

18.7.5.
Sinusoidally Forced Set:

18.7.5.1.
Sinusoidally Forced Undamped Set:

[1/TT]: $\alpha \sin(\omega t) = UTT + \Omega^2 U$

[1/TT] = [α]

Periodic Accelerated Set:

[1/T] : h(t- τ) = sin [Ω (t- τ)]/Ω

And:

0 1 **U(t)=∫ α sin (ω τ) {exp[+ i Ω (t- τ)] + iexp[-iΩ (t- τ)]}/(2i Ω) dτ**

0 1 **U(t) = ∫ α sin (ω τ) sin[Ω_η(t- τ)] /Ω d τ**

Unit-wise:

[1] = [1/TT] T T

Or:

Particular:

$$U_P = a \exp(+ j\, \omega\, t) + b \exp(- j\, \omega\, t) = c \cos(\omega t) + d \sin(\omega t)$$

Substituting:

$$\{\sin(\omega t)\} = \Omega^2 [c \cos(\omega t) + d \sin(\omega t)] + [c \cos(\omega t) + d \sin(\omega t)]^{\bullet\bullet}$$

$$\{\sin(\omega t)\} = [\Omega^2 - \omega^2]\, c \cos(\omega t) + [\Omega^2 - \omega^2]\, d \sin(\omega t)$$

So:

$$0 = [\Omega^2 - \omega^2]\, c \qquad\qquad 1 = [\Omega^2 - \omega^2]\, d$$

Particular: $\Omega^2 \neq \omega^2$

$$U_P = 1/[\Omega^2 - \omega^2] \sin(\omega t)$$

If $\Omega^2 = \omega^2$

$$U_P = d\, t \sin(\omega t)$$

Substituting to get:

$$\{\sin(\omega t)\} / d = \Omega^2 [t \sin(\omega t)] + [\, t \sin(\omega t)]^{\bullet\bullet}$$

$$\{\sin(\omega t)\} / d = \Omega^2 [t \sin(\omega t)] + [\, \sin(\omega t) + t\, \omega \cos(\omega t)]^{\bullet}$$

$$\{\sin(\omega t)\}\, d = \Omega^2\, t \sin(\omega t) + \omega \cos(\omega t) + \omega \cos(\omega t) - t\, \omega^2 \sin(\omega t)]$$

$$\{\sin(\omega t)\} = d \quad \{\, t\, [\Omega^2 - \omega^2\,] \sin(\omega t) + 2\, \omega \cos(\omega t)\, \}$$

If $\Omega^2 = \omega^2$

$$U_P = c\, t \cos(\omega t)$$

Substituting to get:

$$\{\sin(\omega t)\} / c = \Omega^2 [t \cos(\omega t)] + [\, t \cos(\omega t)]^{\bullet\bullet}$$

$$\{\sin(\omega t)\} / c = \Omega^2 [t \cos(\omega t)] + [\cos(\omega t) - t\, \omega \sin(\omega t)]^{\bullet}$$

$$\{\sin(\omega t)\} / c = \Omega^2\, t \cos(\omega t) - \omega \sin(\omega t) - \omega \sin(\omega t) - t\, \omega^2 \cos(\omega t)]$$

$$\{\sin(\omega t)\} = c \quad \{\, t\, [\Omega^2 - \omega^2\,] \cos(\omega t) - 2\, \omega \sin(\omega t)\, \}$$

If $\Omega^2 = \omega^2$

$$U_P = c\, t \cos(\omega t) + d\, t \sin(\omega t)$$

Substituting to get:

$$\{\sin(\omega t)\} = \{c \quad t\, [\Omega^2 - \omega^2\,] \cos(\omega t) - c \quad 2\, \omega \sin(\omega t)\, \} + \{\, d \quad t\, [\Omega^2 - \omega^2\,] \sin(\omega t) + d \quad 2\, \omega \cos(\omega t)\, \}$$

$$\{\sin(\omega t)\} = \{c \quad t\, [\Omega^2 - \omega^2\,] + d \quad 2\, \omega\, \} \cos(\omega t) + \{\, d \quad t\, [\Omega^2 - \omega^2\,] - c \quad 2\, \omega\, \} \sin(\omega t)$$

$$\{\sin(\omega t)\} = \{c \quad t\, [0\,] + d \quad 2\, \omega\, \} \cos(\omega t) + \{\, d \quad t\, [0] - c \quad 2\, \omega\, \} \sin(\omega t)$$

So:

$$0 = 0 + d \quad 2\omega \qquad\qquad 1 = -c \quad 2\omega$$

$$0 = d \qquad\qquad -1/(2\omega) = c$$

So If $\Omega^2 = \omega^2$

$$U_P = -1/(2\omega)\, t \cos(\omega t)$$

Verifying:

$$U_P T = -1/(2\omega)\, [\, \cos(\omega t) - t\, \omega \sin(\omega t)]$$

$$U_P TT = -1/(2\omega)\, [- \omega \sin(\omega t) - \omega \sin(\omega t) - t\, \omega^2 \cos(\omega t)]$$

So:

$$\sin(\omega t) = UTT + \Omega^2\, U$$

Has Particular Solution:

$$\Omega^2 \neq \omega^2 \qquad\qquad \Omega^2 = \omega^2$$

$$U_P = 1/[\Omega^2 - \omega^2] \sin(\omega t) \qquad\qquad U_P = -1/(2\omega)\, t \cos(\omega t)$$

Adding Homogeneous Solution:

If: $\Omega^2 \neq \omega^2$

$$U_s = 1/[\Omega^2 - \omega^2] \sin(\omega t) + C \cos(\Omega t) + D \sin(\Omega t)$$

If: $\Omega^2 = \omega^2$

$$U_s = -1/(2\omega)\, t \cos(\omega t) + C \cos(\Omega t) + D \sin(\Omega t)$$

Periodic Forced Undamped Dynamic:

$$\alpha \ \{\sin(\omega t)\} = + U(t)\, TT + \Omega^2 U(t)$$

Periodic Forced Dynamic:

$$h(t-\tau) = 1/\Omega \sin[\Omega(t-\tau)]$$

And:

$$U(t) = \alpha \int \sin(\omega \tau)\, 1/\Omega \sin[\Omega(t-\tau)] d\tau$$

18.7.5.2.
Sinusoidally Forced Damped Set:

[τTT]: $\{\alpha\} \sin(\omega t) = UTT + 2\,\Omega_\xi\, UT + \Omega^2 U$

Periodic Accelerated Set:

$$h(t-\tau) = 1/\Omega_\eta \exp[-\Omega_\xi(t-\tau)] \sin[\Omega_\eta(t-\tau)]$$

And:

$$U_s(t) = \int \{\alpha\} \sin(\omega\tau) \exp[-\Omega_\xi(t-\tau)] \sin[\Omega_\eta(t-\tau)]/\Omega_\eta\, d\tau$$

:

Particular:

$$U_{sp} = c \cos(\omega t) + d \sin(\omega t)$$

Substituted to get:

$$\{\alpha\} \sin(\omega t) = \ddot{[c \cos(\omega t) + d \sin(\omega t)]} + 2\,\Omega_\xi \dot{[c \cos(\omega t) + d \sin(\omega t)]} + \Omega^2 [c \cos(\omega t) + d \sin(\omega t)]$$

$$\sin(\omega t) = \{[\Omega^2 - \omega^2]\, c + 2\,\Omega_\xi \omega\, d\} \cos(\omega t) + \{-2\,\Omega_\xi\, \omega\, c + [\Omega^2 - \omega^2]\, d\} \sin(\omega t)$$

Then:

$$0 = [\Omega^2 - \omega^2]\, c + 2\,\Omega_\xi\, \omega\, d$$

$$\{\alpha\} = -2\,\Omega_\xi\, \omega\, c + [\Omega^2 - \omega^2]\, d$$

Determinants of:

$$\begin{vmatrix} [\Omega^2 - \omega^2] & +2\,\Omega_\xi\, \omega \\ -2\,\Omega_\xi\, \omega & [\Omega^2 - \omega^2] \end{vmatrix} \quad ([\Omega^2 - \omega^2]^2 + 4\,\Omega_\xi^2\, \omega^2)$$

:

$$\begin{vmatrix} 0 & +2\,\Omega_\xi\, \omega \\ \{\alpha\} & [\Omega^2 - \omega^2] \end{vmatrix} \quad (-2\,\Omega_\xi\, \omega)$$

:

$$\begin{vmatrix} [\Omega^2 - \omega^2] & 0 \\ -2\,\Omega_\xi\, \omega & \{\alpha\} \end{vmatrix} \quad [\Omega^2 - \omega^2]$$

So:

$$(c/\{\alpha\}) = -2\Omega_\xi \omega / \{[\Omega^2 - \omega^2]^2 + 4\,\Omega_\xi^2\, \omega^2\} \qquad (d/\alpha) = [\Omega^2 - \omega^2] / \{[\Omega^2 - \omega^2]^2 + 4\,\Omega_\xi^2\, \omega^2\}$$

Particular:

$$U_P = \{\alpha\} \{-2\,\Omega_\xi\, \omega \cos(\omega t) + [\Omega^2 - \omega^2] \sin(\omega t)\} \ / \ \{[\Omega^2 - \omega^2]^2 + 4\,\Omega_\xi^2\, \omega^2\}$$

When $\xi = 0$ or $\eta = 1$ It becomes Free Undamped Set Where:

$$U_P = \{\alpha\} \sin(\omega t) / [\Omega^2 - \omega^2]$$

$$U_p = \{\alpha\} \{-2\Omega_\xi\, \omega \cos(\omega t) + [\Omega^2 - \omega^2] \sin(\omega t)\}\,] \ / \ \{[\Omega^2 - \omega^2]^2 + 4\Omega_\xi^2\, \omega^2\}$$

If $\Omega^2 = \omega^2$ It is Just A case:

$$U_p = \{\alpha\} \{-2\Omega_\xi \omega \cos(\omega t)\} / \{4\Omega_\xi^2\, \omega^2\} = \{-\cos(\omega t)\} / \{2\Omega_\xi\, \omega\}$$

So:

$$\{\alpha\} \sin(\omega t) = U_d TT + 2\,\Omega_\xi\, U_d T + \Omega^2 U_d$$

It has Particular Solution:

$$U_{sp} = \{\alpha\} \{-2\Omega_\xi\omega \cos(\omega t) + [\Omega^2-\omega^2] \sin(\omega t)\} / \{[\Omega^2-\omega^2]^2 + 4\Omega_\xi^2\omega^2\}$$

Add Homogeneous:

$$U_0 = \{\alpha\} \exp(-\Omega_\xi t) [A \exp(+j\,\Omega_\eta t) + B \exp(-j\,\Omega_\eta t)]$$

So:

$$U_s = \{\alpha\} \{-2\Omega_\xi\omega \cos(\omega t) + [\Omega^2-\omega^2] \sin(\omega t)\} / \{[\Omega^2-\omega^2]^2 + 4\Omega_\xi^2\omega^2\} + \{\alpha\} \exp(-\Omega_\xi t) [A \exp(+j\,\Omega_\eta t) + B \exp(-j\,\Omega_\eta t)]$$

Then:

$$U_s = \{\alpha\} [-2\,\xi\,\beta \cos(\omega t) + (1-\beta^2) \sin(\omega t)] / [(1-\beta^2)^2 + 4\,\xi^2\,\beta^2] + \{\alpha\} \exp(-\Omega_\xi t) [A \exp(+j\,\Omega_\eta t) + B \exp(-j\,\Omega_\eta t)]$$

:

Applied to Periodic Force:

Damped Dynamic:

$$\{\alpha \sin(\omega t)\} = + U_DTT + 2\,\Omega_\xi\, U_DT + \Omega^2 U_D$$

Response to Periodic Forced Dynamic:

$$U(t) = \alpha \int \sin(\omega \tau)\, 1/\Omega_\eta \exp[-\Omega_\xi (t-\tau)] \sin[\Omega_\eta (t-\tau)] d\tau$$

Then:

$$U(t) = \alpha \int \sin(\omega \tau) * \exp[-\Omega_\xi (t-\tau)] * \sin[\Omega_\eta (t-\tau)]/\Omega_\eta\, d\tau$$

18.7.6.
Unit Constant Accelerated Set:

18.7.6.1.
Unit Constant Accelerad Undamped Set

$$1 = UTT + \Omega^2 U$$

:

Accelerated Set:

$$h(t-\tau) = 1/\Omega\, [\exp(-\Omega_\xi)(t-\tau) - \exp(-\Omega_\xi)(t-\tau)] / (2i)$$

And:

$$U(t) = \int \{\exp[(-\Omega_\xi)(t-\tau)] - \exp[(-\Omega_\xi)(t-\tau)]\} / (2i\,\Omega)\, d\tau$$

Let:

$A = -\omega^2 D \exp(+j\,\omega t)$ $\quad$ $V = (+j\,\omega) D \exp(+j\,\omega t)$ $\quad$ $U = D \exp(+j\,\omega t)$

Then:

$$1 = [-\omega^2 + \Omega^2] D \exp(+j\,\omega t)$$

$$1 = (1-\beta^2)\, \Omega^2 D \exp(+j\,\omega t)$$

$$1 = (1-\beta^2)\, \Omega^2 D \equiv (\gamma^2)\, \Omega^2 D \exp(+j\,\omega t)$$

$$[\Omega^2 D \exp(+j\,\omega t)]^{-1} = (1-\beta^2) \equiv (\gamma^2)$$

$$U = D \exp(+j\,\omega t) = (1-\beta^2)^{-1} (\Omega^2)^{-1} \equiv (\gamma^2)^{-1} (\Omega^2)^{-1}$$

$$U = [(\gamma^2)\, \Omega^2]^{-1}$$

$$[\Omega^2 D \exp(+j\,\omega t)]^{-1}$$

$$U = [(\gamma^2)\, \Omega^2]^{-1}$$

$$1-\beta^2 \equiv \gamma^2$$

Φ_D : Flexibility (Form Resilience):[displacement/force]:

$$\Phi_D = U \equiv 1/(\Omega^2 - \omega^2) \equiv 1/[\Omega^2(1-\beta^2)] \equiv 1/[\Omega^2(\gamma^2)]$$

18.7.6.2.

Unit Constant Accelerad Damped Set

$$1 = UTT + 2\,\Omega_\xi\, UT + \Omega^2\, U$$

:

Accelerated Set:

$$h(t-\tau) = 1/\Omega_\eta\,[\exp(-\Omega_\xi + i\,\Omega_\eta)(t-\tau) - \exp(-\Omega_\xi - i\,\Omega_\eta)(t-\tau)]\,/\,(2i)$$

And:

$$U(t) = \int\{\exp[(-\Omega_\xi + i\,\Omega_\eta)(t-\tau)] - \exp[(-\Omega_\xi - i\Omega_\eta)(t-\tau)]\}\,/\,(2i\,\Omega_\eta)\,d\tau$$

Let:

$A = -\,\omega^2\,D\exp(+\,j\,\omega\,t)$ $\quad$ $V = (+\,j\omega)\,D\exp(+\,j\,\omega\,t)$ $\quad$ $U = D\exp(+\,j\,\omega\,t)$

Then:

$$1 = [-\,\omega^2 + 2\,\xi\,\Omega\,(+\,j\,\omega) + \Omega^2]\,D\exp(+\,j\,\omega\,t)$$

$$1 = (1 + i\,2\,\xi\beta - \beta^2)\,\Omega^2\,D\exp(+\,j\,\omega\,t)$$

$$1 = (1-\beta^2 + i\,2\,\xi\beta)\,\Omega^2\,D \equiv (\gamma^2 + i\,2\xi\beta)\,\Omega^2\,D\exp(+\,j\,\omega\,t)$$

$$[\Omega^2\,D\,\exp(+\,j\,\omega\,t)]^{-1} = (1-\beta^2 + i\,2\,\xi\beta) \equiv (\gamma^2 + i\,2\xi\beta)$$

$$U = D\,\exp(+\,j\,\omega\,t) = (1-\beta^2 + i\,2\,\xi\beta)^{-1}\,(\Omega^2)^{-1} \equiv (\gamma^2 + i\,2\xi\beta)^{-1}\,(\Omega^2)^{-1}$$

$$U = [(\gamma^2 + i\,2\xi\beta)\,\Omega^2]^{-1}$$

2ξβ

$[\Omega^2\,D\,\exp(+\,j\,\omega\,t)]^{-1}$

$U = [(\gamma^2 + i\,2\xi\beta)\,\Omega^2]^{-1}$

$1-\beta^2 = \gamma^2$

Φ_D : Flexibility (Form Resilience):[displacement/force]:

$$\Phi_D = U \equiv 1/(\Omega^2 + i\,\Omega 2\xi\Omega - \omega^2) \equiv 1/[\Omega^2(1-\beta^2 + i2\xi\beta)] \equiv 1/[\Omega^2(\gamma^2 + i2\xi\beta)]$$

18.8.
Exponential Accelerated Set:

Given Exponential Accelerated Initial Value:

$\{\alpha\} = \{\alpha_0\}\ \exp(i\omega t_0)$ $\quad$ $U = U_0\ \exp(i\omega t_0)$

18.8.1.
Positive Exponential Accelerated Forms:

Let this Given Exponential Accelerated be in the Same Form:

$F_\omega \equiv M\ \alpha_\omega$ $\quad$ $F_\omega = M\ U_\omega TT + C\ U_\omega T + K\ U_\omega$ $\quad$ $U_\omega = U\exp(+\,j\,\omega t)]$

$F_\omega / M \equiv \alpha_\omega$ $\quad$ $\alpha_\omega = U_\omega TT + 2\,\Omega_\xi\,U_\omega T + \Omega^2\,U_\omega$ $\quad$ $\alpha_\omega = \{\alpha\}\exp(+\,j\omega t)$

$\alpha_\omega = \{\alpha\}\exp(+\,j\omega t)$

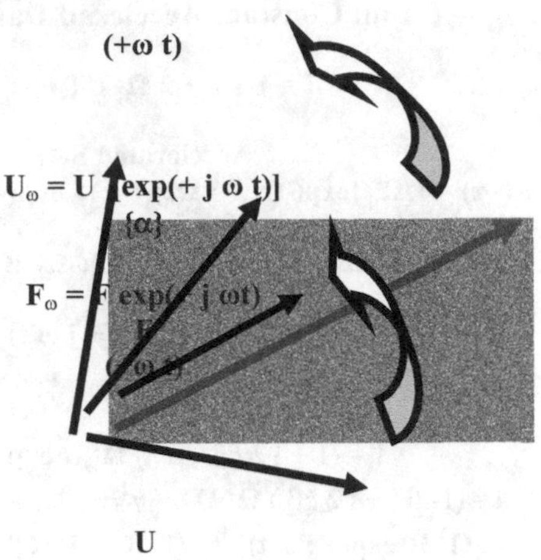

:

α_ω={α}exp(+ jωt) $U_\omega TT=-\omega^2 exp(+ j\omega t)U$ U_ω=(iω)exp(+ jωt)U U_ω=exp(+ jωt)U

18.8.1.1. Positive Exponential Accelerated Damped Set:

Substitute U_ω in Exponential Accelerated Damped Set to Get:

$$\{\alpha\}\ \exp(+ j\ \omega\ t) = U\ [\exp(+ j\ \omega\ t)]\ [- \omega^2 + 2\ \xi\ \Omega\ (+ j\ \omega\ ) + \Omega^2]$$

$$\{\alpha\}\ \exp(+ j\ \omega\ t) = U\ \exp(+ j\ \omega\ t)\ \ \Omega^2\ \rho^2\ \exp(+ j\ \theta)$$

$$\{\alpha_\omega\} = U_\omega\ \ \Omega^2\ \rho^2\ \exp(+ j\ \theta)$$

$$\{\alpha\} = U\ \ \Omega^2\ \rho^2\ \exp(+ j\ \theta)$$

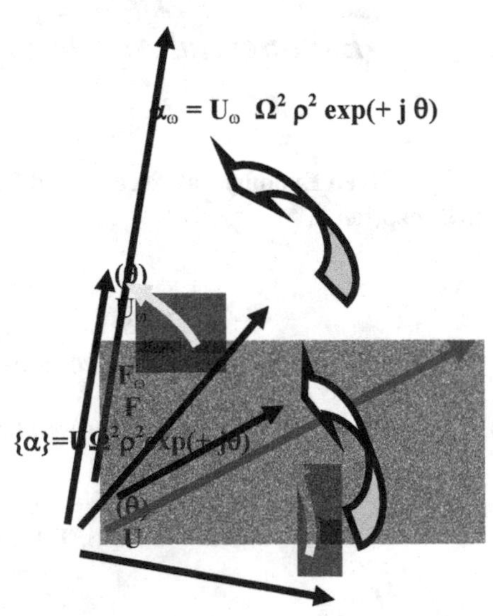

Where:

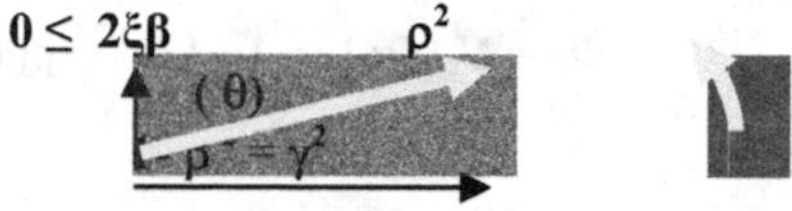

$$[- \omega^2 + 2\,\xi\,\Omega\,(+ j\,\omega) + \Omega^2] = [\Omega^2\,(1 - \beta^2 + i\,2\xi\beta)] = [\Omega^2\,(\gamma^2 + i\,2\xi\beta)]$$

$$(\gamma^2 + i\,2\xi\beta) / \rho^2 \equiv \exp(+ j\,\theta)$$

$$(\rho^4) \equiv (\gamma^4 + 4\,\xi^2\,\beta^2)$$

:

{α} is Inverted to Get Response:

$F_\omega \equiv M\ \alpha_\omega$ $\qquad$ $U_\omega = \alpha_\omega / [\Omega^2\,\rho^2 \exp(+ j\,\theta)]$

$F \equiv M\ \{\alpha\}$ $\qquad$ $U = \{\alpha\} / [\Omega^2\,\rho^2 \exp(+ j\,\theta)]$

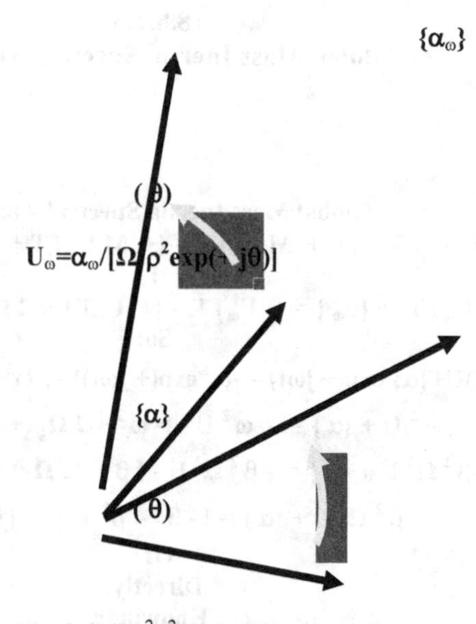

18.8.1.2.
Positive Exponential Accelerated Structure Base:

:

Exponential Accelerated Structure Base:

$$0 = +M\,\{\alpha_\omega\} + M\,U_\omega TT + M\,2\,\Omega_\xi\,U_\omega T + M\,\Omega^2\,U_\omega$$

$$+M\,\{\alpha_\omega\} = - \{+ M\,U_\omega TT + M\,2\,\Omega_\xi\,U_\omega T + M\,\Omega^2\,U_\omega\}$$

:

$$+ \{\alpha_\omega\} = - \{+ U_\omega TT + 2\,\Omega_\xi\,U_\omega T + \Omega^2\,U_\omega\}$$

So:

$$+\{\alpha\}\exp(+j\omega t) = -\{-\omega^2 + 2\,\Omega_\xi(+j\omega) + \Omega^2\}\exp(+j\omega t)\,U$$

$$+\{\alpha\} = -\{-\beta^2 + 2\,\Omega_\xi(+j\beta) + 1\}\,\Omega^2\,U$$

$$+\{\alpha\} = -\{\Omega^2\,\rho^2\exp(+j\theta)\}\,U$$

And:

$$\Gamma_U = \{\alpha\}/U = -\Omega^2\,\rho^2\exp(+j\theta)$$

$$\Phi_U = U/\{\alpha\} = -1/\{\Omega^2\exp(+j\theta)\}$$

Low Frequencies: $\beta^2 \le 1$

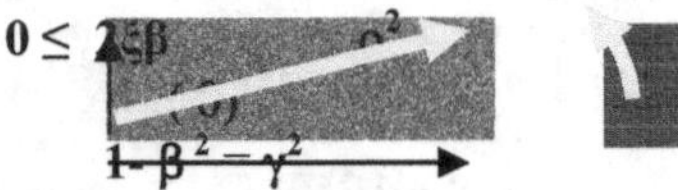

High Frequencies: $1 \le \beta^2$

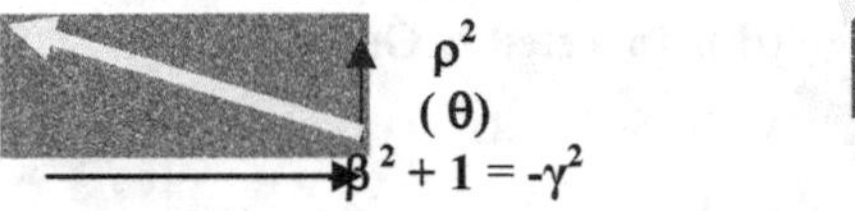

18.8.1.3. Global Mass Inertia Spectral Analysis:

:

Global Mass Inertia Spectral Analysis:

$$+\ M\,U_\omega TT\ + M\{\alpha_\omega\} = +\ M\,U_\omega TT\ - \{+\ M\,U_\omega TT + M\,2\,\Omega_\xi\,U_\omega T + M\,\Omega^2\,U_\omega\}$$

:

$$+\ U_\omega TT\ + \{\alpha_\omega\} = +\ U_\omega TT\ - \{+\ U_\omega TT + 2\,\Omega_\xi\,U_\omega T + \Omega^2\,U_\omega\}$$

So:

$$-\omega^2\exp(+j\omega t)U + \{\alpha\}\exp(+j\omega t) = -\omega^2\exp(+j\omega t)U - \{-\omega^2 + 2\Omega_\xi(+j\omega) + \Omega^2\}\exp(+j\omega t)U$$

$$-\omega^2\,U + \{\alpha\} = -\ \omega^2\,U - \{-\omega^2 + 2\,\Omega_\xi(+j\omega) + \Omega^2\}\,U$$

$$-\beta^2\,\Omega^2\,U + \{\alpha\} = -\ \beta^2\,\Omega^2\,U - \{\beta^2 + 2\,\Omega_\xi(+j\beta) + 1\}\,\Omega^2\,U$$

$$-\beta^2\,\Omega^2\,U + \{\alpha\} = \{-\beta^2 - \rho^2\exp(+j\theta)\}\,\Omega^2\,U$$

Or:

Directly:

Knowing:

$$\{\alpha\} = -\ \{\rho^2\exp(+j\theta)\}\,\Omega^2\,U$$

Adding to Both Sides:

$$-\beta^2\,\Omega^2\,U$$

To Get:

$$-\beta^2\,\Omega^2\,U + \{\alpha\} = \{-\beta^2 - \rho^2\exp(+j\theta)\}\,\Omega^2\,U$$

Define:

$$\{+\beta^2 + \rho^2\exp(+j\theta)\} = +\mu^2\exp(+j\iota)$$

So:

$$-\beta^2\,\Omega^2\,U + \{\alpha\} = -\ \mu^2\exp(+j\iota)$$

Low Frequencies: $\beta^2 \leq 1$

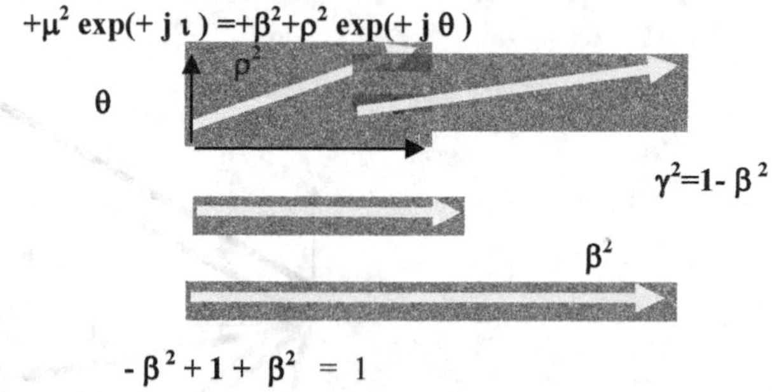

High Frequencies: $1 \leq \beta^2$

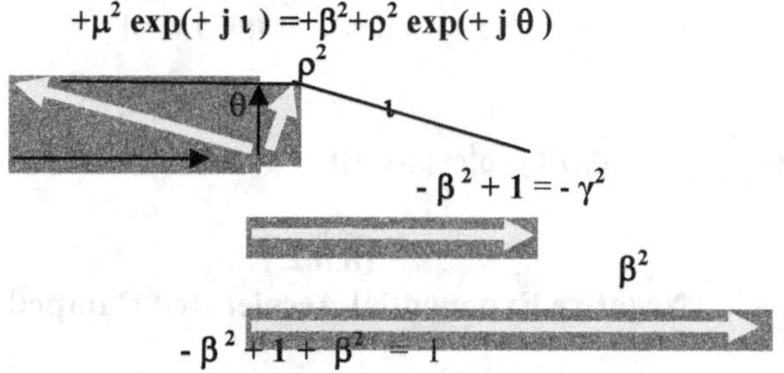

Knowing:

Positive Exponent

$$-\beta^2 \Omega^2 U + \{\alpha\} = - \mu^2 \exp(+ j \iota)$$

Stiffness (Power Resilience):

$$\Gamma_M = \{\alpha\} / \{-\beta^2 \Omega^2 U + \{\alpha\}\} = - \{\alpha\} / \{\mu^2 \exp(+ j \iota) \ \Omega^2\}$$

Flexibility (Form Resilience):

$$\Phi_M = \{-\beta^2 \Omega^2 U + \{\alpha\}\} / \{\alpha\} / = - \{\mu^2 \exp(+ j \iota) \ \Omega^2\} / \{\alpha\}$$

Unitless Global Mass Inertia Stiffness (Power Resilience) Spectral:

$$\Gamma_M \bullet \Omega^2 / \{\alpha\} = - 1 / \{\mu^2 \exp(+ j \iota)\}$$

Unitless Global Mass Inertia Flexibility (Form Resilience) Spectral:

$$\Phi_M \bullet \{\alpha\} / \Omega^2 = - \mu^2 \exp(+ j \iota)$$

Unitary Global Mass Inertia Stiffness (Power Resilience) Spectral:

$$\Gamma_M \bullet \Omega^2 \mu^2 / \{\alpha\} = - 1 / \{\exp(+ j \iota)\}$$

Unitary Global Mass Inertia Flexibility (Form Resilience) Spectral:

$$\Phi_M \bullet \{\alpha\} / \{\mu^2 \Omega^2\} = - \exp(+ j \iota)$$

Global Mass Inertia Stiffness (Power Resilience):

$$\Gamma_M = 1/ \Phi_M = \alpha \ / \{-\beta^2 \Omega^2 U + \{\alpha\}\} = - \{\alpha\} / \{\mu^2 \exp(+ j \iota) \ \Omega^2\}$$

Square of Global Mass Inertia Stiffness (Power Resilience):

$$S_M(\beta) = \Gamma_M^2 = 1/\Phi_M^2 = \alpha^2 / \{-\beta^2 \Omega^2 U + \alpha\}^2 = \alpha^2 / \{\mu^4 \exp(+j 2 \iota) \ \Omega^4\}$$

18.8.2.
Negative Exponential Accelerated Forms:

Let this Given Exponential Acceleration be in the Same Form:

$F_\omega \equiv M \ \alpha_\omega$	$F_\omega = M \ U_\omega TT + C \ U_\omega T + K \ U_\omega$	$U_\omega = U \exp(- j \ \omega t)]$
$F_\omega / M \equiv \alpha_\omega$	$\alpha_\omega = U_\omega TT + 2 \Omega_\xi U_\omega T + \Omega^2 U_\omega$	$\alpha_\omega = \{\alpha\}\exp(-i\omega t)$

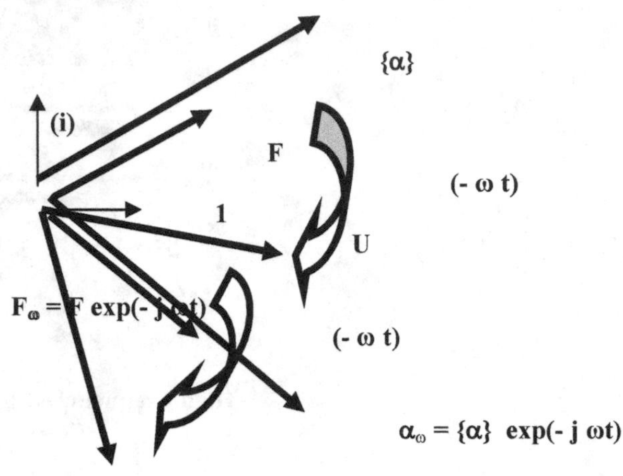

$\alpha_\omega = \{\alpha\}\exp(-i\omega t)$ $\quad U_\omega TT = -\omega^2 \exp(-i\omega t)U$ $\quad U_\omega = (i\omega)\exp(-i\omega t)U$ $\quad U_\omega = \exp(-i\omega t)U$

18.8.2.1.
Negative Exponential Accelerated Damped Set:

Substitute U_ω in Exponential Accelerated Damped Set to Get:

$$\{\alpha\} \ \exp(- j \ \omega \ t) = U \ [\exp(- j \ \omega \ t)] \ [- \omega^2 + 2 \xi \Omega (- j \omega) + \Omega^2]$$

$$\{\alpha\} \exp(- j \ \omega \ t) = U \ \exp(- j \ \omega \ t) \ \Omega^2 \rho^2 \exp(- j \ \theta)$$

$$\{\alpha_\omega\} = U_\omega \ \ \Omega^2 \rho^2 \exp(- j \ \theta)$$

$$\{\alpha\} = U \ \ \Omega^2 \rho^2 \exp(- j \ \theta)$$

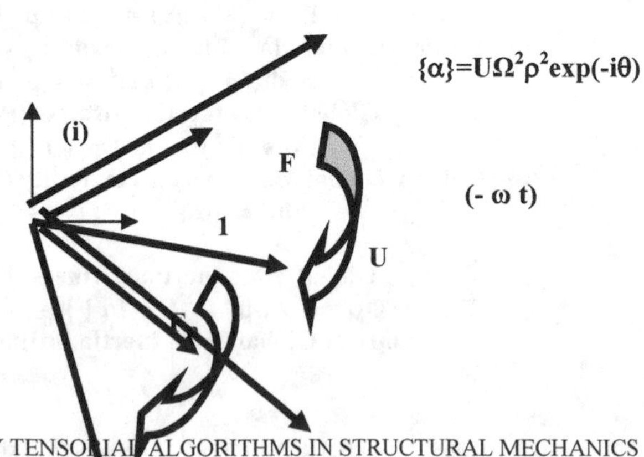

(- ω t)

$$\{\alpha_\omega\}=U_\omega\ \Omega^2\ \rho^2\ \exp(-i\theta)$$

U_ω

U_ω

Where:

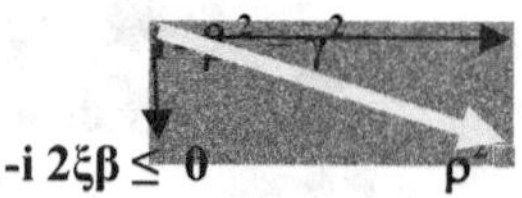

$$[\,-\omega^2 + 2\ \xi\ \Omega\ (-\ j\ \omega\ ) + \Omega^2] = [\Omega^2\ (1 - \beta^2 - i\ 2\xi\beta)] = [\Omega^2\ (\ \gamma^2 - i\ 2\xi\beta)]$$

$$(\ \gamma^2 - i\ 2\xi\beta)\ /\ \rho^2 \equiv\ \exp(-\ j\ \theta)$$

$$(\rho^4) \equiv\ (\ \gamma^4 + 4\ \xi^2\ \beta^2\ )$$

{α} is Inverted to Get Response:

$F_\omega \equiv M\ \alpha_\omega$ $\qquad U_\omega = \alpha_\omega\ /\ [\Omega^2\ \rho^2\ \exp(-\ j\ \theta\ )\ ]$

$F \equiv M\ \{\alpha\}$ $\qquad U\ = \{\alpha\}\ /\ [\Omega^2\ \rho^2\ \exp(-\ j\ \theta\ )\ ]$

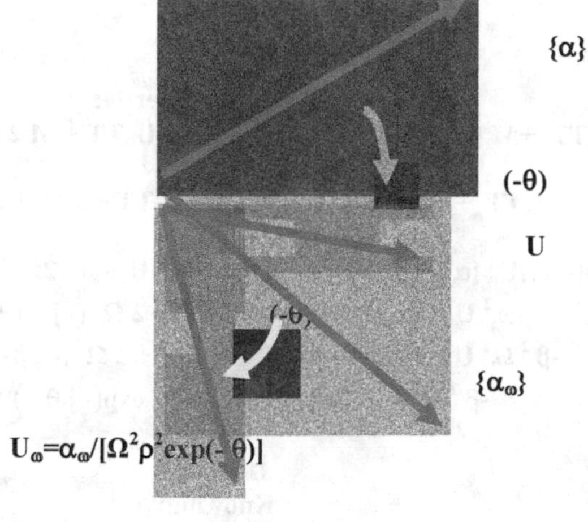

18.8.2.2.
Negative Exponential Accelerated Forms:

Exponential Accelerated Structure Base:

$$0 = +M\,\{\alpha_\omega\} + M\,U_\omega TT + M\,2\,\Omega_\xi\,U_\omega T + M\,\Omega^2\,U_\omega$$

$$+M\,\{\alpha_\omega\} = -\{+ M\,U_\omega TT + M\,2\,\Omega_\xi\,U_\omega T + M\,\Omega^2\,U_\omega\}$$

:

$$+\{\alpha_\omega\} = -\{+ U_\omega TT + 2\,\Omega_\xi\,U_\omega T + \Omega^2\,U_\omega\}$$

So:

$$+\{\alpha\}\exp(-j\,\omega t) = -\{-\omega^2 + 2\,\Omega_\xi(-j\,\omega) + \Omega^2\}\exp(-j\,\omega t)\,U$$

$$+\{\alpha\} = -\{-\beta^2 + 2\,\Omega_\xi(-j\,\beta) + 1\}\,\Omega^2\,U$$

$$+\{\alpha\} = -\{\rho^2\;\Omega^2\exp(-j\,\theta)\}\,U$$

And:

$$\Gamma_U = \{\alpha\}/U = -\Omega^2\;\rho^2\exp(-j\,\theta)$$

$$\Phi_U = U/\{\alpha\} = -1/\{\Omega^2\;\rho^2\exp(-j\,\theta)\}$$

Low Frequencies: $\beta^2 \leq 1$

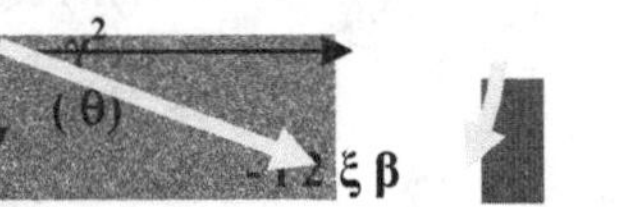

High Frequencies: $1 \leq \beta^2$

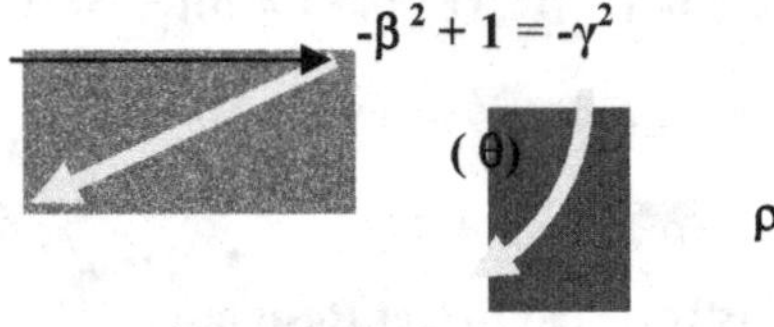

18.8.2.3.
Global Mass Inertia Spectral Analysis:

:

Also:

Global Mass Inertia:

$$+M\,U_\omega TT + M\,\{\alpha_\omega\} = +M\,U_\omega TT - \{+ M\,U_\omega TT + M\,2\,\Omega_\xi\,U_\omega T + M\,\Omega^2\,U_\omega\}$$

:

$$+U_\omega TT + \{\alpha_\omega\} = +U_\omega TT - \{+ U_\omega TT + 2\,\Omega_\xi\,U_\omega T + \Omega^2\,U_\omega\}$$

So:

$$-\omega^2\exp(-i\omega t)U + \{\alpha\}\exp(-i\omega t) = -\omega^2\exp(-i\omega t)U - \{-\omega^2 + 2\Omega_\xi(-i\omega) + \Omega^2\}\exp(-i\omega t)U$$

$$-\omega^2\,U + \{\alpha\} = -\omega^2\,U - \{-\omega^2 + 2\,\Omega_\xi(-j\,\omega) + \Omega^2\}\,U$$

$$-\beta^2\,\Omega^2\,U + \{\alpha\} = -\beta^2\,\Omega^2\,U - \{\beta^2 + 2\,\Omega_\xi(-j\,\beta) + 1\}\,\Omega^2\,U$$

$$-\beta^2\,\Omega^2\,U + \{\alpha\} = \{-\beta^2 - \rho^2\exp(-j\,\theta)\}\,\Omega^2\,U$$

Or:

Directly:

Knowing:

$$\{\alpha\} = -\{\rho^2\exp(-j\,\theta)\}\,\Omega^2\,U$$

Adding to Both Sides:

$$-\beta^2\,\Omega^2\,U$$

To Get:

$$-\beta^2\,\Omega^2\,U + \{\alpha\} = \{-\beta^2 - \rho^2\exp(-j\,\theta)\}\,\Omega^2\,U$$

Define:

$$\{+\beta^2 + \rho^2\exp(-j\,\theta)\} = +\mu^2\exp(-j\,\iota)$$

So:

$$-\beta^2 \Omega^2 U + \{\alpha\} = -\mu^2 \exp(-j\iota)$$

Low Frequencies: $\beta^2 \leq 1$

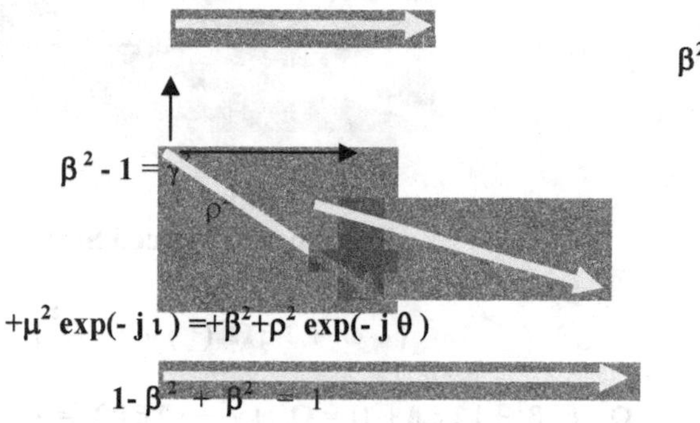

High Frequencies: $1 \leq \beta^2$

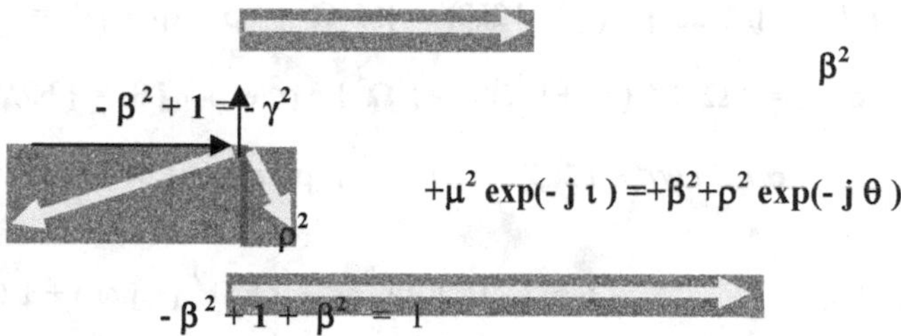

Knowing:

Negative Exponent

$$-\beta^2 \Omega^2 U + \{\alpha\} = -\mu^2 \exp(-j\iota)$$

Stiffness (Power Resilience):

$$\Gamma_M = \{\alpha\} / \{-\beta^2 \Omega^2 U + \{\alpha\}\} = -\{\alpha\} / \{\mu^2 \exp(-j\iota)\ \Omega^2\}$$

Flexibility (Form Resilience):

$$\Phi_M = \{-\beta^2 \Omega^2 U + \{\alpha\}\} / \{\alpha\} / = -\{\mu^2 \exp(-j\iota)\ \Omega^2\} / \{\alpha\}$$

Unitless Global Mass Inertia Stiffness (Power Resilience) Spectral:

$$\Gamma_M \bullet \Omega^2 / \{\alpha\} = -1 / \{\mu^2 \exp(-j\iota)\}$$

Unitless Global Mass Inertia Flexibility (Form Resilience) Spectral:

$$\Phi_M \bullet \{\alpha\} / \Omega^2 = -\mu^2 \exp(-j\iota)$$

Unitary Global Mass Inertia Stiffness (Power Resilience) Spectral:

$$\Gamma_M \bullet \Omega^2 \mu^2 / \{\alpha\} = -1 / \{\exp(-j\iota)\}$$

Unitary Global Mass Inertia Flexibility (Form Resilience) Spectral:

$$\Phi_M \bullet \{\alpha\} / \{\mu^2 \Omega^2\} = -\exp(-j\iota)$$

18.8.3. Stiffness (Power Resilience) And Flexibility (Form Resilience) Spectrals of Exponential Accelerated Set:

Large Frequencies: $1 \leq \beta^2$

$\pi / 2 \leq \theta \leq \pi$

Small Frequencies: $\beta^2 \leq 1$

$0 \leq \theta \leq \pi / 2$

:

Phase Difference Angles with Respect to F:

:

18.8.3.1.
Response of Forced Set:

$$\{-\ \omega^2 + 2\,\xi\,\Omega\ (+\,j\,\omega) + \Omega^2\}$$

$$=$$

$$\Omega^2\ \{-\,\beta^2 + i\,2\,\xi\,\beta + 1\} = \Omega^2\ \{\gamma^2 + i\,2\,\xi\,\beta\} = \Omega^2\ \rho^2\quad \exp(+\,j\,\theta)$$

Then:

$$F = M\ U\ \{-\,\omega^2 + 2\,\xi\,\Omega^1\ (+\,j\,\omega) + \Omega^2\} = K\ U\ \{-\Omega^{-2}\ \omega^2 + 2\,\xi\,\Omega^{-1}\ (+\,j\,\omega) + 1\}$$

$$F\,/\,U = \{M\ \Omega^2\} * (\,\gamma^2 + i\ 2\xi\beta) = \{M\ \Omega^2\} * \rho^2 \exp(+\,j\,\theta) = \{K\}\ \ (\rho^2 \exp(+\,j\,\theta)\,)$$

$$\alpha\,/\,U = \{\,\Omega^2\,\} * (\,\gamma^2 + i\ 2\xi\beta) = \{\ \Omega^2\,\} * \rho^2 \exp(+\,j\,\theta) = \{\,K/M\,\}\ \ (\rho^2 \exp(+\,j\,\theta)\,)$$

$$\alpha\,/\,U\,/\,(\,\gamma^2 + i\ 2\xi\beta) \;=\; \alpha\,/\,U\,/\,\{\,(\rho^2\ \exp(+\,j\,\theta)\,\} = \ \{\,\Omega^2\,\} = \{\,K/M\,\}$$

:

$$F = K\ U\ \{-\Omega^{-2}\ \omega^2 + 2\,\xi\,\Omega^{-1}\ (+\,j\,\omega) + 1\}$$

$$F = \ -\beta^2\ K\ U + \ 2\,\xi\ \ (+\,j\,\beta\,)\ K\ U\ +\ K\ U$$

F = Inertia + Damping + Stiffness

F/M (Power Resilience) = Unit Mass Inertia + Unit Mass Damping + Unit Mass Stiffness

:

$$F/M = \{\alpha\} = \Omega^2\ \rho^2 \exp(+\,j\,\theta\,)\ U \qquad\qquad U = \{\alpha\}/[\Omega^2\ \rho^2 \exp(+\,j\,\theta\,)\,] = \exp(-i\,\theta)\ \{\alpha\}/[\Omega^2\ \rho^2\,]$$

:

Response:		**Angles with Respect to**	
		U	**F**
Displacement Response:	**U**	**0**	$-\theta$
Stiffness (Power Resilience):	**K U**	**0**	$-\theta$
Damping:	$2\,\xi\,\beta\ K\,U$	$+\pi/2$	$-\theta + \pi/2$
Inertia:	$-\beta^2\ K\,U$	**0**	$-\theta$
Inertia & Stiffness (Power Resilience):	$(-\beta^2 + 1)\ K\ U$	**0**	$-\theta$

Seperated Forces :

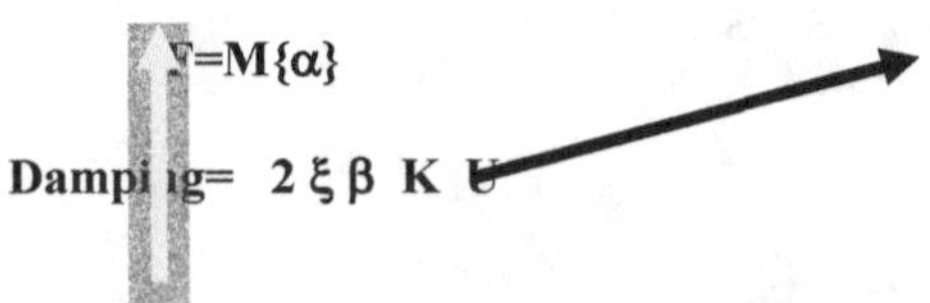

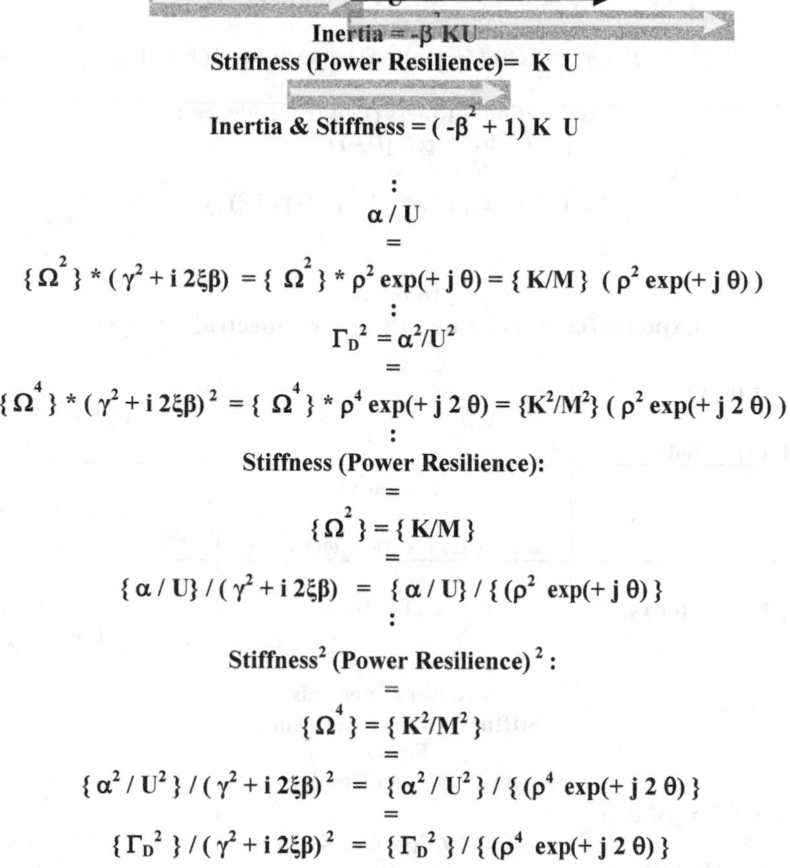

Stiffness (Power Resilience)= K U

Inertia & Stiffness = $(-\beta^2 + 1)$ K U

:

$$\alpha / U$$

=

$$\{\Omega^2\} * (\gamma^2 + i\,2\xi\beta) = \{\Omega^2\} * \rho^2 \exp(+j\,\theta) = \{K/M\}\ (\rho^2 \exp(+j\,\theta))$$

:

$$\Gamma_D{}^2 = \alpha^2/U^2$$

=

$$\{\Omega^4\} * (\gamma^2 + i\,2\xi\beta)^2 = \{\Omega^4\} * \rho^4 \exp(+j\,2\,\theta) = \{K^2/M^2\}\ (\rho^2 \exp(+j\,2\,\theta))$$

:

Stiffness (Power Resilience):

=

$$\{\Omega^2\} = \{K/M\}$$

=

$$\{\alpha / U\} / (\gamma^2 + i\,2\xi\beta) = \{\alpha / U\} / \{(\rho^2 \exp(+j\,\theta)\}$$

:

Stiffness^2 (Power Resilience)^2:

=

$$\{\Omega^4\} = \{K^2/M^2\}$$

=

$$\{\alpha^2 / U^2\} / (\gamma^2 + i\,2\xi\beta)^2 = \{\alpha^2 / U^2\} / \{(\rho^4 \exp(+j\,2\,\theta)\}$$

=

$$\{\Gamma_D{}^2\} / (\gamma^2 + i\,2\xi\beta)^2 = \{\Gamma_D{}^2\} / \{(\rho^4 \exp(+j\,2\,\theta)\}$$

18.8.3.2. Exponential Accelerated Structure Base Spectrals:

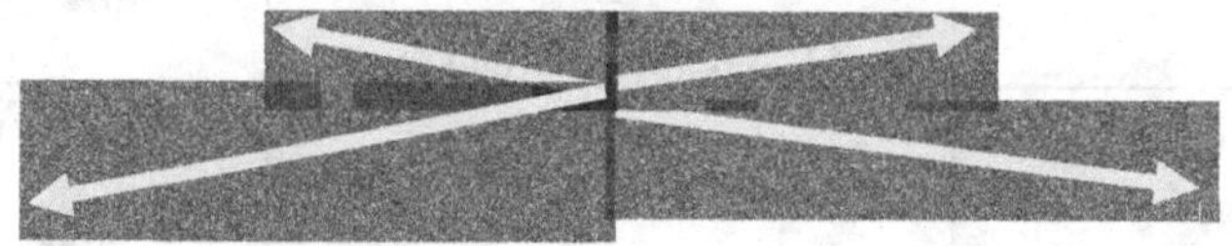

Unit Stiffness (Unit Power Resilience): $[\Gamma_U] = 1/TT$:

$$\alpha / \{U\,\Omega^2\}$$

=

$$(\gamma^2 + i\,2\xi\beta) = \rho^2 \exp(i\,\theta)$$

Unitary Stiffness (Unitary Power Resilience):

$$\alpha / [U\,\rho^2\,\Omega^2]$$

=

$$\rho^{-2}\ (\gamma^2 + i\,2\xi\beta) = \exp(i\,\theta)$$

Stiffness (Power Resilience): $[\Gamma_U] = 1/TT$:

$$\Gamma_U \equiv 1/\Phi_U \equiv \alpha / U$$

=

$$\Omega^2\ (\gamma^2 + i\,2\xi\beta) = \Omega^2\,\rho^2 \exp(i\,\theta)$$

Velocity Stiffness (Power Resilience): $[\Gamma_V] = 1/T$:

$$\Gamma_V \equiv 1/\Phi_V \equiv \alpha \;/ UT \equiv \alpha / U/(+ j\,\omega)$$

=

$$\Omega^2 \;(\gamma^2 + i\,2\xi\beta)/(+ j\,\omega) = \Omega^2 \rho^2 \exp(i\,\theta)/(+ j\,\omega)$$

Acceleration Stiffness (Power Resilience):

$$\Gamma_A \equiv 1/\Phi_A \equiv \alpha \;/ [UTT] \equiv \alpha / U/\omega^2$$

=

$$\Omega^2 \;(\gamma^2 + i\,2\xi\beta)/\omega^2 = \Omega^2 \rho^2 \exp(i\,\theta)/\omega^2$$

18.8.3.3.
Exponential Accelerated Processes Spectral Analysis

$U_\omega TT \equiv (\omega^2)\, U_\omega$ = $(\omega^2)\, U \exp(+ j\omega t)$	$U_\omega TT / U_\omega T = (i\omega)$	$U_\omega TT / U_\omega = (\omega^2)$
$U_\omega T / U_\omega TT = 1/(i\omega)$	$U_\omega T \equiv (\omega^2)\, U_\omega$ = $(i\omega)\, U \exp(+ j\omega t)$	$U_\omega T / U_\omega = (i\omega)$
$U_\omega / U_\omega TT = 1/(\omega^2)$	$U_\omega / U_\omega T = 1/(i\omega)$	U_ω = $U \exp(+ j\omega t)$

Structure Spectrals
Stiffness (Power Resilience)
Equals
{Flexibility (Form Resilience)}$^{-1}$:

$\Gamma_A \equiv \alpha \;/ UTT \equiv 1/\Phi_A$ = $\Omega^2 \;(\gamma^2 + i\,2\xi\beta)/\omega^2$ = $\Omega^2 \rho^2 \exp(i\,\theta)/\omega^2$	$\Gamma_A / \Gamma_V \equiv \Phi_V / \Phi_A$ = $1/(+ j\,\omega)$	$\Gamma_A / \Gamma_U \equiv \Phi_U / \Phi_A$ = $1/(\omega^2)$
	$\Gamma_V \equiv \alpha \;/ [UT]$ = $\Omega^2 \;(\gamma^2 + i\,2\xi\beta)/(+ j\,\omega)$ = $\Omega^2 \rho^2 \exp(i\,\theta)/(+ j\,\omega)$	$\Gamma_V / \Gamma_U \equiv \Phi_U / \Phi_V$ = $1/(+ j\,\omega)$
		$\Gamma_U \equiv \alpha \;/ U$ = $\Omega^2 \;(\gamma^2 + i\,2\xi\beta)$ = $\Omega^2 \rho^2 \exp(i\,\theta)$

:

(Stiffness Spectrals)^2 or (Power Resilience Spectrals)^2
Equals
{Flexibility Spectrals)^2 or (Form Resilience Spectrals)}$^{-2}$:

$S_A \equiv \Gamma_A^2 \equiv \alpha^2/UTT^2 \equiv 1/\Phi_A^2$ = $\Omega^4 \,(\gamma^2 + i\,2\xi\beta)^2/\omega^4$ = $\Omega^4 \rho^4 \exp(2\,i\,\theta)]/\omega^4$	$S_A/S_V \equiv \Gamma_A^2/\Gamma_V^2 \equiv \Phi_V^2/\Phi_A^2$ = $1/\omega^2$	$S_A/S_U \equiv \Gamma_A^2/\Gamma_U^2 \equiv \Phi_U^2/\Phi_A^2$ = $1/\omega^4$
	$S_V \equiv \Gamma_V^2 \equiv \alpha^2/UT^2 \equiv 1/\Phi_V^2$ = $\Omega^4 \,(\gamma^2 + i\,2\xi\beta)^2/\omega^2$ = $\Omega^4 \rho^4 \exp(2\,i\,\theta)/\omega^2$	$S_V/S_U \equiv \Gamma_V^2/\Gamma_U^2 \equiv \Phi_U^2/\Phi_V^2$ = $1/\omega^2$
		$S_U \equiv \Gamma_U^2 \equiv \alpha^2/U^2 \equiv 1/\Phi_U^2$ =

$$\Omega^4 (\gamma^2 + i\, 2\xi\beta)^2 = \Omega^4 \rho^4 \exp(2\, i\, \theta)$$

18.8.3.4.
Global Unit Mass Inertia Spectrals:

Global Mass Inertia Stiffness (Power Resilience):

$$\Gamma_M = 1/\Phi_M = \alpha\ / \{-\beta^2 \Omega^2 U + \{\alpha\}\} = -\{\alpha\} / \{\mu^2 \exp(+j\,\iota)\ \ \Omega^2\}$$

Square of Global Mass Inertia Stiffness (Power Resilience):

$$S_M(\beta) = \Gamma_M^{\,2} = 1/\Phi_M^{\,2} = \alpha^2\ / \{-\beta^2 \Omega^2 U + \alpha\}^2 = \alpha^2 / \{\mu^4 \exp(+j\, 2\,\iota)\ \ \Omega^4\}$$

Unitless:

Global Mass Inertia Stiffness Spectral:

$$\Gamma_M \bullet\ \Omega^2 / \{\alpha\} = -1 / \{\mu^2 \exp(+j\,\iota)\ \}$$

Unitary:

Global Mass Inertia Stiffness Spectral:

$$\Gamma_M \bullet\ \Omega^2 \mu^2 / \{\alpha\} = 1 / \{\exp(+j\,\iota)\ \}$$

$$+\mu^2 \exp(+j\,\iota) = +\beta^2 + \rho^2 \exp(+j\,\theta)$$

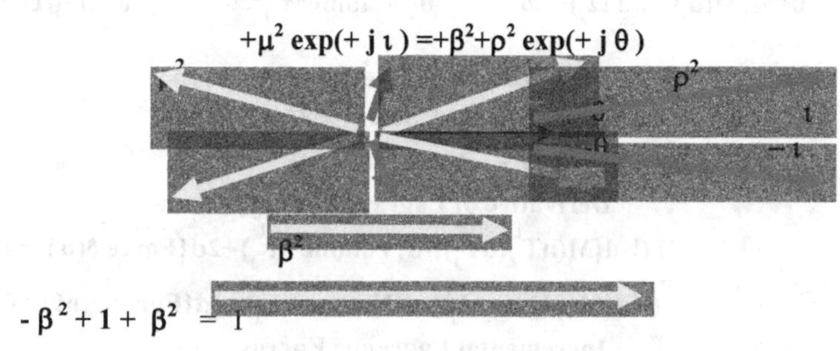

$$-\beta^2 + 1 + \beta^2 = 1$$

Energies:

Lagrange	Kinetic	Potential	Strain	2*(External)
$L \equiv T - V$	T	$V = U - 2W$	U	2W

:

Assembled Energies:

Energy	Kinetic T	- Strain U	2* External W

Derivative of Energies:

(dEnergy/dτ)	$\frac{1}{2}d\{M(uT^j)(uT_j)\}/d\tau$	$-d\{Volume^* \tau^I_{\ J}\}/d\tau$	$2d\{Force^I{}^* uT_J{}^* T\}/d\tau$
(dEnergy/dΩ)	$\frac{1}{2}d\{M(u\Omega^j)(u\,\Omega_j)\}/d\Omega)$	$-d\{Volume^* \tau^I_{\ J}\}/d\Omega)$	$2d\{Force^I{}^* u\Omega_J{}^* \Omega\}/d\Omega)$

Incremental Energies:

(dEnergy)	$\frac{1}{2}d\{M(uT^j)(uT_j)\}/$	$-d\{Volume^* \tau^I_{\ J}\}$	$2d\{Force^I{}^* uT_J{}^* T\}$
(dEnergy)	$\frac{1}{2}d\{M(u\Omega^j)(u\,\Omega_j)\}$	$-d\{Volume^* \tau^I_{\ J}\}$	$2d\{Force^I{}^* u\Omega_J{}^* \Omega\}$

Energies:

$\int dEnergy$ = Energy	$\int(\frac{1}{2}d\{M(uT^j)(uT_J)\})$ = $\frac{1}{2}\{M(uT^j)(uT_J)\}$	$\int(-d\{Volume^*\tau^I_J\})$ = $-Volume^*\tau^I_J$	$\int(2d\{Force^I{}^*(uT_J{}^*T)\})$ = $2Force^I{}^*d(uT_J{}^*T)$
$\int dEnergy$ = Energy	$\int(\frac{1}{2}d\{M(u\Omega^j)(u\,\Omega_J)\})$ = $\frac{1}{2}\{M(u\,\Omega^j)(u\,\Omega_J)\}$	$\int(-d\{Volume^*\tau^I_J\})$ = $-Volume^*\tau^I_J$	$\int(2d\{Force^I{}^*(u\Omega_J{}^*\Omega)\})$ = $2Force^I{}^*d(u\Omega_J{}^*\Omega)$

Average Energies:

$1/T^*\int_{-T/2}^{T/2}$ dEnergy	$1/T^*$ $\frac{1}{2}\{M(uT^j)(uT_J)\}\Big\vert_{-T/2}^{T/2}$	$1/T^*$ $\{-Volume^*\tau^I_J\}\Big\vert_{-T/2}^{T/2}$	$1/T^*$ $2Force^I{}^*d(uT_J{}^*T)\Big\vert_{-T/2}^{T/2}$
$1/\Omega^*\int_{-\Omega/2}^{\Omega/2}$ dEnergy	$1/\Omega^*$ $\frac{1}{2}M(u\,\Omega^j)(u\,\Omega_J)\Big\vert_{-\Omega/2}^{\Omega/2}$	$1/\Omega^*$ $\{-Volume^*\tau^I_J\}\Big\vert_{-\Omega/2}^{\Omega/2}$	$1/\Omega^*$ $2Force^I{}^*d(u\Omega_J{}^*\Omega)\Big\vert_{-\Omega/2}^{\Omega/2}$

Limit of Average Energies:

$limit_{T\to\infty}$ $1/T^*\int_{-T/2}^{T/2}$ dEnergy	$limit_{T\to\infty}$ $1/T^*$ $\frac{1}{2}\{M(uT^j)(uT_J)\}\Big\vert_{-T/2}^{T/2}$	$limit_{T\to\infty}$ $1/T^*$ $\{-Volume^*\tau^I_J\}\Big\vert_{-T/2}^{T/2}$	$limit_{T\to\infty}$ $1/T^*$ $2Force^I{}^*d(uT_J{}^*T)\Big\vert_{-T/2}^{T/2}$
$limit_{\Omega\to\infty}$ $1/\Omega\int_{-\Omega/2}^{\Omega/2}$ dEnergy	$limit_{\Omega\to\infty}$ $1/\Omega^*$ $\frac{1}{2}M(u\,\Omega^j)(u\,\Omega_J)\Big\vert_{-\Omega/2}^{\Omega/2}$	$limit_{\Omega\to\infty}$ $1/\Omega^*$ $\{-Volume^*\tau^I_J\}\Big\vert_{-\Omega/2}^{\Omega/2}$	$limit_{\Omega\to\infty}$ $1/\Omega^*$ $2Force^I{}^*d(u\Omega_J{}^*\Omega)\Big\vert_{-\Omega/2}^{\Omega/2}$

0 < Bounded Energies < ∞:

$0<Energy<\infty$	$0<\frac{1}{2}\{M(uT^j)(uT_J)\}<\infty$	$0<Volume^*\tau^I_J<\infty$	$0<2Force^I{}^*d(uT_J{}^*T)<\infty$
$0<Energy<\infty$	$0<\frac{1}{2}\{M(u\,\Omega^j)(u\,\Omega_J)\}<\infty$	$0<Volume^*\tau^I_J<\infty$	$0<2Force^I{}^*d(u\Omega_J{}^*\Omega)<\infty$

:

Lagrange Energy:

Lagrange = Kinetic - Strain + 2*(External)

$$L \equiv T - V = T - U + 2W$$

Derivative of Lagrange Energy

(dL/dτ) $\quad (\frac{1}{2}d\{M(uT^j)(uT_J)\}-d\{Volume^*\tau^I_J\}+2d\{Force^I{}^*(uT_J{}^*T\})/d\tau$

(dL/dΩ) $\quad (\frac{1}{2}d\{M(u\Omega^j)(u\,\Omega_J)\}-d\{Volume^*\tau^I_J\}+2d\{Force^I{}^*(u\Omega_J{}^*\Omega\})/d\Omega$

Incremental Lagrange Energy

(dL) $\quad \frac{1}{2}d\{M\,(uT^j)(uT_J)\} - d\{Volume^*\,\tau^I_J\} + 2\,d\{Force^I{}^*\,(uT_J{}^*T\}$

(dL) $\quad \frac{1}{2}d\{M\,(u\Omega^j)(u\,\Omega_J)\} - d\{Volume^*\,\tau^I_J\} + 2\,d\{Force^I{}^*(u\Omega_J{}^*\Omega\}$

Lagrange Energy

$\int dL = L$: $\quad \int(\frac{1}{2}d\{M\,(uT^j)(uT_J)\} - d\{Volume^*\,\tau^I_J\} + 2\,d\{Force^I{}^*\,(uT_J{}^*T\})$ = $\frac{1}{2}\{M\,(uT^j)(uT_J)\} - Volume^*\tau^I_J + 2\,Force^I{}^*d(uT_J{}^*T\}$

$\int dL = L$: $\quad \int(\frac{1}{2}d\{M\,(u\Omega^j)(u\,\Omega_J)\} - d\{Volume^*\,\tau^I_J\} + 2\,d\{Force^I{}^*(u\Omega_J{}^*\Omega\})$ = $\frac{1}{2}\{M\,(u\,\Omega^j)(u\,\Omega_J)\} - Volume^*\tau^I_J + 2\,Force^I{}^*d(u\Omega_J{}^*\Omega\}$

Average Lagrange Energy

$1/T\int_{-T/2}^{T/2} dL \quad 1/T(\frac{1}{2}\{M(uT^j)(uT_J)\}-Volume^*\tau^I_J+2Force^I{}^*d(uT_J{}^*T\})\Big\vert_{-T/2}^{T/2}$

$1/\Omega\int_{-\Omega/2}^{\Omega/2} dL \quad 1/\Omega\,(\frac{1}{2}\{M(u\,\Omega^j)(u\,\Omega_J)\}-Volume^*\tau^I_J+2Force^I{}^*d(u\Omega_J{}^*\Omega\})\Big\vert_{-\Omega/2}^{\Omega/2}$

Limit of Average Lagrange Energy

$limit_{T\to\infty}\ 1/T(\frac{1}{2}\{M(uT^j)(uT_J)\}-Volume^*\tau^I_J+2Force^I{}^*d(uT_J{}^*T\})\Big\vert_{-T/2}^{T/2}$

$limit_{\Omega\to\infty}\ 1/\Omega(\frac{1}{2}\{M(u\,\Omega^j)(u\,\Omega_J)\}-Volume^*\tau^I_J+2Force^I{}^*d(u\Omega_J{}^*\Omega\})\Big\vert_{-\Omega/2}^{\Omega/2}$

0 < Bounded Lagrange Energy < ∞

$\int_{-\infty}^{+\infty} dL \quad 0 < (\frac{1}{2}\{M(uT^j)(uT_J)\}-Volume^*\tau^I_J+2Force^I{}^*d(uT_J{}^*T\})\Big\vert_{-\infty}^{+\infty} <\infty$

$\int_{-\infty}^{+\infty}$ **dL** $\quad 0 < (½\{M(u\,\Omega^j)(u\,\Omega_J)\}\text{-Volume}*\tau^I_J+2\text{Force}^I*d(u\Omega_J*\Omega\})\,_{-\infty}^{+\infty} <\infty$

18.9.
Lagrangian Equations of Motion

Energies:

lagrange	Kinetic	Potential	Strain	2*External
			U	2W
		Strain - 2External		
		$V\equiv U-2W\neq V(uT_K)$		
	$T\neq T(u_K)$			

Kinetic-Potential=Kinetic-Strain+2External

$$L \equiv T - V \equiv T - U + 2W$$

$$\{d\,[\,\partial L / \partial uT_K\,] / dt\,\} - \partial[L] / \partial u_K = 0$$

$$\{d[\partial(T-V)/\partial uT_K]/dt\}-\partial[T-V]/\partial u_K=0$$

$$[d\,(\partial T/\partial uT_K)/dt - 0] - (0 - \partial V/\partial u_K)= 0 \qquad \partial T/\partial u_K=0 \qquad \partial V / \partial uT_K = 0$$

$$d\,(\partial T/\partial uT_K)/dt + \partial V/\partial u_K = 0$$

Static:	Static:	Static:
$0 + \partial V/\partial u_K = 0$	$\partial T/\partial uT_K=0$	
$\partial V/\partial u^J = 0$		

Potential	Strain	2*External
$V = U - 2W$	U	2W
	$\int_{volume}\int_0^{\varepsilon} \sigma_J\, d\varepsilon^J\, dVol$	$\int f_J\, u^J\,(dVol) + P_J\, u^J$
$1/2\, u_I\, K^I_J\, u^J - u_I\, B^I - u_I\, P^I$	$\int_{volume} \varepsilon_I\,[1/2 * C^I_J\, \varepsilon^J_q\,]\, dVol$	
$\partial V/\partial u^J = K^I_J\, u^J - R^I$		

Kinetic

$$T= \int_{volume} dT = \int_{volume} 1/2\, uT_I(x,y,z)\ \rho\ uT^I(x,y,z)\ \ (dVolume)$$

$$T = UT_I \int_{volume} 1/2\, B^I_K(x,y,z)\ \rho\ B^K_J(x,y,z)\ \ (dVolume)\,]\ \ UT^J$$

$$M^I_J = \int_{volume} B^I_K(x,y,z)\ \rho\ B^K_J(x,y,z)\ \ (dVolume)$$

$$T = 1/2\, UT_I\, M^I_J\, UT^J$$

$$[\,\partial T / \partial uT_I\,] = M^I_J\ UT^J$$

$$\{d\,[\,\partial T / \partial uT_I\,] / dt\,\} = M^I_J\ UTT^J$$

:

For Static Structural Systems:

$$\partial V/\partial u^J = K^I_J\, u^J - R^I = 0$$

For Dynamic Structural Systems:

$$\{d\,[\,\partial L / \partial uT_I\,] / dt\,\} - \partial[L] / \partial u_I =\{d\,[\,\partial T / \partial uT_I\,] / dt\,\} + \partial[V] / \partial u_I = 0$$

$$M^I_J\ UTT^J + [\,K^I_J\, u^J - R^I] = 0$$

$$M^I_J\ UTT^J + K^I_J\, u^J = R^I$$

Energies are Given in terms of Parameters:

$(u^J), I = 1,\ldots M$

In Finite Elements Parameters are Displacements.

Kinetic Energy T is function of (uT_I) Only **Potential Energy V is function of (u_I) Only**

Such That:

For Static Structural Systems:

$$\partial V/\partial u^J = 0$$

For Dynamic Structural Systems:

$$\{d\,[\,\partial L / \partial uT_K\,] / dt\} - \partial[L] / \partial u_K = 0$$

$$\{d\,[\,\partial (T-V) / \partial uT_K\,] / dt\} - \partial[\,T - V] / \partial u_K = 0$$

$$\{d\,[\,\partial T / \partial uT_K\,] / dt\} + \partial[V] / \partial u_K = 0$$

:

Where:

2*(External Energy)

$$2W = +\int f_J\, u^J\,(dVolume) + P_J\, u^J$$

:

Strain Energy

$$U = \int_{volume} \int_0^{\varepsilon} \sigma_J\, d\varepsilon^J\,(dVolume)$$

:

Linear Strain Energy

$$U = \int_{volume} \{\varepsilon_I\,[1/2 * C^I{}_J\, \varepsilon^J{}_q]\}\,(dVolume)$$

:

Potential Energy V is function of (u_I) Only

$$V(u_I) = U - 2W = \int_{volume} \int_0^{\varepsilon} \sigma_J\, d\varepsilon^J\,(dVolume) - [+\int f_J\, u^J\,(dVolume) + P_J\, u^J]$$

:

Linear Potential Energy

$$V(u_I) = \int_{volume} \{\varepsilon_I\,[1/2 * C^I{}_J\, \varepsilon^J{}_q]\}\,(dVolume) - [+\int f_J\, u^J\,(dVolume) + P_J\, u^J]$$

$$V(u_I) = 1/2\; u_I\, K^I{}_J\, u^J - u_I\, B^I - u_I\, P^I$$

$$V(u_I) = 1/2\; u_I\, K^I{}_J\, u^J - u_I\, R^I$$

And:

$$\partial V/\partial u^J = K^I{}_J\, u^J - R^I$$

:

Kinetic Energy T is function of (uT_I) Only

$$T(uT_I) = \int_{volume} dT = \int_{volume} 1/2\; uT_I(x,y,z)\;\; \rho\;\; uT^I(x,y,z)\;\;(dVolume)$$

$$uT^K(x,y,z) = B^K{}_J(x,y,z)\;\; UT^J \quad \text{where } K=x,y,z$$

$$T(uT_I) = UT_I \int_{volume} 1/2\; B^I{}_K(x,y,z)\;\rho\; B^K{}_J(x,y,z)\;\;(dVolume)\,]\;\; UT^J$$

Define Mass:

$$M^I{}_J = \int_{volume} B^I{}_K(x,y,z)\;\rho\; B^K{}_J(x,y,z)\;\;(dVolume)$$

So That:

$$T = 1/2\; UT_I\; M^I{}_J\;\; UT^J$$

And:

$$[\,\partial T / \partial uT_I\,] = M^I{}_J\;\; UT^J$$

$$\{d\,[\,\partial T / \partial uT_I\,] / dt\} = M^I{}_J\;\; UTT^J$$

:

For Static Structural Systems:

$$\partial V/\partial u^J = K^I{}_J\, u^J - R^I = 0$$

For Dynamic Structural Systems:

$$\{d\,[\,\partial L / \partial uT_I\,] / dt\} - \partial[L] / \partial u_I = \{d\,[\,\partial T / \partial uT_I\,] / dt\} + \partial[V] / \partial u_I = 0$$

$$M^I{}_J\ UTT^J + [\ K^I{}_J\ u^J - R^I] = 0$$

$$M^I{}_J\ UTT^J + K^I{}_J\ u^J = R^I$$

19. Structural Vibrations:

19.1. Random Processes:

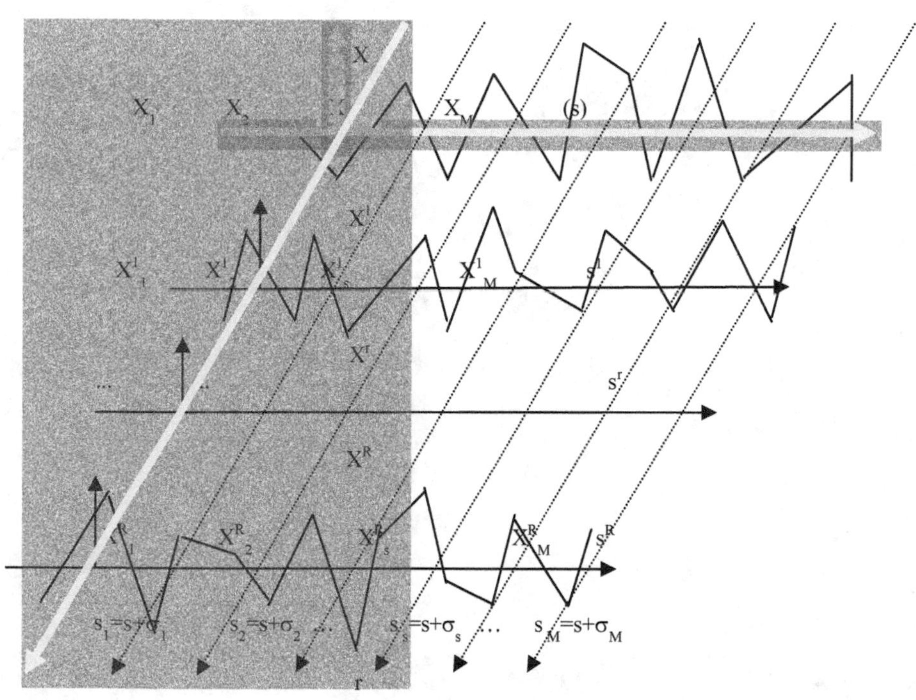

Given:

M Stations, Referenced by Specified Non Negative Values, σ_M:

$$\underline{\sigma} \equiv (\sigma_1,\ \sigma_2,\ \sigma_3,\ ...,\sigma_M)$$

Independent Variable, s:

$$(s+\underline{\sigma}) \equiv (s_1,\ s_2,\ \ldots s_M) \equiv (s+\sigma_1,\ s+\sigma_2,\ s+\sigma_3, s+\sigma_4, ..., s+\sigma_M)$$

Similar Specific Values can be Considered:

(ω or τ) (σ or ε) (τ or ω)

And Corresponding Independent Variable :

(w or t) (s or e) (t or w)

Random Process X is Ensemble Segments Connected at Independent Variable Stations:

$$X^r(s+\underline{\sigma}) \equiv (X^r{}_1, X^r{}_2, \ldots X^r{}_s, \ldots X^r{}_M) \equiv \{\ X^r(s_1), X^r(s_2), \ldots X^r(s_M)\ \} \equiv \{\ X^r(s+\sigma_1),\ X^r(s+\sigma_2),\ \ldots\ X^r(s+\sigma_M)\ \}$$

Two Types of X(s+σ or e+ε) Random Processes are Explored:

Spectral Process
G(w +ω or t+τ)

Temporal Process
F(t+τ or w +ω)

Process Examples:

Acceleration(w)	Acceleration(t)
Velocity(w)	Velocity(t)
Force(w)	Force(t)

Operator H on Random Process Dependent Variable, X: